Animal Behavior

TENTH EDITION

Animal Behavior

AN EVOLUTIONARY APPROACH

TENTH EDITION

JOHN ALCOCK

Arizona State University

Sinauer Associates, Inc. Publishers
Sunderland, Massachusetts U.S.A.

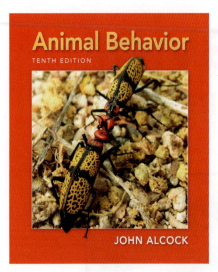

About the Cover

Female animals generally use courtship to evaluate the quality of a potential mate. In the desert blister beetle, *Tegrodera aloga*, the male (above) draws the female's antennae into grooves in the front of his head. In this stance, the female may be able to sense chemical cues that tell her about the ability of her would-be partner to transfer valuable biochemicals to her—should she permit copulation to occur. Photograph by the author.

Animal Behavior: An Evolutionary Approach, Tenth Edition
Copyright © 2013

For information, address:
Sinauer Associates, Inc., P.O. Box 407, Sunderland, MA 01375 U.S.A.
Fax: 413-549-1118
Email: publish@sinauer.com
Internet: www.sinauer.com

Library of Congress Cataloging-in-Publication Data

Alcock, John, 1942-
Animal behavior : an evolutionary approach / John Alcock. -- 10th ed.
  p. cm.
Includes bibliographical references.
ISBN 978-0-87893-966-4 (pbk.)
1. Animal behavior--Evolution. I. Title.
QL751.A58 2013
591.5--dc23

2012012683

Printed in China

10 9 8 7 6 5 4 3 2 1

To all researchers, instructors, and students
who are fascinated by animal behavior

Contents in Brief

CHAPTER 1   An Introduction to Animal Behavior   3

CHAPTER 2   Behavioral Ecology and the Evolution of Altruism   15

CHAPTER 3   The Evolution of Social Behavior   43

CHAPTER 4   The Evolution of Communication   67

CHAPTER 5   Avoiding Predators and Finding Food   101

CHAPTER 6   The Evolution of Habitat Selection, Territoriality, and Migration   139

CHAPTER 7   The Evolution of Reproductive Behavior   171

CHAPTER 8   The Evolution of Mating Systems   217

CHAPTER 9   The Evolution of Parental Care   257

CHAPTER 10   Proximate and Ultimate Causes of Behavior   293

CHAPTER 11   The Development of Behavior   323

CHAPTER 12   Evolution, Nervous Systems, and Behavior   359

CHAPTER 13   How Neurons and Hormones Organize Behavior   391

CHAPTER 14   The Evolution of Human Behavior   423

Contents

CHAPTER 1

An Introduction to Animal Behavior 3

The Behavioral Ecology of a Digger Bee 4
Natural Selection and Infanticide 9
The *Science* of Behavioral Biology 11
The Power of Scientific Logic 11

Summary 13
Suggested Reading 13

CHAPTER 2

Behavioral Ecology and the Evolution of Altruism 15

Explaining Altruism: Intelligent Design? 17
Altruism and For-the-Benefit-of-the-Group Selection 18
Altruism and Indirect Selection 21
Kin Selection and Inclusive Fitness Theory 23
Social Insects and Inclusive Fitness 25
Testing the Haplodiploidy Hypothesis 28

The History of Behavioral Traits 29
The Origin of the Dances of Honey Bees 32
An Initial Hypothesis and Its Test 34
Kin Selection and Social Conflict 37
Summary 41
Suggested Reading 41

CHAPTER 3

The Evolution of Social Behavior 43

The Evolution of Helpful Behavior 45
The Reciprocity Hypothesis 50
Kin Selection and Helpful Behavior 53
Helpers at the Nest: A Darwinian Puzzle 54
Helpers at the Nest: Alternative Hypotheses 56

Helpers at the Burrow: The Case of the Naked Mole Rat 61
Altruism in Vertebrates and Insects: A Comparison 62
Summary 65
Suggested Reading 65

CHAPTER 4

The Evolution of Communication 67

The History of a Strange Display 68

Evolutionary History Occurs via Changes in
Preexisting Traits 69

**Sensory Exploitation and the Origin
of Communication Signals 70**

Sensory Exploitation—or the Retention of
Ancestral Traits? 74

The Panda Principle 76

The Behavioral Ecology of Communication 77

Adaptationist Hypotheses for the Hyena's
Pseudopenis 79

Another Darwinian Puzzle: The Adaptive Value of
Threat Displays 81

The Honest Signal Hypothesis 82

Self-Sacrificing Communication? 86

The Darwinian Puzzle of Deception 88

The Manipulation of Communication Systems 90

One More Darwinian Puzzle: Eavesdropping 93

Summary 97

Suggested Reading 98

CHAPTER 5

Avoiding Predators and Finding Food 101

The Definition of Adaptation 103

Testing Adaptationist Hypotheses 104

The Comparative Method 106

A Cost–Benefit Approach to Social Defenses 110

Game Theory and Social Defenses 113

A Cost–Benefit Approach to Cryptic Behavior 115

The Darwinian Puzzle of Conspicuous Behavior 118

Optimality Theory and Antipredator Behavior 122

Optimality Theory and Foraging Decisions 124

How to Choose an Optimal Mussel 128

Criticisms of Optimal Foraging Theory 129

**Game Theory, Feeding Behavior, and Another
Darwinian Puzzle 132**

Behavioral Variation and Conditional Strategies 134

Summary 136

Suggested Reading 137

CHAPTER 6

The Evolution of Habitat Selection, Territoriality, and Migration 139

**Habitat Selection and Ideal Free Distribution
Theory 140**

When to Invest in Territorial Defense 142

Territoriality and Resource-Holding Power 145

Why Give Up Quickly When Fighting for a
Territory? 146

Contest Resolution via Nonarbitrary Means 148

Resource Value and Payoff Asymmetries 150

The Dear Enemy Effect 151

Dispersal and Migration 153

The History of Migration 156

The Costs of Migration 159

The Benefits of Migration 163

Conditional Strategies and Migration 166

Summary 169

Suggested Reading 169

CHAPTER 7

The Evolution of Reproductive Behavior 171

Sexual Selection and Bowerbird Behavior 172

Sexual Differences Theory 175

Sexual Selection and Parental Investment 177

Testing Sexual Differences Theory 178

Sexual Selection and the Competition for Mates 181

Alternative Mating Tactics 183

Why Settle for Less? 184

The Coexistence of Alternative Mating Strategies 188

Sexual Selection and Sperm Competition 190

Why Stay with a Female after Mating with Her? 193

Sexual Selection and Mate Choice 196

Female Mate Choice for Paternal Males 199

Female Mate Choice without Material Benefits 200

Runaway Sexual Selection 205

Testing Competing Ideas about Mate Choice 207

Sexual Conflict 209

Summary 214

Suggested Reading 215

CHAPTER 8

The Evolution of Mating Systems 217

Is Male Monogamy Adaptive? 218

Male Monogamy in Mammals 222

Male Monogamy in Birds 224

Monogamous Males, Polyandrous Females 227

What Do Females Gain from Polyandry? 230

Polyandry and Good Genes 232

Polyandry and Genetic Compatibility 234

Polyandry and Social Insects 237

Polyandry and Material Benefits 238

Why Are There So Many Kinds of Polygynous Mating Systems? 241

Female Defense Polygyny 241

Resource Defense Polygyny 243

Scramble Competition Polygyny 246

Lek Polygyny 247

Summary 254

Suggested Reading 255

CHAPTER 9

The Evolution of Parental Care 257

The Cost–Benefit Analysis of Parental Care 258
Why More Care by Mothers than by Fathers? 259
Why Are Any Males Paternal? 263
Why Do Male Water Bugs Do All the Work? 264
Discriminating Parental Care 267
Why Adopt Genetic Strangers? 270
The History of Interspecific Brood Parasitism 274

Why Accept a Parasite's Egg? 278
The Puzzle of Parental Favoritism 281
Killer Siblings 284
Parental Behavior in Relation to Offspring Value 287
Summary 290
Suggested Reading 291

CHAPTER 10

Proximate and Ultimate Causes of Behavior 293

Connecting the Four Levels of Analysis 294
The Proximate and Ultimate Causes of Monogamy in Prairie Voles 296
The Proximate Causes of Bird Song 299
Social Experience and Song Development 302
Bird Brains and Bird Songs 304
How the Avian Song Control System Works 305

The Ultimate Causes of Bird Songs 308
The Adaptive Value of Song Learning 311
Adaptive Repertoire Matching 315
Female Preferences and Song Learning 316
Proximate and Ultimate Causes Are Complementary 319
Summary 321
Suggested Reading 321

CHAPTER 11

The Development of Behavior 323

The Nature or Nurture Misconception 324
The Interactive Theory of Development 324
Learning Requires Both Genes and Environment 329
Environmental Differences Can Cause Behavioral Differences 332
Genetic Differences Can Also Cause Behavioral Differences 334

Single Gene Effects on Development 338
Evolution and Behavioral Development 341
Adaptive Developmental Homeostasis 342
Adaptive Developmental Switch Mechanisms 345
The Adaptive Value of Learning 349
Summary 356
Suggested Reading 357

CHAPTER 12

Evolution, Nervous Systems, and Behavior 359

Complex Responses to Simple Stimuli 360

How Moths Avoid Bats 363

Detecting, Processing, and Responding to Ecologically Significant Stimuli 370

Selective Relaying of Sensory Inputs 374

Responding to Relayed Messages 376

The Proximate Basis of Stimulus Filtering 379

Stimulus Filtering via Cortical Magnification 381

The Evolution of Cognitive Skills 384

Summary 389

Suggested Reading 389

CHAPTER 13

How Neurons and Hormones Organize Behavior 391

Neural Command and Control 392

Daily Changes in Behavioral Priorities 395

How Do Circadian Mechanisms Work? 397

Seasonal and Annual Cycles of Behavior 401

Social Conditions and Changing Priorities 409

Hormonal Modulation of Behavior 411

Hormones and Reproductive Behavior 413

The Costs of Hormonal Regulation 416

Summary 420

Suggested Reading 420

CHAPTER 14

The Evolution of Human Behavior 423

Language and the Four Levels of Analysis 424

The History of Human Speech 425

The Neurophysiology of Speech 426

The Adaptive Value of Speech 429

The Evolutionary Analysis of Mate Choice 432

Mate Choice by Women 432

Mate Choice by Men 438

The Evolutionary Analysis of Sexual Conflict 441

Coercive Sex 446

Practical Applications of Evolutionary Theory 449

The Triumph of an Evolutionary Analysis of Human Behavior 452

Summary 454

Suggested Reading 455

Glossary 457

Bibliography 463

Illustration Credits 503

Index 505

Preface

This is my tenth attempt at writing a comprehensive but manageable textbook on animal behavior for undergraduates. Over the years, I have come to believe that there are three really big ideas worth my students' attention. First, the logic of science is powerful, understandable, and beautiful. Students should know how the logic is used to produce hypotheses, predictions, and tests of scientific explanations. Second, my readers ought to learn why evolutionary scientists who study behavioral traits are so dependent on the work of a nineteenth century biologist, Charles Darwin. The Darwinian or adaptationist approach is the foundation for my book for a simple reason—it works and I think students should know why. Third, I am keen to have students understand the distinction between proximate (immediate) and ultimate (evolutionary) causes in biology. I believe that all three goals can be achieved without requiring students to become deeply involved in the formal mathematical and graphical elaborations of modern behavioral biology. Instead, the three big ideas can be discussed in English in ways I hope will make some students want to build on what they have learned.

The recognition that I really had only three primary goals led me to reorganize the text substantially this time around. My textbook has always been largely about one of my main interests, namely, behavioral adaptation and the explanatory beauty of Darwinian theory. An understanding of natural selection theory directs us to puzzles worth solving; in many ways the premier puzzle of this sort is the evolution of altruism. Therefore, instead of tackling the topic of altruism toward the end of the book, I have placed it at the beginning as an introduction to the concept of a Darwinian puzzle and to the kind of scientific research that can help solve such a puzzle.

The chapters that follow continue to direct the reader's attention primarily to the question, what adaptive value, if any, is served by behaviors under examination? After looking at social behavior, anti-predator behavior, mating behavior, and parental behavior, I then present examples of how the developmental and physiological (proximate) mechanisms underlying animal behavior work and why they can also be studied in terms of their adaptive value. I think this sequence puts the proximate details of genes, neurons, and hormones in a more accessible context for students who, initially at least, are likely to be largely interested in whole animals and what they do.

Although the book has been restructured, I have continued to place discussion questions throughout the text in the hope that instructors will use these items or ones of their own making to reinforce the main points outlined above. My questions deal with the logic of science, the meaning of natural selection theory, and the distinction between proximate and ultimate causes in biology. Instructors can also use classroom discussion and out-of-class assignments to emphasize the unfinished nature of behavioral research, which, like all other scientific areas, is constantly expanding and sometimes coming to new conclusions that differ from original ones. I have flagged a few of the many controversies that exist in the discipline; instructors will know of others. These can be exploited to encourage the evaluation of competing ideas, contrary results, and changing conclusions in the discipline.

A full range of classroom activities could be used to encourage students to turn off the academic autopilot and really grapple with the major concepts of the course, and to stop thinking that attractive Powerpoint presentations will teach them all they need to know about the field of animal behavior. Indeed, there is abundant evidence that the old-fashioned 50-minute lecture is exactly that—old-fashioned and relatively ineffective as an educational tool, no matter how entertaining and beautifully delivered. Most instructors know this from personal experience; when we hear others lecturing at departmental seminars and the like, we often stop paying attention within 10 minutes (or at least I do), and instead do some serious daydreaming or doze intermittently, depending on the time of day and our degree of sleep deprivation, even if we are genuinely interested in what the speaker has to say. Our students are no different.

So I suggest that you try breaking your lectures into a series of mini-lectures that lead to student discussions or team problem-solving challenges that require your listeners to engage directly with the major concepts you wish them to absorb. Discussion questions of the sort I have written could be used to this end, and I certainly hope that instructors will take advantage of them or others of their own construction rather than soldiering on in a futile and ultimately counterproductive attempt to "cover the material." We all know that the many details of the courses we ourselves took from lecture-happy college and university instructors have faded into oblivion. The hours I spent memorizing the difference between things like alkanes and alkenes were neither happy nor useful. I do not remember what the difference is nor do I care. So let's not ask our students to memorize for memorization's sake. I would like to think that our students are more likely to learn something important if they have a hand in tackling a key idea that requires them to personally solve a puzzle or to convince a classmate about their view of the answer to a discussion question. I imagine that many instructors in animal behavior are already doing these things and more; if my textbook can help out, great!

Acknowledgments

It is hard for me to believe that I am almost 70. There are very few positive aspects to growing old, very few, but one is that I have had many years to receive help from others in the development of my views on textbook writing and teaching. Gordon Orians and Bob Lockard gave me good advice when I was a youngster teaching my first courses at the University of Washington. Then when I moved to Arizona State University in the early 1970s, I came to appreciate what I could learn from many colleagues, especially Tony Lawson but also Dave Brown, Jim Collins, Stuart Fisher, Dave Pearson, and Ron Rutowski, key members of the lunch bunch. For many years our group met almost every lunch hour for spirited discussions of national and departmental politics as well as general gossip, the cinema, sporting events, assorted trivia, and occasionally issues associated with teaching. We are now much, much older but we still meet, although less regularly, often to hash out why some of us are so disappointed with our current Commander in Chief, whose once glorious promise of "Change" never materialized, but occasionally also to talk about teaching, students, and associated academic matters over which we have slightly more control. These get-togethers are still fun and educational, thank goodness, despite our collective decrepitude.

Each time I have revised my textbook, I have had very generous assistance from colleagues willing to supply me with photographs to illustrate the book. I am grateful to these persons (who are credited in the text) and indeed, to all those behavioral researchers whose ingenuity and scientific resourcefulness has made the field grow ever better over the years. My friend Dave Skryja helps me stay abreast of the field by sending me news items on stories likely to interest me. With assistance from the Web of Science, I can learn more about these and other advances in the field.

The preparation of each new edition has also been helped along by manuscript reviewers whose constructive criticism has resulted in many improvements. (As one always writes at this point, any errors or shortcomings in the final version are mine and mine alone.) In this Tenth Edition, I have benefited from the reviews of David Queller (Chapters 1 and 2), Vittorio Baglione (Chapter 3), Marion East and Diana Hews (Chapter 4), Scott Creel (Chapter 5), Bridget Stutchbury (Chapter 6), Bruce Lyon (Chapter 7), Tom Tregenza (Chapter 8), Naomi Langmore (Chapter 9), Doug Nelson and Bridget Stutchbury (Chapter 10), Gene Robinson (Chapter 11), Ken Catania (Chapter 12), Tom Hahn (Chapter 13), and Tony Little (Chapter 14). Bob Montgomerie helpfully reviewed the entire manuscript, during which time he made great numbers of pertinent editorial comments as well as encouraging me to deal with some much larger issues, such as how to help the reader see the connections between components within and among chapters as well as how to come to grips with controversy in behavioral research. I am very much in his debt as are the readers of this edition.

I have also had much help from the team at Sinauer Associates, my publisher, with whom I have partnered now for over 37 years. During this time I have always had good editors who have deftly handled all the details that have to be dealt with, and there are many. Syd Carroll is my current editor and she has continued the excellent tradition of calmly keeping things moving forward. I have been in touch with Kath Emerson and David McIntyre almost as often as with my editor, and they have made major contributions to the project. Kath Emerson has prepared the manuscript for my superb copy editor, Lou Doucette, and David McIntyre has tracked down photographs for the book's illustrations. I am very lucky to have been linked with these persons and others at Sinauer Associates all these years.

Finally thanks to my wife Sue to whom I have been married for over 4 decades and to my sons, Joe and Nick, who have now both had their 40th birthday, a further reminder, as if I needed any, of my advanced years. Although Joe and his wife Satkirin have decided not to have any children, a decision I applaud at a rational level, Nick and his wife Sara have had two children, Abby and Jake, a decision that I applaud at an emotional level. Grandchildren, especially ones who live nearby as do Abby and Jake, are among the rare positive features of life for us oldsters. I find being with Abby and Jake (about once a week) provides great delight for me (and for Sue). I think I know why from an evolutionary perspective, but this academic knowledge does not detract in the slightest from the pleasure I feel when both kids are snuggled next to me, one on each side, as I read *Go Dogs Go* or *No David* to them. I am fortunate to have lived long enough to have had the experience of being a grandfather as well as having the time to write and rewrite a textbook on a wonderful subject that fascinates me.

Media & Supplements
to accompany *Animal Behavior,* Tenth Edition

eBOOK

Animal Behavior, Tenth Edition is available as an eBook in several formats, all at a substantial discount from the price of the printed textbook. For details on available ebook formats, please visit the Sinauer Associates website at www.sinauer.com, or contact your Sinauer sales representative.

Instructor's Resource Library

Available to qualified adopters of *Animal Behavior,* the Tenth Edition Instructor's Resource Library contains a variety of teaching and laboratory resources. The IRL includes the following elements:

Textbook Figures and Tables

All of the textbook's figures, tables, and photographs are provided in both JPEG (high- and low-resolution) and PowerPoint formats. All of the images have been formatted and optimized for excellent projection quality.

Animal Behavior Video Collection

This collection of video segments brings to life many of the specific behaviors discussed in the textbook. A great many of these high-quality segments were selected by the author from the collection of the Cornell Lab of Ornithology's Macaulay Library. Great for use in class, the segments are short and easy to incorporate into lectures. All segments are provided both as movie files and in ready-to-use PowerPoint presentations.

Teaching Animal Behavior:
An Instructor's Manual to accompany Animal Behavior, *Tenth Edition*

Teaching Animal Behavior provides instructors with several resources to facilitate the preparation of lectures, quizzes, and exams. Contents include:

- Answers to the discussion questions presented in the textbook
- Sample quiz questions with answers
- Sample exam questions with answers
- Descriptions to accompany the collection of animal behavior videos
- A listing of films on animal behavior for use in the classroom

Learning the Skills of Research:
Animal Behavior Exercises in the Laboratory and Field

Edited by Elizabeth M. Jakob and Margaret Hodge

Students learn best about the process of science by carrying out projects from start to finish. Animal behavior laboratory classes are particularly well-suited for independent student research, as high-quality projects can be conducted with simple materials and in a variety of environments. The exercises in this electronic lab manual are geared to helping students learn about all stages of the scientific process: hypothesis development, observing and quantifying animal behavior, statistical analysis, and data presentation. Additional exercises allow the students to practice these skills, with topics ranging from habitat selection in isopods to human navigation. Both student and instructor documentation is provided. Data sheets and other supplementary material are offered in editable formats that instructors can modify as desired.

Animal Behavior

TENTH EDITION

1

An Introduction to Animal Behavior

 The discipline of animal behavior is growing rapidly, thanks to thousands of behavioral biologists who are exploring everything from the genetics of bird song to why women find men with robust chins attractive. A major reason why the field is so active and broad ranging has to do with a book published over 150 years ago, Charles Darwin's *On the Origin of Species*.[8] As soon as it appeared, scientists realized that the evolutionary theory of natural selection provided a revolutionary way of looking at all living things. Darwin's influence continues strongly to this day, which is why this book is entitled *Animal Behavior: An Evolutionary Approach*.

Knowing that animal behavior, like every other aspect of living things, has a history guided by natural selection is hugely important. An understanding of evolutionary theory means that we have a scientific starting point when we set out to determine why animals do the things they do—and why they have the genetic, developmental, sensory, neuronal, and hormonal mechanisms that make these behavioral abilities possible. As the evolutionary biologist Theodosius Dobzhansky said long ago, "Nothing in biology makes sense except in the light of evolution."[11]

I hope to explain why Dobzhansky was right. If I succeed, my readers may come to understand the appeal of the evolutionary approach to animal behavior, which helps scientists identify interesting subjects worthy of explanation, steers them toward hypotheses suitable for testing, and produces solid conclusions about the validity of these hypotheses.

Charles Darwin's study in Down House where he developed the theory of evolution by natural selection, the foundation for the modern study of animal behavior.

The Behavioral Ecology of a Digger Bee

I put an evolutionary approach to animal behavior to work after I moved to Arizona in the early 1970s and began to explore the desert near my hometown of Tempe. At that time, on the advice of a colleague, I drove out to the Blue Point Bridge where a sandy floodplain created by the Salt River had been colonized by a forest of mesquites. As I wandered among the scattered mesquites and paloverdes bordering the river, I came to a large open area where hundreds, perhaps thousands, of large gray bees were cruising noisily close to the ground. Although I had been trained as an ornithologist, I had also learned that insects are delightful, so I stopped to admire the bees as they zoomed this way and that.

Soon I noticed a bee that had landed on the ground and was digging energetically, using its jaws to loosen the sandy soil and its hindlegs to propel the debris away from the depression it was creating (Figure 1.1). The bee paid no attention to me as I came closer but continued to dig as if its life depended on it. I had read about female bees that tunnel into the ground, creating a nest burrow with chambers for their offspring and the provisions that the larvae will feed upon. Therefore I assumed that I had stumbled upon a place where a great many females were searching for nesting sites and that the bee in front of me was a female in the process of burrowing underground.

But my tentative explanation for the bee's behavior received a jolt when the digger bee stopped digging after a few minutes and drew back slightly in the shallow pit that it had dug. Shortly thereafter another bee scrambled out of the opening underneath the waiting gray bee. The ex-digger immediately clambered onto the back of the emerging bee, with which it mated very quickly (Figure 1.2A). The two bees then flew off together, the one above holding on to its partner, before coming to rest on a nearby mesquite (Figure 1.2B). As the lower bee (which was much darker than its mate) clung to a branch tip, the upper bee proceeded to stroke its partner with its middle pair of legs and antennae. When I came closer, I could hear a faint intermittent buzzing that occurred in rhythm with the leg strokes and antennal taps.

I was intrigued. The digger bee was obviously a male, not a female, and he, and others like him that I observed subsequently, clearly possessed the ability to detect emerging females below the soil's surface. How on earth could a

FIGURE 1.1 A bee, *Centris pallida*, digging in the ground. Photograph by the author.

(A)

(B)

FIGURE 1.2   Sexual behavior of _Centris pallida._ (A) A male copulating with a female that he had discovered before she emerged from the ground. (B) After the female was inseminated, the pair flew to a nearby mesquite tree, where the male engaged in postcopulatory courtship with his partner. Photographs by the author.

male bee know exactly where a female was tunneling upward so that he could meet her by digging down through a centimeter or two of hard-packed soil? And why had males of this species (which I later learned was called _Centris pallida_) acquired this extraordinary ability?

With respect to the first question, I guessed that digging males could somehow smell or hear females that had burrowed up close to the surface. You can imagine, perhaps, how I might have determined which of these speculations was correct, research that required me to bury dead female bees in an area being searched by patrolling males.[1]

But although I was interested in the sensory abilities that enabled male bees to find emerging females, I also wanted to learn why male bees of this species searched for and dug down to meet these females. I did not know a lot about male bee behavior at the time, but I did know that males of many bee species mated after finding receptive females at flowers. In "my" bee, males could find adult females that had yet to fully emerge. There is a puzzle here. Since finding newly metamorphosed females is only one of several mate-locating tactics used by sexually motivated male bees, why had males of _Centris pallida_ settled upon the "find buried females" routine?

When behavioral ecologists, like me, ask questions of this sort about the behavioral abilities of an animal species, they are inspired by Darwin's theory of evolution by **natural selection**, which argues that living species are the product of an unguided, unconscious process of reproductive competition among their ancestors. As Darwin explained, if in the past some individuals left more descendants than others because of their distinctive hereditary attributes, then these reproductively dominant individuals would inevitably gradually reshape their species in their image. So, if males of a proto–_Centris pallida_ varied in their mate-finding behavior, and if males could transmit their particular tactic to their descendants, then over time, the one behavior that was most successful in helping individuals pass on their hereditary makeup would come to dominate the species.

The logic of natural selection is such that evolutionary change is inevitable if just three conditions are met:

1. _Variation,_ with members of a species differing in some of their characteristics

2. *Differences in reproductive success,* with some individuals having more surviving offspring than others in their population, thanks to their distinctive characteristics

3. *Heredity,* with parents able to pass on some of their distinctive characteristics to their offspring

If there is hereditary variation within a species (and there almost always is) and if some hereditary variants consistently reproduce more successfully than others, then the increased abundance of living descendants of the more successful types will gradually change the species. The "old" population evolves into one whose members possess the traits that were associated with successful reproduction in the past.

Because the process that causes evolutionary change is natural, Darwin called it natural selection. Darwin not only laid out the logic of his theory clearly but also provided abundant evidence that hereditary variation is common within species and that high rates of mortality are also the rule. Thus, alternative forms within a species are forced into an unconscious competition to be among the relatively few survivors. In other words, the conditions necessary and sufficient for evolutionary change by natural selection are present in all living things, a point that Darwin demonstrated by showing that people could cause dogs and pigeons to evolve by selectively breeding those individuals with hereditary traits that the breeders wanted in future generations of their domesticated animals.

Therefore, *for the purpose of testing evolutionary ideas,* we can assume that whatever trait exists today must have "won" a reproductive competition that took place in the past. If the assumption is wrong, our tests, if they are fair, will reveal this point. If the assumption is correct and the trait did win out over time, then we are dealing with an **adaptation**. Figuring out exactly how a putative adaptation contributes to the reproductive success of individuals is the central goal of behavioral ecologists, some of whom are happy to be known as **adaptationists**. Whatever the label, I believe that Tim Birkhead speaks for behavioral biologists in general when he writes, "The best thing of all about being a behavioural ecologist is that one's enthusiasm for the natural world actually increases over time. The more we discover, the more we discover that there is still more to discover."[4]

Certainly Darwin would have agreed, especially with respect to the hereditary foundation of evolutionary theory. Although Darwin himself knew nothing about genes, we now can reconfigure his argument to deal with selection at the level of the gene. Just as hereditary adaptations that increase the reproductive success of individuals will spread through populations over time, so too the genetic basis for these attributes will increase in frequency. Because **genes** can be present in populations in different forms, known as **alleles**, those alleles that contribute to the development of traits linked to individual reproductive success will become more common over time; those associated with reproductive failure will eventually disappear.

Notice that natural selection, whether acting on individual variation or genetic variation, is not guided by anything or anyone. Selection is not "trying" to do anything. Instead, the better reproducers cause a species to evolve.

Notice also that the only kinds of hereditary characteristics that will become more common in a species are those that promote individual reproductive success, which do not necessarily benefit the species as a whole. Although "for the good of the species" arguments were often made by biologists not so long ago, it is entirely possible for adaptations (and particular alleles) to spread through populations even if they do nothing to "perpetuate the species." Indeed, traits and alleles can be naturally selected that are harmful to group survival in the long run.

Discussion Question

1.1 In order to explain how the blind process of natural selection—a process dependent on random events (mutations)—can generate complex adaptations, Richard Dawkins invites us to imagine that an evolved attribute is like an English sentence—like a line from Shakespeare's *Hamlet*, such as, METHINKS IT IS LIKE A WEASEL.[9] The odds that a monkey would produce this line by tapping at a typewriter are vanishingly small, one in 10,000 million million million million million million (1 in 10^{40}). These are not good odds. But instead of trying to get a monkey or a computer to get the "right" sentence in one go, let's change the rules so that we start with a randomly generated letter set, such as SWAJS MEIURNZMMVASJDNA YPQZK. Now we get a computer to copy this "sentence" over and over, but with a small error rate. From time to time, we ask the computer to scan the list and pick the sequence that is closest to METHINKS IT IS LIKE A WEASEL. Whatever "sentence" is closest is used for the next generation of copying, again with a few errors thrown in. The sentence in this group that is most similar to METHINKS ... WEASEL is selected to be copied, and so on. Dawkins found that this approach required only 40 to 70 runs (generations) to reach the target sentence. What was Dawkins's main point in illustrating what he called cumulative selection? In what sense is this example not a perfect analogy for natural selection?

Natural selection theory helps biologists identify aspects of living things worth studying. If you are aware of the way in which natural selection works, then traits that appear to reduce rather than raise an individual's reproductive success are surprising. We will call these challenges to evolutionary theory **Darwinian puzzles** (and we will call your attention to them with an icon of an ant on a puzzle piece). Biologists deal with these puzzles by developing possible explanations based on natural selection theory for how the surprising trait might actually help individuals reproduce and pass on their genes. For example, rather than proposing ideas about how male digger bee behavior benefits the species as a whole, the behavioral ecologist tries to come up with one or more explanations for how male digging behavior might promote the reproductive success of individual males. Males of *Centris pallida* that could find and dig down to hidden virgin females would seem to have an advantage in mating with those females when they crawled out of their emergence tunnel. The male's advantage would be particularly great if, once having mated, the freshly emerged female was no longer receptive.[1]

Discussion Question

1.2 In experiments in which parent birds are given extra nestlings to rear, the adults usually rear larger numbers of youngsters to fledging than they do naturally.[30] Why does this finding pose a Darwinian puzzle? How might the possibility that egg production and incubation require considerable effort[15,24] help an adaptationist develop hypotheses that may resolve the puzzle?

Darwinian hypotheses are testable because they often can be used to produce predictions that can then be checked against reality. In the case of the bee, we can predict that (1) given the males' enthusiasm for virgins, a *Centris pallida*

FIGURE 1.3   A pair of native bees belonging to the genus *Perdita* copulating on a desert poppy. This species of *Perdita* is just one of many bee species that use flowers as a rendezvous site for mating. Photograph by the author.

female either mates just once in her life or uses the sperm of her first partner to fertilize her eggs. We can also predict that (2) in other bee species in which females remain sexually receptive after mating and do not give male number one a fertilization advantage, males will be far less likely to search for virgin females than males of *Centris pallida*. In addition, we would expect that (3) males of *Centris pallida* should not let fellow males gain access to receptive females if they can prevent it.

Notice that all three predictions or expectations are based on the assumption that males are in a race to inseminate as many females as possible while other males are trying to do the same thing. And, as a matter of fact, (1) females of *Centris pallida* almost certainly do mate just once, judging from the fact that nesting and flower-visiting females are rarely, if ever, pursued or contacted by sexually motivated males. Moreover, (2) in other bee species in which females do mate several times, males often meet their mates at pollen- and nectar-producing flowers rather than at their emergence sites (Figure 1.3). Finally, (3) males of *Centris pallida* are indeed highly aggressive in defense of digging sites and they sometimes lose to other males, often larger ones, that displace them from spots where virgin females are about to emerge (Figure 1.4).

Discussion Question

1.3   Someone proposes that the reason males of *Centris pallida* often fight for access to females is to ensure that only the best males, the largest and most physiologically competent individuals, the ones with the best genes, get to mate. Is this hypothesis based on natural selection theory? Why or why not? How would you test the idea?

FIGURE 1.4   Males of *Centris pallida* fighting. These individuals are struggling to control a spot from which a female is about to emerge. Photograph by the author.

FIGURE 1.5 Hanuman langur females and offspring. Males fight to monopolize sexual access to the females in groups like this one.

Natural Selection and Infanticide

Let us apply natural selection theory to another case, which came to notice in a study of Hanuman langurs. These monkeys live in groups (Figure 1.5) of several females and their offspring accompanied by one or a few adult males. In the course of this research, male Hanuman langurs were seen attacking and even killing the very young infants of females in their own group (Figure 1.6A). The puzzle here is obvious: how can it be adaptive for a male langur to harm the offspring of females in his group, particularly since attacking males can be and sometimes are injured by mothers defending their babies (Figure 1.6B)? Indeed, some primatologists have argued that the very unpleasant

(A)

FIGURE 1.6 Male langurs commit infanticide. (A) A nursing baby langur that has been paralyzed by a male langur's bite to the spine (note the open wound). This infant was attacked repeatedly over a period of weeks, losing an eye and finally its life at age 18 months. (B) An infant-killing male langur flees from an aggressive protective female belonging to the band he is attempting to join. A, photograph by Carola Borries; B, photograph by Volker Sommer, from Sommer.[28]

(B)

FIGURE 1.7 A male lion carrying a cub that he has killed. The conditions under which infanticide occurs in this species are very similar to those associated with male infanticide in Hanuman langurs. Photograph by George Schaller.

infanticidal behavior of these males was not adaptive but was instead the aberrant aggressive response by males to the overpopulation and crowding that occurred when langurs came together to be fed by Indian villagers. According to these observers, overcrowding caused abnormal aggressive behavior.[7] But the behavioral ecologist Sarah Hrdy used natural selection theory to try to solve the puzzle of infanticide in a different way, namely by asking whether the killer males were behaving in a reproductively advantageous manner.[19] By committing infanticide, the males might cause the baby-less females to resume ovulating, which otherwise would not happen for several years in females that retain and nurse their infants.

Hrdy tried to explain how infanticide might have spread through Hanuman langur populations in the past as a reproduction-enhancing tactic for individual males. Hrdy's potential explanation for the evolution of the behavior leads to a number of expectations, of which the most important is the prediction that males will not kill their own progeny but will focus their attacks on the offspring of other males. This prediction in turn generates the expectation that infanticide will be linked to the arrival of a new male or males into a band of females, with the associated ejection of the father or fathers of any baby langurs in the group. In cases of this sort, the new males could father offspring more quickly if they first killed the infants in the band. Females who lose their infants do resume ovulating, and that enables the new males in the band to become fathers of their replacement offspring. Since these predictions have been shown to be correct for this species[5] as well as some other primates,[3,23] various carnivores (Figure 1.7), horses, rodents, and even a bat,[20] we can safely conclude that infanticide as practiced by male langurs is indeed an adaptation, the product of natural selection.

Discussion Question

1.4 Because the reproductive success of female langurs is almost certainly lowered when a newly installed male kills their young infants, selection should favor countermeasures against infanticidal males. In this light, why might already pregnant females mate with a new male soon after a takeover even though they are not ovulating? What significance do you attach to the discovery that when mares are impregnated by stallions at a stable away from their home location, they will also copulate repeatedly with the males in their home stables upon their return?[2]

The *Science* of Behavioral Biology

Let me emphasize that although many persons think that *adaptive* means "good" or "desirable," in evolutionary terms *adaptive* only means "reproductively advantageous for individuals" or "beneficial for the genes that underlie the development of the trait." When Hrdy proposed that infanticide might be adaptive, she was of course not trying to justify in moral terms the behavior of killer males. Thoroughly cruel and ugly behaviors can evolve by natural selection if they happen to enhance the ability of individuals to pass on their genes to the next generation. Infanticidal males that cause some other males' offspring to die are behaving adaptively if they are not themselves seriously injured by the mothers and if the mothers of the deceased offspring come into estrus and mate with the killer males. Often both conditions apply, which is why Hrdy and others have concluded that langur infanticide is an adaptation exhibited by some males under some conditions.

In contrast, an explanation for infanticide based on the claim that overcrowding leads to hyperaggression with incidental attacks on infants produces a prediction that has not been confirmed by additional study. Thus adult male langurs living in natural areas away from villagers that supply them with extra food still exhibit infanticide, which contradicts the idea that infanticide is an aberrant behavior linked with unnatural conditions.

Behavioral ecology and the other equally important components of behavioral biology are scientific disciplines. By this I mean that researchers in these fields use a particular kind of logic to evaluate potential explanations for puzzling phenomena. We used this logic to decide that male digger bees search for and dig down to reach emerging females because this behavior helps them inseminate virgin females, the only kind of females that will mate with them and produce their offspring.

This explanation can be tested by identifying what a researcher must be able to observe if the hypothesis is true. If it is, virgin females will mate with their discoverers but foraging females will not be sexually receptive. When several predictions derived from a hypothesis are shown to be true, the hypothesis gains credibility; if, on the other hand, the expected results do not materialize, the hypothesis loses support. Well-tested hypotheses are the basis for sound scientific conclusions, which take the form of "this idea is probably right" or "this potential explanation is almost surely wrong."

The Power of Scientific Logic

Scientists often pat themselves on the back for using what they believe is a very effective means for solving puzzles about the natural world. But at least on some issues, the general public is far more skeptical. An illustrative example involves the current question of what is causing the Earth's climate to warm up. Almost all climate scientists have claimed that by burning fossil fuels, humans are mainly responsible for the recent rise in the temperature of the atmosphere. But according to a poll by the Pew Research Center (http://pewresearch.org) released on October 22, 2009, only a third of all citizens of the United States believe that human activity is responsible for climate change.

Some skeptics think that climate scientists are fudging the data in order to get grants for themselves and their colleagues. This strategy would, however, require that other climate scientists refrain from commenting on the fraudulent behavior of certain of their fellow scientists. Such restraint is unlikely given that scientists are competitive people who review the ideas of their colleagues at every step of the process, starting right from the beginning when they rate the grant proposals of their fellow scientists. Should a study become funded (and many do not) and should it lead to published research articles, the conclusions presented in these articles might even then be examined skeptically because of the benefits to academics of demonstrating that so-and-so's conclusion is wrong. The successful critic gets a publication or two on this point as well as the social esteem that comes from being known as someone who gets things right. As a result, science tends to be self-correcting, not because scientists are selfless saints but precisely because researchers are competitive individuals who strive for high social status in the community of their peers.

Yes, sometimes the pressures to succeed in academia lead researchers to see what they want to see from their studies. A researcher can mistakenly claim to have secured a result that the data do not support. Or a scientist may ignore an inconvenient finding or reject it outright on spurious grounds. Some scientists, including well-known ones, have even engaged in outright fraud. A fairly recent case involved a stem cell researcher, Dr. Hwang Woo-suk, found guilty by a South Korean court of having manufactured data that appeared in two papers on embryonic human stem cells (see the Reuters online news report [www.reuters.com] entitled "South Korean stem cell scientist accused of fraud"). Before the accusations against Hwang had been aired, these papers had been considered major breakthroughs in the effort to devise methods to treat Alzheimer's disease and the like.

Note, however, that the reason we know that scientists are capable of self-deception and fraud is largely from the work of other scientists, who have gained social rewards from ferreting out mistakes and identifying scientific malfeasance when it occurs. The skepticism of scientists and their willingness to critically examine and reexamine the findings of others means that by the time a scientific consensus is reached, it is likely to have been thoroughly vetted. A scientific consensus has been achieved on the global warming problem.[14] Eventually, perhaps most of us will accept that the Earth is warming up and that people (and their machines) are the culprits. We should, since most of us agree that airplanes are safe to fly, light switches generally work, cornfields tend to produce more corn than ever, computers do what they are supposed to do, medical procedures and drugs actually cure illnesses, and so on—all matters that have been established via scientific consensus. Indeed, almost every aspect of modern life, from agriculture to zoo management, is based on scientific research, which involves many rounds of hypothesis testing and critical review before a position becomes widely accepted.

This book presents a sampler of the conclusions reached by behavioral researchers—along with the test evidence that underlies these conclusions. One of my major goals is to encourage my readers to realize how scientists evaluate hypotheses in ways that are generally considered fair and logical (at least by other scientists). The cases reviewed in the chapters ahead have been selected with this goal in mind. Chapter 2 takes up what many consider to be the premier puzzle for behavioral ecologists, namely the evolution of self-sacrificing behavior in social insect colonies, a puzzle that continues to generate scientific review and debate. Chapter 3 examines the adaptive value of the social behavior of animals other than social insects. Chapter 4 looks at the evolution of communication signals. Chapters 5 and 6 are dedicated to the adaptationist analysis of antipredator behavior, foraging behavior, and habitat selection. Chapters 7, 8, and 9 explore the behavioral ecology of various elements of reproductive and parental behavior. Chapter 10 introduces the concept of the proximate (immediate) causes of behavior and how these causes differ from the ultimate (evolutionary) causes that Chapters 1–9 examine. Chapters 11, 12, and 13 then consider the evolutionary basis of the proximate developmental and physiological mechanisms that underlie adaptive behavior. Finally, Chapter 14 offers an example of how both proximate and ultimate research contribute to an understanding of the causes of our language skills; the chapter also looks at the reproductive strategies of our own species. Let's get started.

Summary

1. Evolutionary theory provides the foundation for behavioral biology, the study of animal behavior.

2. Charles Darwin realized that evolutionary change would occur if "natural selection" took place. This process happens when individuals differ in their ability to reproduce successfully, as a result of their inherited attributes. If natural selection has shaped animal behavior, we expect that individuals will have evolved abilities that increase their chances of passing copies of their genes on to the next generation.

3. Researchers interested in the adaptive value of behavioral traits use natural selection theory to develop particular hypotheses (tentative explanations) on how a specific behavior might enable individuals (not groups or species as a whole) to achieve higher reproductive success than individuals with alternative traits.

4. Adaptationist hypotheses can be tested in the standard manner of all scientific hypotheses by making predictions about what we must observe in nature, or in the outcome of an experiment, if a particular explanation is true. Failure to verify these predictions constitutes grounds for rejecting the hypothesis; the discovery of evidence that supports the predictions means the hypothesis can be tentatively treated as true.

5. The beauty of science lies in the ability of scientists to use logic and evidence to evaluate the validity of competing theories and alternative hypotheses.

Suggested Reading

Three books are essential for all students of behavior: *On the Origin of Species* by Charles Darwin[8] (you will benefit from James Costa's annotated version of this book[6]), *Adaptation and Natural Selection* by George C. Williams,[31] and *The Selfish Gene* by Richard Dawkins.[10] Michael Le Page provides links explaining what is wrong with the common thinking about evolution in his article "Evolution: 24 myths and misconceptions."[21] For a provocative essay on the nature of science itself, read Woodward and Goodstein's article "Conduct, misconduct, and the structure of science."[32] Excellent books by behavioral biologists that capture the pleasures of field research include classics by Niko Tinbergen,[29] Konrad Lorenz,[22] George Schaller,[26,27] and Howard Evans.[12,13] Many delightful accounts of this sort have been written more recently, too many to list here, but for what it is worth, I especially like the work of Bernd Heinrich,[16,17] Craig Packer,[25] and Bert Hölldobler.[18]

2

Behavioral Ecology and the Evolution of Altruism

 Not so long ago, I traveled to the Pantanal, a vast network of marshes and grasslands located largely in southwestern Brazil. As I followed my host and guide, Lucas Leuzinger, he pointed out narrow sandy trails cutting through the scruffy grassland in which we were walking. Hurrying along these straight and narrow paths were large numbers of pale reddish termites, some of which were carrying bits of dried vegetation back to their colony's home, while others were larger, darker in color, and armed with formidable jaws (Figure 2.1). We were looking at two kinds of termites that belonged to the same species. The smaller workers were carrying food items that would provide their fellow termites back at the colony with calories and nutrients. The workers were accompanied by soldiers, which were there to defend the workers from ants and other small predators that attack termites. Amazingly, none of the workers or the soldiers in the colony would ever reproduce. Instead they would spend their entire lives working for others, generally one large reproducing female and her much smaller male partner, the queen and king of the colony, as well as a new generation of brothers and sisters that would eventually try to found colonies of their own.[56]

There are several thousand species of termites, some of which form colonies of millions of individuals and build immense durable structures of packed earth that take your breath away (Figure 2.2). All of these groups, large or small, are composed mostly of sterile individuals that do not reproduce but are willing and able to build the colony's home, to forage for the resources needed to feed their fellow termites, and to care for the queen and king as well as the

These sterile worker weaver ants labor together to make leaf nests for the reproductive benefit of other ants. Why?

FIGURE 2.1 **FIGURE 2.1   A group of worker termites escorted by a single large soldier back to their colony.** None of the termites will ever reproduce—a Darwinian puzzle. Photograph by the author.

eggs laid by the queen, then tend the young immature termites that hatch from those eggs.

The Darwinian puzzle provided by sterile castes of termites should be obvious: how can individuals evolve if they are unable to reproduce given that hereditary differences in individual reproductive success are necessary for natural selection to occur? And this puzzle applies to more than the termites because all ants, some wasps, and some bees also form groups with nonreproducing workers. Among these **eusocial** insects is the familiar honey bee. Indeed, most of us are more familiar with this species than we would prefer; when a honey bee worker stings us in defense of her hive mates, she dies. Her death comes about because her barbed stinger catches in our skin such that when the bee pulls away from us, her sting apparatus remains behind to pump toxins into our body (Figure 2.3).

The honey bee's suicidal sting is a classic example of self-sacrificing behavior that helps individuals other than the bee's own progeny (and honey bee workers almost never produce their own offspring).[37] There are other cases of this sort, such as worker ants that stay outside the nest to seal off the nest entrance in the evening, even though this activity means that they will die before morning.[4,45] Likewise, when colonies of the grenade ant are under attack, the workers break open their abdomens, spilling a gluey substance onto the intruders, a suicidal line of defense for the colony.[28]

In many social insects, the worker caste is composed of females with atrophied ovaries, which make them incapable of reproduction. In cases of this sort, the workers have totally lost the ability to reproduce personally and instead help other members of their colony reproduce (such as the queen or the

FIGURE 2.2   A huge number of tiny sterile termites built this immense home for their colony in Western Australia. The 3-meter tall mound probably contains just one reproducing queen and one king. Photograph by the author.

(A)

(B)

FIGURE 2.3 Sacrifices by social insect workers. (A) In a nasute (pointed-nose) termite colony, soldiers incapable of reproducing attack intruders to their colony at great personal risk and spray these enemies with sticky repellents stored in glands in their heads. (B) When a honey bee stings a vertebrate (but not other insects), she dies after leaving her stinger and the associated poison sac attached to the body of the victim. A, photograph by the author; B, photograph of Bernd Heinrich's knee by Bernd Heinrich.

future kings and queens produced by the colony). Self-sacrifice of this sort is what behavioral ecologists refer to as **altruism**, and it was for a long time very difficult to see how reproductive suicide could spread by natural selection. In fact, for most behavioral biologists the widespread occurrence of altruism represented the premier Darwinian puzzle of them all, which is why we will tackle this phenomenon now.

Explaining Altruism: Intelligent Design?

Because sterile castes are very hard to explain, the colonies of eusocial insects have attracted much attention. Among the large group of scientists interested in social insects are a few adherents of **intelligent design theory**. Advocates of intelligent design assert that the extraordinary complexity of social insect colonies is the kind of thing that cannot be explained by evolutionary theorists. Therefore, so the argument goes, social insect societies, which look as if they had been designed to serve some function, really have been formed by some sort of intelligent entity for the purpose of working together efficiently. According to this view, the several castes of termites cooperate within their colony in much the same way that a watch has many interconnected parts, all of which have to interact in a particular way if the watch is to be an accurate timepiece. A watch requires an intelligent designer, a human; by the same token, a termite or ant colony supposedly also requires an intelligent designer.

Richard Dawkins destroys this analogy in his book *The Blind Watchmaker*.[9] But for our purposes let's focus on the fact that the "intelligent designer" that intelligent design theorists have in mind must be a supernatural being of some sort. We can say this with considerable confidence, in part because a federal judge, John Jones, ruled in 2005 that intelligent design theory is a religious theory, not a scientific one, and as such, intelligent design theory is intended to be taken on faith and not tested in the manner of a scientific theory or hypothesis.[17] Indeed, intelligent design advocates never use their theory to specify exactly what complex adaptations are designed to do. Without a testable hypothesis, science is not possible.

Judge Jones's decision has significance for education policy because many school boards, like the one in Dover, Pennsylvania, where the judge delivered his ruling, have tried to present intelligent design theory in high school science classes. Judge Jones said, "No."[17] Religion is not science and therefore scientists need not concern themselves with intelligent design theory, except as it poses a threat to legitimate science education in our schools.

Discussion Question

2.1 Let's say that intelligent design theorists are correct in claiming that certain features of living things, such as termite societies composed of 20 million cooperating individuals or the long and convoluted biochemical pathways involved in the production of critical cellular compounds, are so complex that they cannot be explained by evolutionary theorists. Why would Darwinists aware of the logic of the scientific method still be justified in rejecting the assertion that these cases in themselves constitute reason to accept intelligent design theory?

Altruism and For-the-Benefit-of-the-Group Selection

The modern scientific analysis of altruism really began in 1962 when V. C. Wynne-Edwards published his book *Animal Dispersion in Relation to Social Behaviour*. Wynne-Edwards did everyone a favor by formally proposing that social attributes, including altruism, had evolved to benefit the group or species as a whole.[58] According to his theory of **group selection**, groups or species with self-sacrificing (altruistic) individuals are more likely to survive than those without altruists, leading to the evolution of group-benefiting altruism.

A classic example of the application of group selection theory to produce a hypothesis involves the territorial behavior of many birds in which territory holders breed but those unable to secure territory do not (Figure 2.4). Wynne-Edwards explained the behavior of nonterritorial individuals as group-benefiting altruism, with these birds holding back in order to avoid depleting the food supply on which the entire species depends for its long-term survival. By much the same token, the sterile castes of social insects might refrain from reproducing in order to help their colonies persist in environments in which resources were difficult to secure.

Soon after Wynne-Edwards's book was published, George C. Williams challenged this for-the-benefit-of-the-group type of selection in *Adaptation and Natural Selection*,[52] arguably the most important book on evolutionary theory written since *The Origin of Species*. Williams showed that the persistence of a hereditary trait, and the genes underlying the development of that trait, were much more likely to be determined by differences in the reproductive success of genetically different individuals than by survival differences among genetically different groups. We can illustrate his point with reference to Hanuman langurs. Imagine that in the past there really were male langurs prepared to risk serious injury, even death, in order to attack infants, an action that would reduce the numbers in their band and conceivably promote the survival of the group. In such a case, group selection would be said to favor male infanticide because the group as a whole would benefit from the removal of excess infants.

But in this species, Darwinian natural selection would also be at work, provided that at any time there were two genetically different kinds of males—

(A)

FIGURE 2.4   The territories of an Australian songbird. (A) The southern emu-wren. (B) A diagram of emu-wren territories over 3 years at one location. Roads appear as dark lines; permanent bodies of water are shown in blue. Although the exact locations of the territories change somewhat from year to year, only pairs with exclusive territories reproduce in this species and most other birds as well. A, photograph by Geoff Gates; B, after Maguire.[26]

(B)

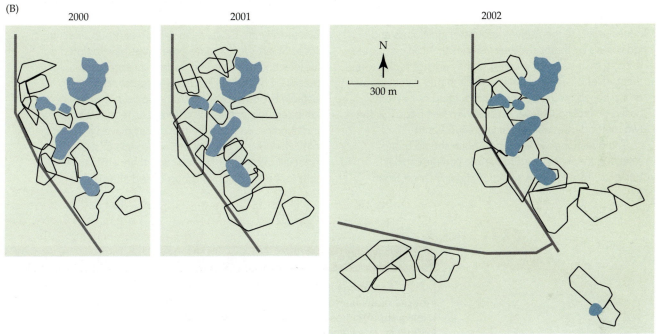

one that practiced infanticide for the good of the group and another that "permitted" other males to pick up the tab for population reduction. If the nonkillers lived longer and reproduced more, which of these two types would become more common in the next generation? Whose hereditary material would increase in frequency over time? Would infanticide long persist in such a population of langurs?

This kind of thought experiment, really a test of the logic of group selection theory, convinced Williams and nearly all his readers that Darwinian selection acting on differences among individuals within a population or species will usually have a stronger evolutionary effect than group selection acting on differences among entire groups. If group selection favors a trait that involves reproductive self-sacrifice while natural selection acts against it, natural selection seems likely to trump group selection,[52] as we have just seen in our hypothetical langur example. There have been attempts to develop more convincing forms of group selection theory, but they have failed to persuade the vast majority of behavioral ecologists (Box 2.1).

BOX 2.1 Altruism and group selection

Ever since the debate between Wynne-Edwards and Williams, some scientists, especially David Sloan Wilson, have argued that it is still possible that group selection might occur under some conditions.[12,53] Although the term *group selection* has been applied to many different evolutionary processes,[51] there is little doubt that competition among genetically different groups can have genetic consequences over time, thereby affecting the course of evolution. So, if you wish, you can calculate the contribution of alleles to the next generation by quantifying the changes in allele frequencies within groups rather than by adding up the genes passed on by the individual members of that species. But only genes replicate themselves; groups of individuals and individuals by themselves are vehicles that can contribute to the transmission of genes, but they are not replicators themselves.[5,10] As a result, it doesn't matter if you use a group-based method of gene accounting or a system based on individual-level kin selection (see page 23); the two methods are mathematically equivalent.[27,39]

The question then becomes, when should group selection theory be used to explain complex social behavior, instead of kin selection theory? David Sloan

Wilson and others (going back to Wynne-Edwards) have argued that because social insects live in colonies, they are ideal candidates for an analysis based on some form of group selection.[55] Here the basic argument is again that colonies with more self-sacrificing individuals will be favored by group selection if groups with more altruists outcompete rival groups and so contribute more genes to the next generation. But most students of social insects and other social animals use the alternative (but mathematically equivalent) perspective derived from kin selection theory.

David Sloan Wilson has complained that the preference for kin selection theory stems from prejudice against the rejected Wynne-Edwardsian version of group selection.[54] But as Stuart West and his colleagues have pointed out,[51] kin selection theory is very widely accepted primarily because it has helped so many researchers develop testable hypotheses for puzzling social behaviors exhibited by creatures as different as termites, bacteria, and humans. Because the theory has played a major role in helping us understand all aspects of sociality, especially altruism,[42] kin selection theory is featured here to the exclusion of group-based alternatives.

Discussion Questions

2.2 Lemmings are small rodents that live in the Arctic tundra. Their populations fluctuate wildly. At high population densities, large numbers leave their homes and travel long distances, during which time many die, some by drowning as they attempt to cross rivers and lakes. One popular explanation for their behavior is that the travelers are committing suicide to relieve overpopulation. By heading off to die, the suicidal lemmings leave shelter and food for those that stay behind. These surviving individuals will perpetuate their species, saving it from extinction. What theory was used to produce this hypothesis? How would George C. Williams use Gary Larson's cartoon (Figure 2.5) to illustrate the logical problem with group benefit selection?

2.3 In some ant species, several unrelated females may join forces to found a colony after they have mated. The females cooperate in digging the nest and producing the first generation of workers, but then they start fighting until only one is left alive. The survival rate of colonies founded by lone females is very low.[1] How might you interpret the behavior of multiple founders as the product of for-the-good-of-the-group selection? If the behavior of the several queens is the product of natural selection, not Wynne-Edwardsian group selection, what prediction can you make about the interactions between the females during the colony establishment phase prior to the fight-to-the-death phase?

Altruism and Indirect Selection

So, if we are to explain the altruism exhibited by sterile worker termites or ants in Darwinian terms, we have to produce a hypothesis in which the apparent self-sacrificing behavior of these workers actually has an evolutionary benefit for them as individuals. But what could this be if workers and soldiers do not reproduce?

In *The Origin of Species*, Darwin confronted several problems for his theory of natural selection illustrated by insect societies, one of which was the puzzle of worker sterility.[38] Darwin saw a fairly simple solution to this puzzle in the form of an analogy with domesticated cattle. Although you cannot breed the tastiest steers, because they are dead, you can let the parents (or other relatives) of the butchered Black Angus or Herefords with desirable attributes remain in the breeding stock even as their offspring are shipped off to the abattoir. Those that are slaughtered are the equivalent of workers, in that they fail to reproduce, but their hereditary attributes (nicely marbled beef, efficient conversion of grass into muscle, and the like) can persist if their parents go on to produce the next generation of breeders as well as the sacrificed animals.

William D. Hamilton reshaped this argument using an understanding of heredity far superior to that available to Darwin.[18] His theory was based on the premise that individuals reproduce with the unconscious goal of propagating their distinctive family genes more successfully than other individuals. Personal reproduction contributes to this ultimate goal in a direct fashion because parents and offspring share genes in common. But helping genetically similar individuals other than offspring—that is, one's nondescendant relatives—survive to reproduce can provide an indirect route to the very same end (Figure 2.6). It does not matter which bodies are carrying certain genes; a trait becomes more common if the allele that promotes the development of that attribute becomes more common.

FIGURE 2.5 Gary Larson's cartoon of presumably suicidal lemmings headed into the ocean. Note the lemming with the inner tube.

FIGURE 2.6 How to achieve indirect fitness. By adopting two nephews and a niece, this woman could be propagating her genes indirectly because relatives share rare family alleles with one another. (Here we assume that the children would not have survived or reproduced as much had they not been adopted.)

To understand why, the concept of the **coefficient of relatedness** is critical. This term refers to the probability that an allele in one individual is present in another because both individuals have inherited it from a recent common ancestor. Imagine, for example, that a parent has the genotype *Aa*, and that *a* is a rare form of the *A* gene, which is to say that *A* and *a* are two alleles of the gene in question. Any offspring of this parent will have a 50 percent chance of inheriting the *a* allele because any egg or sperm that the parent donates to the production of an offspring has one chance in two of bearing the *a* allele. The coefficient of relatedness (*r*) between parent and offspring is therefore 1/2, or 0.5.

The coefficient of relatedness varies for different categories of relatives. For example, a man and his sister's son have one chance in four of sharing an allele by descent because the man and his sister have one chance in two of having this allele in common, and the sister has one chance in two of passing that allele on to any given offspring. Therefore, the coefficient of relatedness for a man and his nephew is $1/2 \times 1/2 = 1/4$, or 0.25. For two cousins, the *r* value falls to 1/8, or 0.125. In contrast, the coefficient of relatedness between an individual and an unrelated individual is 0.

With our knowledge of the coefficient of relatedness between altruists and the individuals they help, we can determine the fate of a rare "altruistic" allele that is in competition with a common "selfish" allele. The key question is whether the altruistic allele becomes more abundant if its carriers forgo reproduction and instead help relatives reproduce. Imagine that an animal could potentially have one offspring of its own or, alternatively, invest its efforts in the offspring of its siblings, thereby helping three nephews or nieces survive that would have otherwise died. A parent shares half its genes with an offspring; the same individual shares one-fourth of its genes with each nephew or niece. Therefore, in this example, personal reproduction yields $r \times 1 = 0.5 \times 1 = 0.5$ genetic units contributed directly to the next generation, whereas altruism directed at three relatives yields $r \times 3 = 0.25 \times 3 = 0.75$ genetic units passed on indirectly in the bodies of relatives. In this case, the altruistic tactic is adaptive because individuals with this characteristic pass on more of the altruism-promoting allele to the next generation than individuals with an alternative hereditary trait.

Discussion Question

2.4   If an altruistic act increases the genetic success of the altruist, then in what sense is this kind of altruism actually selfish? In everyday English, words like *altruism* and *selfishness* carry with them an implication about the motivation and intentions of the helpful individual. Why might everyday usage of these words get us into trouble when we hear them in an evolutionary context? If an individual just happened to help another at reproductive cost to itself, should the behavior be called altruistic under the evolutionary definition of the term?

Another way of looking at this matter is to compare the genetic consequences for individuals who aid others at random versus those who help close relatives. If aid is delivered randomly, then no one form of a gene is likely to benefit the bearer more than any other, and the carrier of an altruism allele pays a price for the help that raises the **fitness**, that is, the number of genes contributed to the next generation, of individuals with other forms of the gene. But if close relatives aid one another preferentially, then any alleles they share with other family members may survive better, causing those alleles to increase in frequency compared with other forms of the gene in the population at large. When one thinks in these terms, it becomes clear that a kind of natural selection can occur

BOX 2.2 Key terms used by evolutionary biologists in the study of altruism

Altruism: Helpful behavior that lowers the helper's reproductive success while increasing the reproductive success of the individual being helped.

Direct selection: The process of natural selection that occurs when hereditarily distinctive individuals differ in the number of surviving offspring they produce or number of genes they pass on to subsequent generations.

Direct fitness: A measure of the reproductive or genetic success of an individual based on the number of its offspring that live to reproduce.

Kin (or indirect) selection: The process that occurs when hereditarily distinctive individuals differ in the number of nondescendant relatives (not their offspring) they help survive to reproduce.

Indirect fitness: A measure of the genetic success of an altruistic individual based on the number of relatives (or genetically similar individuals) that the altruist helps reproduce that would not otherwise have survived to do so.

Inclusive fitness: A total measure of an individual's contribution of genes to the next generation by direct and/or indirect selection.

Group selection: The process that occurs when groups differ in their collective attributes and the differences affect the survival chances of the groups.

Behavioral strategy: An inherited behavior pattern that is in competition with other hereditarily different behavior patterns in ways that have the potential to affect an individual's inclusive fitness. An example of a behavioral strategy is the willingness of individuals to assist close relatives even though their help reduces their direct fitness.

---

when genetically different individuals differ in their effects on the reproductive success of relatives. Jerram Brown called this form of selection **indirect selection** (now often called **kin selection**), which he contrasted with **direct selection** for traits that promote success in personal reproduction (Box 2.2).[7] Although there have been claims that kin (indirect) selection theory should be discarded in favor of an alternative approach,[32] few researchers agree, for reasons spelled out by Andrew Bourke in *Principles of Social Evolution*.[5]

Discussion Question

2.5   We can, if we wish, consider the coefficient of relatedness with respect to those genes on the X chromosome rather than a single gene or the complete genome. All women have two X sex chromosomes. Their sons and daughters each get one X from their mother. An adult son may or may not pass on that X to his children. If he does not and instead gives an offspring his Y chromosome, he will create a grandson (XY) for his mother. That grandson will not carry his grandmother's X chromosome. (Where then did his X chromosome come from?) What about the X chromosomes in the granddaughters of the paternal grandmother? What is the significance of the fact that the survival chances of grandsons with paternal grandmothers living nearby are much less than the odds of survival for granddaughters with paternal grandmothers nearby[15]—(several traditional societies provided the relevant data.) Does the paternal grandmother's effect on the survival of her granddaughters generate indirect selection or direct selection?

Kin Selection and Inclusive Fitness Theory

As I mentioned earlier, Darwin provided a tentative explanation for the evolution of sterile workers in termites and other social insects, namely that selection acting on other (reproductively competent) family members could result in the spread of self-sacrificing traits in the relatives of those reproduc-

FIGURE 2.7   **A foundress female wasp.** If this female paper wasp in the genus *Polistes* is fortunate, she will survive to produce daughters that will become workers, helping her generate still more offspring in the weeks ahead. Photograph by the author.

ers. If we use the modern form of Darwin's argument, namely the Hamiltonian kin selection approach, we can make a prediction about the makeup of termite colonies and those of the other social insects, which is that these groups will generally be composed of parents and offspring (Figure 2.7). This prediction is essentially correct.[41] Even colonies with multiple queens are groups composed of multiple families, so if the sterile workers of one family are helping in ways that benefit the entire colony, the fact that their reproducing relatives are there means that sterile individuals can also gain **indirect fitness** from the enhanced output of their reproducing relatives. (Moreover, queens that join others in the formation of a single colony are almost always relatives, a factor that increases the coefficient of relatedness among their families of offspring.)

Because fitness gained through personal reproduction (**direct fitness**) and fitness achieved by helping nondescendant kin (i.e., siblings) survive and reproduce (indirect fitness) can both be expressed in identical genetic units, we can sum up an individual's total contribution of genes to the next generation, creating a quantitative measure that can be called **inclusive fitness** (see Box 2.2). Note that an individual's inclusive fitness is not calculated by adding up that animal's genetic representation in its offspring plus that in all of its other relatives. Instead, what counts is an individual's own effects on gene propagation (1) directly in the bodies of its surviving offspring that owe their existence to the parent's actions, not to the efforts of others, and (2) indirectly via nondescendant kin that would not have existed except for the individual's assistance. Imagine that a typical diploid organism managed to rear one offspring to reproductive maturity while at the same time helping do the same for three siblings who would otherwise have been unable to raise offspring. In such a case, the direct fitness gained by the individual would be $1 \times 0.5 = 0.5$ whereas that individual's indirect fitness would be $3 \times 0.5 = 1.5$; the total or inclusive fitness of this animal would be $0.5 + 1.5 = 2.0$.

The concept of inclusive fitness, however, is typically used not to secure absolute measures of the lifetime genetic contributions of individuals, but rather to help us compare the evolutionary (genetic) consequences of two alternative hereditary traits.[36] In other words, inclusive fitness becomes important as a means to determine the relative genetic success of two or more competing hereditary behavioral traits or **behavioral strategies** (see Box 2.2). If, for example, we wish to know whether an altruistic strategy is superior to one that promotes personal reproduction, we can compare the inclusive fitness consequences of the two traits. In order for an altruistic trait to be adaptive, the inclusive fitness of altruistic individuals has to be greater than it would have been if those individuals had tried to reproduce personally. In other words, a rare allele "for" altruism will become more common only if the indirect fitness gained by the altruist is greater than the direct fitness it loses as a result of its self-sacrificing behavior. This statement is often presented as **Hamilton's rule**: a gene for altruism will spread only if $r_b B > r_c C$. Spelling this out, we calculate the indirect fitness gained by multiplying the extra number of relatives that exist thanks to the altruist's actions (B) by the mean coefficient of relatedness between the altruist and those extra individuals (r_b); we calculate the direct fitness lost by multiplying the number of offspring not produced by the altruist (C) by the coefficient of relatedness between parent and offspring (r_c). We used this approach when we concluded that altruism would be adaptive in cases in which the altruist gave up one offspring ($1 \times r_c = 1 \times 0.5 = 0.5$ genetic units) as the cost of helping three nephews survive that would have otherwise perished ($3 \times r_b = 3 \times 0.25 = 0.75$ genetic units).

Discussion Question

2.6 Individual amoebae (*Dictyostelium discoideum*) occasionally aggregate into a colony (Figure 2.8). About 20 percent of the amoebae in the colony make up a stalk that supports a ball composed of the remainder of the amoebae (see John Bonner's slime mold movies on YouTube). The stalk-forming individuals die without reproducing, whereas the amoebae in the ball, or fruiting body, give rise to spores that can become new individuals after the spores are carried away from the colony by passing insects and the like. On occasion two genetically different strains of the amoeba cooperate in the formation of a stalked fruiting body. What evolutionary puzzle is posed by this observation? Use kin (indirect) selection theory to make a prediction about the genetic similarity of the strains that work together when forming a colony versus those that do not. You can check your answer by referring to Ostrowski et al.[33] See also Flowers et al.[13]

Social Insects and Inclusive Fitness

Given that most social insects live in family groups, we can see that when these colonies are headed by a queen that will mate with only one male in her lifetime, the offspring of this monogamous mating pair will be full siblings. Sterile workers in these colonies are related to their reproductively competent siblings to the same extent that they would be to their own offspring, were they capable of producing any. Under these circumstances, workers need gain only a tiny indirect fitness advantage from helping their sibs survive to reproduce in order to make helping the adaptive option. Jacobus Boomsma has therefore argued that monogamy must be tightly linked to the evolution of obligately sterile workers (those completely incapable of reproducing) in Hymenoptera and the termites.[3] If Boomsma is correct, the mating system of an insect species can make altruism more or less likely to evolve.

FIGURE 2.8   A colony of *Dictyostelium discoideum*. The stalk of the colony is composed of individuals that have no chance of leaving offspring; the amoebae that make up the round ball at the top of the stalk form spores that can disperse and create new individuals. Photograph by Owen Gilbert.

Discussion Question

2.7   In some species the queen of a colony may have mated once or she may have mated with several different males. If workers of these ants, bees, or wasps retain the ability to lay haploid eggs (see below) that will become their sons upon maturity, what should workers living in a colony headed by a monogamous queen try to do versus workers living in a colony with a queen that has mated with two or more males?

Monogamy in bees, ants, and wasps (the Hymenoptera) plays an important role in the haplodiploidy hypothesis for the evolution of extreme sociality in this group of insects. W. D. Hamilton proposed this hypothesis in 1964 as an offshoot of inclusive fitness theory, and we shall focus on it for historical reasons and especially because the idea illustrates how to engage in "gene thinking" when confronted with a Darwinian puzzle. Remember Hamilton's rule, in which the benefit of helping genetically similar individuals must exceed the cost of sacrificing personal reproduction in order for the altruism to spread through a population. Hamilton realized that the coefficient of relatedness between full sisters in hymenopteran species could be higher than the 0.5 figure that applies to full siblings in most other organisms. A higher coefficient of relatedness among sisters could make adaptive altruism among sisters more likely to evolve, thereby explaining why sterile female workers were especially well represented among the Hymenoptera.[18]

Biologists have long been aware that male ants, bees, and wasps are **haploid**, which is to say they have only one set of chromosomes, not two like you or I have. Therefore, all the haploid sperm a male hymenopteran makes are chromosomally (and thus genetically) identical. So if a female ant, bee, or wasp mates with just one male, all the sperm she receives will have the same set of genes. When the female uses those sperm to fertilize her eggs, all her **diploid** daughters will carry the same set of paternal chromosomes and genes, which make up 50 percent of their total genotype. The other set of chromosomes carried by daughter hymenopterans comes from their mother. The mother's haploid eggs are not genetically uniform, because she is diploid; gamete formation by a parent with two sets of chromosomes involves the production of a cell with just one set drawn at random from those in that parent. An egg made by a female bee, ant, or wasp will share, on average, 50 percent of the alleles carried within her other eggs. Or, if you prefer to focus on a single gene present in two forms in a mother hymenopteran, either two eggs will share the same allele (100 percent similarity) or they will not (0 percent similarity), for an average of 50 percent similarity. An egg contributes half of the genotype of an offspring; half of 50 percent equals 25 percent, the percentage of genetic similarity among daughters of a queen bee, ant, or wasp due to their maternal inheritance. When a queen bee's eggs unite with genetically identical sperm, the resulting offspring share 50 percent of their genes thanks to their father and 25 percent on average because of their mother, for a total r of 0.75 (Figure 2.9).

Under the haplodiploid system of sex determination, hymenopteran sisters may therefore have a higher coefficient of relatedness (ranging from 0.5 to 1.0) than the exact 0.5 figure for a mother hymenopteran and her daughters and sons (or for full siblings in species with the more widespread diplodiploid system of sex determination). As a consequence of this genetic fact, $r_c \times C$ should on average be less than $r_b \times B$ for female Hymenoptera (unlike in diplodiploid animals). If sisters really are especially closely related, indirect selection could more easily favor female hymenopterans that, so to speak, put all their eggs

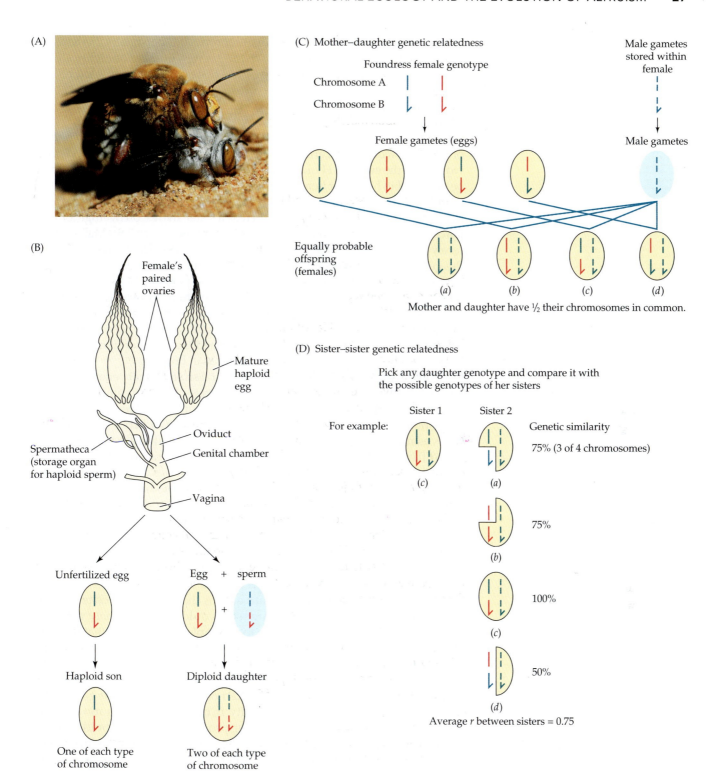

FIGURE 2.9 Haplodiploidy and the evolution of eusociality in the Hymenoptera.
(A) When a haploid male hymenopteran copulates with a diploid female, all the
sperm she receives are genetically identical. (B) Then when she releases a mature
haploid egg from her ovaries, it may or may not be fertilized with sperm from the
spermatheca as the egg passes down the oviduct. Unfertilized eggs become haploid
sons. Fertilized eggs become diploid daughters. (C) The genetic relatedness of mothers that have mated with a single male to their daughters ($r = 0.5$). (D) The average
genetic relatedness between sisters ($r = 0.75$) that are daughters of a monogamous
female. What is the genetic relatedness of a mother bee, ant, or wasp to her sons?

(alleles) in a sister's basket rather than reproducing personally. Perhaps not coincidentally, the Hymenoptera have the greatest number of highly altruistic species, and these species have worker castes that are composed of only females.

Testing the Haplodiploidy Hypothesis

If the haplodiploid system of sex determination contributed to the evolution of altruistic workers in certain species of bees, ants, and wasps, then female workers in the eusocial Hymenoptera (those with sterile helper castes) would be expected to bias their help toward reproductively competent sisters rather than toward male siblings. Why? Because sister hymenopterans share, on average, 75 percent of their genes whereas a sister shares, on average, only 25 percent of her genes with her haploid brother (see Figure 2.9). Males do not receive any of the paternal genes their sisters possess. The remaining half of the genome that sisters and brothers both receive from their mother ranges from 0 to 100 percent identical, averaging to 50 percent; 50 percent of a half means that a sister shares, on average, only one-fourth of her genes with her brothers ($r = 0.25$). In part because sisters are three times more closely related to one another than they are to their brothers (in species whose queen mates only once), Bob Trivers and Hope Hare argued that worker hymenopterans should invest three times as much in sisters as in brothers.[46]

Thus, if female workers employ a behavioral strategy that benefits their sisters as a result of maximizing their indirect fitness, then the combined weight of all the adult female reproductives (a measure of the total resources devoted to the production of females) raised by the colony's workers should be three times as much as the combined weight of all the adult male reproductives. When Trivers and Hare surveyed the literature on the ratio of total weights of the two sexes produced in colonies of different species of ants, they found a 3:1 investment ratio, as predicted by Hamilton's haplodiploid theory.[46]

If queens were in complete control of offspring production, they should ensure that the investment ratio for the two sexes was 1:1 because a queen donates 50 percent of her genes to each offspring, whether male or female. Queens therefore gain no genetic advantage by making a larger total investment in daughters than sons, or vice versa.[52] Imagine a hypothetical population of an ant species in which queens do tend to have more of one sex than the other. In this situation, any mutant queens that did the opposite and had more offspring belonging to the rarer sex would be handsomely repaid in grandoffspring. If males were scarce, for example, then a queen that used her parental capital to generate sons would create offspring with an abundance of potential mates and thus many more opportunities to reproduce than a comparable number of daughters would have. The greater fitness of son-producing queens would effectively add more males to the next generation, moving the sex ratio back toward equality. If, over time, the sex ratio overshot and became male biased, then daughter-producing queens would gain the upper hand, shifting the sex ratio back the other way. When the investment ratio for sons and daughters is 1:1, neither son-producing specialist queens nor daughter-producing queens have a fitness advantage. The fact that most studies (not all) have found an investment ratio skewed toward females in ant, bee, and wasp colonies suggests that when there is a conflict between the fitness interests of queens versus workers, the workers win.[29]

However, if all colonies of eusocial Hymenoptera invest three times as much in females as in males, then the overall investment ratio will be biased 3:1 in favor of females. Because they are relatively rare (investment wise), brothers will therefore provide three times the fitness return per unit of investment relative to females, canceling any indirect fitness gain for workers that put more into making sisters than brothers.[16] For this and several other reasons, it seems unlikely that haplodiploidy per se has been much of a factor, if any, in the evolutionary origin of eusociality in the Hymenoptera. Instead, current thinking leans toward the hypothesis that what counts is strict monogamy on the part of the female haplodiploid ant, bee, or wasp, or diplodiploid termite for that matter. Monogamous "queens" can produce helpers that are on average related to their siblings by a coefficient of 0.5. If the environment is such that a potential helper has even a little less chance of having offspring of her own as opposed to boosting the number of siblings produced by her monogamous mother by an equivalent number, then the indirect route to fitness becomes adaptive for the potential helper.[3,16]

The History of Behavioral Traits

One way to test the significance of monogamy for the evolutionary origin of altruism in the Hymenoptera would be to test the prediction that in the ancestral hymenopterans that gave rise to today's eusocial bees, ants, and wasps, females mated just once. A team led by William Hughes has secured the evidence needed to test this historical prediction.[21] Fortunately, there are ways to reconstruct the history of a behavioral trait with the help of the **theory of descent with modification**, Darwin's other contribution to evolutionary thinking, in addition to natural selection theory. Although Darwin was deeply interested in the process, namely natural selection, that causes adaptive evolutionary changes, he realized that a description of the changes that occurred over time required a different approach.[35] The essence of his "other" theory is that the characteristics of living species are the products of gradual changes in the traits of the ancestors of today's organisms. By identifying a distant ancestral state and then discovering what modifications of that trait occurred over time, we can potentially trace the history of any attribute from past to present.

Evolutionary history can be represented by a **phylogeny** or evolutionary tree of the species that were derived from a common ancestor (Box 2.3). Hughes's team created just such a phylogeny of bees, ants, and wasps based on the molecular and structural similarity of 267 species. Properly constructed, a phylogeny can tell a viewer which modern species or group of species has been least modified over evolutionary time. By overlaying a phylogeny of the social insects with data on whether female colony foundresses mate just once (monogamy) or with more than one male (polyandry), it is possible to determine whether the ancestral species of different lineages was more likely to have been monogamous or polyandrous. Extreme sociality has evolved independently several times in the Hymenoptera, and for all eight lineages for which mating frequency data are available, it appears that the ancestral species at the base of each lineage was monogamous (Figure 2.10), which would have facilitated the evolution of adaptive altruism.[21]

Moreover, in all those lineages that eventually gave rise to polyandrous eusocial species, the workers appear to have lost the capacity to mate. This means that workers are more or less locked into the helping role, so when foundress females mate with several males, they are still largely guaranteed that a sterile worker force will be there to assist in the production of their reproductively competent sisters. In the honey bee, for example, polyandry may have spread secondarily because of the benefits of having high genetic diversity within a colony to promote disease resistance[44] or to make worker specialization and task efficiency more likely.[34] Thus, the highly eusocial lifestyle of the now polyandrous honey bee may currently be maintained by selection pressures that differ from those that were responsible for the

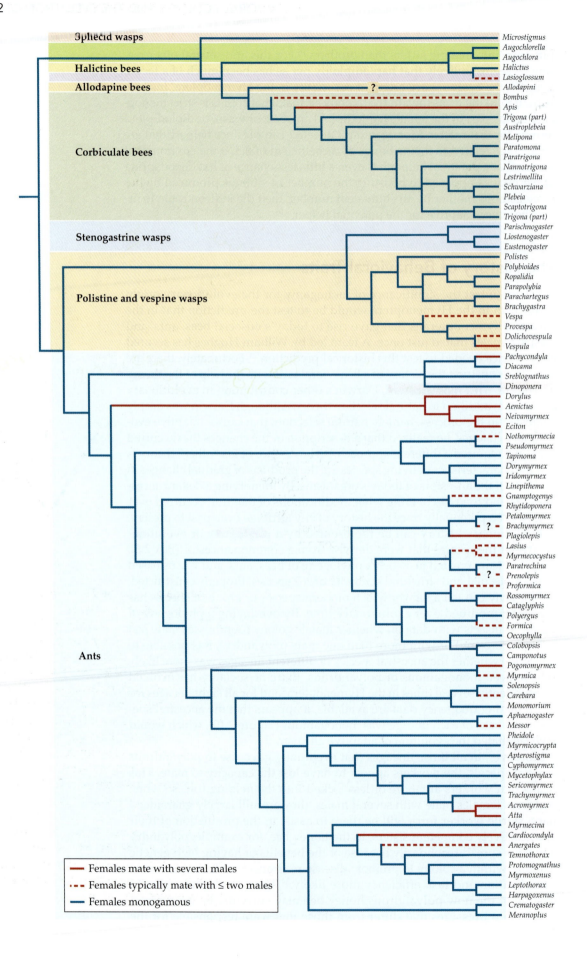

Sphecid wasps — *Microstigmus*
Halictine bees — *Augochlorella*, *Augochlora*, *Halictus*, *Lasioglossum*
Allodapine bees — *Allodapini* ?
Corbiculate bees — *Bombus*, *Apis*, *Trigona (part)*, *Austroplebeia*, *Melipona*, *Paratomona*, *Paratrigona*, *Nannotrigona*, *Lestrimellita*, *Schwarziana*, *Plebeia*, *Scaptotrigona*, *Trigona (part)*
Stenogastrine wasps — *Parischnogaster*, *Liostenogaster*, *Eustenogaster*
Polistine and vespine wasps — *Polistes*, *Polybioides*, *Ropalidia*, *Parapolybia*, *Parachartegus*, *Brachygastra*, *Vespa*, *Provespa*, *Dolichovespula*, *Vespula*
Ants — *Pachycondyla*, *Diacama*, *Sreblognathus*, *Dinoponera*, *Dorylus*, *Aenictus*, *Neivamyrmex*, *Eciton*, *Nothomyrmecia*, *Pseudomyrmex*, *Tapinoma*, *Dorymyrmex*, *Iridomyrmex*, *Linepithema*, *Gnamptogenys*, *Rhytidoponera*, *Petalomyrmex*, *Brachymyrmex* ?, *Plagiolepis*, *Lasius*, *Myrmecocystus*, *Paratrechina*, *Prenolepis* ?, *Proformica*, *Rossomyrmex*, *Cataglyphis*, *Polyergus*, *Formica*, *Oecophylla*, *Colobopsis*, *Camponotus*, *Pogonomyrmex*, *Myrmica*, *Solenopsis*, *Carebara*, *Monomorium*, *Aphaenogaster*, *Messor*, *Pheidole*, *Myrmicocrypta*, *Apterostigma*, *Cyphomyrmex*, *Mycetophylax*, *Sericomyrmex*, *Trachymyrmex*, *Acromyrmex*, *Atta*, *Myrmecina*, *Cardiocondyla*, *Anergates*, *Temnothorax*, *Protomognathus*, *Myrmoxenus*, *Leptothorax*, *Harpagoxenus*, *Crematogaster*, *Meranoplus*

— Females mate with several males
-- Females typically mate with ≤ two males
— Females monogamous

◀ **FIGURE 2.10   Monogamy and the origin of eusociality by kin selection in the Hymenoptera.** In this group, there are many different eusocial species, and among these, polyandry has often evolved independently. However, the phylogeny reveals that the ancestral species of the different eusocial groups (each lineage is presented with a different background color) were in every case monogamous, the condition necessary for sister hymenopterans to be unusually closely related. After Hughes et al.[21]

origin of eusociality in this species. The key point, however, is that monogamy appears to have been essential for the evolution of sterile workers in the Hymenoptera.[21]

Discussion Question

2.8 If a female of a monogamous social wasp species could help to produce more sisters with an *r* of 0.75, why would she ever reproduce personally, given that reproducers are related to their offspring by just 0.5?

BOX 2.3   How are phylogenetic trees constructed and what do they mean?

The diagram in this box is a phylogenetic tree that represents the evolutionary history of three animal species (X, Y, and Z) that are living today and their links to two ancestral species (A and B). To create a phylogeny of this sort, it is necessary to determine which of the three living species are more closely related and thus which are descended from the more recent common ancestor. Phylogenetic trees can be drawn up on the basis of anatomical, physiological, or behavioral comparisons among species, but more and more often, molecular comparisons are used. The molecule DNA, for example, is very useful for this purpose because it contains so many "characters" on which such comparisons can be based, namely, the specific sequences of nucleotide bases that are linked together to form an immensely long chain. Each of the two strands of that chain has a base sequence that can now be read by an automated DNA-sequencing instrument. Therefore, one can, in theory and in practice, compare a cluster of species by extracting a specific segment of DNA from either the nuclei or mitochondria in cells from each species and identifying the base sequences of that particular segment.

For the purposes of illustration, here are three made-up DNA base sequences that constitute part of a particular gene found in all three hypothetical species of animals:

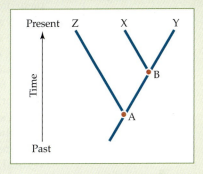

Position	1	2	3	4	5	6	7	8	9	10	11	12	13	14	15
Species X	A	T	T	G	C	A	T	A	T	G	T	T	A	A	A
Species Y	A	T	T	G	C	A	T	A	T	G	G	T	A	A	A
Species Z	G	T	T	G	T	A	C	A	T	G	T	T	A	A	T

These data could be used to claim that species X and Y are more closely related to each other than either is to species Z. The basis for the claim is that the base sequences of species X and Y are nearly identical (differing by a single change in position 11 of the chain), whereas species Z differs from the other two by four and five base changes. The shared genetic similarity between X and Y can be explained in terms of their history, which must have featured the recent common ancestor (species B in the phylogenetic tree shown here). Species B must have split so recently into the two lineages leading to living species X and Y that there has not been sufficient time for more than one mutation to become incorporated in this segment of DNA. The lesser, but still substantial, similarities among all three species can be explained in terms of their more ancient common ancestor, species A. The interval between the time when A split into two lineages and the present has been long enough for several genetic changes to accumulate in the different lineages, with the result that species Z differs considerably from both species X and species Y.

Discussion Question

2.9   In termites both males and females are diploid and both sexes make up the worker caste. Moreover in at least one species, two small colonies may amalgamate without conflict except between the reproductive members of the same sex, which may try to kill one another.[22] The members of merged colonies cooperate with one another. Some individuals in these colonies also have the ability to become replacement kings or queens upon the demise of one of the original surviving reproductives. What might these termites tell us about haplodiploidy and the origin of eusocial behavior? Evaluate the possibility that unrelated workers (with the capacity to become reproductives) may have helped unrelated queens as an early stage in the evolution of eusocial colonies.

The Origin of the Dances of Honey Bees

Here I shall digress for a bit to present another example of the value of Darwin's other theory, the theory of descent with modification. The *origin* of sterile workers in the Hymenoptera is a historical puzzle; the origin of the dances of honey bees is yet another. These dances are performed by workers when they return to their hives after finding good sources of pollen or nectar.[47] As the dancers move about on the vertical surface of the honeycomb in the complete darkness of the hive, they attract other bees, which follow them around as they move through their particular routines. Researchers watching dancing bees in special observation hives have learned that their dances contain a surprising amount of information about the location of a new food source (such as a patch of flowers). If the bee executes a round dance (Figure 2.11), she has found food fairly close to the hive—say, within 50 meters of it. If, however, the worker performs a waggle dance (Figure 2.12), she has found a nectar or pollen source more than 50 meters away. The longer the waggle-run portion lasts, the more distant the food.

Moreover, by measuring the angle of the waggle run with respect to the vertical, an observer can also tell the direction to the food source. Apparently, a foraging bee on its way home from a distant but rewarding flower patch notes the angle between the flowers, hive, and sun. The bee transposes this angle onto the vertical surface of the comb when she performs the waggle-run portion of the waggle dance. If the bee walks directly up the comb while waggling, the flowers will be found by flying directly toward the sun. If the bee waggles straight down the comb, the flower patch is located directly away from the sun. A patch of flowers positioned 20 degrees to the right of a line between the hive and the sun is advertised with waggle runs pointing 20 degrees to the right of the vertical on the comb. In other words, when outside the hive, the bees' directional reference is the sun, whereas inside the hive, their reference is gravity.

The conclusion that the dances of honey bees contain information about the distance and direction to good foraging sites was reached by Karl von Frisch after years of experimental work.[47] His basic research protocol involved training bees (which he daubed with dots of paint for identification) to visit feeding stations, which he stocked with concentrated sugar solutions. By watching the dances of these trained bees, he saw that their behavior changed in highly predictable ways depending on the distance and direction to a feeder. More importantly, his dancing bees were able to direct other bees to a feeder they had found (Figure 2.13), leading him to believe that bees use the information in the dances

FIGURE 2.11 Round dance of honey bees. The dancer (the uppermost bee) is followed by three other workers, which may acquire information that a profitable food source is located within 50 meters of the hive. After von Frisch.[47]

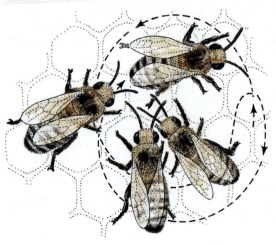

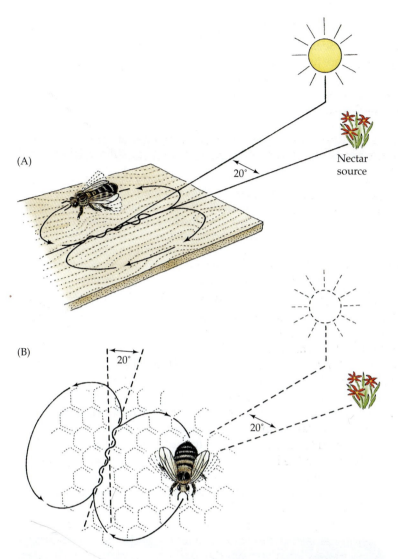

FIGURE 2.12 Waggle dance of honey bees. As a worker performs the waggle-run portion of the dance, its shakes its abdomen from side to side. The duration and the orientation of the waggle runs contain information about the distance and direction to a food source. In this illustration, workers attending to the dancer learn that food may be found by flying 20 degrees to the right of the sun when they leave the hive. (A) The directional component of the dance is most obvious when it is performed outside the hive on a horizontal surface in the sunlight, in which case the bee uses the sun's position in the sky to orient its waggle runs directly toward the food source. (B) On the comb, inside the dark hive, dances occur on vertical surfaces, so they are oriented with respect to gravity; the deviation of the waggle run from the upward vertical equals the deviation of the direction to the food source from a line between the hive and the sun.

FIGURE 2.13 Testing directional and distance communication by honey bees. (A) A "fan" test to determine whether foragers can convey information about the direction to a food source they have found. After training scout bees to come to a feeding station at F, von Frisch collected all newcomers that arrived at seven feeding stations with equally attractive sugar water. Most new bees arrived at the feeder in line with F. (B) A test for distance communication. After training scouts to come to a feeding station 750 meters from the hive, von Frisch collected all newcomers arriving at feeders placed at various distances from the hive. In this experiment, 17 and 30 newcomers were captured at the two feeders closest to 750 meters, far more than were caught at any other feeder. After von Frisch.[47]

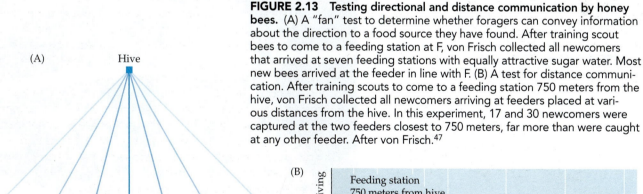

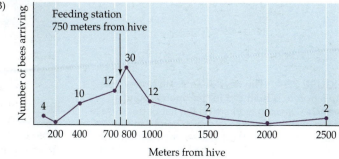

of their hive mates to find good foraging sites. Many years later, Jacobus Biesmeijer and Thomas Seeley were able to show that more than half the young worker bees that were just beginning their careers as pollen or nectar gatherers spent some time following dancing bees before launching their collecting flights. This finding suggests that dance information really is useful to bees about to start foraging for food.[2]

Discussion Question

2.10   Wolfgang Kirchner and Andreas Grasser[23] found that when the hive was turned on its side, bees continued to dance in the darkness, but on a horizontal surface, not a vertical one. Under these conditions, recruitment at distant feeders (more than 100 meters from the hive) that had been visited by dancing bees was very poor. When, however, the hive was returned to an upright position and the comb surface on which recruiters danced was vertical (as it would be in natural hives), most recruits appeared at the feeders that the scouts had visited. What bearing do these results have on the argument about whether recruits derive information from the dances of their colony mates? What prediction can you make about relative rates of recruitment to sites less than 50 meters from the hive when it is turned on its side as opposed to when it is upright?

An Initial Hypothesis and Its Test

Martin Lindauer was the first to propose a hypothetical historical scenario for the extraordinary dances of honey bees.[25] He began his work by comparing three other members of the genus *Apis*, which he found to perform dance displays identical to those of the familiar honey bee (*Apis mellifera*) except that in one species, *A. florea*, the bees dance on the horizontal surface of a comb built in the open over a tree branch (Figure 2.14). To indicate the direction to a food source, a

FIGURE 2.14   The nest of an Asian honey bee, *Apis florea*, is built out in the open around a branch. Dancing workers on the flat upper surface of the nest (two nests are shown here) can run directly toward the food source when performing the waggle dance. Photograph by Steve Buchmann.

worker of this species simply orients her waggle run directly toward the food's location. Because this is a less sophisticated maneuver than the transposed pointing done in the dark on vertical surfaces by *A. mellifera*, it may resemble a form of dance communication that preceded the dances of *A. mellifera*.

Lindauer then looked to stingless tropical bees in other genera for recruitment behaviors that might provide hints about the steps preceding waggle dancing. These comparisons led him to suggest the following historical sequence.

A possible first stage: Workers of some species of stingless bees in the genus *Trigona* run about excitedly, producing a high-pitched buzzing sound with their wings, when they return to the nest from flowers rich in nectar or pollen. This behavior arouses their hive mates, which detect the odor of the flowers on the dancers' bodies. With this information, the recruits leave the nest and search for similar odors. The same kind of behavior also occurs in the bumblebees, which form small colonies with "dancing" scouts that do not provide signals containing directional or distance information but that do alert their fellow bees to the existence of resource-rich flowers in the neighborhood of the colony.[11]

A possible intermediate stage: Workers of other species of *Trigona* do convey specific information about the location of a food source. In these species, a worker that makes a substantial find marks the area with a pheromone produced by her mandibular glands. As the bee returns to the hive, she deposits pheromone on tufts of grass and rocks every few meters. Inside the hive entrance, other bees wait to be recruited. The successful forager crawls into the hive and produces buzzing sounds that stimulate her companions to leave the hive and follow the scent trail that she has made (Figure 2.15).

A still more complex pattern: A number of stingless bees in the genus *Melipona* (a group related to *Trigona* bees) convey distance and directional information separately. A dancing forager communicates information about the distance to a food source by producing pulses of sound; the longer the pulses, the farther away the food. In order to transmit directional information, she leaves the nest with a number of followers and performs a short zigzag flight that is oriented toward the food source. The scout returns and repeats the flight a number of times before flying straight off to the food source, with the recruited bees in close pursuit.

FIGURE 2.15 Communication by scent marking in a stingless bee. In this species, workers that found food on the side of a pond opposite from their hive could not recruit new foragers to the site until Martin Lindauer strung a rope across the pond. Then the scouts placed scent marks on the vegetation hanging from the rope and quickly led others to their find. Photograph by Martin Lindauer.

Lindauer was not suggesting that some stingless bees behaved in a more adaptive manner than others. He simply used the diversity of existing traits in this group to provide possible clues about the behaviors of now extinct bees whose communication systems were modified in species derived from these ancestral bees. On the basis of what modern species do, he hypothesized that communication about the *distance* to a food source was initially encoded in the degree of activity by a returning food-laden worker.[31] Subsequently, selection acting in some species may have favored standardization of the sounds and movements made by successful foragers, which set the stage for the round and waggle dances of *Apis* bees.[25,56]

Lindauer's comparisons indicated that communication about the *direction* to a food source might have originated with a worker guiding a group of recruits directly to a nectar-rich area. This behavior has evolved gradually to contain a less and less complete performance of the guiding movements, beginning with partial leading (as in some *Melipona*) and later consisting of pointing in the proper direction with a waggle run on a horizontal surface (as in *A. florea*). From antecedents like this came the transposed pointing of *A. mellifera*, in which the direction of flight relative to the sun is converted into a signal (the waggle run) oriented relative to gravity.

FIGURE 2.16 Evolutionary history of the honey bee dance communication system. (A) A phylogeny of four closely related groups, one of which (the Apini) includes *Apis mellifera*. According to this diagram, the ancestor of the four modern groups was an ancient member of the Bombini, a bumblebee, probably a weakly social bee with rudimentary information exchange between foragers. The Meliponini and Apini are shown as close relatives, so an ancestral species of both groups might have elaborated on the simpler communication system of a distant bumblebee ancestor. Descendants of this one ancestral species later incorporated distance and direction information into their recruitment signals. The honey bee still more recently evolved dance attributes that built upon the signals of a fairly recent ancestor. (B) A newer phylogeny in which the closest relative of the Apini is the Euglossini (the orchid bees, which are not social), not the Meliponini (the social tropical stingless bees). This phylogeny indicates that complex dance communication has evolved independently in the Meliponini and Apini. After Cardinal and Danforth.[8]

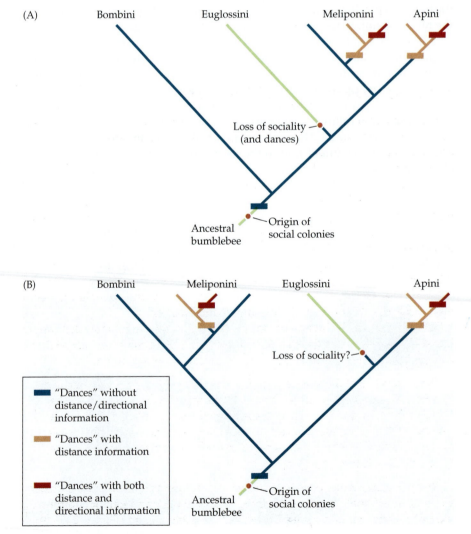

The historical sequence of events outlined by Martin Lindauer can be tested if we have an accurate phylogeny of four closely related groups of bees: the bumblebees, the orchid bees, the tropical stingless bees, and the *Apis* bees. In the previous edition of my book, I presented a phylogeny (Figure 2.16A) based on molecular and other evidence, suggesting that the tropical stingless bees (the Meliponini) and the honey bees and the other members of the genus *Apis* were derived fairly recently from a common ancestor. This phylogeny indicated that the dances of *Apis mellifera* were the modified products of descent from ancestors of the Meliponini.

Recently, however, Sophie Cardinal and Bryan Danforth have developed a more likely phylogeny (Figure 2.16B) that makes use of a very large data set of molecular similarities and differences among the four groups of bees as well as sophisticated statistical procedures.[8] Their newer phylogeny does *not* support the hypothesis that the honey bee's dance behavior was derived with descent by modification from bees that gave rise to today's tropical stingless bees. Why not? (If you can answer this question, you will demonstrate that you understand how phylogenies can contribute to an understanding of the evolutionary history behind complex behavioral traits.)

Discussion Question

2.11 In what way does the replacement of one view of the evolutionary history behind honey bee dances with another view illustrate one major difference between science and religion?

Kin Selection and Social Conflict

Honey bee workers perform beautifully in recruiting others to collect ephemeral supplies of nectar and pollen, which are used to rear additional workers and new queens and male reproductives (drones). The origin of the workers' dances may lie in the behavior of some now extinct, weakly social members of the Euglossini, the closest relatives of the honey bees. Alternatively, the initial stage and all subsequent ones may have occurred in ancestral species within the Apini lineage itself. The origins of the honey bee workers' extreme altruism probably can be linked to the monogamous mating system of a very distant ancestor, which long ago made indirect fitness gains possible for workers that sacrificed personal reproduction in order to help full siblings ($r = 0.5$) achieve reproductive age. Kin selection almost certainly underlies the evolution of eusociality in all its manifestations among the Hymenoptera and elsewhere (see Foster et al.,[14] Strassmann and Queller[41]).

To demonstrate the breadth of application of kin selection theory, let's apply it to the converse of altruism, the selfish conflicts that occur within social insect colonies. And these conflicts do occur, as W. D. Hamilton noted:

> Every schoolchild, perhaps as part of religious training, ought to sit watching a *Polistes* wasp nest for just an hour.... I think few will be unaffected by what they see. It is a world human in its seeming motivations and activities far beyond all that seems reasonable to expect from an insect: constructive activity, duty, rebellion, mother care, violence, cheating, cowardice, unity in the face of a threat—all these are there.[19]

Rebellion and violence in a paper wasp nest take any number of forms: for example, one female lunging at another or biting her opponent or climbing on top of the other female. Should a worker wasp lay an egg (Figure 2.17), then

FIGURE 2.17   Workers and the queen monitor the reproductive behavior of others. Eggs laid by individuals other than the queen will usually be eaten by other colony members. Eggs can be seen in several of the cells to the upper right side of this nest; larvae occupy the other cells and are being fed by the workers. Photograph by the author.

another member of the group, the queen or another worker, may well discover the haploid egg and eat it.[37] Egg destruction is a common form of "policing" in social insects. Negative interactions can be more elaborate, as in the ant *Dinoponera quadriceps*, where the dominant reproducing female (the queen) smears a potential competitor with a chemical from her stinger, after which lower-ranking workers immobilize the unlucky pretender queen for days at a time (Figure 2.18A).[30] Likewise, workers of *Harpegnathos saltator* punish nest mates that are developing their ovaries by holding the offenders in a firm grip (Figure 2.18B), preventing them from doing anything and thereby inhibiting further development of the immobilized ant's ovaries.[24]

Tom Wenseleers and Francis Ratnieks tested whether sanctions imposed on workers that try to reproduce can make it more profitable for the sanctioned workers to behave altruistically. Using quantitative data from 20 species of social insects on (1) the effectiveness of policing efforts within colonies and (2) the proportion of workers that laid unfertilized eggs in their colonies, Wenseleers and Ratnieks were able to show that when the eggs of workers were destroyed, a smaller proportion of workers developed ovaries in the colony (Figure 2.19).[50] In other words, the policing behavior of colony mates can mean that workers have little chance of boosting their inclusive fitness directly—thus, the indirect route to inclusive fitness is superior.

Another way of looking at this issue is to consider what percentage of the males in a colony of social insects are the sons of workers, a figure that varies from 0 to 100 percent. If worker policing is responsible for the cases

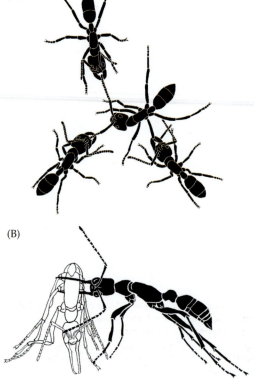

(A)

(B)

FIGURE 2.18   Conflict within ant colonies in which workers that are about to reproduce are detected by colony mates and physically restrained. (A) In this species, three workers grasp the would-be reproductive that has been marked by the queen, preventing her from moving. (B) In another species, the ant in black has grabbed and immobilized a nest mate whose ovaries had begun to develop. After holding her nest mate captive for 3 or 4 days, that worker may turn her prisoner over to another worker to continue the imprisonment. A, from Monnin et al.[30]; B, drawing by Malu Obermayer, from Liebig et al.[24]

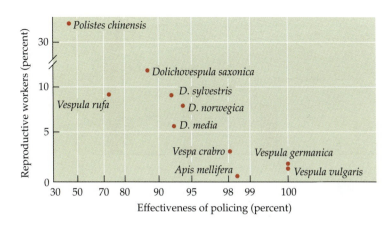

FIGURE 2.19 **A test of the effectiveness of policing** on the likelihood that workers will reproduce in social insect colonies. The better that workers are at destroying the eggs of their fellow workers, the less likely workers are to try to reproduce. After Wenseleers and Ratnieks.[50]

in which few males are the offspring of workers, then these cases should involve species in which the workers are more closely related to the queen's sons than to the sons of other workers. This prediction is also correct (Figure 2.20).

In the *Melipona* bees, female larvae are sealed within brood cells that are the same size for both workers and queens and so are fed equally by other workers. Under these circumstances, female larvae control their own fate and can become either workers or queens, depending upon what option confers the higher inclusive fitness. In different species, different proportions of the female larvae opt to develop into queens, with as many as 25 percent of all immature larvae becoming queens. Evidently it pays to help, but it can pay even more to *be helped*. The production of so many queens ready to take advantage of worker assistance reduces the size of the workforce because these individuals lack the modifications of the hindlegs that workers use to collect pollen for the brood pots. The excess mature queens (only one is needed to head up a colony) are quickly killed by the workers, but even so, the fact that they are produced lowers the reproductive output of the colony as a whole.

Wenseleers and Ratnieks predicted that the extent of queen overproduction in *Melipona* bees would be a function of how many reproductively competent males in the colony were the sons of workers.[48] The greater the proportion of workers' sons, the greater the kin-selected cost to a female larva for opting to develop into a queen. The coefficient of relatedness between a female and her

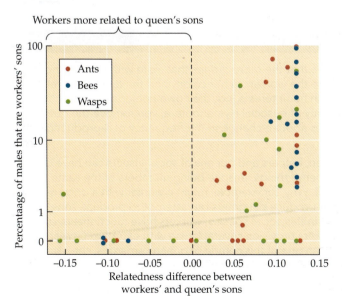

FIGURE 2.20 **The proportion of males produced by workers varies among ants, social bees, and social wasps.** The greater the difference in the relatedness of workers' and the queen's sons (for values greater than zero), the greater the proportion of males that are produced by the colony's workers. After Wenseleers and Ratnieks.[49]

(A)

(B)

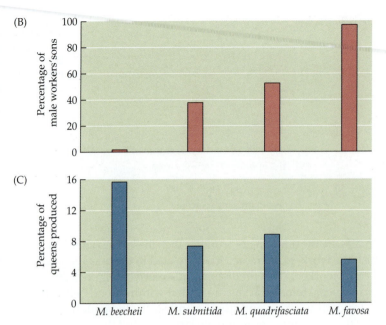

(C)

FIGURE 2.21   Colony kin structure is linked to queen production in eusocial *Melipona* bees including (A) *M. beecheii*. (B) The percentage of males produced by workers varies enormously among four species of *Melipona*. (C) To the extent to which workers are reproducing, creating nephews ($r = 0.375$) of their sisters, females are less likely to become queens (which cannot carry out worker tasks). As predicted, the higher the percentage of workers' sons in the colonies of different species, the less likely immature females are to try to become queens. B and C, after Wenseleers and Ratnieks.[48]

sisters' sons is $0.75 \times 0.5 = 0.375$ (the *Melipona* mother of these females mates only once, resulting in the high r for sisters). In contrast, the coefficient of relatedness for a female and her brothers, the sons of the queen, is only 0.25 (as discussed earlier). If worker sisters are producing many sons, a new queen harms these nephews by removing herself from the colony workforce. Harm done to relatives reduces an individual's inclusive fitness in proportion to the similarity between these kin. And in fact, as predicted, in a species in which many workers are laying haploid eggs (which develop into their sons), most females "choose" not to develop into queens. In a species in which workers' sons are relatively rare, larval females are more likely to opt to develop into queens (Figure 2.21).[48] The fact that kin selection theory (or inclusive fitness theory, if you wish) has been used to make testable predictions about the kinds of conflicts that occur within colonies explains why this approach is so strongly favored by most scientists who study the social behavior of colonial insects (and other highly social animals).[6,41]

Discussion Questions

2.12   Why might it be that the more drones that mate with a queen honey bee, the more loyal are the workers in her colony and the longer she retains a workforce that acts in her interests, not their own personal reproductive interests?[40] If in a colony of eusocial Hymenoptera, you remove the queen, what prediction follows about a change in the frequency of reproduction by workers?[49]

2.13   As we have seen, helpers and queens of social insects can have disputes about any number of things, despite being members of the same family (e.g., Heinze et al.[20]). Why, for example, might a worker with a once-mated mother let her mother produce daughters but attempt to produce sons herself (assuming that in this social species, workers have functional ovaries)? In the honey bee, workers provision several larvae with extra food, creating future queens. But when one of this coterie of queens emerges, she usually kills her sister queens. Why is this puzzling? Why is the behavior adaptive?

Summary

1. The greatest puzzle identified by Darwin was the evolution of altruism. Because self-sacrificing altruists help other individuals reproduce, one would think that this behavior should be naturally selected against and so should disappear over time.

2. Yet extreme altruism is common in eusocial (caste-forming) insect colonies where workers rarely, if ever, reproduce and instead help their colony mates survive to reproduce. Many solutions to this puzzle have been offered, including the claim that group selection favors colonies that contain altruists because these colonies produce more new colonies than groups without altruists.

3. However, most students of animal social behavior employ kin selection theory because it has repeatedly proven to be useful for scientific investigation of sociality. Specifically, kin selection theory has been used to show that altruism can spread through a population if the cost to the altruist in terms of a reduction in the number of offspring produced (multiplied by the coefficient of relatedness between the altruist and those offspring) is less than the increase in the number of related individuals helped by the altruist (multiplied by the average coefficient of relatedness between the altruist and the helped relatives).

4. Kin (indirect) selection can result in an increase in the number of genes transmitted indirectly by an individual to the next generation in the bodies of relatives that exist because of the altruist's help. An individual's total genetic contribution to the next generation has been called its inclusive fitness.

5. Kin selection theory underlies modern "gene-centered thinking," with researchers now aware that individuals should behave in ways that boost their inclusive fitness whether this is achieved through self-sacrificing cooperation or through self-serving conflict with others, even close relatives.

Suggested Reading

V. C. Wynne-Edwards's *Animal Dispersion in Relation to Social Behaviour*[58] is loaded with fascinating natural history, which alone makes it well worth reading, and it also presents the original version of group selection theory. George C. Williams explains why group selection theory has not fared well.[52] For a summary of "new" group selection theory see Wilson and Wilson,[55] but then read an article by Stuart West and colleagues.[51] W. D. Hamilton wrote the revolutionary papers on altruism, evolution, and inclusive fitness.[18] Richard Dawkins provides a popular account of the evolution of altruism by kin selection in *The Selfish Gene*,[10] as does E. O. Wilson in *Sociobiology*.[57] Articles by Joan Strassmann and David Queller provide cogent analyses of kin selection in relation to the evolution of altruism.[41,43]

3 The Evolution of Social Behavior

 I have just been watching a YouTube video of a pack of wolves hunting buffalo in the winter in Alberta, Canada. The wolves tested the herd, pursuing them first one way and then another before launching a co-ordinated attack on a single individual that they managed to isolate from the rest of the herd. They quickly swarmed over their prey and pulled it down. The wolves all had a big meal as a result. No single wolf could have accomplished what the pack did together.

The social skills exhibited by hunting wolves are perhaps not as stunning as those of the social insects we discussed in the preceding chapter, but they are testimony to the fact that vertebrates as well as insects have evolved sophisticated social behavior. In some of these social groups, individuals appear to sacrifice their reproductive chances to help others, the same sort of Darwinian puzzle that was the focus of the preceding chapter. In wolves, for example, subordinate females rarely try to reproduce, instead deferring to the dominant female in the pack[39] and indeed helping the head female by sharing the costs of hunting for large prey. Puzzles of this sort will be the main focus of this chapter.

These immature leaf-footed coreid bugs cluster together even though they face heightened competition for food and a greater risk of contracting a communicable disease. What then is the adaptive value of their social behavior?

FIGURE 3.1 Cliff swallows are highly social. Pairs nest right next to one another, a tactic that has costs as well as possible benefits. Photograph by Charles Brown.

But first, although the idea that social behavior is superior to solitary living is popular, this view is rarely held by those aware of the cost-benefit approach of behavioral ecology. The fact is that living with others comes with substantial fitness costs to individuals. For example, because cliff swallows nest side by side in colonies (Figure 3.1), the birds are subject to parasitic bugs that make their way from nest to nest. Moreover, the larger the colony of swallows, the more the site attracts transient birds, and with these visitors come more bugs, some of which slip off their carriers and onto resident swallows and their offspring. As a result, the larger the colony, the higher the infestation rate of the parasite.[9] When the blood-sucking parasites attack swallow chicks, the price for avian sociality becomes evident in the form of stunted nestlings (Figure 3.2).[7] In order for social life to be adaptive for cliff swallows, some benefit(s) must exist that more than compensates the birds for the risk of having bug-infested offspring. For the cliff swallow, a major benefit appears to be improved foraging success for adult birds searching in groups for flying insects.[8] But for many other species, social individuals do not gain sufficient benefits to replace the solitary types over evolutionary time.[1]

Discussion Question

3.1   Figure 3.3 shows a phylogeny (see page 31) of many bee species in the genus *Lasioglossum*, with the social system of the species superimposed on the evolutionary tree.[20] What is the minimum number of times that eusociality (i.e., colonial species with obligately sterile workers) has evolved in this group? How many times has eusociality been lost, either completely or partially? (Partial loss of eusociality has occurred in those species that are labeled "polymorphic," which means that in these cases some populations form colonies with sterile workers and others do not.) How might this phylogeny be used to criticize the common value judgment that complex social systems are generally superior, and more recently evolved, than simpler ones?

FIGURE 3.2 Effect of parasites on cliff swallow nestlings. The much larger nestling on the right came from an insecticide-treated nest; the stunted baby of the same age on the left occupied a nest infested with swallow bugs. From Brown and Brown.[7]

FIGURE 3.3 Social behavior has an evolutionary history. Among the many species of bees belonging to the genus *Lasioglossum*, some are eusocial, others are solitary, and still others are polymorphic (some populations are eusocial, others solitary). The social parasitic species exploits established colonies of other bees. The photograph shows two females of *L.* (*Sphecodogastra*) *oenotherae*, a solitary species whose closest relatives are mostly eusocial bees. After Danforth et al.;[20] photograph by Bryan Danforth.

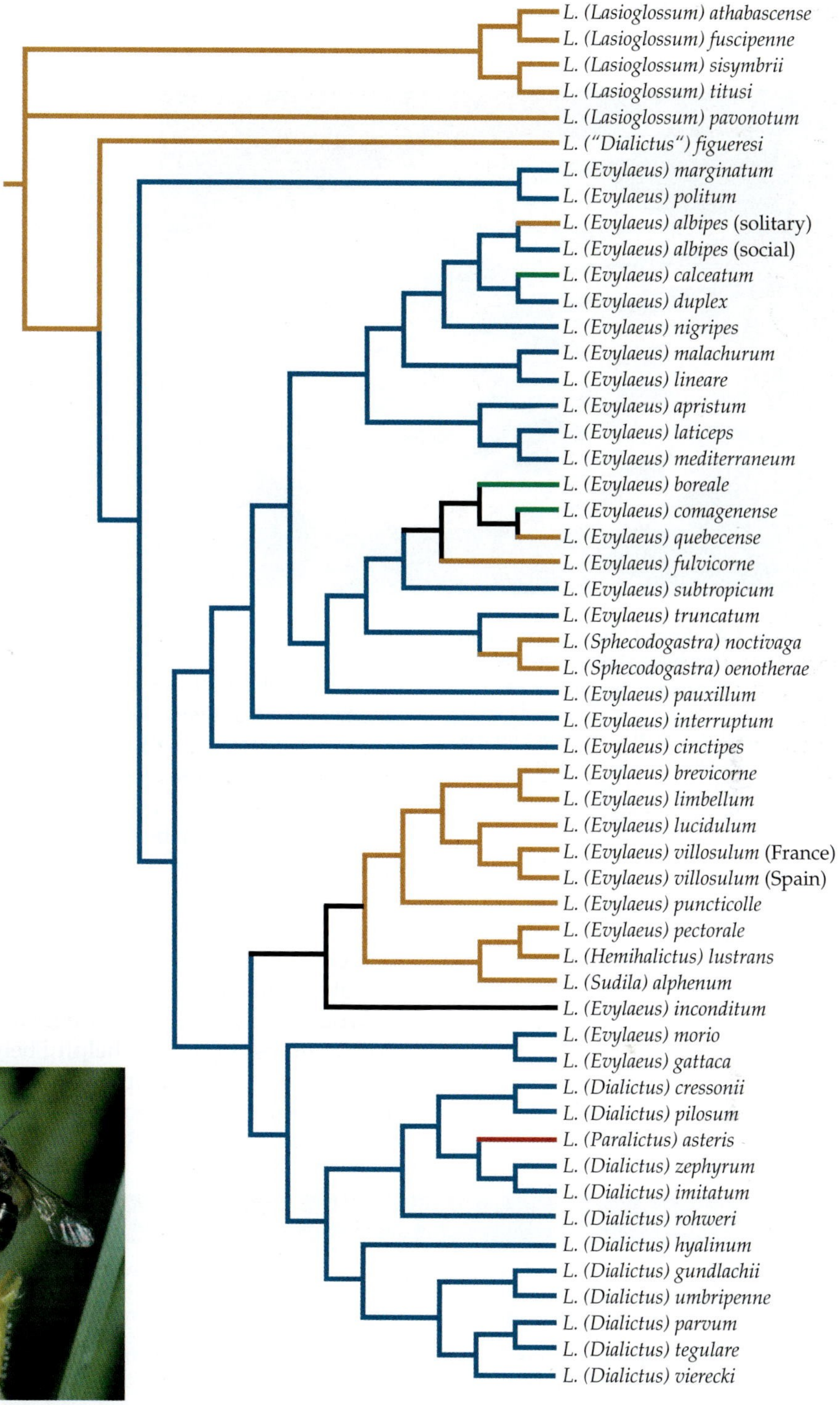
L. (Lasioglossum) athabascense
L. (Lasioglossum) fuscipenne
L. (Lasioglossum) sisymbrii
L. (Lasioglossum) titusi
L. (Lasioglossum) pavonotum
L. ("Dialictus") figueresi
L. (Evylaeus) marginatum
L. (Evylaeus) politum
L. (Evylaeus) albipes (solitary)
L. (Evylaeus) albipes (social)
L. (Evylaeus) calceatum
L. (Evylaeus) duplex
L. (Evylaeus) nigripes
L. (Evylaeus) malachurum
L. (Evylaeus) lineare
L. (Evylaeus) apristum
L. (Evylaeus) laticeps
L. (Evylaeus) mediterraneum
L. (Evylaeus) boreale
L. (Evylaeus) comagenense
L. (Evylaeus) quebecense
L. (Evylaeus) fulvicorne
L. (Evylaeus) subtropicum
L. (Evylaeus) truncatum
L. (Sphecodogastra) noctivaga
L. (Sphecodogastra) oenotherae
L. (Evylaeus) pauxillum
L. (Evylaeus) interruptum
L. (Evylaeus) cinctipes
L. (Evylaeus) brevicorne
L. (Evylaeus) limbellum
L. (Evylaeus) lucidulum
L. (Evylaeus) villosulum (France)
L. (Evylaeus) villosulum (Spain)
L. (Evylaeus) puncticolle
L. (Evylaeus) pectorale
L. (Hemihalictus) lustrans
L. (Sudila) alphenum
L. (Evylaeus) inconditum
L. (Evylaeus) morio
L. (Evylaeus) gattaca
L. (Dialictus) cressonii
L. (Dialictus) pilosum
L. (Paralictus) asteris
L. (Dialictus) zephyrum
L. (Dialictus) imitatum
L. (Dialictus) rohweri
L. (Dialictus) hyalinum
L. (Dialictus) gundlachii
L. (Dialictus) umbripenne
L. (Dialictus) parvum
L. (Dialictus) tegulare
L. (Dialictus) vierecki

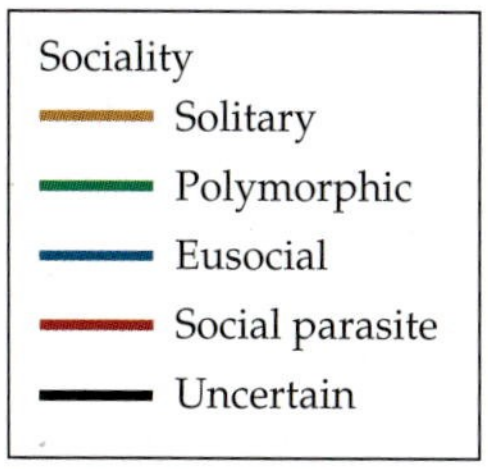
Sociality
Solitary
Polymorphic
Eusocial
Social parasite
Uncertain

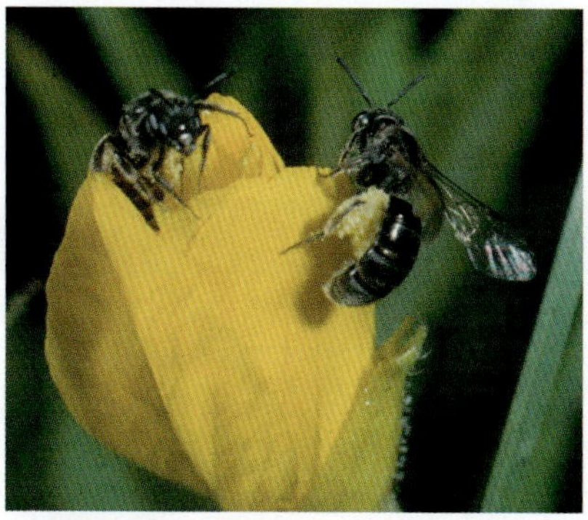

Lasioglossum oenotherae

The Evolution of Helpful Behavior

One of the potential benefits of social life is the possibility of cooperation among the members of the group, as in that pack of wolves hunting buffalo. As we saw in the preceding chapter, members of social insect colonies can gain by helping related members of their group; those that receive help also benefit if they reproduce more than they would otherwise. Here we shall expand our evolutionary analysis of helpful behavior to social organisms other than insects, a collection of animals that includes bacteria[79] and amoebae,[55] har-

FIGURE 3.4 Social life has costs and benefits. Why have these little catfish living on a coral reef near Sulawesi joined forces to form a school of fish? Does each individual gain direct fitness from its decision? Do some individuals make sacrifices for relatives? Photograph by Roger Steene.

vestmen,[49] spiders,[4] and aphids[69] as well as fish,[33] lizards,[21] birds,[15] and of course mammals.[35]

We begin by categorizing the kinds of social interactions that can take place between two individuals of the same species, whether these are amoebae or zebras, with respect to the payoffs for the two interactors (Figure 3.4, Table 3.1). When two individuals both benefit from helping one another, their behavior can be placed in the category of **cooperation.** Thus, when one wolf helps its fellow wolves pull down a buffalo, the cooperative hunter will usually get some meat, even if it did not apply the actual killing bite. The same applies to brown-necked ravens that hunt large lizards in teams, with two members of the group standing by the lizard's burrow entrance while others attack and kill an animal that they have surprised outside its retreat; only after the victim is dead do the ravens that blocked the lizard's escape route leave to get their share of the food.[85] Likewise, if several male bluegills succeed in fending off a bullhead catfish that has entered their part of the nesting colony (Figure 3.5), the eggs in all the males' nests are more likely to survive to hatch. When all cooperators derive roughly the same fitness gains from their interactions, their cooperation requires no special evolutionary explanation.

Nevertheless it can be challenging to identify whether benefits are available to both participants in a relationship believed to be cooperative. For example, consider the social paper wasp, *Polistes dominulus*, in which several females may band together to build and provision a nest in the spring. At any one moment, only one wasp in a group of females acts as a queen while the others are subordinate to the leader. You might think that these subordinate workers would all be sisters or close relatives of the queen wasp (thereby gaining indirect fitness from their helpful behavior), but you would be wrong; some 15 to 35 percent of the helper wasps are completely unrelated to the dominant female.[47] Under these circumstances, there is no possibility that the subordi-

TABLE 3.1 Fitness consequences of social interactions between two individuals[a]

Category	Individual A (a helper)	Individual B
Cooperation (A and B help each other right now)	Gains direct fitness (immediately)	Gains direct fitness
Postponed cooperation (A eventually gains access to a resource controlled by B because of its prior help)	Gains direct fitness (after delay)	Gains direct fitness
Reciprocity (B *directly* pays A back for its prior aid)	Gains direct fitness (after delay)	Gains direct fitness
Maladaptive altruism (A sacrifices its lifetime inclusive fitness in order to help B)	Loses inclusive fitness	Gains direct fitness
Adaptive altruism (The initial direct fitness sacrifice made by A leads to indirect fitness gains for A)	Gains indirect fitness	Gains direct fitness
Spite (A reduces its reproductive success in order to harm B)	Loses inclusive fitness	Loses inclusive fitness
Deceit and manipulation (B exploits or manipulates A in ways that harm A but benefit B)	Loses inclusive fitness	Gains inclusive fitness

[a]The distinctions between categories are often subtle but revolve around such key factors as whether the helpful individual is repaid by boosting its direct fitness or its indirect fitness.

FIGURE 3.5   Mutual defense in a society of bluegills. Each colonial male gains from its social behavior because of the mutual benefits derived from the collective response to predators such as bass (above), bullhead catfish (left), snails, and pumpkinseed sunfish (right foreground), which roam the colony in search of eggs. Drawing courtesy of Mart Gross.

nates are engaged in kin-directed altruism, in which their sacrifices—in terms of lowered production of their own offspring—are compensated by increases in the number of relatives that exist because of their helpful behavior.

However, we might account for their apparent self-sacrifice by demonstrating that these unrelated subordinates derive a future reproductive benefit by helping now. A team led by Ellouise Leadbeater showed that social subordinates had a chance of inheriting a well-protected nest from the dominant queen upon the queen's demise. If so, the successor to the initial queen would have the assistance of the remaining subordinates at the nest. Because of the considerable reproductive success of the inheritors, subordinates on average have higher direct fitness than solitary nesting females, which run a high risk that they will be killed and/or their nests will be destroyed by the predatory birds and mammals that plague weakly defended nests of paper wasps (Figure 3.6).[47] Thus by helping a dominant female, the subordinate receives a ticket from the dominant that permits the helper to stay with the colony now, thereby gaining a chance to inherit the nest at some later date.

Behavioral ecologists have uncovered many other cases of individuals helping unrelated members of their species in ways that eventually result in

(A)

(B)

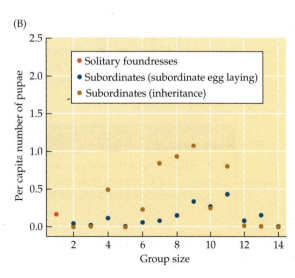

FIGURE 3.6   A direct fitness benefit for helping in a paper wasp. Females that help an unrelated foundress female have a chance of inheriting a colony from the dominant foundress. (A) The photograph shows an early-stage colony with two foundress females. One is dominant to the other, but the subordinate female has a chance to inherit the nest should the dominant disappear. (B) Inheritors often acquire a well-defended colony with helpers, and as a result, they generally reproduce more successfully than solitary nest foundresses in *Polistes dominulus*. A, photograph by Ellouise Leadbeater; B, after Leadbeater et al.[47]

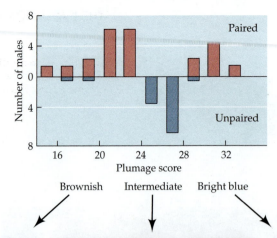

Brownish   Intermediate   Bright blue

FIGURE 3.7   Cooperation among competitors. Yearling male lazuli buntings range in color from dull brownish to bright blue and orange. (Their plumage scores range from less than 16 to more than 32.) Bright yearling males permit dull males, but not males of intermediate brightness, to settle on territories neighboring their own. As a result, brownish males often pair off with females in their first year, whereas yearling males of intermediate plumage typically remain unpaired. After Greene et al.;[27] photographs courtesy of Erick Greene.

heightened personal reproductive success for the helper, on average. Some of these cooperative acts are more byzantine than the one based on the chance of inheriting a paper wasp nest. Consider, for example, the strange cooperation that links subordinate yearling male lazuli buntings, which have dull brown plumage, with brightly colored, dominant yearling males (Figure 3.7). The dominant males tolerate dull-plumaged birds as neighbors, letting them settle in superior habitat next door while aggressively driving away other males that are more brightly colored. One reproductive benefit for the brightly colored males from this tactic is that they get to mate with the subordinate males' females, which presumably find the brightly colored males more appealing than their own dull-colored social partners, as is true for some other birds in which both dull and brightly colored males compete for mates.[76] Erick Greene and his coworkers found that dull-plumaged males regularly care for between one and two extra-pair young, which are presumed to be the genetic offspring of their more attractive neighbors.[27]

The real Darwinian puzzle has to do with the benefit, if any, that subordinate males secure from living next to such sexually attractive males. Greene and his fellow researchers found that subordinate buntings do generally rear a few offspring of their own as a result of possessing high-quality territories that appeal to females. In contrast, males of intermediate plumage brightness are often pushed by dominant rivals into habitat so poor that no female will join them, with the result that they must wait an entire extra year before trying to reproduce again.[27]

Discussion Question

3.2   Given the differences in reproductive success for the three categories of male lazuli buntings, how can we account for the evolutionary persistence of males with dull and, especially, intermediate plumage?

Because both dull-colored male buntings and their brightly plumaged neighbors gain some fitness from their interactions, this social arrangement

FIGURE 3.8 Cooperative courtship of the long-tailed manakin. The two males are in the cartwheeling portion of their dual display to a female, which is perched on the vine to the right.

can be categorized as cooperation. But is the same true for male partnerships among long-tailed manakins? In this bird (and some other manakins[48,63]), unrelated males form pairs that exhibit extraordinary coordinated displays in an attempt to attract females.[26,51] Females that respond to the cooperative callers may land on the males' display perch, often a horizontal section of liana that lies a foot or so above the ground. In response, the two males dart in and land close to the prospective mate before performing their astonishing cartwheel display (Figure 3.8). After a series of these moves, the males may then flutter slowly back and forth in front of the female, showing off their beautiful plumage in the "butterfly flight." Should a female visitor start jumping excitedly on the perch in response to these displays, one member of the duo discreetly departs, while the remaining male stays to copulate with the visitor. When the female then flies off, the mated male calls for its display partner, which hurries back to resume its cooperative but apparently unrewarded duties.

By marking the males at display perches, David McDonald and his manakin watchers were able to identify an alpha male manakin, which did all the mating, not leaving anything for its display companions, not even for the alpha's favorite colleague, the beta male, which in turn is dominant to any other part-time cooperators.[50] How can it be adaptive for a celibate beta subordinate to help a sexually monopolistic alpha male year after year? McDonald proposed that by conceding all receptive females to the alpha male, a beta manakin establishes his claim to be next in line, if and when the alpha male disappears, keeping other (mostly somewhat younger) birds at bay. When an alpha male does disappear, the beta male moves up in status and then usually gets to mate with many of the same females that copulated with the previous alpha (Figure 3.9).[50] Thus, beta males work for unrelated partners because this is the only way to join a queue to have any chance of becoming a reproducing alpha male—eventually. Note that there is no guarantee that a beta male will become an alpha, only a better than average chance of doing so.

FIGURE 3.9 Cooperation with an eventual payoff. After the death of his alpha male partner, the beta male long-tailed manakin (now an alpha) copulates about as frequently as his predecessor did, presumably because the females attracted to the duo in the past continue to visit the display arena when receptive. After McDonald and Potts.[50]

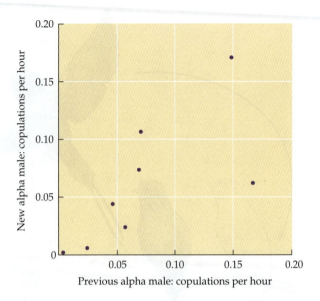

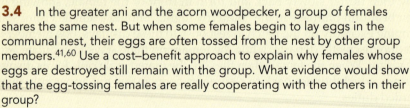

Discussion Questions

3.3   Lions live in prides composed of a variable number of females and males. Each pride defends its own territory.[52] If lion social groups form because of the benefits of communal defense of a territory, what prediction follows about the interactions between prides of different sizes? What additional predictions can you make about the size of prides and (1) the quality of the defended habitat, (2) the survival of female pride members, and (3) the fitness of females? Would you be surprised to learn that as pride size grows, the foraging efficiency of the group does not increase? How can you reconcile this finding with the existence of relatively large prides? When lion prides get large, one subgroup may move away and take up residence in a nearby territory. Prides that have recently split tend to interact with relatively little aggression; why?

3.4   In the greater ani and the acorn woodpecker, a group of females shares the same nest. But when some females begin to lay eggs in the communal nest, their eggs are often tossed from the nest by other group members.[41,60] Use a cost–benefit approach to explain why females whose eggs are destroyed still remain with the group. What evidence would show that the egg-tossing females are really cooperating with the others in their group?

The Reciprocity Hypothesis

The study of long-tailed manakins reveals that some superficially self-sacrificing actions actually advance the reproductive chances of at least some patiently helpful individuals that survive long enough to graduate to alpha (mating) status. Cases of postponed cooperation (see Table 3.1) are very similar to instances of reciprocity in which the helpful individuals also receive delayed compensatory repayment *directly* from the helped individual, rather than waiting to inherit a resource. Robert Trivers called this social arrangement **reciprocal altruism** because individuals that are helped generally return the favors they receive—eventually.[73] But because the helpers are not really sacrificing direct fitness over the long haul, perhaps reciprocal altruism should be called just plain **reciprocity**. Whatever the label, if the initial direct fitness

FIGURE 3.10 Reciprocity occurs in primates that groom one another. Here two females are cleaning the fur of their partner and protector, a male hamadryas baboon that will keep other males from sexually harassing his groomers in the future. Photograph by Mat Pines.

cost of helping is modest but the delayed direct fitness benefit from receiving the returned favor is on average greater, then selection can favor being helpful in the first case. A classic example is the grooming behavior of assorted primates in which individuals pick through the fur of a companion, a helpful act that may be repaid at a later date when the groomee becomes the groomer or performs some other useful behavior for the animal that had been its helper (Figure 3.10).[64]

Although determining that an apparent case of reciprocity is legitimate can be difficult,[58] reasonably good experimental evidence for reciprocity exists. For example, a convincing case involves pied flycatchers, which are among the many species of small songbirds that will mob (i.e., harass by flying around or even assaulting) a predator, such as a hawk or an owl, especially one that appears near their nests. In one experiment, pairs of pied flycatchers participated in an experiment in which a test pair and a neighboring pair were confronted with a stuffed tawny owl placed near both their nests. Both pairs mobbed the pseudo-owl by flying around it and dive-bombing the "predator." Some time later, after a third, captured pair of flycatchers had been released (after having been held in captivity during the initial part of the experiment), two stuffed owls were placed, one near the nest of this previously captured pair, which had been unable to join the mobbers previously, and one near the nest of the pair that had helped the test pair. The test pair observed the other flycatchers and had a choice to make, namely, which of the other two pairs to help by joining in the mobbing of their "owl." In 30 of 32 trials, the test pair chose to help the neighbors that had been able to assist them an hour earlier.[45] Thus, pied flycatchers appear to remember who has helped them and who has not, and they use this information to pay back those that have been helpful, while ignoring those that have not been cooperative.

Although pied flycatchers and some primates appear to be capable of reciprocity, direct payback behavior is not particularly common,[18,80] perhaps in part because a population composed of reciprocal helpers would generally be vulnerable to invasion by individuals happy to accept help but likely to "forget" about the payback. "Defectors" reduce the fitness of helpers in such a system, which ought to make reciprocity less likely to evolve. The problem can

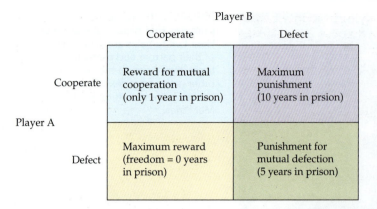

Player B

	Cooperate	Defect
Cooperate	Reward for mutual cooperation (only 1 year in prison)	Maximum punishment (10 years in prsion)
Defect	Maximum reward (freedom = 0 years in prison)	Punishment for mutual defection (5 years in prison)

Player A

FIGURE 3.11   The prisoner's dilemma. The diagram lays out the payoffs for player A that are associated with cooperating or not cooperating with player B. Defection is the adaptive choice for player A given the conditions specified here (if the two individuals will interact only once).

be illustrated with a game theoretical model called the **prisoner's dilemma** (Figure 3.11). The dilemma can be illustrated by imagining that a crime has been committed by two persons who have agreed not to squeal on each other if caught. The police have brought them in for interrogation and have put them in separate rooms. The cops have enough evidence to convict them both on lesser charges but need to have the criminals implicate each other in order to jail them for a more serious crime. The police therefore offer each suspect freedom if he will squeal on his pal. If suspect A accepts the offer ("defect") while B maintains their agreed-upon story ("cooperate"), A gets his freedom (the maximum reward) while B gets hit with the maximum punishment—say, 10 years in prison (the "sucker's payoff," as shown in the upper right panel of Figure 3.11). If together they maintain their agreement (cooperate + cooperate), then the police will have to settle for conviction of both on the lesser charges, leading to, say, a 1-year prison term for each suspect. And if each one fingers the other, the police will use this evidence against both and renege on their offer of freedom for the snitch, so both A and B will be punished quite severely with, say, a 5-year prison sentence each.

In a setting in which the payoffs for the various responses are ranked "defect while other player cooperates" > "both cooperate" > "both defect" > "cooperate while other player defects," the optimal response for suspect A is always to defect, never to cooperate. Under these circumstances, if suspect B maintains their joint innocence, A gets a payoff that exceeds the reward he could achieve by cooperating with a cooperative B; if suspect B squeals on A, defection is still the superior tactic for A, because he suffers less punishment when both players defect than if he cooperates while his companion squeals on him. By the same token, suspect B will always come out ahead, on average, if he defects and points the finger at his buddy.

This model predicts, therefore, that reciprocal cooperation should never evolve. How, then, can we account for the cases of reciprocity that have been observed in nature (and are also common features of human social behavior[80])? One answer comes from examining scenarios in which two players interact repeatedly, not just once. When this condition applies, individuals that use the simple decision rule "do unto individual X as he did unto you the last time you met" can reap greater overall gains than cheaters who accept assistance but never return the favor.[2] When multiple interactions are possible, the rewards for back-and-forth cooperation add up, exceeding the short-term payoff from a single defection. In fact, the potential accumulation of rewards can even favor individuals who "forgive" a fellow player for an occasional defection because that tactic can encourage maintenance of a long-term relationship with its additional payoffs.[68,77]

Vampire bats appear to meet the required conditions for adaptive multiple-play reciprocity. These animals must find scarce vertebrate victims from which to draw the blood meals that are their only food. After an evening of foraging, the bats return to a roost where individuals known to one another regularly assemble. A bat that has had success on a given evening can collect a large amount of blood, so much that it can afford to regurgitate a life-sustaining amount to a companion that has had a run of bad luck. Under these circumstances, the cost of the gift to the donor is modest, but the potential benefit to the recipient is high, since vampire bats die if they fail to get food three nights in a row. Thus, a cooperative blood donor is really buying insurance against starvation down the road. Individuals that establish durable give-and-take relationships with one another are better off over the long haul than those cheaters that accept one blood gift but then renege on repayment, thereby ending a potentially durable cooperative arrangement that could involve many more meal exchanges.[81] However, it remains to be determined whether the reciprocating bats are often relatives, in which case their behavior might well fall under the category of kin-directed altruism.[22]

Kin Selection and Helpful Behavior

As we saw in Chapter 2, kin, or indirect, selection has been valuable in analyzing cases in which helpful behavior seems unlikely to ever provide *direct* fitness benefits for helpers, as in the case of sterile worker insects. One of the first persons to consider the possible effects of kin selection on social interactions in a vertebrate species was Paul Sherman in his study of the Belding's ground squirrel.[66] This medium-size North American rodent produces a staccato whistle (Figure 3.12) when a coyote or badger approaches. The sound of one Belding's ground squirrel whistling sends other nearby ground squirrels rushing for safety.

Sherman discovered that calling squirrels are tracked down and killed by weasels, badgers, and coyotes at a higher rate than noncallers, which indicates that callers pay a survival—and thus reproductive—price for sounding the alarm. However, this price might be more than offset by an increase in the survival of the caller's offspring, which if true would boost the caller's direct fitness. The fact that adult females generally live near their offspring and are more than twice as likely to give the alarm than adult males, which move away from their progeny and so cannot alert offspring to danger, is consistent with a direct fitness explanation for alarm calling.

But there may be more than a probable direct fitness benefit in some cases. Sherman observed that adult female squirrels with **nondescendant kin** (relatives other than offspring) living nearby were significantly more likely to yell when a predator appeared than were adult females without sisters, aunts, and nieces as neighbors. This result is what we would predict if alarm callers sometimes help relatives other than offspring survive to pass on shared family genes, resulting in indirect fitness gains for those that warn their relatives of danger. Sherman's work therefore indicates that both direct and indirect selection contribute to the maintenance of alarm-calling behavior in this species.[66]

FIGURE 3.12 An alarmed Belding's ground squirrel gives a warning call after spotting a terrestrial predator.

Discussion Question

3.5 Siberian jays that find perched owls or hawks in the forests where they live often call to others to join them in harassing these predators with noisy vocalizations and swooping attacks.[28] This behavior could have direct fitness benefits for a mobber if its actions induced a dangerous predator to move out of the home range of the jay and its offspring. But the behavior could also have indirect fitness benefits for a bird whose behavior helped nondescendant kin by driving a predator away. Figure 3.13 presents data on the mean number of all mobbing calls given by members of groups of Siberian jays, some of which were composed of kin and others of which were made up of nonkin. How do these data help us reach conclusions about the two hypotheses?

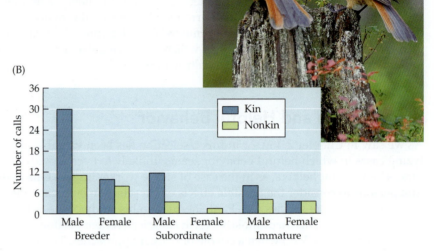

(A)

FIGURE 3.13 Mobbing and kinship in groups of Siberian jays. Male members of kin groups gave many more mobbing calls upon discovering a predator in their territory than males in groups composed of nonrelatives. B after Griesser.[28]

(B)

Helpers at the Nest: A Darwinian Puzzle

Uli Reyer showed that both direct and indirect selection played a role in the evolution of helping by male pied kingfishers.[59] These attractive African birds nest in tunnels in the banks of large lakes and rivers. Year-old males that are unable to find mates become primary helpers that bring fish to their mothers and their nestlings while attacking predatory snakes and other nest enemies. Are these males propagating their genes as effectively as possible by helping to raise their siblings? They do have other options: they could help unrelated nesting pairs in the manner of secondary helpers, which are unrelated to the pairs that they are assisting, or they could simply sit out the breeding season, waiting for next year in the manner of delayers, which do not help rear any offspring.

To learn why primary helpers help, we need to measure the costs and benefits of their actions. Primary helpers work harder than delayers and the more laid-back secondary helpers (Figure 3.14). The greater sacrifices of primary helpers translate into a lower probability of their surviving to return to the breeding grounds the next year (just 54 percent return) compared with second-

FIGURE 3.14 Altruism and relatedness in pied kingfishers. Primary helpers deliver more calories per day in fish to a nesting female and her offspring than do secondary helpers, which are not related to the breeders they assist. After Reyer.[59]

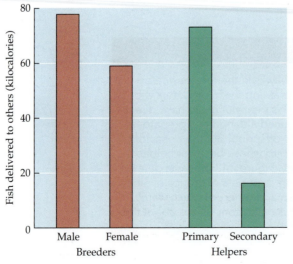

Pied kingfisher

ary helpers (74 percent return) or delayers (70 percent return). Furthermore, only two in three surviving primary helpers find mates in their second year and reproduce, whereas 91 percent of returning secondary helpers go on to breed. Many one-time secondary helpers bond with the females they helped the preceding year (10 of 27 in Reyer's sample), suggesting that improved access to a potential mate is the ultimate payoff for their helpful behavior.

These data enable us to calculate the direct fitness cost to the altruistic primary helpers in terms of reducing their own reproduction in their second year of life. For simplicity's sake, we shall restrict our comparison to solo primary helpers that help their parents rear siblings in the first year and then breed on their own in the second year, if they survive and find a mate, versus secondary helpers that help nonrelatives with no other helpers present in the first year and then reproduce on their own in the second year, if they survive and find a mate.

Primary helpers throw themselves into helping their parents produce offspring at the cost of being less likely to survive and thus to reproduce themselves in the next year. True, primary helpers do better than delayers in the second year (0.41 versus 0.29 units of direct fitness), but secondary helpers do better still in year 2 (0.84 units of direct fitness) because they are more likely to survive to year 2, at which time they usually secure a partner (Table 3.2).

But is the cost to primary helpers of 0.43 lost units of direct fitness (0.84 – 0.41 = 0.43) in the second year offset by a gain in indirect fitness during the first year? To the extent that these males increase their parents' reproductive success, they create siblings that would not otherwise exist, indirectly propagating their genes in this fashion. In Reyer's study, the parents of a primary helper gained an extra 1.8 offspring, on average, when their son was present. Some primary helpers assisted their genetic mothers and fathers, in which case the extra 1.8 siblings were full brothers and sisters, with a coefficient of relatedness of 0.5. But in other cases, one parent had died and the other had remated, so the offspring produced were only half siblings ($r = 0.25$). The average coefficient of relatedness for sons helping a breeding pair was thus between one-fourth and one-half ($r = 0.32$). Therefore, the average gain for primary helper sons was 1.8 extra sibs $\times$ 0.32 = 0.58 units of indirect fitness in the first year, a figure higher than the mean direct fitness loss experienced in their second year of life.

Note that Reyer used Hamilton's rule (see page 24) to establish that primary helpers sacrifice future personal reproduction in year 2 in exchange for increased numbers of nondescendant kin in year 1.[59] Because these added siblings carry some of the helpers' alleles, they provide indirect fitness gains that more than cover the loss in direct fitness that primary helpers experience in their second year relative to secondary helpers.

TABLE 3.2 Calculations of inclusive fitness for male pied kingfishers

Behavioral tactic	First year			Second year				
	y	r	f_1	o	r	s	m	f_2
Primary helper	1.8 $\times$ 0.32 = 0.58			2.5 $\times$ 0.50 $\times$ 0.54 $\times$ 0.60 = 0.41				
Secondary helper	1.3 $\times$ 0.00 = 0.00			2.5 $\times$ 0.50 $\times$ 0.74 $\times$ 0.91 = 0.84				
Delayer	0.3 $\times$ 0.00 = 0.00			2.5 $\times$ 0.50 $\times$ 0.70 $\times$ 0.33 = 0.29				

Source: Reyer[59]

Symbols: y = extra young produced by helped parents; o = offspring produced by breeding ex-helpers and delayers; r = coefficient of relatedness between the male and y, and between the male and o; f_1 = fitness in the first year (indirect fitness for the primary helper); f_2 = direct fitness in the second year; s = probability of surviving into the second year; m = probability of finding a mate in the second year.

FIGURE 3.15 Purple-crowned fairy wrens are similar to pied kingfishers. Here an adult breeding bird is surrounded by helpers. In this species, some helpers are the offspring of the adult birds they assist, while others are unrelated to those they help. Photograph by Michelle Hall.

Discussion Questions

3.6   Given the results of our calculations of inclusive fitness for male pied kingfishers, isn't it maladaptive to be a delayer? Why aren't there any completely sterile helper male pied kingfishers?

3.7   In the purple-crowned fairy wren (Figure 3.15), most subordinate helpers at the nest feed full siblings or half siblings, but some helpers are unrelated to the nestlings they assist.[38] What Darwinian puzzle is created by these findings and how might you solve the problem by using inclusive fitness theory as Uli Reyer did in his kingfisher study? What predictions follow from the explanation(s) that you have proposed?

Helpers at the Nest: Alternative Hypotheses

In the pied kingfisher, primary helpers raise their fitness *indirectly* through their increased production of nondescendant kin, whereas secondary helpers raise their fitness *directly* by increasing their future chances of reproducing personally. Primary helpers reproduce less over a lifetime but are compensated genetically for their helpful behavior via a gain in indirect fitness; secondary helpers *temporarily* reproduce less but sometimes inherit territories and sexual partners as a result of their helpful behavior, thereby securing direct fitness benefits. Thus, this one species offers support for two different hypotheses on the evolution of helping at the nest. The indirect and direct fitness hypotheses can be tested for other cases of helpers, which are found in some other birds as well as certain other vertebrates.[10,36,72]

However, helping could also be a nonadaptive side effect of other fitness-enhancing traits. As Ian Jamieson has pointed out, perhaps helping occurs simply because it is adaptive for young adults to delay their dispersal from their natal territory.[34] If these stay-at-home birds were exposed to the nestlings being cared for by their parents, the begging behavior of the baby birds might trigger parental behavior in the young nonbreeding adults. Thus, delayed dispersal and the tendency to care for one's own offspring might lead to helping behavior, even if feeding someone else's young reduced the fitness of nonbreeding helpers.

This nonadaptive **by-product hypothesis** assumes that selection could not eliminate the helper's tendency to feed its parents' offspring without also destroying the capacity of the helper to care for its own nestlings at a later date. This proposition is testable. One of its key predictions is that the underlying mechanisms of parental care should be no different in species with helpers than in species without helpers. The group of birds known as jays provides the necessary comparative test. In the Mexican jay (*Aphelocoma ultramarina*) and Florida scrub jay (*Aphelocoma coerulescens*), some nonbreeding birds help their parents rear additional siblings (Figure 3.16). In contrast, the western scrub jay (*Aphelocoma californica*) is a member of the same genus but lacks helpers at the nest. In this species, only breeding individuals have high prolactin levels, while nonbreeders have low levels of this hormone, which appears to promote parental care in many birds. But nonbreeding helpers in the Mexican jay and Florida scrub jay have prolactin levels that match those of their breeding parents.[65,75] Moreover, the prolactin in nonbreeding Mexican jays rises to peak levels *before* there are young to feed in the nest (Figure 3.17), suggesting that

FIGURE 3.16 Cooperation among scrub jay relatives. In the Florida scrub jay, helpers at the nest provide food for the young, defense of the territory, and protection against predators. Based on a drawing by Sarah Landry, from Wilson.[82]

FIGURE 3.17 Seasonal changes in prolactin concentrations in (A) breeders and (B) nonbreeding helpers at the nest in the Mexican jay. Nonbreeding birds in a group exhibit the same pattern of increased prolactin production prior to the hatching of eggs as do breeding adults. After Brown and Vleck.[11]

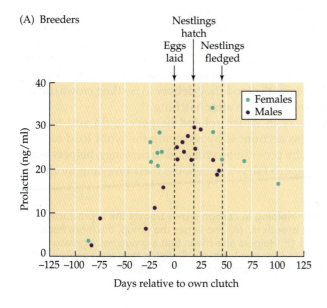

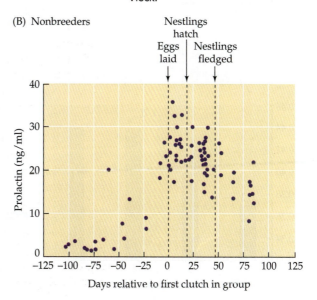

TABLE 3.3 Effect of Florida scrub jay helpers at the nest on the reproductive success of their parents and on their own inclusive fitness

	Parents without breeding experience[a]	Parents with breeding experience
Average number of fledglings produced with no helpers	1.03	1.62
Average number of fledglings produced with helpers	2.06	2.20
Increase in reproductive success due to help	1.03	0.58
Average number of helpers	1.70	1.90
Indirect fitness gained per helper	0.60	1.30

Source: Emlen[24]

[a]Includes pairs in which one parent has reproduced, which is why some pairs in this category acquire a helper at the nest.

Carrion crow

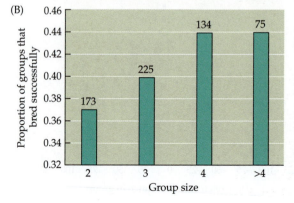

selection has favored nonbreeding individuals of this species that become hormonally primed to rear their siblings.[75] These results are at odds with the nonadaptive by-product explanation for helping.

But if helping is adaptive, do helpers derive inclusive fitness gains via the direct or indirect route, or both? In the Mexican jay and Florida scrub jay, some stay-at-home helpers inherit their natal territories from their parents in much the same way that some paper wasp helpers do when they inherit nests from unrelated dominant females. The jays that inherit good territories derive direct fitness from helping, and so do some adult male carrion crows that leave their natal territory to join a group of cooperative breeders nearby where they may sire offspring.[3]

However, indirect fitness benefits are also possible for nonreproducing helpers if their parent or parents rear more offspring than other adults that don't have helpers at the nest (Table 3.3).[10,84] In fact, the vast majority (estimated at over 90 percent) of cooperative-breeding bird species are composed of family or kin groups with helpers that are generally the offspring of breeding pairs.[31] In the carrion crow, for example, kin that are helpers derive an indirect fitness benefit from their work by increasing the number of full and half siblings that survive to fledge from the nest (Figure 3.18). Their positive effect on nestling survival stems in part from the food they deliver to nestling crows during the critical first few days after hatching.[13]

Another way in which related helpers could add to their indirect fitness account is by reducing the work load of their parents.[37] So, by providing extra food to nestlings, helpers at the nest in the superb fairy-wren enable their breeding mother to

FIGURE 3.18 Helping in carrion crows boosts the reproductive success of the breeders. (A) The larger the group, the more fledglings produced. (B) Helpers also mean that a group has a greater probability of producing at least one fledgling than pairs working on their own. Numbers above the columns show the numbers of nests observed. After Canestrari et al.[13]

make lighter eggs that contain significantly less fat and protein than those laid by a breeding female without a helpful retinue.[61] By cutting back on their egg investments per breeding attempt, females with helpers may live longer and, thus, have more lifetime opportunities to reproduce. If so, helpers may boost the number of siblings created by their mother, adding to the helpers' indirect fitness if they are related to the adult female that they are assisting.[17]

Nonetheless, we have to consider the possibility that any apparent benefits supplied by helpers may actually be the effect of occupying superior territories, which provide more food or superior nesting sites. Thus, differences in the numbers of offspring fledged by parents with versus without helpers might be due to differences in the quality of the territories held by pairs in the two categories, not because one set had helpers at the nest and the other did not. Ronald Mumme tested this hypothesis by capturing and removing the nonbreeding helpers from some randomly selected breeding pairs of Florida scrub jays, while leaving others with helpers untouched. The experimental removal of helpers reduced the reproductive success of the experimental pairs by about 50 percent, as measured by the number of offspring known to be living 60 days after hatching (Figure 3.19). Helpers apparently really do help, at least in this species.[54]

In fact, helper scrub jays also improve the chances that their parents will live to breed again another year, as has been suggested for the superb fairywren. Improved parental survival means that the scrub jay helpers are responsible for still more siblings in the future; these extra siblings yield an average of about 0.30 additional indirect fitness units for each helper.[53] The total indirect fitness gained from helping at the nest can potentially exceed its costs in terms of lost direct fitness, especially if young helpers have almost no chance of reproducing personally. When very few openings are available for dispersing youngsters, helping is more likely to be the adaptive option.

Whether saturated nesting habitats contribute to the maintenance of facultative (optional) helping at the nest is also testable. If young birds remain on their natal territories because they cannot find suitable nesting habitat, then yearlings given an opportunity to claim unoccupied territories of high quality should promptly exercise the option to become breeders. Jan Komdeur did the necessary experiment with the Seychelles warbler, a small brown bird that has played a big role in testing evolutionary hypotheses about helping at the nest. When Komdeur transplanted 58 birds from one island (Cousin) to two other nearby islands with no warblers, he created vacant territories on Cousin, and helpers at the nest there immediately stopped helping in order to move into the open spots and begin breeding. Since the islands that received the transplants initially had many more suitable territorial sites than warblers, Komdeur expected that the offspring of the transplanted adults would also leave home promptly in order to breed on their own. They did, providing further evidence that young birds help only when they have little chance of making direct fitness gains by dispersing.[43]

Moreover, the sophisticated internal mechanisms that control the dispersal decisions made by young Seychelles warblers are sensitive to the quality of their natal territory. Breeding birds occupy sites that vary in size, plant cover, and edible insects. By using these variables to divide warbler territories into categories of low, medium, and high quality, Komdeur showed that young helpers on good territories were likely to survive there while also increasing the odds that their parents would reproduce successfully. Young birds whose parents had prime sites often stayed put, securing both direct and indirect fitness gains in

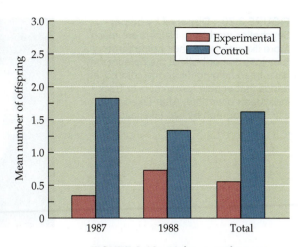

FIGURE 3.19 Helpers at the nest help raise more siblings. In the Florida scrub jay, fewer offspring were alive after 60 days in experimental nests from which the helpers had been removed versus unmanipulated control nests during a 2-year experiment. After Mumme.[54]

FIGURE 3.20 **Dispersal by helper Seychelles warblers.** Helpers are much more likely to leave their home territory if they lose one or both of the parents they have been helping. Note the colored leg bands that enable researchers to track individuals over time. After Eikenaar et al.;[23] photograph by Cas Eikenaar.

Seychelles warbler

the process. In contrast, young birds on poor natal territories had little chance of helping to boost the reproductive success of their parents. Instead, they left home and tried to find breeding opportunities of their own.[42] The probability that a warbler would disperse was also influenced by whether both of its genetic parents were alive and in control of the family territory or whether one or both had been replaced by new stepparents (Figure 3.20).[23] Thus, the dispersal of helpers became more likely if opportunities to help close kin had declined.

Discussion Question

3.8 Helpers at the nest have been found in only about 9 percent of all bird species.[16] One attribute of this minority of birds that has often been linked to the evolution of helping behavior is the delayed dispersal of juveniles, as we have just illustrated for Florida scrub jays and Seychelles warblers. But another factor that might have promoted the evolution of helping is a very low adult mortality rate. These two ideas have sometimes been presented as competing hypotheses, but how might they both reflect the same ecological pressure that makes helping at the nest an adaptive temporary option for young birds?

The behavioral flexibility exhibited by Seychelles warblers is not unique to that species. Consider how young female white-fronted bee-eaters make adaptive decisions about reproducing (Figure 3.21). This African bird nests in loose colonies in clay banks. Like male pied kingfishers, young female white-fronted bee-eaters can choose to breed, or to help a breeding pair at their nest burrow, or to sit out the breeding season altogether. If an unpaired, dominant, older male courts a young female, she almost always leaves her family and natal territory to nest in a different part of the colony, particularly if her mate has a group of helpers to assist in feeding the offspring they will produce. Her choice usually results in high direct fitness payoffs. But if young, subordinate males are the only potential mates available to her, the young female will usually refuse to set up housekeeping. Young males come with few or no helpers, and when they try to breed, their fathers often harass them, trying to force their sons to abandon their mates and return home to help rear their siblings.[25]

A female that opts not to pair off under unfavorable conditions may choose to slip an egg into someone else's nest or to become a helper at the nest in her natal territory—provided that the breeding pair there are her parents. If one or both of her parents have died or moved away, she is unlikely to help rear the chicks there, which are at best half siblings, and instead will simply wait, conserving

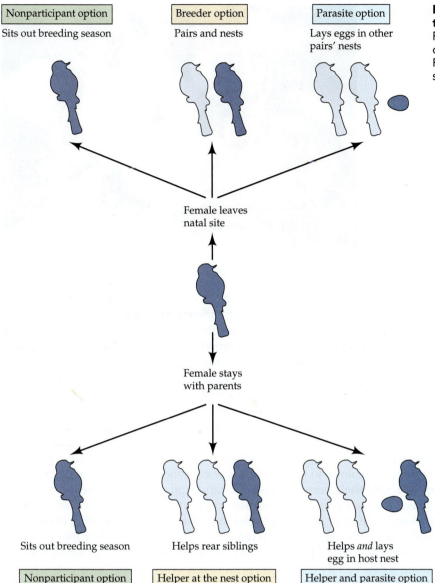

Nonparticipant option	Breeder option	Parasite option
Sits out breeding season	Pairs and nests	Lays eggs in other pairs' nests

Female leaves natal site

Female stays with parents

Sits out breeding season

Helps rear siblings

Helps *and* lays egg in host nest

Nonparticipant option	Helper at the nest option	Helper and parasite option

FIGURE 3.21 Conditional reproductive tactics of female white-fronted bee-eaters. Females of this species have many options, of which helping at the nest is only one. Females select a given tactic suited for their special circumstances. After Emlen et al.[25]

her energy for a better time in which to reproduce. Thus, although daughter bee-eaters have the potential to become helpers at the nest, they choose this option only when the indirect fitness benefits of helping are likely to be substantial.[25]

Helpers at the Burrow: The Case of the Naked Mole Rat

One of the strangest social mammals is the naked mole rat.[12,67] This little hairless, sausage-shaped creature (Figure 3.22) forms colonies of up to 200 individuals living in a complex maze of burrows under the African plains. The impressive size of their subterranean home stems from extraordinary cooperation among chain gangs of colony members, which work together to move tons of earth to the surface each year while burrowing about in search of edible tubers. Yet when it comes to reproducing, breeding is restricted to a single big "queen" and several "kings" that live in a centrally located nest chamber. Females other than the queen do not even ovulate. Instead, they serve as sterile helpers at the nest, consigned to specialized support roles for the queen and kings, as do most of the males in the colony.[46]

FIGURE 3.22   A mammal with an effectively sterile caste. Naked mole rats live in large colonies made up of many workers that serve a queen and one or a few breeding males. Three pups are shown here nursing from their mother, the queen mole rat.

The altruism seen in this eusocial, caste-forming vertebrate appears to stem at least in part from policing by the queen mole rat in particular. The chief policewoman shoves other members of the colony around, inducing high levels of stress in subordinate females and males. (Note that aggressive interactions also occur in groups of birds where there are helpers at the nest, and these involve such things as destruction of the eggs of some females and forced eviction from the group's territory.[32])

Bullying by the queen mole rat suppresses the production of sex hormones in her underlings, so they become incapable of reproducing. Thus, the altruism shown by subordinate naked mole rats occurs at one level in part because they are forced to forgo reproduction. At that juncture, their only options are to leave the colony to try to reproduce elsewhere (a very risky proposition) or to accept their nonreproductive status and assist the queen mother sufficiently to be permitted to remain within the safety of their group. Nonetheless, at another level, the decision to stay with the bullying queen may well be adaptive because naked mole rat workers help produce the occasional reproductively capable sibling, a plump brother or sister that leaves home to found a new colony elsewhere, presumably with an unrelated individual of the opposite sex from another group.[6]

Discussion Question

3.9   In meerkats, another cooperatively breeding African mammal, helpers supply insects to youngsters, and this provisioning fattens up the juveniles more quickly. In light of this result, why did a research team supply one group of young meerkats with supplemental food while refraining from doing so with another equivalent group (Figure 3.23A)?[62] And why did researchers collect the data shown in Figure 3.23B?[86]

Altruism in Vertebrates and Insects: A Comparison

Self-sacrificing behavior clearly occurs in animals other than the eusocial bees, wasps, ants, and termites. But almost all helper vertebrates retain the capacity to reproduce. Instead of working selflessly for others for a whole lifetime, they typically assist their kin only for a limited time and then attempt to reproduce

(A)

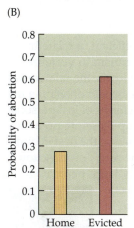

Meerkats

(B)

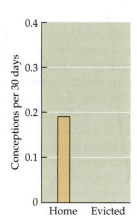

FIGURE 3.23 Why do meerkats live in groups? (A) The effect of supplemental food on the probability that young females (light green) and young males (dark green) would achieve some reproductive success during their lifetimes. (B) Subordinate females that were temporarily driven from their band were more likely to abort their fetuses and less likely to conceive during the period of eviction than those females that avoided being evicted. A, after Russell et al.;[62] B, after Young et al.[86]

personally. In other words, highly social vertebrates engage in **facultative altruism**, not **obligate altruism**, and appear to maximize their inclusive fitness via a mix of indirect and direct fitness.

With respect to that period in a social mammal or helper bird's life when it is assisting others, kin selection theory provides a powerful tool for understanding the helper's behavior. As we have seen, this theory is the basis for predictions that altruists will direct their assistance to relatives, usually close ones, and that such assistance will boost the reproductive success of the individuals they help. Helpers at the nest in birds do indeed tend to provide help toward others in relation to the degree to which they are related.[29]

Furthermore, Hamilton's rule suggests that in cases of adaptive altruism, helpers are giving up relatively little in the way of direct fitness, because their chances of reproducing personally are very low. Among the factors that reduce the cost of forgoing reproduction is an environment saturated by competitors so that few territories or other resources are available to a would-be disperser. If the cost of not reproducing is low for this reason or simply because leaving a relatively safe home base is very dangerous, then the downside of giving up a small chance to reproduce is likely to be outweighed by the indirect fitness benefits derived from boosting the fitness of close relatives, such as siblings. These conditions apply with equal strength to termites, Seychelles warblers, and meerkats.[14]

Another factor in the evolution of altruism in vertebrates and insects alike is the role played by the animal's mating system. As noted in Chapter 2, monogamous insect queens, whether hymenopterans or termites, produce offspring

FIGURE 3.24 Cooperative breeding in birds is linked to monogamy. In 267 species of birds for which data exist on the degree to which females are promiscuous (i.e., mate with more than one male), species with promiscuous females are significantly less likely to have helpers at the nest and the like. After Cornwallis et al.[19]

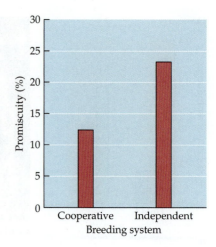

that are full siblings with a relatively high coefficient of relatedness. Monogamy would also promote adaptive altruism in vertebrates. The more partners taken by a female bird, the lower the relatedness of siblings to one another and the less likely the benefit of helping relatives will overcome the costs to the altruist in terms of lost personal reproduction. Charlie Cornwallis and his group therefore predicted that monogamy would be associated with helpers at the nest in birds. Although there are definitely exceptions to the rule, "promiscuous" species (those in which there was a relatively large percentage of nests that contained young sired by more than one male) were, as predicted, less likely to engage in cooperative breeding with helpers at the nest than were monogamous bird species (Figure 3.24).[19] Thus, kin selection theory helps us understand complex sociality in vertebrates as well as in insects.

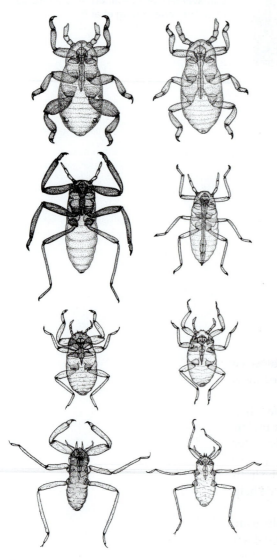

FIGURE 3.25 Soldier aphids (on the left) compared with the aphids they live with and protect (on the right). Soldiers are larger, more robust—and sterile. The species were drawn at different scales by Christina Thalia Grant. After Stern and Foster.[69]

Discussion Question

3.10 Some organisms form clonal societies, such as the aphids that are parthenogenetically (asexually) produced by their mother.[70] These carbon copy individuals live together, often in hollow plant galls. Some individuals in the clone develop into larger-bodied soldiers that fight with intruders that attempt to enter the gall and feed on the aphids within (Figure 3.25).[57] Likewise, certain sea anemones sometimes form clusters of genetically identical individuals that live on the ocean floor. When two such populations come into contact, they may "fight" with one another for control of the local sea bottom.[74] In what ways would you expect these creatures to behave more like the honey bee and other eusocial insects than like cooperative breeding birds? Why? Use gene thinking to predict how clonal organisms might still differ from the eusocial insects in the extent and nature of their altruism.

Summary

1. Animals other than eusocial insects engage in various kinds of social interactions, including cooperation from which both parties derive immediate gains in direct fitness (aka reproductive success).

2. In addition, some cases of reciprocity occur in which a helper provides assistance to another animal from which it later receives repayment that on average increases the helper's reproductive success enough to overcome the cost of its earlier helpfulness.

3. Altruism in which a helper loses direct fitness as it helps another individual reproduce is more common than reciprocity, perhaps because individuals that help now in return for later reciprocity run the risk of never being repaid. In contrast, individuals that provide carefully calibrated altruistic assistance to close relatives will definitely be repaid in the form of indirect fitness gains.

4. However, helpers in some birds and other animals can inherit a valuable resource, such as a territory or a mate, as a result of their helpful behavior. Natural selection can therefore promote the evolution of behavior that has the appearance of altruism but in reality generates gains in direct fitness for the helpful individual.

5. The loss of personal reproductive success experienced by some helpful individual birds and other vertebrates almost always falls into the category of facultative (optional) altruism. Kin selection can result in the evolution of this kind of helping, which boosts an individual's indirect fitness. Facultative altruists retain the capacity to engage in personal reproduction, unlike the obligate helpers seen in so many eusocial insects, which are generally locked into their role for life.

Suggested Reading

Some recent books on evolution and social behavior include those by Bourke[5] and Székely et al.[71] Although Jerry Brown's *Helping and Communal Breeding in Birds*[10] focuses on birds (obviously), the book also helps readers understand the kinds of selection that affect the evolution of social behavior in all groups. All of these books owe much to W. D. Hamilton's contribution to our understanding of social behavior.[30] See also Richard Alexander,[1] Mary Jane West-Eberhard,[78] and Steve Emlen et al.[25]

There have been many excellent studies of social behavior of birds. For examples, see Uli Reyer's paper on the social pied kingfisher,[59] Jan Komdeur and colleagues' work on the Seychelles warbler,[42,44] and Walt Koenig and Ron Mumme's review of acorn woodpecker sociality.[40] Mammalian social behavior has also attracted much attention: see Packer et al.[56] on lions and Wolff and Sherman[83] on rodents, as well as Sherman, Jarvis, and Alexander[67] on naked mole rats.

4 The Evolution of Communication

 If you were fortunate enough to go on an African safari in search of large mammals, you might see a spotted hyena endowed with what looks like an erect penis approach another hyena, which would then inspect the engorged organ of its companion with apparent interest. The inspector, which would also have what looks like an enlarged penis, might also permit the other hyena to examine its phallus (Figure 4.1). Each hyena appears to be signaling something to the other animal in this odd interchange. If this display truly is an evolved communication system, then both the *signaler* and the *receiver* must benefit from providing the signal and from responding to it.[91]

Why? A thought experiment is helpful here. Imagine that the costs to a penis presenter were greater than the benefits of the behavior. Over time, if there were some individuals with a hereditary aversion to penis presentation, the distinctive alleles of these hyenas would be favored by natural selection, with the result that eventually penile signaling would disappear from the population. Likewise, if receivers were in some way harmed by responding to the signal provided by the penis, then hyenas with a hereditary predisposition to ignore penis presentations would have higher fitness, and eventually responders would no longer exist. Thus, both signalers and receivers are required if a communication system is to evolve.

Males of many frogs and toads, including this calling gray tree frog, participate in evolved communication systems from which both male signalers and female receivers derive fitness benefits.

(A)

(B)

FIGURE 4.1 The pseudopenis of the female spotted hyena can be erected (A), in which condition it is often presented to a companion (B) in the greeting ceremony of this species. A, photograph courtesy of Steve Glickman; B, photograph by Heribert Hofer.

So let's assume that because spotted hyenas regularly engage in penis presenting and penis sniffing, we are dealing with a cooperative interaction (see Table 3.1), an evolved communication system with benefits to both parties. Many aspects of this system are fascinating, not the least of which is that many of the penis presenters are actually female spotted hyenas, not males! So our very first challenge is to explain how females of this species evolved a structure that looks very much like a male hyena's penis.

The History of a Strange Display

We turn again to Darwin's theory of descent with modification to try to deal with this challenge. In 1939, L. Harrison Matthews proposed that the pseudopenis might have originated as the developmental result of high levels of male sex hormones in female hyenas while they were still in their mother's uterus.[82] Much later, Stephen Jay Gould introduced this hypothesis to a large audience in one of his popular essays in the magazine *Natural History*.[51] As Matthews and Gould knew, the male penis and the female clitoris of mammals develop from the same embryonic tissues. The general mammalian rule is that if these tissues are exposed very early on to male sex hormones (testosterone and other androgens), as they almost always are in male embryos, a penis is the end result. If the same target cells do not interact with androgens, as is the case for the typical embryonic female mammal, then a clitoris develops. But when a female embryo of most mammals comes in contact with testosterone, either in an experiment or because of an accident of some sort, her clitoris becomes enlarged and rather penis-like.[42] This effect has been observed in our own species in the unlucky daughters of pregnant women who received medical treatment that inadvertently exposed their offspring to extra testosterone,[86] as well as in females whose adrenal glands produce more testosterone than normal.[93]

Given the general pattern of development of mammalian external genitalia, which surely appeared long ago in an ancient proto-mammal, Gould thought it likely that female embryos in the spotted hyena must be exposed to unusual concentrations of male androgens. In support of this descent with modification hypothesis, he pointed to a paper written in 1979 in which P. A.

Racey and J. D. Skinner[98] reported that female spotted hyenas had circulating levels of testosterone equal to those of males, unlike other hyenas and unlike mammals generally. This discovery appeared to confirm the hypothesis that an unusual hormonal environment led to the origin of the spotted hyena's false penis, setting the stage for the origin of the dual-sex penis-sniffing greeting ceremony.

Work on the extra androgen hypothesis continued, with some researchers checking on androgen levels in wild female spotted hyenas (not a simple task). These workers found to their surprise that testosterone levels in free-living females are actually lower than in adult males.[53] Nonetheless, *pregnant* female hyenas do have higher testosterone levels than lactating females (Figure 4.2),[30] although measuring hormone levels in hyenas is far from easy or certain. If pregnant females do indeed produce more testosterone than lactating females, then the door remains slightly open for the extra androgen hypothesis,[33] as does the finding that females of another mammal with a penile clitoris, the ring-tailed lemur, also boost the concentration of androgens in their blood at the onset of pregnancy.[35] Because the placenta of a pregnant spotted hyena converts certain androgens to testosterone, her female embryos could conceivably be exposed to masculinizing levels of this hormone.[48] However, during the time that embryonic clitoral development is likely to be most sensitive to androgens, the mother hyena's placental cells are producing substantial amounts of an enzyme that inactivates these hormones. In humans and most other mammals, the same enzyme prevents masculinization of the genitalia of embryonic females.[20]

The extra androgen hypothesis also leads to the prediction that experimental administration of anti-androgenic chemicals to pregnant adult hyenas should abolish the pseudopenis in their subsequent daughters while also feminizing the external genitalia of their newborn sons. In reality, however, when anti-androgens are administered to pregnant hyenas, the daughters of the treated females retain the elaborate pseudopenis, albeit in somewhat altered form.[49] Results like this are at odds with the view that early exposure to male sex hormones is sufficient in and of itself to cause female embryos of the spotted hyena to develop their unusual genital equipment.[32,94]

Evolutionary History Occurs via Changes in Preexisting Traits

Although our understanding of the history of hormonal and developmental changes that resulted in the female hyena's pseudopenis is far less certain than it once seemed,[34] the most important point for our purposes here is that the solution to this problem surely lies in identifying how changes to *preexisting* developmental and sensory mechanisms might have provided the impetus for the first pseudopenis. Given the widespread mammalian processes for the development of secondary sexual characteristics, it was reasonable to propose (although still not verified) that this weird (to us) structure is the result of the exposure of female embryos to a male sex hormone that activates a developmental program that evolved in both sexes prior to its use in producing a pseudopenis in females.

Another example of this sort comes from the "ears" of male and female whistling moths, *Hecatesia exultans*, that are capable of detecting ultrasonic signals generated by male moths (Figure 4.3). Persons interested in the history of these unusual structures have argued that they represent modifications of ears that were used by the ancestors of this species to detect bats. Predatory bats locate and track moth prey by listening to the ultrasonic echoes from the high-frequency calls bats produce when out hunting.[45] If the ancestor of today's whistling moths could hear bat cries and take defensive action, it could also have heard the ultrasound produced when the first perched male whistling

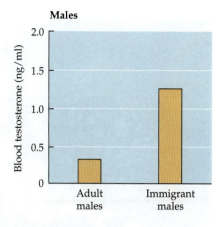

Males

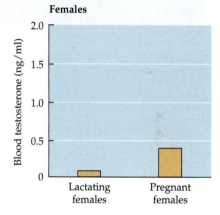

Females

FIGURE 4.2 Concentrations of testosterone in male and female spotted hyenas. Testosterone was measured in the blood of four categories of free-living hyenas: adult males born in the group where they live, adult males that had immigrated into a group, lactating females, and pregnant females. Levels are shown as nanograms of testosterone per milliliter of blood. After Dloniak et al.[30]

(A)

(B)

FIGURE 4.3   Ultrasonic communication by a moth. (A) Male whistling moths produce ultrasounds by striking together the two "castanets" on their forewings; the castanet on the left forewing appears against the yellow background provided by an acacia flower. (B) A male whistling moth has been attracted by the recorded calls of another male played from a small speaker in his territory, confirming that this species communicates via ultrasound. Photographs by the author.

moth beat its wings together in a way that generated an ultrasonic signal. If true, a bat-detecting system in moths could have been co-opted as a communication system with mate-searching females (receivers) locating acoustical males (signalers) by hearing moth-produced ultrasound.[1]

This hypothesis has been tested with females of another species of moth whose males also use ultrasonic courtship signals to induce sexual receptivity in potential partners. Tapes of these signals and tapes of simulated bat calls did equally well in triggering female receptivity, as one would predict from the hypothesis that a sensitivity of females to bat ultrasound preceded and facilitated the evolution of ultrasonic courtship by male moths.[88]

Sensory Exploitation and the Origin of Communication Signals

The argument that communication signals originate in actions that activate sensory abilities and biases of receivers that are already in place is often called **sensory exploitation**.[4,56] Thus, courting male water mites appear to have tapped into the preexisting sensory abilities of females of this predatory species.[95] While a female is waiting for an edible copepod to make the mistake of bumping into her, the water mite adopts a particular pose, called the net stance. A male that encounters a female in this position may vibrate a foreleg in front of her, a behavior that may generate vibrational stimuli in the water similar to those provided by an approaching copepod, according to Heather Proctor. The female water mite, in turn, may grab the male mite, using the same response that she uses to capture her prey, although she will release the captured male unharmed. He may then turn around and place spermatophores (packets of sperm) near the female, which she will pick up in her genital opening if she is receptive (Figure 4.4).[95,96]

(A)

(B)

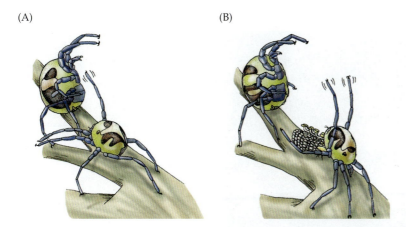

FIGURE 4.4 Sensory exploitation and the evolution of a courtship signal in the water mite *Neumania papillator*. (A) The female (on the left) is in her prey-catching position (the net stance). The male approaches and waves a trembling foreleg in front of her, setting up water vibrations similar to those a copepod might make. The female may respond by grabbing him, but she will release him unharmed. (B) The male then deposits spermatophores on the aquatic vegetation in front of the female before waving his legs over them. After Proctor.[95]

If males trigger the prey-detection response of females, then unfed, hungry female water mites held in captivity should be more responsive to male signals than well-fed females. They are, providing support for the contention that once the first ancestral male happened to use a trembling signal, the male's behavior and its hereditary basis spread because it effectively activated a pre-existing prey-detection mechanism in females[95,96] (but see Borgia[8]).

Discussion Question

4.1 Females of many African cichlid fish lay their eggs on lake bottoms in depressions made by males. The females brood their eggs and young fry in their mouths. A female picks up her orange eggs almost as quickly as she lays them. As this happens, the male cichlid that made the "nest" may move in front of her and spread his anal fin (Figure 4.5), which in many species is decorated with a line of large orange spots. The female may try to pick up the objects on the fin.[119] As she does, the male releases his sperm, some of which swim into the female's mouth, where they fertilize her eggs. If sensory exploitation explains the evolutionary origin of the female's behavior, what prediction can you make about how female fish of a related species will respond to the normally unspotted anal fins of males that have been painted with colorful egg-like spots? Check your prediction against Egger et al.[40] Was the first male to use this signal exploiting his mate in the sense of reducing her fitness to benefit himself?

FIGURE 4.5 Sensory exploitation in spawning by a cichlid fish? (A) The male cichlid's anal fin has large orange spots on it. (B) The female approaches the anal fin closely soon after the male has released his sperm.

(A)

(B)

(A)

(B)

(C)

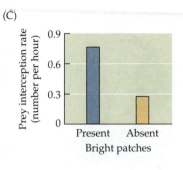

FIGURE 4.6 **The bright spots and stripes** in the color pattern of the spider *Nephila pilipes* appear to attract prey to the predator at night. (A) The upper side of the spider. (B) The underside of the spider. (C) The number of prey captured per hour in the web of another brightly colored nocturnal spider when the spider was present (blue bar) and when the spider was absent (orange bar). A, photograph by Chih-Yuan Chuang; B, photograph by Jin-Nan Huang; both photographs courtesy of I. Min-Tso; C, after Chuang et al.[19]

Discussion Question

4.2   The body of the giant wood spider is extremely colorful (Figure 4.6). When the bright patches are painted black, the rate at which moths and other nocturnal prey fly into the spiders' webs declines sharply, especially at night.[18,19] What relevance does this research have for persons interested in the sensory exploitation hypothesis for the evolution of courtship signals in animals?

The argument that has been developed for water mite courtship has also been applied to the signals and responses of male and female guppies. Females in some populations of this small fish prefer to mate with males that have bright orange markings (Figure 4.7).[55] Male guppies cannot synthesize the orange pigments that go into their body coloration but instead have to

(A)

(B)

FIGURE 4.7 **Food, carotenoids, and female mate preferences in the guppy.** (A) Males have to acquire orange pigments from the foods they eat, like this *Clusia* fruit that has fallen into a stream where guppies live. Males that secure sufficient amounts of carotenoids incorporate the chemicals into ornamental color patches on their bodies. (B) Females (like the larger fish on the right) find males with large orange patches more sexually appealing than males without them. Photographs by Greg Grether.

acquire these carotenoids from the plants they eat. Those that get enough carotenoids from their food become attractive to females, but why?

One hypothesis on the origin of the female preference for orange-spotted males suggests that when this mating preference first appeared, it was a by-product of a sensory preference that had evolved in another context. This hypothesis has received support from observations of female guppies feeding avidly on rare but nutritionally valuable orange fruits that occasionally fall into the Trinidadian streams where guppies live.[102] Thus, it is possible that females evolved visual sensitivity to orange stimuli because of the foraging benefits associated with this ability, not because of any fitness benefits from selective mate choice. If so, then female guppies can be predicted to respond more strongly to orange stimuli than to other colors when feeding. In fact, female guppies do approach and try to bite an inedible orange disc far more often than discs of any other color. Moreover, the strength of the mate choice preference of females from different populations was matched by the relative rates at which females pecked at orange discs presented to them (Figure 4.8).

If sensory exploitation is a major factor in the origin of effective signals, then it should be possible to create novel experimental signals that trigger responses from animals that have never encountered those stimuli before.[103] To test this prediction, researchers have played sounds to frogs that contain acoustical elements not present in the species' natural calls,[107,108] attached strips of yellow plastic to the shorter tails of male platyfish,[4] supplied male sticklebacks with bright red spangles to add to their nests,[89] and added feather crests to the crestless heads of least auklets[65] (Figure 4.9) as well as to the heads of some crestless Australian finches (Figure 4.10).[14] In all of these cases, researchers found that the artificial attributes elicited stronger reactions from females than the natural ornaments (see Figure 4.9).

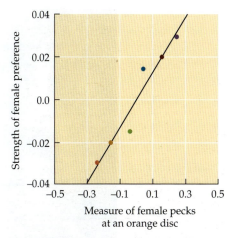

FIGURE 4.8 Sexual preferences for orange spots match foraging preferences for orange foods by female guppies. The strength of the female response to orange spots on males (vertical axis) varies from population to population and is proportional to the strength of the foraging response to an orange disc. Each point represents a different guppy population living in a different stream. After Rodd et al.[102]

(A)

(B)

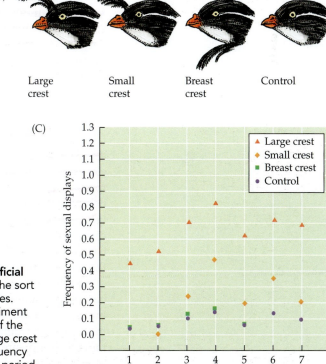

FIGURE 4.9 The response of least auklets to three novel artificial signals. (A) A stuffed male least auklet with an artificial crest of the sort used in an experiment on female sexual preferences in this species. (B) A diagram of some of the heads of models used in the experiment (from right to left): the control (which lacks a crest, as do males of the least auklet), a breast crest model, a small crest model, and a large crest model. (C) The models with large crests elicited the highest frequency of sexual displays by the females during a standard presentation period. A, photograph by Ian Jones; B and C, after Jones and Hunter.[65]

(A)

(B)

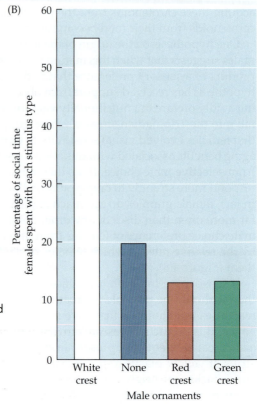

FIGURE 4.10 Mate preferences for a novel ornament. (A) A male long-tailed finch (left) and a male zebra finch (right) have been outfitted with bizarre white plumes. (B) The "white-plumed" male zebra finches were more attractive to females than were control males without plumes or those given headdresses of red or green feathers. A, photograph by Kerry Clayman, courtesy of Nancy Burley; B, after Burley and Symanski.[14]

Sensory Exploitation—or the Retention of Ancestral Traits?

The experiments described above suggest that novel signals could originate when signalers happened to possess mutant attributes that turned on preexisting sensory preferences that had evolved for other purposes. However, some artificial signals may elicit responses because the ancestors of the tested species used similar signals during courtships in the past and today's descendants have retained the sensory preferences of their ancestors.[106] This conjecture is plausible, given the numerous cases in which elaborate male traits used in courtship and aggression have been lost after having once evolved. In these cases, the close relatives of the ornament-free species possess the ornaments in question (which suggests that their mutual common ancestor did, as well).[120]

An example is provided by the lizard *Sceloporus virgatus*, which lacks the large blue abdominal patches of many other members of its genus. These other species use their blue patches in the male threat posture, which involves the elevation of the lizard's body and its lateral compression, making the abdomen visible. But although the signal has apparently been lost in *S. virgatus*, the behavioral response to the signal has not, as shown when some male lizards had blue patches painted on them. Lizards that observed a rival giving a threat display experimentally enhanced by the blue paint were far more likely to back off than were lizards that saw a displaying opponent that did not have the added blue patches (Figure 4.11).[97]

So what are we to make of the fact that the fish *Priapella olmecae* does not have an elaborate elongate swordtail and yet when Alexandra Basolo endowed males of *Priapella* with an artificial yellow sword, females found males with this novel trait highly attractive; furthermore, the longer the sword, the greater the female's desire to stay close to the altered male. Basolo concluded that a sensory preference for long tails must have preceded the eventual evolution of swordtails in some relatives of *P. olmecae* belonging to the genus *Xiphophorus*,

Sceloporus virgatus

FIGURE 4.11   Receivers can respond to an ancestral signal not present in their species. Lizards of a species whose relatives have blue patches on the abdomen are more likely to abandon a conflict when confronted by a conspecific rival with blue patches painted on its abdomen than when seeing the threat display of an unmanipulated control individual or one that has had white patches or black dots painted on the abdomen. After Quinn and Hews;[97] photograph by Paul Hamilton.

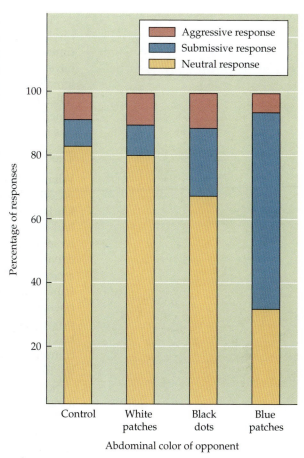

in which the male's swordtail currently plays an important role in courtship (Figure 4.12).[5] Just what adaptive aspect of the female's sensory system was responsible for its original swordtail preference is not known for sure, but Basolo suggests that, for example, the female's sensory bias might have evolved in the context of recognizing conspecific males (which have a large, vaguely sword-like gonopodium for the transfer of sperm to their mates).

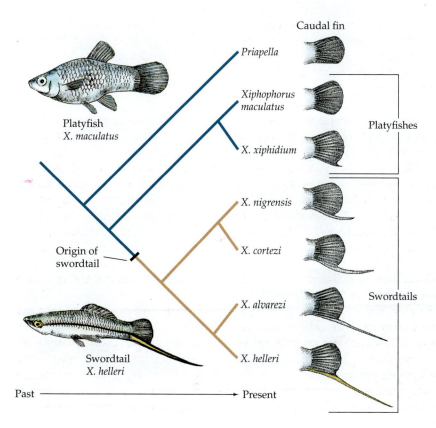

FIGURE 4.12   Sensory exploitation and swordtail phylogeny. The genus *Xiphophorus* includes the swordtails, which have elongated caudal fins, and the platyfishes, a group without tail ornaments. Because the closest relatives of the platyfishes and swordtails belong to a genus (*Priapella*) in which males lack long tails, the ancestor of *Xiphophorus* probably also lacked this trait. The long tail apparently originated in the evolutionary lineage that diverged from the platyfish line. Even so, females of the platyfish *X. maculatus* find males of their species with experimentally lengthened tails more attractive, suggesting that they possess a sensory bias in favor of elongate tails. After Basolo.[4]

However, Axel Meyer and his colleagues argue on the basis of a detailed phylogeny of *Xiphophorus* that swords have evolved and been lost several times in this genus.[84] Therefore, even though *P. olmecae* does not currently have a swordtail, it is conceivable that some of its ancestors did and that the female preference for this trait is an evolutionary holdover from the past, a reflection of the tendency for sexual signals to be gained and lost rapidly over time.[8] The point is that it can be tricky to test the sensory exploitation hypothesis for the evolution of communication signals.

The Panda Principle

Whatever its merits in particular cases, the sensory exploitation hypothesis is derived from the argument that what has already evolved influences what kinds of additional changes are likely and which are not. If natural selection had been given the job of coming up with a jet plane, for example, it would have had to start with what was already available, presumably a propeller-driven plane, changing this "ancestral" structure piece by piece, with each modified airplane flying better than the preceding one, until a jet was built.[29] In nature, evolutionary transitions must occur this way because selection does not start from scratch but instead acts on what already exists. As a result, the products of evolution often appear to be jury-rigged devices that might have been designed by Rube Goldberg, a cartoonist famous for his ridiculously overcomplicated inventions designed for everyday tasks. (Goldberg died in 1970, but his wacky cartoons can still be seen at the official Rube Goldberg website.)

So, for example, the panda's thumb is not a "real" finger at all but, instead, a highly modified wrist bone (the panda's hand has the same five digits that

FIGURE 4.13 The panda principle. Descent with modification is responsible for the sexual behavior of a parthenogenetic whiptail lizard. In the left column, a male of a sexual species engages in courtship and copulatory behavior with a female. In the right column, two females of a closely related parthenogenetic species engage in very similar behavior. From Crews and Moore.[22]

most vertebrates have plus a small thumb-like projection from the radial sesamoid bone, which is present, although as a much smaller bone, in the feet of bears, dogs, raccoons, and the like).[52] Why do pandas have their own special thumb? According to Stephen Jay Gould, pandas evolved from carnivorous ancestors whose first digit had become an integral part of a foot used in running. As a result, when pandas evolved into herbivorous bamboo eaters, the first digit was not available to be employed as a thumb in stripping leaves from bamboo shoots. Instead, selection acted on variation in the panda's radial sesamoid bone, which is now used in a thumb-like manner by the bamboo-eating panda.

What Gould called "the panda principle," Darwin labeled "the principle of imperfection." No matter the title, the phenomenon is widespread. Consider, for example, the persistence of sexual behavior in species of whiptail lizards that are composed entirely of parthenogenetic (asexual) females. In these species, a female may be "courted" and mounted by another female (for reasons that are not fully understood); females subjected to pseudomale sexual behavior are much more likely to produce a clutch of eggs than if they do not receive this sexual stimulation from a partner (Figure 4.13).[22] The effect of courtship on fecundity in unisexual lizards obviously exists because these reptiles are derived from sexual ancestors. The parthenogenetic females retain characteristics, such as an acceptance of courtship, that their nonparthenogenetic ancestors possessed—characteristics that a biological engineer would surely eliminate if he or she could play God by designing an all-female asexual species on paper and then creating it in one go. Natural selection, however, cannot play God, because it is a blind process with no goal in mind and no means to get to a predetermined end point.

FIGURE 4.14 A mating pair of empid flies. The female is feeding on a fly that her mate gave to her.

Discussion Question

4.3 In studying the courtship behavior of the empid flies, E. L. Kessel was amazed to find a species, *Hilara sartor*, in which males gather together to hover in swarms, carrying empty silken balloons, which females accept prior to mating with balloon-carrying males.[66] Use the concept of a co-opted, preexisting sensory system to explain how this sort of behavior could have originated. You should know that many empid flies are predatory (Figure 4.14) and that males often offer their mates a prey item as an inducement to mate. In one species of this sort, some males supply their mates with an inedible dandelion-like tufted seed.[70]

The Behavioral Ecology of Communication

This chapter's focus now shifts from describing the history of communication signals, with special emphasis on how new signals originated, to an attempt to identify why natural selection resulted in the spread of some signal changes over time. Even though the evolutionary origin of the female spotted hyena's pseudopenis is still not known, we can use natural selection theory to explore the adaptive value of the trait. This task is challenging, however, given that the early forms of the pseudopenis probably had substantial fitness costs. In our own species, females "experimentally" exposed to much higher than normal amounts of androgens as embryos not only develop an enlarged clitoris but also may be sterile as adults. Indeed, pregnant women who have naturally high levels of circulating testosterone produce babies of low birth weight, a lifetime handicap for the infants so affected.[16] If developmental costs of this sort were experienced by hormonally distinctive ancestors of the spotted hyena, how could the genes for a pseudopenis spread through the species by natural selection?[44,87]

One solution might be that the pseudopenis is *not* adaptive per se but instead developed as a by-product of a change that had other kinds of positive fitness effects. Perhaps the pseudopenis is a developmental side effect of the hormones present in newly born spotted hyenas that promote extreme aggression between siblings, which enables them to compete for control of their mother's parental care. Baby spotted hyenas are born with eyes that open and erupted teeth, which they often use on one another. Infant males have a large penis and infant females have a large pseudopenis, which could be the by-product of whatever hormonal features produce rapid development, the foundation for neonatal aggression, which leads in some cases to the death of one of the two sibs.[36,50]

Alternatively, a hereditary change of some sort that resulted in an enlarged clitoris might have spread because it helped to make adult females larger and more aggressive, but not because it helped them acquire a pseudopenis.[49] By all accounts, spotted hyena females are indeed unusually aggressive, at least by comparison with immigrant males, which are usually subordinate and deferential to the opposite sex. Females not only keep these males in their place, sometimes with threats or attacks, but also compete with other females in their clan for high rank, with winners gaining priority of access to the wildebeests and zebras that their clan kills or steals from lions (Figure 4.15).[43,116] Moreover, with high rank comes reproductive success. The sons and daughters of high-ranking females grow faster, are more likely to survive, and are more likely to become high-ranking individuals themselves than are the offspring of subordinate hyenas (Figure 4.16). Many of the advantages enjoyed by the offspring of dominant females come from the support they receive from their mother and those hyenas in her coalition of helpers.[38,62] Even after leav-

FIGURE 4.15 Competition for food is fierce among **spotted hyenas**, which may favor highly aggressive individuals. A hyena clan can consume an entire giraffe in minutes.

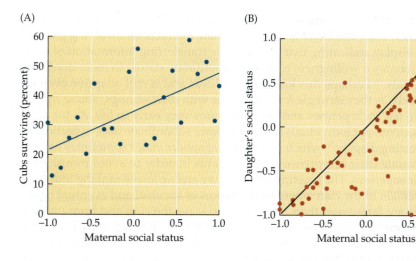

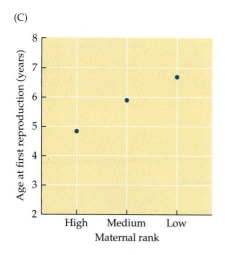

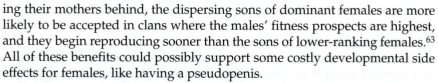

FIGURE 4.16 Dominance greatly advances female reproductive success in the spotted hyena. A mother's social status is directly linked to (A) cub survival to age 2 years and (B) the dominance rank of her daughters, as determined by observation of interactions between pairs of hyenas. (C) Maternal social status (rank) also affects when her sons begin to reproduce. A and B, after Hofer and East;[62] C, after East.[38]

ing their mothers behind, the dispersing sons of dominant females are more likely to be accepted in clans where the males' fitness prospects are highest, and they begin reproducing sooner than the sons of lower-ranking females.[63] All of these benefits could possibly support some costly developmental side effects for females, like having a pseudopenis.

But there are problems here. For one thing, if the pseudopenis is strictly a liability, why hasn't selection favored mutant alleles that happened to reduce the negative clitoral effects that androgens (or other developmentally potent biochemicals) have on female embryos? So let's consider the possibility that the pseudopenis might have adaptive value in and of itself, despite the apparent costs associated with its development.

Adaptationist Hypotheses for the Hyena's Pseudopenis

At some point, a female ancestor of today's spotted hyenas must have been the first female ever to greet a companion by offering her a chance to inspect her "penis," perhaps during a bout in which the receiver was analyzing chemical signals from the anogenital region of the signaler. All four modern species of hyenas engage in chemical communication via anal scent glands and the like.[69,116] Female spotted hyenas appear to use the anal gland secretions of other individuals to evaluate their social rank.[13] In a predecessor of spotted hyenas, females added inspection of the erect penises of their fellow hyenas, male and female alike, when inspecting a companion. Could the female's enlarged clitoris have been so useful in this encounter as to outweigh any reproductive handicaps imposed by the structure?

One answer to this question comes from considering the possibility that sensory exploitation played a role in the origins of the display. Imagine that the female-dominant, male-subordinate relationship was well established before the first pseudopenis appeared. Furthermore, imagine that one of the cues that sexually motivated, subordinate male hyenas supplied to dominant females during courtship was an erect penis. Presentation of this organ to the female would clearly signal the suitor's male identity and intentions and thus his subordinate, nonthreatening status. Such a signal might have encouraged females on occasion to accept the male's presence rather than rejecting or attacking him.

Once this kind of male–female communication system was in place, mutant females with a pseudopenis would have been able to tap into an already established system to signal subordination or willingness to bond with the other individual, or both.[116] Evidence in support of the submission hypothesis comes from the observation that subordinate females and youngsters are far more likely than dominant animals to initiate interactions involving pseudo-penile display, indicating that they gain by transferring information of some sort (probably a willingness to accept subordinate status) to more dominant individuals.[36] Perhaps this kind of signal was and is still adaptive even for hyena cubs as they struggle for dominance with their siblings. Young hyenas begin to use their penises and pseudopenises very early in life. As for adult hyenas, dominant individuals sometimes force subordinate ones to participate in a greeting display,[36] which suggests that subordinates sometimes have to demonstrate their acceptance of lower status in order to avoid punishment by their superiors. The greeting display may enable lower-ranking individuals to remain in the clan where their chances of survival and eventual reproduction are far greater than if they had to leave the clan and hunt on their own. Permission to remain could be more readily granted if the dominant female could accurately assess the physiological (e.g., hormonal) state of a subordinate female by inspecting its erect, blood-engorged clitoris. Perhaps truly submissive subordinates are signaling that they lack the hormones (or some other detectable chemicals) needed to initiate a serious challenge to the dominant individual, which can then afford to tolerate these others because they do not pose a threat to her.

The social-bonding hypothesis is similar to the submission hypothesis in that this odd mode of greeting could promote the formation of cooperative coalitions within the clan, an effect that requires the willingness of dominant individuals to accept somewhat lower-ranking individuals as associates. Coalition members cooperate with one another in ways that appear to raise the direct and indirect fitness of these individuals.[111] Moreover, coalitions within a clan of spotted hyenas are constantly shifting, which might help explain why the animals so often engage in their bizarre greeting ceremonies.[112] In addition, this social-bonding hypothesis is congruent with the observation that after a conflict, some individuals initiate a greeting ceremony to reconcile their differences with previous opponents.[61]

There are still other hypotheses on the possible adaptive value of the pseudopenis. For example, because the vagina is reached through the clitoris in spotted hyenas, copulation is a tricky and delicate matter, requiring complete cooperation by the female. Thus, the pseudopenis gives the female spotted hyena a great degree of control in the selection of a sexual partner, which could contribute to the maintenance of this feature.[37] Perhaps the most important point to be derived from this case is that much more remains to be learned about spotted hyenas, even though this animal has been carefully studied for decades.[39] Those of my readers who are students today should realize that many animal behavior puzzles remain to be explored by persons who know how to use hypotheses to generate testable predictions, in order to reach defendable scientific conclusions.

Discussion Question

4.4 To test adaptationist hypotheses on the spotted hyena's greeting ceremony, how would you take advantage of information on the behavior of the other species of hyenas (see Mills,[85] Owens and Owens,[92] and Watts and Holekamp[116]). In addition, could you make use of information on another highly social mammal, the naked mole rat (see Figure 3.22 and pages 61–62)? Then there are some mammals in which females have an enlarged clitoris.[90,93] What might these species tell us about the adaptive value of the spotted hyena's behavior?

Another Darwinian Puzzle: The Adaptive Value of Threat Displays

The discipline of animal behavior continues to grow as researchers tackle the many remaining challenges in the field. One worthy Darwinian puzzle that has already been extensively explored is why animals so often resolve key conflicts with harmless threat displays. For example, when one hyena threatens another by approaching with tail raised while walking in a stiff-legged manner, the interaction often ends without any physical contact between the two animals, let alone an all-out fight.[41] Instead, one hyena usually runs off or crouches down submissively, in effect conceding whatever resource stimulated one or both individuals to initiate the threat display. Moreover, the "winning" individual often accepts signals of submission by a rival and does not follow up with a physical attack on the other animal.

The use of noncontact threat displays to resolve conflicts is common in the animal kingdom, although violent interactions are also widespread. But rather than assaulting a rival, males of many bird species sometimes settle their disputes over a territory or a mate with much singing and plumage fluffing, without ever so much as touching a feather of the opponent. Even when genuine fighting does occur in the animal kingdom, the "fighters" often appear to be auditioning for a comic opera. After a body slam or two, a subordinate elephant seal generally lumbers off as fast as it can lumber, inchworming its blubbery body across the beach to the water, while the victor bellows in noisy, but generally harmless, pursuit.

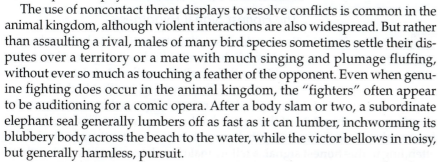

Discussion Question

4.5   One sometimes hears that the reason why so many species resolve their contests via mostly harmless threat signals is to reduce the number of injuries and thereby protect the breeding adults needed to produce the next generation of offspring. What's the problem with this hypothesis?

But if threat displays and the typical response to them are the result of natural selection, it should be possible to demonstrate that the losers, those signal receivers that run off or give up without a fight, benefit from their reaction to the signals of others. This task can be difficult. Consider the European toad, *Bufo bufo*, whose males compete for receptive females. When a male finds another male mounted on a female, he may try to pull his rival from her back. The mounted male croaks as soon as he is touched, and often the other male immediately concedes defeat and goes away, leaving his noisy rival to fertilize the female's eggs. How can it be adaptive for the signal receiver in this case to give up a chance to leave descendants simply on the basis of hearing a croak?

Because European toad males come in different sizes, and because body size influences the pitch of the croak produced by a male, Nick Davies and Tim Halliday proposed that males can judge the size of a rival by his acoustical signals. If a small male can tell, just by listening, that he is up against a larger opponent, then the small male ought to give up without getting involved in a fight he cannot win. If this hypothesis is correct, then deep-pitched croaks (made by larger males) should deter attackers more effectively than higher-pitched ones (made by smaller males).[27] To test this prediction, the two researchers placed mating pairs of toads in tanks with single males for 30 minutes. Each mounted male, which might be large or small, had been silenced by a rubber band looped under his arms and through his mouth. Whenever a single male touched a pair, a tape recorder supplied a 5-second call of either low or high pitch. Small paired males were much less frequently attacked if

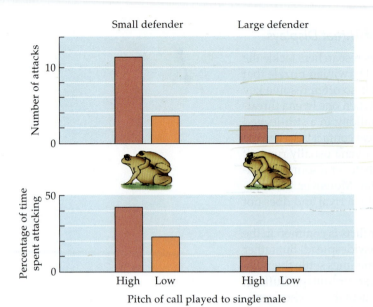

Small defender   Large defender

Number of attacks

Percentage of time spent attacking

High   Low   High   Low
Pitch of call played to single male

FIGURE 4.17   Deep croaks deter rival toads. When a recording of a low-pitched croak was played to them, male European toads made fewer contacts with, and interacted less with, silenced mating rivals than when a higher-frequency call was played. Tactile cues also play a role, however, as one can see from the higher overall attack rate on smaller toads. After Davies and Halliday.[27]

the interfering male heard a deep-pitched croak (Figure 4.17). Thus, deep croaks do deter rivals to some extent, although tactile cues also play a role in determining the frequency and persistence of attacks.

The Honest Signal Hypothesis

So why don't small males pretend to be large by giving low-pitched calls? Perhaps they would, if they could, but they can't. A small male toad simply cannot produce a deep croak, given that body mass and the unbendable rules of physics determine the dominant frequency (or pitch) of the signal that he can generate. Thus, male toads have evolved a warning signal that accurately announces their body size. By attending to this **honest signal**, a fellow male can determine something about the size of his rival and thus his probability of winning an all-out fight with him. The same is true for the barking gecko (Figure 4.18)[60] and some other animals that employ acoustical signals in conflicts with others[11] (but as is so often the case, there are exceptions to the rule—e.g., Reichert and Gerhardt[99]).

In paper wasps, females signal their fighting ability not by an acoustic signal but via the markings on their faces, which differ between dominant and subordinate wasps. So why don't subordinate females produce the kind of faces that signal dominance, given that dominance is associated with higher reproductive success? As Elizabeth Tibbetts and Amanda Izzo have shown, the females on a paper wasp nest test each other from time to time, and in so doing, they can detect cheaters whose facial signals are not in synchrony with their real fighting ability. Individuals whose faces have been painted to signal dominance but whose capacity for combat is actually low have been attacked more often by their colony mates than dominant females whose faces have been painted but not altered in appearance (Figure 4.19).[114] In this system, dishonest signalers are punished by their companions and so pay a special price for the mismatch between facial markings and behavior, and this outcome in turn creates selection for the evolution of accurate signaling.

FIGURE 4.18   Honest acoustical signals of size in the barking gecko. The larger the male, the lower the frequency of its calls. After Hibbitts et al.[60]; photograph by Tony Hibbits.

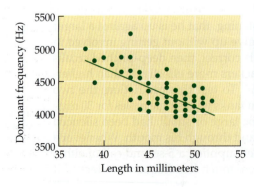

Dominant frequency (Hz)
Length in millimeters

Barking gecko

FIGURE 4.19 **Honest visual signals of dominance in the paper wasp *Polistes do-minulus.*** In this species, dominance is linked to dark patches on the face of the wasp. Photographs courtesy of Elizabeth Tibbetts.[114]

Likewise, males of certain bizarre antlered flies convey honest information about body size to their opponents by standing directly in front of each other in a position that enables the two flies to compare the relative size of their antlers. The size of these structures is nicely correlated with the fly's body size, and smaller (weaker?) males quickly abandon the field of battle to the larger, more powerful males.[31,122] Antler size is also linked to body size in roe deer, just one of many vertebrates in which opponents can determine their size relative to a rival by examination of his weapons.[115] For male collared lizards, an honest signal of fighting ability is provided by the ultraviolet-reflecting patches near the opponent's mouth;[72] the larger the patch, the harder the bite that the lizard can deliver to a rival (Figure 4.20).

(A)

(B)

(C)

FIGURE 4.20 **Honest visual signals of size and of strength.** (A) Males of this Australian antlered fly in the genus *Phytalmia* confront each other head to head, permitting each fly to assess his own size relative to the other's size. (B) Antler size in fallow deer bucks provides information about male fighting ability. (C) The larger the patch of ultraviolet-reflecting skin by the open mouth of a gape-displaying collared lizard, the harder the reptile can bite an opponent. The UV-reflecting skin is outlined in the image on the right. A, photograph by Gary Dodson; C, photograph from Lappin et al.[72]

Discussion Questions

4.6   Baby songbirds usually produce fairly loud vocalizations in response to the arrival of a parent bird with food at the nest. This begging behavior could be an honest signal of the need for food by each nestling. Alternatively, the vocalizations could be an honest signal of the nestling's "quality," its likelihood of achieving high fitness, which could provide a parent with information needed to invest more in offspring with the potential for higher fitness. If the first hypothesis is correct, what prediction follows about the begging intensity of well-fed nestling songbirds when placed in a nest with food-deprived youngsters of the same age?

4.7   Birds are not the only animals in which offspring have special signals that appear to provide information to a parent. Develop an evolutionary analysis of crying by human infants in which you consider the two hypotheses outlined in Discussion Question 4.6 for begging by baby birds. Then employ the following data in evaluating your hypotheses: (1) young infants expend considerable energy when crying; (2) the growth rate for typical infants is highest during the first 3 months of life, with smaller and smaller portions of the energy budget thereafter going to growth as opposed to maintenance; (3) consumption of breast milk peaks at 3 to 4 months of age and then declines; (4) crying peaks at about 6 weeks of age and occurs progressively less often after 3 months of age, except when the child is being weaned; (5) babies who are carried everywhere and nursed on demand (as in traditional societies) cry far less than babies in Western societies; and (6) the high-pitched cries of unhealthy babies are considered especially unpleasant by adult listeners.[46,79,117]

When smaller toads or geckos or antlered flies or collared lizards withdraw upon receiving an honest signal of size or fighting power from another individual, both the smaller and larger competitors gain: small males do not waste time and energy or risk injury in a battle they are unlikely to win, and large males save time and energy that they would otherwise have to spend brushing aside annoying smaller opponents. To understand the value of a quick concession, imagine two kinds of aggressive individuals in a population: one that fought with each opponent until he was either victorious or physically defeated, and another that checked out the rival's aggressive potential and then withdrew as quickly as possible from fighters that were probably superior. If this argument is correct, then in a species in which there are nondisplaying individuals, such as the silent males that occur in some cricket species, when two such individuals fight each other, the level of aggression should be higher than when two signal-producing males engage in an aggressive encounter. This prediction has been confirmed for a cricket species, some of whose males provide information-bearing acoustical signals while others do not.[78]

In species in which combatants provided honest signals of competitive ability, any "fight no matter what" types would eventually encounter a powerful opponent that would administer a serious thrashing. The "fight only when the odds are good" types would be far less likely to suffer an injurious defeat at the hands of an overwhelmingly superior rival.[83,118]

Further, imagine two kinds of superior fighters in a population: one that generated signals that other, lesser males could not produce, and another whose threat displays could be mimicked by smaller males. As mimics became more common in the population, natural selection would favor receivers that ignored the easily faked signals, reducing the value of producing them. This, in turn, would lead to the spread of the genetic basis for an honest signal that could not be devalued by deceitful signalers.

(A)

(B)

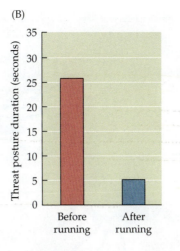

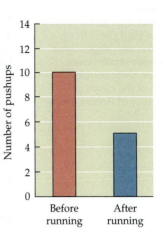

FIGURE 4.21 Threat displays require energy. (A) Males of the side-blotched lizard (B) that have been run on a treadmill to lower their endurance are not able to maintain their threat posture for as long as they can when they are not tired; they also generate fewer push-up displays. A, photograph by Paul Hamilton; B, after Brandt.[10]

If honest signal theory is correct, we can predict that male threat displays should be relatively easy for large males to perform but more difficult for small, weak, or unhealthy males to imitate. As predicted, both the duration of a male side-blotched lizard's display and the number of push-ups he performs fall markedly after he has been forced to run on a treadmill for a time (Figure 4.21). Conversely, males that first perform a series of displays before being placed on the treadmill run for a shorter time than when they have not been displaying. In other words, a male's current condition is accurately reflected by his push-up performance,[10] which enables him to advertise his true fighting capacity to rival males.

Discussion Questions

4.8 When males of the Australian slender crayfish compete aggressively with one another (Figure 4.22), they begin by displaying their enlarged front claws. The larger the claw, the more likely the male is to dominate his rival, which may leave without grappling with the larger-clawed crayfish. However, the muscles in the claws of males generate only half the force of the claw muscles of females. In addition, the actual strength of the claw has no bearing on which male is dominant.[123] Much the same applies to male fiddler crabs in which males with a regenerated large claw quickly defeat rivals, despite the fact that replacement claws are relatively weak.[71] Are males with large but weak claws dishonest signalers? But how could a mindless crab be dishonest?

4.9 Honey bee queens produce a complex olfactory signal, the queen pheromone, which they use either to control the workers, forcing them to keep working for the queen, or to provide accurate information about the reproductive state of the queen, making it adaptive for workers to be altruists (by helping them determine that the queen is capable of producing many siblings of the workers). Which of these hypotheses is more congruent with the logic of natural selection theory? Which hypothesis is supported by the discovery that components of the queen pheromone are related to the volume of sperm stored in the queen and her ovarian function? For a full and fair evaluation of the two competing hypotheses, see Kocher and Grozinger.[68]

FIGURE 4.22 A dishonest signal of strength? Males of the slender crayfish display their large front claws to opponents, and males with larger claws dominate those with smaller ones. However, claw size is not correlated with the strength of the claw's grip. Photograph courtesy of Robbie Wilson, from Wilson et al.[123]

Self-Sacrificing Communication?

When Bernd Heinrich heard a mob of ravens calling loudly next to a dead moose killed by a poacher in a Maine forest, he was both surprised and intrigued (Figure 4.23).[59] Ravens would seem to have nothing to gain by calling loudly and attracting still more consumers to a food bonanza that, if not shared, might last for most of the winter. So why do they yell?

Heinrich could see that this Darwinian puzzle had much in common with alarm calling by Belding's ground squirrel (see page 53), an apparently helpful behavior now known to generate both direct and indirect fitness benefits for the callers. Perhaps ravens are similar, in which case the costs of sharing a deceased moose with others are outweighed by the benefits of attracting offspring to the feast (thereby boosting the caller's direct fitness) or by alerting nondescendant relatives to the carcass (thereby increasing the caller's indirect fitness). Heinrich felt, however, that offspring and/or other relatives were unlikely to make yelling adaptive, given that pairs of ravens produce a maximum of six offspring per breeding season whereas several dozen birds often assemble at a food source advertized by yellers. Subsequently, DNA fingerprinting studies showed unequivocally that feeding flocks were indeed composed of unrelated individuals.[59]

All right, but perhaps ravens yell because the signals get the attention of a bear or coyote, which can open up a tough-skinned moose, eventually provid-

FIGURE 4.23 A group of ravens feeding on a carcass. Most of the birds were attracted to the food by yelling companions. Photograph by Bernd Heinrich.

ing the caller with access to meat that it could not get in any other way. To test this hypothesis, Heinrich hauled a 150-pound goat carcass through the Maine woods, taking it out to various sites during the day and storing the none-too-sweet-smelling goat in his cabin at night to prevent its loss to a nocturnal scavenging bear. Ravens occasionally approached the dead goat, though only after making Heinrich wait for hours in cramped and bitterly cold observation blinds. But contrary to his prediction, the birds never yelled when they found the bait. Moreover, Heinrich sometimes observed ravens yelling at carcasses that had already been ripped open. These findings forced him to abandon the attract-a-carcass-opener hypothesis.

Instead, he switched his attention to an alternative explanation, an idea stimulated by seeing how cautiously some ravens approached carcasses when they first found them. Perhaps, he suggested, a carcass discoverer yells to draw in other ravens so that if a predator lurks nearby, the other birds will provide possible targets, reducing the yeller's risk of being taken by a hiding coyote or fox. The incoming birds would be attracted because they would gain a wealth of food in return for taking a small risk of being the unlucky victim of a predator. However, this hypothesis generates the prediction that once a group has assembled and feeding has begun, the birds should shut up to avoid attracting still more ravens, which would be unwanted competitors as well as unnecessary for safety purposes. The observation that yelling continued at baits that had already acquired a mob of actively feeding birds convinced Heinrich to discard the diluted-risk-of-predation hypothesis.

As Heinrich continued stoically lugging dead goats into the Maine woods, he came to realize that whenever he saw a single bird or a pair at a feeding site, those ravens were quiet. Yelling occurred only when three or more ravens were present, and it was then, and only then, that large numbers of other ravens came to the area. Heinrich knew that older adult ravens form pairs that defend a territory year-round. Unmated young birds usually travel solo over great distances in search of food. If a singleton attempts to feed in a resident pair's territory, the solo bird is attacked by the pair. Heinrich wondered if yelling could be a signal given by nonterritorial intruders, a signal that attracts other unmated wanderers to a food bonanza that they can exploit if they can overwhelm the defenses of the resident pair.

This gang-up-on-the-territorial-residents hypothesis leads to a number of predictions: (1) resident territory owners should never yell, (2) nonresident ravens should yell, (3) yelling should facilitate a mass assault on a carcass by nonresident ravens, (4) resident pairs should be unable to repel a communal assault on their resources, and (5) a food bonanza should be eaten either by a resident pair alone or by a mob of ravens. Heinrich collected data that supported all of these predictions (Figure 4.24).[58] He concluded that when a young raven yells, the consequences of providing information ("food

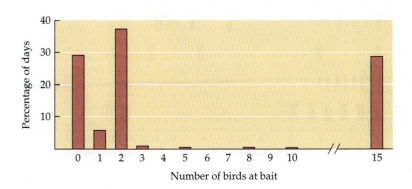

FIGURE 4.24 **Yelling is a recruitment signal.** The graph shows the percentage of days on which carcass baits were visited by various numbers of ravens. Carcasses were exploited either by non-yelling territorial singletons and pairs or by large groups of ravens, many of them yelling, most of them young, nonterritorial birds. After Heinrich.[58]

bonanza here") to other nonterritorial birds can include benefits (personal access to food for the yeller) that outweigh the energetic costs of yelling, as well as the risk of attack by the resident ravens guarding the carcass.[59] Both yelling and the response to yelling therefore have the properties of adaptations, because these actions can yield net fitness benefits to both signalers and receivers under the appropriate circumstances. Puzzle solved.

The Darwinian Puzzle of Deception

Having made the case that senders and receivers in evolved communication systems must both benefit, how are we to account for the evolution of deception? There are many such cases, which include predatory margay cats imitating the calls of their prey, juvenile tamarin monkeys;[15] an octopus that can change its appearance to look like a toxic marine flatfish;[64] and a butterfly whose larvae are raised in ant colonies where the caterpillars mimic the sounds that ant queens make in order to receive preferential care from ant workers.[3]

Firefly "femme fatales" provide another example of this phenomenon.[75] Females of these nocturnal predatory insects in the genus *Photuris* respond to firefly flashes given by males in the genus *Photinus* as these males try to locate and attract conspecific mates (Figure 4.25).[74] The predatory female answers her prey's signal with a flash of light, perfectly timed to resemble the response given by the right *Photinus* female. Some *Photuris* females can supply the differently timed answering flashes of three different *Photinus* species.[77] Typically, a male *Photinus* that sees a receiver's signal of the appropriate type flies toward the light flash. But if he encounters a deceptive *Photuris* female, he may well be grabbed, killed, and eaten by the predator (Figure 4.26), which recycles the toxic chemical defenses of her victims for her own benefit.[74]

Much the same sort of thing has been documented for a predatory katydid that quickly answers the loud calls of a variety of cicadas with acoustical clicks

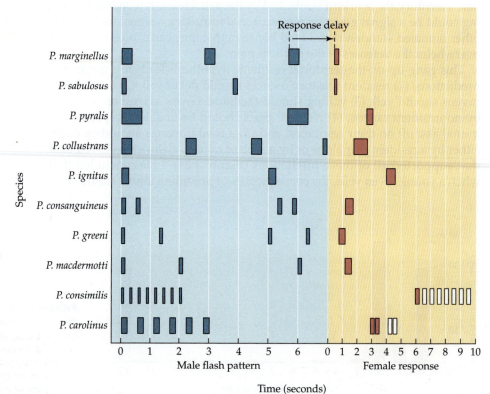

FIGURE 4.25 The diversity of communication flashes by different species of fireflies. On the left, the timing (in seconds) of male flashes is shown for ten North American species of *Photinus* fireflies. On the right, the diagram shows when the female provides her flash response to the last pulse of light from a male of her species. The differences among species lead to reproductive isolation of the various species. Based in part on the research of Jim Lloyd (e.g., Lloyd[76]). After Lewis.[74]

that mimic the come-hither sounds made by receptive female cicadas. Male cicadas that track the sound to its source may be captured and eaten by the katydid (Figure 4.27).[81]

The male firefly or cicada that pays attention to the wrong signal loses any future chances to reproduce. Why then do some males of these insects do the wrong thing? When a behavior is clearly disadvantageous, evolutionary biologists have recourse to two main possibilities (see also Table 5.1):[21]

1. *Novel environment theory*: The maladaptive behavior of the receiver occurs because the current environment is so different from that in which the behavior evolved that there has not been sufficient time for advantageous mutations to occur that would "fix the problem."

2. *Net benefit theory*: The maladaptive response of the receiver is caused by a sensory mechanism that may result in fitness losses for some individuals under some circumstances but does not erase the fitness gain that signal receivers derive on average for reacting to a signal giver in a particular way.

Novel environments seem unlikely to account for the nature of interactions between *Photinus* males and their predatory relatives or between male cicadas and the katydid that lures them to their doom. This theory more often applies to cases in which very recent human modifications of the environment appear responsible for eliciting maladaptive behavior (see Figure 12.6). Instead, instances in which prey respond maladaptively to their predator's signals have been explained by arguing that, on average, the response of a male *Photinus* to certain kinds of light flashes increases his fitness, even though one of the costs of responding is the (small) chance that he will be devoured by a *Photuris* femme fatale. Males that avoided the deceptive *Photuris* signals might well live longer, but they would probably ignore females of their own species as well, in which case they would leave few or no descendants to carry on their cautious behavior.

FIGURE 4.26   A firefly femme fatale. This female *Photuris* firefly has killed a male *Photinus* firefly that she lured to her by imitating the timed flash response given by females of his species. Photograph by Jim Lloyd.

FIGURE 4.27   A deceptive killer katydid. This katydid uses acoustical sexual deception to attract male cicadas, the better to capture and eat them. Photograph by Dave Marshall; from Marshall and Hill.[81]

Discussion Question

4.10 An assassin bug that captures and kills orb-weaving spiders sometimes approaches a web and plucks at the silken strands it encounters. The resident spider sometimes responds by moving across the web to the place where the assassin bug is at work. The spider's behavior can lead to its death. Use net benefit theory to explain why spiders respond to web vibrations in this way. Use your hypothesis to make predictions about the response of spiders to web movements caused by the predatory bug, prey struggling in a web, leaves falling into the web, and male spiders that have moved onto the web to court the female web builder. Check your predictions against Wignall and Taylor.[121]

The Manipulation of Communication Systems

Hypotheses on deception based on net benefit theory highlight the definition of adaptation employed by most behavioral biologists, which is that an adaptation need not be perfect, but it must contribute more to fitness *on average* than other alternative traits. The male firefly that responds to the signals of a predatory female of another species possesses a mechanism of mate location that clearly is not perfect but is better than those alternatives that improve a male's survival chances at the cost of making him a sexual dud. In effect, the evolved communication system involving male *Photinus* signalers and female *Photinus* receivers has been taken advantage of by predatory fireflies in another genus. These **illegitimate signalers** presumably reduce but do not eliminate the net fitness benefits of participating in a generally advantageous communication system.[91] Cases of this sort are usually ones in which a deceptive signaler of one species exploits a receiver of another, but there are possible examples of intraspecific deception (e.g., see Figure 4.22).

If this adaptationist hypothesis is correct, then deception by an illegitimate signaler should exploit a response that has clear adaptive value under most circumstances.[17] Consider the interaction between male thynnine wasps and the elbow orchid of Western Australia. A wasp in search of a mate will sometimes fly to the bizarre lip petal of an elbow orchid flower, which it will then grasp and attempt to fly away with. The lip petal is jointed, and so any wasp that flies while holding onto the object at the end of the petal moves upward in an arc. This trajectory quickly brings the male into contact with the column, the part of the flower where packets of pollen are located. By this time, the male will have released the lip petal, but the orchid does not let him get away immediately; instead it traps him because the wasp's wings slip into the two hooks on the underside of the column. There the wasp is held temporarily while his thorax touches the sticky pollen packets of the orchid. When the wasp manages to struggle free from his floral captor, he is likely to carry with him the orchid's pollen (Figure 4.28). Should he be deceived again, the wasp will transfer the pollen packets on his back to the pollen-receiving surface of the column of another plant. Thus, the wasp acts as a pollinator for the elbow orchid thanks to the illegitimate signals offered by the elbow orchid flower.[2] The benefits of this interaction accrue entirely to the orchid, although at least the orchid does not kill the wasp, unlike a firefly femme fatale. However, the pollinating wasp does waste his time and energy when interacting with the elbow orchid, as is true for many other species of wasps that are fooled by other species of orchids.[109] In at least one wasp–orchid system, the male thynnine finds the orchid's female decoy so stimulating that he also wastes his sperm by ejaculating after grasping the lip petal.[47]

(A)

(B)

FIGURE 4.28 **Sexual deception by the elbow orchid.** (A) This small brownish Australian orchid has flowers with lip petals that are highly modified to resemble a female wasp. This plant has three female decoys. (B) Male thynnine wasps are attracted by the odor and appearance of female decoys, which they may grasp and attempt to carry off. In so doing, a male thynnine travels up in an arc with the decoy and is trapped by the orchid column's "wings," where it remains long enough to pick up the pollen packets of the orchid for later transfer to another deceptive elbow orchid. A, photograph by the author; B, photograph by Bert Wells.

Discussion Question

4.11   Charles Darwin loved both orchids and sundews, the latter because of their carnivory.[25,26] As is true for certain orchids, sundews engage in deceptive signaling (Figure 4.29). The fluids exuded by the plants attract insects that become trapped when they alight upon the sticky, glue-like droplets, a prelude to their death and digestion by the plant. Analyze the evolutionary basis for this case of deception, using the approach we have outlined above.

Why are male thynnine wasps prone to the fitness-reducing deception practiced by elbow orchids and other Australian orchids? To answer this question, we need to observe the interactions between female and male wasps. The

FIGURE 4.29 **Plants can engage in deception.** (A) Sundews exude droplets that look like sugary fluids but are instead sticky glues, which (B) trap insects deceived into landing on the sticky droplets. Two very different species of Australian sundews are shown here. Photographs by the author.

(A)

(B)

(A)   (B)   (C)

FIGURE 4.30 Why the elbow orchid's sexual deception works. (A) A female thynnine wasp releasing a sex pheromone from her perch. (B) A male of the same species grasps the calling female. (C) A copulating pair of this thynnine on a eucalyptus flower. Note the orchid pollinia attached to the thorax of the male, which has been deceived at least once before securing the real McCoy. Photographs by the author.

wingless females of this species come to the surface of the ground after searching underground for beetle larvae to paralyze with a sting before laying eggs on their victims. Once aboveground, females crawl upward a short distance on a plant or twig before perching and releasing a sex **pheromone**, a chemical signal that females use to attract males (Figure 4.30). Males patrol areas where females may appear, and they rush to calling females, which they grab and carry away for a mating elsewhere.[101] The elbow orchid unconsciously (obviously) exploits the males' ability to respond to the olfactory and visual cues associated with the right kind of female wasp. The lip petal offers odors similar enough to those released by calling females to attract sexually motivated males to the plant. Once nearby, the flying wasp can see the lip petal, which looks vaguely like a wingless female wasp of his own species. The male wasp that pounces is deceived into becoming a potential pollinator.[2]

Let us imagine what would happen if there were mutant males that tried to discriminate between deceptive orchids and the real McCoy. These careful males almost certainly would be slower to approach elbow orchids—and females of their own species. They might not be deceived by elbow orchids very often, but they would also be somewhat less likely to reach a calling female wasp before other males. Competition for females is stiff among thynnine wasps, and the first male to grab a legitimate signaler is very likely to win the chance to inseminate his mate; those arriving even a few seconds later wind up with nothing. So, as is true for *Photuris* fireflies, the elbow orchid shows that illegitimate signalers can evolve if there is a legitimate communication system to exploit.

In a way the plant is truly exploiting a preexisting sensory bias in male wasps, which first evolved a sensitivity to the appearance and odor of females of their species, which the orchid later mimicked via the evolution of female decoys that look and smell like female wasps. Even plants that feed rather than deceive their pollinators appear to have evolved their come-and-get-it signals *after* the sensory capacities of their nectar- and pollen-consuming pollinators had evolved. These perceptual skills originated earlier for some other purpose, further evidence in support of the importance of preexisting biases in the evolution of communication signals.[110]

4.12   The fork-tailed drongo, an African bird that often perches in trees, sometimes gives alarm calls that warn of terrestrial predators when it is accompanying flocks of the pied babbler, a bird that forages on the ground. Upon hearing the drongo's alarm call, pied babblers dash to cover, sometimes leaving behind recently captured insect prey.[100] If we hypothesize that these alarm calls are often deceptive, what prediction can we make about the kind of alarm call produced by drongos vocalizing in the absence of babblers? Why might it be adaptive for babblers to react to drongo warning calls if some, or even most, are false alarms? Does the same argument apply to the case of male topi antelopes that produce false alarm signals when associating with groups of females on their mating territories?[12]

One More Darwinian Puzzle: Eavesdropping

Just as light-signaling male *Photinus* fireflies and acoustical-signaling Australian cicadas may come to an unhappy end when they are "overheard" by **eavesdropping** predators, so too calling túngara frogs sometimes have the great misfortune to attract a fringe-lipped bat rather than a female of their own species.[104] Predators that take advantage of acoustical, visual, and olfactory (pheromonal) signals provided by their prey are **illegitimate receivers**, listening in on individuals that lose fitness by having their message reach the wrong target.[73,125] When predatory illegitimate receivers succeed in using their prey's signals to secure a meal, the consumed signal givers lose all future opportunities to reproduce, which one would think would select strongly against signaling.

Indeed, Marlene Zuk and her colleagues have documented that on the Hawaiian island of Kauai, almost no males of the field cricket *Teleogryllus oceanicus* now sing to attract mates.[127] By remaining silent, these males are less likely to be visited by a fatal parasitoid fly, *Ormia ochracea*, which uses the chirps of crickets as a guide to their location.[54] On finding a cricket, the female fly then stealthily deposits larvae on the unfortunate creature, which the immature flies eventually consume from the inside out. In most areas, field crickets are noisy, but on Kauai, where the lethal fly has only recently arrived, silent males now make up a large proportion of the cricket population, a change that has taken only 20 generations to evolve.[127] Still, in most places with *O. ochracea*, calling male crickets are not uncommon, just as the presence of the fringe-lipped bat has not resulted in the evolution of silent male túngara frogs. Why do so many animals provide their enemies with cues that can result in their death?

The same general argument that we examined when discussing illegitimate signalers can help explain why some animals provide illegitimate receivers with information that can be used against them. When an evolved communication system contributes to the fitness of legitimate signalers and legitimate receivers, the system can be exploited by outsiders. When predators or parasites are eavesdropping, they impose costs on the legitimate signalers nearby. If these costs are very high, signaling can be all but eliminated, as shown by the case of field crickets on fly-blown Kauai. But if the reproductive costs do not exceed the benefits of signaling on average, legitimate signalers will persist in the population.

Male túngara frogs that fail to call might well live somewhat longer on average than callers, but any quiet males incur a reproductive disadvantage. Indeed, one of the things that Mike Ryan and his associates have learned about

(A)

(B)

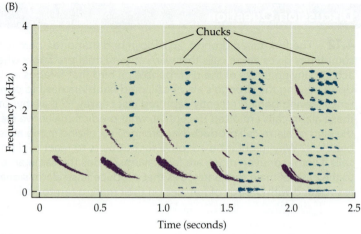

FIGURE 4.31 Illegitimate receivers can detect the signals of their prey. (A) A calling male túngara frog may inadvertently attract an illegitimate receiver—the fringe-lipped bat, a deadly predator. (B) The risk of attack is greater if the male's call includes one or more chucks (blue in the sonograms) as well as the introductory whine (purple in the sonograms). A, photograph by Merlin D. Tuttle, Bat Conservation International; B, sonograms courtesy of Mike Ryan.

túngara frogs is that many males that produce longer, louder calls composed of an introductory whine followed by a series of chucks are more likely to attract females[6]—just as males of some fireflies that produce their flashes at a relatively high rate are more likely to be answered by females of their species.[74] However, both male frogs and fireflies that generate the most attractive calls are also likely to attract predators, namely eavesdropping bats (Figure 4.31)[105] and *Photuris* fireflies,[74] respectively. The tradeoff between attracting females and attracting predators has had an effect on the male túngara frog's calling strategy (and probably on vulnerable male fireflies too). Because frogs calling in small assemblages are at greater danger of becoming a bat's meal (see page 110), males in small groups are more likely to drop the chucks from their calls than are males in large groups.[7,104]

The behavior of other signalers has also been shaped by the risk of attracting eavesdropping predators. For example, nestling songbirds produce loud cheeps and peeps that could give a listening hawk or raccoon information on the location of a nest full of edible nestlings. In fact, when tapes of begging tree swallows were played at an artificial swallow nest containing a quail's egg, the egg in that "noisy" nest was taken or destroyed by predators before the egg in a nearby quiet control nest in 29 of 37 trials.[73] The raccoon that listens in on the communication taking place between a brood of baby tree swallows and their parents uses information produced by the nestlings to the detriment of these individuals and their parents.

The begging calls of birds vary in the degree to which they can be exploited by listening predators. For example, ground-nesting warblers would seem to be more exposed to raccoons, skunks, and the like than those species that nest in relative safety high in trees.[57] The young of ground-nesting warblers produce begging cheeps of higher frequencies than their tree-nesting relatives (Figure 4.32A). These higher-frequency sounds do not travel as far and so should better conceal the individuals that produce them. David Haskell tested this prediction by creating artificial nests with clay eggs, which he placed on the ground by a tape recorder that played the begging calls of either tree-nesting or ground-nesting warblers. The eggs advertised by the tree nesters' begging calls were found and bitten significantly more often than the eggs associated with the ground nesters' calls (Figure 4.32B).[57]

(A)

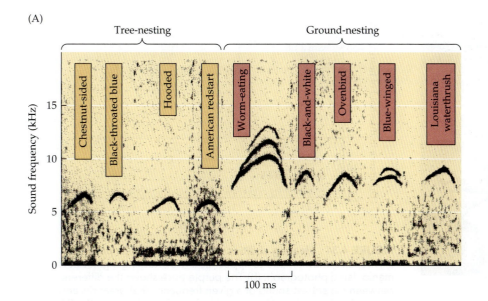

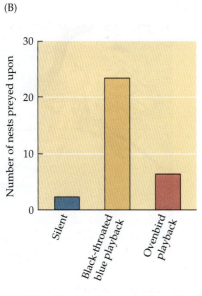

FIGURE 4.32 Predation risk has affected the evolution of begging calls in warblers. (A) The nestling begging calls of ground-nesting warbler species are higher in sound frequencies than the calls of young tree-nesting warblers. (B) Experimental playback of the begging calls of the black-throated blue warbler (a tree-nesting species) at artificial nests placed on the ground resulted in higher rates of discovery by predators than did playback of begging calls of the ground-nesting ovenbird. After Haskell.[57]

Discussion Question

4.13 Males of a fish called the northern swordtail attract mates with flashy displays that show off their elaborate tails. But this species shares its habitat with a lethal predatory fish, the Mexican tetra. The bodies of male northern swordtails, especially their tails, reflect a considerable amount of ultraviolet radiation.[23] When researchers put male and female swordtails in tanks with and without ultraviolet filters, they found that female swordtails were more attracted to males when the ultraviolet channel was available to them. Given that these researchers were interested in how swordtails might reduce the risk of eavesdropping by Mexican tetras, what other experiment would they have to do as well? Outline the science behind this research, starting with the causal question and ending with the possible conclusions the researchers might have reached.

The risk of exploitation by an illegitimate receiver may also be responsible for the differences between the rasping mobbing call and the "seet" alarm call of the great tit (Figure 4.33).[80] These small European songbirds sometimes approach a perched hawk or owl and give a loud mobbing call whose dominant frequency is about 4.5 kHz. This easily located acoustic signal helps other birds find the mobbers and join in the harassment of their mutual enemy (see page 104). If, however, a great tit spots a flying hawk, it gives a much quieter

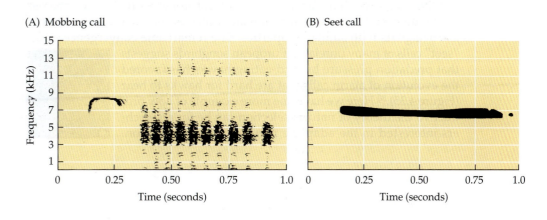

FIGURE 4.33 Great tit alarm calls. Sonograms of (A) the mobbing call and (B) the "seet" alarm call. Note the lower sound frequencies in the mobbing signal. A, courtesy of William Latimer; B, courtesy of Peter Marler.

(A)

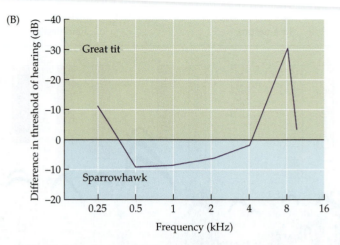

(B)

FIGURE 4.34 **Hearing abilities of a predator and its prey.**
(A) A sparrowhawk made to appear as if it is attacking a great tit in this manipulated photograph. (B) The purple trace shows the difference between the softest sound of a given frequency that great tits and sparrowhawks can hear. A sparrowhawk can hear sounds in the 0.5 to 4 kHz range that are fainter (5 to 10 dB lower in intensity) than those that great tits can hear. But great tits can detect an 8 kHz sound (in the range of the "seet" call) that is fully 30 dB fainter than any 8 kHz sound that a sparrowhawk can detect. After Klump et al.[67]

"seet" alarm call, which appears to warn mates and offspring of possible danger while making it harder for the predator to track down the signaler.

Interestingly, in the Japanese population of the great tit, parents use a harsh signal rather like the mobbing call to warn their chicks of an approaching rat snake, which induces the nestlings to leap out of the nest; the adults give a much softer call in a narrow range of sound frequencies when a jungle crow is nearby, which induces the nestlings to crouch down in the nest cavity, thereby making the baby birds harder for a crow to extract.[113]

The European great tit's "seet" call's dominant frequency lies within 7 to 8 kHz, so the sound attenuates (weakens) after traveling a much shorter distance than the mobbing signal. The rapid attenuation of the "seet" call compromises its effectiveness in reaching distant legitimate receivers, but it also lowers the chance that a dangerous predator on the hunt will be able to tell where the caller is. Moreover, the frequencies of the "seet" call lie outside the range that hawks can hear best, while falling within the range of peak sensitivity of the great tit (Figure 4.34). As a result, a great tit can "seet" to a family member 40 meters away but will not be heard by a sparrowhawk unless the predator is less than 10 meters distant.[67]

If the "seet" call of the great tit has evolved properties that reduce the risk of detection by its enemies, then unrelated species should also have evolved alarm signals with similar properties. The remarkable convergence in the "seet" calls of many unrelated European songbirds suggests that predation pressure by bird-eating hawks has favored the evolution of alarm calls that are hard for hawks to hear (Figure 4.35).[80] Here we have one more example of how productive the theory of natural selection can be for researchers interested in the adaptive nature of communication.

FIGURE 4.35 Convergent evolution in a signal. The great tit's high-pitched "seet" alarm call (see Figure 4.33B) is very similar to the alarm calls given by other, unrelated songbirds when they spot an approaching hawk. After Marler.[80]

Reed bunting

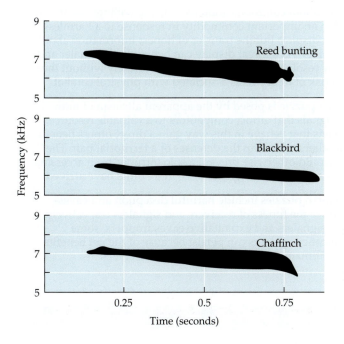

Blackbird

Chaffinch

Summary

1. If Darwin's theory of natural selection is correct, communication systems can persist only when both signalers and receivers gain fitness benefits from their interactions. If just one party benefits, then selection should act against participation in the system by the other party.

2. Darwin's other evolutionary idea, the theory of descent with modification, tells us that the adaptive traits of an animal living today (including its communication system) are the products of past changes layered on still older changes that occurred during the history of the species.

3. One result of descent with modification has been given the label "sensory exploitation." The term is applied to cases in which an individual produces a signal that happens to activate a preexisting sensory mechanism already present in a receiver. Sensory exploitation theory argues that behaviors originate as a result of what has already evolved; in the case of communication systems, legitimate signalers may tap into the sensory biases of legitimate receivers.

4. Changes will spread through populations by selection when individuals with the modified signal (or response) gain fitness from their altered behavior. Awareness of this rule helps researchers identify Darwinian puzzles, namely attributes that seem to reduce rather than increase the individual's fitness, such as female spotted hyenas' use of a false penis (an enlarged clitoris) in greeting displays with other members of their clan.

5. Darwinian puzzles about communication include the willingness of individuals to abandon an important commodity in response to a harmless (noncontact) threat display. But the transfer of honest information about relative fighting ability via a threat display can benefit both signaler and receiver when they resolve conflicts quickly and accurately without the time, energy, and risk of injury costs associated with physical combat.

6. Another Darwinian puzzle is posed by the apparent altruism of individuals that produce signals that attract competitors to a valuable resource. However, the ravens that behave in this way gain a direct fitness benefit from drawing in others to swamp the defenses of a territorial pair. The territory owners would otherwise prevent a solo intruder from accessing food that becomes available to a signaler once more birds have arrived.

7. Additional Darwinian puzzles include harmful deception and eavesdropping that reduce the fitness of receivers and signalers respectively. Despite deceptive signalers and exploitative eavesdroppers, some evolved communication systems can persist because of the net benefits they provide to legitimate signalers and legitimate receivers.

Suggested Reading

The vast field of animal communication has been reviewed in an updated edition of *Principles of Animal Communication* by Jack Bradbury and Sandra Vehrencamp.[9] Darwin discussed "descent with modification" in *On the Origin of Species*, especially in Chapter 13;[24] see also Gould.[52] Bernd Heinrich's *Ravens in Winter* illustrates how an adaptationist uses multiple hypotheses to solve a Darwinian puzzle in communication.[59] For an evolutionary analysis of manipulation and deceit by communicating animals, read Dawkins and Krebs;[28] for the classic paper on honest signaling, see Zahavi.[126] Tristram Wyatt focuses on the role of pheromones in animal communication and provides many examples of olfactory eavesdropping and deceit.[124]

5

Avoiding Predators and Finding Food

 During a trip to New Zealand long ago, I visited a coastal nature reserve where hundreds of pairs of silver gulls were nesting on the stony seaside. While I was watching the gulls from a distance, a young researcher walked down to the shore carrying a scale for weighing gull chicks and a clipboard for recording the results. As she walked toward the colony, the gulls took notice, and soon those closest to her began to fly up, calling raucously. By the time she came within a few meters of the first nests, the colony was in an uproar, and many of the adult gulls were swooping about, some diving at the intruder like kamikaze pilots, others yapping loudly, a few attempting to splatter the researcher with liquid excrement (Figure 5.1).

It is not hard to guess why gulls become upset when potential predators get close to their nests—and their progeny. If the parents' assaults distract predators from their offspring, then mobbing gulls may increase their reproductive success, enabling them to pass on the hereditary basis for joining other gulls in screaming at, defecating on, and even assaulting those who might eat their eggs and youngsters. Indeed, the distraction hypothesis motivated Hans Kruuk, at that time a student of Niko Tinbergen, to investigate group mobbing in the black-headed gull, another ground-nesting, colony-forming species (Figure 5.2) but one that lives in Europe rather than New Zealand.[64] Because Kruuk was interested in studying the adaptive basis of mobbing, he did what behavioral ecologists do to solve problems of this sort—he considered the costs and benefits of the behavior. His working hypothesis was based on the notion that the costs to the mobbers (such as the time and energy expended by gulls) were outweighed by the fitness benefits to the birds from

The great potoo provides a wonderful example of an animal that uses a camouflaged color pattern and cryptic behavior to avoid detection by its predators.

FIGURE 5.1 **Mobbing behavior of colonial, ground-nesting gulls.** Silver gulls reacting to a trespasser in their breeding colony in New Zealand. Photograph by the author.

their social harassment of egg and chick predators (such as an improvement in the survival chances of their offspring). Note that when behavioral ecologists speak of costs, they are talking about **fitness costs**, the negative effect of a trait on the number of surviving offspring produced by an individual or a reduction in the number of copies of its alleles that it contributes to the next generation. When they speak of **fitness benefits**, they are referring to the positive effects of the trait on reproductive (and genetic) success.

FIGURE 5.2 **A nesting colony of black-headed gulls in Great Britain.**

The Definition of Adaptation

Kruuk and all other behavioral ecologists study traits that they believe are adaptations, an assumption that they make in order to test specific hypotheses about the possible **adaptive value** (contribution to fitness) of the characteristic of interest. So what is an adaptation? Here we shall use the following definition: an **adaptation** is a hereditary trait that has spread or is spreading by natural selection and has replaced or is replacing any alternative traits in the species. Such an attribute has a better fitness-benefit-to-fitness-cost ratio than the alternative forms of this trait that have occurred in the species.

This definition does *not* require that the naturally selected trait be "perfect." Selection does not produce perfect traits, for any number of reasons (Box 5.1).[23,77] For one thing, selection cannot create the best of all possible genes for a particular task but has to wait for mutations to occur by chance; only when differences exist in the reproductive success of the different allele carriers can selection winnow out the less effective alleles, increasing the frequency of the form that does the best job in promoting reproduction but only in the environment in

---

BOX 5.1   Why a trait may not be perfectly adaptive

The adaptive value of a trait is the contribution it makes to an individual's inclusive fitness. Not every trait has adaptive value. For example, some traits persist when they are no longer adaptive, some traits are due to genes that have both a positive and a negative adaptive value, and some traits are slow to develop. Mutations, pleiotropy, and coevolution all constrain the adaptive value of traits.

Constraint 1: Failure of appropriate mutations to occur

If the appropriate mutation does not occur, selection cannot keep up with environmental change. Thus, maladaptive or nonadaptive traits can persist, especially in environments only recently invaded by a species. So, for example, some arctic moths live in regions where bats are absent, but the moths still cease flying in response to an ultrasonic stimulus.[96] Likewise, arctic ground squirrels react defensively upon experimental exposure to snakes, even though there are no snakes living in the Arctic.[19] (Because these behaviors are not used by these species in nature, they presumably are not very costly, but the underlying physiological systems that make the behavior possible probably do require developmental energy.)

Man-made changes in the environment are especially likely to lead to inappropriate expression of previously adaptive traits.[98] For example, some moths are so strongly attracted to artificial lights that bats visit lights in order to make some easy kills.[37] Likewise, male buprestid beetles (see Figure 12.6) may die while persistently attempting to mate with beer bottles,[51] and sea turtles sometimes consume plastic bags floating in the water that the turtles mistake (fatally) for edible

jellyfish.[9,65] The current obesity epidemic in Western societies may well be caused in part by the once adaptive desire of humans to consume calorie-rich foods that were rare in the past but are now easily obtained.[76]

Constraint 2: Pleiotropy (in which a gene has multiple effects on development)

Pleiotropy occurs when a gene has more than one developmental effect, some of which may be positive while others are negative. If the negative consequences of a gene are outweighed by the positive ones, the less than perfect effects can persist in populations. For example, misdirected parental care is not uncommon in nature, a result of the intense drive to care for offspring, which usually is adaptive but in relatively rare instances can lead an adult to provide assistance to a youngster of a genetic stranger.[54] The costs of these rare events may be so small that the gene persists in the population.

Constraint 3: Coevolution (which occurs when individuals interact in ways that affect each other's fitness)

If what is adaptive for one individual is maladaptive for another, then evolutionary stability may never be reached. Instead, as the members of one group or species change in response to selection pressure imposed by another group or species, selection may favor counterresponses, as in the coevolutionary arms races between predator and prey. The inability of selective processes to generate an immediate effective solution to an altered environmental problem means that some less than perfect traits can remain for a time within a species.

which the trait evolved. As a result, although a trait that has spread as a result of natural selection should be better than the alternatives it has replaced, the adaptation need not work perfectly, and indeed, more than one black-headed gull has made a lethal miscalculation when dive-bombing an unusually agile fox that has leapt up and caught the gull in midair. After all, any hereditary improvement in a gull's ability to thwart a certain predator would tend to select for improvement in that predator's ability to circumvent the bird's defense. As a result of this kind of **arms race** between predator and prey, we might well find that neither one would be able to thwart the other completely.[32]

Discussion Questions

5.1   Many people think that an adaptation is a trait that improves the survival chances of an organism. Under what circumstances would a survival-enhancing attribute be selected against?

5.2   Stephen Jay Gould and Richard Lewontin claimed that adaptationists—a group that contains most of the world's behavioral ecologists—often make the elementary mistake of believing that every characteristic of living things is a perfected product of natural selection,[43] when in reality many attributes of living things are not adaptations (see Box 5.1). One of the effects of making this mistake is, according to Gould and Lewontin, the tendency of adaptationists to invent fables as absurd as the fictional "just so stories" of Rudyard Kipling, who made up amusingly silly explanations for the leopard's spots and the camel's hump. How might adaptationists defend themselves against these charges? Do adaptationists have the means to discover whether an evolutionary explanation for a particular trait is wrong?

Testing Adaptationist Hypotheses

Once Kruuk produced his adaptationist hypothesis, he wanted to test it, which he did in the standard manner by using the hypothesis to make a prediction, namely that mobbing gulls should force distracted predators to expend more searching effort than they would otherwise. This expectation can be tested simply by watching gull–predator interactions.[64] Kruuk saw that egg-hunting carrion crows have to continually face gulls diving at them and so, while being mobbed, they cannot look around comfortably for their next meal. Because distracted crows are probably less likely to find their prey, Kruuk established that a probable benefit exists for mobbing. Moreover, the benefit of mobbing crows plausibly exceeds the costs, given that crows do not attack or injure adult gulls.

However, the predator distraction hypothesis yields much more demanding predictions. Because adaptations are better than the traits they replace, we can predict that the benefit experienced by mobbing gulls in protecting their eggs should be directly proportional to the extent that predators are actually mobbed. Kruuk used an experimental approach to test this prediction.[64] He placed ten chicken eggs as stand-ins for gull eggs, one every 10 meters, along a line running from the outside to the inside of a black-headed gull nesting colony. The eggs placed outside the colony, where mobbing pressure was low, were much more likely to be found and gobbled up by carrion crows and herring gulls than were eggs placed inside the colony, where the predators were communally harassed by the many parents whose offspring were threatened by the presence of a predator (Figure 5.3).

Thus, Kruuk assembled both observational and experimental evidence that supports the hypothesis that mobbing is an adaptation that helps adult black-headed gulls protect their eggs and youngsters. Note that behavioral ecologists

Black-headed gull

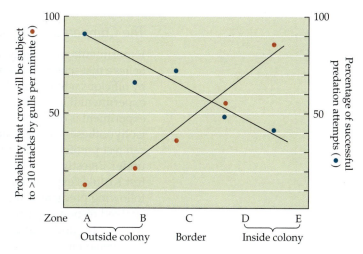

FIGURE 5.3   Does mobbing protect eggs? When chicken eggs were placed outside and inside a black-headed gull nesting colony, crows searching for the eggs within the colony were subject to more attacks by mobbing gulls (red circles), and as a result, they discovered fewer hen eggs (blue circles). After Kruuk.[64]

often have to settle for an indicator or correlate of reproductive success or genetic success when they attempt to measure fitness. In the chapters that follow, *fitness* or *reproductive success* or *genetic success* is often used more or less interchangeably with such indicators as egg survival (Kruuk's measure), the number of young that survive to fledging, the number of mates inseminated, or even more indirectly, the quantity of food ingested per unit of time, the ability to acquire a breeding territory, and so on. These proxies for an individual's **reproductive success** (i.e., the total number of his or her offspring—or grandoffspring—that reach the age of reproduction) are pretty clearly related to reproductive success but only as an approximation.

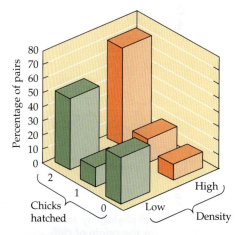

Discussion Questions

5.3   As is true for many issues in biology, scientists have proposed a number of competing definitions for *adaptation* (see Harvey and Pagel;[53] Lauder et al.;[66] Reeve and Sherman[91]). For some, the term must be reserved for a characteristic that provides "current utility to the organism and [has] been generated historically through the action of natural selection for its current biological role.[1]" What do you think *current utility* means? If a trait originated for function X and later took on a different, but still adaptive, biological role Y, does that mean it is not an adaptation? Track down the evolutionary history of the flight feathers of modern birds (see Prum and Brush[88]). Where did these feathers come from, and what function did their predecessor feathers exhibit? If you go back far enough in time, will the ancestral form of any current trait have the same function that it does now?

5.4   The arctic skua, a close relative of gulls, also nests on the ground and mobs colony intruders, including another relative, the great skua, a larger predator that eats many arctic skua eggs and chicks. In one study, hatching success and post-fledging survival were greater for arctic skuas that nested in dense colonies than for those in low-density groups (Figure 5.4); the number of near neighbors was, however, negatively correlated with the growth rate of their chicks.[82] Rephrase these findings in terms of the reproductive costs and benefits of communal nesting and mobbing by the arctic skua. If adaptation meant a perfect trait, would communal mobbing by arctic skuas be labeled an "adaptation"?

Arctic skua

FIGURE 5.4   Benefit of high nest density for the arctic skua. In one population studied in 1994, skuas nesting with relatively many nearby neighbors were more likely to rear two chicks than were individuals nesting in areas with a lower density of breeding pairs. After Phillips et al.[82]

The Comparative Method

Experiments are highly valued in science, so much so that most persons believe that scientific research can be performed only in high-tech laboratories by officious white-coated researchers. As Kruuk's work shows, however, good experimental science can be done in the field. Moreover, the manipulative experiment is only one of several ways in which predictions from adaptationist hypotheses can be tested. For example, users of the **comparative method** can test predictions about which other species should have evolved the trait under investigation, and which should not.[16] We can use this approach to test Kruuk's explanation for mobbing by black-headed gulls by checking the following prediction: if mobbing by ground-nesting black-headed gulls is an evolved response to predation on gull eggs and chicks, then other gull species whose eggs and young are at low risk of predation should not exhibit mobbing behavior.

The rationale behind this prediction is as follows: the various costs of mobbing (such as the increased risk of being killed by a predator) will be outweighed only if there are sufficient benefits derived from distracting predators. If predators have not posed a problem, then the odds are that the costs of mobbing (such as the energy spent dive-bombing a nonpredator) would be greater than the benefits.

There is good reason to believe that the ancestral gull was a ground-nesting species with many nest-hunting predators against which the mobbing defense would have been effective. Of the 50 or so species of gulls living today, most nest on the ground and mob enemies that hunt for their eggs and chicks.[107] These similarities among gulls, which share many other features, are believed to exist in part because all gulls are the descendants of a relatively recent common ancestor, from which they all inherited the genetic package that predisposes today's many gull species to develop similar traits. Thus, all other things being equal, we expect all gulls to behave similarly.

However, some things are not equal. For example, a few gull species nest on cliff ledges or on trees rather than on the ground. Perhaps these species are all the descendants of a more recent cliff-nesting gull species that evolved from a ground-nesting ancestor. The alternative possibility, that the original gull was a cliff nester, requires the cliff-nesting trait to have been lost and then regained, which produces an evolutionary scenario that requires more changes than the competing one (Figure 5.5). Many evolutionary biologists, although not all,[92] believe that simpler scenarios involving fewer transitions are more probable than more complicated alternatives. In this, the majority accepts a commonly held philosophical principle, known as Occam's razor or the principle of par-

FIGURE 5.5 Gull phylogeny and two scenarios for the origin of cliff-nesting behavior.   (A) Hypothesis A requires just one behavioral change, from ground nesting to cliff nesting. (B) Hypothesis B requires two behavioral changes, one from the ancestral cliff nester to ground nesting, and then another change back to cliff nesting.

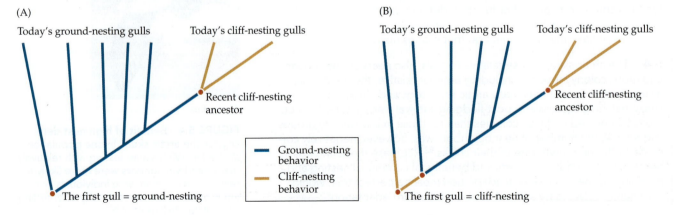

(A)

(B)

FIGURE 5.6   Not all gulls nest on the ground.
(A) Steep cliffs are utilized for nesting by kittiwake gulls, which appear in the lower half of this photograph. (B) Kittiwakes are able to nest on extremely narrow ledges; their youngsters crouch in the nest, facing away from the cliff edge. A, photograph by Bruce Lyon; B, photograph by the author.

simony, which holds that simpler explanations are more likely to be correct than complex ones—all other things being equal. (But as Einstein pointed out, "Everything should be made as simple as possible, but not simpler.")

In any event, cliff-nesting gulls currently have relatively few nest predators because it is hard for small mammalian predators to scale cliffs in search of prey, while predatory birds have a difficult time maneuvering near cliffs in turbulent coastal winds. Thus, a change in nesting environment surely led to a reduction in predation pressure, which should have altered the cost–benefit equation for mobbing by these gulls. If so, cliff-nesting gulls are predicted to have modified or lost the ancestral mobbing behavior pattern.

The black-legged kittiwake nests on nearly vertical coastal cliffs (Figure 5.6), where its eggs are relatively safe from predators.[70] The relatively small size of the kittiwakes may also make the adults themselves more vulnerable to attack by some nest predators, making the cost-to-benefit ratio for mobbing still less favorable. As predicted, groups of nesting adult kittiwakes do not mob their predators, despite sharing many other structural and behavioral features with black-headed gulls and other ground-nesting species. The kittiwake's distinctive behavior provides a case of **divergent evolution** and supports the hypothesis that mass mobbing by black-headed gulls evolved in response to predation pressure on the eggs and chicks of nesting adults.[26] Take away this pressure, and a gull will lose its mobbing behavior.

Another comparative approach is that species from different evolutionary lineages that live in similar environments should experience similar selection pressures and thus can be predicted to evolve similar traits, resulting in **convergent evolution**. If true, these species will adopt the same adaptive solution to a particular environmental obstacle to reproductive success, despite the fact

FIGURE 5.7 **The logic of the comparative method.** Members of the same evolutionary lineage (e.g., gull species of the family Laridae) share a common ancestry and therefore share many of the same genes, and thus they tend to have similar traits, such as mobbing behavior, which are widespread among ground-nesting gulls. But the effects of shared ancestry can be overridden by a novel selection pressure. A reduction in predation pressure has led to divergent evolution by the cliff-nesting kittiwake, which no longer mobs potential enemies. The other side of the coin is convergent evolution, which is illustrated by the mobbing behavior of some colonial swallows, even though gulls and swallows are not related, having come from different ancestors long ago. These colonial swallows and gulls have converged on a similar antipredator behavior in response to shared selection pressure from predators that have fairly easy access to their nesting colonies. What kind of evolution is responsible for the difference between bank swallows, which mob their enemies, and rough-winged swallows, which do not? This difference constitutes evidence in support of what evolutionary hypothesis about mobbing?

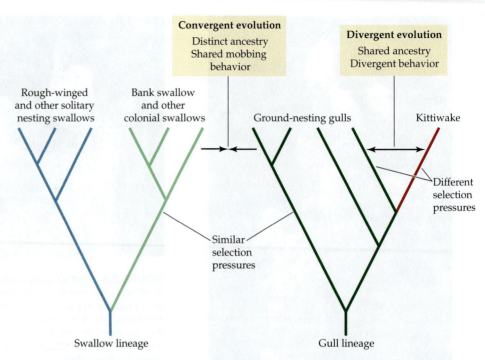

that their different ancestral species had very different genes and attributes (Figure 5.7). All other things being equal, unrelated species ought to behave differently—unless they have been subjected to the same selection pressures.

As predicted, mobbing behavior has evolved convergently in many other birds only distantly related to gulls.[103] Among these species is the bank swallow, which also nests in colonies where predatory snakes and blue jays come to eat swallow eggs and nestlings.[56] The common ancestor of bank swallows and gulls occurred long, long ago, with the result that the lineages of the two groups have evolved separately for eons—a fact recognized by the taxonomic placement of gulls and swallows in two different families, the Laridae and the Hirundidae. Despite their evolutionary and genetic differences from gulls, bank swallows behave like gulls when they are nesting. As they swirl around and dive at their predators, they sometimes distract hunting jays or snakes that would otherwise destroy their offspring.

Even some colonially nesting mammals have evolved mobbing behavior.[80] Adult California ground squirrels, which live in groups and dig burrows in the ground, react to a hunting rattlesnake by gathering around it and shaking their infrared-emitting tails vigorously, a signal to the snake to encourage it to depart before the ground squirrels kick sand in its face.[94] Rattlers molested in this fashion cannot hunt leisurely for nest burrows to enter in search of vulnerable young ground squirrels (Figure 5.8).

Because mobbing behavior has evolved independently in several unrelated species whose adults can sometimes protect their vulnerable offspring by distracting predators, mobbing appears to be an antipredator adaptation. However, this conclusion would be less certain if for every group-living species in which mobbing occurred under the expected conditions there were two in which the behavior was absent. Therefore, researchers increasingly require that the comparative method be used in a statistically rigorous fashion.[53] For our purposes, however, the point is that one can, in principle, test hypotheses about the adaptive value of a behavior by predicting that particular cases of divergent or convergent evolution will have occurred—a prediction about the past that can be checked by making disciplined comparisons among species living today.

FIGURE 5.8   Colonial California ground squirrels mob their snake enemies. One squirrel kicks sand at a rattlesnake, while others give a variety of alarm signals. Courtesy of R. G. Coss and D. F. Hennessy.

Discussion Question

5.5   The ability of one species of noctuid moth to hear ultrasound is considered an antipredator adaptation because it apparently enables individuals to hear and avoid nocturnal, ultrasound-producing bats. Imagine that you wished to test this hypothesis via the comparative method. Identify the utility of each of the following lines of evidence about the hearing abilities of other insect species. Specify whether these cases involve convergent evolution, divergent evolution, or neither.

1. Almost all other species of noctuid moths also have ears that respond to ultrasound.

2. Almost all the species in the evolutionary lineage that includes the noctuid moths and several other groups also have ears that respond to ultrasound.[118]

3. Some diurnal noctuid moths have ears but are largely or totally incapable of hearing ultrasound.[38]

4. Almost all butterflies, which belong to the same large evolutionary grouping as the noctuids, are usually active during the day and they lack ears to detect ultrasound.[36]

5. Six species of noctuid moths found only on the Pacific Ocean islands of Tahiti and Mooréa have ears and can hear ultrasound but do not react to this stimulus with antibat responses, such as diving for cover.[39]

6. Members of one small group of nocturnal butterflies have ears on their wings and can hear ultrasound; they respond to ultrasonic stimulation by engaging in unpredictable dives, loops, and spirals.[118]

7. Lacewings and praying mantises fly at night and have ears that detect ultrasound and lead to antibat defensive behavior.

(A)

(B)

FIGURE 5.9   Why do *Pogonomyrmex* ants mate in groups? (A) Harvester ants form huge but brief mating aggregations on hilltops. So many individuals come to these rendezvous sites that the local predators, like (B) this dragonfly (which has captured one unlucky ant) cannot possibly consume them all. Photographs by the author.

Discussion Question

5.6   Some persons might say that the fact that most noctuid moths have ultrasound-sensitive ears is "simply" a reflection of their shared ancestry, a holdover from the past, and therefore that ultrasound sensitivity is not an adaptation in these species.[6] Others disagree, arguing that it makes no sense to define adaptations in a way that limits them to just those traits that have diverged from the ancestral pattern.[91] Who is right?

A Cost–Benefit Approach to Social Defenses

Communal mobbing behavior is an example of a social defense against predators; it has costs and benefits, as is true for all behavioral traits. Consider the harvester ants that come together to mate on only a few days each year when huge numbers of individuals briefly form a dense aggregation (Figure 5.9). Dragonflies and birds can locate these swarms, where they capture and eat some ants, but the odds that any one ant will be targeted (the fitness cost of gathering together in conspicuous masses) are small given the ratio of ants to predators (and there are fitness benefits in being part of a mating aggregation).

This **dilution effect** argument may explain why many butterflies create tightly packed groups when drinking from mud puddles, where they may be securing fluids high in nitrogen or sodium.[74] One would think that these groups are likely to be spotted by butterfly-eating birds (but see Ioannou et al.[59]). However, any costs imposed on mud-puddling butterflies arising from increased conspicuousness of the groups may be offset by a dilution in the chance that any one individual will be killed by an attacker.[109] Imagine that only a few predators (say, five birds) occupy the space that contains a mud-puddling aggregation of butterflies. Further imagine that each of the five predators kills two prey per day there. Under these conditions, the risk of death for a member of a group of 1000 butterflies is 1 percent per day, whereas it is ten times higher for members of a group of 100 butterflies. The dilution effect has been confirmed by Joanna Burger and Michael Gochfeld, who showed any butterfly puddling by itself or with only a few companions would be safer if it moved to an even slightly larger group (Figure 5.10).[11]

This possibility has also been tested for mayflies in a population that synchronizes the change from aquatic nymph to flying adult such that most indi-

FIGURE 5.10   The dilution effect in butterfly groups.
(A) A group of butterflies drinking fluid from a Brazilian mud puddle. (B) Individual butterflies that "mud-puddle" in large groups experience a lower risk of daily predation than those that suck up fluids from the ground by themselves or in small groups.
A, photograph by the author; B, after Burger and Gochfeld.[11]

viduals emerge from the water during just a few hours on a few days each year.[106] This temporal bottleneck ought to limit the access of predators to the mayflies. If so, the higher the density of emerging individuals from a stream, the lower the risk to any one mayfly of being inhaled by a trout as the mayflies make the risky transition to adulthood, which is followed by mating. To check this prediction, Bernard Sweeney and Robin Vannote placed nets in streams to catch the cast-off skins of mayflies, which molt on the water's surface as they change into adults, leaving their discarded cuticles to drift downstream on the current. Counts of the molted cuticles revealed how many adults emerged on a particular evening from a particular segment of stream. The nets also caught the bodies of females that had laid their eggs and then died a natural death; a female's life ends immediately after she drops her clutch of eggs into the water, provided a nighthawk or a whirligig beetle does not consume her first. Sweeney and Vannote measured the difference between the number of molted cuticles of emerging females and the number of intact corpses of spent adult females that washed into their nets on different days. The greater the number of females emerging together on a given day, the better the chance each mayfly had to live long enough to lay her eggs before expiring, as predicted from the dilution effect hypothesis (Figure 5.11).[106]

Of course, there are other possible advantages of social defenses,[14] including improved or shared vigilance[108] and communal defense, as illustrated by various animals but especially the bees, ants, and wasps (Figure 5.12), as well as the simple ability to use others to hide behind.[52] Imagine, for example, a

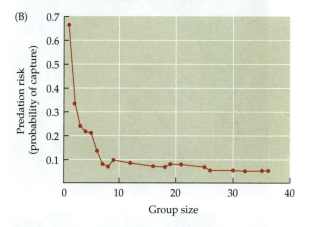

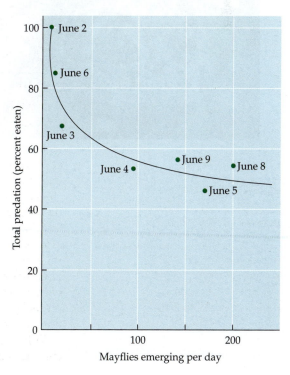

FIGURE 5.11   The dilution effect in mayflies. The more female mayflies that emerged together on a June evening, the less likely any individual mayfly was to be eaten by a predator. After Sweeney and Vannote.[106]

(A)

(B)

(C)

FIGURE 5.12   Fighting back by groups of terns and wasps.
(A) Several royal terns confront a gull trying to steal a tern egg.
(B, C) A colony of *Polybia* wasps (B) just before and (C) just after a nest was tapped by an observer. These wasps will leave their nest en masse to assault any predator foolish enough to continue disturbing them. A, photograph by Bruce Lyon; B and C, photographs by Bob Jeanne.

population of antelope grazing on an African plain in which all individuals stay well apart, thereby reducing their conspicuousness to their predators. Now imagine that a mutant individual arises in this species, one that approaches another animal and positions itself so as to use its companion as a living shield for protection against attacking predators. The social mutant that employs this tactic would incur some costs; for example, two animals may be more conspicuous to predators than one and so attract more attacks than scattered individuals, as has been demonstrated in some cases.[113] But if these costs were consistently outweighed by the survival benefit gained by social individuals, the social mutation could spread through the population. If it did, then eventually all the members of the species would be aggregated, with individuals jockeying for the safest position within their groups, actively attempting to improve their odds at someone else's expense. The result would be a **selfish herd**, whose members would actually be safer if they could all agree to spread out and not try to take advantage of one another. But since populations of solitary individuals would be vulnerable to invasion by an exploitative social mutant that takes fitness from its companions, social behavior could spread through the species—a clear illustration of why we define adaptations in terms of their contribution to the fitness of individuals relative to that of other individuals with alternative traits.

Discussion Questions

5.7   In my front yard, I sometimes find several hundred male native bees clustering in the evening on a few bare plant stems (Figure 5.13). Assassin bugs are bee killers that sometimes prey on my front yard bees as they are settling down for the night. Devise at least three alternative hypotheses on the possible anti-assassin bug value of these sleeping clusters, and list the predictions that follow from each hypothesis.

5.8 When Scott Creel and his colleagues studied elk behavior in Montana,[21,22] they found that elk tended to form larger groups when foraging in the open far away from forest cover. Why might this result lead us to interpret large group formation by elk as an antipredator response? Creel and company noted that the elk aggregated only on days when wolves were *absent*. In the presence of wolves, elk remained in small herds. What is the significance of these observations for the antipredator hypothesis for the tendency of elk to group together under some conditions? In scientific terms, what label should be given to these observations: hypothesis, prediction, test data, or scientific conclusion? What is the significance of this work for studies of mud-puddling by butterflies?

Game Theory and Social Defenses

W. D. Hamilton's examination of social evolution (see page 24, Chapter 2) provides an example of how **game theory** can be used to study behavioral adaptations. Users of this theory treat individuals as participants in a contest in which the success of one competitor is dependent on what its rivals are doing. Under this approach, decision-making is treated as a game, just as it is by the economists who invented game theory in order to understand the choices made by people as they compete with one another for consumer goods or wealth. Game theoretical economists have shown that individuals using one strategy, which works well in one situation, may come up short when matched against other individuals using another strategy. (Remember that in evolutionary biology, strategy designates an inherited behavioral trait, such as a decision-making ability of some sort, not the consciously adopted game plans of the sort humans often employ.[28])

The fact that all organisms, not just humans interested in how to spend their money or get ahead in business, are really competing with one another in a reproductive sweepstakes means that the game theory approach is a natural match for what goes on in the real world. The fundamental competition of life revolves around getting more of one's genes into the next generation than one's fellows. Winning this game almost always depends on what those other individuals are doing, which is why game theory is so popular with evolutionary biologists and behavioral ecologists as they develop hypotheses to explain the behaviors they observe.

FIGURE 5.13 A group of sleeping bees. In this species of bee, males spend the night together in sleeping clusters. An assassin bug sometimes attacks sleeping bees, just as it captures and kills other bees during the day. Photograph by the author.

Discussion Question

5.9 Why might game theory be useful in analyzing decision making by Adélie penguins as they wait to enter the water by the edge of their breeding colonies (Figure 5.14)? The penguins gather in groups before diving into the ocean where a huge predator, a leopard seal, may be lurking. How might one individual's decision to dive in be dependent on what other penguins are doing? You can see a relevant video of penguins and leopard seals in action on YouTube at http://www.youtube.com/watch?v=r3RElQWsh34, or search for "emperor penguins versus leopard seal."

Hamilton's selfish herd hypothesis generates testable predictions that can be applied to any prey species that forms groups. Thus, we expect that there will be competition among group members for the safest positions within the

FIGURE 5.14   Are these penguins members of a selfish herd? Adélie penguins often wait at the ice's edge before plunging into the water. They have formidable predators, such as this leopard seal, which wait for penguins to leave the safety of the ice. But while the seal is dispatching one penguin, others can dive in and make their way to the open ocean. Photograph by Gerald Kooyman/Hedgehog House.

Redshank

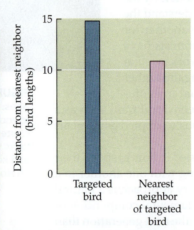

selfish herd, which generally means that individuals will jockey for a place in the center of the group.[75] In bluegill nesting colonies that form on the bottoms of lakes and ponds, the larger, more dominant individuals do indeed come to occupy the central positions, forcing smaller, subordinate males to build their nests on the periphery, where these unfortunates are subject to more frequent attacks by predatory fish.[49]

Likewise, if the feeding flocks of redshanks, which are European sandpipers, constitute selfish herds, we would expect that those individuals targeted by predatory sparrowhawks should be relatively far from the shelter provided by their companions. John Quinn and Will Cresswell collected the data needed to evaluate this prediction by recording 17 attacks on birds whose distance to a nearest neighbor was known. Typically, a redshank selected by a sparrowhawk was about five body lengths farther from its nearest companion than that companion was from its nearest neighbor (Figure 5.15).[90] Thus, birds that moved a short way from their fellow redshanks, presumably to forage with less competition for food, put themselves at greater risk of being singled out for attack.

Redshank flock

FIGURE 5.15   Redshanks form selfish herds. Redshanks that are targeted by hawks are usually standing farther from their nearest neighbors in the flock than those birds are standing from their nearest neighbors. After Quinn and Cresswell.[90]

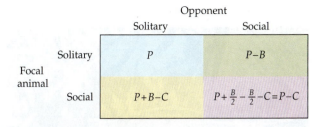

Discussion Question

5.10   The concept of a selfish herd can be illustrated with a game theory matrix that shows the fitness of individuals that adopt different strategies (Figure 5.16) (with thanks to Jack Bradbury). In a population entirely composed of solitary prey, the fitness payoff to individuals is *P*. But then mutant individuals arise that use others as living shields. When a solitary individual is found and used by a social type for protection, the solitary animal loses some fitness (*B*) to the social type. The costs (*C*) to social individuals include the time spent searching for another individual to hide behind. When two social types interact, we will say that they each have one chance in two of being the one that happens to hide behind the other when a predator attacks. If *B* is greater than *C*, what behavioral type will come to predominate in the population over time? Now compare the average payoff for individuals in populations composed entirely of solitary versus social types. If the average fitness of individuals in a population of social types is less than that of individuals in a population composed of solitary types, can hiding behind others be an adaptation?

FIGURE 5.16   A game theoretical model in which the fitness gained by a solitary or social animal is dependent on the behavior of its opponent, which may be either a solitary or a social individual.

A Cost–Benefit Approach to Cryptic Behavior

Although we have focused on the analysis of social defenses against predators, there are vast numbers of solitary creatures, and they too have much to gain by staying alive longer than the average for their species if their greater life span translates into greater reproductive success. As a result, solitary prey often evolve camouflaged color patterns along with behavioral tactics that enhance their concealment. One potential tactic of this sort is for the cryptically colored animal to select the kind of resting place where it is especially hard to see (Figure 5.17). A classic example involves the peppered moth, *Biston betularia*. The melanic (black) form of this moth, once extremely rare, almost completely replaced the once abundant whitish salt-and-pepper form in the period from about 1850 to 1950 in Great Britain.[44] Most biology undergraduates have heard the standard explanation for the initial spread of the melanic form (and the genetic basis for its mutant color pattern): as industrial soot darkened the color of forest tree trunks in urban regions, the whitish moths living in these places became more conspicuous to insectivorous birds, which ate the whitish form and thereby literally consumed the allele for this color pattern.

Despite some claims to the contrary,[20] this story remains valid,[45,93] particularly with respect to the conclusions of H. B. D. Kettlewell.[62] That British researcher pinned fresh specimens of the black and the whitish forms of the moth on dark and pale tree trunks. When he checked later, he found that whichever form was more conspicuous to humans had been taken by birds much more often than the other form. So, for example, paler individuals were at special risk of attack when they were placed on dark backgrounds. Michael Majerus repeated Kettlewell's experiments but with living specimens and vastly more of them over a 7-year period. His data confirm that the form more conspicuous to human observers is also at significantly higher risk of attack and capture by birds, nine species of which were seen taking the moths as they perched on trees during the daytime.[18]

(A)

(B)

FIGURE 5.17   Cryptic coloration depends on where the animal chooses to rest.   (A) The Australian thorny devil is camouflaged, but only when the lizard is motionless in areas littered with bark and other varicolored debris, not on roads. (B) The famous moth *Biston betularia* comes in two forms, the "typical" salt-and-pepper morph and the melanic type, each of which is well hidden on a different kind of background. A, photographs by the author; B, photographs by Michael Tweedie.

In nature, *B. betularia* tends to settle within the shaded patches just below where branches meet the trunk. If the moth's actual perch site selection is an adaptation, then we can predict that individuals resting underneath tree limbs should be better protected against predators than if they were to choose an alternative location. Majerus and a colleague attached samples of frozen moths to open trunk areas and to the undersides of limb joints. They found that birds were particularly likely to overlook moths on shaded limb joints (Figure 5.18).[58] Here is another demonstration that a putative behavioral adaptation almost certainly does provide a survival benefit—enabling the moths to reproduce more.

Discussion Question

5.11   Consider the following finding: In the years since 1950, pollution controls have reduced the amount of soot deposited on tree trunks, and the melanic form of *B. betularia* has correspondingly become increasingly scarce in Europe[5,17] as well as in North America[46] where the species also occurs. Put this statement in the context of a scientific investigation into whether the typical salt-and-pepper coloration of some members of this species constitutes an adaptation. Begin with a research question and proceed through hypothesis, prediction, test, and conclusion.

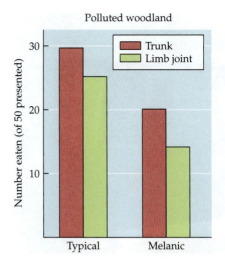

Polluted woodland

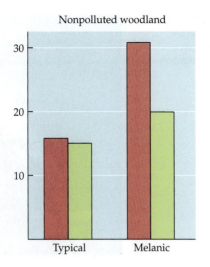

Nonpolluted woodland

FIGURE 5.18   Predation risk and background selection by *Biston betularia*. Specimens of typical and melanic forms of the moth were attached to tree trunks or to the undersides of limb joints. Birds were more likely to find and remove moths on tree trunks than on limb joints, and this was true for both types of moths. But overall, melanic forms were less often discovered by birds in polluted (darkened) woods, while typical forms "survived" better in unpolluted woods. After Howlett and Majerus.[58]

Discussion Question

5.12   Some persons have suggested that the real cause of the (temporary) increase in the frequency of the melanic form of *B. betularia* was selective predation by bats. If, however, Kettlewell was correct about the role of visual predation by birds, and if it were possible to examine how bats responded to the two forms at night, what would you predict about the mortality rates of the two forms of the moth? Check Majerus[69] for the relevant data.

Moths other than *B. betularia* also can make decisions about where to perch during the day. For example, the whitish moth *Catocala relicta* usually perches head-up, with its whitish forewings over its body, on white birch and other light-barked trees (Figure 5.19). When given a choice of resting sites, the moth selects birch bark over darker backgrounds.[97] If this behavior is truly adaptive, then birds should overlook moths more often when the insects perch on their favorite substrate.

To evaluate this prediction, Alexandra Pietrewicz and Alan Kamil used captive blue jays, photographs of moths on different backgrounds, and **operant conditioning** techniques (Figure 5.20).[84] They trained the blue jays to respond to slides in which cryptically colored moths were shown positioned on an appropriate background. When a slide flashed on a screen, the jay had only a short time to react. If the jay detected a moth in the picture, it pecked at a key, received a food reward, and was quickly shown a new slide. But if the bird pecked incorrectly when a slide appeared with a scene without a moth, it not only failed to secure a food reward but had to wait a minute for the next chance to evaluate a slide and get some food. The caged jays' responses demonstrated that they saw the moth 10 to 20 percent less often when *C. relicta* was pinned to pale birch bark than when it was placed on darker bark. Moreover, the birds were especially likely to overlook a moth when it was oriented head-

FIGURE 5.19   Cryptic coloration and body orientation. The orientation of a resting *Catocala* moth determines whether the dark lines in its wing pattern match up with the dark lines in birch bark. Photograph by H. J. Vermes, courtesy of Ted Sargent, from Sargent.[97]

FIGURE 5.20 Does cryptic behavior work? Images of moths on different backgrounds and in different resting positions are shown to a captive blue jay, which is rewarded for detecting a moth. Photograph by Alan Kamil.

up on birch bark. Thus, the moth's preference for white birch resting places and its typical perching orientation appear to be antidetection adaptations that thwart visually hunting predators such as blue jays.

The Darwinian Puzzle of Conspicuous Behavior

Cryptic behavior comes with its costs, notably the time and energy that individuals expend in finding the right background on which to perch, as well as the time spent immobile during the day when visual predators might spot them if they were to move. Given the apparent advantages of remaining hidden, it is surprising that some camouflaged insects do things that make themselves conspicuous. For example, the walnut sphinx caterpillar, a handsome big green species that rests on green leaves, can produce a whistling squeak up to 4 seconds long by forcing air out of one pair of its respiratory spiracles.[10] Likewise, the peacock butterfly, which hibernates during the winter in sheltered sites, flicks open its wings repeatedly when approached by a potential predator, like a mouse, and in so doing, generates hissing sounds with ultrasonic clicks.[78]

Although these actions announce the presence of otherwise cryptic insects, and so would seem costly, the two prey species only do so when under attack or about to be attacked. The sounds they produce frighten at least some of their enemies. Thus, yellow warblers that pecked the sphinx caterpillar departed in a hurry when the caterpillar whistle-squeaked in response; likewise, when infrared-sensitive cameras were used to film mice approaching peacock butterflies at night, the predators were often seen to back off promptly as soon as the butterflies began their wing flicking. The unexpected sounds made by these edible prey species appear to startle their predators into retreat.

Of course, many animals are not camouflaged at all but instead seem to go out of their way to make themselves obvious to their predators (Figure 5.21). Take the monarch butterfly, for example—a species whose orange-and-black wing pattern makes it very easy to spot. Bright coloration of this sort is correlated with greater risk of attack in some cases.[105] How can it be adaptive for monarchs to flaunt their wings in front of butterfly-eating birds?

In order to overcome the costs of conspicuous coloration, the trait ought to have some very substantial benefits, and in the case of the monarch, it does. These benefits appear to be linked to the ability of monarch larvae to feed on poisonous milkweeds, from which they sequester an extremely potent plant poison in their tissues.[8] In dealing with this highly toxic species, you would

(A)

(B)

FIGURE 5.21   Warning coloration and toxins. Animals that are chemically defended typically are conspicuous in appearance and behavior. (A) Monarch butterflies that feed on toxic milkweeds as caterpillars store the cardiac glycosides acquired from their food in their bodies and wings when they become adults. (B) Blister beetles, which have blood laced with cantharidin, a highly noxious chemical, often mate conspicuously for hours on flowering plants. B, photograph by the author.

therefore be wise to ignore the recommendation of the lepidopterist E. B. Ford, who wrote, "I personally have made a habit, which I recommend to other naturalists, of eating specimens of every species which I study."[35] Any bird that makes the mistake of accepting the Ford challenge with respect to monarchs usually finds the experience most unpleasant, albeit highly educational (Figure 5.22). After vomiting up a noxious monarch just once, a surviving blue jay will avoid this species religiously thereafter.[7,8]

If you have absorbed the essence of the **cost–benefit approach**, with its assumption that the fitness benefits of an evolved trait must exceed its fitness costs, you will now be asking how an eaten monarch could gain fitness by inducing vomiting, given that regurgitated monarchs rarely fly off into the sunset. Is this case an exception to the rule that dead animals cannot pass on their genes? Probably not, although the possibility exists that toxicity has evolved via indirect selection (see page 21), provided that the dead individuals educate the potential predators of their close relatives, thereby helping these genetically similar monarchs to pass on shared genes to the next generation.

FIGURE 5.22   Effect of monarch butterfly toxins on a predator. After eating a toxic monarch butterfly, a blue jay vomits a short while later. Photographs by Lincoln P. Brower.

FIGURE 5.23 Why behave conspicuously? This tephritid fly (left) habitually waves its banded wings, which gives it the appearance of a leg-waving jumping spider (right). When the spiders wave their legs, they do so to threaten one another. The fly mimics this signal to deter attack by the spider. Photographs by Bernie Roitberg, from Mather and Roitberg.[71]

A direct selection hypothesis, however, is that monarchs recycle poisons from their food plants to make themselves so bad tasting that most birds will release any monarch immediately after grabbing it by the wing.[8] In fact, when caged birds are offered a frozen but thawed monarch, many pick it up by the wing and then drop it at once, evidence that in nature living monarchs could benefit personally from being highly unpalatable, a fact advertised by their color pattern.

Bright coloration is not the only means by which a prey species can make itself conspicuous. Take the tephritid fly that habitually waves its banded wings as if trying to catch the attention of its predators. This puzzling behavior attracted two teams of researchers, who noticed that the wing markings of the fly resemble the legs of jumping spiders, important consumers of flies. The biologists proposed that when the fly waves its wings, it creates a visual effect similar to the aggressive leg-waving displays that the spiders themselves use (Figure 5.23).[47,71] Thus, the fly's appearance and behavior might activate escape behavior on the part of the predator.

In order to test this signal deception hypothesis (see page 88), one group of researchers became expert fly surgeons. Armed with scissors, Elmer's glue, and steady hands, they exchanged wings between clear-winged houseflies and pattern-winged tephritid flies. After the operation, the tephritid flies behaved normally, waving their now plain wings and even flying about their enclosures. But these modified tephritids with their housefly wings were soon eaten by the jumping spiders in their cages. In contrast, tephritids whose own wings had been removed and then glued back on repelled their enemies in 16 of 20 cases. Houseflies with tephritid wings gained no protection from the spiders, showing that it is the combination of leg-like color pattern and wing movement that enables the tephritid fly to deceive its predators into treating it as a dangerous opponent rather than a meal.[47]

Some vertebrates also behave in ways that paradoxically make them easier to spot. Take the Thomson's gazelle. When pursued by a cheetah or lion, this antelope may leap several feet into the air while flaring its white rump patch (Figure 5.24). Any number of possible explanations exist for this behavior, which is called stotting. Perhaps a stotting gazelle sacrifices speed in escaping from one detected predator in order to scan ahead for other as-yet-unseen enemies, like lions, lying in ambush.[85]

FIGURE 5.24 An advertisement of unprofitability to deter pursuit? The springbok, a small antelope, leaps into the air when threatened by a predator, just as Thomson's gazelles do.

TABLE 5.1 Predictions derived from four alternative hypotheses on the adaptive value of stotting by Thomson's gazelle

Prediction	Alternative hypotheses			
	Alarm signal	Social cohesion	Confusion effect	Signal of unprofitability
Solitary gazelle stots	No	Yes	No	Yes
Grouped gazelles stot	Yes	No	Yes	No
Stotters show rump to predator	No	No	Yes	Yes
Stotters show rump to gazelles	Yes	Yes	No	No

The antiambush hypothesis predicts that stotting will not occur on short-grass savanna but will instead be reserved for tall-grass or mixed grass-and-shrub habitats, where predator detection could be improved by jumping into the air. But gazelles feeding in short-grass habitats do stot regularly, so we can reject the antiambush hypothesis and turn to some others.[12,13]

Table 5.1 lists some predictions that are consistent with each of these hypotheses. In checking these predictions, Tim Caro learned that a single gazelle will sometimes stot when a cheetah approaches, an observation that helps eliminate the alarm signal hypothesis (if the idea is to communicate with other gazelles, then lone gazelles should not stot) and the confusion effect hypothesis (because the confusion effect can occur only when a group of animals flee together). We cannot rule out the social cohesion hypothesis on the grounds that solitary gazelles stot, because there is the possibility that solitary individuals stot in order to attract distant gazelles to join them. But if the goal of stotting is to communicate with fellow gazelles, then stotting individuals, solitary or grouped, should direct their conspicuous white rump patch toward other gazelles. Stotting gazelles, however, orient their rumps toward the predator. Only one hypothesis is still standing: gazelles stot to tell a predator that they have seen it and that they have plenty of energy and so will be hard to capture. Cheetahs get the message, since they are more likely to abandon hunts when the gazelle stots than when the potential victim does not perform the display.[13]

In order for stotting to be an evolved communication system, the signaler (the gazelle) and the receiver (the cheetah) must both benefit from their interaction (see page 67). If so, stotting is an honest signal (see Chapter 4) indicating that the stotter will be hard to capture, which therefore makes it advantageous for cheetahs to accept the attack deterrence signal and to call off the hunt to avoid wasting valuable time and energy. This proposition has been formally tested in the case of an *Anolis* lizard that performs a series of push-up displays when it spots a predatory snake coming its way. To conduct the test, Manuel Leal and an assistant first counted the number of push-ups each individual performed in the laboratory when exposed to a model snake. They then took each lizard to a circular runway, where they induced it to keep running by lightly tapping it on the tail. The total running time sustained by tapping was proportional to the number of push-ups the lizard performed in response to the model of its natural predator (Figure 5.25).[67] Thus, as the attack deterrence hypothesis requires, predators could derive accurate information about the physiological state of a potential prey by observing its push-up performance. Since this anole sometimes does escape when attacked, it could pay predatory snakes to make hunting decisions based on the signaling behavior of potential victims.

FIGURE 5.25 **Are push-up displays an honest signal of a lizard's physiological condition?** (A) The lizard *Anolis cristatellius* performs a push-up display when it spots an approaching snake. (B) The time a lizard spent running until exhaustion was positively correlated with the number of push-ups performed by that individual under perceived threat from a model snake. A, photograph by Manuel Leal; B, after Leal.[67]

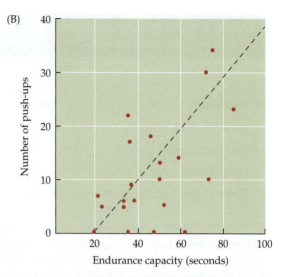

Discussion Question

5.13 If snakes do tend to avoid lizards that perform a series of push-ups, why don't weaker lizards "cheat" by boosting the number of push-ups they perform upon spotting a predator?

Optimality Theory and Antipredator Behavior

Behavioral ecologists have then a variety of ways of testing their adaptationist explanations for behaviors of interest. Much work of this sort in the past has been dedicated to demonstrating that the behavior under examination provides fitness benefits of some sort for individuals. Other cost–benefit approaches, however, can potentially yield precise quantitative predictions, rather than the more general qualitative ones that we have been discussing up to this point. This information may permit a researcher to determine whether or not a current trait is likely to have been "better" than alternatives over evolutionary time. One technique of this sort relies on **optimality theory**, the notion that adaptations have greater benefits-to-costs ratios than the putative alternatives that have been replaced by natural selection. We shall illustrate the theory by examining the costs (C) and benefits (B) of four alternative hereditary behavioral traits in a hypothetical species (Figure 5.26). Of these four traits (W, X, Y, and Z), only one (Z) generates a net loss in fitness (C > B) and is therefore obviously inferior to the others. The other three all generate net fitness gains, but just one, trait X, is an adaptation, because it produces the greatest net benefit of the four alternatives. (What if trait X never arose because the necessary mutation never appeared? Would W or Y be considered an adaptation?)

Given the occurrence of X in a population, the alleles associated with this trait will spread at the expense of the alternative alleles underlying the development of the alternative traits in this population. Behavior X can be considered optimal because the difference between benefit and cost is greatest for this trait, and because this trait will spread while all others are declining in frequency (as long as the relationship between their costs and benefits remains the same).

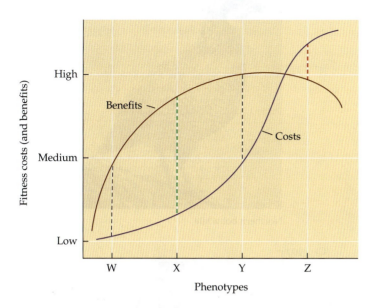

FIGURE 5.26 An optimality model. If one can measure the fitness costs and benefits associated with four alternative behavioral phenotypes in a population, then one can determine which trait confers the greatest net benefit on individuals in that population. In this case, phenotype X is the adaptation—an optimal trait that would replace competing alternatives, given sufficient evolutionary time.

If it were possible to measure B and C for a set of reasonable alternatives, one could predict that the characteristic with the greatest net benefit would be the one observed in nature. Unfortunately, it is often difficult to secure precise measures of B and C in the same fitness units, although some studies of foraging behavior have been able to measure the calories gained from collected foods (the benefits) and calories expended in collecting those foods (the costs). But optimality theory has been applied to some antipredator behaviors as well.

For example, in the midwestern United States, northern bobwhite quail spend the winter months in small groups, called coveys, which range in size from 2 to 22 individuals, but most are composed of intermediate size, with about 11 individuals. These groups almost certainly form to gain protection against predators. For one thing, members of larger coveys are safer from attack, judging from the fact that overall group vigilance (percentage of time that at least one member of the covey has its head up and is scanning for danger) increases with increasing group size and then levels off around a group size of 10. Moreover, in aviary experiments, members of larger groups reacted more quickly than members of smaller ones when exposed to a silhouette of a predatory hawk.[116]

The antipredator benefits of being in a large group are almost certainly offset to some degree by the increased competition for food that occurs in larger groups. This assumption is supported by evidence that relatively large groups move more each day than coveys of 11. (Small groups also move more than groups of 11, probably because these birds are searching for other groups with which to amalgamate.) The mix of benefits and costs associated with coveys containing different numbers of individuals suggests that birds in intermediate-size coveys have the best of all possible worlds, and in fact, daily survival rates are highest during the winter in coveys of this size (Figure 5.27).[116] This work demonstrates not just that the quail form groups to deal with their predators effectively, deriving a benefit from their social behavior, but that they attempt to form groups of the optimal size. If they succeed in joining such a group, they will derive a greater net benefit in terms of their survival than by joining a smaller or larger group.

(A)

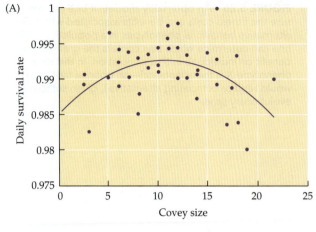

(B)

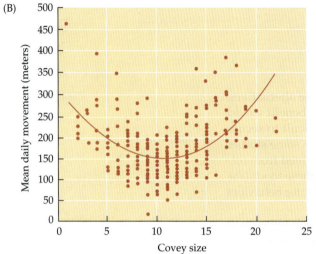

FIGURE 5.27 **Optimal covey size for northern bobwhite quail** is a function of the costs and benefits of belonging to groups of different sizes. (A) The probability that an individual would survive on any given day, that is, its survival rate, which is a benefit, is highest for birds in coveys of about 11 birds. (B) Distance traveled, a cost, is lowest for individuals in coveys of about 11 birds. (C) The most common covey size was about 11 in this bobwhite population. After Williams et al.[116]

Northern bobwhite

(C)

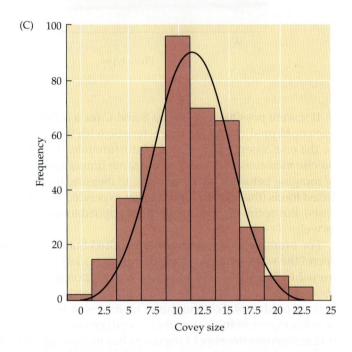

Discussion Question

5.14 Several factors might influence optimal flock size in redshanks. Included in Figure 5.28 are the relative risks for the sandpipers in different flock sizes as functions of the rate of attacks by sparrowhawks and the overall risk of predation per individual member of a flock. During this study, the average redshank flock size was a little more than 18. Most birds were in flocks of about 50, although when large numbers of redshanks were in the marsh, some flocks contained about 80 individuals.[24] Did the redshanks in this study form flocks of optimal size in relation to predation risk from sparrowhawks?

Optimality Theory and Foraging Decisions

Animals use all manner of ingenious methods to secure the food they need to survive and reproduce, ranging from a blood-consuming leech that specializes on the hippopotamus (in whose rectum the leeches assemble to find mates)[79] to a number of Australian hawks that consume the tongues of road-killed cane

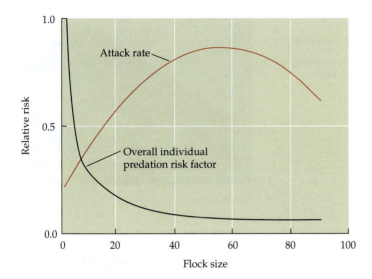

FIGURE 5.28 Redshank flock size and optimality theory. The risk that a flock would be attacked was much higher for larger groups. But the overall risk to an individual bird from predation was lowest for birds in flocks of more than 40 redshanks, thanks to a combination of the dilution effect and the reduced attack success of hawks attempting to capture individuals in larger flocks. After Cresswell and Quinn.[24]

toads (the safest part of this generally toxic nonnative animal).[2] Whatever the tactics that animals use to get enough food to survive and reproduce, the various behaviors all come with costs and benefits. It is not surprising, therefore, that both optimality and game theoretic approaches have helped behavioral ecologists produce and test hypotheses on the possible adaptive value of the foraging decisions of assorted animals.

Consider a northwestern crow out hunting for food on a tidal flat in British Columbia. When the bird spots a clam, snail, mussel, or whelk, it sometimes picks it up, flies into the air, and then drops its victim onto a hard surface. If the mollusk's shell shatters on the rocks below, the bird flies down to the prey and plucks out the exposed flesh. The adaptive significance of the bird's behavior seems straightforward: It cannot use its beak to open the extremely hard shells of certain mollusks. Therefore, it opens the shells by dropping them on rocks. This seems adaptive. Case closed. But Reto Zach was much more ambitious in his analysis of the crow's foraging decisions.[119] When a hungry crow is searching for food, it has to decide which mollusk to select, how high to fly before dropping the prey, and how many times to keep trying if the mollusk does not break on the first try.

Zach made several observations while watching foraging crows:

1. The crows selected and flew up with only large whelks about 3.5 to 4.4 centimeters long.

2. The crows flew up about 5 meters to drop their chosen whelks.

3. The crows kept trying with each chosen whelk until it broke, even if many flights were required.

Zach sought to explain the crows' behavior by determining whether the birds' decisions were optimal in terms of maximizing whelk flesh available for consumption per unit of time spent foraging.[119] First he had to check whether the birds' behavior was efficient in terms of generating broken whelks. If so, then the following predictions should apply:

1. Large whelks should be more likely than small ones to shatter after a drop of 5 meters.

2. Drops of less than 5 meters should yield a reduced breakage rate, whereas drops of much more than 5 meters should not greatly improve the chances of opening a whelk.

3. The probability that a whelk will break should be independent of the number of times it has already been dropped (or better still, the breakage odds should increase the more times the whelk is dropped).

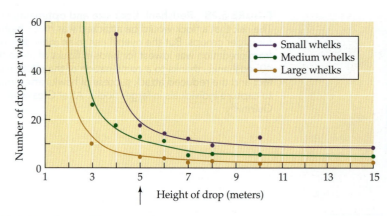

Northwestern crow

FIGURE 5.29 Optimal foraging decisions by northwestern crows when feeding on whelks. The curves show the number of drops at different heights needed to break whelks of different sizes. Northwestern crows pick up only large whelks, increasing the calories available to them, and they drop their whelks from a height of about 5 meters, thereby minimizing the energy they expend in opening their prey. After Zach.[119]

Zach checked each of these points in the following manner: He erected a 15-meter pole on a rocky beach and outfitted it with a platform whose height could be adjusted and from which whelks of various sizes could be pushed onto the rocks below. He collected samples of small, medium, and large whelks and dropped them from different heights (Figure 5.29). He found, first, that large whelks required significantly fewer 5-meter drops before they broke than either medium-size or small whelks. Second, the probability that a large whelk would break improved sharply as the height of the drop increased—up to about 5 meters—but going higher did little to improve the breakage rate. Third, the chance that a large whelk would break was not affected by the number of previous drops and was instead always about one chance in four on any given drop. Therefore, a crow that abandoned an unbroken whelk after a series of unsuccessful drops would not have a better chance of breaking a replacement whelk of the same size on its next attempt. Moreover, finding a new prey would take time and energy.

Zach then calculated the average number of calories required to open a large whelk (0.5 kilocalories), a figure he subtracted from the food energy present in a large whelk (2.0 kilocalories), for a net gain of 1.5 kilocalories. In contrast, medium-size whelks, which require many more drops, would yield a net loss of 0.3 kilocalories; trying to open small whelks would have been even more disastrous.[119] Thus, the crows' rejection of all but large whelks was adaptive, assuming that fitness is a function of energy gained per unit of time. (Note, however, that another group of researchers has argued that the crows would have been better off calorically if they had dropped their prey from greater heights.[86] If true, why might Zach's crows have chosen to fly up only 5 meters rather than maximize caloric intake?)

Discussion Question

5.15 In some places, American crows open walnuts by dropping them on hard surfaces. Unlike northwestern crows opening whelks, American crows reduce the height from which they drop walnuts from about 3 meters on the first drop to about 1.5 meters on the fifth drop. If this tendency is adaptive, what prediction follows about a difference between whelks and walnuts in the likelihood of breaking on successive drops? In addition, American crows tend to drop walnuts from lower heights when other crows are present. If this trait is an adaptation, what do you predict about the likelihood of breakage over a series of drops? Check your answers against data in Cristol and Switzer.[25]

The approach behind Zach's research included the simple premise that the caloric cost of gathering food had to be less than the caloric benefits gained by opening whelks. The same premise has been examined for blue whales, which feed by diving underwater for up to 15 minutes in the search for krill, shrimp-like animals that form huge swarms in Antarctic waters. A feeding blue whale surges into a shoal of krill, engulfing huge amounts of water before pushing the water out through the baleen plates in its mouth, an action that traps the edible krill contained in the water. Whale biologists had assumed that open-mouthed blue whales paid a heavy price to force their way through the water at a high enough speed to catch enough krill to pay for their foraging movements. When a team of blue whale experts estimated the energy expended on a dive in which three to four underwater lunges occurred, they came up with a figure of about 60,000 kilojoules for a typical dive. (The whales they studied had been outfitted with hydrophones, pressure sensors, and accelerometers, which provided the data needed to estimate the cost of a lunge.)

So what does a blue whale secure in return for its expenditure of metabolic and mechanical energy? Of course, the answer depends on how many krill a whale can capture, which is a function of how much water it engulfs and the density of krill in a given volume of water. Based upon the size of blue whale jaws and published data on krill densities, the team estimated that a single underwater lunge could yield anywhere from 30,000 to almost 2 million kilojoules. Given that during a typical dive, a blue whale makes several lunges, which in total yield about 5 million kilojoules on average, krill-hunting whales come out way ahead, with the food ingested paying for a standard dive nearly 80 times over.[42]

Both northwestern crows and blue whales gain more energy from their foraging activities than they expend in the attempt to secure food. But do these and other animals achieve maximum reproductive success by maximizing the number of calories ingested per unit of time, a key assumption of the standard optimality model? This is an assumption that deserves to be investigated. The relationship between whelk-opening efficiency and fitness has not been established for crows, nor do we know if blue whale foraging efficiency translates into maximum reproductive gain. However, in an experiment with captive zebra finches given the same kind of food under different feeding regimes so that some birds had higher foraging costs than others, the individuals with the highest daily net caloric gains survived best and reproduced the most,[68] as well as beginning to reproduce sooner than their less fortunate compatriots.[115] In another experiment, male red crossbills were placed on both sides of a divided cage that contained either a branch with pinecones from which most of the edible seeds had been removed or a branch whose pinecones were unaltered. Female crossbills that observed the two categories of males foraging associated preferentially with males they had seen securing many seeds (Figure 5.30). Female crossbills can evidently assess the feeding rate of male foragers, and they use this information to choose potential mates accordingly; male foraging efficiency leads to reproductive opportunities.[102]

FIGURE 5.30   Red crossbills feeding on seeds hidden in a pinecone. Thanks to their crossed bills, these birds are capable of twisting open the closed elements of pinecones to remove seeds, one at a time, from within the cones. Here a male (the redder individual) pauses while a female (the greenish bird) probes the pinecone with her specialized beak. Photograph by Craig Benkman.

Discussion Question

5.16 The Cape gannet, a seabird, normally feeds on oceanic fishes such as sardines, but during the nonbreeding season, the birds consume large quantities of fishery waste discarded by fishing vessels that process catch at sea. Despite the fact that the adult birds do fine on a mixed diet of discards and sardines, when there are young chicks to feed, gannets try to provide their youngsters with whole fish caught at sea rather than giving them the easily retrieved odds and ends thrown out of fishing boats. What prediction do you have about the development of gannet chicks fed fish guts and the like versus whole fish? Check Grémillet et al.[48]

How to Choose an Optimal Mussel

The Eurasian oystercatcher is a shorebird whose foraging decisions can also be matched against predictions taken from optimality models. Two Belgian researchers, P. M. Meire and A. Ervynck, developed a calorie maximization hypothesis to apply to oystercatchers feeding on mussels.[73] Like Reto Zach, they calculated the profitability of different-size prey, based on the calories contained in the mussels (a fitness benefit) and the time and energy required to open them (a fitness cost). Even though mussels over 50 millimeters long require more time to hammer or stab open, they provide more calories per minute of opening time than smaller mussels. Therefore, the model predicts that oystercatchers should focus primarily on the largest mussels. But in real life, the birds do not prefer the really large ones (Figure 5.31). Why not? Here are two hypotheses.

> **HYPOTHESIS 1:** The profitability of very large mussels is reduced because some cannot be opened at all, reducing the average return from handling these prey.

In their initial calculations of prey profitability, the researchers had considered only those prey that the oystercatchers actually opened (Figure 5.32, model A). As it turns out, oystercatchers occasionally select some large mussels that they find impossible to open, despite their best efforts. The handling time wasted on these large, impregnable mussels reduces the average payoff for dealing with

Oystercatcher

FIGURE 5.31   Available prey versus prey selected. Foraging oystercatchers tend to choose fairly large mussels, but they do *not* concentrate on the very largest mussels. After Meire and Ervynck.[73]

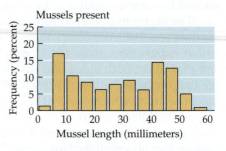
Mussels present

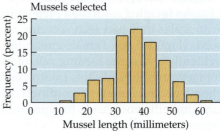
Mussels selected

FIGURE 5.32 Two optimal foraging models yield different predictions because they calculate prey profitability differently. Model A calculates the profitability of a mussel based solely on the energy available in opened mussels of different sizes divided by the time required to open these prey. Model B calculates profitability with one added consideration, namely, that some very large mussels must be abandoned after being attacked because they are impossible to open. After Meire and Ervynck.[73]

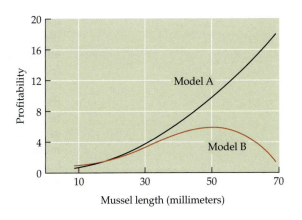

this size class of prey. When this factor is taken into account, a new optimality model results, yielding the prediction that the oystercatchers should concentrate on mussels 50 millimeters in length, rather than the very largest size classes (Figure 5.32, model B). The oystercatchers, however, actually prefer mussels in the 30- to 45-millimeter range. Therefore, time wasted in handling large, invulnerable mussels fails to explain the oystercatchers' food selection behavior.

> **HYPOTHESIS 2:** Many large mussels are not even worth attacking because they are covered with barnacles, which makes them impossible to open.

This additional explanation for the apparent reluctance of oystercatchers to feast on large, calorie-rich mussels is supported by the observation that oystercatchers never touch barnacle-encrusted mussels. The larger the mussel, the more likely it is to have acquired an impenetrable coat of barnacles, which eliminates these prey from consideration. According to a mathematical model that factors in prey-opening time, time wasted in trying but failing to open a mussel, and the actual size range of realistically available prey, the birds should focus on 30- to 45-millimeter mussels—and they do. Note that these researchers used optimality theory to produce an initial hypothesis, which they rejected on the basis of the evidence they collected. They then modified their model and subjected it to a new test, which is a productive scientific approach. As a result, they gained an improved understanding of oystercatcher feeding behavior.

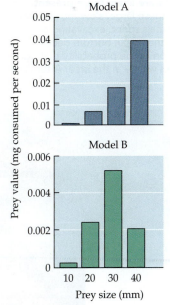

Pike cichlid

Discussion Question

5.17   The pike cichlid is a predatory fish that tends to attack and consume relatively large individual guppies, its primary prey in the rivers in Trinidad where they both live. A Swedish research team measured the time it took for pike cichlids to detect, approach, stalk, and attack guppies of four size classes (10, 20, 30, and 40 millimeters long). They also recorded the capture success rate for each size class, as well as noting the time it took to handle and consume those prey that were actually captured. With these data, they constructed two models of prey value as measured by mass of prey consumed per unit of time.[60] Model A considered only the time to attack and the capture rate, whereas model B incorporated these two factors plus the post-capture handling time in calculating the weight of food consumed per unit of time (Figure 5.33). Which of the two models strikes you as the most realistic optimal foraging model, and why? In light of the two models, what evolutionary issues are raised by the actual preference of pike cichlids for 40-millimeter-long guppies?

Criticisms of Optimal Foraging Theory

By developing and testing optimality models, researchers have concluded that northwestern crows and European oystercatchers, among other species, choose prey that tends to provide the maximum caloric benefit in relation to time spent

FIGURE 5.33 Two optimal foraging models. These models make different assumptions about the value of hunting guppies of different sizes by a predatory fish, the pike cichlid. After Johansson et al.[60]

foraging. But some scientists have criticized the use of optimality theory on the grounds that animals apparently do not always select or hunt for food as efficiently as possible. Consider, for example, the tropical jumping spider that feeds largely on the specialized leaf tips of certain acacias, a food high in fiber but low in nitrogen, a valuable substance found in much larger quantities in the insects preyed upon by thousands of other jumping spiders.[72] How can this be the most efficient way for this one spider to secure large amounts of nitrogen? This question raises an interesting problem, but remember that optimality models are constructed not to make statements about perfection in evolution but rather to make it possible to test whether one has correctly identified the variables that have shaped the evolution of an animal's behavior. As we have also seen, the factors included in an optimality model have a large effect on the predictions that follow. If an oystercatcher is assumed to treat every mussel in a tidal flat as a potential prey item, then it will be expected to make different foraging decisions than if the modeler assumes that oystercatchers simply ignore all barnacle-covered mussels. If the predictions of an optimality model fail to match reality, researchers will reject that model.

If ecological factors other than caloric intake affect oystercatcher foraging behavior, for example, then a caloric maximization model will fail its test, as it should. And for most foragers, foraging behavior does indeed have consequences above and beyond the acquisition of calories and certain nutrients. If you suspected, for example, that predators had shaped the evolution of an animal's foraging behavior, then the kind of optimality model you might choose to construct and test would not focus solely on calories gained versus calories expended. If foraging exposes an animal to the risk of sudden death, then when that risk is high, we would expect foragers to sacrifice short-term caloric gain for long-term survival[61,112] (but see Urban[110]).

A sacrifice of this sort has been demonstrated for dugongs, which are large, relatively slow-moving marine mammals that are preyed upon by tiger sharks in Shark Bay, Western Australia. There the dugongs feed on sea grasses in two different ways: cropping, in which the herbivores quickly strip leaves from standing sea grasses, and excavation, in which the foragers stick their snouts into the sea bottom to pull up the sea grasses by their "roots," which are called rhizomes. Dugongs that eat sea grass with rhizomes attached secure more energy per time spent feeding. But when the animals have their heads partly

FIGURE 5.34 Dugongs alter their foraging tactics when dangerous sharks are likely to be present. The time these large, slow-moving sea mammals spend excavating sea grass from the ocean bottom declines in relation to the probability that tiger sharks will be in the area. After Wirsing et al.[117]

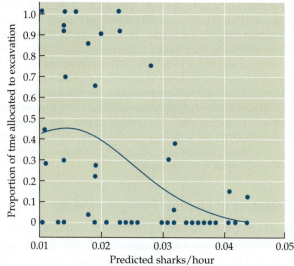

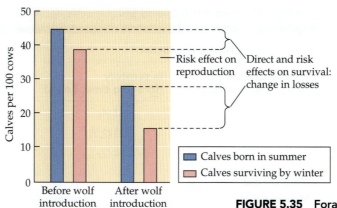

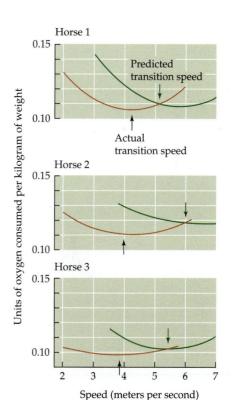

FIGURE 5.35 Foraging decisions affected by predator pressure. After wolves were reintroduced into the Yellowstone ecosystem, elk spent more time hidden in woodlands rather than feeding in exposed meadows. This change reduced the production of calves by elk cows (blue bars) and decreased the survival of the calves they did have (pink bars). After Creel and Christianson.[22]

buried in the sandy ocean floor, they cannot see well, whereas when they are cropping sea grass, they can more easily scan for approaching enemies, reducing the risk of shark attack. Dugongs in Shark Bay stop excavating sea grass rhizomes when tiger sharks are relatively common in the area, as measured by the shark catches made by local fishermen (Figure 5.34).[117] An optimality model that failed to consider the trade-offs between foraging success and predation risk would fail to predict the dugongs' behavior accurately.

Dugongs are not the only animals to alter their behavior in the face of predation risk in ways that have costs as well as benefits for individuals. For example, elk living in Yellowstone National Park have changed their foraging behavior considerably following the reintroduction of wolves into the area. Instead of feeding comfortably in open meadows where their preferred foods are present in abundance, elk now leave the grasslands when wolves arrive and move into wooded areas where they are less easily spotted and chased by their enemies.[21] Although safer in forests, elk pay a price for altering their foraging behavior. In winter, analysis of fecal samples suggests that elk in areas with wolves are eating a quarter less than elk in wolf-free sites.[15] In the summer, the probability that a given cow elk will give birth to a calf has fallen sharply, and the likelihood that her calf will still be with her by the time winter rolls around has also declined (Figure 5.35). In order to reduce the risk of being killed, the animals lower their energy intake, which increases the chances that they will be malnourished while reducing the odds for females of producing and tending a calf. If we were to fail to consider the fatal consequences of ignoring the presence of wolves, we might conclude that elk in some places were foraging in a suboptimal manner.

Discussion Question

5.18   When a horse switches from slower trotting to faster galloping, it might change gaits to minimize the energetic expense of locomotion, if we assume that animals able to minimize the energetic costs of getting from point A to point B will enjoy greater reproductive success than individuals that engage in less efficient locomotion. The energy minimization hypothesis was tested by Claire Farley and Richard Taylor with the help of three cooperative horses willing to run on a treadmill while outfitted to provide data on their oxygen consumption, a factor directly related to energy use.[33] What scientific conclusion is justified on the basis of Figure 5.36? Does this study support those who claim that optimality theory is not useful because the assumptions underlying particular hypotheses are often oversimplified and incorrect?

FIGURE 5.36 Energy consumption by three horses in relation to two locomotory modes: trotting (red line) versus galloping (green line). Energy consumption was measured in terms of milliliters of oxygen consumed per kilogram of weight for every meter covered. After Farley and Taylor.[33]

TABLE 5.2 Benefits of territoriality to golden-winged sunbirds under different conditions

Nectar production[a]		Hours of rest gained by territorial birds[b]	Calories saved[c]
In undefended site (foraging time, hours)	In territory (foraging time, hours)		
1 (8)	2 (4)	8 − 4 = 4	2400
2 (4)	3 (2.7)	4 − 2.7 = 1.3	780
4 (2)	4 (2)	2 − 2 = 0	0

Source: Gill and Wolf[40]

[a]Measured as microliters of nectar produced per blossom per day

[b]Calculated from the number of hours in foraging time needed to meet daily caloric needs, which depends on nectar production:

Nectar production (microliters/blossom)	1	2	3	4
Foraging time (hours)	8	4	2.7	2

[c]For each hour spent resting instead of foraging, a sunbird expends 400 calories instead of 1000.

Discussion Question

5.19   The golden-winged sunbird feeds exclusively on nectar from certain flowers during the winter in South Africa. Sometimes the birds are territorial at some patches of flowers, but at times they are not. Frank Gill and Larry Wolf devised a way to measure the rate of nectar production per bloom at a given site. With this information and previously published data on the caloric costs of resting, flying from flower to flower, and chasing intruders, they were able to calculate when it would be advantageous to be territorial (Table 5.2). Gill and Wolf assumed that the nonbreeding birds' goal was to collect sufficient nectar to meet their daily survival needs.[40,41] (Why was this assumption reasonable?) Based on Table 5.2, how many minutes of territorial defense would be worthwhile if a territory owner had access to a 2-microliters-per-blossom-per-day site while other flower patches were producing nectar at half that rate?

Game Theory, Feeding Behavior, and Another Darwinian Puzzle

Earlier in this chapter, game theory was used to analyze the possible anti-predator function of certain social behaviors. Here we shall return to game theory in order to identify the adaptive basis of certain foraging behaviors in the little wormlike larvae of fruit flies. The larvae come in two forms, active rovers and sedentary sitters, which differ in how far they maneuver through fruit fly medium in the laboratory in search of food. The odds that rovers and sitters will secure exactly the same net caloric benefits per unit of time spent foraging seem vanishingly small. If one type of larva did even slightly better on average than the other, the genes specifically associated with that trait should spread and replace any alternative alleles linked to the less effective food-acquiring behavior. So here is a puzzle, namely, why are both types still reasonably common in some places?

In the fruit fly case, the two kinds of larvae differ genetically,[29] and so game theorists would say that they employ two different strategies (hereditary behaviors), which means that if one strategy confers higher fitness than the other in a population, eventually only the superior strategy will persist. But under some special circumstances, two strategies can coexist indefinitely, thanks to the effects of **frequency-dependent selection**. This kind of selection occurs when the fitness of one strategy is a function of its frequency relative to the other inherited behavioral trait. When the fitness of one type increases as that type becomes rarer, then that type will become more frequent in the population—but only until such time as it has the same fitness as individuals playing the other strategy. Frequency-dependent selection will act against either type if it becomes even a little more common, pushing the proportion of that form back toward the equilibrium point at which both types have equal fitness. At equilibrium, the two types of individuals will coexist indefinitely.

In the case of rover and sitter fruit flies, experiments have shown that when food resources are scarce, the odds that an individual of the rarer phenotype will survive to pupation (a probable correlate of fitness) are greater than the corresponding odds for the more common type in the population (Figure 5.37).[34] The effect of this kind of selection is to lead to an increase in the frequency of the rarer behavioral type, which keeps it in the population.

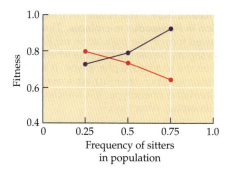

FIGURE 5.37 Frequency-dependent selection. When resources are scarce for fruit fly larvae, the fitness of a sedentary sitter (red line) versus an active rover (purple line) depends on which of the two types is the rarer. After Fitzpatrick et al.[34]

Discussion Questions

5.20 Imagine a population of 1000 fruit fly larvae in which there are two hereditarily distinct foraging phenotypes, rover and sitter. Imagine that there are 195 rovers and 805 sitters. Let's say that both types survive to adulthood equally well and both types have 1.2 surviving offspring on average. What were the frequencies of the two behavioral types in the parental generation? What will the frequencies be in the generation composed of their offspring? What would happen if rovers had 1.1 surviving offspring on average, whereas sitters had 0.9? What's the point of this question?

5.21 When the frequency of sitters is 0.75 (see Figure 5.37), the fitness of rovers is much higher than that of the sitters. So why don't rovers quickly and completely eliminate the sitters in this population?

Another famous example of frequency-dependent selection in action involves an African cichlid fish, *Perissodus microlepis*. This cichlid comes in two forms, one with the jaw twisted to the right and the other with the jaw twisted to the left. The fish makes a living, believe it or not, by darting in to snatch scales from the bodies of other fish in Lake Tanganyika. Adults with the jaw twisted to the right always attack the prey's left flank, while the other kind hits the right side (Figure 5.38).[57] Right-jawed parents usually produce offspring that inherit their jaw shape and behavior; ditto for left-jawed parents, indicating that the difference between the two forms is hereditary, although some recent studies have raised questions about this conclusion, including the possibility that the degree of asymmetry in jaw shape is enhanced by the attack behavior of individuals.[81]

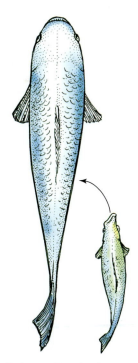

"Right-jawed" *Perissodus* attack prey from the left rear side

"Left-jawed" *Perissodus* attack prey from the right rear side

FIGURE 5.38 Two hereditary forms of an African cichlid fish. Right-jawed and left-jawed cichlids have asymmetrical mouths that they employ to snatch scales from the left and right sides, respectively, of the fish they prey on.

FIGURE 5.39 **The results of frequency-dependent selection in *Perissodus microlepis*.** The proportion of the left-jawed form in the population oscillates from slightly above to slightly below 0.5 because whenever it is more common than the alternative phenotype, it is selected against (and becomes less numerous); whenever it is rarer than the alternative phenotype, it is selected for (and becomes more numerous). After Hori.[57]

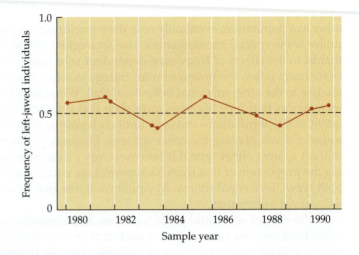

Nonetheless, here is the classic response to the question of why both kinds of predators occur in this species. Michio Hori proposed that the prey this fish attacks could learn to expect a raid on their scales from the left if most attacks were directed at the prey's left flank. Thus, in a population of predators in which the right-jawed form predominated, a lefty would have an advantage because its victims would be less vigilant with respect to scale snatchers darting in on the right flank. This advantage would translate into higher reproductive success for the rarer phenotype and an increase in its frequency until left-jawed fish made up half the population.[57] With a 50:50 split, the equilibrium point would have been reached. Can you use this same argument to explain why in species in which males and females are equally costly to produce, the sex ratio is typically 1:1?

If Hori's hypothesis is correct, the frequency of any one phenotype should oscillate around the equilibrium point. Hori confirmed that this prediction was true by measuring the relative frequencies of the two types over a decade (Figure 5.39).

Behavioral Variation and Conditional Strategies

But not every example of the coexistence of multiple foraging types has been attributed to frequency-dependent selection. Consider the ruddy turnstone, a small sandpiper with many different ways of finding prey items on beaches, ranging from pushing strands of seaweed to one side, to turning stones over, to probing in mud and sand for little mollusks. Some individuals specialize in one foraging method, while others prefer a different technique, but individual turnstones are rarely committed to just one way of finding food.[114] This observation suggests that the differences between the sandpipers are not caused by genetic differences but instead are due to an environmental difference of some sort. Philip Whitfield wondered if that environmental factor might be the dominance status of the foragers. Turnstones often hunt for food in small flocks, and the individuals in these groups form a pecking order. The birds at the top can displace subordinates merely by approaching them, thereby keeping them from exploiting the richer portions of their foraging areas. Dominants use their status to monopolize patches of seaweed on beaches, which they push about and flip over; subordinates keep their distance and are often forced to probe into the sand or mud instead of feasting on the invertebrates contained within seaweed litter.[114]

The turnstones exhibit flexibility in their foraging behavior, as individuals are apparently capable of adopting feeding methods in keeping with their ability to control different sources of food on the beach. The capacity to be flexible is provided by what game theoreticians have labeled a **conditional strategy**,[27,30,50] which is an inherited mechanism that gives the individual the ability to alter its behavior adaptively in light of the conditions it confronts (such as having to deal with socially dominant competitors on a beach). Unlike the left- and right-jawed cichlids, which are locked into one particular behavioral phenotype, turnstones can almost certainly switch from one feeding tactic (or option) to another. But subordinate turnstones tend to stick with the probing technique because if they were to challenge stronger rivals for seaweed patches, the low-ranking birds would probably lose, in which case they would have wasted their time and energy for nothing while also running the risk of being injured by an irate dominant. Instead, by knowing their place, these subordinates make the best of a bad situation and presumably secure more food than they would if they tried without success to explore seaweed patches being searched by their superiors. It can be adaptive to concede the better foraging spots to others if you are a turnstone with more powerful rivals.

Discussion Question

5.22   Here are two examples of curious behaviors that might help individuals either avoid predators or capture food: (1) some burrowing owls place mammalian dung near their burrows, and (2) some orb-weaving spiders add thick silk to their webs, creating conspicuous white lines or crosses on the webs (Figure 5.40). Why are these traits Darwinian puzzles? Use selection theory to produce alternative explanations for these behaviors, and then generate predictions from your hypotheses that would enable you to test your ideas. Check your answers by referring to the modest literature on burrowing owls[100,101] and the much larger literature on spider stabilimenta.[3,4,31,55,104,111]

FIGURE 5.40   Web ornament of an orb-weaving spider. The large female spider has added four thick, conspicuous zigzagging lines of ultraviolet-reflecting silk to her web, which radiate out from her central resting point. Photograph by William Eberhard.

Summary

1. Behavioral ecologists can test hypotheses about the adaptive value of antipredator behaviors and foraging tactics via the comparative method. If a behavior of a given species is indeed the product of selection associated with a particular environmental pressure, then a comparison of other closely related species subject to different pressures should reveal that they have evolved different behavioral characteristics via divergent evolution. In contrast, unrelated species subject to the same pressures are predicted to have evolved similar behavioral characteristics, producing examples of convergent evolution.

2. Behavioral ecologists also use several offshoots of natural selection theory in order to develop hypotheses on the possible adaptive value of behavioral traits. One of these is the cost–benefit approach in which traits are analyzed in terms of their possible fitness costs and fitness benefits to determine whether the benefits are likely to exceed the costs to individuals in terms of their inclusive fitness.

3. In addition, the game theory approach views behavioral decisions as a game played by competitors. Here the better strategy (an inherited behavior) for making decisions about how to achieve reproductive success is one that takes into account the competing strategies of other individuals.

4. The third approach employs optimality theory, which is based on the premise that optimal traits are characteristics with a better benefit-to-cost ratio than any alternative traits that have arisen during the evolution of a species.

5. Certain antipredator and foraging behaviors occur in two or more forms, which raises a major Darwinian puzzle: how can alternative behaviors coexist within a species or population given that natural selection should eliminate any alternative that yields lower individual fitness? The puzzle of coexisting alternatives can be explained in some instances as a product of frequency-dependent selection, in which the rarer of two hereditary attributes confers higher fitness on individuals, thereby maintaining both traits in a population. Other cases of variation exist because animals possess a single conditional strategy that enables them to be flexible in their choice of a tactic. A conditional strategy can spread when, for example, the ability to make the best of a bad situation (such as being dominated by rivals) helps individuals secure more genetic success than if they were to persist in futilely employing the tactic of their more powerful rivals.

Suggested Reading

S. J. Gould and R. C. Lewontin's attack on adaptationism[43] is worth read-ing—critically, with the help of David Queller.[89] Bernie Crespi succinctly reviews the various definitions of *adaptation* and the several reasons for the occurrence of maladaptations in nature[23] (see also Randolph Nesse[77]). The comparative method for testing adaptationist hypotheses is the subject of a book by Paul Harvey and Mark Pagel.[53] Game theory as applied to behavior is clearly explained in *The Selfish Gene* by Richard Dawkins.[28] Frequency-dependent selection is the subject of a review by Barry Sinervo and Ryan Calsbeek,[99] while the theory of conditional strategies is outlined by Mart Gross.[50]

A modern and comprehensive review of antipredator coloration and behav-ior has been written by Graeme Ruxton, Thomas Sherratt, and Michael Speed.[95] See also Rod and Ken Preston-Mafham.[87] You will find a general analysis of predator foraging decisions in a chapter by John Krebs and Alejandro Kacelnik in *Behavioural Ecology*.[63] Reto Zach has written a paper on optimal foraging in crows that is a model of clarity.[119] For a critique of optimality models, see a paper by G. J. Pierce and J. G. Ollason[83] and the much different view of Krebs and Kacelnik.[63]

6 The Evolution of Habitat Selection, Territoriality, and Migration

 When I was a teenager, my father and I went on a birding marathon with some fellow bird-watchers one weekend every May. As we searched through the fields and forests near our home in southeastern Pennsylvania, we usually found about 100 species, including yellow warblers singing high in the sycamore trees by the White Clay Creek, blue-winged warblers flitting among the little saplings growing in abandoned farm fields, and common yellowthroats (also a species of warbler) skulking in the marshy spots. The male warblers calling loudly from their perches not only occupied a well-defined space but also were prepared to defend their turf against others of their species, as they demonstrated by dashing over to chase off intruders whenever males of their species had the audacity to enter their territories. In addition to the costly investments each bird had made in defending a territory, the warblers we saw had traveled peacefully in mixed-species flocks of migrants from wintering grounds as far away as South America. After spending a few months in an exclusive breeding territory in our neighborhood, they would head south again on another great journey, which their offspring would undertake as well.

This chapter focuses on decisions that animals make about where to live, whether to defend a patch of real estate, and whether to stay put in a particular location or to leave it (perhaps to return in another season). The decisions about the use of space all come with obvious and substantial costs, which is why they illustrate Darwinian puzzles. This chapter will explore these problems and demonstrate how several theories derived from natural selection theory have been used by behavioral ecologists to solve these puzzles.

Habitat Selection and Ideal Free Distribution Theory

The rule that certain species live in particular places applies to all kinds of animals, not just warblers, perhaps because in so many cases, the opportunities for successful reproduction by members of any given species are much better in habitat A than in habitat B. This fact helps explain why habitat destruction is generally considered the main cause of species' declines everywhere[78] and why climate change, which really is a form of habitat destruction, is likely to continue the trend toward reduced populations of many animals and plants, leading to heightened rates of extinction.[57]

Given the importance of the "right" habitat for successful reproduction, we might expect an animal's preferred habitat to be the one where its breeding success is the greatest. So, for example, male blackcap warblers, a species of European songbird, settle first along stream edges in deciduous forests, demonstrating a preference for that particular streamside habitat. Males that settle later occupy mixed coniferous woodlots away from water. But the average number of offspring produced by blackcaps in those two habitats is the same.[104] Why don't warblers in the preferred environment do better than those that breed in the second-choice habitat?

One answer has been provided by Steve Fretwell and his colleagues, who used game theory to develop what they called **ideal free distribution theory**.[36] You may recall from the preceding chapter that game theory is a kind of optimality approach in which behavioral decisions are treated as if the individuals involved in the "game" are attempting to maximize their reproductive success (or proxies for that success, such as number of mates secured or amount of food ingested). Ideal free distribution theory enables behavioral ecologists to predict what animals should do when choosing between alternative habitats of different quality in the face of competition for space and food. Fretwell and Lucas demonstrated mathematically that if individuals were free to distribute themselves spatially in relation to resource quality and the intensity of competition from others of their species, then as the density of resource consumers in the superior habitat increased, there would come a point at which an individual could gain higher fitness by settling for a lower-ranked habitat that had fewer occupants of the same species.[36] In line with this approach, the Czech ornithologist Karel Weidinger found that the density of nesting blackcaps was four times higher in the preferred stream-edge habitat than in the second-ranked habitat away from streams.[104] Thus, these birds must make habitat selection decisions based not just on the nature of the vegetation and other indices of insect productivity but also on the intensity of intraspecific competition as reflected by the density of nesting birds in a location.

FIGURE 6.1 Annual changes in the average density of breeding black-throated blue warblers and the average number of offspring fledged per territory. After Rodenhouse et al.[88]

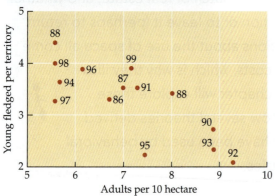

Discussion Question

6.1 A team of ornithologists led by Nicholas Rodenhouse[88] measured the average number of young fledged by pairs of black-throated blue warblers in areas where the average number of breeding adults ranged, over the years, from a little over five to a little less than ten per 10 hectares (Figure 6.1). They also found that the average number of young fledged per territory occupied on all 3 years of the study was about twice as high as the value for territories that were occupied on only 1 or 2 years. What do these data tell us about the roles of competition and food resources in habitat selection by the warblers? Are the birds behaving in the manner predicted from ideal free distribution theory?

Ideal free distribution theory requires that individuals move about in order to evaluate the quality of different habitats. The key prediction is that individuals will settle on the sites where their reproductive success is maximized. Gordon Orians used this approach in his study of red-winged blackbirds, a polygynous species in which a male can attract up to a dozen females to settle on his territory. Orians predicted female blackbirds would choose territories of lower quality when better territories with safer nesting sites were densely occupied by other females. In so doing, a new resident in a low-quality territory would have a better nest site or access to more food or more male parental care than she would have secured if she had chosen to live elsewhere with a larger number of competitors.[75]

The same theory has been used to determine whether wintering red knots, a medium-size shorebird, move about among seven large tidal areas in the United Kingdom, the Netherlands, and France in ways that equalize the amount of food eaten per individual.[84] Red knots feed on a small snail that occurs in varying densities in the mud. By combining data on the numbers of knots present in seven wintering sites with information on the density of mud snails over 5 years, a European research team was able to show that the birds were not distributed uniformly over the seven areas. Instead, as predicted by ideal free distribution theory, individual birds achieved equal food intakes by shifting from site to site depending on the availability of mud snails and the density of their fellow red knots in the area (Figure 6.2).

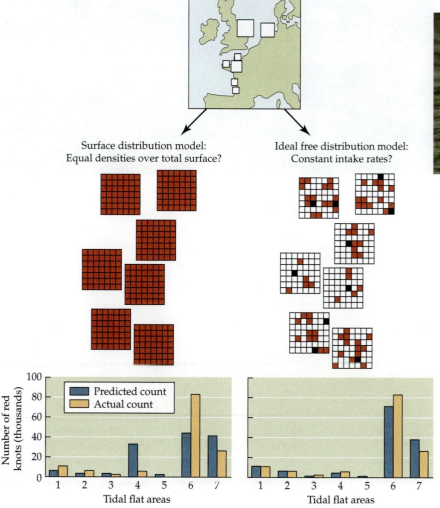

FIGURE 6.2 Habitat selection and ideal free distribution theory. Different models were developed to explain the distribution of the shorebird, the red knot, at seven coastal locations in Europe. One of the models was based on the assumption that the knots would move about to achieve equal densities over the total surfaces provided by seven tidal flats in countries from Great Britain to coastal France. Another model, the ideal free distribution model, was based on the assumption that individual knots were free to move about so as to achieve a constant intake of prey from the seven tidal flats where snails ranged in local abundance from low (white) to moderate (red) to high (black). This second model made predictions that more accurately matched the actual numbers of knots counted at the seven locations. After Quaintenne et al.[84]

When to Invest in Territorial Defense

In some cases, like the red knot example, individuals can distribute themselves freely across their environment. But in others, such as those involving breeding blackcap warblers and red-winged blackbirds, free movement is "discouraged" by **territorial** males that attempt to monopolize high-quality sites, forcing losers to look elsewhere for breeding sites. The benefits of territorial behavior include the ability to utilize the resources on a territory without interference from others. Thus, young male collared lizards that acquire a territory upon the death of an older territory holder have significantly more opportunities to court females than those males that lack the good fortune that comes with the demise of a territorial neighbor.[5]

The potential benefits of possessing a territory can also be seen in our nearest relatives, the chimpanzees. Males patrol the boundaries of their territo-

(A)

(B)

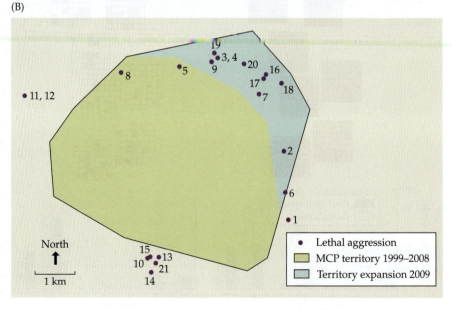

FIGURE 6.3 Male chimpanzees patrolling their territory. (A) This group of males will march along the boundaries of their band's territory (in Uganda, Africa), where they will attack and even kill the members of other groups. (B) The effect of this behavior, if successful, is the expansion of the attacking band's territory, labeled MCP by the research term. A, photograph by John Mitani; B, after Mitani et al.[68]

rial living space in African forests. If a group on patrol encounters a smaller number of chimps from another territory during one of these forays, they will attack and even kill their neighbors. Males from one such band killed 21 fellow chimpanzees over a 10-year period, so depleting the size of one next-door group that the aggressors were eventually able to expand their territory by more than 20 percent, giving them access to food resources once controlled by their neighbors (Figure 6.3).[68]

However, in order to be adaptive, the fitness benefits of territorial expansion (added feeding grounds and presumably more offspring for the killers) have to be weighed against the very real costs of the aggression, which include the time and energy expended by males that also risked injury when violently attacking their neighbors. In general, the costs of trying to control a territory are substantial, a point that Catherine Marler and Michael Moore made experimentally in their studies of Yarrow's spiny lizard.[61,62] Marler and Moore made some nonterritorial male lizards territorial by inserting small capsules containing testosterone beneath the skin of lizards they captured in June and July, a time of year when the reptiles are only weakly territorial. These experimental animals were then released back into a rock pile high on Mount Graham in southern Arizona. The testosterone-implanted males patrolled more, performed more push-up threat displays, and expended almost a third more energy than controls (lizards that had been captured at the same time and given a chemically inert implant). As a result, the hyperterritorial males used up their energy reserves and died sooner than males with normal concentrations of testosterone (Figure 6.4).

FIGURE 6.4 Energetic costs of territoriality. (A) Males of Yarrow's spiny lizard that received an experimental implant of testosterone spent much more time moving about than did control males. (B) Testosterone-implanted males that did not receive a food supplement disappeared at a faster rate than did control males. Testosterone-implanted males that received a food supplement (mealworms) survived as well as or better than controls; thus the high mortality experienced by the unfed group probably stemmed from the high energetic costs of their induced territorial behavior. After Marler and Moore;[61,62] photograph by the author.

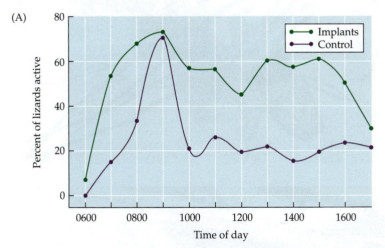

(A)

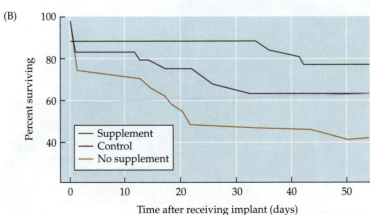

(B)

Yarrow's spiny lizard

Because territoriality is costly, we can predict that peaceful coexistence on an undefended living space or **home range** should evolve when the benefits of owning a valuable resource do not outweigh the costs of monopolizing that resource. (You may wish to ask yourself why red knots on their wintering grounds are willing to feed peacefully with others.) In Yarrow's spiny lizard, for example, males do live together in relative harmony in June and July, when females are not sexually receptive, but when fall comes and females are ready to mate, the males become highly aggressive, as do many other male animals that are competing for mates.

The adaptive capacity of individuals to adjust their aggressiveness up or down in relation to the costs and benefits of territoriality is also found in the tiny tropical pseudoscorpion *Cordylochernes scorpioides*. The life cycle of this arthropod involves periods of colonization of recently dead or dying trees, followed in a few generations by dispersal to fresh sites from the now older trees. Dispersing pseudoscorpions make the trip under the wing covers of a huge harlequin beetle (Figure 6.5), disembarking when the beetle touches

(A)

(B)

FIGURE 6.5 Dispersal by territorial beetle-riding pseudoscorpions. (A) Two huge male harlequin beetles battle for position on a tree trunk. (B) Individuals of the pseudoscorpion *Cordylochernes scorpioides* compete for space on the beetle's back before it flies from one tree to another. Photographs by David and Jeanne Zeh.

down on an appropriate tree. The pseudoscorpions mate both on trees and on the backs of beetles, but much-larger-than-average male pseudoscorpions attempt to control real estate only when they occupy a beetle's inch-long abdomen, which can be defended in its entirety at less expense than a larger swath of tree trunk.[119] A male riding on a harlequin beetle may be joined by a harem of receptive females traveling with him to their new destination, much to the male's reproductive benefit.[118]

Territoriality and Resource-Holding Power

The cost–benefit approach to territoriality also predicts that if individuals vary in their territory-holding abilities, namely their fighting ability and social dominance, then those with superior competitive ability should be found in the highest-quality habitat. A case in point is the American redstart. These warblers compete for territories during the nonbreeding season, when they are on their tropical wintering grounds in the Caribbean and Central America. In Jamaica, males tend to occupy black mangrove forests along the coast, while females are more often found in second-growth scrub inland. Field observations reveal that older, heavier males in mangrove habitat attack intruding females and younger males, apparently forcing these birds into second-rate habitats.[63] If older, dominant males derive a net benefit from their investment in territorial aggression, then there should be some survival (and thus reproductive) advantages for individuals that succeed in occupying the favored habitat. In fact, birds living in mangroves retained their weight over the winter, whereas redstarts in the apparently inferior scrub habitat generally lost weight.[63]

Probably because they have more energy reserves, territory holders in mangroves leave their wintering grounds for their distant breeding sites far to the north sooner than birds living in second-growth scrub (Figure 6.6). In many migrant songbird species, early male arrivals secure a reproductive advantage by claiming the best territories and gaining quick access to females when

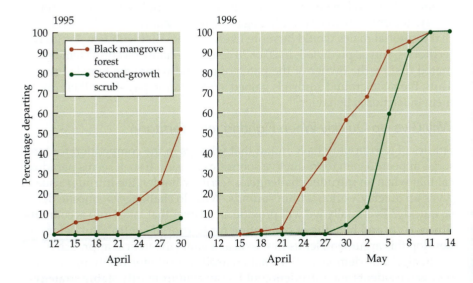

American redstart (female)

American redstart (male)

FIGURE 6.6 Habitat quality and the date of departure from Jamaican wintering grounds by American redstarts. Birds occupying the preferred black mangrove habitat are able to leave sooner for the northern breeding grounds than redstarts forced into second-rate, second-growth scrub. After Marra and Holmes.[63]

FIGURE 6.7 Redstart winter territoriality affects reproductive success. The estimated number of fledglings produced by redstarts in relation to where they spent the winter. Birds that wintered in the most productive habitat, wet forest, are believed to enjoy the highest reproductive success. After Norris et al.[73]

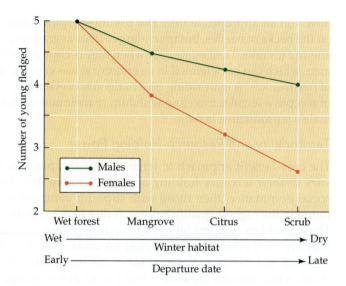

they arrive.[41] Female redstarts also have something to gain by reaching their breeding grounds early (and in good condition), because they will have only a few months to raise their offspring. Early birds get the most fledglings in this species (Figure 6.7).[73]

As we have seen, the benefits of territoriality can include access to mates. In the arctic ground squirrel, for example, males compete with one another for control of patches of meadows in the Canadian Arctic. Female ground squirrels also live in these meadows, and at the start of the breeding season, they mate with one or several males. Given that a territorial male does not necessarily monopolize sexual access to the females living within his plot of grassy meadow, it seems odd that he invests considerable time and energy keeping other males away from his home ground. But Eileen Lacey and John Wieczorek knew that even though a female arctic ground squirrel regularly mates with several males, the male that is first to copulate with her is almost always the only one to fertilize her eggs. With this knowledge, they predicted that territorial males would be more likely to mate first with females in their territories, and as predicted, in 20 of 28 cases, a sexually receptive female was mated first by the owner of the patch of ground where the female resided. Thus, the male ground squirrels gained a fertilization payoff for their investment in territorial behavior.[51]

Why Give Up Quickly When Fighting for a Territory?

Studies of territoriality have found that winners in the competition for territories appear to gain reproductive success as a result. If so, it seems paradoxical that when an intruder challenges a territory holder, the intruder usually gives up quickly—often within seconds. Why do challengers concede defeat so quickly? Eventually we will provide an answer that takes into account the fitness costs and benefits of competing for territorial control. But before we develop this argument, let us first look at a game theoretical approach based on an algebraic demonstration that an arbitrary rule for resolving conflicts between residents and intruders could be an **evolutionarily stable strategy**— that is, one that cannot be replaced by an alternative strategy. One simple arbitrary rule is, the resident always wins. If all competitors for territories were to adopt the resident-always-wins rule, so intruders always gave up and residents never did, a mutant with a different behavioral strategy would not spread by natural selection. For example, a mutant that always challenged a resident would sometimes pick on a better fighter and be injured during the contest. Intruders that immediately gave up would never be damaged by a tougher resident. Therefore, the resident-always-wins strategy could persist indefinitely.[65]

When the resident-always-wins version of the arbitrary conflict resolution rule was first introduced, it was thought that males of the speckled wood butterfly provided a supportive example. Territorial males defend small patches of sunlight on the forest floor, where they encounter receptive females. Males that successfully occupy sun-spot territories mate more frequently than those that do not, judging from an experiment in which males and females were released into a large enclosure. Under these conditions, sun-spot males secured nearly twice as many matings as their nonterritorial rivals.[10] Based on these results, you might think that there would be stiff competition for sun spots, but Nick Davies had previously found that territorial males always quickly defeated intruders, which invariably departed rather than engaging in lengthy territorial conflicts.[28]

The butterflies were capable of prolonged fights, as Davies showed by capturing and holding territorial males in insect nets until new males had arrived and claimed the seemingly empty sun-spot territories. When a prior resident was released, he returned to "his" territory, only to find it taken by a new male. This new male, having claimed the site, reacted as if he were the resident, with the result that the two males had what passes for a fight in the butterfly world. The combatants circled one another, flying upward, occasionally clashing wings, before diving back to the territory, sometimes repeating this maneuver. Eventually the previous resident gave up and flew away, leaving the site under the control of the replacement resident, evidence that the "resident always wins" (Figure 6.8).[28]

However, although Davies's results were consistent with the hypothesis that male butterflies use an arbitrary rule to decide winners of territorial contests, Darrell Kemp and Christer Wiklund decided to repeat the experiment but without subjecting the captured initial resident to the potentially traumatic effects of being held in an insect net before release. Instead, they put their captives in a cooler, retrieving them after replacements had taken up residence in

FIGURE 6.8 Does the resident territory owner always win? An experimental test (now known to be flawed) of the hypothesis that territorial residents always win conflicts with intruders was performed with males of the speckled wood butterfly. When one male ("White") is the resident, he always defeats intruders (1, 2). But when White is temporarily removed (3), permitting a new male ("Black") to settle on his sun-spot territory (4), then Black always defeats White upon his return after release from captivity (5, 6). After Davies.[28]

Speckled wood butterfly

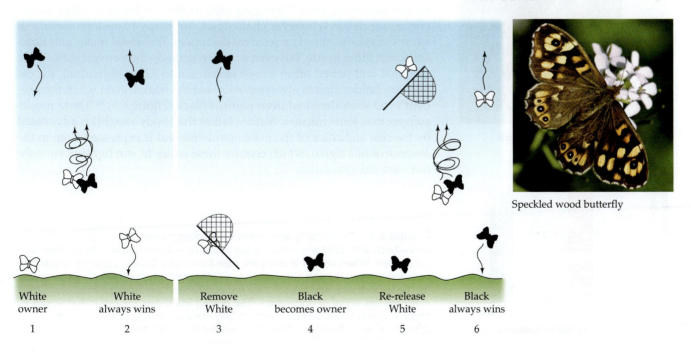

White owner	White always wins	Remove White	Black becomes owner	Re-release White	Black always wins
1	2	3	4	5	6

their sun-spot territories for about 15 minutes. After placing the original resident on the ground near his old territory, they tossed a wood chip over him. Butterflies pursue visual stimuli of this sort, so the new male could be steered toward his rival. When the perched replacement male saw the incoming initial resident, he reacted in the manner of a territorial speckled wood butterfly by engaging his opponent in a spiral flight. But these flights lasted much longer than the ones that Davies had observed, and their outcomes were strikingly different, with the original sun-spot holders winning 50 of 52 contests with new residents.[49] Kemp and Wiklund's study shows just how important experimental design can be. Thanks to their work, we now know that the current resident speckled wood butterfly does *not* always win, a result that eliminates the arbitrary rule hypothesis for this species, which demonstrates the scientific value of revisiting the research of others. (Kemp and Wiklund believe that in Davies's experiment, the original resident males that were held in nets until their release just wanted to escape rather than fight for their old territories, hence the relatively short interactions that led to victory by the new residents.)

Contest Resolution via Nonarbitrary Means

Although the arbitrary contest resolution hypothesis was interesting and mathematically viable, work on the speckled wood butterfly and other species has convinced behavioral researchers that territorial winners in the animal kingdom are rarely, if ever, decided on arbitrary grounds. Instead, a currently more popular hypothesis is one in which winners are said to be decided on the basis of having an edge in physical combat. In fact, in species ranging from rhinoceroses[85] to fiddler crabs[10] to wasps,[74] territory holders are relatively large individuals, apparently stronger than their smaller rivals, and therefore capable of securing the territorial resource against weaker opponents.

Nevertheless, being larger is not the key to territorial success in every species. In the red-shouldered widowbird, an African bird that looks remarkably like the North American red-winged blackbird, males with bigger and redder shoulder patches are more likely to hold territories than are rivals with smaller, duller epaulettes. The less gaudy males become nonterritorial "floaters," hanging around the territories of other males, conceding defeat whenever challenged, but ready to assume control of vacant territories should a resident disappear. Territorial males and floaters do not differ in size, but even so, when males in the two categories were captured and paired off in a cage unfamiliar to either individual, the ex-territory holder almost always dominated the floater, even when the resident's red shoulders had been painted black (Figure 6.9).[82] These results suggest that some intrinsic feature (other than body weight) is advertised by the size and color of the male epaulettes and is expressed even in the absence of this signal, which enables these males to win fights even without their red epaulettes.

Red-shouldered widowbird

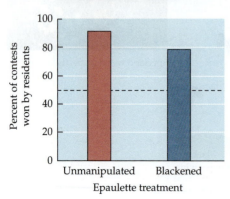

FIGURE 6.9   Territoriality and resource-holding power. Male red-shouldered widowbirds with territories have higher resource-holding power than nonterritorial floaters. When residents compete with floaters for food in captivity, residents usually win, even though they have been removed from their breeding territories for the experiment, and even if their red epaulettes have been painted black to eliminate this signal of dominance. The dotted line shows the result that would occur if residents and floaters were equally likely to win dominance contests. After Pryke and Andersson;[82] photograph by Alan Manson.

Discussion Question

6.2 In some songbirds, a nonterritorial pair sets up housekeeping unobtrusively within the territory defended by another pair of the same species. Develop a game theoretical hypothesis to analyze why there are any nonterritorial pairs of this sort. Come up with a two-strategies hypothesis and a conditional strategy hypothesis. What predictions follow from your two hypotheses? What significance do you attach to the findings that nonterritorial birds either do not breed at all[99] or, if paired, fledge fewer offspring on average than the owners of the territories they live in?[32]

Just as is true for red-shouldered widowbirds, males of some damselflies can be divided into territory holders and floaters. Although territorial male damselflies are not necessarily larger or more muscular, those that win the lengthy back-and-forth aerial contests that occur over streamside mating territories almost always have a higher fat content than the males they defeat (Figure 6.10).[58,80] In these insects, contests do not involve physical contact but instead consist of what has been called a war of attrition, with the winner outlasting the other, and the ability to continue flying is related to the individual's energy reserves.

Differences in **resource-holding power**, however, whether based on size or on energy reserves, cannot account for every case in which territory owners have a huge territorial advantage. For example, older female Mediterranean fruit flies, which are almost certainly not as physically fit as younger flies, can keep younger females away from the fruits in which the insects lay their

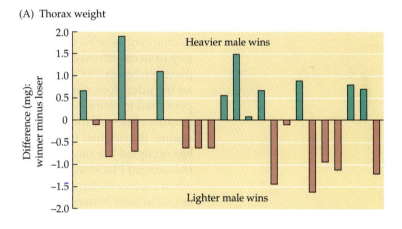

(A) Thorax weight

FIGURE 6.10 Fat reserves and resource-holding power. Fat reserves determine the winner of territorial conflicts in black-winged damselflies. In this species, males sometimes engage in prolonged aerial chases that determine ownership of mating territories on streams. (A) Larger males, as measured by dry thorax weight, do not enjoy a consistent advantage in territorial takeovers. (B) Males with greater fat content, however, almost always win. Photograph by the author; A and B, after Marden and Waage.[59]

Black-winged damselfly

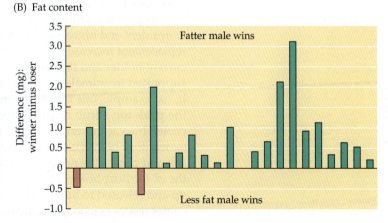

(B) Fat content

(A)

(B)

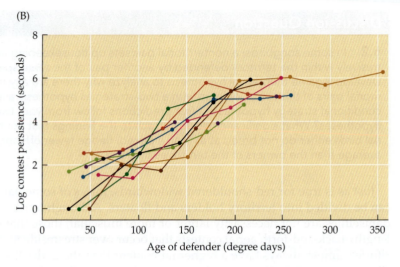

FIGURE 6.11 Motivation affects territorial success. Older males fight harder to control territories in the eggfly butterfly. (A) An old male missing large chunks of his hindwings but still defending his perching territory in Queensland, Australia. (B) Results of a series of encounters in the laboratory between a hand-reared male of known age and a fresh young rival placed in his cage. In the ensuing aerial clashes, the resident persisted longer as he grew older. Each color represents a different individual resident. A, photograph by Darrell Kemp; B, after Kemp.[48]

eggs.[77] By the same token, older males of the song sparrow respond more intensely to simulated intrusion of their territories via song playback than younger ones,[46] and some older male butterflies are more willing to persist in their reproductive attempts than their younger rivals (Figure 6.11).[34,48] Findings of this sort do not support the hypothesis that territorial winners in these species are in better condition or more physically imposing than losers.

Resource Value and Payoff Asymmetries

A different kind of explanation for why some males win and others lose derives from a realization that two individuals might value the same resource differently, with the resident typically deriving a greater payoff from maintaining control of his territory than a newcomer could gain by taking the property from its current owner. This hypothesis suggests that the value a resident places on his territory is linked to the male's familiarity with a location, which an intruder does not have. If so, we can predict that when a newcomer is permitted to claim a territory from which the original resident has been temporarily removed, the likelihood of the replacement resident's winning a fight against the original resident will be a function of how long the replacement has occupied the site. This experiment has been done with birds, such as the red-winged blackbird, as well as with some insects and fishes. In red-wings, when captive ex-territory holders are released from an aviary and allowed to return to their home marsh to compete again for their old territories, they are more likely to fail if the new males have been on the experimentally vacated territories for some time.[7]

This payoff asymmetry hypothesis also predicts that contests between an ex-resident and his replacement will become more intense as the tenure of the replacement increases, because longer tenure boosts the value of a site to the current holder and thus his motivation to defend it. This prediction has been supported in animals as different as tarantula hawk wasps[1] and songbirds.[50] If, for example, one removes a male tarantula hawk wasp (Figure 6.12) from the peak-top shrub or small tree that he is defending and pops him into a cooler, his vacant territory will often be claimed within a few minutes. If the ex-territory holder is quickly released, he usually returns promptly to his old site and chases the newcomer away in less than 3 minutes on average. But if the ex-territory holder is left in the cooler for an hour, then when he is warmed up and released to hurry back to his territory, a battle royal ensues. The newcomer resists eviction, and the two males engage in a long series of ascending

FIGURE 6.12   Residency gives a territorial advantage. A male tarantula hawk wasp (*Hemipepsis ustulata*) perched on his territory, a large creosote bush, where he waits for arriving females. Intruder males almost never oust the resident. Photograph by the author.

flights; the two males climb rapidly up into the sky side by side for many meters before diving back down to the territory, only to repeat the activity again and again until finally one male—usually the replacement wasp—gives up and flies away. The mean duration of these contests is about 25 minutes, and some go on for nearly an hour.[1]

Although examples of this sort support the payoff asymmetry hypothesis to explain why residents usually win, it is possible that lengthier contests occur after replacements have been on territories for some time because the removed residents have lost some resource-holding power while in captivity. This possibility was checked in a study of European robins by taking the first replacement away and permitting a second replacement to become established for a short time before releasing the resident. In this way, Joe Tobias was able to match 1-day replacements against residents that had been held captive for 10 days. Under these circumstances, the ex-territory holders always won, despite their prolonged captivity.[101] In contrast, when ex-residents that had been caged for 10 days went up against replacements that had been on territories for 10 days, the original territory holders always lost. Therefore, contests between replacements and ex-residents were decided by how long the replacement had been on the territory, not by how long the ex-resident had been in captivity, a result that is derived from the payoff asymmetry hypothesis.

The Dear Enemy Effect

One of the reasons why established territory holders may have more to gain by hanging on to their real estate than younger intruders is that boundary disputes with neighbors usually get settled with the passage of time, with the result that neighbors treat familiar rivals as "dear enemies."[35] For example, territorial males of an African lizard charge after an intruder when he is far away, whereas they permit familiar neighbors to come much closer. Should the resident chase a neighbor, the pursuit covers only a few centimeters, whereas a territory holder dashes after an unfamiliar intruder for a meter and a half on average.[111]

Thus, once a territory owner and his neighbors have learned who is who, they no longer need to expend time and energy in lengthy chases—just one of the several advantages associated with becoming familiar with one's living space.[79] If an established resident is ousted, the new territory owner will have to fight intensely for a time with his unfamiliar neighbors in order to settle who owns what. The original resident therefore has more to lose should he be ousted than the new intruder can secure by acquiring his territory, given the expenditures associated with being a new boy on the block.

The hypothesis that familiar enemies save much time and energy compared with those that have to work things out with newcomers has been tested in a novel way by conservation biologists interested in moving groups of Stephens' kangaroo rats to new nature reserves in California. These highly territorial rodents are endangered, and so are in need of assistance. Yet the initial attempts to establish new breeding populations failed. Thinking that perhaps the failures stemmed from the absence of dear enemies in the reconstituted groups, Debra Shier and Ron Swaisgood moved 99 kangaroo rats, permitting about half to retain their familiar neighbors and mixing unfamiliar pairs in the remainder. The group composed of dear enemies fought less and reproduced much more than those kangaroo rats that were saddled with unfamiliar opponents.[97] Here is a wonderful example of how basic research in behavioral ecology can help conservation biologists do their work more effectively (see also Caro[21]).

Discussion Questions

6.3 The dear enemy effect has been explained in terms of familiarity (individuals learn who their neighbors are, and as they become familiar with these others, they become less aggressive toward them). An alternative explanation can be labeled the threat level hypothesis, which states that the dear enemy effect results from the reduced threat to the fitness of a territory holder offered by neighbors that no longer challenge the territory owner next door. Could both of these hypotheses be right? The banded mongoose is a group-living, territorial mammal in which individuals react more aggressively to members of neighboring bands than to strangers.[70] If the threat level hypothesis is right, what prediction can you make about the nature of interactions between two neighboring bands versus the band and an intruding stranger?

6.4 The gall-forming cottonwood poplar aphids studied by Tom Whitham[108–110] compete for spots near the midribs of leaves. Figure 6.13 shows the reproductive consequences for asexual females that secure places near the base of a leaf versus those that form their galls farther along the midrib. Table 6.1 shows how well female aphids do on leaves of different sizes. Given these data, where do you predict the most intense or lengthiest fights for territory will occur? What is the importance of the finding that the females farther along the midrib of a medium-size leaf (which they have chosen after another female has settled on the basal portion of the leaf) do as well as females that have a small leaf all to themselves?

TABLE 6.1 Effect of leaf size and position of the gall on the reproductive success of female poplar aphids

Number of galls per leaf	Mean leaf size (cm)	Mean number of progeny produced by:		
		Basal female	Second female	Third female
1	10.2	80		
2	12.3	95	74	
3	14.6	138	75	29

Source: Whitham[110]

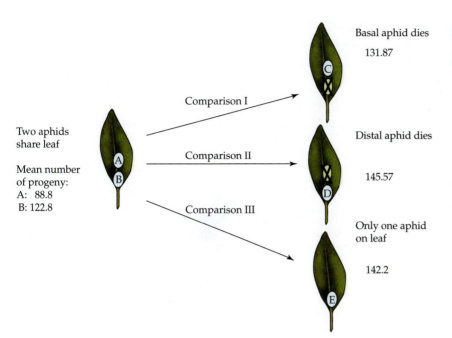

Two aphids
share leaf

Mean number
of progeny:
A: 88.8
B: 122.8

Comparison I

Comparison II

Comparison III

Basal aphid dies

131.87

Distal aphid dies

145.57

Only one aphid
on leaf

142.2

FIGURE 6.13   Territories and reproductive success. The average number of progeny produced by gall-forming poplar aphids that succeed in monopolizing an entire poplar leaf versus those that are forced to share a leaf of the same size with a rival. After Whitham.[110]

Dispersal and Migration

The cost–benefit approach has helped us make sense of the behavioral diversity exhibited by animals as they select places in which to settle and as they defend these areas against intruders—or leave potential rivals alone. We shall employ the same approach in the analysis of why animals often leave areas where they have been living (dispersal), only to come back again, sometimes to the same place, after an interval of months (migration). Although some animals, such as clown fishes,[19] Australian sleepy lizards,[18] and terrestrial salamanders,[64] spend all their lives in one spot, it is common for a young animal to leave its birthplace and go elsewhere. Moreover, adults may also move around, as noted for the red knots that shift from one European mudflat to another in winter before migrating to their high Arctic breeding grounds in spring.

Dispersal is costly in as much as individuals, while moving, not only have to secure extra energy for their travels but may also run the risk of falling prey to predators in the unfamiliar area through which they are moving, a possibility that James Yoder and his coworkers examined in their study of ruffed grouse. They followed birds that had been captured and outfitted with radio transmitters, which enabled them to map the movements of individuals precisely and to find any grouse whose transmitter signaled that the bird had not moved for 8 daytime hours, a very good indication that the bird was dead. Some birds stayed for months near the place where they had been captured, while others moved at considerable intervals from one location to another (Figure 6.14). Being in a new area boosted the risk of being killed by a hawk or other predator at least threefold compared with birds that stayed in locations with which they had become familiar.[116]

So why are so many animals willing to leave home? Users of the cost–benefit approach begin by thinking how the social behavior and ecology of the species might affect the balance between the advantages and disadvantages of dispersal. Consider that the typical young male Belding's ground squirrel travels about 150 meters from the safety of his mother's burrow before he settles in a new burrow, whereas a young female usually goes only 50 meters

Ruffed grouse

FIGURE 6.14 **Two patterns of move-ment by radio-tracked ruffed grouse.** (A) This bird stayed pretty much within the same fairly small home range for many months. (B) Another individual alternated bouts of staying put with substantial dispersal movements through unfamiliar terrain, a risky business for a ruffed grouse. After Yoder et al.[116]

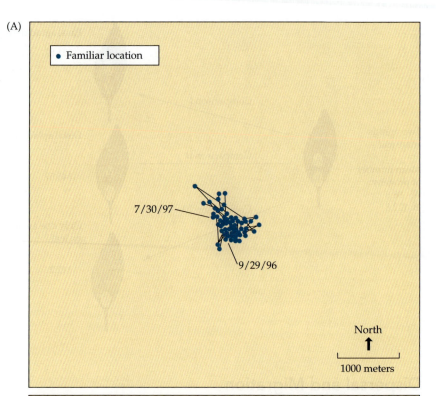

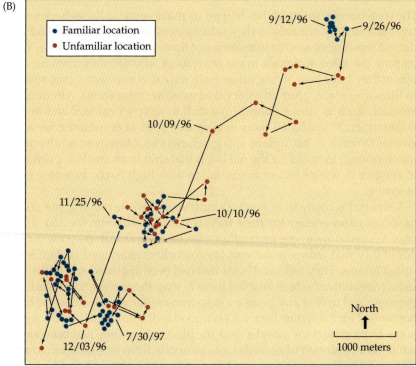

or so from where she was born (Figure 6.15).[44] Why should young male Belding's ground squirrels disperse farther than their sisters?

According to one argument, dispersal by juvenile animals of many species reduces the chance of inbreeding, which often affects fitness negatively.[83] When two closely related individuals mate, the offspring they produce are more likely to carry damaging recessive alleles in a double dose than are off-

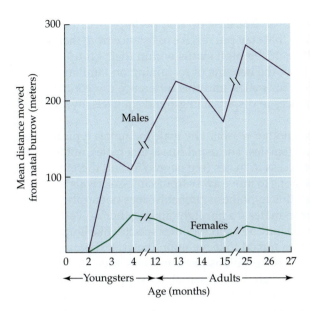

FIGURE 6.15 Distances dispersed by young male and female Belding's ground squirrels. Males searching for a new home go much farther on average from their natal burrows than females. After Holekamp.[44]

Belding's ground squirrel

spring produced by unrelated individuals. The risk of associated genetic problems should in theory reduce the average fitness of inbred offspring, and it does (Figure 6.16).[47,60,86] The prairie vole is an example of a species in which females appear to have evolved a preference for unfamiliar partners with which to produce a litter, a preference that should prevent them from reproducing with littermates, thereby increasing the odds of adaptive outbreeding.[54] Likewise, in the spotted hyena, most young males leave their natal clans because females prefer sexual partners that have either recently joined the group from another clan or have been with the group for a very long time, and are therefore not likely to be their close relatives. As a result, females generally avoid males that were born in their clan, making it adaptive for these males to leave to search for unrelated females elsewhere.[45]

If avoidance of inbreeding is the primary benefit of dispersing, then one might expect as many female as male ground squirrels (or hyenas) to travel

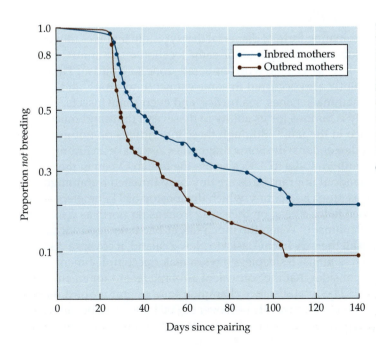

Oldfield mouse

FIGURE 6.16 Inbreeding depression in oldfield mice. Inbred females appear to reproduce later than outbred females. After Margulis and Altmann;[60] photograph by Mike Groutt, U.S. Fish and Wildlife Service, courtesy of Rob Tawes.

away from their natal burrows. But they do not, perhaps because female ground squirrels (or hyenas) that stay put receive assistance from their mothers and so may eventually achieve high status within their groups and all the attendant benefits.[38] If so, the benefits of remaining on familiar ground are greater for females than for males, and this difference can contribute to the evolution of sex differences in dispersal in ground squirrels, hyenas, and other mammals.[44]

There may, however, be another reason why male mammals typically disperse greater distances than females. The usual rule is that males, not females, fight with one another for access to mates (see page 175), and therefore loser males may find it advantageous to move away from same-sex rivals that they cannot subdue.[69] Although this hypothesis probably does not apply to Belding's ground squirrels, since males do not fight among themselves around the time of dispersal, the idea may be valid for other species, like lions. Young males in lion prides are often forced to leave upon the arrival of new older males that violently displace the previous pride masters, then chase off the subadult males as well.[83] Although these observations support the mate competition hypothesis for male dispersal, if young males are not evicted after a pride takeover, they often leave anyway, without any coercion from adult males and without ever having attempted to mate with their female relatives. Inbreeding avoidance may therefore be the cause of voluntary dispersal by subadult male lions.[40]

Discussion Question

6.5   In one study of brown bears (grizzlies) in Sweden, 15 of 16 males left their mothers and natal territories behind while only 13 of 32 females dispersed. Older and heavier females were less likely to be part of the dispersing cohort.[117] So here, as in Belding's ground squirrels and many other mammals, males disperse while most females remain on or near their natal territories. Given the information above, do the explanations given for the pattern of ground squirrel dispersal also apply to brown bears? What other information would be useful in order to evaluate these hypotheses?

The History of Migration

Migration is a fascinating form of dispersal that typically involves movement away from and subsequent return to the same location on an annual basis, a phenomenon strongly associated with temperate-zone songbirds.[27] However, migration apparently also occurred in some extinct dinosaurs[37] and is seen today in many mammals, fishes, sea turtles, and even some insects.[23,25] The monarch butterfly is famous for its ability to fly several thousand miles from Canada to Mexico in the fall;[20] although these individuals do not make the return trip, their descendants do. Some dragonflies also make an impressively long trip from southern India to Africa across the Indian Ocean in the fall.[4]

Among the standout migrant birds are tiny ruby-throated hummingbirds, weighing about as much as a penny, which fly nonstop 850 kilometers across the Gulf of Mexico twice a year, on their way from as far north as Canada to as far south as Panama.[106] When the much larger bar-tailed godwit flies in autumn nonstop all the way from Alaska to New Zealand, it covers a distance of 11,000 kilometers in just 8 sleepless days, a major accomplishment, especially when you consider that the equivalent airborne record for an aircraft is less than 4 days.[42] The total distance champion for a migratory bird may be the sooty shearwater, which travels well over 60,000 kilometers per annum in a figure-eight journey over the whole Pacific Ocean (Figure 6.17). These shear-

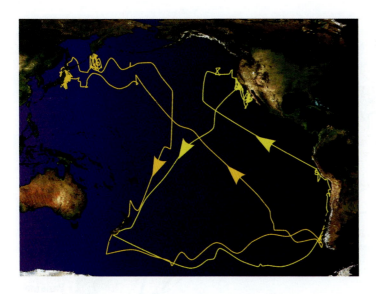

FIGURE 6.17 A very long migratory route. The sooty shearwater travels around the whole of the Pacific Ocean each year. Shown here are the routes taken by two shearwaters outfitted with electronic tracking tags as they left their breeding place in New Zealand and traveled across the Pacific Ocean to South America. One bird then went north to California, where it spent considerable time foraging before flying back to New Zealand; the other bird went northwest to Japan to its summer foraging ground before returning to nest in New Zealand. From Shaffer et al.,[96] courtesy of Scott Shaffer.

waters breed in New Zealand and then move north to feeding areas off Japan, or Alaska, or California, only to loop back down to their breeding grounds in time for another breeding season.[96]

Discussion Question

6.6 Data on the travel routes of migrating birds today largely come from extremely lightweight radios, geolocators, and satellite tags, which can be safely attached to individual birds.[13] The use of geolocator backpacks has revealed that migrant wood thrushes and purple martins take much longer to travel from Pennsylvania to the Amazon basin in fall than they do when going in the opposite direction in the spring.[100] (One martin completed its spring migration in about 2 weeks, which required an average trip of 600 kilometers each day.) Why the difference in speed of travel between the fall and spring migrations?

How did migration originate and evolve? Here is another historical problem that requires Darwin's theory of descent with modification. If we assume that sedentary species were ancestral to migratory ones, as they probably were, then we have to show how gradual modifications of a sedentary pattern could lead eventually to the evolution of a species that travels thousands of kilometers each year between two points. One possible start in this process may be exhibited by some living tropical bird species that engage in fairly short-range "migrations" of dozens to hundreds of miles, with individuals moving up and down mountainsides or from one region to another immediately adjacent one. The three-wattled bellbird, for example, has an annual migratory cycle that takes it from its breeding area in the mid-elevation forests on the mountains of north-central Costa Rica to lowland forests on the Atlantic side of Nicaragua, then to coastal forests on the Pacific side of southwestern Costa Rica, from which the bird returns to its mountain breeding area (Figure 6.18).[81] The distances between any two locations visited by migrating bellbirds are substantial (up to 200 kilometers apart), but not breathtaking.

Douglas Levey and Gary Stiles point out that short-range migrants occur in nine families of songbirds believed to have originated in the tropics. Of these nine families, seven also include long-distance migrants that move thousands of kilometers from tropical to temperate regions. The co-occurrence of

(A)

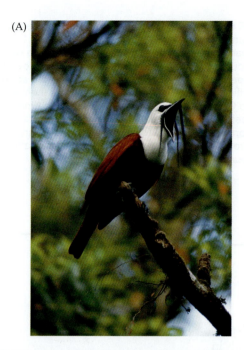

(B)

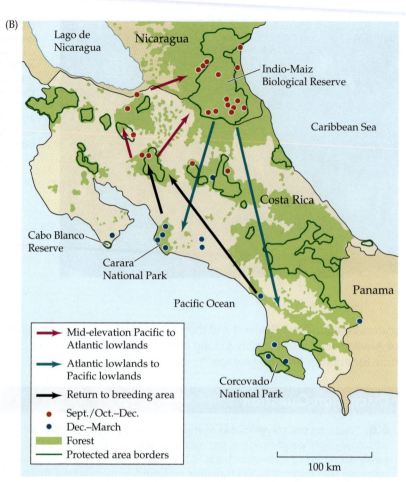

FIGURE 6.18 Short-range migration in the three-wattled bellbird. (A) A male three-wattled bellbird calling from a perch in a Costa Rican forest. (B) After breeding in the mountains of north-central Costa Rica, bellbirds first head to the north and east, then go south and west to reach forests on the Pacific coast before returning north to the mountains. A, photograph by Michael and Patricia Fogden; B, after Powell and Bjork.[81]

short-range and long-distance migrants in these seven families suggests that short-range migration preceded long-distance migration, setting the stage for the further refinements needed for the impressive migratory trips of some species.[53] Thus, long-distance migrants are probably descended from species that moved far less on an annual basis.

One genus of birds, the *Catharus* thrushes, may shed some light on this theory of avian migration. This genus contains 12 species, 7 of which are year-round residents in areas from Mexico to South America; the other 5 are migratory species that travel between breeding areas in northern North America and wintering zones far to the south, especially in South America (Figure 6.19). These observations suggest that the ancestors of the migratory species lived in Mexico or Central America. Moreover, the most parsimonious interpretation of a phylogeny of this genus is that migratory behavior has evolved three times, with subtropical or tropical resident species giving rise to migratory lineages each time (Figure 6.20).[76] Thus, the history of this genus supports the hypothesis that migratory species evolved from tropical nonmigratory ancestors.

On the other hand, a phylogeny of the wood warblers, which belong to the family Parulidae, indicates that the ancestral species in this family was a migrant species that shuttled between northern North America and the Neotropics. This evolutionary tree indicates that migratory behavior has been lost repeatedly during the radiation of the wood warblers, with sedentary species having been derived from migratory ones. Of course, the first ancestral migrant parulid may have evolved from a sedentary tropical species, and if so,

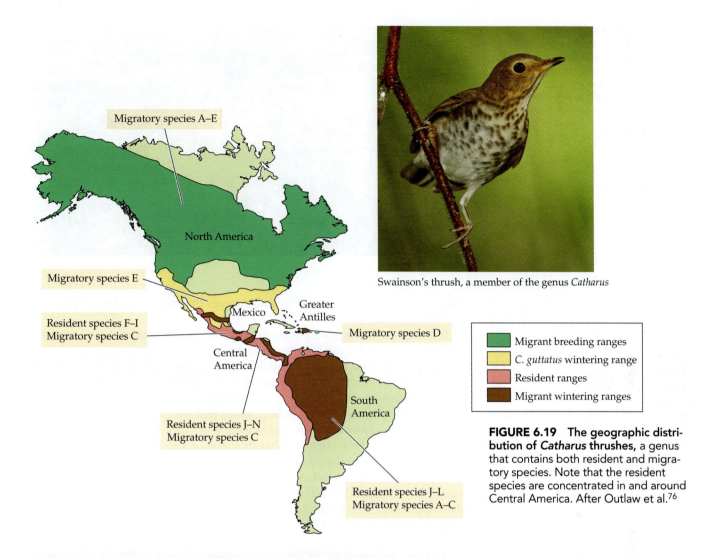

Migratory species A–E

North America

Migratory species E

Resident species F–I
Migratory species C

Mexico

Central
America

Greater
Antilles

Migratory species D

Resident species J–N
Migratory species C

South
America

Resident species J–L
Migratory species A–C

Swainson's thrush, a member of the genus *Catharus*

🟩	Migrant breeding ranges
🟨	*C. guttatus* wintering range
🟧	Resident ranges
🟫	Migrant wintering ranges

FIGURE 6.19 The geographic distribution of *Catharus* thrushes, a genus that contains both resident and migratory species. Note that the resident species are concentrated in and around Central America. After Outlaw et al.[76]

the hypothesis that migration can be traced back to a nonmigratory bird may still be true.[113]

The Costs of Migration

Whatever the history of migratory behavior for a given species, the trait will only be maintained if its fitness benefits exceed its fitness costs, which are not trivial. For birds, those costs include the extra weight that the migrants have to gain in order to have the energy reserves they will need to fly long distances. The storage of fuel is only one of a battery of costly physiological changes that make migration possible for birds. Among these changes are the temporary atrophy of reproductive organs, increases in muscle contraction efficiency, and an altered metabolism that enables the bird to process stored fats quickly.[103]

The red knot illustrates the importance of securing the extra energy reserves (Figure 6.21). This bird, which migrates in both the New and Old World, can add over 70 grams to its initial base weight of about 110 grams after arriving at Delaware Bay from South America, provided that there are plenty of horseshoe crab eggs on the beaches of that staging area. The knots need these fat-rich eggs to fuel the long trip to their Arctic breeding grounds

FIGURE 6.20 The long-distance migratory trait overlain on the phylogeny of *Catharus* thrushes. The phylogenetic tree was constructed on the basis of similarities among the species with respect to mitochondrial DNA. There appear to have been three independent origins of migratory behavior in this cluster of species.[114] After Outlaw et al.[76]

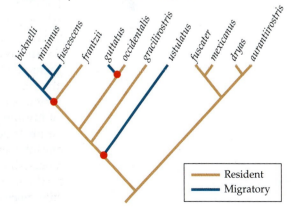

bicknelli *minimus* *fuscescens* *frantzii* *guttatus* *occidentalis* *gracilirostris* *ustulatus* *fuscater* *mexicanus* *dryas* *aurantiirostris*

▬	Resident
▬	Migratory

FIGURE 6.21   Red knots on migration at a critically important staging area. The Delaware Bay once had huge numbers of horseshoe crabs whose eggs fed red knots at a key point along their migratory route. The current scarcity of horseshoe crabs threatens the well-being of migrant red knots.

and to provide for the physiological changes that make egg laying possible after they do arrive.

If horseshoe crab eggs are the essential fuel for migrant knots, we can predict that a decline in crab eggs should have negative effects on the population of red knots that use Delaware Bay during their migration. In fact, as crab egg levels fell by 90 percent due to rampant overharvesting of horseshoe crabs as bait by fishermen, the red knot population at this site decreased by 75 percent.[72] As the population was falling, knots that left Delaware Bay for the far north at relatively low body weights were less likely to be recaptured in the following year, which suggests that they were more likely to die than well-fed birds.[6] Thanks to their knowledge of the basic biology of knot migration through Delaware Bay, behavioral ecologists have helped develop management plans designed to rebuild the horseshoe crab population and sustain the population of red knots that migrates from southern South America north along the eastern seaboard.[66]

Discussion Question

6.7   Figure 6.22 shows the heart rates and wing beat frequencies of great white pelicans flying in V-formation as opposed to flying alone. Birds traveling behind others can take advantage of updrafts created by the wing beats of their companions, which enables them to cut their energetic costs by about 10 percent. Perhaps this is why so many large birds, like Canada geese, typically fly in V-formation when migrating. But what is there about Figure 6.22 that constitutes a Darwinian puzzle? See the paper by Henri Weimerskirch et al.,[105] where they explain how they managed to get these big birds to divulge information on their heart rates and wing beat frequencies.

Another cost of migration is the risk of dying during the trip, a special factor for those songbirds that attempt long crossings over water. Perhaps the survival advantages of reducing the overwater component of a migratory trek explain why so many small songbirds travel east to west across all of Europe before crossing the Mediterranean at the narrow point between southern Spain and northern Africa.[11] Although this route greatly lengthens the total journey for birds headed to central Africa, it may prevent some individuals from drowning at sea.

If this hypothesis is true, then we would expect other songbirds to make migratory decisions that decrease the risks of mortality during the trip. Red-

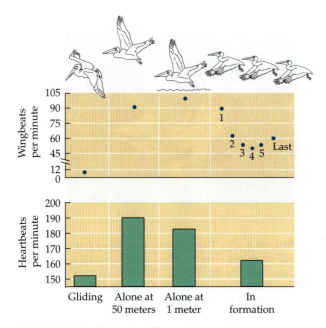

FIGURE 6.22 **Flying in V-formation can be an energy saver.** Data on wing beat frequency and heart rate are presented for various flight options available to the great white pelican. Pay special attention to the upper right portion of the figure. After Weimerskirch et al.[105]

FIGURE 6.23 **A funnel cage used in migration experiments.** As the captive bird tries to begin its nighttime flight during the migratory season, it jumps up from an inkpad floor and leaves ink marks on the lining of the funnel cage that reveal the direction in which it intended to travel. Photograph by Jonathan Blair.

eyed vireos migrating in the fall from the eastern United States to the Amazon basin of South America must either cross a large body of water, the Gulf of Mexico, or stay close to land, moving in a southwesterly direction along the coast of Texas to Mexico and then south. The trans-Gulf flight is shorter, but vireos that cannot make it all the way to Venezuela are dead ducks, so to speak.

In light of this danger, Ronald Sandberg and Frank Moore predicted that red-eyed vireos that happened to have low fat reserves (for whatever reason) would be less likely than those with considerable body fat to risk the long journey due south across the Gulf of Mexico. They captured migrating vireos in the fall on the coast of Alabama, classified each individual as lean or fat, and placed the birds in orientation cages (Figure 6.23). Vireos with less than about 5 grams of body fat showed a mean orientation at sunset toward the west-northwest, whereas vireos that had been classified as having more fat tended to head due south, just as Sandberg and Moore had predicted (Figure 6.24).[93]

Red-eyed vireo

FIGURE 6.24 **Body condition affects the migratory route chosen by red-eyed vireos.** (A) Birds with low fat reserves do not head south toward the Gulf of Mexico but instead head west (sunset symbol) as if to begin an overland flight toward Mexico. (B) Birds with ample energy reserves orient due south. The central arrow shows the mean orientation for the birds tested in each group. After Sandberg and Moore.[93]

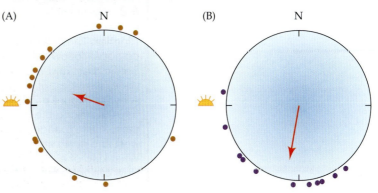

Blackpoll warbler

FIGURE 6.25   Transatlantic migratory path of black-poll warblers. The warblers travel across the Atlantic Ocean as they fly from southeastern Canada and New England to their South American wintering grounds. Courtesy of Janet Williams.

Given that many songbirds appear to avoid long overwater journeys, it is very surprising that some blackpoll warblers appear to voluntarily make a nonstop flight of 3000 kilometers from Canada to South America over the Atlantic Ocean (Figure 6.25).[112] Surely the little warblers should take the safer passage along the coast of the United States and down through Mexico and Central America. However, the blindly courageous blackpoll warbler that manages this overwater trip has substantially reduced some of the costs of getting to South America. First, the sea route from Nova Scotia to Venezuela is about half as long as a land-based trek, although admittedly it requires an estimated 50 to 90 hours of continuous flight. Second, there are very few predators lying in wait in mid-ocean or on the islands of the Greater Antilles that the transoceanic blackpolls reach. Third, the birds leave the Canadian coast only when a west-to-east-traveling cold front can push them out over the Atlantic Ocean for the first leg of their journey, after which the birds use the westerly breezes typical of the southern Atlantic to help them reach an island.[52,67]

Discussion Question

6.8   Swainson's thrush (*C. ustulatus*), one of the species in the genus *Catharus* mentioned earlier, breeds in a large region right across North America. Those birds that live in the northwestern part of North America do not all follow the same migratory route. Some birds go right down the Pacific coast and winter in Central America. But others travel all the way to the eastern part of North America before flying south to winter in South America (Figure 6.26).[91] One hypothesis to account for the behavior of the thrushes taking the long way south is that these birds are descendants of those that expanded the species' range from the East Coast far out to the west and north after the retreat of the glaciers about 10,000 years ago.[92] What kind of evolutionary hypothesis is this? Is the behavior maladaptive? How can you account for the persistence of the trait?

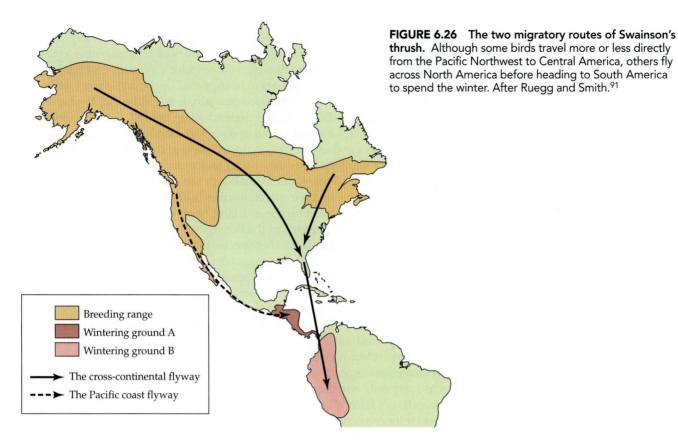

FIGURE 6.26   The two migratory routes of Swainson's thrush. Although some birds travel more or less directly from the Pacific Northwest to Central America, others fly across North America before heading to South America to spend the winter. After Ruegg and Smith.[91]

Legend:
- Breeding range
- Wintering ground A
- Wintering ground B
- → The cross-continental flyway
- ---► The Pacific coast flyway

The Benefits of Migration

For all their navigational and meteorological skills, migrating blackpolls and other birds cannot eliminate the costs of their travels altogether. What ecological conditions might elevate the benefits of migration enough to outweigh these costs, leading to the spread and maintenance of migratory abilities by natural selection? One answer for many songbirds in the Americas may lie in the immense populations of protein-rich insects that appear in the northern United States and Canada in the summer, when long days fuel the growth of plants on which herbivorous insects feed.[17] Moreover, the many hours of summer daylight mean that breeding migrant songbirds can search for food longer each day than tropical bird species, which have only about 12 hours each day to harvest prey for their offspring. But a summertime food bonanza cannot be the only factor favoring migration, given that many migrants abandon areas where food is still plentiful in order to winter elsewhere.[9]

Discussion Question

6.9   For some whale species that migrate from the Arctic or Antarctic Ocean to give birth in warmer water nearer the equator, food cannot provide a benefit, since the adults do not feed on the calving grounds. Therefore, other hypotheses for whale migration have been advanced, such as the idea that whale calves can gain weight more quickly in subtropical waters, where they need to invest less energy in keeping warm. Alternatively, some scientists have suggested that infant whales in these waters are less likely to be attacked by predators, especially killer whales.[24,26] How would you test these hypotheses, given the practical difficulties of directly measuring the metabolic costs of thermoregulation by whale calves or of actually observing killer whale attacks in any environment?

Resources other than food can also vary in availability seasonally, making migration adaptive. In the Serengeti National Park of Tanzania, over a million wildebeests, zebras, and gazelles move from south to north and back again each year. The move north appears to be triggered by the dry season, while the onset of the rains sends the herds south again. It might be that the herds are tracking grass production, which is dependent on rainfall. Eric Wolanski and his colleagues have, however, tested another hypothesis, namely that a decline in water supplies and an increase in the saltiness of the water in drying rivers and shrinking water holes is the critical factor underlying migration. If one knows the salinity of the water available to the great herds, one can predict when they will leave on their march north,[115] although the precise route they follow north is influenced by the availability of vegetation, which in turn is influenced by rainfall patterns in the Serengeti.[71]

The monarch butterfly is another species that does not migrate to find food. When monarchs fly in fall from the eastern half of North America, they head for central Mexico, where they will spend the winter roosting (not feeding) in Oyamel fir forests high in the mountains northwest of Mexico City.[14,90,102] Whereas red knots use expensive flapping flight to get where they are going, monarchs use favorable winds to help them glide and soar relatively cheaply toward their destination. True, as they make their long journey south, monarchs must find enough flower nectar to fuel their flight, but unlike red knots that create and expend large fat reserves on their migration, migrating monarchs carry only small quantities of lipids for much of their trip. Only when monarchs get fairly close to the Oyamel forests do they collect large amounts of nectar for conversion to the lipid energy reserves they will need for the long months of cold storage in their winter roosts.[15]

But why go to the trouble of flying up to 3600 kilometers to reach a tree in the cold high mountains of Mexico? Even if the butterflies keep the costs of the journey relatively low by migrating primarily on days when the winds facilitate inexpensive soaring flight, still one would think that they could spend the winter roosting in places much closer to the milkweed-producing areas that female monarchs visit to produce their offspring the next spring and summer.

But perhaps not, since killing freezes occur regularly at night throughout eastern North America during winter. In contrast, freezes are very rare in the Mexican mountain refugia used by the monarchs. In these forests, at about 3000 meters elevation, temperatures rarely drop below 4°C, even during the coldest winter months. Occasionally, however, snowstorms do strike the mountains, and when this happens, as many as 2 million monarchs can die in a single night of subfreezing temperatures. The risk of freezing to death could be completely avoided in many lower-elevation locations in Mexico. But William Calvert and Lincoln Brower note that monarchs would quickly use up their water and energy reserves in warmer and drier areas. By remaining moist and cool—without freezing to death—the butterflies conserve vital resources, which will come in handy when they start back north after their 3 months in the mountains.[20]

Discussion Question

6.10   Imagine that there are other places in Mexico or the southern United States where monarchs could overwinter even more safely and successfully than they do in their current Mexican wintering grounds. Imagine that the only reason the butterflies do not utilize these locations is that the mutations that are needed to change their migratory route have not occurred. In other words, the accidents of monarch history are responsible for their present migratory choices. Does this mean that the adaptationist approach to monarch migration is flawed in this instance because the migratory behavior of today's monarch butterflies could be more efficient than it is?

(A)

FIGURE 6.27 Monarch butterfly habitat selection. (A) Vast numbers of monarchs spend the winter resting in huge clusters on fir trees in a few Mexican mountain sites. (B) Habitat quality correlates with survival of overwintering monarchs. Protection from freezing in the high Mexican mountains depends on a dense tree canopy that reduces wetting of the butterflies by rain or snow and their exposure to open sky. A, photograph by Lincoln P. Brower; B, after Anderson and Brower.[3]

(B)

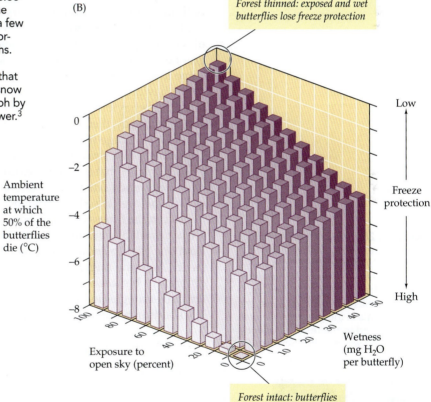

Forest thinned: exposed and wet butterflies lose freeze protection

Forest intact: butterflies protected from freezing

The hypothesis that the stands of Oyamel fir used by the monarchs provide a uniquely favorable microclimate that promotes winter survival is being unwittingly tested in an unfortunate manner. Even in supposedly protected reserves, an alarming amount of woodcutting and logging has occurred.[87] Brower and his associates believe that timber removal causes butterfly mortality, even when some roosting trees are left in place. Opening up the forest canopy increases the chances that the butterflies will become wet, which increases the risk that they will freeze (Figure 6.27). Thus, even partial forest cutting may destroy the conditions needed for the survival of monarch aggregations. If the loss of a relatively small number of Oyamel firs causes the local extinction of overwintering monarch populations, it will be a powerful but sad demonstration of the value of a specific habitat for this migratory species[3] as well as illustrating once again how the world is afflicted by the destructive activities of humans, which some believe threaten the very discipline of behavioral ecology.[22]

Discussion Question

6.11 Recently Sonia Altizer and her colleagues have found that the monarch butterflies that travel the longest distances have the lowest levels of infection by a protozoan parasite, whereas the prevalence of these parasites is highest in the nonmigratory populations of the butterfly.[2] They suggest that because badly infected individuals cannot reach distant destinations, migration has the beneficial effect of culling parasite carriers, which keeps the species healthier than it would be otherwise. According to this hypothesis, who benefits from the removal of infected monarchs? From a theoretical perspective, why does this matter?

Conditional Strategies and Migration

Both migratory and nonmigratory individuals occur in some species, like the European blackbird (actually a thrush, like the American robin), some of which migrate in fall while others remain to overwinter on the breeding grounds.[95] We have dealt with this kind of puzzle previously and have used game theory to explain why selection has not resulted in the replacement of one behavioral trait by the other. If migratory behavior is one strategy (an inherited behavioral trait—see page 23) and the nonmigratory alternative is another strategy, then individuals of the two types would have to enjoy equal fitness on average if the two strategies were to persist together over time. Therefore, a two-strategies hypothesis for blackbird behavior generates two predictions: (1) that the lifetime fitness of the two types should be the same, on average, and (2) that the differences between migratory and nonmigratory individuals are caused by differences in their genes.

Data on the fitness consequences of being migratory versus being a resident are not available, but individual blackbirds regularly change their behavior from one year to another (Figure 6.28). The fact that an individual bird can switch from the migrant strategy to the resident strategy indicates that the differences between the two behavior patterns are not hereditary. Because the two-strategies hypothesis seems unlikely (see page 132), perhaps all the blackbirds in a population share the same conditional strategy that provides

FIGURE 6.28 A conditional strategy controls the migratory behavior of European blackbirds. (A) Birds that were residents in the preceding winter tended to be nonmigratory the next winter as well. (B) In contrast, birds that were migrants in the preceding winter often switched to the resident option the following winter. After Schwabl.[95]

European blackbird

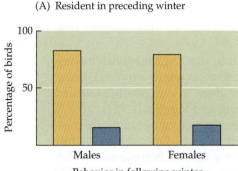

(A) Resident in preceding winter

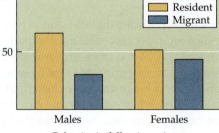

(B) Migrant in preceding winter

them with the flexibility to choose whether to migrate or stay put, depending conceivably on their social status. Socially dominant individuals should be in a position to select the better of the two options under the control of a conditional strategy, forcing subordinates to make the best of a bad situation by adopting the option with a lower reproductive payoff (but one that confers higher fitness than they could get from futile attempts to behave like dominants). For example, perhaps an area can support only a few resident blackbirds during the winter. Under such circumstances, subordinates faced with a cadre of more powerful residents might do better to migrate away from the competition, returning in the spring to occupy territories made vacant by the deaths of some rivals over the winter.

Given the logic of this hypothesis, we can make several predictions: (1) blackbirds should have the ability to switch between **tactics** (available behavior options), rather than being locked into a single behavioral response; (2) socially dominant birds should adopt the superior tactic; and (3) when choosing freely between tactics, individuals should choose the option with the higher reproductive payoff. In light of these predictions, it is significant that migratory European blackbirds head off in the fall at times when dominance contests are increasing in frequency.[55] Moreover, when blackbirds do drop the migratory option in favor of staying put, they do so typically when they are older and presumably more dominant.[56]

Discussion Question

6.12   Some males of the white-ruffed manakin of Costa Rica move temporarily to lower areas while others skip migration to stay in the higher-elevation breeding locations. Birds that stay have greater mating success than those that leave.[12] So why do any males migrate? Use conditional strategy theory to make some predictions about the age and condition of the birds that leave the breeding area.

The migratory component of a species can also be composed of subgroups that vary in where they breed and overwinter, as in the case of the Swainson's thrush.[91] Black-throated blue warblers migrate to islands in both the western and eastern Caribbean. The feathers of the warblers in the western Caribbean region are chemically distinct from the feathers of birds of the same species that winter in the eastern Caribbean. The chemical differences in the two populations have been traced back to where the feathers were produced in the United States. (Keith Hobson discusses the use of isotopic technology in establishing differences of this kind.[43]) As a result, we know that the western Caribbean black-throated blue warblers breed in the northeastern United States while their counterparts in the eastern Caribbean come from the mountains of the southern United States.[89]

The occurrence of migratory variation of this sort poses the same kind of evolutionary question that we discussed in the context of the coexisting resident and migrant strategies of blackbirds: why do members of the same species vary in their choice of breeding and wintering sites, as is true for six different populations of the fox sparrow in western North America (Figure 6.29).[8] The members of each population are sufficiently different in appearance to warrant placing them in six different subspecies (and in fact, some ornithologists believe that the "fox sparrow" is actually several different species). One subspecies (A) in the northwestern United States and southwestern Canada is essentially sedentary, but members of the other five populations migrate. They do not, however, select the same destination. Instead, the birds that breed in

FIGURE 6.29 The leapfrog migratory pattern of western North American populations of the fox sparrow. The populations that breed farthest to the north migrate farthest to the south in the fall. (Population A is composed of birds that live year-round in the same area.) After Bell.[8]

Fox sparrow

southern Alaska (subspecies D, E, and F) travel great distances (probably over the open ocean) to winter in Southern California. Other populations (B and C) travel many fewer kilometers, from southern Canada to wintering grounds in central to northern California. How can we account for this pattern?

One largely untested adaptationist hypothesis (the multistrategy hypothesis) is that birds in the different populations achieve the same fitness by adopting the optimal migration patterns for their particular breeding sites. Thus, the birds that form the southern Alaskan populations (D, E, and F) benefit by moving to Southern California to exploit the surge of food available there in the mid to late spring, which enables these individuals to put on the large fat reserves needed for their long flight north. If sparrows from a breeding population in southern Canada (B or C) also went to Southern California for the winter, they would have to compete for food with the Alaskan birds that were there, and they would be late in getting back to Canada, where the breeding season begins sooner than in Alaska. Late arrival in southern Canada would damage their chances in territorial competition with other fox sparrows that had migrated earlier over shorter distances. Thus, members of the southern Canadian breeding population spend the winter closer to Canada than Alaskan fox sparrows, and they start their migration back earlier in the year as well.[8]

Discussion Questions

6.13 As an exercise, apply conditional strategy theory to the western fox sparrow case in the same way that we did with European blackbirds. Use your hypothesis to make predictions about what decisions individuals of different competitive abilities would make about remaining in an area year-round versus migrating various distances. For example, if your hypothesis were correct, what should happen if a bird improved its condition from one year to the next? What information presented above permits you to evaluate your predictions? Is it useful to know that in the white-crowned sparrow, resident and migrant individuals have almost identical annual survival rates, or that those sparrow species that migrate relatively long distances do not have lower survival rates than other species that travel shorter distances?[94]

6.14 Conditional strategy theory is one of many evolutionary theories that we have examined thus far. Others include natural (direct) selection theory, kin (indirect) selection theory, inclusive fitness theory, group selection theory, optimality theory, game theory, ideal free distribution theory, and evolutionarily stable strategy theory. Some of the theories listed above can be considered part of another theory (for example, ideal free distribution theory is derived from game theory). Produce an organizational scheme that ranks theories in terms of their relationships. How would you define a scientific theory? Why are theories so important to scientific research?

Summary

1. In choosing where to live, many animals actively select certain places over others. Ideal free distribution theory deals with the surprising observation that some animals may occupy less favored areas rather than competing for prime habitat. Hypotheses based on ideal free distribution theory predict that animals that choose areas where the competition is less may have the same fitness as those that have joined many others in using a favored habitat.

2. Territorial behavior poses another puzzle for the adaptationist because of the clear costs to individuals that attempt to monopolize a patch of real estate. Defense of living space evolves only when individuals can gain substantial benefits from holding a territory—such as access to food or mates. As predicted, individuals abandon their territories when the costs exceed the benefits.

3. Territorial contests are usually won quickly by owners. The competitive edge held by territorial residents over intruders may stem from superior physical strength or energy reserves (which give them higher resource-holding power), or it may exist because residents have more to lose than intruders can gain (a payoff asymmetry) thanks to the dear enemy effect (in which familiar neighbors stop fighting with one another over territorial boundaries).

4. In many species, young individuals leave the territories defended by their parents, and adults regularly abandon sites they have invested much time and energy in controlling. Dispersal, like territoriality, comes with fitness costs. Major benefits can include the avoidance of inbreeding with close relatives, escape from aggressive competitors, and the ability to find and exploit resources in short supply in the natal territory or home range.

5. One form of dispersal involves migration between two well-separated areas, with the dispersers returning eventually to the place they left. The ability to migrate very long distances probably originated in populations that had acquired the capacity for short-range migrations. The behavior may have adaptive value because migrants keep the costs of travel low while gaining major benefits from their long-distance journeys, such as access to superior breeding sites that are available only during certain seasons.

6. The coexistence of both resident and migratory individuals (or both territorial and nonterritorial members of a species) is puzzling because one would think that the behavior associated with greater fitness should have replaced the alternative behavioral trait. Conditional strategy theory provides a solution by explaining why individuals with the behavioral flexibility to choose from two or more options can have higher fitness than those that are locked into a single behavioral trait.

Suggested Reading

Richard Dawkins[30,31] and Mart Gross[39] explain what an evolutionarily stable strategy is. The related ability of individual organisms to develop or behave in different ways depending on their environment has been explored in depth by Mary Jane West-Eberhard.[107] See Shuster and Wade for a different view of conditional strategy theory.[98] The cost–benefit approach to territoriality was first presented by Jerram Brown[16] and has been applied subsequently with special skill by Nick Davies and Alistair Houston.[29] For more on game theory as it applies to conflict resolution, see Dugatkin and Reeve.[33] The use of stable isotope technology, an important new tool in the study of migration, is covered by Dustin Rubenstein and Keith Hobson;[90] see also Stutchbury et al.[100]

7

The Evolution of Reproductive Behavior

 I was thrilled the first time I saw a male satin bower-bird fly down to his bower in an Australian forest. His handiwork looked more like something a precocious child might have built than the construction of a bird not much larger than a robin (Figure 7.1). The arriving male held a blue rubber band in his beak, which he dropped among the far more attractive blue parrot feathers that he had strewn about his bower. Although I did not stay to see a female satin bowerbird visit the bower and inspect the feathers and rubber bands there, Gerald Borgia tells me that when a female arrives, the male begins with a preamble of chortles and squeaks, followed by an elaborate court-ship that has the male dancing across the entrance to his bower while opening and closing his wings in synchrony with a buzzing trill. This dance may be followed by another in which the male bobs up and down while imitating the songs of several other species of birds. (See for yourself at http://www.life.umd.edu/biology/borgialab.) Yet despite the apparent elaborateness of the male's **displays** (stereo-typed actions used to communicate with others), most courtships end with the abrupt departure of a seemingly indifferent female.[25]

In fact, female satin bowerbirds initially visit several bowers, scattered through the Australian forest, but just to look, not yet to mate with any bower builder.[200] After the first round, the female takes a break of several days to construct a nest before returning to a number of bowers, during which time she usually observes the full courtship routines of several males. Finally, after several weeks, the female chooses one male and enters his bower, where she is court-ed again before she crouches down to invite the male to copulate. Afterwards she flies off and will have no further contact with her

These three male chim-panzees are all staring at a fertile female climbing in a tree overhead. You can probably guess why they are so interested. Photo-graph by John Mitani.

FIGURE 7.1 Bowerbird courtship revolves around the bower. A male satin bowerbird, with a yellow flower in his beak, courts a female that has entered the bower that he has painstakingly constructed and ornamented with blue feathers. Photograph by Bert and Babs Wells.

partner, incubating her eggs and rearing her nestlings all by herself. Her mate stays at or near his bower for most of the 2-month breeding season, courting other females that come to inspect his creation and copulating with any that are willing.

Thus, not only do the two sexes of satin bowerbirds look different (see Figure 7.1), but their reproductive tactics are very dissimilar. Nor are satin bowerbirds unusual, because in most animals, males do the courting and females do the choosing, whether we are talking about bowerbirds or belugas, aardvarks or zebras. This pattern is so widespread that biologists ever since Darwin have tried to provide an evolutionary explanation for it. This chapter reviews what we now know about why the sexes differ in the way they propagate copies of their genes.

Sexual Selection and Bowerbird Behavior

What does a male satin bowerbird gain by spending so much time constructing his bower, gathering decorations (often taken from other males' bowers), and defending his display site from rival males while also destroying the bowers of his neighbors? Here we turn to Darwin's theory of evolution by sexual selection to get at the possible adaptive value of male bowerbird behavior. **Sexual selection** theory[50] was Darwin's solution to a major evolutionary puzzle, namely why extraordinarily extravagant courtship behaviors and ornaments had evolved. Darwin realized that bower building and things like the huge tails of peacocks (Figure 7.2) must surely make the males with these costly traits more likely to die young. I myself have seen male digger bees so focused in trying to dig up virgin females that thrashers and woodpeckers could walk or hop over to the bees and dispatch them with a single peck. Natural selection eliminates males that die while trying obsessively to find females or to court them with bizarre displays or stunning bowers. So Darwin invented the theory of sexual selection to help him and others explain why individuals, usually males, had evolved traits that lowered the survival chances of individuals. Darwinian sexual selection could favor such traits, provided that these attributes helped one sex, usually males, gain an advantage over others of the same sex in acquiring mates, an end that could

FIGURE 7.2 Elaborate costly traits exhibited by male birds. These characteristics evolved despite their probable negative effect on the survival of individuals. (A) This displaying male bird of paradise is showing off his extraordinary plumage to a visiting female of his species. (B) The sage grouse male can hold his own with other examples of birds that possess conspicuous feather ornaments. (C) The elongate tail of the male quetzal complements the bird's exquisite body plumage. B, photograph by Marc Dantzker; C, photograph by Bruce Lyon.

be achieved either because the ornamented or aggressive males intimidated their rivals or because these males were especially attractive to females. If males lived shorter lives but reproduced more, then the ornaments, displays, and behaviors that reduced longevity could nevertheless spread over time.

Discussion Question

7.1   Darwin defined sexual selection as "the advantage which certain individuals have over others of the same sex and species, in exclusive relation to reproduction."[50] Today most evolutionary biologists consider sexual selection for traits that promote success in acquiring mates to be a form of natural selection. Why? Compare the conditions that cause the process of natural selection to occur with the conditions that must cause sexual selection to occur. Is the factor "differences among individuals in age at death" on your list? Does sexual selection theory have elements in common with game theory?

If sexual selection is responsible for the evolution of bower building, perhaps male satin bowerbirds invest in these structures because they make the builders sexually attractive to discriminating females. If this hypothesis is true, then we expect male mating success in modern satin bowerbirds to be correlated with some features of the bower that vary from male to male, such as the skill with which the bower has been constructed and decorated, or perhaps the number of blue feathers (stolen or not) littering the bower entrance.[212] In fact, even humans can detect differences among the bowers built by different males. Some contain neat rows of twigs (see Figure 7.1), lined up to create a tidy, symmetrical bower, while others are obviously messier, less professionally assembled. Bowers also differ markedly in how well they are decorated with feathers and the like. Female bowerbirds evidently notice these differences too, because they tend to remain calmer at well-constructed bowers, a response that often leads the female to mate with the bower builder.[148,149] Female behavior helps explain why bower quality and the number of bower decorations correlate with male mating success in this[24] and other bowerbird species.[127,128] In the satin bowerbird, male mating success translates directly into male genetic success, because females typically mate with only one male, whose sperm fertilize all her eggs.[167]

Perhaps attractive, well-decorated bowers are built by males that are superior in some way to those birds that cannot construct a top-flight bower. If true, good bower builders could be healthier birds, unlikely to infect their mates with parasites or pathogens and more likely to possess sperm with genes for disease resistance that could be passed on to their offspring. In keeping with this proposal, Stéphanie Doucet and Bob Montgomerie found that males that build better bowers do indeed have fewer ectoparasitic feather mites than those that make less appealing display structures[60] (see also Borgia et al.[26]).

Another idea along these lines is that the bower's quality is in some sense an indicator of the developmental history of the male. For example, birds that had plenty of food as they matured should have well-constructed brains and should therefore be able to excel at the many demanding manipulative tasks needed to assemble a bower. Joah Madden recognized that if this hypothesis were correct, then the brains of bower-building species of bowerbirds should be proportionally larger than those of bird species that do not build bowers—a group that includes a few bowerbird species, such as the green catbird, which only clears off a display court but builds nothing on it. Madden went to the British Museum to X-ray the skulls of a series of stuffed bowerbirds and some of their relatives. Sure enough, bower builders are unusually brainy birds (Figure 7.3).[126] When another research group looked at the relationship between the complexity of bowers built by different species and the sizes of their brains, they found no correlation between bower sophistication and overall brain size. (Note again that contradictory findings are not uncommon

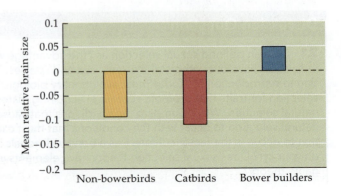

FIGURE 7.3   Bower building may be an indicator of brain size. Bowerbirds that build bowers have relatively large brains compared with other bowerbirds, the catbirds, that merely construct cleared display courts. The mean brain size (as determined by comparing brain cavity volumes against a measure of body size) of the bower-building bowerbirds also far exceeds that of a sample of other, unrelated bird species. After Madden.[126]

in behavioral ecology.) This team did, however, discover a correlation between bower complexity and the size of one component of the bowerbird brain, the cerebellum.[54] Yet another research team tried to measure relative cognitive abilities in individual male satin bowerbirds to examine the validity of the prediction that females should prefer males with greater cognitive skills. They evaluated the ability of male satin bowerbirds in nature to solve several problems (e.g., whether they would be able to cover up an unremovable red item experimentally inserted in the bower, given that red objects are not tolerated at the bowers of this species). This team found that the average cognitive ability of the bowerbirds tested on five such problems was highly correlated with the males' mating success (Figure 7.4).[105] Thus, there is some evidence that brainier bowerbirds experience greater reproductive success than their less with-it rivals (but see Healy and Rowe[92]).

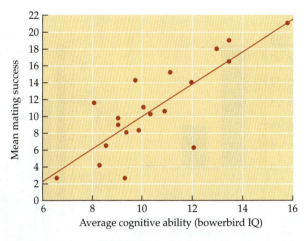

FIGURE 7.4 Male mating success in the satin bowerbird is higher for birds with higher cognitive ability. Males were tested on six different cognitive tasks. An integrated measure of the birds' performances was highly correlated with male mating success. After Keagy et al.[105]

Discussion Questions

7.2   Female satin bowerbirds appear to favor males that give very intense, even aggressive, courtship displays. Perhaps that is why females are often "jumpy" when in the presence of a displaying male. Males differ in how they adjust their displays in response to the female's reaction. Females that often flinch as the male displays tend to leave the site without mating, whereas females that crouch down in the male's bower are more likely to stay and mate with the displaying male. Why might these observations have led Gerald Borgia and his coworkers to hypothesize that the bower enables females to protect themselves against forced copulation attempts by displaying males? How could males gain by making it harder for themselves to force females to mate? See Patricelli et al.[147]

7.3   Male rats, sheep, cattle, rhesus monkeys, and humans that have copulated to satiation with one female are speedily rejuvenated if they gain access to a new female. This phenomenon is called the Coolidge effect, supposedly because when Mrs. Calvin Coolidge learned on a tour that roosters copulate dozens of times each day, she said, "Please tell that to the President." When the President was told, he asked his guide, "Same hen every time?" Upon learning that roosters select a new hen each time, he said, "Please tell that to Mrs. Coolidge." Provide a sexual selectionist hypothesis for the evolution of the Coolidge effect. Use your hypothesis to predict what kinds of species should lack the Coolidge effect.

Sexual Differences Theory

Because most female bowerbirds mate with just one male—often the same individual that is popular with other females—male reproductive success in any given breeding season is variable (Figure 7.5).[24,200] This fact of reproductive life has great significance for a key puzzle that we have yet to address: Why do male bowerbirds, not females, build courtship display structures, and why do females evaluate male performance instead of the other way round? Indeed, it is common throughout the animal kingdom for males to try to mate with many females, while the objects of their desire are content with one or a few matings, albeit with a male or males they have carefully chosen.

This widespread pattern is almost certainly related to a truly fundamental difference between the sexes, which is that males produce small sperm and females produce large eggs.[174] Indeed, in sexual species, eggs are by definition larger than sperm, which are usually just big enough to contain the male's

(A)

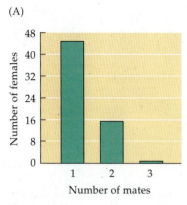

(B)

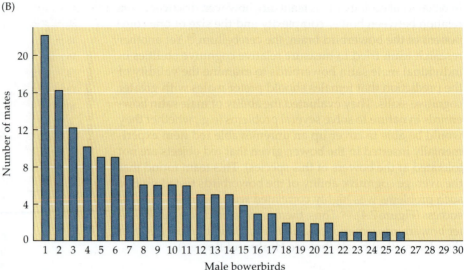

Male bowerbirds

FIGURE 7.5 **Variance in reproductive success is greater for males than for females in the satin bowerbird.** (A) Very few female bowerbirds have more than two mates per breeding season, and few, if any, use the sperm of more than one male to fertilize their eggs. (B) Some male bowerbirds, however, mate with more than 20 females in a single season, while others do not mate at all. After Uy et al.[200]

DNA and enough energy to fuel the journey to an egg. Even in species in which males produce oversize sperm—such as a fruit fly whose males make sperm that are (when uncoiled) nearly 6 centimeters long, or 20 times the length of the flies[21]—the mass of an egg is still vastly greater than that of a sperm, as is typical (Figure 7.6). A single bird egg may constitute up to about 20 percent of the female's body mass.[114] By way of contrast, a male splendid fairy-wren, a very small bird indeed, may have as many as 8 billion sperm in his testes at any given moment.[198] The same pattern applies to coho salmon, whose males shower about 100 billion sperm on a typical batch of 3500 eggs, according to Bob Montgomerie.[38] Likewise, a woman has only a few hundred cells that can ever develop into mature eggs,[49] whereas a single man could theoretically fertilize all the eggs of all the women in the world, given that just one ejaculate contains on the order of 350 million minute sperm.

FIGURE 7.6 **Male and female gametes differ greatly in size.** A hamster sperm fertilizing a hamster egg (magnified 4000×) illustrates the trivial contribution of materials to the zygote by the male. Photograph by David M. Phillips.

The critical point is that small sperm usually vastly outnumber the many fewer large eggs available for fertilization in any population. This sets the stage for competition among males to fertilize those eggs.[111] A male's contribution of genes to the next generation generally depends directly on how many sexual partners he has: the more mates, the more eggs fertilized, the more descendants produced, and the greater the male's fitness relative to less sexually successful individuals—and the greater the effect of sexual selection on the evolution of male attributes.

Whereas male animals usually try to have many sexual partners, females generally do not, because their reproductive success is typically limited by the number of eggs they can manufacture, not by any shortage of willing mates. Eggs are costly to produce because they are large, which means that females have to secure the resources to make them. Furthermore, after one batch of eggs has been fertilized, a female may spend still more time and energy caring for the resultant offspring. Thus, during the breeding season of the satin bowerbird, females are much more likely to spend their time foraging, building a nest, or caring for young than looking for mates, whereas adult male bowerbirds devote most of their time to building or guarding their bowers and sneaking off to destroy the bowers of other males. A female's reproductive success is far more dependent on the quality of her partner (and her own ability to produce eggs and care for offspring) than on the number of males she mates with.

FIGURE 7.7 A clone of aphids. Photograph by the author.

Discussion Question

7.4 Gene-centered thinking tells us that sexual reproduction is a Darwinian puzzle in and of itself. Why? (Consider the fitness consequences for asexual (i.e., parthenogenetic) females in competition with sexual ones in a given population in which both types produce equal numbers of offspring.) When you have dealt with this problem, you should be surprised to learn that in many aphids, females are perfectly capable of cloning themselves (Figure 7.7), which occurs in one generation after another before one group of females produces a generation of both sons and daughters; these males and females then engage in sexual reproduction before their descendants resume cloning themselves. What is the Darwinian puzzle here? Read Moran and Dunbar[138] for one possible solution.

Sexual Selection and Parental Investment

But why are females much more likely than males to spend their time and energy caring for the offspring that come from their fertilized eggs? On the plus side, putting resources into big gametes and helping offspring become adults increases the probability that an existing offspring will live long enough to reproduce and pass on the parental genes to the next generation. On the negative side of the equation, what a parent supplies to one offspring cannot be used to make additional offspring down the road. Males typically try to excel in the mating game, whereas females typically put their resources into the parental game.

The parental game is dominated by expenditures of time and energy and risks taken by a parent to help existing offspring at the cost of reducing future opportunities to reproduce,[195] a phenomenon that Robert Trivers labeled **parental investment** (Figure 7.8). Given that there are two sexes in sexually reproducing species, females are more likely than males to derive a net benefit from taking care of existing offspring. For one thing, the offspring that a female

FIGURE 7.8 Parental investment takes many forms. (Clockwise from top left) A male frog carries his tadpole offspring on his back. A male katydid gives his mate an edible spermatophore containing orange carotenoid pigments that will be incorporated into her eggs. The nutritious coiled spermatophore of a male *Photinus* firefly (shown here greatly enlarged) is donated to the female along with the male's sperm during copulation. A parental eared grebe protects its young by letting them ride on its back. Photographs by Roy McDiarmid; Klaus Gerhard-Keller; Sara Lewis; Bruce Lyon.

cares for are extremely likely to carry her genes. In contrast, a male's paternity is often less certain, given that females of many species are inseminated by more than one male. In addition, males have less incentive to be parental when paternal males lose fertilization opportunities. If a male can mate with several females, it pays him to do so, particularly if he has attributes that give him an edge in the race to fertilize eggs.[163] Although the average reproductive success of males will be the same as that of females (since every offspring has one father and one mother), some nonparental males still do extremely well in the mating game, which favors low parental investment by all males that have a chance to be highly successful in this arena. Because females that have already mated often have nothing to gain by copulating again, there are typically many fewer sexually active females than males at any given time, creating a male-biased **operational sex ratio** (the ratio of sexually active males to sexually receptive females).[69] Thus, key behavioral differences between the sexes have apparently evolved in response to the difference in what they produce (in size and number of gametes), a difference that is often amplified by a difference in the degree to which a female and male provide parental care for their putative offspring.

Testing Sexual Differences Theory

We have reviewed a theory of sex differences that focuses on the role of gametic differences and inequalities between the sexes in parental investment (Figure 7.9). We can test this theory by finding unusual cases in which males make the larger parental investment or engage in other activities that cause the operational sex ratio to become reversed, so there are more sexually active females than males. For example, in some species, males make contributions other than sperm toward the welfare of their offspring (or their mates) because if they do not, they may not get a chance to fertilize any eggs at all, a precondition for male reproductive success. For species of this sort, we can predict female competition for mates and careful mate choice by males—in other words, a **sex role reversal** with respect to which sex competes for mates and which does the choosing. Such a reversal occurs in the mating swarms of certain empid flies, in which the operational sex ratio is heavily female biased because most males are off hunting for insect prey to bring back to the swarm as a mating inducement.[189] When a male enters the swarm, bearing his **nuptial gift**, he may be able to choose among females advertising themselves with (depending on the species) unusually large and patterned wings or decorated legs[89] or bizarre inflatable sacs on their abdomens (Figure 7.10).[77]

Likewise, males of some fish species offer their mates something of real value, namely, a brood pouch in which the female can place her eggs. For example, in the pipefish *Syngnathus typhle*, "pregnant" males provide nutrients and oxygen to a clutch of fertilized eggs for several weeks, during which time the average female produces enough eggs to fill the pouches of two males. Females evidently compete for the opportunity to donate eggs to parental males, which pay a price when they are pregnant, because they feed and grow less while brooding eggs. As a result, at the start of the breeding season, males that are given a choice (in an aquarium experiment) between allocating time to feeding or responding to potential mates actually show more interest in feeding than mating.[17] In the lab and in the ocean, large males with free pouch space actively choose among mates, discriminating against small, plain females in favor of large, ornamented ones, which can provide selective males with larger clutches of eggs to fertilize.[16,170] (What disadvantage might afflict males that acquired a full batch of eggs from several smaller females over some period of time?)

Another example of a species in which females sometimes compete for males is the flightless Mormon cricket, which despite its common name has no religious affiliation and is a katydid, not a cricket. When male Mormon crickets mate, they transfer to their partners another kind of nuptial gift, an enormous edible spermatophore that gives a female nutrients for the production of more eggs (Figure 7.11).[87] Given that the spermatophore constitutes 25 percent of the male's body mass, most male Mormon crickets probably cannot mate more than once. In contrast, some females are able to produce several clutches of eggs, but they have to persuade several males to mate with them if all their eggs are to be fertilized.

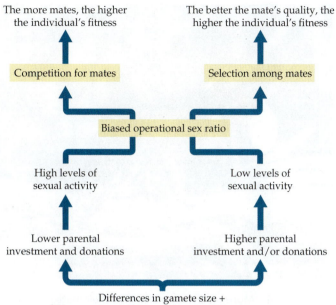

FIGURE 7.9   **A theory of sex differences.** The sexual behavior of males and females may differ because of differences in parental investment that affect the rate at which individuals can produce offspring. The sex that can potentially leave more descendants gains from high levels of sexual activity, whereas the other sex does not. An inequality in the number of receptive individuals of the two sexes leads to competition for mates within one sex, while the opposite sex can afford to be choosy.

FIGURE 7.10   **A sex role reversal.** In the long-tailed dance fly, *Rhamphomyia longicauda*, females fly to swarms where they advertise themselves while waiting for the arrival of a gift-bearing male. The female's inflated abdomen and dark, hairy legs make her appear as large as possible to discriminating males. Photograph by David Funk, from Funk and Tallamy.[77]

FIGURE 7.11 Mormon cricket males give their mates an edible nuptial gift. Here a mated female carries a large white spermatophore just received from her partner. Photograph by the author.

FIGURE 7.12 A katydid with flexible sex roles. Female *Kawanaphila* katydids only compete to mate with males when food is scarce, making the male's donated nutritious spermatophore especially valuable. In this photograph, the female perches on a pollen-poor flower while eating a spermatophore that she received from her mate. The female's two thin antennae and shorter, thicker ovipositor point downward. Photograph by Darryl Gwynne.

Female competition for mates in Mormon crickets is evident during the times when huge numbers of these katydids leave home and march across the countryside, devouring farmers' crops (and one another).[184] In a band of Mormon crickets, a male may announce his readiness to mate, which he does by producing an acoustical signal that causes females to come running, jostling with one another for the opportunity to climb onto the stridulating male, which is the prelude to insertion of the male's genitalia and transfer of a spermatophore. Males, however, may refuse to transfer spermatophores to lightweight females. A choosy male that rejects a 3.2-gram female in favor of one weighing in at 3.5 grams fertilizes about 50 percent more eggs as a result.[87] So here, too, when there are more receptive females than males, competition for mates takes place among females.

The theory of sex role differences also predicts that if the operational sex ratio were to change over the course of the breeding season, the sexual tactics of males and females should change as well. A test of this prediction was provided by a study of a skinny Australian katydid (Figure 7.12) whose food supply varies greatly over the course of the breeding season. When these katydids are limited to pollen-poor kangaroo paw flowers, the male's large spermatophore is both difficult to produce and valuable to females as a nuptial gift. Under these conditions, sexually receptive males are scarce and they are choosy about their mates, whereas females fight with one another for access to spermatophore-offering males. But when pollen-rich grass trees start to flower and males can produce spermatophores much more rapidly, the operational sex ratio can become male biased since the production of eggs by females is limited by the speed at which they can turn pollen into gametes. At this time, sex roles switch to the more typical pattern, with males competing for access to females and females rejecting some males.[88]

Discussion Question

7.5 In the two-spotted goby, a fish that lives on rocky shorelines in northern Europe, males provide parental care for one or more clutches of eggs. Males initially compete fiercely for territories (nest sites in seaweed or mussel shells) during the short breeding season (May to July). But over the course of the breeding season, males become scarce, and females begin to behave aggressively toward one another and to court males.[72] Do the data in Table 7.1 permit you to test the theory outlined above for why the sexes differ in their reproductive tactics?

TABLE 7.1 Seasonal changes in the reproductive biology of the two-spotted goby

	May	June	Late June	Late July
Territorial males per square meter	0.56	0.32	0.13	0.07
Females ready to spawn	0.15	0.39	0.29	0.33
Nest space (number of additional clutches that each nest could accommodate)	2.98	2.53	1.49	0.99

Source: Forsgren et al.[72]

Sexual Selection and the Competition for Mates

In most species, males, with their minute sperm, can potentially have a great many offspring, but if they are to achieve even a fraction of their potential, they must deal both with other males, which are attempting to mate with the same limited pool of receptive females, and with the females themselves, which often have a great deal to say about which male fertilizes their eggs. Let's first consider those male traits that appear to be adaptations that help males compete effectively with other males. We mentioned that male bowerbirds dismantle each other's bowers when they have a chance.[161] Because males with destroyed bowers lose opportunities to copulate, males have been sexually selected to keep a close eye on their display territories and to be willing to fight with rival intruders. Outright fighting among males is one of the most common features of life on Earth (Figure 7.13) because winners of male–male competition generally mate more often, something that is true for everything from giraffes[158,183] to digger bees[6] (see Figure 1.4).

Sexual selection for fighting ability often leads to the evolution of large body size because larger males tend to be able to beat up smaller ones. In keeping with this hypothesis, when males regularly fight for access to mates, they tend to be larger than the females of their species.[22] In addition, males in animals ranging from spiders to dinosaurs, and from beetles to rhinoceroses, have evolved weapons in the form of horns, tusks, antlers, clubbed tails, and enlarged spiny legs, which they use when fighting with other males over females (Figure 7.14).[68]

Although males of many species battle for females, conflicts among males of some species are not about an immediate mating opportunity but instead have to do with the male **dominance hierarchy**. Once individuals have sorted themselves out from top dog to bottom mutt, the alpha male need only move toward a lower-ranking male to have that individual hurry out of the way or otherwise signal submissiveness. If the costly effort to achieve high status (and the priority of access to useful resources that is associated with dominance) is adaptive, then high-ranking individuals should be rewarded reproductively. In keeping with this prediction, among mammalian species, dominant males generally mate more often than subordinates.[45]

FIGURE 7.13 Males of many species fight for females. Here two male giraffes slam each other with their heavy necks and clubbed heads. Note that the long necks of giraffes may be as much a product of sexual selection as of natural selection for access to treetop food. Photograph by Gregory Dimijian.

FIGURE 7.14   Convergent evolution in male weaponry. Rhinoceros-like horns have evolved repeatedly in unrelated species thanks to competition among males for access to females. The species illustrated here are 1, narwhal (*Monodon monoceros*); 2, chameleon (*Chamaeleo* [*Trioceros*] *montium*); 3, trilobite (*Moroccomites malladoides**); 4, unicornfish (*Naso annulatus*); 5, ceratopsid dinosaur (*Styracosaurus albertensis**); 6, horned pig (*Kubanochoerus gigas**); 7, protoceratid ungulate (*Synthetoceras* sp.*); 8, dung beetle (*Onthophagus raffrayi*); 9, brontothere (*Brontops robustus**); 10, rhinoceros beetle (*Allomyrina* [*Trypoxylus*] *dichotomus*); 11, isopod (*Ceratocephalus grayanus*); 12, horned rodent (*Epigaulus* sp.*); 13, giant rhinoceros (*Elasmotherium sibiricum**); (*extinct species). Courtesy of Doug Emlen.

FIGURE 7.15   Dominance and mating success in savanna baboons. (A) Male baboons fight for social status. (B) In a Kenyan reserve in which many different troops were followed over many different breeding seasons, the relation between male dominance rank and ability to form consortships with fertile females generally yielded strongly positive correlation coefficients, which measure the association between the two variables. The closer the correlation coefficient is to +1, the closer the match between male rank and courtship success. A, photograph by Joan Silk; B, after Alberts et al.[4]

The relationship between dominance and sexual access to mates has been especially well studied in groups of savanna baboons, where males compete intensely for high status. Opponents are willing to fight to move up the ladder (Figure 7.15A), and as a result, individual males get bitten about once every 6 weeks, an injury rate nearly four times greater than that for females.[62] So, does it pay males to risk damaging bites and serious infections in order to secure a higher ranking? When Glen Hausfater first attempted to test this proposition, he counted matings in a troop of baboons and found that, contrary to his expectation, males of low and high status were equally likely to copulate.[91] Hausfater subsequently realized, however, that he had made a dubious assumption, which was that any time a male mated, he had an equal chance of fathering an offspring. This assumption would be wrong if matings really only "count" when the females have recently ovulated. When Hausfater reexamined the timing of the copulations, he found that dominant males had indeed monopolized females during the few days when they were fertile. The low-ranking males had their chances to mate, but typically only when females were in the infertile phase of their estrous cycles.

(A)

(B)

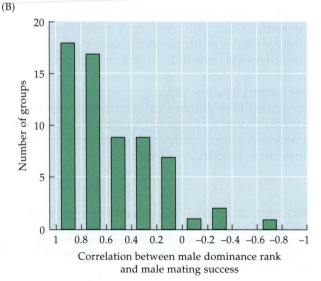

Since Hausfater's pioneering research, others have followed up with more such studies. A summary of data gathered on troops of savanna baboons observed in a Kenyan reserve over many years reveals that male dominance is almost always positively correlated with male copulatory success with fertile females (Figure 7.15B).[5] A male that copulates with an ovulating female while keeping all other males away from her would seem all but certain to be the father of any offspring that his partner subsequently produces. But we no longer need to guess about paternity in these cases, thanks to developments in molecular technology, including microsatellite analysis, which make it possible to determine with nearly complete certainty whether a given male has indeed fathered a given baby (see Box 8.1). And, yes, a male's dominance predicts not only his mating success but also his genetic success.[5] Dominant male baboons sired more offspring than subordinate ones by virtue of their ability to identify receptive females that were highly likely to conceive,[79] and they also kept other males away from these fertile females,[5] as is true for other primate species.[209]

Discussion Question

7.6   In some species, like spotted hyenas and meerkats, females form dominance hierarchies in competition with others of their sex. In species of this sort, variance in reproductive success can be greater for females than for males because dominant females are able to suppress reproduction by subordinate females in their groups.[39,40] Sexual selection theory is usually applied to males rather than females, but what about status striving and aggressiveness among female spotted hyenas and meerkats? Are these sexually selected traits in the Darwinian sense, or do we need a new theory for the evolution of attributes like these, one that focuses on the effects of social competition over resources rather than competition for mates? A theory of social selection exists;[207] check it out, as well as a paper that explains why sexual selection can be considered a form of social selection.[125]

Alternative Mating Tactics

Although research on baboons and many other animals has confirmed that high dominance status yields paternity benefits for males able to secure the alpha position, the benefits are often not as large as one would expect (Figure 7.16).[4] As it turns out, socially subordinate baboons can compensate to some

FIGURE 7.16   Dominant male baboons do not control fertile females as completely as one would expect if the top-ranked male always had priority of access to estrous females at times when only one was available in the troop. (A) A male guarding an estrous female. (B) Males of high rank did not remain close to estrous females for as much time as expected based on male dominance status alone. A, photograph by Joan Silk; B, after Alberts et al.[4]

(A)

(B)

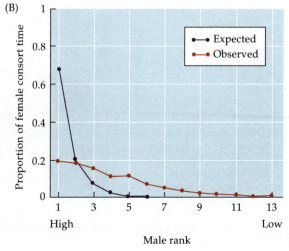

extent for their inability to physically dominate others in their group. For one thing, lower-ranking males can and do develop friendships with particular females, relationships that depend less on physical dominance than on the willingness of a male to protect a given female's offspring (see Figure 9.14A).[143] Once a male, even a moderately subordinate one, has demonstrated that he will protect a female and her infant, that female may seek him out when she enters estrus again.[187]

Male baboons also form friendships with other males. Through these alliances, they can sometimes collectively confront a stronger rival that has acquired a partner, forcing him to give her up, despite the fact that the high-ranking male could take out any one of his opponents mano a mano. Thus, for example, in one troop of savanna baboons that contained eight adult males, three low-ranking males (fifth through seventh in the hierarchy) regularly formed coalitions to oppose a single higher-ranking male when he was accompanying a fertile female. In 18 of 28 cases, the threatening gang of subordinates forced their higher-ranking but lone opponent to relinquish the female.[140]

Yet another alternative tactic that may help less successful males do a little better than they would otherwise has evolved in the famous marine iguana, which lives in dense colonies on the Galápagos Islands. Male iguanas vary greatly in size, and when a small male runs over to mount a female, a larger male may arrive almost simultaneously to remove him unceremoniously from her back (Figure 7.17). Because it takes 3 minutes for a male iguana to ejaculate, one would think that small males whose matings were so quickly interrupted would not be able to inseminate their partners. However, the little iguanas have a solution to this problem; they have already released sperm but have retained them prior to any copulation attempt. When, in the course of mounting a female, the small male everts his penis from an opening at the base of his tail and inserts it into the female's cloaca, these "old" sperm immediately begin to flow down a penile groove, so even if he does not have time to achieve another ejaculation, the small iguana can at least pass some viable sperm to his mate during their all-too-brief time together.[211]

Why Settle for Less?

Although the little iguanas' ability to inseminate without ejaculation does help them reproduce, odds are that the big boys, which are able to copulate and inseminate at their leisure, reproduce more successfully. So here we have yet another example of behavioral variation within a species in which some

FIGURE 7.17 Small males of the marine iguana must cope with sexual interference from larger rivals. Two small males are mounted on the same female; a much larger male is trying to pull them off her back. Photograph by Martin Wikelski, from Wikelski and Baurle.[211]

(A)

(B)

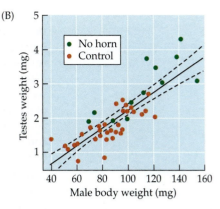

FIGURE 7.18 **Large horns means smaller testes.** When males of (A) the horned scarab beetle *Onthophagus nigriventris* are experimentally induced to develop as hornless males, (B) they tend to grow larger than horned controls and invest in relatively large testes for their body size. The regression line for both data sets combined is shown, along with the dotted 95% confidence limit lines. A, photograph by Doug Emlen; B, after Simmons and Emlen.[182]

individuals employ a behavior with a high fitness payoff while others appear to be making the best of a bad situation, which helps individuals that might otherwise be complete losers gain at least some reproductive success. This pattern is the sort associated with conditional strategies (see page 134), which evolve when selection favors behaviorally flexible individuals that can opt for the tactic that provides them with the best they can do, given their standing with others.

Applying this approach to the competitively disadvantaged little iguanas, we can argue that these males do better by trying to inseminate females on the fly rather than fighting futilely for mates with larger, more powerful individuals. Likewise, males of the horned scarab beetle *Onthophagus nigriventris* make both developmental and behavioral decisions that reflect the reality that smaller individuals will invariably lose fights with males endowed with bigger horns. Therefore, at some point during its larval growth, when a male's developmental mechanisms sense that his body is likely to be relatively small, presumably because the beetle larva is poorly nourished, the male shifts his investment of resources away from growing horns and into what will become his sperm-producing testes. Such a "minor" male has small to nonexistent horns as an adult but larger testes than his bigger opponents. In keeping with his body plan, the minor male sneaks into a burrow where a female is being guarded by a big-horned major male and attempts to inseminate the female on the sly, passing large amounts of sperm to her should he be successful in evading detection by her consort.[100] Large males cannot build both large weapons and large testes (Figure 7.18),[182] just as it has been suggested that male bats cannot afford both large testes and large brains.[155] A large-horned male *O. nigriventris* with his relatively small testes can lose egg fertilizations to a smaller rival, as the smaller male's large ejaculate can swamp the sperm the female has received from her more heavily armed consort (see Sexual Selection and Sperm Competition below).

In species with conditional strategies, the ability of disadvantaged individuals to switch to an alternative tactic secures a higher payoff for these individuals than if they were to behave like their dominant opponents (Figure 7.19). In the horseshoe crab, for example, some males patrol the water off the beach, finding and grasping females heading toward shore to lay their eggs there. Other males are Johnny-come-latelies that swim onto the beach alone and crowd around paired couples there. As it turns out, an attached male fertilizes at least 10 percent more eggs than any competing **satellite male**.[30,31] Although satellite males do not do as well as attached horseshoe crabs, they surely do better than they would if they tried to attach themselves to a female at sea only to be pushed aside or displaced by males that are in better physical condition.

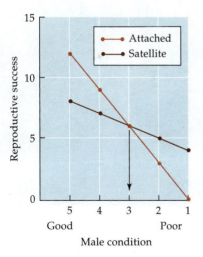

FIGURE 7.19 **A conditional strategy model of the relationship between male condition and reproductive success in the horseshoe crab.** This model predicts that when male condition has fallen to 3 on a 5-point scale, males will switch tactics because they will then gain more reproductive success from the satellite option than by trying to attach themselves to females. "Reproductive success" and "male condition" are given in arbitrary units. After Brockmann.[32]

To test this proposition, Jane Brockmann removed samples of attached and unattached males from a beach and returned them to the sea with plastic bags over their claws so that they could not grasp females in the water. Before releasing the males, she marked each of her subjects and scored its body condition based on such things as whether the carapace was worn or smooth and whether the animal's eyes were covered by marine organisms or free from obstruction. She found that worn, fouled males were more likely to show up again on the beach as unattached individuals. In other words, males in poorer condition were prepared to accept the lower-payoff tactic of coming to shore without a female to search for satellite opportunities, while males in good condition evidently remained at sea, attempting to find a female to grab and defend, which they could not do because of their covered claws.[32] In general, the role of a satellite (Figure 7.20) is an adaptive making-the-best-of-a-bad-job tactic if individuals that select this option gain more than they would by persisting in hopeless attempts to employ another tactic.

Discussion Questions

7.7   Satellite and attached male horseshoe crabs do not have the same reproductive success. Why hasn't sexual selection eliminated the low-payoff satellite option if males exercising this option do not leave as many descendants as attached males do?

(A)

(B)

(C)

FIGURE 7.20   Satellite male mating tactics. (A) A satellite male Great Plains toad crouches by a singing male (note the inflated throat pouch) whose signals he may be able to exploit by intercepting females attracted by the calls. (B) Six subordinate male bighorn sheep trail after a dominant male, which stands between them and the female with which he will mate at intervals. (C) Two satellite male horseshoe crabs wait by a female on either side of a male that has attached himself to his mate. A, photograph by Brian Sullivan; B, photograph by Jack Hogg; C, photograph by Kim Abplanalp, courtesy of Jane Brockmann.

7.8   Roosters compete with one another for social dominance, and not surprisingly, dominant males have greater copulatory success than subordinate males. Use sexual selection theory to account for these differences among the two categories of males: dominant males release more sperm per ejaculate than subordinates, and dominants transfer more and better (faster-moving) sperm to females with large red combs on their heads, whereas subordinates provide all their mates with sperm of the same quality (the same velocity).[43] In addition, use conditional strategy theory to predict how males should behave if two dominant males were placed together until one became subordinate.

Male horseshoe crabs appear to evaluate their body condition, and they use this information to adopt one or another mating tactic, evidence that they possess a conditional strategy. Let's apply the same theory to three alternative mating tactics in a *Panorpa* scorpionfly (Figure 7.21) in which (1) some males aggressively defend dead insects, a food resource highly attractive to receptive females; (2) other males secrete saliva on leaves and wait for occasional females to come and consume this nutritional gift; and (3) still others offer females nothing at all but instead grab them and force them to copulate.[192] In experiments with caged groups of ten male and ten female *Panorpa*, Randy Thornhill showed that the largest males monopolized the two dead crickets placed in the cage, which gave these males easy access to females and about six copulations on average per trial. Medium-size males could not outmuscle the largest scorpionflies in the competition for the crickets, so they usually produced salivary gifts to attract females but gained only about two copulations each. Small males were unable to claim crickets, nor could they make salivary presents, so these scorpionflies instead forced some females to mate but averaged only about one copulation per trial.

Thornhill proposed that in this case, all the males, large and small alike, possessed a conditional strategy that enabled each individual to select one of three options based on his social standing. This hypothesis predicts that the differences between the behavioral phenotypes are environmentally caused, not based on hereditary differences among individuals, and that males should switch to a tactic yielding higher reproductive success if the social conditions they experienced made the switch possible.

FIGURE 7.21   A male *Panorpa* scorpionfly with its strange scorpion-like abdomen tip, which it can use to grasp females in a prelude to forced copulation, one of three mating tactics available to males of this species. Photograph by Jim Lloyd.

FIGURE 7.22 Hereditary differences in male mating strategies. In the ruff, some males fight fiercely for small display courts, as these two individuals are doing, while others associate with court holders without attempting to take the sites from their companions. The behavioral differences between these kinds of males are hereditary.

To test these predictions, Thornhill removed the large males that had been defending the dead crickets. When this change occurred, some males promptly abandoned their salivary mounds and claimed the more valuable crickets. Other males that had been relying on forced copulations hurried over to stand by the abandoned secretions of the males that had left them to defend dead crickets. Thus, a male *Panorpa* goes with whichever tactic gives him the highest possible chance of mating, given his current competitive status. These results clinch the case in favor of a conditional strategy as the explanation for the coexistence of three mating tactics in *Panorpa* scorpionflies.[192]

The Coexistence of Alternative Mating Strategies

Although most cases of alternative mating tactics can be explained via conditional strategy theory[52,63,86] (but see Shuster and Wade[180]), exceptions to this general rule exist. The ruff, for example, is a sandpiper whose males occupy very small display courts near the courts of other males. The males can be categorized as (1) territorial "independents" or (2) subordinate satellites that join an independent on his territory or (3) female mimics. Since males pass on these traits to their male offspring, we can conclude that the differences among them are strategies that are hereditary[116] rather three conditional tactics. Satellites differ in their plumage from territorial independents and are tolerated on the display courts (Figure 7.22), where they sometimes get to mate when the resident male's attention is elsewhere. The rare female mimics look like females of their species, which may make territory holders less likely to attack them.[104] The extra-large testes of female mimics may enable these males to swamp any other sperm their occasional mates may have received from other males.

The marine isopod *Paracerceis sculpta* is another species with three coexisting reproductive strategies. This creature vaguely resembles the more familiar terrestrial sow bugs and pill bugs that live in moist debris in suburban backyards, but this species resides in sponges found in the intertidal zone of the Gulf of California. If you were to open up a sufficient number of sponges, you would find females, which all look more or less alike, and an assortment of males, which come in three dramatically different sizes: large (alpha), medium (beta), and small (gamma) (Figure 7.23), each with its own behavioral phenotype.

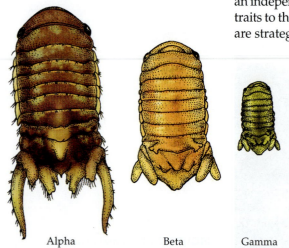

Alpha   Beta   Gamma

FIGURE 7.23 Three different forms of the sponge isopod: the large alpha male, the female-size beta male, and the tiny gamma male. Each type not only has a different size and shape but also uses a different hereditary strategy to acquire mates. After Shuster.[179]

The big alpha males attempt to exclude other males from the interior cavities of sponges where one or several females live. If a resident alpha encounters another alpha male in a sponge, a battle ensues that may last hours before one male gives way. Should an alpha male find a tiny gamma male, however, the larger isopod will simply grasp the smaller one and throw him out of the sponge. Not surprisingly, gammas avoid alpha males as much as possible while trying to sneak matings from the females living in their sponges.[177] When an alpha and a medium-size beta male meet inside a sponge cavity, the beta behaves like a female, and the male courts his rival ineffectually. Through female mimicry, the female-size beta males coexist with their much larger and stronger rivals and thereby gain access to the real females that the alpha males would otherwise monopolize.

Discussion Question

7.9   Female mimicry by males occurs in many other species as well. For example, in the Augrabies flat lizard, some males have the brown coloration of females while others are far more colorful.[208] Female mimics do secure some matings in the territories of their larger, more colorful rivals. Why are the bigger males (and those of the marine isopod) ever fooled into tolerating a female mimic? Why do female mimics occur in any species if the mating success of these individuals is lower than that of the bigger territorial males?

In the marine isopod, therefore, we have three different types of males, and one type has the potential to dominate others in male–male competition. If the three types represent three distinct *strategies*, then (1) the differences between them should be traceable to genetic differences, and (2) the mean reproductive success of the three types should be equal, a requirement based on frequency-dependent selection (see Chapter 5). If, however, alpha, beta, and gamma males use three different *tactics* resulting from the same conditional strategy, then (1) the behavioral differences between them should be the developmental result of different environmental conditions, not different genes, and (2) the mean reproductive success of males using the alternative tactics need not be equal.

Steve Shuster and his coworkers collected the information needed to check the predictions derived from these two hypotheses.[178,179] First, they showed that the size and behavioral differences between the three types of male isopods are the hereditary result of differences in a single gene represented by three alleles. Second, they measured the reproductive success of the three types in the laboratory by placing various combinations of males and females in artificial sponges. The males used in this experiment had special genetic markers—distinctive characteristics that could be passed on to their offspring—enabling the researchers to identify which male had fathered each baby isopod that each female eventually produced. Shuster found that the reproductive success of a male depended on how many females and rival males lived with him in a sponge. For example, when an alpha male and a beta male lived together with one female, the alpha isopod fathered most of the offspring. But when this male combo occupied a sponge with several females, the alpha male could not control them all, and the beta male outdid his rival, siring 60 percent of the resulting progeny. In still other combinations, gamma males outreproduced the others. For each combination, it was possible to calculate an average value for male reproductive success for alpha, beta, and gamma males.

Shuster and Michael Wade then returned to the Gulf of California to collect a large random sample of sponges, each one of which they opened to

(A)

(B)

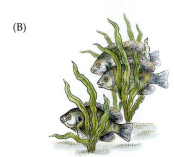

(C)

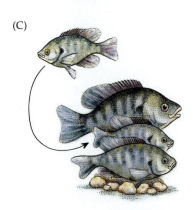

FIGURE 7.24 **Three different egg fertilization behaviors coexist in the bluegill sunfish.** (A) A territorial male guards a nest that may attract gravid females. (B) Little sneaker males wait for an opportunity to slip between a spawning pair, releasing their sperm when the territory holder does. (C) A slightly larger satellite male with the body coloration of a female hovers above a nest before slipping between the territorial male and his mate when the female spawns.

count the isopods within.[178] Knowing how often alphas, betas, and gammas lived in various combinations with competitors and females enabled Shuster and Wade to estimate the average reproductive success of the three types of males, given the laboratory results gathered earlier. When the mathematical dust had settled, they estimated that alpha males in nature had mated with 1.51 females on average, while betas checked in at 1.35 and gammas at 1.37 mates. Since these means were not significantly different statistically, Shuster and Wade concluded that the three genetically different types of males had essentially equal fitnesses in nature and, thus, the requirements for a three-distinct-strategies explanation had been met.

Discussion Question

7.10 Figure 7.24 shows three different tactics used by male bluegill sunfish for fertilizing eggs: the territorial nest defense tactic used by large males, the sneaker option used by small males, and the female mimic option.[85] How would you test the competing three-different-strategies and one-conditional-strategy hypotheses to account for the existence of alternative mating tactics in this species? For test data, see DeWoody et al.[56]

Sexual Selection and Sperm Competition

The sponge isopod alpha male and beta male differ in how they go about securing copulatory partners. But the reproductive competition need not stop there. When females mate with more than one male in a short time, the males may not divvy up a female's eggs evenly. In the case of bluegill sunfish, for example, older males nesting in the interior of the colony (see page 47) produce ejaculates with more sperm, and their sperm swim faster than those of an average bluegill, suggesting that older males should generally enjoy a fertilization advantage.[38] On the other hand, when both a sneaker male bluegill and a guarding territorial male release their sperm over a mass of eggs (see Figure 7.24), the sneaker male fertilizes a higher proportion of the eggs than the nest-guarding male, in part because the sneaker gets closer to the egg mass before spawning.[76] What we have here is evidence of competition among males with respect to the fertilization success of their sperm. **Sperm competition** is a very common phenomenon in the animal kingdom, no matter whether fertilization is external (as in bluegills and many other fishes) or internal (as in some fishes, insects, birds, and mammals). If the sperm of some males have a consistent advantage in the race to fertilize eggs, then counting up a male's spawnings or copulatory partners will not measure his fitness accurately.[19]

Discussion Question

7.11 In the European frog, *Rana temporaria*, some males find and grasp egg-laden females and then release their sperm as the female deposits a batch of eggs in a pond. Some other males locate floating egg masses soon after they have been laid. While grasping the clutch as if it were a female, these after-the-fact males release their sperm on the eggs, with the result that more than 80 percent of the clutches in one pond had multiple paternity.[203] Some researchers have argued that the egg-mass-copulating males are ensuring that the maximum amount of genetic diversity is passed on to the next generation, given that the sex ratio is heavily male-biased. Devise another evolutionary explanation and evaluate the two alternatives.

Sperm wars occur in most animal groups, including insects.[144,181] In *Calopteryx maculata*, the black-winged damselfly of eastern North America, the male tries to win the sperm war by physically removing rival gametes from his mate's body before transferring his own,[204] a common mechanism of sperm competition in damselflies and dragonflies.[42] Male black-winged damselflies defend territories containing floating aquatic vegetation, in which females lay their eggs. When a female flies to a stream to lay her eggs, she may visit several males' territories and copulate with each site's owner, laying some eggs at each location. The female's behavior creates competition among her partners to fertilize her eggs,[144] and the resulting sexual selection has endowed males with an extraordinary penis.

To understand how the damselfly penis works, we need to describe the odd manner in which damselflies (and dragonflies) copulate. First, the male catches the female and grasps the front of her thorax with specialized claspers at the tip of his abdomen. A receptive female then swings her abdomen under the male's body and places her genitalia over the male's sperm transfer device, which occupies a place on the underside of his abdomen near the thorax (Figure 7.25A). The male damselfly then rhythmically pumps his abdomen up and down, during which time his spiky penis acts as a scrub brush (Figure 7.25B),

(A)

(B)

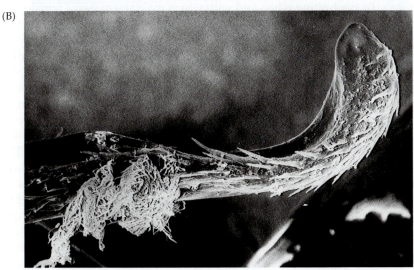

FIGURE 7.25 Copulation and sperm competition in the black-winged damselfly. (A). The male (on the right) has grasped the female with the tip of his abdomen; the female bends her abdomen forward to make contact with her partner's sperm-removing and sperm-transferring penis. (B) The male's penis has lateral horns and spines that enable him to scrub out a female's sperm storage organ before passing his own sperm to her. A, photograph by the author; B, photomicrograph by Jon Waage, from Waage.[204]

FIGURE 7.26   Sperm competition may require female cooperation. A male dunnock pecks at the cloaca of his partner after finding another male near her; in response, she ejects a droplet of sperm-containing ejaculate just received from the other male. After Davies.[51]

catching and drawing out any sperm already stored in the female's sperm storage organ. Jon Waage found that a copulating male *C. maculata* removes between 90 and 100 percent of any competing sperm before he releases his own gametes, which he earlier transferred from his testes on the tip of his abdomen to a temporary storage chamber very near his penis. After emptying the female's sperm storage organ, he lets his own sperm out of storage and into the female's reproductive tract, where they remain for use when she fertilizes her eggs—unless she mates with yet another male before ovipositing, in which case his sperm will be extracted in turn.[204]

If we took the black-winged damselfly as our guide, we might conclude that sperm competition is basically something males do to one another. In reality, females often have an active role in deciding which of their partners' sperm will win the egg fertilization contest.[181] Even a female of the black-winged damselfly can arrange for the immediate removal of some sperm simply by mating with a second male after copulating with an individual that she does not favor. In other cases, females do not have to rely on males for sperm removal but can expel sperm themselves (Figure 7.26).[51,156] For example, hens eject a larger proportion of the semen received from low-ranking roosters, indicating that they are able to bias the fertilization of their eggs by giving an advantage to the dominant males.[55]

Although in many species the female bird lives and mates with a social partner, she may also engage in **extra-pair copulations** with other males. Sperm is stored in female birds in tiny tubules where the sperm can survive for up to a month (Figure 7.27). Females of the collared flycatcher, for example, are able to use their storage system to bias male fertilization chances by copulating in ways that give one male's sperm a numerical advantage over another's. By controlling when and with which male she mates, a female flycatcher can manipulate the number of sperm from different males within her reproductive tract. A female that stopped mating with her social partner for a few days and then mated with an attractive neighbor would have five times as many of her extrapair mate's sperm available for egg fertilizations as she retained from earlier matings with her social partner. This imbalance would give the extrapair male a big fertilization advantage. In fact, a female collared flycatcher paired with a male with a small white forehead patch is likely to secure sperm around the time of egg laying by slipping away for a tryst with a nearby male sporting a larger white patch. Therefore, females of this species seem to play a major role in determining whose sperm will fertilize their eggs,[135] a conclusion that may apply to most animal species.[65] Here, as is so often the case, not everyone agrees that females manage extra-pair matings to their advantage; the alternative view is that they are essentially forced to mate more often than they would otherwise by sexually motivated males that are the fitness beneficiaries of these matings.[13] A recent paper by Akçay et al.[3] is relevant to this argument; check it out.

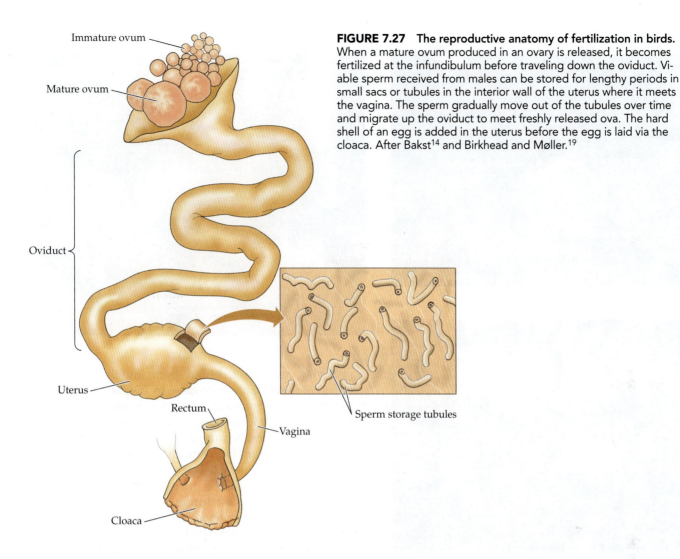

Immature ovum

Mature ovum

Oviduct

Uterus

Rectum

Vagina

Cloaca

Sperm storage tubules

FIGURE 7.27 The reproductive anatomy of fertilization in birds. When a mature ovum produced in an ovary is released, it becomes fertilized at the infundibulum before traveling down the oviduct. Viable sperm received from males can be stored for lengthy periods in small sacs or tubules in the interior wall of the uterus where it meets the vagina. The sperm gradually move out of the tubules over time and migrate up the oviduct to meet freshly released ova. The hard shell of an egg is added in the uterus before the egg is laid via the cloaca. After Bakst[14] and Birkhead and Møller.[19]

Why Stay with a Female after Mating with Her?

Males of the bee *Centris pallida* quickly inseminate their partners but remain with them for a number of minutes afterwards, during which time the pair flies to a nearby tree or bush where the male strokes his partner vigorously with his legs and antennae (see Figure 1.2). Here is a Darwinian puzzle because while the male is engaged in these postcopulatory activities, he might miss finding another virgin female whose eggs he could fertilize. Yet despite this kind of cost, males of many species stay with partners after copulation, sometimes for hours or days, rather than resuming the search for additional mates. In some of these cases, mated males deceptively lure new suitors away from their sexual partners,[70] while in others males seal their mates' genitalia with various secretions[57,157] or they induce their mates to fill their genital openings themselves.[1,2] The most extreme commitments of this sort occur when males sacrifice themselves after copulating,[136,199] as happens, for example, when the male of an orb-weaving spider expires after inserting both pedipalps (the sperm-transferring appendages) into a female's paired genital openings. The dead male is then carried about by the female spider as a kind of morbid chastity belt.[71] Somewhat less extreme examples of this phenomenon in spiders include cases in which the male amputates his own sperm-transferring pedipalp during copulation so that it blocks the female's genital opening. The emasculated male nonetheless can gain fitness when the sperm-containing appendage not only acts as a mat-

FIGURE 7.28 Mate guarding occurs in many animals. (A) A red male damselfly grasps his mate in the tandem position so that she cannot mate with another male. (B) The male blueband goby, an Indonesian reef fish, closely accompanies his mate wherever she goes. (C) A male tropical harvestman (right) stands guard by a female ovipositing in his territory.[36] A, photograph by the author; C, photograph courtesy of Glauco Machado.

ing plug but continues to pump sperm into the female, even if the rest of the male has been eaten by his predatory mate.[119]

Another cost to males of their post-insemination associations is the time lost while the male guards his companion and fights off other males. Because guarding males of the western Mexican whiptail lizard forage much less and fight much more, they pay an energetic price to keep other males at bay.[7] For the New Zealand stitchbird, the price to a male is about 4 percent of his body weight lost during the period when he chases other males away from a previous sexual partner.[121] Therefore, if remaining with an inseminated partner is an adaptation, the behavior must provide fitness benefits to males that outweigh the several costs that they pay. The key benefit appears to be a reduction in the probability that a female will copulate with another male, thereby diluting or removing the sperm the guarding male already donated to his partner (Figure 7.28).[9]

Many studies have demonstrated that guarding males do decrease the odds that a mate will attract an additional sexual partner. For example, Catherine Crockford and her colleagues found that when "bachelor" baboon males heard a taped copulation call of a female about 25 meters to one side and the grunt of her consort partner about 25 meters to their other side, they took notice and often approached the hidden speaker from which the female's call had come. They evidently recognized both individuals by their calls and deduced that the grunting male was no longer guarding his recent partner, whose copulation call indicated that she had been mounted by another male. The listening male hurried over to see if perhaps she would be available to him too.[46] Dominant males must guard their fertile partners closely and constantly if they wish to maintain exclusive access to their females.

If **mate guarding** is adaptive, then the costs of lost opportunities to mate with other females must be less than the benefits of monopolizing one female and her egg(s). Janis Dickinson measured this opportunity cost in the blue milkweed beetle, in which the male normally remains mounted on the female's back for some time after copulation. When Dickinson pulled pairs apart, about 25 percent of the separated males found new mates within 30 minutes. Thus, remaining mounted on a female after inseminating her carries a considerable cost for the guarding male, since he has a good chance of finding a new mate elsewhere if he just leaves his old one after copulating. On the other hand, nearly 50 percent of the females whose guarding partners were plucked from their backs acquired a new mate within 30 minutes. Since mounted males cannot easily be displaced from a female's back by rival males, a mounted male reduces the probability that an inseminated partner will mate again, giving his sperm a better chance to fertilize her eggs. Dickinson calculated that if the last male to copulate with a female fertilized even 40 percent of her eggs, he would gain fitness by giving up the search for new mates in order to guard his current one.[58]

In general, the benefits of mate guarding increase with the probability that unguarded females will mate again and use the sperm of later partners to fertilize their eggs. But how do you figure out what unguarded females would do in a species in which all the available females are guarded? Dickinson's technique involved the simple removal of a male from a female. Other studies have checked the prediction that experimental removal of the male would lead to higher rates of copulation by his mate with other males.[102,214] Jan Komdeur and his colleagues tricked male Seychelles warblers into ending their guarding prematurely by placing a false warbler "egg" in a nest a few days before the male's partner was due to lay her one and only egg. The male warblers used the false egg cue to stop mate guarding at a time when their partners were still fertile. In short order, many of these unguarded females copulated with neighboring males (Figure 7.29)[112] and used the extra-pair sperm to fertilize

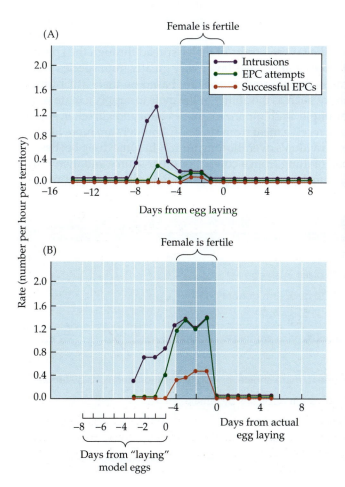

Seychelles warbler

FIGURE 7.29 Adaptive mate guarding by the Seychelles warbler. The graphs show the rate of intrusions and extra-pair copulations (EPCs) by males other than the female's social mate in relation to the female's fertile period (shaded area). (A) Control pairs, in which the female's mate was present throughout her fertile period. (B) Pairs in which the female's mate was experimentally induced to leave her unguarded by the placement of a false egg in the nest. A and B, after Komdeur et al;[113] photograph by Cas Eikenaar.

their eggs. Indeed, the probability that a nestling would be sired by a male other than the female's social partner increased in relation to the number of days that her "mate" neglected to guard her during her fertile period.[113] Mate guarding provides clear fitness benefits for male Seychelles warblers. In chaffinches the male stops associating closely with his mate when the female's fertile period has ended,[18] and the same is true for many other birds.[19]

Discussion Questions

7.12   The digger bee's "postcopulatory courtship" consists of elaborate tactile stimulation that the male provides his partner *after* she has accepted his sperm. Why is this behavior a Darwinian puzzle, and what might its adaptive value be?

7.13   Mate guarding should be common in species in which females retain their receptivity after mating and are likely to use the sperm of their last mating partners when fertilizing their eggs. But there are species, including some crab spiders, in which males remain with immature, unreceptive females for long periods and fight with other males that approach these females.[59] How can "guarding" behavior of this sort be adaptive? Produce sexual selectionist hypotheses and some predictions derived from them.

Sexual Selection and Mate Choice

Sexual selection arises not just as a result of competition for mates among the members of one sex (usually males). In addition, discriminating mate choice (usually by females) can also enable members of the opposite sex with certain hereditary characteristics to reproduce more successfully than others (Table 7.2).[205,216] As a result, the favored traits of males spread through the species.

The attributes of males that females favor vary greatly from species to species, but in some cases, female preferences are based on male attributes of obvious practical utility, such as the ability of the male to supply a mate with a good meal. So, for example, in chimpanzees, males hunt and kill other, smaller primates; females that receive meat from successful hunters are more likely to copulate with these individuals over the long term.[82] In some dung beetles, the male constructs a ball of dung from a cowpat or other source and then rolls it away to a burrow; a female may accompany him to the burrow (Figure 7.30),

---

TABLE 7.2 Ways in which females and males attempt to control reproductive decisions

Key reproductive decisions controlled primarily by females

Egg investment: What materials, and how much of them, to place in an egg

Mate choice: Which male or males will be granted the right to be sperm donors

Egg fertilization: Which sperm will actually fertilize the female's eggs

Offspring investment: How much maintenance and care goes to each embryo and offspring

Ways in which males influence female reproduction

Resources transferred to female: May influence egg investment, mate choice, or egg fertilization decisions by female

Elaborate courtship: May influence mate choice or egg fertilization decisions by female

Sexual coercion: May overcome female preferences for other males

Infanticide: May overcome female decisions about offspring investment

Source: After Waage[205]

FIGURE 7.30 Female mate choice and male contributions. In some dung beetles, the female mates with the male after he has made a dung ball and has rolled it to a distant burrow where he turns the present over to a female. She may accompany him on his journey, as in this species. Photograph by the author.

in which mating will probably occur, with the female rewarding her partner for his contribution of food to her or her offspring.[173]

An analogous case involving insects comes from Randy Thornhill's study of the black-tipped hangingfly, an insect in which female acceptance of a male depends on the nature of his nuptial gift (Figure 7.31A). In this species, a male that tries to persuade a female that an unpalatable ladybird beetle is a good mating present is out of luck. Even males that transfer edible prey items to their mates will be permitted to copulate only for as long as the meals last. If the nuptial gift is polished off in less than 5 minutes, the female will separate from her partner without having accepted a single sperm from him. When, however, the nuptial gift is large enough to keep the copulating female feeding for 20 minutes, she will depart with a full complement of the gift giver's sperm (Figure 7.31B).[191] Males of many other animals provide food presents before or during copulation,[201] including male fireflies that add packets of protein to the

(A)

(B)

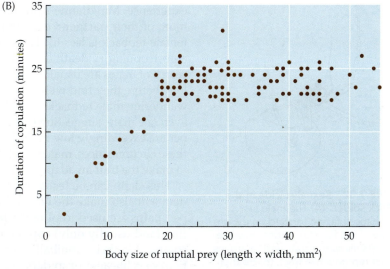

FIGURE 7.31 Sperm transfer and the size of nuptial gifts. (A) A male hangingfly has captured a moth, a material benefit to offer his copulatory partner. (B) In black-tipped hangingflies, the larger the nuptial gift, the longer the mating, and the more sperm the male is able to pass to the female. A, photograph by the author; B, after Thornhill.[191]

ejaculates that they donate to their mates. In at least one firefly species, females evaluate mates on the basis of the duration of their light flashes, which correlates with the size of the spermatophore the male will give to his partner.[118]

(A)

(B)

(C)

Discussion Question

7.14   Although the number of cases of mate choice by females dwarfs the known examples of choosy males, that rarer form of mate choice does occur and may be more widespread than often appreciated.[67] In this light, why might it be adaptive for male jungle fowl to enhance the speed with which their sperm can travel when roosters inseminate attractive versus unattractive females?[44] Why might male potbellied seahorses (see Figure 8.4) strongly prefer to court large females, whereas females show no such preference for large males?[131] And why might male black widow spiders bias courtship in favor of well-fed females as opposed to starved ones?[103]

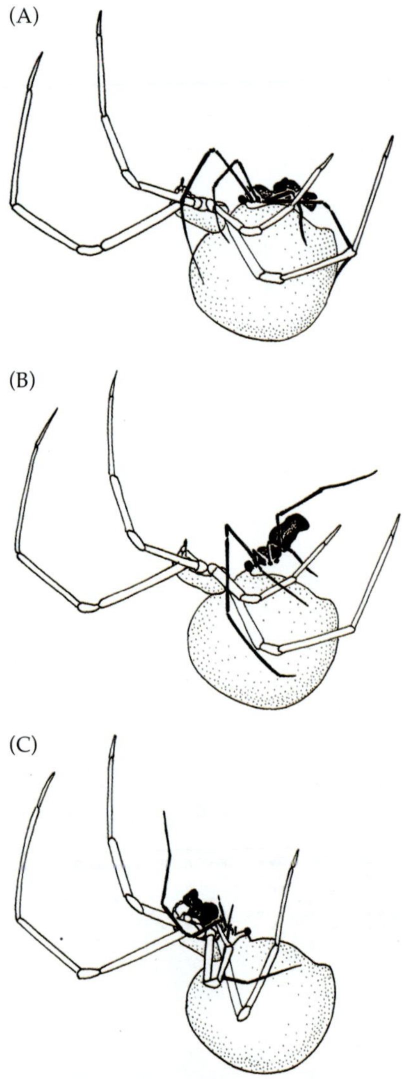

Some researchers have suggested that a special class of nuptial gift givers should be recognized, namely, those males—mainly mantids and spiders—that wind up being eaten by their mates (see Figure 13.2). Indeed, it could be adaptive, under some special circumstances, for a male to wrap up a copulation by turning himself into a meal for his sexual partner.[35,160] It is true that males of several species of mantids avoid or very cautiously approach poorly fed females, suggesting that males of these species typically do not benefit from sacrificing themselves to a mate.[117,132] On the other hand, male redback spiders often do make it easy for their mates to eat them.[186] After the male has transferred sperm into both of the female spider's sperm receptacles, which he has blocked with his two sperm-transferring organs, the male redback performs a somersault, literally throwing his body into his partner's jaws (Figure 7.32). About two-thirds of the time, the female accepts the invitation and devours her sexual companion.[10]

Since a male redback weighs no more than 2 percent of what a female weighs, he does not make much of a meal for a cannibalistic partner. But once eaten, for whatever reason, the deceased male redback does in fact derive substantial benefits from what might seem like a genuinely fitness-reducing experience. Maydianne Andrade showed that cannibalized males fertilized more of their partners' eggs than uneaten males did, partly because a fed female redback is less likely to mate again promptly.[10] Moreover, the cost of a postcopulatory death is very low for male redbacks. Young adult males in search of mates are usually captured by predatory ants or other spider hunters long before they find webs with adult females. The intensity of predation on wandering male redbacks is such that fewer than 20 percent manage to locate even one mate, suggesting that the odds that a male could find a second partner are exceedingly low even if he should survive his initial mating.[11] Under these circumstances, males need very little benefit from sexual suicide in order to make the trait an adaptive option.

This claim can be tested by predicting that males of other, unrelated spiders will also make the ultimate sacrifice in those species in which the male pedipalps are used to plug the female's genital openings during the male's first (and usually only) copulation. Jeremy Miller has found that males are complicit in their cannibalism or die of "natural causes" soon after mating in five or six lineages of spiders. In all but one of these, the male's genitalic pedipalps become broken or nonfunctional in the course of an initial mating,[136] providing support for the hypothesis that sexual suicide occurs when males have almost no chance of mating again in the future.

FIGURE 7.32   Sexual suicide in the redback spider. (A) The male first aligns himself facing forward on the underside of the female's abdomen while he inserts his sperm-transferring organs into her reproductive tract. (B) He then elevates his body and (C) flips over into the jaws of his partner. She may oblige by consuming him while sperm transfer takes place. After Forster.[73]

Female Mate Choice for Paternal Males

Although redbacks and some other spiders offer themselves up as nuptial gifts, males of other animals may care for the female's offspring, and in these species we would expect females to prefer to mate with males that provide more paternal care than average. In keeping with this expectation, females of the fifteen-spined stickleback, a small fish with nest-protecting males, associate more with males that shake their bodies relatively frequently when courting. Males that can behave in this way also perform more nest fanning after courtship is over and eggs have been laid in their nests (see Figure 9.4). Nest fanning sends oxygenated water flowing past the eggs, which increases gas exchange and, ultimately, egg hatching success.[142]

The female fifteen-spined stickleback that evaluates the courtship display of a male sees him behaving in ways that are linked to his parental capacities. Males of a related species, the three-spined stickleback, also appear to offer their females cues of paternal helpfulness—in the form of a colorful belly. As a rule, males with redder bellies are more attractive to potential mates.[171] The reddish pigment that colors a male's skin comes from the carotenoids he has consumed. Males with red bellies are for some reason able to fan their eggs for longer periods under low-oxygen conditions than paler males whose diets are short in carotenoids.[152] Thus, a male stickleback's appearance advertises his ability to supply oxygen for the eggs he will brood.

Could the carotenoid pigments in a male bird's plumage or bill also be an indicator of his capacity for paternal behavior? Certainly females of many animal species are especially attentive to the reds and yellows in male color patterns,[83,133] which might reveal something about the health of a male if, as has been argued, the quality of an individual's immune system is enhanced by a diet rich in carotenoids (or if only healthy males can use their dietary carotenoids to make bright feathers[94]). In either event, a healthy male might well be a better provider for his offspring.

When zebra finch males were provided with extra carotenoids in their food, they had more carotenoids in their blood, brighter red beaks, and stronger immune responses.[134] Moreover, female zebra finches found experimentally carotenoid-enhanced males more attractive than those that received normal diets.[23] Female finches even prefer males that have been given red leg bands over those with green bands; female zebra finches reward red-banded males by laying eggs with greater mass than those they produce for the less attractive males.[81] In nature, female zebra finches that acquired more brightly colored mates might benefit by investing more in their offspring because having a healthy mate might mean that he would provide superior care for these nestlings.

Much the same thing may be happening in the blue tit, another small songbird with a carotenoid-based plumage ornament, a bright yellow breast. Male blue tits collect and deliver food for their nestlings, usually in the form of carotenoid-containing caterpillars. If the amount or quality of food supplied by a male is related to how bright his yellow feathers are, then the offspring of brilliantly yellow males should be larger and healthier when they fledge than the youngsters of less brightly colored males. Indeed, the offspring of brighter parents are in better condition and have stronger immune systems than those of less yellow parents.[93]

But wait a minute. This same result might occur if bright yellow males were themselves large and healthy at fledging, thanks to their genetic makeup, in which case their offspring would simply inherit these traits from their parents. Because a team of Spanish behavioral ecologists recognized this problem, they wisely controlled for heredity through a cross-fostering experiment. They took complete clutches of eggs and transferred them between nests, moving the offspring of one set of parents to another pair's nest. The foster parents were willing to rear these genetic strangers, and when their adopted chicks had reached the age of independence, the size of the fledglings was a function of the brightness of their foster father's yellow plumage, not the color of their genetic father. Bright yellow foster males produced larger fledglings. If the foraging capabilities of bright yellow males really are greater than those of duller individuals, as they appear to be,[78] females could benefit by choosing their mates on the basis of the males' plumage.[176]

Discussion Question

7.15 Male barn swallows have thin outer tail feathers that are somewhat longer than those possessed by females. When Anders Møller analyzed the effect of tail length on male mating success in the barn swallow in Europe, he did an experiment in which he made some males' tail feathers shorter by cutting them and made other males' tail feathers longer by gluing feather sections onto their tails.[137] But he also created a group in which he cut off parts of the males' tail feathers and then simply glued the fragments back on to produce a tail of unchanged length. What was the point of this group? And why did he randomly assign his subjects to the shortened, lengthened, and unchanged tail groups? And why did a team of Canadian biologists repeat Møller's experiment on another continent?[185] And why did yet another team of British ornithologists study the effect of the tail "streamers" on the maneuverability of male swallows, given their interest in female mate choice?[29]

Female Mate Choice without Material Benefits

Some of the examples that we have just presented are consistent with **good parent theory**, which explains aspects of male color, ornamentation, and courtship behavior as sexually selected indicators of a male's capacity to provide parental care, a factor of obvious significance for female reproductive success. We have not, of course, presented every possible case in which choosy females gain material benefits from their choices. For example, in a Japanese damselfly, females apparently evaluate potential mates on the basis of cues linked to the male's body temperature. The thorax of the male is warmer in the damselfly whose streamside territory is exposed to sun, which not only enables these males to raise their thoracic temperatures but to beat their wings more rapidly during their courtship flights.[197] Because females prefer hotter males, their eggs are laid in relatively sunny, warmer places where they may develop more rapidly than in shady territories. If true, the female damselfly's choice of mates based on their temperatures could help her choose males that provide better care of their offspring—indirectly.

However, in the satin bowerbird, as well as other species, the males do not provide food or any other material benefit to their mates or their offspring. Even so, female bowerbirds prefer males with more ornaments (in their bowers) and the ability to court more intensely.[24,148] The same is true for many other animals.[110]

For example, a female canary's choice of a mate appears to be heavily influenced by his ability to sing a certain portion of the male song, the "A phrase," composed of many two-note syllables (Figure 7.33). Estradiol-primed females that hear an A-phrase trill that packs many syllables into a second of song

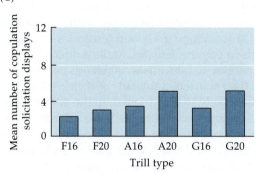

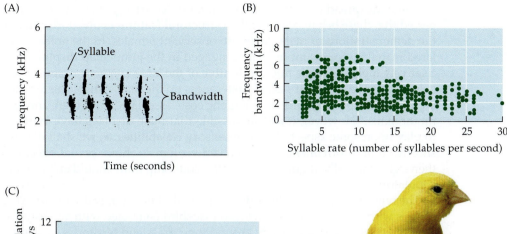

FIGURE 7.33 Mate choice based on a physiologically challenging task. (A) Male canaries sing songs that contain a special trill that is composed of a series of syllables, each composed of two notes that require the coordinated action of the syrinx and the respiratory system to produce. Trill syllables have frequency bandwidths that range from small (from 2 to 4 kHz, for example) to large (from 2 to 6 kHz, for example). The rate at which the syllables are produced also can vary. (B) There is an upper limit on the rate at which males can sing syllables of a given frequency bandwidth. (C) Females prefer trills that are composed of broad-bandwidth syllables sung at a very fast rate. Tapes of three trill types were played to females: F had the narrowest bandwidth, and G the greatest. Each trill type was played at two rates, 16 syllables per second and 20 syllables per second. Females responded with significantly more copulation solicitation displays to A20 and G20 trills, those with many syllables per second. After Drăgăniou et al.[61]

readily adopt the precopulatory position (because estradiol is the hormone that induces female sexual receptivity).[202]

Passing a female canary's song test requires that males not only generate a rapid trill but also make the individual syllables in the trill cover a relatively wide range of sound frequencies (the bandwidth of the trill). We know this because of the responses of females to tapes of artificial trills, including some that were impossibly exaggerated versions of the A phrase. The most extreme versions of the trill elicited the most copulation solicitation displays from listening female canaries. Because there is an upper limit for male canaries with respect to how rapidly they can sing syllables of a given bandwidth, the female preference in effect favors males able to sing at the upper limit of their ability.

Much the same is true for Alston's singing mouse. Here too mechanical constraints mean that the faster a male produces his trills, the lower the range of frequencies he can generate in each note. When researchers produced artificial songs with exaggerated trill rates and broad bandwidths and female mice were given a choice between approaching a speaker playing the "enhanced" song and another playing a "normal" song, the subjects preferred the enhanced song that males could never actually sing.[146]

Experiments of this sort indicate that the greater the sensory stimulation provided by a courting male, the more attractive he becomes to females. This point has been established for golden-collared manakins, small tropical birds with an extraordinarily complex courtship display whose maneuvers are carried out at high speed. The faster the male performs the components of his jump-snap display, the more likely he is to acquire mates; very small differences among males translate into large differences in male reproductive success[15] (see a video of the male's behavior at http://www.huffingtonpost.com/2011/06/07/golden-collared-manakin-dance-video-manacus-candei_n_872645.html). Indeed, throughout the animal kingdom, courtship displays by males are characterized by the vigorous performance of motor skills to potential partners.[37]

The widespread occurrence of demanding courtship displays has suggested that these behaviors and allied structural features evolved to enhance the demonstrations of male quality to observant females. In the golden-collared manakin, for example, males engaged in the jump-snap display have relatively high metabolic rates, indicative of the large energetic costs of their behavior. By basing their sexual preferences on the males' displays, females secure males able to afford the energetic costs associated with their unusual motor skills. (However, it is also possible that male displays have evolved to play upon some aspect of female sensory capabilities, as we shall discuss shortly.) If male ornaments have evolved via female choice sexual selection, then experimentally augmenting a male's courtship traits should enhance his copulatory success.

The relevant experiment was done first with the long-tailed widowbird, a species with the body of a red-winged blackbird combined with an absurdly long tail (Figure 7.34). Males fly around their grassland territories in Kenya, displaying their magnificent tails to passing females. Malte Andersson took advantage of the wonders of superglue to perform an ingenious experiment. He captured male widowbirds, then shortened the tails of some of them by removing a segment of tail feathers, only to glue them onto the tails of other males, thereby lengthening those ornaments.[8] The tail-lengthened males were far more attractive to females than those that had much-reduced ornaments. Moreover, the tail-lengthened males also did much better than controls, whose tails had been cut and then put back together without changing their length.

The elements of courtship that females may use to assess the quality of potential mates vary across the animal kingdom. Sexy songs and displays of extreme ornaments work for some species, but in others the male's performance during or even immediately after copulation might constitute a kind of test of male quality, an idea originally proposed by Bill Eberhard[65] (see also Hosken and Stockley[98]). The various tactile signals provided by male digger bees mounted on a female that they have already inseminated (see Figure 1.2) affect her readiness to mate again, an example of postcopulatory courtship. During copulation itself, females could evaluate partners on the basis of how they performed; if true, then in species whose females mate with several males

FIGURE 7.34   A sexually selected ornament. The extraordinary tail feathers of the long-tailed widowbird are displayed to choosy females while the male flies above his grassland territory.

(A)   (B)   (C)

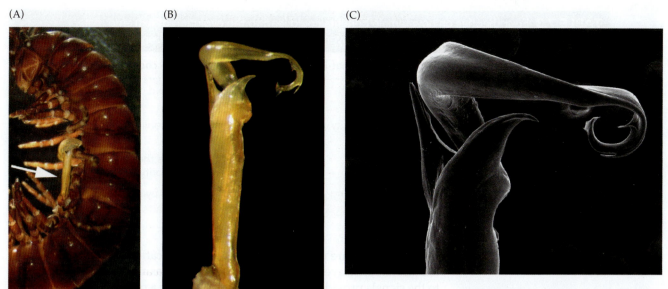

FIGURE 7.35 Cryptic female choice and male copulatory performance. The elaborate structure of the gonopod (sperm delivery organ) of this male millipede may provide his partner with copulatory stimuli that determine his success in fertilizing her eggs. (A) The location of the gonopod (arrow) on the body of the millipede. (B and C) The gonopod at close range, showing its remarkable architecture. Photographs courtesy of Janine Wojcieszek and Leigh Simmons.

in a breeding cycle, the mating success of a male may be linked to the structure of his sperm delivery organ or its use during mating, as seems to be the case for one species of millipede (Figure 7.35).[213] **Cryptic female choice** (choice hidden from view of human observers) has been demonstrated for male oriental beetles, which have a spiny hook on the aedeagus (penis); the larger the hook, the greater the proportion of eggs fertilized by the male following an initial copulation with a female that mates twice.[206]

Discussion Question

7.16 Peacocks and long-tailed widowbirds have truly extravagant ornaments that could be fairly recently evolved (that is, derived) from an ancestral pattern in which male plumage was not nearly so extreme. In the peacock-pheasants, six species in a genus related to that of the peafowl, there is considerable variation in the degree of male plumage ornamentation. In four species, males have highly elaborate plumage featuring large eyespots, but in two species they do not. Draw two phylogenies, one in which elaborate plumage is the derived trait and another in which a reduction in ornamentation is the derived condition. Then test your hypotheses with data available in Kimball et al.[107]

DOES MALE COURTSHIP HELP FEMALES IDENTIFY QUALITY MATES? No matter what the basis for female preferences for one male's courtship or copulatory skill over another, a key question is, does female choosiness translate into fitness gains *for females*? After all, males rejected by females sometimes turn nasty; sexual conflict often arises over such matters,[145] as we noted when discussing infanticide by langurs.

Now, it could be that females generally do not benefit by being choosy but instead are being manipulated by showy males, which gain if they are persuasive, while the females that select them actually lose fitness as a result. This possibility is rendered less likely if we apply the thinking that has convinced most behavioral ecologists that communication must benefit signal receivers as well signal producers if it is to persist (see page 67). If courtship and copu-

TABLE 7.3 Four theories on how female benefits could lead to the evolution of extreme male ornamentation and striking courtship displays even if males provide no material benefits to their mates

Theory	Females prefer trait that is:	Primary adaptive value to choosy females
Healthy mates	Indicative of male health	Females (and offspring) avoid contagious diseases and parasites
Good genes	Indicative of male viability	Offspring may inherit the viability advantages of their father
Runaway selection	Sexually attractive	Sons inherit trait that makes them sexually attractive; daughters inherit the majority mate preference
Chase-away selection	Exploitative of preexisting sensory biases	Females derive no benefits by being choosy

latory behavior are part of an evolved communication system, then choosy females should leave more surviving descendants as a result of their responses to the signals provided by courting males. Here is a prediction worth testing.

One of several elements of male quality that could affect female fitness is the health of a sexual partner (Table 7.3). Unhealthy males could give their mates any of a variety of unpleasant sexually transmitted diseases, making it advantageous for females to avoid them. According to the **healthy mates theory**, female preferences are focused on a potential sexual partner's health or parasite load as indicated by his courtship displays and appearance.[166] Females could use these traits to mate with males that are less likely to carry lice, mites, fleas, or bacterial pathogens, any one of which could harm the females or their future offspring. As we noted earlier, male bowerbirds with high-quality bowers are less likely to carry transmissible ectoparasites in their feathers.[60]

Of course, females that mate with healthy males may not only steer clear of contagious parasites and diseases, but they may also acquire sperm that supply the genetic basis for good health in their offspring. The **good genes theory** proposes that preferences for certain male ornaments and courtship displays enable females to choose partners whose genes will help their offspring develop physiological mechanisms to combat infection and disease. In some species, for example, females might be able to evaluate (unconsciously) the strength of a male's immune system by his courtship displays. One possible example involves the cricket *Teleogryllus oceanicus*, whose females prefer to approach artificial male songs that have been manipulated to sound like those sung by males with strong immune systems, as opposed to songs that sound like those sung by males with weak immune systems.[194]

Discussion Question

7.17   Males of the Houbara bustard (Figure 7.36) differ in the time spent displaying their remarkable ornamental feathers to females. The more time a male spends in displaying during the early years of adulthood, the faster the quality of male sperm declines over the bird's lifetime.[159] What does this finding say about the cost of sexually selected traits for males? What is puzzling about the discovery that males with high rates of display early in adulthood continue displaying at a relatively high rate later in life?

In other species, a male's appearance could be correlated with his hereditary resistance to parasites, a valuable attribute to pass on to offspring. Indeed W. D. Hamilton and Marlene Zuk predicted that selection for honest signals

FIGURE 7.36 **A male Houbara bustard,** a species in which producing mating displays eventually affects male sperm quality.

of noninfection would lead bird species with numerous potential parasites to evolve strikingly colored plumage. They argued that brightly colored feathers are difficult to produce and maintain when a bird is parasitized, because parasitic infection causes physiological stress. Hamilton and Zuk found the predicted correlation between plumage brightness and the incidence of blood parasites in a large sample of bird species, supporting the view that males at special risk of parasitic infection engage in a competition that signals their condition to choosy females.[90]

In addition, "good genes" derived from males could be involved in the development of other fitness-advancing traits besides resistance to parasites and diseases. For example, if females had a way to avoid genetically similar males or to identify males with high levels of heterozygosity, these choosy females might well help their offspring avoid the developmental problems that can occur when individuals have two copies of damaging recessive alleles, a common result of inbreeding. Female preferences for males with relatively heterozygous genomes have been documented for a number of bird species (e.g., house sparrows,[84] sedge warblers,[129] and superb starlings [see page 236]).

Runaway Sexual Selection

A quite different view of what constitutes "good genes" is embodied in **runaway selection** theory.[12,53,162] This theory is based on the fact that female choice creates a genetic link between mate choice by females and the male trait and, because of this correlation, female choice then begins to select on itself, leading to the evolution of preferences for ever more extreme traits. As a result, a sexual preference for the elaborate song of male canaries could be adaptive for females if their sons inherit the capacity to sing attractive songs, even if they are costly to produce, because these songsters may be especially appealing to females in the next generation.

Because the runaway selection alternative is the least intuitively obvious explanation for extreme male courtship displays, let us sketch the argument underlying the mathematical models of Russell Lande[115] and Mark Kirkpatrick.[108] Imagine that a slight majority of the females in an ancestral population had a preference for a certain male characteristic, perhaps initially because the preferred trait was indicative of some survival advantage enjoyed by the male. Females that mated with preferred males would have produced offspring that inherited the genes for the mate preference from their mothers and the genes for the attractive male character from their fathers. Sons that expressed the preferred trait would have enjoyed higher fitness, in part simply because they possessed the key cues that females found attractive. In addition, daughters that responded positively to those male cues would have gained by producing sexy sons with the trait that many females liked.

Thus, female mate choice genes as well as genes for the preferred male attribute could be inherited together. This pattern could generate a runaway process in which ever more extreme female preferences and male attributes spread together as new mutations affecting these traits occurred. The runaway process would end only when natural selection against costly or risky male traits balanced sexual selection in favor of traits that appealed to females. Thus, if bowerbird females originally preferred males with well-decorated bowers because such males could forage efficiently, they might later favor males with extraordinary bowers because this mating preference had taken on a life of its own, resulting in the production of sons that were exceptionally attractive to females and the production of daughters that would choose these extraordinary males for their own mates. Note that the original connection between female mate choice and male foraging skill or any other useful trait would have disappeared.

In fact, the Lande–Kirkpatrick models demonstrate that, right from the start of the process, female preferences need not be directed at male traits that are utilitarian in the sense of improving survival, feeding ability, and the like. Any preexisting preference of females for certain kinds of sensory stimulation (see page 70) could conceivably get the process under way. As a result, traits opposed by natural selection because they reduced viability could still spread through the population by runaway selection.[108,115] Instead of mate choice based on genes that promote the development of useful characteristics in offspring, runaway selection could yield mate choice for arbitrary characters that are a burden to individuals in terms of survival, a disadvantage in every sense except that females mate preferentially with males that have them!

Discussion Questions

7.18 Many have felt that the sails of extinct reptiles in the genus *Dimetrodon* (Figure 7.37) evolved to dissipate unwanted heat. Produce another hypothesis based on runaway sexual selection. Read Tomkins et al.[193] to see how to test this kind of sexual selectionist explanation.

7.19 Richard Prum argues that researchers need a "null model" of the effects of sexual selection in order to determine whether in fact mate choices lead to adaptive outcomes, with female preferences for certain male attributes providing the choosy females with better genes or superior parents.[162] For Prum, the Lande–Kirkpatrick models provide the null in which preferences have no utilitarian adaptive effect but instead are the arbitrary products of a nonselectionist process. He argues that we need to test the predictions from the null model first, rather than focusing on predictions from adaptationist hypotheses about such things as signal honesty and fitness benefits of female preferences. Why have so many researchers employed the adaptationist approach? Do we also need a null hypothesis for behavioral traits that adaptationists have assumed (for the purposes of hypothesis development and testing) promote survival by defeating predators?

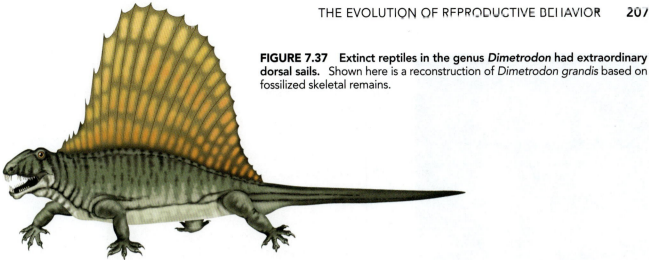

FIGURE 7.37 Extinct reptiles in the genus *Dimetrodon* had extraordinary dorsal sails. Shown here is a reconstruction of *Dimetrodon grandis* based on fossilized skeletal remains.

Testing Competing Ideas about Mate Choice

Discriminating among these three alternative explanations for elaborate male courtship displays and female choice—the healthy mate, good genes, and runaway selection theories—has proved very difficult, in part because hypotheses derived from them are not mutually exclusive. As just mentioned, female preferences and male traits that originated through a good genes process could then be caught up by runaway selection. Note, too, that males with hereditary resistance to certain parasites (good genes benefits) would also be less likely to infect their partners with those parasites (healthy mate benefits). Moreover, if at the end of a period of runaway selection, males had evolved extreme ornaments and elaborate displays, then only individuals in excellent physiological condition would be able to develop, maintain, and deploy their ornaments in effective displays. Males in such superb physiological condition would probably have to be highly effective foragers (good genes benefits) as well as parasite-free (healthy mate benefits), in which case females mating with such males would be unlikely to acquire parasites, and their daughters might well get some survival-benefiting genes while their sons received the pure attractiveness genes of their fathers.

Given the overlap among these three ideas, let's just pick one and try to test it. As an example, Marion Petrie applied good genes theory to peacocks and derived the following predictions: (1) males should differ genetically in ways related to their survival chances, (2) male behavior and ornamentation should provide accurate information on the survival value of the males' genes, (3) females should use this information to select mates, and (4) the offspring of the chosen males should benefit from their mothers' mate choice.[150,151] In other words, males should signal their genetic quality in an accurate manner, and females should pay attention to those signals because their offspring would derive hereditary benefits as a result.[109,215]

Petrie studied a captive but free-ranging population of peacocks in a large forested English park, where she found that males killed by foxes had significantly shorter tails than their surviving companions. Moreover, she observed that most of the males taken by predators had not mated in previous mating seasons, suggesting that females could discriminate between males with high and low survival potential.[150] The peahens' preferences translated into offspring with superior survival chances, as Petrie showed in a controlled breeding experiment. She took a series of males with different degrees of ornamentation from the park and paired each of them in a large cage with four females chosen at random from the population. The young of all the males were reared under identical conditions, weighed at intervals, and then eventually released back into the park. The sons and daughters of males with larger

Peacock

FIGURE 7.38 Do male ornaments signal good genes? Peacocks with larger eyespots on their tails produced offspring that survived better when released from captivity into an English woodland park. After Petrie.[151]

eyespots on their ornamented tails weighed more at day 84 of life and were more likely to be alive after 2 years in the park than the progeny of males with smaller eyespots (Figure 7.38).[151]

Similar studies have been conducted on the role of eyespot *number* as opposed to eyespot size but with conflicting results.[48,90,122] But to the extent that Petrie's earlier studies hold up, we may tentatively conclude that peahens appear to be choosing males on the basis of criteria that are tied to offspring survival. If so, female mate choice in this species could be of the good genes variety. But as noted above, perhaps healthy mates benefits are involved as well if, for example, peahen mate choice helps them avoid parasitized males and so reduce their risk of acquiring nasty parasites that they would pass on to their offspring. Moreover, a demonstration of current selective advantages associated with female preferences and male traits does not rule out the possibility that these attributes originated as side effects of runaway selection in the manner described above.

Yet another complication comes from the discovery that females of some species, like zebra finches,[81] invest more resources in offspring that are sired by preferred males than in those fathered by less attractive males. In some species, the female adjusts the amount of testosterone she contributes to an egg in relation to the attractiveness of her mate.[80,123,175] Likewise, female mallards make larger eggs after copulating with unusually attractive males,[47] while female black grouse produce and lay more eggs subsequent to mating with top-ranked males.[169] All of these effects might easily be attributed to the "good genes" of the male, when in reality they arise from the female's manipulation of her own parental investment, not the genetic contribution of her partner. Untangling the contributions of the two parents to the welfare of their offspring is challenging business.

Sexual Conflict

Although cooperation between the sexes is often the case in sexually reproducing species, particularly in those in which both males and females rear the young together, sexual conflict also appears to be common.[145] Consider that

(A)

(B)

FIGURE 7.39 Rejection behavior by females. (A) A female cabbage butterfly lifts her abdomen up and away from a male trying to copulate with her, a mate refusal behavior studied by Obara et al.[141] (B) A female cane toad inflates her body to make it difficult for a male to grasp her properly so that he can fertilize her eggs when she releases them. A, photograph by Y. Obara; B, photograph by Crystal Kelehear, courtesy of Rick Shine.

females of many species often turn down sexually motivated males (Figure 7.39), as seen in the typical response of female bowerbirds to displaying males.[34]

Much less frequently, males may reject potential sexual partners, as happens, for example, in the African topi. Choosy male topi occupy the popular central positions in their leks (see page 248), where they can become sperm depleted after copulating with one female after another. Central males therefore may refuse to mate again with females they have already inseminated, saving sperm to donate to new partners. A jilted female sometimes responds by attacking the male and disrupting his efforts to mount a newcomer (Figure 7.40).[28]

Sexual conflict can become even more unpleasant when, for example, males kill the infants of females to cause them to become sexually receptive again (see page 9) or to force apparently unwilling females to mate. In these cases, it is difficult (but not impossible) to believe that the male's behavior is somehow

FIGURE 7.40 Sexual conflict. Male topi that are sperm depleted may refuse to mate with some females. In response, a rejected female may attack the male. Photograph by Jakob Bro-Jørgensen.

FIGURE 7.41 Forced copulation in a bird. Males of the New Zealand stitchbird sometimes assault females on neighboring territories and force them to the ground, where the male transfers sperm to his "partner" while she is on her back. Note the tail of the pinned female in the lower left. Photograph by Matt Low.

reproductively advantageous to the female. When a male hangingfly grabs a female by a wing and mates with her without providing a food gift, she loses a meal.[192] When a female wolf spider is pushed to the ground and bitten by a coercive male, she may lose some of her hemolymph from the wound.[101] When a male stitchbird chases a female paired to another male and eventually drives her into the ground for a face-to-face copulation (Figure 7.41), a most unusual pattern in birds, the female has at the very least expended time and energy for what certainly looks like an unnecessary copulation.[120] Likewise, when orangutan females finally copulate with the sexually eager young males that have been harassing them for days on end, it appears that those females are mating against their will. Given a choice, female orangutans seek out and copulate with huge, older adult males that protect them from those annoying smaller males.[74] Similar preferences are exhibited by female chimpanzees, which do not synchronize their estrous cycles with other band members, perhaps to ensure that at least one dominant male will be available to guard them against sexual harassers of lower rank.[130] Nonetheless, dominant and subordinate male chimpanzees alike often sexually assault fertile females.[139] Some violent copulations are fatal for females, as is true for our own species as well.[33]

Discussion Question

7.20 When forced copulation results in the death of the female, neither sex benefits. Explain why males may behave in ways that result in the demise of their sexual partners.

CHASE-AWAY SELECTION   Sexual conflict plays a central role in a fourth general theory for why males have evolved extreme ornaments and elaborate courtship displays (see Table 7.3).[96] According to Brett Holland and Bill Rice, these traits could be the result of **chase-away selection**, a process that begins when a male happens to have a mutation for a novel display trait that manages to tap into a preexisting sensory bias (see page 70 on sensory exploitation) that affects female mate preferences in his species. Such a male might induce individuals to mate with him even though he might not provide the material or genetic benefits offered by other males of his species.[172] The resulting spread over time of the traits of such truly exploitative males would create selection on females favoring those that were psychologically resistant to the purely attractive display trait. As females with a higher threshold for sexual responsiveness

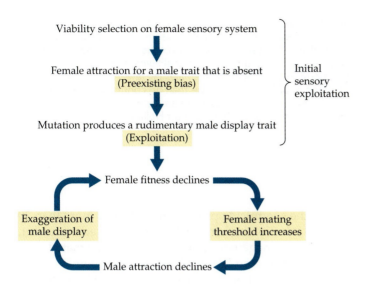

FIGURE 7.42 Chase-away selection theory. The evolution of extreme male ornaments and displays may originate with exploitation of females' preexisting sensory biases. If sensory exploitation by males reduces female fitness, the stage is set for a cycle in which increased female resistance to male displays leads to ever greater exaggeration of those displays. After Holland and Rice.[96]

to the exploitative trait spread, selection would then favor males able to overcome female resistance, which might be achieved by mutations that further exaggerated the original male signal. A cycle of increasing female resistance to, and increasing male exaggeration of, key characteristics could ensue, leading gradually to the evolution of costly ornaments of no real value to the female and useful to the male only because without them, he would have no chance of stimulating females to mate with him (Figure 7.42). According to this theory, females gain neither material nor genetic benefits by mating with males with exceptionally stimulating courtship moves or stunning ornaments.

Chase-away selection theory illustrates how far we have come from the once popular view of sexual reproduction as a gloriously cooperative enterprise designed to perpetuate the species or to maximize the fitness of both participants. Instead, many behavioral biologists now see reproduction as an activity in which the two sexes battle for maximum genetic advantage, even if one member of a pair loses fitness as a result.

For example, males of many insects transfer chemicals along with their sperm that increase the delay in remating by their partners.[95] Although receiving these chemicals may be advantageous to females in some instances, in other cases females are damaged by male seminal fluid. When males of the fruit fly *Drosophila melanogaster* copulate, they transfer to their mates a protein (Acp62F)[124] that boosts male fertilization success (perhaps by damaging rival sperm) at the expense of females, whose lives are shortened, whose sleep patterns are disrupted,[106] and whose fecundity is lowered.[154] Despite the negative long-term effects that toxic protein donors have on their mates, males still gain because they are unlikely to mate with the same female twice. Under these circumstances, a male that fertilizes a larger proportion of one female's current clutch of eggs can derive fitness benefits even though his chemical donations reduce the lifetime reproductive success of his partner.

If male fruit flies really do harm their mates as a consequence of their success in sperm competition, then placing generations of males in laboratory environments in which no male gets to mate with more than one female should result in selection against donors of damaging seminal fluid. Under these conditions, any male that happened not to poison his mate would reap a fitness benefit by maximizing his single partner's reproductive output. In turn, a reduction in the toxicity of male ejaculates should result in selection for females that lack the chemical counteradaptations to combat the nega-

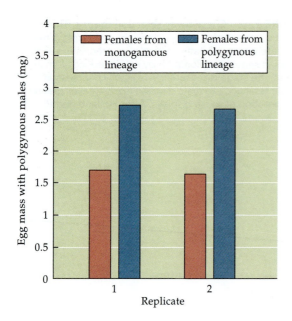

FIGURE 7.43 **Sexual selection and the evolution of male traits harmful to females.** Females from an experimental monogamous lineage of fruit flies have lost much of their biochemical resistance to the damaging chemicals present in the seminal fluids of polygynous male fruit flies. Therefore, monogamous females lay fewer eggs when mated with males from a polygynous lineage than do control females that evolved with those males. The results of an experiment repeated twice are shown here as replicates 1 and 2. After Holland and Rice.[97]

tive effects of the spermicide protein. Indeed, after more than 30 generations of selection in a one male–one female environment, females from the monogamous population that were mated once with spermicide-donating "control" males taken from a typical multiple-mating population laid fewer eggs and died earlier than females that had evolved with polygynous, spermicide-donating males (Figure 7.43).[97]

Discussion Question

7.21   Stuart Wigby and Tracey Chapman formed three populations of fruit flies with different sex ratios (female biased, even sex ratio, and male biased). Not surprisingly, the frequency with which females mated increased from female-biased to male-biased populations. After 18 and 22 generations of selection, fresh females from the three selected lines were taken from their environments and placed in cages with equal numbers of males. The mortality rate of females from the male-biased line was less than that of females from the even-sex-ratio line, and much less than that of females from the female-biased line.[210] What do these results tell us about the evolutionary consequences of sexual conflict between the sexes in this species?

In the light of chase-away selection theory, it is revealing that female fruit flies actually reduce their fitness by preferring to mate with larger males, which they choose either because they find large body size an attractive feature in and of itself or because large body size is correlated with some other attractive feature, such as more persistent courtship. Whatever the reason for their preference, mate choice by females based on this characteristic lowers their longevity and also reduces the survival of their offspring.[75] These negative effects on female fitness may arise from the physiological costs of dealing with increased rates of courtship by "preferred" mates and possibly the increased quantities of toxins received from "attractive" males.[154] These findings can be taken as evidence that, for the moment, large males are ahead in a chase-away arms race between the sexes in the fruit fly.

Another kind of arms race may be taking place in bedbugs[164] and some other creatures[99,168] that employ traumatic insemination or its equivalent. In the bedbug, this method of sperm transfer takes place when the male uses his knifelike intromittent organ to stab a female in her abdomen before injecting his sperm directly into her circulatory system (Figure 7.44).[188] This odd method of insemination presumably originated as a result of males injecting their sperm into sexually resistant females that had already acquired sperm from earlier partners via the traditional, less damaging mode of insemination. If so, then traumatic insemination ought to be costly to females, and it is, given

(A)

(B)

FIGURE 7.44 A product of conflict between the sexes? (A) Male bedbugs have evolved a saber-like penis that they insert directly into the abdomen of a mate prior to injecting her with sperm. The trait may have originated when males benefited by employing traumatic insemination to overcome resistance to mating by unwilling females. (B) The white box shows the site on the female's abdomen that males penetrate with the penis. A, photograph by Andrew Syred; B, photograph by Mike Siva-Jothy.

that females mating at high frequencies live fewer days and lay fewer eggs.[188] On the other hand, females that mated for a minute with three males per week laid eggs at a higher rate for a longer period than females that were permitted just one 60-second copulation per week.[165] Something in the male ejaculate appears to boost female fitness in bedbugs, at least for those that experience traumatic insemination in moderation.

The bedbug case illustrates the complexity of interactions between males and females. But let us accept that in at least some instances, there are mating costs to females, such as a reduction in longevity or the developmental damage done to a female's daughters as a result of having genes that are beneficial only when expressed in males.[153] But these and other costs could be outweighed by the exceptional reproductive success of a female's adult sons.[145] If so, the females that permitted the fathers of these sons to mate with them might be more than compensated by the increased reproduction of their male offspring. Females might even accept a reduction in their lifetime reproductive success if they produced sons that inherited effective manipulative, or even coercive, tactics from their fathers—tactics that might make their sons unusually successful reproducers, thereby endowing their mothers with extra grandoffspring.[41] Apparent physical conflict among males and females may even be the way that females judge the capacity of males to supply them with sons that will also be able to overcome the resistance of females in the next generation.[66] Alternatively, aggressive court- ship by males may be the way in which females choose males whose sons will do well in aggressive competition with other males.[27] These kinds of hypotheses suggest that females may be "winning by losing" when they mate with males that appear to be forcing them to copulate or to be blocking their apparent preference for other males.

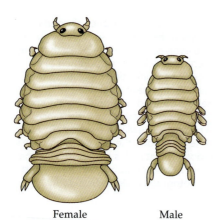

Female   Male

FIGURE 7.45 A mutually cannibalistic species: the ultimate in sexual conflict. Either the larger female or the smaller male of the freshwater isopod *Ichthyoxenus fushanensis* may kill and consume its partner. Drawing courtesy of C.-F. Dai.

Discussion Question

7.22   Males of the parasitic isopod *Ichthyoxenus fushanensis* (Figure 7.45) pair off with females in cavities they construct inside their victims, freshwater fish. At times, sexual cannibalism occurs early in the breeding season, when females eat their partners; males may eat their mates later in the season.[196] Replacement partners of both sexes are readily available. In addition, males can transform themselves into females, a phenomenon that occurs only when males have achieved a fairly large size. The larger the female, the more fecund she is. The smaller the difference in size between male and female, the fewer the number of offspring produced by a pair. Why has conflict between the sexes reached such an extreme state in this species? What kinds of males are females expected to eat? What kinds of males are expected to eat their partners? How can you account for the difference in the timing of cannibalism by males and females?

Summary

1. Darwin's sexual selection theory explains why certain traits can spread through populations even though these attributes appear to lower individual survival, a result that would seem disfavored by natural selection. But sexual selection can overcome natural selection for improved survival if the sexually selected characteristics promote success in securing mates.

2. Typically males compete with each other for mates while trying to convince females to mate with them. These standard sex roles arise because males usually make huge numbers of very small gametes and often try to fertilize as many eggs as possible, while providing little or no care for their offspring. In contrast, females make fewer, larger gametes and often provide parental care as well. As a result, receptive females are scarce, becoming the focus of male competition while ensuring that females have many potential partners to choose among.

3. The competition-for-mates component of sexual selection has led to the evolution of many characteristic male reproductive behaviors, including a readiness to fight over females or for social dominance. In addition, this form of sexual selection has resulted in the evolution of mating tactics that enable some males to make the best of bad situations created by their rivals as well as the ability of males to give their sperm an edge in the competition for egg fertilizations, especially by guarding their mates for some time after copulation.

4. In a typical species, females exercise mate choice because they control egg production, egg fertilization, and the care of offspring. Males of some species seek to win favor with females by offering them material benefits, including nuptial gifts or parental care. In these cases, males with better gifts or superior paternal capabilities often produce more descendants.

5. Mate choice by females occurs even in some species in which males provide no material benefits of any sort. Female preferences for elaborate ornaments may arise because males with these attributes are healthy and parasite-free, and so less likely to transmit disease or parasites to them (healthy mates theory). Or choosy females may gain by securing genes that enhance the viability of their offspring (good genes theory). On the other hand, extravagant male features could spread through a population in which even arbitrary elements of male appearance or behavior became the basis for female preferences. Exaggerated variants of these elements could be selected strictly because females preferred to mate with individuals that had them (runaway selection theory). A fourth possibility is that the extreme ornaments of males evolve as a result of a cycle of conflict between the sexes, with males selected for ever-improved ability to exploit female perceptual systems and females selected to resist those males ever more resolutely (chase-away selection theory). The relative importance of these possible mechanisms of sexual selection remains to be determined.

6. Interactions between the sexes can be viewed as a mix of cooperation and conflict as males seek to win fertilizations in a game whose rules are set by the reproductive mechanisms of females. Conflict between the sexes is widespread and includes sexual harassment and the transfer of damaging ejaculates from males to females, demonstrating that what is adaptive for one sex may be harmful to the other.

Suggested Reading

Charles Darwin introduced the theory of sexual selection in *The Descent of Man and Selection in Relation to Sex*,[50] a book that is still relevant today. For a modern review of all aspects of sexual selection, read Malte Andersson's comprehensive book on the subject.[9] The male–male competition component of sexual selection includes sperm competition; the classic paper on this subject was written by Geoff Parker[144] (see also Simmons[181] and Birkhead and Møller[19]). In addition, Tim Birkhead has written an enjoyable popular account of sperm competition and its consequences.[20] The female choice component of sexual selection is at the heart of several of Bill Eberhard's books,[64,65] which explore cryptic mate choice by females, among other things. The June 2009 issue (volume 106; supplement 1) of the *Proceedings of the National Academy of Sciences* presents many papers that look at various modern developments in sexual selection theory.

8

The Evolution of Mating Systems

 Male satin bowerbirds, as we saw in the previous chapter, are capable of copulating with many females in a single breeding season, although they rarely have the good fortune to do so.[14] The successful males are **polygynous**, and polygyny applies also to males of the digger bee that locate and mate with several females, sometimes in a single morning. In contrast, female satin bowerbirds (and female digger bees) are almost always **monogamous**, typically mating with just one male per nesting attempt. But the combination of potentially polygynous males interacting with monogamous females is only one of a variety of arrangements found in the animal kingdom. In fact, different mating systems can be found even among the bowerbirds. For example, both males and females of the monogamous green catbird, a close relative of the polygynous satin bowerbird, pair off one by one for a given breeding season before rearing offspring together.[59] In some other birds, such as the spotted sandpiper, **polyandry** is the order of the day: females copulate with two or three males in a breeding season,[105] sometimes using the sperm of more than one male to fertilize the eggs that they lay during this period. A still more extreme version of this pattern is exhibited by the honey bee, whose young queens fly out from their hives into aerial swarms of drones that pursue, capture, and mate with them in midair. The average queen is highly polyandrous, coupling with many males and using the sperm of perhaps a dozen or so males during her lifetime of egg laying.[140] In contrast, drones never mate with more than one queen, because a drone violently propels his genitalia into his first and only mate, a suicidal act that ensures that he is both monogamous and, shortly thereafter, dead.[164]

This ruff, a male shorebird, participates in an unusual mating system in which males display their extraordinary plumage to females that visit groups of performing males before selecting one as a mate.

The diversity of mating systems offers a rich banquet of Darwinian puzzles for the evolutionist, including three that we address here: (1) why are males ever voluntarily monogamous, (2) why do females of some species practice polyandry, and (3) why do males of different species differ in the tactics they use to achieve polygyny? These puzzles are the central focus of this chapter.

Is Male Monogamy Adaptive?

Although examples of monogamy by males are not common, they do occur, as we have just indicated. But why should a male honey bee (Figure 8.1) or a monogamous male prairie vole or the male of any other species restrict himself to a single mate? The standard rule, which was first discovered in a now classic study of *Drosophila* fruit flies,[11] is that the more females inseminated, the more eggs fertilized and thus the greater the reproductive success of a male (see page 176). As a result, males typically compete for mates, and the winners are polygynists that are likely to gain higher fitness than their monogamous rivals. For example, male house wrens that attract two females father about nine fledglings per year on average, whereas monogamous male wrens have fewer than six.[137] Viewed in these terms, males that voluntarily restrict themselves to one sexual partner are a mystery.

In the previous chapter we discussed mate guarding by males (see page 194), which results in monogamy for certain spiders—the male gives its all, quite literally, to its first mate. Those males that commit sexual suicide by breaking off their genital appendages in a partner's reproductive tract or by feeding themselves to a partner[3,97] may increase their fertilization success with that one female. This gain from what might be called the male's posthumous mate guarding may exceed the cost of his actions if both (1) his mate has the potential to remain receptive after one mating and (2) the male's probability of finding a second female is extremely low. These conditions appear to be met in some spiders with male monogamy[4,97] and in some other species as well.[51]

Mate guarding also helps explain why males of the beautiful clown shrimp, *Hymenocera picta* (Figure 8.2), spend weeks in the company of one female.[159] As expected, the ratio of searching males to receptive females is highly male-biased since females are widely dispersed and will mate only during a short period every 3 weeks or so. Therefore, a male that encounters a potential mate guards her closely until she is willing to copulate.

FIGURE 8.1 The monogamous honey bee drone dies after mating. (A) An intact male. (B) A queen with the yellow genitalia of a deceased partner attached to the tip of her abdomen. Photographs by Christal Rau, courtesy of Nikolaus Koeniger.

(A)

(B)

FIGURE 8.2 A monogamous mate-guarding shrimp. When a male clown shrimp encounters a potential mate, he remains with her because receptive females are scarce and widely distributed. Here, a couple feeds on the severed arm of a starfish. Photograph by Stephen Childs.

A different explanation for male monogamy has been labeled the **mate-assistance hypothesis**, which proposes that a male remains with a single female in order to help her in various ways but generally because paternal care and protection of offspring are especially advantageous.[51] Of course, guarding a mate does not necessarily preclude helping her, and in fact, a study of crickets in a Spanish field (with video cameras monitoring each burrow) showed that guarded females were much less likely to fall victim to predators, mostly birds, than unguarded ones. Why? Because when a predator approached, the male permitted his mate to dash into the burrow first, a chivalrous and self-sacrificing act that increased his odds of being killed (Figure 8.3).[119] But

FIGURE 8.3 Mate guarding and mate assistance in the cricket, *Gryllus campestris*. (A) A male that has acquired a mate remains with her in his territory. If the pair is attacked, the male permits the female to rush into the burrow first, at great cost to his personal safety. (B) Solitary territorial males are much more likely to survive an attack. A female that lives alone is *much less* likely to survive than one paired with a self-sacrificing partner. A, photograph by Rolando Rodríquez-Muñoz, courtesy of Tom Treqenza; B, after Rodríquez-Muñoz et al.[119]

(A)

(B)

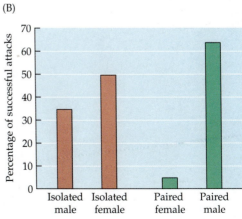

**FIGURE 8.4   Mate-assistance mono-
gamy in a seahorse.** A pregnant male
is giving birth to his single partner's
offspring. Two youngsters can be seen
emerging from their father's brood
pouch.

because the male had copulated frequently with the female, most of the offspring that she produced carried his genes.

In the case of the crickets and other species in which males make sacrifices to help their mates, the additional offspring that are produced or survive
because of the male's efforts more than compensate the male for giving up
chances to reproduce with other females. Readers will be most familiar with
male parental care in birds, but the phenomenon occurs in other groups as
well. Males of the seahorse *Hippocampus whitei*, for example, even take on
the responsibility of "pregnancy," carrying a clutch of eggs in a sealed brood
pouch for about 3 weeks (Figure 8.4). Each male has a durable relationship
with one female, which provides him with a series of clutches. Pairs even greet
one another each morning before moving apart to forage separately; they will
ignore any others of the opposite sex they happen to meet during the day.[147]
Since a male's brood pouch can accommodate only one clutch of eggs, he gains
nothing by courting more than one female at a time. And, in fact, in another
species of *Hippocampus*, genetic data indicate that males do not accept eggs
from more than one female (even though in this species, groups of females are
sometimes seen courting single males).[161]

A male seahorse may not benefit by switching mates if his long-term mate
can supply him with a new complement of eggs as soon as one pregnancy is
over. Many females apparently do keep their partners pregnant throughout
the lengthy breeding season but cannot produce batches of eggs so quickly
that they have some to give to other males.[148] By pairing off with one female,
a male might be able to match his reproductive cycle with that of his partner. When the two individuals were in sync, the male would complete one
round of brood care just as his partner had prepared a new clutch. Therefore,
he would not need to spend time waiting or searching for alternative mates
but could immediately secure a new clutch from his familiar mate. If true,
the experimental removal of a female partner should lengthen the interval
between spawning for the male, which it does. Males of a paternal pipefish
that had to change mates took 8 days extra to acquire replacement partners
with clutches of mature eggs, compared with males permitted to retain their
mates from one clutch to the next.[136]

In other species, although males might gain by acquiring several mates, a
female may block her partner's polygynous intentions in order to monopolize
his parental assistance, leading to **female-enforced monogamy**. This hypothesis is similar to the mate-guarding hypothesis, but with the female as the
guardian of monogamy in this instance. So, for example, paired females of
the burying beetle *Nicrophorus defodiens* are aggressive toward intruders of
their own sex. In this species, a mated male and female work together to bury
a dead mouse or shrew, which will feed their offspring once they hatch from
the eggs the female lays on the carcass. But once the carcass is buried, the
male may climb onto an elevated perch and release a sex pheromone to call
a second female to the site. If another female added her clutch of eggs to the
carcass, her larvae would compete for food with the first female's offspring,
reducing their survival or growth rate. Thus, when the paired female smells
her mate's pheromone, she hurries over to push him from his perch. These
attacks reduce his ability to signal, as Anne-Katrin Eggert and Scott Sakaluk
showed by tethering paired females so that they could not suppress their
partners' sexual signaling.[49] Freed from controlling spouses, the experimental
males released scent for much longer periods than control males that had to
cope with untethered mates (Figure 8.5). Thus, male burying beetles may often
be monogamous not because it is in their genetic interest but because females
make it happen—another example of the kinds of conflicts that occur between
the sexes (see Chapter 7).

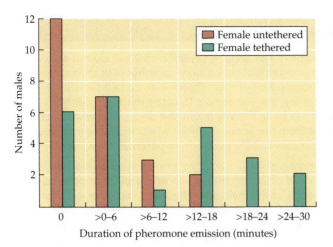

Burying beetle

FIGURE 8.5 Female-enforced monogamy in burying beetles. When a paired female beetle is experimentally tethered so that she cannot interact with her partner, the amount of time he spends releasing sex pheromones rises dramatically. After Eggert and Sakaluk;[49] photograph courtesy of C. F. Rick Williams.

Likewise, hostility toward other females by a dominant female emerald coral goby may force her partner to be monogamous even though in this species a number of females live with a single male in their coral retreat. The large dominant female fish in the group suppresses reproduction by the subordinate members, which accept their nonreproductive status because they may die if chased from their safe coral shelter. If they live long enough, they may eventually assume dominant status themselves. When they do, they will prevent other group members from breeding and will also attack intruder females, especially large mature females, which they usually force to leave the area (Figure 8.6).[162] By keeping mature rivals from her home base, a female coral goby keeps her male partner all to herself.

Discussion Question

8.1 We began this chapter with a mention of male monogamy in the honey bee. Try to explain the male's suicidal mating behavior in light of the alternative hypotheses on male monogamy outlined above. Also include in your list one hypothesis based on group selection theory. What predictions follow from the different explanations you have considered? What data are required to resolve the issue?

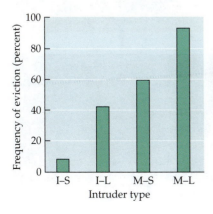

Emerald coral goby

FIGURE 8.6 Female-enforced monogamy may involve aggression by a breeding female toward other females. Breeding females of the emerald coral goby evict female intruders that are experimentally introduced into their coral territory, especially if the intruders are mature, large (M–L) individuals capable of reproducing. Immature, small (I–S) females are permitted to remain. After Wong et al.;[162] photograph by João Paulo Krajewski.

Discussion Questions

8.2   In the starling, some males acquire several mates but do not assist them, whereas other, monogamous males work together with their sole partners to rear their broods. Sometimes when there are two females nesting on a male's territory, the first female to settle there pierces her companion's eggs with her beak.[115] Why might she do so, and what kind of monogamy could result from her actions? Under what circumstances would the female's behavior actually qualify as a form of parental investment?

8.3   The mimic poison frog of Peru is a truly monogamous animal, as revealed by genetic tests of offspring that are cared for by both parents.[25] The male fertilizes the eggs of a female after these are laid on a leaf. He then remains by the fertilized eggs until they hatch into tadpoles, after which he carries each tadpole on his back to its own small pool of water in vegetation high in a tree. His partner remains nearby so that when he calls to her, she comes to lay an unfertilized, edible egg in each pool for her tadpole to eat. What kind of mating system is exhibited by this species?

Male Monogamy in Mammals

Although monogamy is not common in any group, this mating system is exceptionally rare in mammals, a group notable for the size of female parental investment in offspring. Sexual selection theory suggests that male mammals, which cannot become pregnant and do not offer milk to infants (with the exception of one species of bat[58]), should usually try to be polygynous—and they do. However, exceptions to the rule are useful for testing alternative hypotheses on male monogamy in this group.

For example, if the mate-assistance hypothesis for monogamy applies to mammals, then males of the unusual mammalian species that exhibit paternal behavior[163] should tend to be monogamous. One monogamous mammal with paternal males is the Djungarian hamster, whose males actually help deliver their partners' pups (Figure 8.7).[84] Male parental care contributes to offspring survival in this species and also in the monogamous California mouse, pairs of which were consistently able to rear a litter of four pups under laboratory

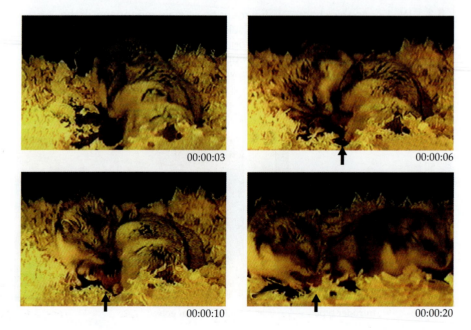

00:00:03

00:00:06

00:00:10

00:00:20

FIGURE 8.7   An exceptionally paternal rodent. A male Djungarian hamster may pull newborns from his mate's birth canal and then clear the infants' airways by cleaning their nostrils, as shown in these photographs (the male is the hamster on the left; arrows point to the pink newborn). From Jones and Wynne-Edwards.[84]

conditions, whereas single females did not do nearly as well.[29] The relationship also held under natural conditions, with the number of young reared by free-living California mice falling when a male was not present to help his mate keep the pups warm (Figure 8.8).[68]

In some mammals, one helpful thing a paternal male might do is to drive other males away from his brood, thereby blocking potential infant killers. In the mate-guarding prairie vole, males can and do defend their offspring against infanticidal rivals, so both mate guarding and mate assistance may contribute to the tendency toward monogamy in this species.[61] Infanticide may also favor monogamy in the white-handed gibbon, given that there was a much lower risk of infant disappearance in groups composed of a monogamous pair living with their offspring as opposed to polyandrous groups in which adult females associated with several adult males.[16]

However, a systematic comparative test of the proposition that mammalian monogamy goes hand in hand with male parental care produced completely negative results for primates, rodents, and all other groups.[88] In fact, the only mammalian pattern revealed by that comparative analysis is that males tend to live with females in two-adult units more often when females live well apart from one another in small territories. Arrangements of this sort favor mate guarding by males. The monogamous rock-haunting possum of northern Australia is a case in point. A female, her mate, and their young live along the edges of rock outcrops in territories of about 100 meters by 100 meters (Figure 8.9), an area about one-sixth of that occupied by other herbivorous mammals of equivalent weight.[121] In these small territories, a male possum can effectively monitor the activities of one female without expending a great deal of energy. Any male that attempted to move back and forth between the home ranges of several females would run the risk that intruder males would visit and mate with those mates that he had temporarily left behind. Thus, ecological factors

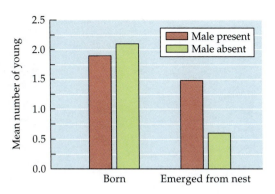

FIGURE 8.8   Male care of offspring affects fitness in the California mouse. The mean number of offspring reared by female mice falls sharply in the absence of a helpful male partner. After Gubernick and Teferi.[68]

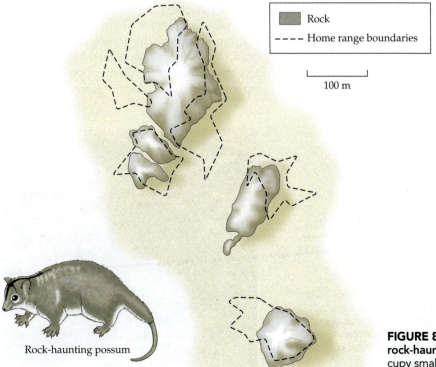

Rock-haunting possum

FIGURE 8.9   Mate-guarding monogamy in the rock-haunting possum. Females of this species occupy small, discrete home ranges along the edges of rock outcrops in northern Australia. After Runcie.[121]

that enable females to live in small, defensible territories tilt the cost–benefit equation toward mate guarding, which then leads to male monogamy.

Discussion Questions

8.4 In a small African antelope called Kirk's dik-dik, most males and females live in monogamous pairs.[24] Evaluate alternative hypotheses for monogamy in this species in light of the following evidence: the presence of a male does not affect the survival of the offspring; the male conceals the female's estrous condition by scent-marking over all odors deposited by his mate in the pair's territory; the male sires his social partner's offspring; a female left unaccompanied wanders from the pair's territory; some territories contain five times the food resources of others; the few polygynous groups observed do not occupy larger, or richer, territories than monogamous pairs of dik-diks.

8.5 In their classic paper on mating systems, Steve Emlen and Lew Oring suggested that two ecological factors could promote the evolution of monogamy: a high degree of synchrony in reproductive cycling within a population, and a highly dispersed distribution of receptive females.[51] Try to reconstruct the logic of these predictions, and then make counterarguments to the effect that synchronized breeding could facilitate acquisition of multiple mates while a relatively dense population of receptive females might actually promote monogamy.

Male Monogamy in Birds

In the vast majority of birds, males and females form partnerships for one or more breeding seasons.[89] In keeping with the mate-assistance hypothesis, the males in these partnerships often contribute in a big way to the welfare of the offspring produced by their mates.[104] In the yellow-eyed junco, for example, the male takes care of his mate's first brood of fledglings while the female incubates a second clutch of eggs. The paternal help provided is essential for the survival of these young "bumblebeaks," which are at first highly inept foragers.[152] The value of male assistance has also been documented for starlings. In a population in which some males helped their mates incubate their eggs and others did not, the clutches with biparental attention stayed warmer (Figure 8.10) and so could develop more rapidly. Indeed, 97 percent of the eggs

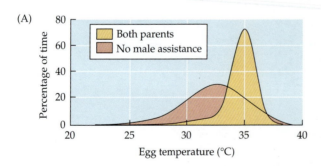

(A)

(B)

FIGURE 8.10 Paternal male starlings help their mates. Males keep their clutches warmer by sharing in incubation duties. (A) As the graph shows, eggs that were incubated by both parents were kept at about 35°C most of the time, whereas eggs incubated by the female alone were often several degrees cooler. (B) Both male and female starlings incubate the eggs and in this nest, all the eggs hatched successfully. After Reid et al.[115]

Spotless starling

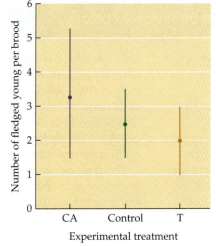

FIGURE 8.11 Paternal care boosts reproductive success in spotless starlings. Males whose testosterone levels were reduced by the anti-androgen cyproterone acetate (CA) provided more food for their broods and had the highest fledging rates per brood. Males given extra testosterone (T) provided less food and had the lowest fledging rates. Untreated controls were intermediate with respect to both feeding and fledging rates. After Moreno et al.[98]

that had been incubated by both parents hatched, compared with 75 percent of those that had been cared for by mothers alone.[115]

Demonstrations of the importance of male parental care include some studies in which females have been experimentally "widowed" and left to rear their broods on their own. Widowed snow buntings, for example, usually produce three or fewer young, whereas control pairs often fledge four or more.[94] When male parental care is reduced rather than eliminated altogether, this too can have a negative effect on reproductive success. Give a male spotless starling some extra testosterone, and he becomes less willing to feed nestlings, whereas males that receive an anti-androgenic chemical, which blocks the effects of naturally circulating testosterone, feed their offspring at increased rates. The mean number of fledged young per brood was lowest for starlings with extra testosterone and highest for those with the testosterone blocker (Figure 8.11).[98]

Thus, we can safely conclude that paternal, pair-bonded males of at least some bird species really do increase the number of offspring their mates can produce in a breeding attempt. But have these paternal males actually fathered the offspring of their mates? Because the fitness gained by these males will increase only to the extent that they care exclusively for their own genetic offspring, we expect monogamous males to sire all the offspring of their mates. This prediction has been confirmed for some species, such as the common loon: DNA fingerprinting of 58 young from 47 loon families revealed that all were the genetic offspring of the pair that raised them.[111] A similar study of Florida scrub jays (see Figure 3.16) also showed that nestlings were always the offspring of the adults believed to be their parents.[114]

But we now know, thanks to advances in molecular genetics (Box 8.1), that loons and scrub jays are exceptions to the rule. In most other birds, social monogamy (the pairing of male and female) does not equate with sexual monogamy (which happens when pairs produce and rear only their own genetic offspring). In nearly 90 percent of all bird species, some females engage in extra-pair copulations with males other than their social partners and use the sperm they acquire in this manner to fertilize some or all of their eggs.[66,158] In other words, socially monogamous males in most bird species run a grave risk of helping to raise offspring other than their own, which clearly reduces the benefits of rearing the offspring of a social partner.

BOX 8.1 Microsatellite analysis and behavioral ecology

The discovery that social monogamy is not synonymous with genetic monogamy in birds came about primarily through the use of multilocus DNA fingerprinting technology.[26,27,79] However, DNA fingerprinting has now been largely replaced by microsatellite analysis[113,155] because mutations in the noncoding microsatellite regions of DNA (see below) are selectively neutral, which means that they stay in populations indefinitely rather than being eliminated by natural or sexual selection.[9] As a result, individuals usually differ greatly in their microsatellites, which enables researchers to identify more readily whether a female has used the sperm from males other than her social partner to fertilize some of her eggs. Many studies of avian paternity now exist, demonstrating that extra-pair paternity is common.[66]

Microsatellite analysis takes advantage of the fact that scattered through the chromosomes are stretches of repeated sequences of DNA, such as

AAT AAT AAT AAT AAT

Here AAT is a sequence that is repeated five times, and therefore this microsatellite would be written as $(AAT)_5$. The number of times such a sequence is repeated at a given location on a chromosome often varies greatly, so one individual might carry five copies of AAT on one chromosome, whereas the other chromosome of the pair might have eight copies—$(AAT)_8$. Still other individuals might have other microsatellite alleles, such as $(AAT)_{10}$ or $(AAT)_{21}$. To perform microsatellite analyses, researchers first identify the nucleotide sequences in individuals of interest. Imagine that a female with the microsatellite genotype $(AAT)_5(AAT)_8$ mates with two males, one of which has the genotype $(AAT)_{10}(AAT)_{21}$

while the other has the genotype $(AAT)_9(AAT)_{33}$. Each offspring will carry a microsatellite allele from its mother, either $(AAT)_5$ or $(AAT)_8$, and a microsatellite allele present in the DNA of the sperm from its father. One need only establish an offspring's microsatellite genotype, usually for several loci, in order to determine which is the real father with a high level of certainty.

The table below shows the results of a microsatellite analysis of eight white-winged fairy-wrens with respect to four different arbitrarily labeled microsatellite loci (Mcy3 to Mcy7). For each microsatellite locus, there are two columns. The numbers in each column are the alleles (labeled by size) of that DNA repeating sequence or locus possessed by the individuals being studied. The male and female at the top of the table were social partners that cared for three offspring in their nest. The nest was also attended by two helper males, which assisted the social pair in rearing the young. The territory of this group was adjacent to another territory, one of whose occupants was the "neighbor male." The microsatellite genotypes of baby 1 and baby 2 show that these were almost certainly the offspring of the social pair. The mother was homozygous for the allele of a particular size (274) of Mcy3, and so all of her offspring had to possess one maternally derived copy of this allele. The real father donated one copy of each microsatellite gene to each of his genetic offspring, so the female's social partner was the sire of babies 1 and 2. (Check out the alleles present in the male, female, baby 1, and baby 2 in the other microsatellites analyzed.) But baby 3 had an allele (266) at the Mcy3 locus that could not have been donated by either member of the pair at this nest. This allele was present in helper 2 and in the neighbor male, so either of these males could

(continued)

Bird	Microsatellite locus							
	Mcy3		Mcy4b		Mcy5		Mcy7	
Male	274	258	168	168	130	98	106	104
Female	274	274	195	195	98	96	146	104
Baby 1	274	274	195	168	96	96	104	104
Baby 2	274	274	195	168	98	96	104	104
Baby 3	**274**	**266**	**195**	**140**	**102**	**96**	**112**	**104**
Helper 1	274	258	168	195	98	96	146	106
Helper 2	274	266	184	184	98	96	106	104
Neighbor male	256	**266**	140	**140**	**102**	98	**112**	106

BOX 8.1 Microsatellite analysis and behavioral ecology (*continued*)

conceivably have been the donor of the haploid sperm that fertilized the egg that produced baby 3. Why then did the complete analysis of this data set rule out helper 2 as the father of baby 3 and rule in the neighbor male as the genetic father of this youngster?

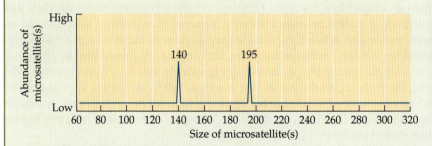

White-winged fairy-wren

The paternity of the male white-winged fairy-wren can be determined via microsatellite analysis. Individuals can be captured and a tiny drop of blood taken for microsatellite analysis. The idealized chromatogram (actual data are not quite so tidy) provides a visual record made by a DNA sequencer used in microsatellite analyses. The data shown here represent one individual found in the table. What is that individual's genotype? Which one has that genotype? Which was that individual's father? Data for the table and chromatogram courtesy of Bob Montgomerie.

Monogamous Males, Polyandrous Females

That social monogamy by males often coexists with genetic polyandry by their mates greatly surprised ornithologists who had always assumed that males helped their mates care for their mutual offspring, not for the progeny of other males. We shall deal with the puzzles associated with socially monogamous females that accept and use sperm from more than one male in the next section, but for the moment let's focus on this Darwinian puzzle: why does a male attach himself to a single female only to have her fertilize some or all of her eggs with sperm from one or more extra-pair partners? One answer is that the male may be able to take advantage of the opportunity for extra-pair copulations with females other than his social partner. A polygynous male of this sort may be able to fertilize the eggs of his social partner while also inseminating other females whose offspring will receive care, but not from him. These males, which often exhibit ornaments and displays thought to be indicative of superior physical condition,[66] avoid some or most of the costs of monogamy. Therefore, it falls to the truly monogamous males, which are more likely to be cuckolded by their mates, to suffer the disadvantages of monogamy. The question remains, why do these truly monogamous males fail to engage in fitness-boosting extra-pair copulations? One answer is that females have something to say about who gets to mate with them. If female mate choice revolves around parental contributions from a social partner and honest indicators of quality from extra-pair partners, then males that are unwilling to be paternal and are also unable to offer cues of good condition will likely find themselves between a rock and a hard place. They could drop out of the reproductive competition altogether but would obviously fail to leave descendants for as long as they remained on the sidelines. The only real option for these monogamous males is to help social partners that may fertilize at least some eggs with their sperm.

The existence of constraints imposed by choosy females and competitors may also help explain the evolution of mating systems in which most breeding females form social bonds with several males, rather than having just one social partner and a variable number of extra-pair mates. The mating system of the Galápagos hawk, for example, ranges from monogamy to extreme polyandry.[53] Polyandry appears to be associated with a scarcity of suitable territories, which leads to a highly male-biased operational sex ratio since males outnumber the limited number of territorial, breeding females. The intense competition for these females and the territories on which they live has favored males capable of forming a cooperative defense team to hold an appropriate site. A breeding female may acquire as many as eight mates prepared to pair-bond with her for years, forming a male harem that will help her rear a single youngster per breeding episode. In this hawk and another odd bird, the pukeko, all the males that associate with one female appear to have the same chance of fertilizing her eggs.[78] If females do give all their mates an equal chance at fertilizations, the mean fitness of each mate is certainly higher than that of males that fail to be part of a team.

In some other polyandrous species, the several mates of a female each receive a clutch of eggs from her, which forces those males to share the reproductive output of the female with the other members of her harem. In the wattled jacana, for example, aggressive females fight for territories that can accommodate several males. These males mate with the territory owner, which then supplies each of them with one clutch of eggs, which will be cared for exclusively by that male. Males that pair monogamously with a polyandrous female are also likely to care for offspring sired by other males, since 75 percent of the clutches laid by polyandrous female jacanas are of mixed paternity.[52] The same sort of thing happens in the red phalarope, a shorebird in which females are both larger and more brightly colored than males (Figure 8.12). Female phalaropes fight for males, and a male provides all the parental care for the clutch deposited in his nest, whether or not the eggs have been

FIGURE 8.12 Female red phalaropes are polyandrous. After securing one male partner, a female phalarope may attract another, donating a clutch of eggs to each male in turn. The more brightly colored bird (on the left) is the female. Photograph by Bruce Lyon.

fertilized by other males.[36] Cases of this sort illustrate the disadvantages to males that are monogamously bonded with a polyandrous female, but male jacanas and phalaropes may have to accept the eggs their partners lay if they are to leave any descendants at all.

However, a monogamous male under the control of a potentially polyandrous female may be able to do some things that at least reduce the probability that he will care for eggs fertilized by other males. For example, males of the red-necked phalarope, like those of its relative the red phalarope, care for broods on their own. Females of this species produce two clutches of eggs in sequence. Polyandrous females almost always draw their second mates from those males that have lost a first clutch to predators. Such a male, however, favors his original sexual partner over a novel female more than nine times out of ten.[125] By copulating with this female again and then accepting her eggs, the male reduces the risk that the eggs he receives will have been fertilized by another male's sperm.

The same applies to the spotted sandpiper, whose females also behave like males in many ways.[105] In addition to taking the lead in courtship, females are the larger and more combative sex, and they arrive on the breeding grounds first, whereas in most migrating birds, males precede females. Once on the breeding grounds, females fight with one another for territories (Figure 8.13). A female's holdings may attract first one and then later another male. The first male mates with the female and gets a clutch of eggs to incubate and rear on his own in her territory, which she continues to defend while producing a new clutch for a second mate, which may not fertilize all of the eggs he broods if his mate uses sperm stored from male number 1 to fertilize some of her second clutch. In this species, a few females achieve higher reproductive success than the most successful males,[107] an atypical result for animals generally (see page 178).

An understanding of the mating system of spotted sandpipers is advanced by recognizing that in all sandpiper species, females never lay more than four eggs at a time, presumably because five-egg clutches cannot be incubated properly.[106] Indeed, egg addition experiments have shown that sandpipers given extra eggs to incubate sometimes damage their clutches inadvertently and lose eggs more often to predators as well.[6] If female spotted sandpipers are "locked into" a four-egg maximum, then they can capitalize on rich food resources only by laying more than one clutch, not by increasing the number of eggs laid in any one clutch. To do so, however, they must acquire more than one mate to

FIGURE 8.13 **Polyandrous female spotted sandpipers fight for males.** Two female spotted sandpipers (A) about to fight and (B) fighting for a territory that may attract several monogamous, paternal males to the winner. Photographs by Stephen Maxson.

(A)

(B)

care for their sequential clutches, making this a rare case in which female fitness is limited more by access to mates than by production of gametes.

Male spotted sandpipers may be forced into monogamy by the confluence of several unusual ecological features.[90] First, the adult sex ratio is slightly biased toward males. Second, spotted sandpipers nest in areas with immense mayfly hatches, which provide superabundant food for females and for the young when they hatch. Third, a single parent can care for a clutch about as well as two parents, in part because the young are precocial—that is, they can move about, feed themselves, and thermoregulate shortly after hatching. The combination of excess males, abundant food, and precocial young means that female spotted sandpipers that desert their initial partners can find new ones without harming the survival chances of their first broods. Once a male has been deserted, he is stuck. Were he also to leave the nest, the eggs would fail to develop, and he would have to start all over again. If all females are deserters, then a male single parent presumably experiences greater reproductive success than he would otherwise, even if his partner acquires another mate to assist her with a second clutch. Furthermore, the first male to mate with a female spotted sandpiper may provide her with sperm that she stores and uses much later to fertilize some or all of her second clutch of eggs, as noted above.[108] Once again, the males that get the short end of the stick are the less competitive individuals, the ones slower to arrive on the breeding grounds. They must deal with polyandrous females if they are to have any chance of reproducing, but the female's ability to control the reproductive process puts them at a major disadvantage.

What Do Females Gain from Polyandry?

In a massive field study of crickets whose activities were monitored with video cameras, the research team who carried out this work found that both males and females produced more offspring when they acquired more mating partners.[73] The fact that males do better when they have more mates is not surprising, but adaptive cricket polyandry is unexpected given that females receive a great many sperm from a single male. Mating with "extra" males exposes the female cricket and most other polyandrous females to a host of costs, including the time and energy spent searching for and mating with several males, the risk of being killed by a predator during these forays and when mating, and the chance of acquiring a sexually transmitted disease from some mates.

> ### Discussion Question
>
> **8.6** Earlier in the text we noted that termites, which are eusocial insects, are almost always monogamous.[70] Use kin selection theory to propose a hypothesis on why polyandry is so rare in this group of insects.

If the risk of sexually transmitted diseases is greater for polyandrous females, then we would expect the immune systems of polyandrous species to be stronger than those of related species that have a greater tendency toward monogamy. One research team led by Charles Nunn tested this hypothesis using comparative data from primates rather than birds. The team took advantage of the fact that there are extremely polyandrous primates, such as the Barbary macaque, whose females mate with as many as ten males in one day. At the other end of the spectrum, there are monogamous species, such as the gibbons. Nunn and his colleagues measured immune system strength

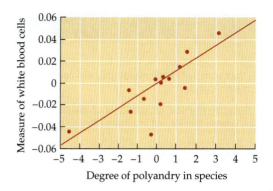

FIGURE 8.14 Polyandry has fitness costs. In primates, the number of mates accepted by females is correlated with the investment made in white blood cells, a component of an animal's immune system. (The higher the measure of white blood cells, the stronger the immune system.) This finding suggests that the more polyandrous the species, the greater the challenge to the female's immune system. Each data point represents a different primate species. After Nunn et al.[102]

across the spectrum of mating systems by looking at data on white blood cell counts from large numbers of adult females of 41 primate species, most of which were held in zoos, where their health is regularly monitored. (Naturally, only healthy specimens were included in the samples.) White blood cell counts were indeed higher (but within the normal range) in females of the more polyandrous species (Figure 8.14).[102]

Discussion Question

8.7 Why did Nunn and company also look at the relationship between white blood cell counts and the mean sizes of groups in the species they studied, as well as considering the extent to which each species was terrestrial as opposed to arboreal?

Given the potential costs of extra-pair copulations for females, it is not too surprising that in at least some species, females attempt to avoid insemination by males other than their social partners. Female mallard ducks, for example, resist every copulatory attempt by males other than their long-term partners.[35] Their favored partners typically appear to be individuals of well-above-average condition, having molted into breeding plumage sooner than others. The ability to overcome the severe energetic demands of the molt relatively early in the year indicates that these males are healthier and thus less likely to transmit venereal diseases to their mates (and waterfowl are at risk of sexually transmitted disease).[132]

Although female mallards always try to say no to unsolicited copulatory attempts, females of other species accept, or even invite, matings from males other than their primary partners.[12] For example, 85 percent of female bluethroats whose social mates had been made infertile by outfitting them with a kind of avian condom went on to lay fertile eggs, demonstrating unequivocally that they were mating with other males.[56] Just what bluethroats and other socially monogamous songbirds might gain from extra-pair copulations initially puzzled behavioral biologists, but many hypotheses are now available (Table 8.1). The possible benefits to polyandrous females have been categorized as genetic (or indirect) versus material (or direct). One genetic benefit, for example, is that extra-pair fertilizations could reduce the risk to a female of having an infertile partner as a social mate.[67] The **fertility insurance hypothesis** is supported by the observation that the eggs of polyandrous female red-winged blackbirds are somewhat more likely to hatch than the eggs of monogamous females;[64] in Gunnison's prairie dogs, polyandrous females become pregnant 100 percent of the time, whereas only 92 percent of monogamous females achieve this state.[76]

TABLE 8.1 Why do females mate voluntarily with more than one male?	
Genetic, or indirect, benefits of polyandry	
Fertility insurance hypothesis	Mating with several males reduces the risk that some of the female's ova will remain unfertilized because some of her partners are infertile.
Good genes hypothesis	A female mates with more than one male because her social partner is of lower genetic quality than her extra-pair partner, whose genes will improve offspring viability or sexual attractiveness.
Genetic compatibility hypothesis	Mating with several males increases the genetic variety of the sperm available to the female, which boosts the chance that the female will receive some sperm whose DNA is unusually compatible with hers.
Material, or direct, benefits of polyandry	
More resources hypothesis	More mates means more resources received from the sexual partners of a female.
More care hypothesis	More mates means more caregivers to help rear the female's offspring.
Better protection hypothesis	More mates means more protectors to keep other males from sexually harassing the polyandrous female.
Infanticide reduction hypothesis	More mates means greater uncertainty about paternity of an infant when it is born and thus fewer males with no stake in the welfare of the polyandrous female's offspring.

Polyandry and Good Genes

Although fertility insurance may have contributed to the evolution of polyandry in some cases, a far more frequently mentioned genetic benefit linked to this mating system has to do with improved offspring quality. In an African tree frog, the tadpoles of polyandrous females whose clutches had been fertilized by at least 10 males were significantly more likely to survive to metamorphosis than those of females that had had just a single mate (Figure 8.15A).[28] This result suggests a genetic benefit from polyandry. The same factor may be at work in a wild guinea pig (the yellow-toothed cavy), whose females will copulate with more than one male when given the opportunity. The payoff for polyandry in this species appears to be a reduction in stillbirths and losses of babies before weaning (Figure 8.15B).[75]

But polyandry may provide for more than just improved early survival for the offspring of females that mate with several males. In the dark-eyed junco, a common songbird of North America, not only are the genetic offspring of females and their extra-pair mates more likely to survive to reproduce, but when they actually do reproduce, they do better than the offspring of females and their social partners.[60] The sons of extra-pair liaisons appear to boost their lifetime reproductive success by mating with females outside the pair-bond; the daughters of these arrangements are more fecund, perhaps because they lay more eggs. Because extra-pair male juncos provide only genes to their sexual partners, the fitness benefits that females gain from polyandry appear to be due to those genes.

There is, however, another explanation for effects of this sort, namely that when a female songbird mates with an extra-pair male, she subsequently invests more in the offspring of such a relationship. For example, females can, and sometimes do, make larger than usual eggs to be fertilized by the sperm of their extra-pair mates. The greater supply of nutrients can give extra-pair offspring an early developmental boost, enabling these youngsters to become sexy sons and fecund daughters.[146]

(A)

(B)

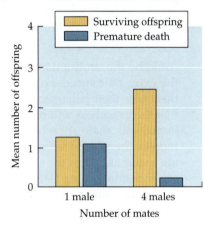

FIGURE 8.15 Polyandry has fitness benefits. (A) Females of an African tree frog lay clutches of eggs that may be fertilized by anywhere from 1 to 12 or more males. The offspring of polyandrous females are more likely to survive than those of females mated to only one male. (B) A male yellow-toothed cavy, whose very large testes speak to the intense sperm competition in this species, which occurs because females usually mate with several males. When females were experimentally restricted to a single mate, they had fewer surviving offspring than females allowed to mate freely with four males. A, photographs by Philip Byrne; courtesy of Martin Whiting; B, after Hohoff et al.;[75] photograph by Matthias Asher, courtesy of Norbert Sachser.

In other words, in some species with polyandrous females, the production of fitter offspring is due to the contributions of the mother to her progeny, not because females have acquired good genes from one or more males. Having said that, the superb fairy-wren is another animal for which the good genes hypothesis is a plausible explanation for polyandry. In this gorgeous little Australian wren, a socially bonded pair lives on a territory with a number of subordinate helpers, usually males. When the breeding female is fertile, she regularly leaves her mate and auxiliary helpers before sunrise and travels to another territory, where she often mates with a dominant male before returning home.[47] Tiny radio transmitters attached to the females revealed what they were up to, and genetic analyses of offspring demonstrated that the females' social partners often lost paternity to distant rivals. When social mates did well, it was generally because they sired extra-pair young with females other than their supposed "mates" (Figure 8.16).[156] For example, 10 of the 14 fledglings fathered by one dominant male were the result of extra-pair copulations; these youngsters were cared for by fairy-wrens other than their father.

Thus, for both superb fairy-wrens and dark-eyed juncos, a major component of the fitness benefits gained by females through extra-pair copulations

FIGURE 8.16 Extra-pair paternity and male reproductive success. In the superb fairy-wren, males vary greatly in their reproductive success in part because of differences in the number of offspring they have with females other than their social partners. Shown are the number of young produced per year by dominant males with social partners and by auxiliary helper males that live in groups with a "breeding" pair. After Webster et al.[156]

Male superb fairy-wren

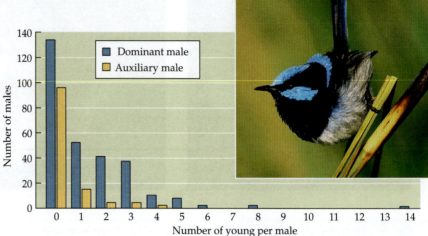

could be the production of sons that are so attractive that they repeat the extra-pair success of their fathers.[151] If they inherit the very traits that made their fathers sexually appealing, "sexy sons" increase the odds that their mothers will have many grand-offspring. As required by this hypothesis, male attractiveness is hereditary in at least some species.[141] Indeed, the sons of sexually successful field crickets were more appealing to females than the sons of males that failed to acquire a mate in an experimental setting (Figure 8.17).[157]

The idea that extra-pair copulations enable females to acquire good genes for their offspring is also compatible with the discovery that older males appear to have greater success in extra-pair matings in some songbirds.[44,82] Older males have demonstrated an ability to stay alive, which might be due in part to their genotype (but remember that the early development of these individuals could have been helped along by their mothers). If some males have the good genes for survival or sexiness, the offspring of such males might live longer and be more attractive as well, something that has been documented in the polyandrous blue tit.[86,87]

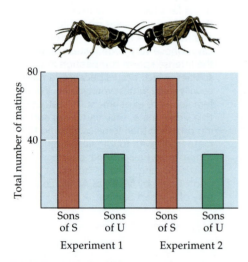

FIGURE 8.17 A father's mating success can be transmitted to his sons. In experiment 1, two male field crickets were given an opportunity to compete for a female; one male (S) mated successfully, while the other male (U) was unsuccessful. When the sons of male S were placed in competition for a female with the sons of male U (which had been given a female to mate with after failing to win the initial competition), the sons of S were about twice as likely to mate with the female as the sons of U. In experiment 2, a male that had won a mating competition was later allocated a female at random for breeding, as was a male that had lost the competition. The sons of the two males were then placed in an arena with a female, and as before, the sons of S were much more likely to mate with the female than were the sons of U. After Wedell and Tregenza.[157]

Discussion Question

8.8 In a study of song sparrows, a team led by Rebecca Sardell assembled data on various indirect measures of fitness of female and male birds that were fathered by the social partner of a female (within-pair young) and of those that were sired by an extra-pair male (extra-pair young).[123] Based upon the good genes theory for polyandry, what did they expect the data to reveal? Are their results (Figure 8.18) consistent with that expectation? What conclusion did the research team reach?

Polyandry and Genetic Compatibility

Females might engage in polyandry not to hit the genetic jackpot with one especially sexy or long-lived male but rather to increase the odds of receiving genetically complementary sperm *and* to then have that special

(A)

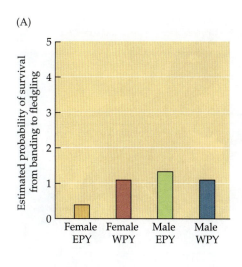

(B)

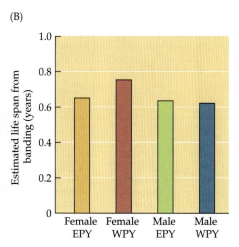

FIGURE 8.18 A test of the hypothesis that female birds receive good genes from extra-pair mates. In the song sparrow, females can produce young with extra-pair males (extra-pair young, or EPY) or with their social partners (within-pair young, or WPY). Shown are (A) the probabilities of survival to fledging for the two categories of sons and daughters and (B) the estimated life span of the two kinds of offspring. After Sardell et al.[123]

sperm fertilize their eggs. When gametes with especially compatible genotypes unite, they can result in highly viable progeny that, for example, have high levels of heterozygosity.[165,167] Individuals with two different forms of a given gene often enjoy an advantage over homozygotes (see Fossøy et al.[57]). If genetically similar pairs are in danger of producing inbred offspring with genetic defects, then we can predict that females socially bonded with genetically similar individuals will be prime candidates to mate with other males that are either (1) genetically dissimilar or (2) highly heterozygous in their own right (which means that these males are likely to carry rare alleles that, when donated to offspring, increase the odds that these individuals will also be heterozygous).[145]

The **genetic compatibility hypothesis** has been tested by examining the number of surviving offspring produced by polyandrous pseudoscorpions versus females that were experimentally paired with lone males (Figure 8.19).[101,166] When Jeanne Zeh compared the numbers of offspring different females produced from matings with the *same* male, she found no correlation, which means that male pseudoscorpions cannot be divided into studs and duds. Instead, the effect of a male's sperm on a female's reproductive success depended on the match between the two gametic genomes, as predicted by the genetic compatibility hypothesis. Because a female's chances of securing genetically compatible sperm increase with the number of her partners, we

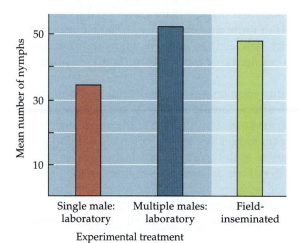

Pseudoscorpion

FIGURE 8.19 Polyandry boosts female reproductive success in a pseudoscorpion. In laboratory experiments, female pseudoscorpions restricted to lone partners produced fewer nymphs than females that mated with several males. Paternity tests of the offspring of wild-caught females confirmed that females usually mate with several males under natural conditions. After Zeh;[166] photograph by Jeanne Zeh.

FIGURE 8.20 The effect of extra-pair matings on the genetic heterozygosity of offspring. (A) The comparison here shows that the heterozygosity of the mated pair's offspring (orange bars) is lower than that of the extra-pair offspring (red bars) created when a bird in one group mated with a bird from another group altogether. (B) In contrast, the offspring of a mated pair were no less heterozygous than those produced by extra-pair copulations with another member of the same group of starlings. A and B, after Rubenstein;[120] photograph by Dustin Rubenstein.

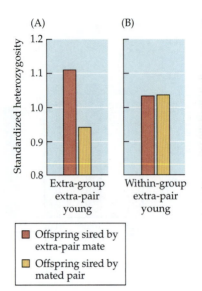

Superb starling

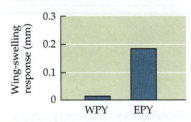

Bluethroat

FIGURE 8.21 Extra-pair matings can boost the immune responses of offspring in the bluethroat. The mean wing-swelling response (an indicator of immune system strength) of youngsters sired by their mother's social partner (within-pair young, or WPY) was less than that of young birds sired by their mother's extra-pair partner (extra-pair young, or EPY) in broods of mixed paternity. After Johnsen et al;[83] photograph by Bjørn-Aksel Bjerke.

can predict that females of this pseudoscorpion should prefer to mate with new males rather than with previous partners (provided that the sperm of these new mates are more likely to fertilize her eggs). Indeed, when a female was given an opportunity to mate with the same male 90 minutes after an initial copulation, she refused to accept his spermatophore in 85 percent of the trials. But if the partner was new to her, she would usually accept his gametes,[168] behavior reminiscent of the Coolidge effect (see Discussion Question 7.3).

Likewise, when Dustin Rubenstein studied the choice of extra-pair partners by female superb starlings, he found that when females went looking for partners outside their social groups, they tended to have social mates that were less heterozygous than they were.[120] The offspring fathered by the outsiders were more heterozygous on average than those produced by the females' social partners (Figure 8.20).

In the polyandrous songbird the bluethroat, extra-pair offspring were also more heterozygous than those sired by the females' social partners.[57] Extra-pair progeny also enjoyed stronger immune systems, a point established by injecting a foreign substance into the wing webs of nestling bluethroats and measuring the swelling at the injection sites (Figure 8.21). But females that engaged in extra-pair copulations were not acquiring universally good genes, because the extra-pair males and their own primary mates produced youngsters with only average levels of immunocompetence.[83] Thus, it must have been the especially good match between the genes of a polyandrous female and her extra-pair mate that resulted in a more robust immune system for her offspring.

Male mandrills are stunning monkeys with bright red and blue facial skin. Whether this coloration affects female mate choice is not yet known, although dominant males sire the largest proportion of the offspring in mandrill troops.[45] But it has been established that here, too, females tend to reproduce with genetically compatible males, namely those whose MHC (major histocompatibility complex) genes are dissimilar to their own, the better to produce heterozygous offspring with stronger immune systems.[131] The MHC genes code for proteins that contribute to immune system functioning, so mate choice by mandrills appears to have converged on adaptive criteria similar to those used by bluethroats.

Discussion Questions

8.9 In the polyandrous brown anole lizard, females produce more sons than daughters when paired with males in good condition (i.e., with considerable fat reserves).[34] The bias could stem either from paternal effects based on the genes in sperm received from males in good condition or from maternal "decisions" made after females have mated with these males. If this sex ratio biasing is adaptive for mother lizards, what do you predict about the fitness of sons in relation to the condition of their fathers?

8.10 The gray mouse lemur is a species in which females come into estrus and mate with several males on one night each year. Females accept any and all partners, but a genetic analysis of offspring produced by female gray mouse lemurs indicates that mothers and actual fathers differed with respect to their MHC genes to a greater degree than mothers and males selected at random from the population.[129] Use these results to evaluate the various hypotheses on the indirect genetic benefits of polyandry. What do they suggest about the mechanism that enables a sexually promiscuous female mammal to exercise mate choice?

Polyandry and Social Insects

Some members of the genus *Apis*, a group to which the familiar honey bee belongs, are highly polyandrous, with females of *A. florea* accepting and using sperm from around 10 males on average while *A. dorsata* females acquire sperm from an average of about 63 males.[130] As mentioned earlier, virgin queens of this group fly out from the hive on one or more nuptial flights, during which time they are "captured" and mated by a number of males in quick succession. What makes this business particularly puzzling is the finding that in most bees other than those in genus *Apis*, including some from large social colonies, the queens mate just once.[139] In the typical bee, a single mating suffices to supply the female with all the sperm she will need for a lifetime of egg fertilizations (queens of stingless bees can live for several years), thanks to her ability to store and maintain the abundant sperm she receives from her single partner. The widespread nature of bee monogamy indicates that this trait was the ancestral mating pattern, with polyandry in the genus *Apis* a more recently derived characteristic. What selective pressures might have led to the replacement of monogamy by extreme polyandry in a relatively few bee species?

Benjamin Oldroyd and Jennifer Fewell were able to track down a full dozen potential answers to this question, ranging from the idea that polyandry guarantees that long-lived queens do not run out of sperm to hypotheses such as ones in which polyandry is said to enhance the disease or parasite resistance of a colony by increasing the genetic diversity within a colony. Alternatively, genetic diversity among the workers could promote colony-level productivity by increasing the range of skills exhibited by the workforce.[103]

Support for the anti-disease hypothesis includes the results of an experiment in which some honey bee colonies were set up with queens that had been artificially inseminated just once and some colonies were set up with queens that had been supplied with the sperm of ten drones. Both types of colonies were then infected with a bacterium that causes the disease American foulbrood, a killer of honey bee larvae. The colonies with polyandrous queens not only had milder forms of foulbrood but also had larger populations of workers on average than the colonies headed by once-mated queens.[130]

The experimental results described here are supportive of an anti-disease function associated with queen polyandry in the honey bee. However, the

FIGURE 8.22 Polyandry provides genetic benefits in honey bees. Genetically diverse colonies (the queens had mated with several males—shown in green) on average produced more comb than did workers in genetically uniform colonies (whose queens had mated with only one drone—shown in red). After Mattila and Seeley.[95]

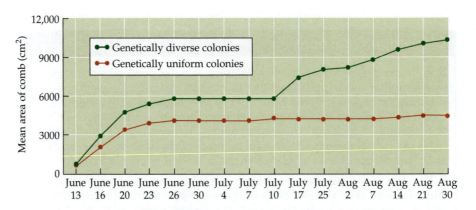

advantages shown by the colonies with polyandrous queens might have in part or in whole arisen from the ability of a more genetically diverse worker force to carry out the full range of worker tasks more efficiently than a less diverse worker force. To test this hypothesis, Heather Mattila and Thomas Seeley again used artificial insemination techniques to create two categories of colonies: those headed by single-insemination queens and those whose queens had mated with many males. The colonies were kept free from diseases through antibacterial and antiparasitic treatments to eliminate this variable from the experiment. The genetically uniform and genetically diverse colonies were also monitored to measure such things as the total weight of the bees, the foraging rates of workers, and the mean area of comb in the hive (where offspring are reared and food is stored) (Figure 8.22).[95] None of the genetically uniform colonies made it through the winter, whereas about a quarter of the colonies led by polyandrous queens survived. These and other similar findings demonstrate that the colony's genetic diversity, which is boosted when the queen mates with several males, elevates the fitness of the queen. The puzzle of polyandry for the honey bee and its relatives is at least partly solved by the demonstration that a genetically diverse work force is better than one of limited genetic variety.

Polyandry and Material Benefits

Our focus thus far on genetic (indirect) gains for polyandrous females should not obscure the possibility that females sometimes mate with several males in order to secure certain direct benefits, often in the form of useful resources, rather than genes alone from these males. Thus, female red-winged blackbirds may be allowed to forage for food on the territories of males with which they have engaged in extra-pair copulations, whereas truly monogamous females are chased away.[65] Similarly, females of some bees must copulate with territorial males each time they enter a territory if they are to collect pollen and nectar there (Figure 8.23).[1] It is even possible that males of some insects pass sufficient fluid in their ejaculates to make it worthwhile for water-deprived females to copulate in order to combat dehydration.[48]

Female hangingflies and other insects also have an incentive to mate several times in order

FIGURE 8.23 Polyandry can yield material benefits. By mating with many males, females of this megachilid bee gain access to pollen and nectar in those males' territories. Photograph by the author.

to receive **material benefits** in the form of food presents or nutritious spermatophores from their partners. The spermatophores of highly polyandrous butterfly species contain more protein than the spermatophores of generally monogamous species.[13] Males of polyandrous butterfly species could bribe females to mate with them by providing them with nutrients that can be used to make more eggs (in these species, the male's spermatophore can represent up to 15 percent of his body weight, making him a most generous mating partner). If this hypothesis is true, then the more polyandrous a female, the greater her reproductive output should be. Christer Wiklund and his coworkers used the comparative method to check this prediction. They took advantage of the fact that a number of closely related butterfly species in the same family (the Pieridae) differ substantially in the number of times they copulate over their lifetimes. They used the mean number of copulations for females of each species as a measure of its degree of polyandry. To quantify reproductive output, they measured the mass of eggs produced by a female in the laboratory and divided that figure by the body mass of the female, to control for differences among the species. As predicted, across these eight species, the more males a female mated with on average, the more spermatophores she received, and the greater her production of eggs (Figure 8.24).[160]

In some butterflies, polyandry means more eggs produced. In some other animals, polyandry enables females to get more parental assistance from their several mates. Thus, in the dunnock, a small European songbird, a female that lives in a territory controlled by one (alpha) male may actively encourage another, subordinate male to stay around by seeking him out and copulating with him when the alpha male is elsewhere. Female dunnocks are prepared to mate as many as 12 times per hour, and hundreds of times in all, before laying a complete clutch of eggs. The benefit to a polyandrous female of distributing her copulations between two males is that both sexual partners will help her rear her brood—provided that they have both copulated often enough with her. Female dunnocks ensure that this paternal threshold is reached by actively soliciting matings from whichever male, alpha or beta—usually the latter—has had less time in their company (Figure 8.25).[39] Likewise, a female superb starling that is pair-bonded with one male sometimes mates with another unpaired male in her flock; the second male then sometimes helps rear the brood of his mate.[120] By copulating with the extra male and using his sperm to fertilize an egg or two, the polyandrous female has in effect made it advantageous for the "other male" to invest parentally in her offspring.

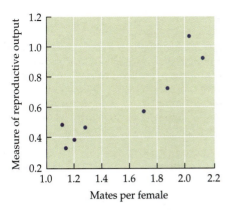

FIGURE 8.24 Polyandry boosts reproductive output in pierid butterflies. The mean number of mates during a female's lifetime and the mean reproductive output of females (measured as the cumulative mass of eggs produced divided by the mass of the female) is shown for eight pierid species. After Wiklund et al.[160]

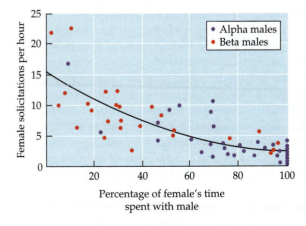

Dunnock

FIGURE 8.25 Adjustment of copulation frequency by polyandrous female dunnocks. A female living in a territory with two males solicits copulations more often from the male that has spent less time with her, whether that is the alpha or the beta male. After Davies.[39]

8.11 In the traditional cultures of the Amazon, many women had extra-marital affairs. In these cultures, it was standard practice to accept these affairs. Often the men involved believed that they shared in the biological paternity of the children a woman had during their relationships.[150] Explain how a belief in multiple paternity may have benefited women in these cultures. Use at least one of your hypotheses to make a prediction about the parental behavior of the additional partners. If this culturally promoted belief reduced male fitness, what should have happened over time in these cultures?

It is also possible that by mating with several males, a female may encourage all her sexual partners to leave her next newborn alone. Potentially infanticidal males of the Hanuman langur generally do ignore a female's baby if they have mated with the mother prior to the birth of the infant.[15] The fact that females will copulate with more than one male even when they are not ovulating—in fact, even when they are pregnant—suggests that polyandry promotes the female langur's interests by lowering the risk of infanticide.[72]

8.12 The average proportion of extra-pair offspring in a brood varies among bird species from 0.0 to almost 0.8.[80] Consider how variation in the benefits and costs of extra-pair paternity might be responsible for variation in the willingness of females to engage in extra-pair copulations.[66] For example, how might low variation in male genetic quality in a species affect the extra-pair paternity figure? What about differences among species in the risk of venereal disease, or in the likelihood that partners will detect and punish the sexual infidelity of a mate? Furthermore, if mating with extra-pair males provides females of a given species primarily with the opportunity to trade up to a genetically or materially superior partner,[31] what is the predicted relationship between extra-pair copulations and "divorce" among songbirds?

8.13 In many animals, the female mates repeatedly with the same male, typically her social partner. Perform a cost–benefit analysis of this kind of mating pattern by females, and contrast it with that applied to polyandrous matings.

Having just now read the accounts that seem to indicate that females can gain any or all of an extremely wide range of indirect (genetic) benefits and/or direct (material) benefits by mating with more than one male, you may find it disconcerting to hear that some biologists have concluded that none of these potential benefits actually amounts to much. Göran Arnqvist and Mark Kirkpatrick used the admittedly small number of appropriate bird studies to measure the strength of selection on female infidelity. They found that indirect (good genes) benefits were close to zero (see also Townsend et al.[144]), while selection via the direct benefits side of the equation was actually negative.[7] In other words, female birds may not gain by mating with males other than their social partners. The fact that female birds of so many species do indeed copulate outside the pair-bond can be explained, according to Arnqvist and Kirkpatrick, as the result of positive selection on males (not females) to inseminate females other than their primary partners. Selection of this sort could lead males to do things that are not in the best interests of their primary mates, such as copulating with other females, because these actions are in the best interests of the sexually demanding males' genes. If true, perhaps there should be a mating system called "convenience polyandry" to account for cases in which females apparently do not benefit from mating with several males—except that it costs them less to mate with unwanted sexual harassers than to fight them off.[43,143]

Simon Griffith has disputed the conclusions of Arnqvist and Kirkpatrick,[67] but his critique has in turn been challenged,[8] generating an example of how scientists argue in print. I suspect that most behavioral biologists working in this area find it hard to believe that female birds typically gain nothing by participating in extra-pair copulations, but Arnqvist and Kirkpatrick's position is one that has to be considered by those trying to solve the persistent puzzle of female polyandry.[50]

Why Are There So Many Kinds of Polygynous Mating Systems?

We have illustrated what is interesting about male monogamy and female polyandry. Let's now look at polygyny. Although a male's attempts to be polygynous are easily understood in terms of sexual selection theory, what is puzzling is the great variation in the tactics employed by males to achieve polygyny. Consider the behavior of bighorn sheep, black-winged damselflies, and satin bowerbirds. Even though some males of all three species acquire many mates by defending territories of one sort or another, bighorn rams go where potential mates are (see Figure 7.20B), and they then fight with other males to monopolize females there (**female defense polygyny**). In contrast, male black-winged damselflies wait for females to come to them (see Figure 7.25), defending territories that contain the kind of aquatic vegetation in which females prefer to lay their eggs (**resource defense polygyny**). Male satin bowerbirds, on the other hand, defend territories containing only a display bower (see Figure 7.1), not food or nesting sites or any other resources that females might use to promote their reproductive success (**lek polygyny**). Moreover, males of many other polygynous species skip combative territoriality altogether and instead try to outrace their rivals to receptive females (**scramble competition polygyny**). The differences in mating systems may stem from the degree to which females are clustered in defendable groups as a result of predation pressure or food distribution, a point that we shall develop in the following pages.[51]

Female Defense Polygyny

The theory that differences in female distribution patterns underlie the diversity of mating systems generates many predictions, among them the expectation that when receptive females occur in defensible clusters, males will compete directly for those clusters and female defense polygyny will result. As predicted, social monogamy in mammals never occurs when females live in groups.[88] Instead, in animals as different as bighorn sheep and gorillas, groups of females, which have formed in part for protection against predators, attract males that compete to control sexual access to the whole group.[69] Likewise, male lions fight over prides of females, which have formed in part for defense of permanent hunting territories and protection against infanticidal males.[109] Given the existence of female groups, female defense polygyny is the standard male tactic (Figure 8.26).[96] Thus, males of the Montezuma oropendola, a tropical blackbird, try to control clusters of nesting females, which group their long, dangling nests in certain trees. The dominant male at a nest tree may

FIGURE 8.26   Female defense polygyny in the greater spear-nosed bat. The large male (on the bottom right) guards a roosting cluster of smaller females. A successful male may sire as many as 50 offspring with his harem females in a year.

secure up to 80 percent of all matings by driving subordinates away.[153] These dominant males shift from nest tree to nest tree, following females rather than defending a nesting resource per se, demonstrating that they are employing the female defense tactic when the level of competition makes this possible (Figure 8.27).[154]

(A)

(B)

(C)

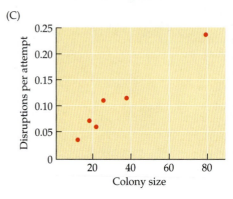

(D)

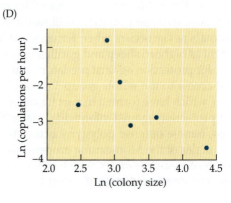

FIGURE 8.27   Female defense polygyny in a bird, the Montezuma oropendola. (A) Males of this species attempt to monopolize females in (B) small colonies of nesting females. But as colony size increases, (C) mating attempts are often disrupted by rivals, and as a result, the (D) frequency of copulations per hour at the colony site decreases. After Webster and Robinson.[154]

(A)

(B)

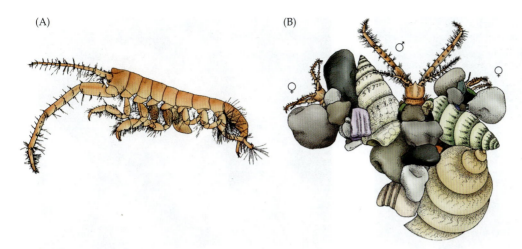

FIGURE 8.28 Female defense polygyny in a marine amphipod. (A) A male without his house. (B) A male that has glued the houses of two females to his house. Drawings courtesy of Jean Just.

Groups of receptive females also occur in insects and other invertebrates, and female defense polygyny has evolved here as well.[143] For example, nests of *Cardiocondyla* ants produce large numbers of virgin queens, leading to lethal combat among the males of the colony, the survivors of which get to copulate with dozens of freshly emerged females.[71] But one does not have to be hyper-aggressive in order to practice female defense polygyny. Thus, males of tiny siphonoecetine amphipods construct elaborate cases composed of pebbles and fragments of mollusk shells found in the shallow ocean bays where they live. They move about in these houses and capture females by gluing the females' houses to their own, eventually creating an apartment complex containing up to three potential mates (Figure 8.28).[85]

Resource Defense Polygyny

In many species, females do not live permanently together, but a male may still become polygynous if he can control a rich patch of resources that females visit on occasion. Some male black-winged damselflies, for example, defend floating vegetation that attracts a series of sexually receptive females, each of which mates with the male and then lays her eggs in the vegetation he controls.[149] Likewise, male antlered flies fight for small rot spots on fallen logs or branches of certain tropical trees because these locations attract receptive, gravid females, which mate with successful territory defenders before laying their eggs (Figure 8.29).[46]

A safe location for eggs constitutes a defensible resource for a host of animal species; the more of this resource a territorial male holds or the longer he holds it, the more likely he is to acquire several mates. This rule applies to a Brazilian harvestman, a relative of the more familiar daddy long-legs. The males of the Brazilian species occupy and defend nest burrows in muddy banks that attract gravid, sexually receptive females, particularly if there are already eggs present being protected by a guarding male. In effect, the male's parental care, in addition to the safe burrow that he defends, constitutes a resource for receptive females, enabling some nest- and egg-guarding males to be polygynists.[100]

For an example of resource defense polygyny in a vertebrate, we turn to an African cichlid fish, *Lamprologus callipterus*, in which a female deposits a clutch of eggs in an empty snail shell and then pops inside to remain with the eggs and hatchlings until they are ready to leave the nest. Territorial males of this species, which are much larger than

FIGURE 8.29 Resource defense polygyny in an Australian antlered fly. Males compete for possession of egg-laying sites that can be found only on certain species of recently fallen trees. Photograph by Gary Dodson.

FIGURE 8.30   Resource defense polygyny in an African cichlid fish. (A) A territorial male bringing a shell to his midden. The more shells, the more nest sites are available for females to use. (B) The tail of one of the territorial male's very small mates can be seen in a close-up of the shell (lower left) that serves as her nest. Photographs by Tetsu Sato.

their tiny mates, not only defend suitable nest sites but collect shells from the lake bottom and steal them from the nests of rival males, creating middens of up to 86 shells (Figure 8.30). Because as many as 14 females may nest simultaneously in different shells in one male's midden, the owner of a shell-rich territory can enjoy extraordinary reproductive success.[124]

If the distribution of females is controlled by the distribution of key resources and if male mating tactics are in part dictated by this fact of life, then it ought to be possible to alter the mating system of a species by moving resources around, thereby altering where females are located. This prediction was tested by Nick Davies and Arne Lundberg, who first measured male territories and female foraging ranges in the dunnock. In this drab little songbird, females hunt for widely dispersed food items over such large areas that the ranges of two males may overlap with those of several females, creating a polygynous–polyandrous mating system. However, when Davies and Lundberg gave some females supplemental oats and mealworms for months at a time, female home ranges contracted substantially.[38] Those females with the most reduced ranges had, as predicted, fewer social mates than other females whose ranges had not diminished in size (Figure 8.31). In other words, as a female's home range contracted, the capacity of one male to monitor her activities increased, with the result that female polyandry tended to be replaced

FIGURE 8.31   A test of the female distribution theory of mating systems. When food supplements reduce the size of a female dunnock's home range, a male can monopolize access to that female and reduce the number of males with which she interacts. After Davies and Lundberg.[38]

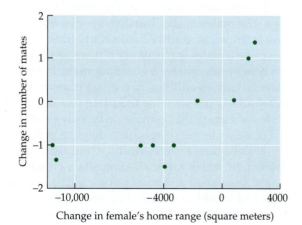

by female monogamy. These results support the general theory that males attempt to monopolize females within the constraints imposed on them by the spatial distribution of potential partners, which in turn may be affected by the distribution of food or other resources.

Discussion Question

8.14 The alpine accentor is a songbird with a mating system rather like that of its close relative, the dunnock, in that two males, an alpha and a beta, may live in the same area and copulate with the same female(s). However, in contrast to the dunnock, alpha male accentors reduce the amount of parental assistance they give to females as their mating share decreases, whereas the beta male does not make his paternal care donations dependent on his past mating frequency.[41] If female accentors have evolved a mating strategy designed to increase the parental assistance they receive from males, how should their polyandry differ from that of female dunnocks?

What should a female do when the selection of an already mated male means that she must share the resources under his control with other females? As we noted in Chapter 7, there is a point at which a female can gain more by mating with a polygynist on a good territory than by pairing off with a single male on a resource-poor, or predator-vulnerable, territory.[104] If females really are sensitive to this threshold, then we can predict that females paired with polygynous and monogamous males should have about the same fitness. Michael Carey and Val Nolan checked this prediction by counting the number of young fledged in a population of indigo buntings, a small songbird whose parental males may attract either one or two mates to their territories in overgrown fields.[30] A monogamous female whose relationship with a male lasted the whole breeding season had only slightly greater reproductive success (an average of 1.6 young fledged) than a female that participated in a polygynous arrangement (1.3 young fledged). Therefore, females choosing an already occupied territory were not penalized heavily, if at all.

Stanislav Pribil and William Searcy put the **polygyny threshold hypothesis** to an experimental test. They took advantage of their knowledge that female red-winged blackbirds in an Ontario population almost always choose unmated males over mated ones. Their usual refusal to mate with a paired male was adaptive, judging from the fact that on those rare occasions when females did happen to select mated males, they produced fewer offspring than females that have their monogamous partners' territories all to themselves. Pribil and Searcy predicted that if they could experimentally boost the quality of territories held by already mated males while lowering the value of territories controlled by unmated males, then mate-searching females should reverse their usual preference. They tested this prediction by manipulating pairs of red-wing territories in such a way that one of the two sites contained one nesting female and some extra nesting habitat (added cattail reeds rising from underwater platforms) while the other territory had no nesting female but had supplemental cattail reed platforms placed on dry land. Female redwings in nature prefer to nest over water, which offers greater protection against predators. For 14 pairs of territories, the first territory to be settled by an incoming female was the one where she could nest over water, even though this meant becoming part of a polygynous male's harem. Females that made this choice reared almost twice as many young on average as latecomers that had to make do with an onshore nesting platform in a monogamous male's territory.[112]

Thus, when there is a free choice between a superior territory and an inferior one, it can pay a female to pick the better site even if she has to share it with another female. But why, then, do some female red-wings, and females of other birds as well, ever accept a second-rate territory, as this usually means that the second female rears fewer offspring than the first female? Svein Dale and his colleagues found that unmated female pied flycatchers visit many males and do not rush into a paired relationship, suggesting that when they choose an already mated male, they do so voluntarily and with full knowledge.[37] Perhaps they choose these males anyway because of the high costs of finding other potential partners[138] or because the remaining unmated males have extremely poor territories.[142] If so, the options for some females may be either to accept secondary status or to not breed at all, just as late-arriving female red-wings may have to make do with what is available in their neighborhood, even if it means making the best of a bad job.

Scramble Competition Polygyny

Although female defense and resource defense tactics by competitive males make intuitive sense when females or resources are clumped in small, defensible areas, in many other species, receptive females and the resources they need are widely dispersed. Under these conditions, the costs of mating territoriality usually increase; when the costs exceed the benefits, males may simply try to find scarce receptive females before other males do. Flightless females of a *Photinus* firefly, for example, can appear almost anywhere over wide swaths of Florida woodland. Searching males of this species make no effort to be territorial; instead, they fly, and fly, and fly some more. When Jim Lloyd tracked flashing males, he walked 10.9 miles in total, following 199 signaling males, and saw exactly two matings. Whenever Lloyd spotted a signaling female, a firefly male also found her in a few minutes.[92] Mating success in this species almost certainly goes to those searchers that are the most persistent, durable, and perceptive, not the most aggressive.

Male thirteen-lined ground squirrels behave like fireflies, searching widely for females, which become receptive for a mere 4 to 5 hours during the breeding season. The first male to find an estrous female and copulate with her will fertilize about 75 percent of her ova, even if she mates again. Given the widely scattered distribution of females and the first-male fertilization advantage, the ability to keep searching should greatly affect a male's reproductive success. In addition, male fitness may depend on a special kind of intelligence,[55] namely, the ability to remember where potential mates can be located. After visiting a number of females near their widely scattered burrows, searching males often return to those places on the following day. When researchers experimentally removed several females from their home sites, returning males spent more time searching for those missing females that had been on the verge of estrus. Moreover, the males did not simply inspect places that the females had used heavily, such as their burrows, but instead biased their searches in favor of the spots where they had actually interacted with about-to-become-receptive females.[127,128] Individuals with superior spatial memory can probably relocate potential mates efficiently (Figure 8.32).

Another form of scramble competition polygyny, the **explosive breeding assemblage**, occurs in species with a highly compressed breeding season. One such species is the horseshoe crab, whose females lay their eggs on just a few nights each spring and summer. Males are under the gun to be near the egg-laying beaches at the right times and to accompany females to shore, where egg laying and fertilization occur.[23] A race to find and pair off with a female also occurs in the wood frog, another species in which the opportunity to acquire a mate is restricted to one or a few nights each year. On that night

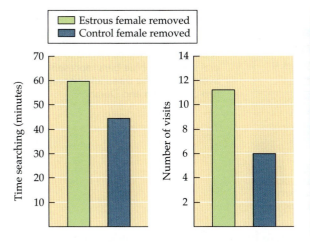

Thirteen-lined ground squirrel

FIGURE 8.32 Scramble competition polygyny selects for spatial learning ability. Male thirteen-lined ground squirrels remember the locations of females that are about to become sexually receptive. When males returned to an area where such a female had been the previous day but then had been experimentally removed, they spent more time searching for her than for a removed female not near estrus, and they returned to the near-estrus female's home range more often. After Schwagmeyer.[126]

or nights, most of the adult males in the population are present at the ponds that females visit to mate and lay their eggs. Just as in horseshoe crabs, the high density of rival males raises the cost of repelling them from a defended area. And because females are available only on this one night, a few highly aggressive territorial males cannot monopolize a disproportionate number of mates. Therefore, male wood frogs eschew territorial behavior and instead hurry about trying to encounter as many egg-laden females as possible before the one-night orgy ends (Figure 8.33).[21] They then have everything to gain by moving far from the breeding pond where predators come to kill and eat the frogs.[118]

Lek Polygyny

In some species, males do not search for mates, nor do they defend groups of females or resources that several females come to exploit. Instead, the males fight to control a very small area that is used only as a display arena; these

FIGURE 8.33 An explosive breeding assemblage. A male wood frog grasps a female (upper left) that he has found before rival males, two of which are near the mating pair. Numerous fertilized egg masses float in the water around the frogs. Photograph by Rick Howard.

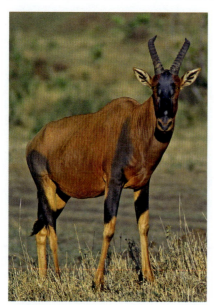

Topi male

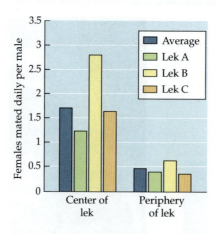

FIGURE 8.34   Mating success at leks. Each topi male in a central position at his lek mates with more females per capita than do males forced to peripheral sites. The blue bars represent the averages for the three leks shown. After Bro-Jørgensen and Durant.[22]

mini-territories may be clustered in a traditional display ground, or **lek**, or they may be somewhat scattered, forming a dispersed lek, as is true for satin bowerbirds.[18] Despite the fact that the males' territories do not contain food, nesting sites, or anything else of practical utility, females come to the leks anyway. When white-bearded manakin females arrive at a lek in a Trinidadian forest, they may find as many as 70 sparrow-sized males in an area only 150 meters square. Each male will have cleared the ground around a little sapling rising from the forest floor. The sapling and cleared court are the props for his display routine, which consists of rapid jumps between perches accompanied by loud sounds produced by snapping his clubbed primary wing feathers together. When a female is around, the male jumps to the ground with a snap and immediately back to the perch with a buzz, and then back and forth "so fast he seems to be bouncing and exploding like a firecracker."[135] If the female is receptive and chooses a partner, she flies to his perch for a series of mutual displays followed by copulation. Afterward, she leaves to begin nesting, and the male remains at the lek to court newcomers. Alan Lill found that in a lek with ten manakins, where he recorded 438 copulations, one male achieved nearly 75 percent of the total; a second male mated 56 times (13 percent), while six other males together accounted for a mere 10 matings.[91] Preferred males tend to occupy sites near the center of the manakin lek, and they engage in more aggressive displays than less successful males.[133]

Huge inequalities in male mating success are standard features of lekking species. Thus, in the topi antelope of the African savannas, it is generally the older males that occupy the center of a lek, and they copulate much more often than younger rivals forced to the periphery (Figure 8.34).[22] Likewise, just 6 percent of the males in a lek of the bizarre West African hammer-headed bat (Figure 8.35) were responsible for 80 percent of the matings recorded by Jack Bradbury. In this species, males gather in groups along riverbanks, each bat defending a display territory high in a tree, where he produces loud cries that sound like "a glass being rapped hard on a porcelain sink."[17] Receptive females fly to the lek and visit several males, each of which responds with a paroxysm of wing-flapping displays and strange vocalizations (note the similarity of these displays to those of satin bowerbirds).

Why do male hammer-headed bats, bowerbirds, and topi antelope behave the way they do? Bradbury has argued that lekking evolves only when other mating tactics do not pay off for males, thanks to a wide and even distribution of females.[18] Female satin bowerbirds and hammer-headed bats do not live in permanent groups but instead travel over great distances in search of widely scattered sources of food of unpredictable availability, especially figs and other tropical fruits. A male that tried to defend one tree might have a long wait before it began to bear attractive fruit, and when it did, the large amount of

FIGURE 8.35 A lek polygynous mammal: the hammer-headed bat.

food would attract hordes of consumers, which could overwhelm the territorial capacity of a single defender. Thus, the feeding ecology of females of these species makes it hard for males to monopolize them, directly or indirectly. Instead, males display their merits to choosy females that come to leks to inspect them.

HOTSPOTS, HOTSHOTS, AND FEMALE PREFERENCES   But you may recall that the absence of defensible clusters of females was invoked as an explanation for nonterritorial scramble competition polygyny as well as for leks. Why some species subject to these conditions simply search for mates while others form elaborate leks is not at all clear. Nor is it known for certain why males of some lekking species congregate in small areas as opposed to displaying solitarily in a dispersed lek (as in the satin bowerbird), although one possibility is that congregated males may be relatives that assist one another in the attraction of mates,[33] and another is that predation favors adoption of a safety-in-numbers tactic.[63] Any reductions in the cost of displaying help explain why lekking males congregate, but we still need to know what benefit a male derives from defending a little display territory at a group lek. Here we shall review three ideas: (1) the **hotspot hypothesis**, according to which males cluster in places (hot spots) where the routes frequently traveled by receptive females intersect;[20] (2) the **hotshot hypothesis**, according to which subordinate males cluster around highly attractive males in order to have a chance to interact with females drawn to these "hotshots";[19] and (3) the **female preference hypothesis**, according to which males cluster because females prefer sites with large groups of males, where they can more quickly, or more safely, compare the quality of many potential mates.[18]

To test these hypotheses, Frédéric Jiguet and Vincent Bretagnolle managed to create artificial leks populated by plastic decoys painted to resemble males and females of the little bustard, a bird that exhibits a dispersed-lek mating system. The two researchers placed different numbers of decoys of the two sexes in different fields and then, over some time, counted the number of living little bustards attracted to their experimental leks. They found that female decoys failed to draw in males, leading to the rejection of the hotspot

(A)

(B)

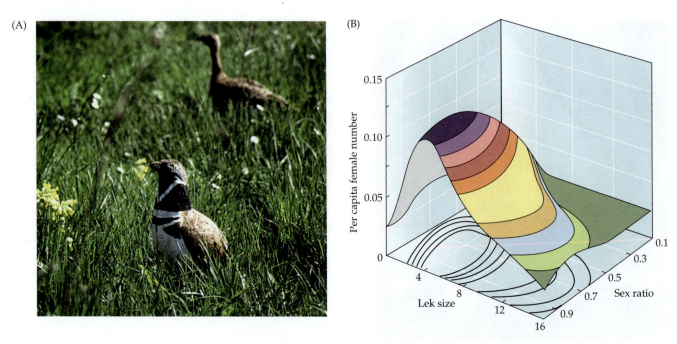

FIGURE 8.36   An experimental test of alternative hypotheses for lek formation.
(A) A female little bustard visiting a decoy made to look like a male of her species.
(B) More females visit groups of four decoys than smaller (or larger) groups, as shown
by the peak in the graph of the number of females visiting per decoy when there were
four decoys in the lek. The sex ratio represents the number of male decoys divided by
the total number of decoys. A, photograph courtesy of Frédéric Jiguet; B, after Jiguet
and Bretagnolle.[81]

hypothesis. On the other hand, male decoys regularly attracted both females
and males, particularly if those decoys had been painted to resemble individuals with highly symmetric plumage patterns. Therefore, the hotshot hypothesis may apply to little bustards. The fact that more females per decoy were
attracted to clusters of four decoys than to smaller groups (Figure 8.36) is
consistent with the female preference hypothesis as well, although more than
four decoys did not draw in additional females.[81]

   Another way of testing the hotspot versus hotshot hypotheses is by temporarily removing males that have been successful in attracting females. If
the hotspot hypothesis is correct, then removal of these successful males from
their territories will enable others to move into the favored sites. But if the hotshot hypothesis is correct, removal of attractive males will cause the cluster of
subordinates to disperse to other popular males or to leave the site altogether.

   In a study of the great snipe, a European sandpiper that displays at night,
removal of central dominant males caused their neighboring subordinates to
leave their territories. In contrast, removal of a subordinate while the alpha
snipe was in place resulted in his quick replacement on the vacant territory by
another subordinate. At least in this species, the presence of attractive hotshots,
not the real estate per se, determines where clusters of males form.[74] Likewise,
in the unrelated black grouse, although relatively large, centrally located display
sites are associated with higher mating success,[116] the exact location of the most
successful territory can change somewhat from year to year, suggesting that
a popular male, rather than a particular spot, most influences the behavior of
other males (Figure 8.37).[117]

   Although the hotshot hypothesis seems likely in some cases, the hotspot
hypothesis has received support in others (Figure 8.38).[62] For example, a site at

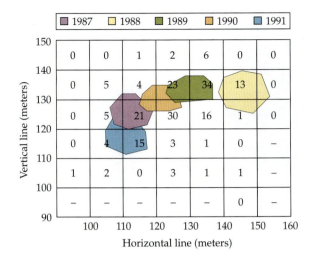

Black grouse

FIGURE 8.37 **Hotspots or hotshots?** Researchers divided a black grouse lek into 100-square-meter sectors and recorded the total number of copulations in each sector over a 5-year period from 1987 to 1991. The irregular polygons show the location of the top territory for each of the 5 years. The shifts in the preferred territory suggest that male attractiveness, rather than the territory itself, plays the key role in reproductive success in this species, as required by the hotshot hypothesis. After Rintamäki et al.[117]

which fallow deer bucks gathered to display shifted when logging activity altered the paths regularly followed by fallow deer does.[5] Likewise, peacocks tend to gather near areas where potential mates are feeding (hotspots); the removal of some males has no effect on the number of females visiting leks of this species, as one would predict if females are inspecting those areas because of the food they provide, not because of the males there.[93] In addition, the leks of certain tropical manakins are located in areas relatively rich in fruits, the food of these birds, which is consistent with the idea that males gather near places with resources attractive to females.[122]

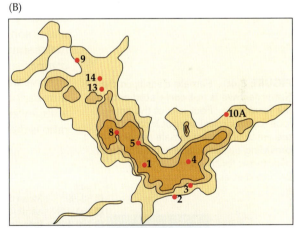

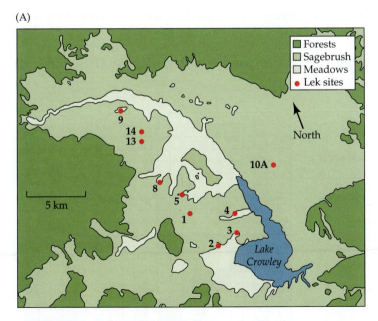

FIGURE 8.38 **A test of the hotspot hypothesis.** (A) The position of sage grouse leks (numbered red circles) in relation to sagebrush, meadows, forests, and a lake. (B) The distribution of nesting females in relation to the leks where males gather to display. The darker the shading, the more females present. After Gibson.[62]

Sage grouse

FIGURE 8.39 The mating success of male marine iguanas at a site in the Galápagos Islands in 1987, 1988, 1994, and 1995. Territories are numbered; color indicates number of copulations at the site. After Partecke et al.[110]

Marine iguana

Number of copulations

	0
	1–10
	11–20

Discussion Question

8.15 In the marine iguana of the Galápagos Islands, studies were done over 4 years measuring male mating success at a site with 22 reliably occupied territories. The territories with the most matings in a given year were identified (Figure 8.39). Are these data consistent with the hotspot or hotshot hypothesis for mate choice at a lek?

The hotspot hypothesis cannot, however, apply to those ungulates in which ovulating females leave their customary foraging ranges to visit groups of males some distance away, perhaps to compare the performance of many males simultaneously (Figure 8.40).[10] A female preference for quick and easy comparisons might make it advantageous for males of the Uganda kob, an African antelope, to form large groups. If so, those leks with a relatively large number of males should attract disproportionately more females than leks with fewer males. Contrary to this prediction, however, the operational sex ratio is the same for leks across a spectrum of sizes (Figure 8.41), so males are

FIGURE 8.40 Female density in the Kafue lechwe is not correlated with lek formation. Four leks of the antelope (open circles) are not located in the areas of highest female density, providing evidence against the hotspot hypothesis for this species. After Balmford et al.[10]

	Mean density of
<49	adult females
50–99	(per square kilometer)
100–149	
150–199	
200–249	

Kafue lechwe

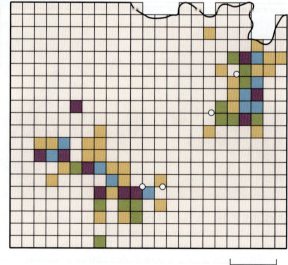

4 km

(A)
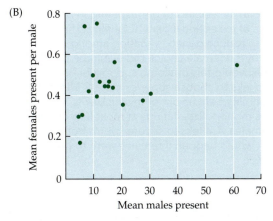

(B)

Uganda kob

FIGURE 8.41 **Female Uganda kob do not aggregate disproportionately at leks with large numbers of males.** (A) Female attendance at leks is simply proportional to the number of males displaying there. (B) As a result, the female-to-male ratio does not increase as lek size increases. After Deutsch.[42]

no better off in large groups than in small ones.[42] For this species at least, the female preference hypothesis can be rejected. The same is true for the barking treefrog. Here too, the more males chorusing in a pond, the more receptive females show up on a given night (Figure 8.42). But preventing males from coming to the pond to call does not reduce the number of females arriving, which is not what one would expect if a large number of calling males is needed to attract a large number of choosy females.[99]

Thus, no one hypothesis on why lekking males form groups holds for every species. Nevertheless, the interactions among males on a lek usually seem to enable individuals of high physiological competence to demonstrate

Female barking treefrog

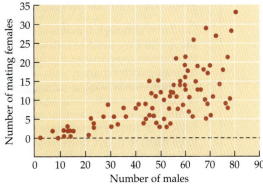

FIGURE 8.42 **The correlation between mating females and calling males of the barking treefrog.** The more female barking treefrogs mating at a pond on a given night, the more males found chorusing there. But the correlation stems not from the ability of large numbers of males to attract more females with their calls but from the fact that both sexes respond similarly to a set of environmental variables, including temperature and rainfall. After Murphy.[99]

their superior condition to their fellow males and to visiting females.[54] Whatever the basis for lek formation, lekking males are forced to compete in ways that separate the men from the boys, making it potentially advantageous for females to come to a lek to compare and choose, the better to select a partner of the highest quality. Indeed, one of the main themes of mating system theory is that in the vast majority of species, female reproductive tactics create the circumstances that determine which competitive and display maneuvers will provide payoffs for males. As a result, male mating systems are an evolved response to female mating systems—and to the ecological factors that determine the spatial distribution of females.

Discussion Question

8.16 One of the possible benefits to females of participating in lek mating systems is the selection of mates of exceptional genetic quality. This form of sexual selection should tend to reduce genetic variation among males of lekking species over time. (Why?) If female choice did eliminate genetic variation among males over time, would this outcome support or undercut the argument that females go to leks to choose superior mates? Might the phenomenon of genetic compatibility help us out of this pickle? Is, however, the finding that a very few males monopolize matings at most leks consistent with the suggestion that females visit leks to secure sperm from genetically compatible partners? What would happen if females were to prefer different traits from year to year?[32]

Summary

1. Mating systems can be defined in terms of the number of sexual partners an individual acquires during a breeding season. Males can be monogamous (mating with a single female) or polygynous (mating with several females). Likewise, females can be monogamous (having a single partner) or polyandrous (mating with several males).

2. Monogamy by males is a Darwinian puzzle because a male that restricts himself to a single female would seem unlikely to have more descendants than a male that successfully mates with many females. Monogamous males can gain fitness, however, if there are large payoffs to those individuals that prevent their sole mates from accepting sperm from other males or to those males that invest paternally in their sole mates' offspring. Alternatively, conflict between females and males may thwart male attempts to be genetically polygynous, even if it would be advantageous from the male perspective.

3. Similarly, although females of most birds and other animals can secure all the sperm they need to fertilize their eggs from a single mate, polyandry is common, even in socially monogamous birds. What do polyandrous females gain by mating with several males, given the costs of superfluous copulations? Polyandrous females may secure any of several genetic or material benefits, ranging from superior genes for their offspring to greater amounts of parental assistance from their partners. But perhaps females of some species mate with several males because those males force them to do so. Both female choice and conflict between the sexes probably play a part in the evolution of mating systems.

4. The diversity of polygynous mating systems may have evolved in response to different patterns of female distribution that affect the profitability of different kinds of territorial mating tactics for males. When females or the resources they need are clumped in space, female defense or resource defense polygyny becomes more likely. If, however, females are widely dispersed or male density is high, males may engage in non-territorial scramble competition for mates, or they may acquire mates by displaying at a lek. Many aspects of lek polygyny are still not completely understood, such as why it is that males of most lekking species gather in groups to display to females.

Suggested Reading

Steve Emlen and Lew Oring's now classic paper changed the way we look at the evolution of mating systems.[51] Their view is developed by Malte Andersson in his book on sexual selection.[2] For an especially thorough examination of a single species with several mating systems, read *Dunnock Behaviour and Social Evolution*.[40] Steve Shuster and Michael Wade offer their own take on mating system evolution in *Mating Systems and Strategies*.[134] The at-one-time surprising discovery that birds are not nearly as monogamous as they seem to be is the subject of reviews by Griffith et al.[66] and Westneat et al.[158]

9

The Evolution of Parental Care

 Although my children are now middle-aged adults, I can remember when they were young and in need of parental care. My recollection, no doubt exaggerated, is that I did my share back then. Even if I am giving myself more credit than is my due, which is likely, it is true that the human species is one in which biparental care is the norm. In contrast, in a great many animal species neither the mother nor the father lifts a finger for their offspring, while in others just one parent—either the female or, much more rarely, the male—takes full responsibility for her or his progeny. You may recall that nest building and the provisioning of brood cells is strictly the province of the female digger bees, whereas males do not spend a minute helping their offspring.

The key to explaining the diversity in parental behavior lies with the cost–benefit approach of behavioral ecology. The benefits of caring for one's offspring typically include the improved survival of the assisted progeny. But what is given to one infant is rarely available for the production of additional offspring, a future cost that is difficult to measure but can potentially overwhelm any immediate benefit.[107] We shall therefore consider both sides of the parental equation, beginning with a study of songbird parenting. In this group, both sexes often bring food to their all-but-helpless offspring in a nest. But parent songbirds adjust their devotion to their young in line with the costs and benefits of their feeding visits, as we shall see shortly.

A male frog of a newly discovered species protecting eggs on a leaf in Papua New Guinea. Why has paternal care evolved in this frog?

(A)

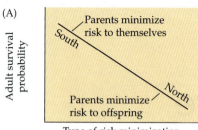

(B) Nest predator

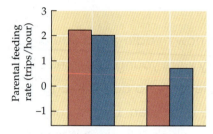

(C) Adult predator

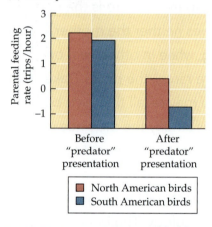

■ North American birds
■ South American birds

FIGURE 9.1 Parental care has fitness costs and benefits. (A) The reactions of parent birds to predatory threats to themselves and their offspring should vary in relation to the annual mortality rates of adults, which differ between South and North America. Shorter-lived North American birds are predicted to do more to reduce risks to their offspring; longer-lived South American birds are predicted to do more to promote their own survival. (B) The rate at which parents feed their nestlings falls more sharply in North American birds than in South American birds in response to an apparent risk of nest predation by jays, which can find nests by watching where parent birds go. (C) The rate at which parents feed their nestlings falls more sharply in South American birds than in North American birds after the adults see a hawk capable of killing them. After Ghalambor and Martin.[24]

The Cost–Benefit Analysis of Parental Care

By feeding their young, adult birds increase the survival chances of their nestlings. But there are risks attendant upon the parents' feeding trips. Predators can use parental comings and goings to find the nest and feast on its occupants; alternatively, predators can lurk near the nest to intercept and eat the parents as they return to their offspring with food. How does a parental bird balance the pluses and minuses of its activities?

Cameron Ghalambor and Tom Martin predicted that parent birds should adjust their provisioning behavior adaptively in accordance with two key factors: the nature of the predator (whether it consumes nestlings or adults) and the annual mortality rate for breeding adults.[24] In birds with a generally low adult mortality rate, parents ought to minimize the risk of getting themselves killed by a predator, because they will probably have many more chances to reproduce in the future—if a predator does not get them now. In breeding birds with a high annual mortality rate, however, parents should be less concerned with their own safety and more sensitive to the risks that their nestlings may confront from nest-raiding predators; these parents will have relatively few chances to reproduce in the future, so they can gain by doing more for their current brood.

Ghalambor and Martin knew that birds that breed in North America tend to have shorter lives and produce larger clutches of eggs than their close relatives that breed in South America. So they matched up five pairs of these relatives, including, for example, two members of the same genus, the short-lived North American robin and the longer-lived Argentinian rufous-bellied thrush. When the ornithologists played tapes of Steller's jays to Arizonan robins and tapes of plush-capped jays to their Argentinian counterparts, the robins and thrushes both greatly reduced their visits to their nests for some time, which under natural conditions would help hide their nests from these predators. However, the robins curtailed their activity around the nest more strongly than the thrushes, presumably because they had more to gain by protecting their current crop of nestlings from keen-eyed jays, given their relatively low probability of reproducing in subsequent years.

When the ornithologists placed a stuffed sharp-shinned hawk, a killer of adult birds, near active nests and played recorded sharp-shin calls, the parent birds in the sampled species again reduced their visits to their nests for some time. In this round of tests, however, the potentially long-lived Argentinian birds delayed their return to the nest longer than the corresponding Arizonan species (Figure 9.1).[24] This case is merely one of many in which the parental strategies of North American birds appear to differ from those of South American species, probably because of differences in predation pressure on nesting adults in the two regions.[64] The costs of taking care of offspring, not just the benefits, have fine-tuned the evolution of parental behavior in these birds.

Discussion Question

9.1 If songbirds can evaluate the risk of nest predation, and make adaptive adjustments in the length of incubation periods, what predictions follow for New Zealand bellbirds in (1) mainland populations where introduced nonnative predators (rats, cats, possums, and the like) are common versus (2) island populations that have never been affected by exotic predators? If incubating females do differ behaviorally in the two locations, and they do, what does this say about the costs of engaging in relatively long bouts of incubation? What are the conservation implications of finding that bellbirds have adjusted their parental behavior in relation to the presence of exotic predators, which were introduced after AD 1300?[66]

(A)

(B)
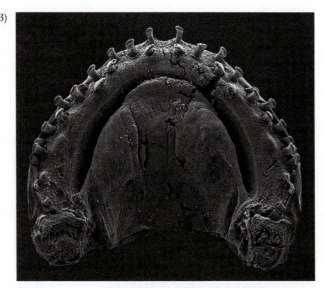

FIGURE 9.2 Extreme maternal care in caecilian amphibians. (A) Mother caecilians live with their offspring and permit them to remove and feed upon their nutritious skin. (B) Nestling caecilians have curved teeth for the very purpose of stripping the edible skin from their mothers. See Wilkinson et al.;[111] photographs by Carlos Jared.

Why More Care by Mothers than by Fathers?

Although in many species of birds, including robins and thrushes, both fathers and mothers help their progeny survive, care by just one adult is common in the animal kingdom. You might think that about half of all cases of uniparental care would involve the mother and the other half would involve the father. If so, you would be wrong. Females are much more likely than males to be the sole parent in charge in those species with uniparental care. Some maternal sacrifices are quite striking, as in the readiness of certain mother caecilians (burrowing worm-like amphibians) to let their youngsters feed on their lipid-rich epidermis, which the infant caecilians strip from their mother's body (Figure 9.2).[111] Then there are the females of a spider, *Stegodyphus lineatus*, that not only feed their spiderlings regurgitated food but eventually permit their brood to cannibalize them completely,[88] a phenomenon also observed in some pseudoscorpions.[22]

The maternal sacrifices made by some species of treehoppers are less dramatic but still substantial as the females, never the males, stand watch over their eggs day and night to protect them against predatory or parasitic insects that would destroy their broods. In some cases, females even stay on to protect their nymphs until they have become adults. Chung-Ping Lin and his coworkers explored the evolutionary history of egg-guarding parental care by mapping the trait on a phylogeny of the treehopper subfamily, the Membracinae, that was derived from molecular comparisons among the genera in this group.[52] Their work indicates that while maternal care has probably originated three different times in the Membracinae (Figure 9.3), paternal care has never evolved in these insects.

Discussion Question

9.2   On the basis of Figure 9.3, how many times has maternal care been lost after evolving in the Membracinae? If you were to employ the comparative method to examine the evolved function of maternal care in this group, what genera would be of special interest to you, and why?

One intuitively appealing explanation for the female domination of parental caregiving, as illustrated by the treehoppers, is that because females (unlike

FIGURE 9.3 Only females provide parental care in the treehoppers. The genera of Membracinae that exhibit maternal care have been placed on a molecular phylogeny of this subfamily of treehoppers. The circles at the bases of the lineages represent ancestral species; the proportion of a given circle that is blue reflects the probability that females of this ancestral species cared for their eggs. The numbers mark the three probable origins of maternal care. The photograph shows a female treehopper in the genus *Guayaquila* standing guard over a cluster of eggs covered by a secretion applied by the female. After Lin et al.;[52] photograph by Chung-Ping Lin.

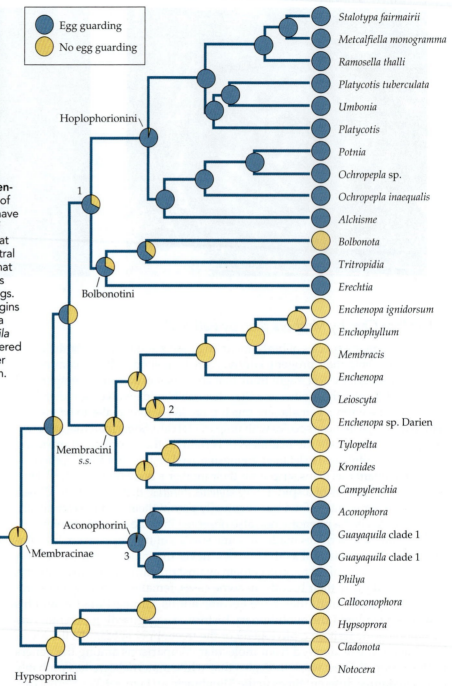

males) have already invested so much energy in making eggs, they have a special incentive to make sure that their large initial gametic investment is not wasted, and so they come to the aid of their offspring more often than males. This argument comes to grief, however, when we observe that females of a substantial number of species, including spotted sandpipers and many fishes, abruptly end their parental activities after laying their large and costly eggs, which they leave totally in the care of their male partners (Figure 9.4). These species show that a considerable initial investment in offspring does not automatically make it advantageous for females to invest still more in their broods. Instead, both the benefits and costs of each additional increment of maternal care will determine whether giving still more is adaptive.[20]

(A)

(B)

FIGURE 9.4　Paternal care in fishes and opportunities for polygyny. (A) A male Randall's jawfish holds his mate's eggs in his mouth. Mouth brooding limits a male to one clutch at a time. (B) In contrast, a male stickleback caring for a nest with a clutch of eggs can attract additional females, which add their eggs to the nest. This male is aerating the eggs in his nest, at the base of the aquatic plant, by drawing water through the nest.

That maternal behavior costs the caregiver something can be illustrated by examining the effects of brood tending on female European earwigs, which often stay with clutches of eggs laid in burrows in order to feed their larval offspring after the eggs have hatched (Figure 9.5). Females can help their little earwigs survive, an obvious benefit of their tending behavior, but they pay a price as well. For maternal females, the interval between laying one clutch of eggs and the next is a week longer than for females that do not stay around to help one group of offspring get a good start in life.[45] The same cost applies to females of the harvestman *Serracutisoma proximum*, which lay more clutches and have nearly a 20 percent higher lifetime fecundity if they are prevented from standing guard over their eggs, compared with those females that tend their clutches.[9]

Because there is no guarantee that females will always derive a net benefit from an extra dollop of parental care, how can we explain the general pattern of female-only parental care? David Queller has come to the rescue (see also page 177) by pointing out that if the costs of parental care are usually lower for females than for males, as they may well be, then this could tip the scales so that females would be expected to provide more care than males.[84]

FIGURE 9.5　A maternal female earwig guards her clutch of eggs, which she protects against predators. In nature, as opposed to the laboratory, most females lay their eggs under leaf litter or some other shelter. Photograph by Mathias Kölliker.

Let's assume for the sake of simplicity that one standard unit of parental care invested in a current offspring reduces the future reproductive output of a male and a female by the same amount. Let's also assume that we are looking at a polyandrous species in which females sometimes mate with more than one male in a breeding season. In this case, the average benefit to a male from caring for a brood of offspring will be reduced to the extent that some of "his" offspring were actually fathered by other males. For example, if his paternity is, on average, 80 percent, then for every five offspring assisted, the male's investment can yield at most only four descendants, whereas all five youngsters could advance the female's reproductive success. In other words, when cuckolded paternal males provide parental care to nonrelatives, the benefits of being a good parent decrease for them.

Not only are the benefits from paternal care likely to be less than the benefits from a comparable amount of maternal care, but the costs of parental care are likely to be greater for males than for females. As we noted when discussing sexual selection theory in Chapter 7, males that acquire many mates generally leave many descendants. Potential Casanovas would pay a steep price if they were to care for a few offspring at the cost of missing some additional mates. Imagine a lek of black grouse in which the top male fertilizes most of the eggs of the 20 or so females that come to the lek to mate. Because regular attendance at the lek is one of the main correlates of male mating success, a grouse with a reasonable chance of becoming an alpha male would lose fitness if he took time off from lekking to incubate a batch of eggs.[84] The same rule surely applies to sexually attractive males of many other species.

Discussion Questions

9.3 Here's another way of looking at the effect of differences between males and females on the probability that their putative offspring are truly their genetic offspring. If a female lays a fertilized egg or gives birth to a baby, this offspring will definitely have 50 percent of her genes. In contrast, a male that mated with this female may or may not be the father of that offspring. Thus, the argument goes, males have less to gain by parental behavior. But imagine a hypothetical species in which males have, say, a 40 percent chance of siring the offspring of any given mate. Further, imagine that there are two hereditarily different male phenotypes in the population, a paternal type and a nonpaternal type. The average paternal male mates with two females (each with an average of 10 eggs), whereas the average nonpaternal male mates with five females (which enables him to fertilize 40 percent of 50 eggs in all). In addition, let's say that the paternal male boosts the survival chances of the eggs under his care to 50 percent, whereas the unprotected offspring of nonpaternal males have a 10 percent survival rate. Which behavior is adaptive here? Show your math. What point does this example make about the evolution of male parental care?

9.4 Among certain monkeys and apes with prolonged parental care, females live longer than males in species in which females do most or all of the parenting, but males live longer than females in species in which males make the major contribution to offspring care.[2] In other words, adults of the parental sex tend to live longer than the nonparental sex. Does this finding indicate that parental care improves the survival chances of caretakers? Imagine that someone claims that the longer life span of the parental sex has been selected for because primate young are very slow to develop, and therefore parents must live long enough to get their offspring to the age of independence in order to maintain a stable population. Do you agree? Do you have an alternative explanation for the observed pattern?

Why Are Any Males Paternal?

Given the clear logic of Queller's analysis of the costs of parental care, you might think that paternal care should be exceptionally rare, but in fact, male-only parental care has evolved in a number of animals and is actually quite common among fishes (see Figure 9.4). Because male fish are loaded with sperm and could therefore potentially have many more offspring than the most fecund female of their species, they would seem, at first glance, to have much to lose by taking time and energy away from mating effort to be good parents. Upon reflection, however, we can see that there does not have to be a trade-off between parental care and mate attraction in mating systems in which females are drawn to egg-guarding, parentally committed males.

In fact, stickleback males can care for up to ten clutches of eggs in one nest over the 2 weeks or so that it takes the eggs to hatch. In contrast, an average female stickleback can produce only seven clutches of eggs during this period, even without taking time out to guard her eggs.[14] Parenting would be far less advantageous for stickleback females than for males, given that a female would be brooding only one clutch at a time. Furthermore, while the female was so engaged, she would not be able to forage freely and so would not grow as rapidly as she might otherwise, an especially damaging outcome for those species in which female fecundity increases exponentially with increasing body size. Males that are parental also grow more slowly than they would otherwise, but since they must remain in a territory in any case if they are to attract mates, the decrease in their growth resulting exclusively from parental care is negligible.

Discussion Question

9.5 Males of the golden egg bug are sometimes chosen by females to receive their eggs, which are glued to the males' backs. Why do males accept this burden? Either males carry eggs to attract gravid females, which may then copulate with them, or males carry eggs (of their mates) to decrease the risk that their offspring will be afflicted by parasites. Given these alternatives, what significance do you attach to the following three results: (1) males from an area where egg parasites are numerous are much more likely to carry eggs than those from a region where egg parasites are essentially absent; (2) eggs laid on plants, an alternative for egg-laden females, are up to ten times more likely to be destroyed by parasites than eggs laid on male bugs; and (3) when females were given a choice between mating with an egg-bearing male versus one unencumbered by eggs, they did not choose the egg bearers significantly more often than those without eggs.[26]

The costs of parental behavior for the two sexes have been directly measured for a mouth-brooding cichlid, St. Peter's fish, in which either the male or the female may care for their young by orally incubating the fertilized eggs. Both sexes lose weight while mouth brooding, not surprisingly since it is difficult to eat with a mouth full of eggs or baby fish. Furthermore, the interval between spawnings increases for parental fish of both sexes compared with individuals whose clutches are experimentally removed (Figure 9.6). However, parental females wait 11 more days between spawnings than nonparental females, whereas brooding males pay a smaller price—only 7 extra days between spawnings compared with nonparental males. Moreover, parental females produce fewer young in the next clutch than do nonparental females,

St. Peter's fish

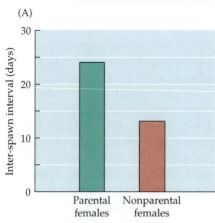

(A)

Inter-spawn interval (days)

Parental females — Nonparental females

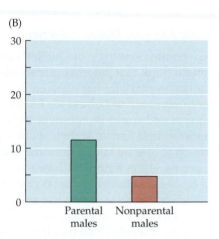

(B)

Parental males — Nonparental males

FIGURE 9.6 Parental care costs female St. Peter's fish more than it costs males. (A) Females that have cared for a clutch of eggs are much slower to produce a new batch of mature eggs than are females that have not provided care to their previous clutch. (B) Parental males also spawn less often than nonparental males, but the difference between the two groups is less than for females. After Balshine-Earn.[5]

whereas parental males are just as able to fertilize complete clutches of eggs in their next spawning as are nonparental males. Therefore, although both sexes pay a price for being parental, the costs of brood care are greater for females than for males.[5] This factor makes it more likely that paternal males will have higher fitness than nonpaternal individuals.[30]

Why Do Male Water Bugs Do All the Work?

Although exclusive paternal care of young is common among fishes, the trait is, as expected, rare among other animals, vertebrates and invertebrates alike.[13,105] The exceptions to the rule include some harvestmen, such as *Iporangaia pustulosa*, whose males sometimes spend months standing near eggs laid on the underside of leaves (Figure 9.7). In a male-removal experiment lasting 12 days, clutches that were left unattended were attacked in almost every instance, whereas half of the control clutches with a guarding male were intact at the end of the study.[85] In species of harvestmen with paternal care, females appear to be attracted to partners with eggs in their care, suggesting that the males' unusual behavior has evolved via sexual selection, just as is often true for fishes with highly paternal males.[63,77]

Among the paternal insects are large male water bugs in the genus *Lethocerus*, which guard and moisten clutches of eggs that females glue onto the stems of aquatic vegetation above the waterline (Figure 9.8A).[101] In some other water bug genera (e.g., *Abedus* and *Belostoma*), males permit their mates to lay eggs directly on their backs (Figure 9.8B). A brooding male *Abedus* spends hours perched near the water surface, pumping his body up and down to keep well-aerated water moving over the

FIGURE 9.7 A male harvestman (*Iporangaia pustulosa*) protecting eggs that he has surely fertilized. In this tropical species, males are the guardians of offspring; without paternal care, egg survival is seriously compromised. Photograph by Glauco Machado.

(A)

(B)

FIGURE 9.8   Male water bugs provide uniparental care. (A) In the genus *Lethocerus*, males stand watch over eggs their mates lay on vegetation out of water. (B) In other genera of belostomatids, males remain in water to brood eggs glued onto their backs by their mates. A, photograph by Robert Smith; B, photograph by the author.

eggs. Clutches that are experimentally separated from male attendants do not develop, demonstrating that male parental care is essential for offspring survival in this case.

Bob Smith has explored both the history and the adaptive value of these unusual paternal behaviors.[102] Since most insects, including some close relatives of the Belostomatidae, lack male parental care, it is probable that species with egg-brooding males evolved from nonpaternal ancestors (Figure 9.9).

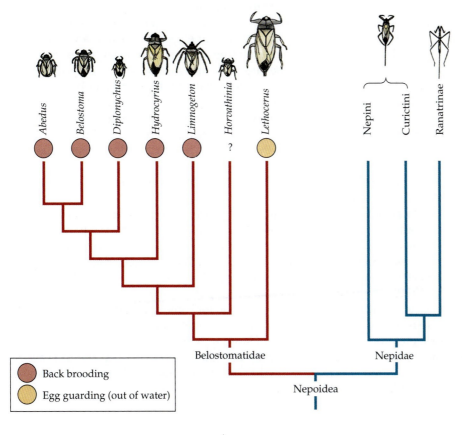

Abedus
Belostoma
Diplonychus
Hydrocyrius
Limnogeton
? *Horvathinia*
Lethocerus

Nepini
Curictini
Ranatrinae

Belostomatidae

Nepidae

Nepoidea

Back brooding

Egg guarding (out of water)

FIGURE 9.9   Evolution of brood care by males in the *Nepoidea*, the group that includes the belostomatid water bugs. The illustrations, drawn to scale, show the largest representatives of each group. Paternal care is widespread throughout the family Belostomatidae, but no species in the family Nepidae exhibits this trait. After Smith.[102]

Whether out-of-water brooding and back brooding evolved independently from such an ancestor, or whether one preceded the other, is not known, although when *Lethocerus* females cannot find suitable exposed vegetation for their eggs, they sometimes lay their eggs on the backs of other individuals, male or female. This unusual behavior indicates how the transition from out-of-water brooding to back brooding might have occurred.

But why do the eggs of water bugs require costly brooding? Huge numbers of aquatic insects lay eggs that do just fine without a caretaker of either sex. However, Smith notes that belostomatid eggs are much larger than the standard aquatic insect egg, with a correspondingly large requirement for oxygen needed to sustain the high metabolic rates underlying embryonic development. But the relatively low surface-to-volume ratio of a large aquatic egg leads to an oxygen deficit inside the egg. Since oxygen diffuses through air much more easily than through water, laying eggs out of water can solve that problem. But this solution creates another problem, which is the risk of desiccation that the eggs face when they are high and dry. The solution, brooding by males that moisten the eggs repeatedly, sets the stage for the evolutionary transition to back brooding at the air–water interface.

Wouldn't things be simpler if belostomatids simply laid small eggs with large surface-to-volume ratios? To explain why some water bugs produce eggs so large that they need to be brooded, Smith points out that water bugs are among the world's largest insects, an advantage when it comes to grabbing and stabbing fish, frogs, and tadpoles. Water bugs, and all other insects, grow in size only during the immature stages. After the final molt to adulthood, no additional growth occurs. As an immature insect molts from one stage to the next, it acquires a new flexible cuticular skin that permits an expansion of size, but no more than a 50 or 60 percent increase per molt. One way for an insect to grow large, therefore, would be to increase the number of molts before making the final transition to adulthood. However, no member of the belostomatid family molts more than six times. This observation suggests that these insects are locked into a five- or six-molt sequence, just as spotted sandpipers evidently cannot lay more than four eggs per clutch. If a water bug is to grow large enough to kill an adult frog, then the first instar (the nymph that hatches from the egg) must be large, because it will get to undergo only five or so 50-percent expansions. In order for the first-instar nymph to be large, the egg must be large. In order for large eggs to develop quickly, they must have access to oxygen, which is where male brooding comes into play. Thus, male brooding is an ancillary evolutionary development whose foundation lies in selection for a body size that enables a bug to take down relatively large prey.[101] The water bug story revolves around the "panda principle" (see page 76) in that evolutionary modifications of body size were layered on what had already evolved in this lineage.

Female water bugs, however, could conceivably provide care for their own eggs after laying them on exposed aquatic vegetation. Why is it that the males do the brooding, never the females? Here we have a story that parallels the fish case closely. First, male water bugs with one clutch of eggs sometimes attract a second female, perhaps because a male bug with a partial load of eggs is effectively advertising his capacity for parental care, just like a male stickleback at his nest.[105] Second, just as is true for some fishes, the costs of parental care may be disproportionately great for females in terms of lost fecundity. In order to produce large clutches of large eggs, female belostomatids require far more prey than males do. Because brooding limits mobility and thus access to prey, parental care probably has greater fitness costs for females than for males, biasing selection in favor of male parental care.

Discriminating Parental Care

No matter which parent provides care, an adaptationist would not expect the caring parent or parents to provide assistance freely to young animals that are not their own genetic offspring. But sometimes parents might have a hard time identifying their youngsters. Consider the Mexican free-tailed bat, which migrates to certain caves in the American Southwest, where pregnant females form colonies in the millions. After giving birth to a single pup, a mother bat leaves her offspring clinging to the roof of the cave in a crèche that may contain 4000 pups per square meter (Figure 9.10). When the female returns to nurse her infant, she flies back to the spot where she last nursed her pup and is promptly besieged by a host of hungry baby bats.[67] Given the swarms of pups, early observers believed that mothers simply provided milk on a first-come, first-served basis.

But are Mexican free-tailed bat mothers really indiscriminate parents? To find out, Gary McCracken captured female bats and the pups nursing from them and took blood samples from both.[67] He then analyzed the samples using starch gel electrophoresis, a technique that can show whether two individuals have the same variant form of a given enzyme, and thus the same allele of the gene that codes for that enzyme. If female bats are indiscriminate care providers, then the enzyme variants of females and the pups they nurse should often be different. But if females tend to nurse their own offspring, then females and pups should tend to share the same alleles. McCracken focused on the gene for superoxide dismutase, an enzyme represented by six different forms in the population he sampled. Despite the chaotic conditions within the bat colony, the enzyme data indicated that females found their own pups at least 80 percent of the time, a conclusion confirmed and extended by more recent evidence that females use the vocal and olfactory signals of pups to recognize their own pups before letting them nurse.[4,68]

FIGURE 9.10 **Mexican free-tailed bat mothers recognize their pups,** despite the fact that they leave their infants in a dense mass of baby bats when they leave their caves to forage outside. When a female returns to nurse her pup, she almost always relocates her son or daughter among thousands of other individuals.

Discussion Questions

9.6   McCracken found that although female Mexican free-tailed bats usually feed their own pups, they do make occasional "mistakes," which they could have avoided if each pup were left in a spot by itself instead of in a crèche with hundreds of other babies.[67] Does this mean that the parental behavior of this species is not adaptive? Use a cost–benefit approach to develop alternative hypotheses to account for these "mistakes."

9.7   In some cases, males or females do care for young other than their own, as when certain male fish take over and protect egg masses being brooded by other males or when female ducks acquire ducklings that have just left someone else's nest (Figure 9.11). Devise alternative hypotheses to explain this phenomenon. Under what circumstances might adoptions actually raise the caregiver's reproductive success? Under what other circumstances might adopters help nongenetic offspring as a cost of achieving some other goal?

If the ability to recognize one's genetic offspring is advantageous in proportion to the risk of misdirected parental care, then we would expect convergent evolution of parent–offspring recognition in colonial mammals other than Mexican free-tailed bats. Tests of this prediction have been positive in species as different as the degu, a roly-poly little rodent whose females rear young together in communal burrows,[40] and the subantarctic fur seal, whose females give birth on crowded island beaches. Female seals remain with their pups for about a week before leaving for an oceanic fishing trip that can last as long as 3 weeks. Playback experiments demonstrate that a baby seal takes no more than 5 days to learn its mother's call, while the females are also quick learners. When a mother seal returns to the beach, she calls out, and her infant calls back, usually leading to their reunion in less than 15 minutes.[10]

Yet another comparative test of the hypothesis that offspring recognition functions to prevent misdirected care takes advantage of variation among swallows in the risk of making parental mistakes. Although both the bank swallow and the rough-winged swallow nest in clay banks, the bank swallow is colonial, whereas the rough-winged swallow nests by itself. Individual fledglings of the colonial bank swallow produce highly distinctive vocalizations, giving their parents a reliable cue to use when making decisions about which individuals to feed and which to repel (such as the fledglings that sometimes wind up in the wrong nests begging for food). Bank swallow parents rarely make mistakes, despite the high density of nests in their colonies.[6,69]

FIGURE 9.11   Adoption by a female goldeneye. The duck escorts a number of ducklings, some of which are not her offspring. Photograph by Bruce Lyon.

The solitary rough-winged swallow, on the other hand, never has a chance in nature to feed another's fledglings and so would not be expected to have evolved sophisticated offspring recognition mechanisms. Indeed, rough-winged swallow chicks produce calls that sound much more alike than those of bank swallows, a reflection of the fact that young rough-wings need not communicate their identity to their parents.

Two other swallow species, the highly colonial cliff swallow and the less social barn swallow, also differ in their chick recognition attributes. As

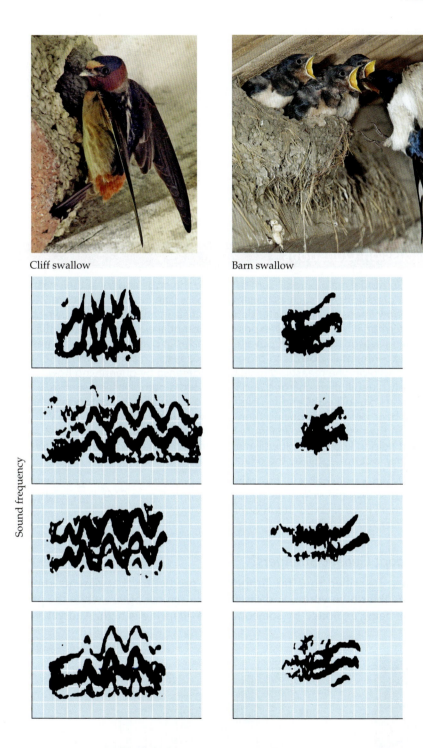

Cliff swallow   Barn swallow

Sound frequency

FIGURE 9.12 Recognition of offspring evolves when the risk of misdirected parental care is high. Chicks of cliff swallows, a colonial species, produce highly structured and distinctive calls, helping their parents recognize them as individuals. The calls of barn swallow chicks, a less colonial species, are much less structured and more similar. Sonograms of the sort shown here (and elsewhere) provide a visual record of the sound frequencies produced by the signalers over time. The call frequencies of both species lie between 1 and 6 kHz; the durations of the calls range from 0.7 to 1.3 seconds for the cliff swallows and from 0.4 to 0.8 seconds for the barn swallows. After Medvin et al.[70]

expected, cliff swallow chicks produce calls containing about 16 times as much variation as the corresponding calls of barn swallow chicks (Figure 9.12).[70] Therefore, it should be easier for cliff swallows to recognize their young than for barn swallows to discriminate among barn swallow chicks. In operant conditioning experiments that required adults of both species to discriminate between pairs of chick calls, cliff swallows reached 85 percent accuracy significantly faster than barn swallows. These results suggest that the acoustical perception systems of the cliff swallow, as well as its calls, have evolved to promote accurate offspring recognition,[70] just as is true of Mexican free-tailed bats and subantarctic fur seals.

FIGURE 9.13   Reactions of nest-defending bluegill males to potential egg and fry predators under two conditions. Experimental males had been exposed to clear containers holding smaller male bluegills, mimicking the presence of rivals who might fertilize some of the eggs in the defenders' nests; control males were not subjected to this treatment. "Parental care" was quantified using a formula based on the number of displays and bites directed at a plastic bag holding a predatory pumpkinseed sunfish. After Neff.[78]

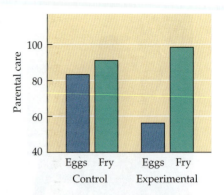

Bluegill

Discussion Questions

9.8   Territorial male bluegill sunfish defend the eggs and fry in their nests against predatory fish such as largemouth bass (see Figure 3.5). Figure 9.13 shows how intensely males defended their nests in an experiment in which some territorial bluegills were exposed to potential cuckolders during the spawning season. Bryan Neff put sneaker males in plastic jars near the nests of his experimental subjects to provide the cues associated with a high risk of cuckoldry; he measured male brood defense by quantifying how intensely bluegill dads charged and threatened a pumpkinseed sunfish, which eats bluegill eggs and fry. Neff placed the sunfish predators in clear plastic bags before introducing them next to bluegill nests.[79] How do you interpret the results shown in Figure 9.13? What is puzzling about them? Does it help to know that bluegill males can apparently evaluate the paternity of fry, but not eggs, by the olfactory cues they offer?

9.9   When a male baboon intervenes in disputes between two juveniles, he tends to take the side of his genetic offspring (Figure 9.14). How might researchers have secured this information?

Why Adopt Genetic Strangers?

The cases described thus far support the prediction that parents should recognize their own young and discriminate against others when the probability of being exploited by someone else's offspring is high. And yet some

(A)

(B)

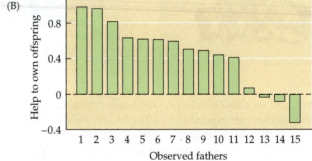

FIGURE 9.14   Male baboons intervene on behalf of their own offspring when young baboons start fighting with one another. (A) The adult male cradles his offspring, protecting it against an aggressive youngster. (B) Of 15 fathers whose behavior was monitored, 12 helped their own offspring rather than unrelated juveniles. A, photograph by Joan Silk; B, after Buchan et al.[8]

colonial, ground-nesting gulls occasionally adopt unrelated chicks. Here we have another challenging Darwinian puzzle. Although researchers initially reported that adults consistently rejected older, mobile chicks when they were experimentally transplanted between nests,[71] attacks by adults on these transferred youngsters apparently occurred because of the frightened behavior of the displaced chicks.[27] When juveniles voluntarily leave their natal nests—which they sometimes do if they have been poorly fed by their parents (Figure 9.15)—they do not flee from potential adopters but instead beg for food and crouch submissively when threatened. These youngsters have a good chance of being adopted, even at the ripe old age of 35 days.[36] If they are taken in, they are more likely to survive than if they had remained with the genetic parents that were failing to supply them with enough food.[7]

When parents apparently fail to act in the best interests of their genes, we had better think about the costs, not just the benefits, of what might appear to be better for those genes. And learned recognition of offspring carries costs as well as benefits, notably the risk of making a mistake by not feeding, or even attacking and killing, one's own offspring. Rather than erring on the side of harming their genetic offspring, gulls have evolved a readiness to feed any chicks in their nest that beg confidently when approached by an adult.[82] Sometimes this rule of thumb permits a genetic outsider to steal food from a set of foster parents by slipping into a nest with other youngsters of its age and size.[44] When adoption occurs, the adoptive parents lose about 0.5 chicks

Ring-billed gull and chick

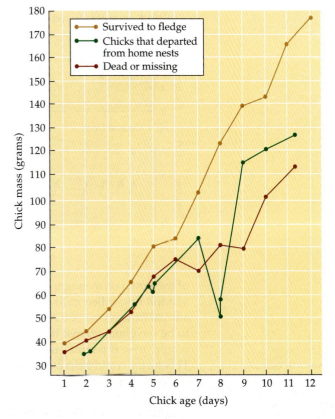

FIGURE 9.15 Why seek adoptive parents? Gull chicks that abandoned their natal nests in search of foster parents weighed much less than average for chicks their age. But these potential adoptees were sometimes adopted by nonrelatives, and as a result they weighed more on average at day 11 than the subset of unadopted (poorly fed) gull nestlings that later died or disappeared. After Brown.[7]

FIGURE 9.16 The European cuckoo chick's begging call matches that of four baby reed warblers. (A) A cuckoo chick begging for food from its reed warbler foster parent. The calls shown below the photograph are those of (B) a single reed warbler chick, (C) a brood of four reed warblers, and (D) a single cuckoo chick. A, photograph by Roger Wilmshurst; B–D, after Davies et al.[18]

(A)

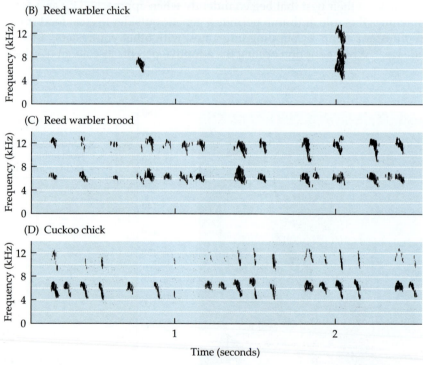

(B) Reed warbler chick

(C) Reed warbler brood

(D) Cuckoo chick

Time (seconds)

of their own on average; however, adoption is rare, with fewer than 10 percent of adult ring-billed gulls taking in a stranger in any year.[7] The modest average annual fitness cost of a rule of thumb that results in occasional adoptions has to be weighed against the cost of rejecting one's own genetic offspring that would arise if parent gulls were more reluctant to feed chicks in their nests. This factor could help explain the gullibility of adoptive parent gulls.

Specialized **brood parasites** can also insinuate themselves into a family other than their own by employing deceptive signals that trigger parental care by their hosts. So it is that a parasitic European cuckoo mimics an entire brood of reed warbler chicks, stimulating their caregivers to bring them as much food as they would to several of their own chicks (Figure 9.16). Horsfield's bronze-cuckoo chicks also produce begging calls very much like their hosts' nestlings, in this case superb fairy-wrens (Figure 9.17), presumably as

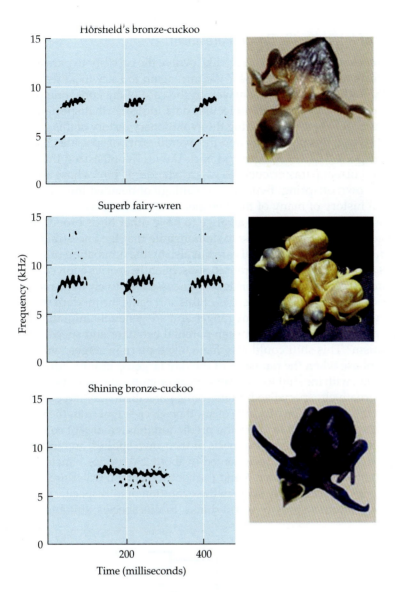

Horsfield's bronze-cuckoo

Superb fairy-wren

Shining bronze-cuckoo

Frequency (kHz)

Time (milliseconds)

FIGURE 9.17   A product of an evolutionary arms race? Chicks of Horsfield's bronze-cuckoo, a specialist brood parasite on Australian fairy-wrens, mimic the calls of their hosts' chicks very closely, which may help them overcome the defenses of fairy-wren host parents. In contrast, the chicks of another bronze-cuckoo species, which rarely parasitizes fairy-wren nests, not only do not look like fairy-wren chicks but lack a good facsimile of the begging calls of fairy-wren nestlings. After Langmore et al.[48]

an evolved response to their hosts' ability to discriminate against illegitimate signalers. In support of this conclusion, Naomi Langmore and her colleagues point to the fact that superb fairy-wrens always abandon nests that have been parasitized by another species, the shining bronze-cuckoo, whose nestlings produce vocalizations very different from those of the host chicks.[48] (I suspect that you can guess how often shining bronze-cuckoos make the mistake of laying an egg in a superb fairy-wren nest.)

Discussion Question

9.10   Interspecific brood parasitism is very rare in birds, an ability of only about 1 percent of all species.[47] Make a prediction about which group of birds, those with precocial young or those with altricial young, would be more likely to evolve into specialist brood parasites. (In altricial species, the eggs are small in relation to parental body weight, but the hatchlings are initially completely dependent on food supplied to them by parents. In precocial species, the eggs are relatively large, but the youngsters can move about and feed themselves shortly after hatching.) Check your prediction against Lyon and Eadie.[57]

The History of Interspecific Brood Parasitism

But how did cuckoos, cowbirds, and the like evolve the capacity to parasitize the parental care of other bird species? We turn again to Darwin's theory of descent with modification for an answer. In the case of cuckoos, one phylogenetic reconstruction based on molecular data indicates that specialized parasitism has arisen three times during the evolutionary history of this group (Figure 9.18).[3] Given this phylogeny and the overall rarity of specialist brood parasitic bird species, Oliver Krüger and Nick Davies hypothesized that the ancestor of the current parasitic cuckoos was a "standard" bird whose adults cared for their own offspring. By taking advantage of detailed information on the natural history of many of the 136 species of cuckoos worldwide, of which 83 are nonparasitic while 53 arrange to have other birds brood their young, Krüger and Davies were able to demonstrate that the ancestral state was represented not only by parental care but also by the occupation of small home ranges and the absence of migration. Subsequently, species evolved that provided parental care for their offspring but possessed relatively large ranges and a tendency to migrate. From lineages of this sort came the modern brood parasites, which also have large home ranges and are generally migratory.[46]

But what about the transition between parental behavior and specialized brood parasitism? This shift could also have taken place in stages, with an intermediate phase when the parasites of the day targeted nesting adults of their own species, with the shift to members of one or more other species occurring later. Alternatively, specialized interspecific parasitism may have arisen abruptly with the exploitation of adults of another species right from the start. The first, gradualist scenario leads to the prediction that females of some of today's birds should lay their eggs in nests of their own species, and indeed, parasitism of this sort has been documented in about 200 species,[62] including birds as different as the zebra finch [90] and the coot, a common waterbird.[58]

A possible hint of a very early stage in the evolution of intraspecific brood parasitism comes from a study of wood ducks. In this species, suitable nest

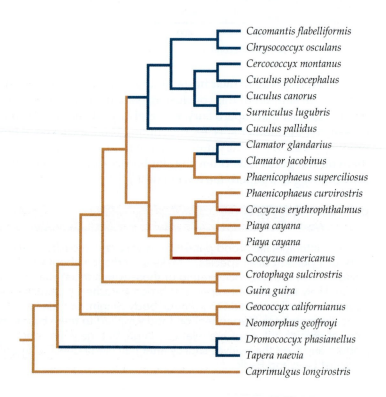

FIGURE 9.18 Multiple origins of brood parasitism by cuckoos. In this family of birds, brood parasitism has evolved three times. The blue branches indicate those lineages whose modern descendants are specialist parasites. The red branches are for two species that are occasional parasites. After Aragón et al.[3]

cavities in trees are scarce, with the result that two females sometimes lay eggs in the same nest before one duck evicts the other. The "winner" then cares for the eggs of the other female along with her own, having made the "loser" an involuntary parasite in the process.[95] If the losers were to produce more off-spring than they would have if they had retained the nest site, then selection could favor variants that voluntarily deposited and abandoned their eggs in the nests of others—a behavior that is fairly common in another hole-nesting duck, the Barrow's goldeneye.[23]

In the coot, "floater" females without nests or territories of their own lay their eggs in the nests of other coots, apparently in an effort to make the best of a bad situation, since they cannot brood their own eggs themselves. But some fully territorial females with nests of their own also regularly pop surplus eggs into the nests of unwitting neighbors, indicating that several hypotheses can apply to intraspecific brood parasitism.[62] Since there are limits to how many young one female coot can rear with her partner, even a territorial female can boost her fitness a little by surreptitiously enlisting the parental care of other pairs.[58] The exploitative nature of this behavior is revealed by the finding that older, larger females select younger, smaller ones to receive their eggs, presumably because this kind of host cannot easily prevent a larger female from gaining access to her nest.[60]

Pressure of this sort has apparently shaped the evolution of coot behavior, judging from the fact that parasitized females tend either to bury the eggs of others or to keep their own eggs in the better brooding position in the center of the nest.[61] Eggs in the outer part of the nest tend to hatch later than the eggs in the nest center. Coots can discriminate between their own chicks and those of a parasite *if* members of their own brood hatch first. By increasing the odds that parasitic chicks hatch later, the adult coots have a chance to learn the cues associated with their own offspring, which enables them to use this informa-tion to reject later-hatched chicks, if these are not their own—a nice example of how animals avoid misdirecting their valuable parental care.[96]

In any event, the gradual shift hypothesis for the evolution of parasitism among species yields the prediction that when intraspecific brood parasites first began to exploit other species as hosts, they should have selected other related species with similar nestling food requirements. Currently, most specialized brood parasites take advantage of species that are not closely related to them, but perhaps most brood parasites have been evolving for many millions of years since the onset of their interspecific parasitic behavior. Thus, to check this pre-diction, we need to find brood parasites that have a relatively recent origin. The familiar cowbirds of the Americas make up one such group, its parasitic species having originated "only" 3 to 4 million years ago, whereas the parasitic cuckoos evolved about 60 million years earlier.[19] The living cowbird species believed to be closest to the ancestral brood parasite does indeed parasitize a single host species that belongs to the same genus that it does; the next closest species to the ancestral parasite parasitizes other birds belonging to its own family (Figure 9.19).[51] These data, if they have been interpreted properly, provide support for the gradual shift hypothesis.

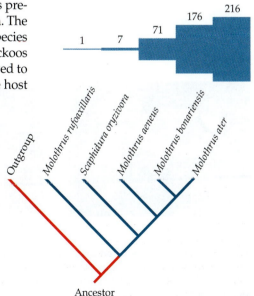

FIGURE 9.19 Evolution of brood parasitism among cowbirds. The phylogeny depicts the evolutionary relationships among cowbirds as deter-mined by molecular genetic analyses. Above the phylogeny, the number of host species parasitized by each current living species of cowbird is shown. (The "outgroup" is a species of cowbird that does not engage in brood parasitism.) The pattern suggests that the first brood parasitic cowbird victimized only a single closely related host species, with increasingly gener-alized brood parasitism evolving subsequently. After Lanyon.[51]

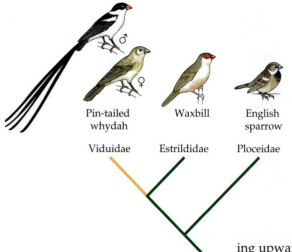

Pin-tailed
whydah

Waxbill

English
sparrow

Viduidae   Estrildidae   Ploceidae

FIGURE 9.20   Widowbirds parasitize closely related species. Widowbirds (in the family Viduidae) parasitize nests of finches in the family Estrildidae (the family most closely related to the Viduidae). Adult male widowbirds, like the male pin-tailed whydah (*Vidua macroura*) shown here, look nothing like their hosts, but female pin-tailed whydahs and their nestlings do resemble adult and nestling waxbills (*Estrilda astrild*), an estrildid finch parasitized by this widowbird.

Similarly, female widowbirds, in the family Viduidae, parasitize finch species that belong to a closely related family, the Estrildidae (Figure 9.20), which may be why both parasites and hosts share a number of important features in common, especially bright white eggs and an unusual form of begging behavior by nestlings, in which the baby birds turn the head nearly upside down and shake it from side to side, rather than stretching upward in the manner of most other nestling songbirds. Assuming that the ancestral parasitic widowbird also possessed these attributes, sensory exploitation (see Chapter 4) could account for the success the offspring of the original brood parasite had after hatching in the nest of an estrildid finch host[104] (but see Hauber and Kilner.[32]).

On the other hand, the very large majority of living brood parasites take advantage of unrelated species much smaller than they are,[97] a finding that could be explained if the ancestral parasites made an abrupt shift from normal parental care to exploiting one or more unrelated smaller host species. Such a shift might well have been more likely to succeed, given that, as already noted, brood parasite nestlings that become larger than their hosts' offspring are more likely to be fed, another form of sensory exploitation that works to a parasite's advantage.

The importance of the size disparity between host and parasite has been demonstrated experimentally. When Tore Slagsvold shifted blue tit eggs into great tit nests, the experimentally produced brood-parasitic blue tit nestlings, which are smaller than those of great tits, did very poorly. In the reciprocal experiment, however, most of the great tit chicks cared for by blue tit parents survived to fledge (Figure 9.21).[97] These findings suggest that unless the original mutant interspecific brood parasites happened to deposit their eggs in the nests of a smaller host species, the likelihood of success (from the parasite's perspective) was not great.

Yoram Yom-Tov and Eli Geffen have applied the comparative method to determine which historical scenario for the evolution of obligatory avian

Great tit

Blue tit

FIGURE 9.21   The size of an experimental "brood parasite" nestling relative to its host species determines its survival chances. Larger great tit nestlings survived well when transferred to the nests of smaller blue tits, whereas blue tits did poorly in great tit nests. After Slagsvold.[97]

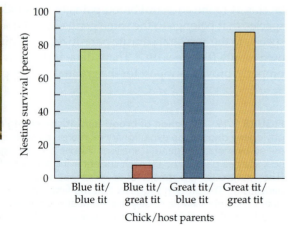

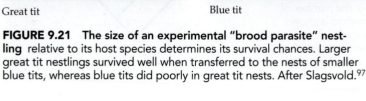

parasites is more likely—the indirect, or gradual, pathway with intraspecific parasitism as an intermediate stage, or the direct pathway in which standard parental behavior was quickly replaced by obligate parasitism (Figure 9.22). Their analysis indicates that for a large group of altricial birds the probability was much greater that the ancestor of today's obligate parasites was a bird that did not engage in intraspecific parasitism and instead took advantage of members of an entirely different species.[113] Thus, proponents of both evolutionary scenarios for interspecific brood parasitism have supportive evidence to which they can point, which leaves the resolution of this issue up in the air.

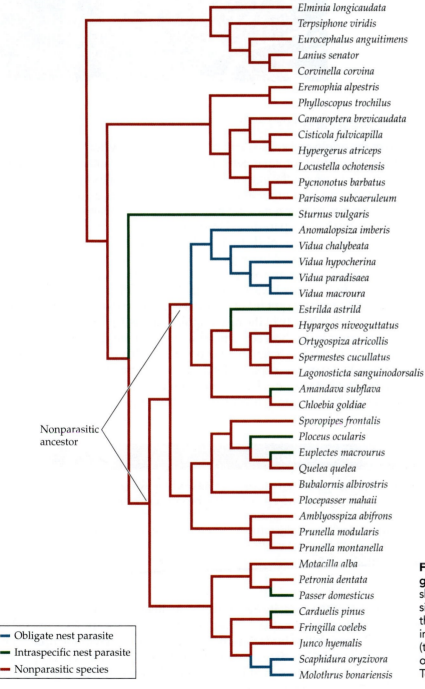

FIGURE 9.22 **Abrupt transitions to obligate parasitism.** Here an avian phylogeny shows that the two clusters of obligate parasites (those that exclusively lay their eggs in the nests of other species) had ancestors that in all likelihood were completely nonparasitic (the birds did not lay eggs in the nests of other members of their species). After Yom-Tov and Geffen.[113]

FIGURE 9.23   Why do many birds accept a cowbird's egg? This nest contains three eggs of the chipping sparrow host and one very different egg laid by a cowbird. The puzzle is, why doesn't the chipping sparrow recognize that its nest has been parasitized? Photograph by Bernd Heinrich.

Why Accept a Parasite's Egg?

Whatever the origin and subsequent history of interspecific brood parasitism, it is true that in order for a parasite nestling to take advantage of a host species' parental decisions, the egg containing the parasite has to hatch. Because some host species take immediate action against the eggs of parasites (by burying a foreign egg, or removing it from the nest, or abandoning the nest), we can ask why don't all parasitized bird species reject unwanted eggs (Figure 9.23)[112]? However, parent birds that made incubation of eggs dependent on their learned recognition of the eggs they had laid might sometimes abandon or destroy some of their own eggs by mistake, treating them as if they were foreign eggs. Indeed, reed warblers sometimes do throw out some of their own eggs while trying to get rid of cuckoo eggs.[16] If brood parasites victimize only a very small minority of the host population, then even a small chance of recognition errors by the host could make accepting parasite eggs the adaptive option.[54] Indeed, this prediction has been tested with reed warblers in Britain, where the risk of parasitic exploitation is low while the chance of recognition errors is modest but not zero. Unless the warblers had seen a (stuffed) cuckoo near their nest, they generally accepted model cuckoo eggs placed in their nest by experimental biologists.[17]

Discussion Question

9.11   The mangrove gerygone, a small warbler-like songbird of Australia, sometimes incubates the egg of a little bronze-cuckoo, a specialized parasitic species. When the parasitic cuckoo hatches, the gerygone may pick the nestling up and physically remove it from its nest, a very unusual response of bird hosts to brood parasites.[29,106] What Darwinian puzzles are associated with this case?

Acceptance of the parasite egg is also more likely to be adaptive when the host is a small species unable to grasp and remove large cowbird or cuckoo eggs.[86] Such small-billed birds have two choices: abandon the clutch, either by leaving the site or by building a new nest on top of the old one, or stay put and continue brooding the clutch along with the parasite egg. The abandonment

Prothonotary warbler

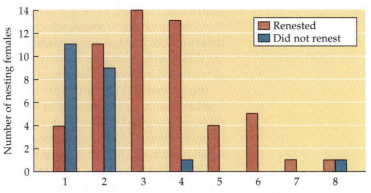

Number of potential nest sites in each female's territory

FIGURE 9.24 The probability that a female prothonotary warbler will nest again in her territory is a function of the number of potential nest sites there. When only a few good nest holes are present, a female rarely makes a second nesting attempt in her territory; if relatively many sites are available, a female usually does renest there. After Petit.[81]

option imposes heavy penalties on the hosts, which must at a minimum build a new nest and lay a new clutch of eggs. The costs of this option are especially high for hole-nesting species because acceptable tree holes are generally rare. Indeed, lack of suitable alternative nest sites often leads prothonotary warbler females to tolerate cowbirds in their nests (Figure 9.24).[81] Likewise, yellow warblers tend to accept foreign eggs when their nests are parasitized near the end of the breeding season, when little time remains to rear a new brood from scratch.[92]

Even if host birds could throw out or cover up a brood parasite's eggs without making mistakes, the parasite might make this option unprofitable by returning to the nest to check whether its egg had been harmed; if so, the parasite might retaliate against the host (Figure 9.25). This "mafia hypothesis" has been tested by examining the interactions between parasitic great spotted cuckoos and a host species, the European magpie.[103] Magpie nests from which cuckoo eggs had been ejected suffered a significantly higher rate of predation than nests with accepted cuckoo eggs (87 percent versus 12 percent in one sample). Furthermore, when researchers removed the cuckoo egg from a nest that was apparently being checked by a cuckoo and replaced the magpie's eggs as well with plasticine imitations, the cuckoo approached the nest after the researchers

FIGURE 9.25 Cowbirds, such as this female visiting a host's nest, are among the parasitic birds that may punish hosts that fail to incubate the parasite's eggs.

had finished and left its beak marks on the false magpie eggs. Magpies that lose their clutches have to renest, which exposes them to all the negative effects that delays in breeding have in a seasonal environment. In this light, it is not surprising that acceptor magpies actually have somewhat higher reproductive success, as measured in terms of fledglings produced, than do ejectors of cuckoo eggs.[103]

Brown-headed cowbirds are also avian mafiosi, as Jeffrey Hoover and Scott Robinson demonstrated in a study of prothonotary warblers. The ornithologists worked with a large sample of warblers that had built their nests in boxes on top of greased poles, which made them immune to snakes and small mammals that often prey upon the eggs and young of prothonotary warblers. But these nests were still vulnerable, at least initially, to cowbirds, which could enter a nest and deposit an egg there. Nests parasitized in this fashion were then divided into three groups: (1) the cowbird egg was removed but the nest opening was left as it was—wide enough to permit the reentry of a cowbird, (2) the cowbird egg was not touched and the nest entrance was not modified, and (3) the cowbird egg was removed and the nest entrance was made smaller so that only warblers, not cowbirds, could then enter the nest box. The results in the groups were as follows: (1) When the cowbird egg was removed, often the nest was subsequently visited by an avian predator, almost certainly the cowbird whose egg had been taken away, and the warbler eggs were destroyed. (2) When the parasite's egg was not ejected by the experimenters, the warbler's eggs were much less likely to disappear, even though adult cowbirds still had free access to the nests in this category. (3) When cowbirds were prevented from revisiting nests that they had parasitized, the loss of warbler eggs did not occur (Figure 9.26). These results strongly suggest that cowbird females often come back to nests they have parasitized, which enables them to destroy the eggs of any host bold enough to get rid of an unwanted egg. This tactic may cause the host to renest, a decision that may enable the parasite to strike again with greater success.[37]

The approach we have taken thus far is to examine the proposition that the costs of refusing to incubate a brood parasite's egg or feed a parasitic nestling after it hatched could actually outweigh the benefits of these actions. Such a conclusion comes from viewing the interaction between host and parasite from the perspective of evolutionary arms race theory. Whenever there are two parties in conflict with each other, they exert reciprocal selection pressure on each other, with an adaptive advance made by one often leading eventually to an adaptive counter-response by the other, something that has been seen in host–parasite interactions in birds and in social insects as well.[42]

The arms race approach helps us make sense of the interaction between the parasitic Horsfield's bronze-cuckoo and one of its primary hosts, the superb fairy-wren.[49] If breeding fairy-wrens find an egg in their nest before they have

FIGURE 9.26 The mafia hypothesis as tested with parasitic cowbirds and prothonotary warblers. (A) In treatment 1, a cowbird laid an egg in the nest, which was then removed by the experimenter. Subsequently, the warbler eggs in most nests in this treatment were destroyed, presumably by the thwarted cowbirds. In treatment 2, all nests were parasitized but the cowbird eggs were left in the nests, which were largely untouched by predators thereafter. In treatment 3, the cowbird eggs were removed from the parasitized nests, which were then made inaccessible to cowbirds; none of these nests was harmed after removal of the parasite's egg. (B) The warblers produced more offspring under treatments 2 and 3 than under treatment 1. After Hoover and Robinson.[37]

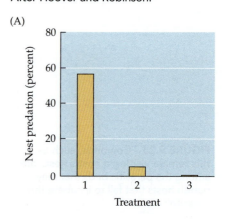

(A)

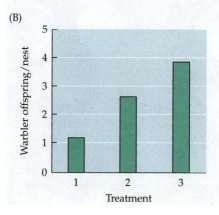

(B)

Female brown-headed cowbird

begun to deposit their own eggs there, they almost always build over the intruder's egg, and they abandon nests altogether if a cuckoo drops an egg in after the wrens have begun to incubate their own complete clutch. So the wrens have defenses against cuckoos. Adult female cuckoos, however, have evolved an answer; they are very good at slipping in to lay an egg when the host nest contains only a partial clutch of fairy-wren eggs. In such a case, the cuckoo egg is almost always accepted and incubated along with the hosts' eggs. But when the cuckoo chick hatches and kills its wren nest mates by pushing them out of the nest, the adult fairy-wrens abandon the nest about 40 percent of the time, leaving the baby cuckoo to die as well. In the remainder of cases, the fairy-wrens continue to care for the sole occupant of their nest, a baby cuckoo, and so waste their time and energy rearing the killer of their offspring.

Discussion Questions

9.12   The fairy-wren is one of the very few bird species that abandon the chicks of brood parasites. Why don't other victimized species do the same? After all, many other birds exploited by cuckoos can identify and take action against the parasite's eggs, which they do by learning the distinctive visual features of their own eggs and then rejecting those that do not match. But the same species that are extremely good at learning to recognize egg features usually completely fail to recognize a cuckoo chick.[53] However, consider the costs of learned chick recognition for birds that are successfully parasitized in their first year of breeding by a single cuckoo chick that takes over the nest and eliminates all the hosts' youngsters, as is the cuckoos' habit. How would the host adults respond to their own chicks in the next breeding attempt? To find out how superb fairy-wrens reduce the costs of learned errors of recognition, see Langmore et al.[50]

9.13   About 15 to 20 percent of all nestling cuckoo parasites are abandoned and left to die by their reed warbler hosts after about 2 weeks of foster parent care. Tomas Grim suspected that reed warblers had evolved a means to avoid helping a parasite, namely a time limit on parental care for a brood.[28] In order to test this idea, Grim performed experiments in which he manipulated broods of reed warbler chicks so as to extend the period of parental care needed for the young to fledge. He created experimental broods by transferring younger (and older) chicks between nests. How did he expect the parent warblers to respond, if the time limit hypothesis was correct? Why might it be advantageous for reed warblers to use the time limit system rather than learning what kind of nestlings to care for and which ones to reject?

The Puzzle of Parental Favoritism

Even when parents invest only in their own progeny, they rarely distribute their care in a completely equitable fashion, even when the coefficient of relatedness between parents and all their genetic offspring is the same (0.5 as a result of sexually reproducing parents placing 50 percent of their genes in each egg or sperm). Consider the biased parental decisions made by the red mason bee, *Osmia rufa*, whose females nest in hollow stems and supply pollen and nectar for a series of brood cells, provisioning one after another. Initially, when the adult females are young and in good condition, they tend to give the first few offspring large amounts of food.[94] These initial offspring are the products of fertilized eggs and so will develop into daughters of the red mason bee mothers (see page 26). But then as the season progresses and the females get older, their physiological condition declines, making it more and more diffi-

(A)

(B)

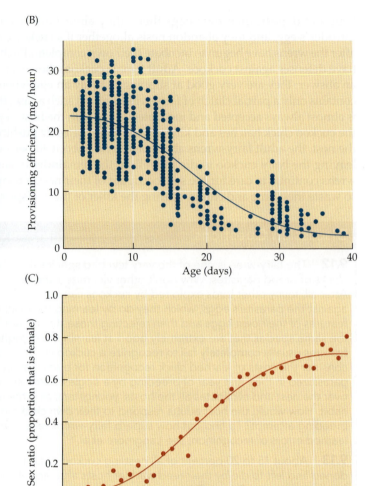

(C)

FIGURE 9.27   Adjustment of investment in sons and daughters by (A) the red mason bee, *Osmia rufa*. (B) When females are young at the start of the breeding season, their provisioning efficiency is high, and (C) under these conditions, the sex ratio of their offspring is biased toward daughters. Small sons are produced when the females are older and their efficiency in filling their brood cells is low. Provisioning efficiency is measured in terms of the average increase in the mass of a larva per hour of bringing food to the nest. A, photograph by Nicolas Vereecken; B and C, after Seidelmann.[93]

cult for them to forage efficiently. As this happens, females provide much less food per brood cell and they lay unfertilized eggs in these cells, which develop into sons that weigh much less than their sisters (Figure 9.27). Because females of this (and other) bee species are able to control both the sex of an egg and the amount of brood provisions the offspring will receive, they can invest more in a daughter than a son. In this way mothers give their daughters the resources needed to make their energy-demanding eggs and to do the hard work of foraging for their offspring. Sons can afford to be smaller, because they make tiny sperm and spend their time searching for receptive females, presumably less demanding endeavors than those tackled by females.

Other animals may lack the mechanisms needed to control the sex of their offspring as precisely as bees, ants, and wasps, but they can still give more to some progeny than others. Adults of the burying beetle *Nicrophorus vespilloides* cooperate by burying a dead mouse or vole, removing the hair from the deceased animal, and fashioning a lump of flesh from the remains. The female then lays eggs near this brood ball. When the larvae hatch, they can feed themselves from the prepared carcass, or they can receive processed carrion regurgitated by a parent. The beetle grubs can differ markedly in size because some hatch out sooner than others, and under these circumstances, their parents give more food to the earlier-hatched (senior) larvae than to later-

(A)

(B)

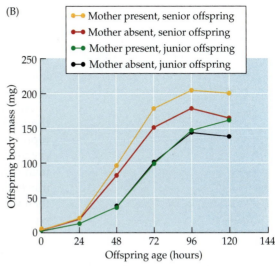

FIGURE 9.28 **Discriminating parental care by the burying beetle *Nicrophorus vespilloides*.** (A) An adult beetle inspecting larvae feeding on a ball of carrion prepared by the parents for their offspring. (B) When the mother is present, the senior (older, larger) grubs are fed more and grow to a larger size than when the mother is absent. No such effect applies to the junior (younger, smaller) offspring. A, photograph by C. F. Rick Williams; B, after Smiseth et al.[98]

hatched (junior) grubs (Figure 9.28). Given that only so many progeny can be supported by one mouse carcass, parents may gain by helping those offspring most likely to achieve adulthood, especially since the absolute amount of food needed to reach maturity is less for larvae that are farther along the road to adult metamorphosis.[99]

Discussion Question

9.14   Burying beetles also feed on the carcass themselves, but older parents take proportionally less while allocating more to their offspring.[15] Why might this decision be adaptive? What does this case have in common with the one discussed in Chapter 6 (see page 150), in which older male eggfly butterflies invest more in territorial contests than younger males?[41] And what about the discovery that single women are less likely to abort a pregnancy as they grow older?[56] You may find it helpful to read "Life History Theory" on *Wikipedia*, where you will be introduced to the terms *residual reproductive value* and *terminal investment*.

Burying beetle adults sometimes distribute food unevenly to their offspring by directly feeding some larvae while refusing to provision some others, and the same thing occurs in the common earwig. Mother earwigs respond to the distinctive chemical cues emanating from the cuticle of well-fed nymphs by foraging more actively specifically for these likely-to-survive youngsters.[65] Just as offspring condition can sometimes influence parental donations to their brood, so too parental condition is predicted to affect how adults care for their progeny. Robert Trivers and Dan Willard have argued that a parental bias toward one sex or another is expected to evolve in response to variation in parental condition *or* to variation in how well one sex converts extra parental investment into more of their own offspring.[108]

Although there has been debate about the robustness of the evidence in support of the Trivers–Willard theory, some researchers have found that, in our own species, women with access to abundant food resources are slightly more likely to have sons than women with less to eat. This outcome reflects the fact that sons have the potential in most human societies to have many children by securing multiple wives or many sexual partners. But sons are more likely to achieve this happy state if they have been developmentally advantaged and so are capable of competing successfully with rival males for women. Mothers with high-quality diets can give their male embryos the good start in life that they will need to become potent competitors as adults. In keeping with this argument, high-ranking women in polygynous marriages in Rwanda are more likely to have sons than are subordinate low-ranking, later-arriving wives, who appear to experience reduced access to resources (and more stress).[83]

Additional evidence in support of the idea that in our species parents bias their investment in sons when they can help sons achieve especially high reproductive success comes from data on inheritance rules in relation to the potential for polygyny. John Hartung showed that in societies in which some men had many wives because of culturally sanctioned polygyny, sons were much more likely to receive a disproportionate share of the family inheritance than were daughters. In these groups, sons can potentially convert inherited wealth into multiple wives, providing many more grandchildren for their (deceased) parents than can daughters, whose reproductive success will of course generally be less than that of men with more than one wife.[31,100]

Discussion Question

9.15 In one species, the eclectus parrot, mothers sometimes take parental favoritism to extreme lengths by killing their sons, but never their daughters. Sex-specific infanticide occurs more often at nests that can be flooded during the rainy season. Sons spend longer in the nest than daughters; link this factor to why it might be adaptive for a parent parrot to kill a son in a vulnerable nest occupied by offspring of both sexes.[34]

Killer Siblings

Some animal species do not, at first glance, appear to favor some of their offspring over others. Nesting great egrets, for example, bring small fish back to the nest and drop them in front of their brood. They do not intervene when their sons and daughters fight for possession of these fish (Figure 9.29).[72] Fighting among siblings, however, can escalate, with a dominant nestling bludgeoning a brother or sister to death or pushing it out of the nest. You might think that the parent should intervene to prevent the loss of even one chick from a brood of three or four.

It is conceivable that **siblicide** may have evolved only because of the fitness advantages enjoyed by offspring able to dispose of siblings that were competing for the same supply of food. Imagine parents with two offspring, each of which will eventually produce three surviving youngsters of its own, on average. Now imagine that if one of the two eliminates its sibling, the siblicidal offspring boosts its output to five surviving offspring, thanks to its ability to hog the food its parents supply after the death of its brother or sister. Although the killer loses three nephews or nieces, which its sib would have produced were it not for its premature death, this indirect fitness cost is more than compensated for, from the vantage point of the siblicidal individual, by the two extra offspring of its own. (On average, nephews and nieces share one-fourth of their genes with their uncles and aunts, whereas parents and offspring have one-half of their genes in common [see page 22]. You can do the math.) But the parent of a murderous offspring loses because instead of six grandoffspring, it has just the five as a result of the successful siblicidal offspring's reproduction. This situation is predicted to lead to **parent–offspring conflict**, a concept also developed by Bob Trivers after he realized that some actions can advance the fitness of an offspring while reducing the reproductive success of its parent, and vice versa.[109]

FIGURE 9.29 Sibling aggression in the great egret. Two chicks fight viciously in a battle that may eventually result in the death of one of them. The adult egret yawns while the fight goes on. Photograph by Douglas Mock, from Mock et al.[74]

FIGURE 9.30 Early siblicide in the brown booby. A very young chick is dying in front of its parent, which continues to brood the larger, siblicidal chick that has forced its younger brother or sister out of the nest and into the sun. Photograph by the author.

In some animals in which siblicide occurs, parents can and apparently do resist their progeny's me-first behavior. Evidence for this claim comes from studies of seabirds called boobies, some of which exhibit "early siblicide," in which an older chick A disposes of a younger chick B in the first few days of B's short and unhappy life.[55] Chick A's ability to kill its brother or sister stems in part from the pattern of egg laying and incubation in boobies. In these birds, females lay one egg (A), begin incubating it at once, and then some days later lay a second egg (B). Because the first egg hatches sooner than the second, chick A is relatively large by the time chick B comes on the scene. In early siblicidal species, chick A immediately begins to manhandle chick B, soon forcing it out of the nest scrape, where it dies of exposure and starvation (Figure 9.30).

Early siblicide is standard practice for the masked booby but not for the blue-footed booby, in which chicks engage in siblicide less often and generally later in the nesting period. If, however, you give a pair of blue-footed booby chicks the chance to be cared for by masked booby parents, which tolerate early sibling aggression, chick A often quickly kills chick B under the vacant gaze of one of its substitute parents. In contrast, blue-footed booby parents appear to keep their chick A under control during its initial days with its sibling. If so, then when masked booby chicks are given to blue-footed booby parents, the foster parents should sometimes be able to prevent them from immediately killing their siblings. As predicted, they do (Figure 9.31),[55] providing evidence that parents can interfere with lethal sibling rivalries, should it be in their interest to do so.

FIGURE 9.31 Parent boobies can control siblicide to some extent. The rate of early siblicide by masked booby (MB) chicks declines when they are placed in nests with intervention-prone blue-footed booby (BFB) foster parents. Conversely, the rate of early siblicide by blue-footed booby chicks rises when they are given laissez-faire masked boobies as foster parents. After Lougheed and Anderson.[55]

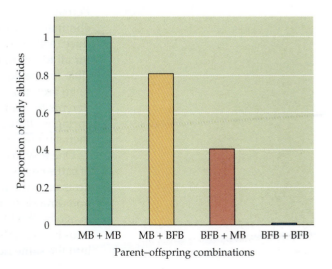

In egrets, however, parental intervention does not occur when two juveniles go at one another. Indeed, lethal sibling battles are actually promoted by earlier parental decisions about when to begin incubating eggs. Thus, as soon as a female egret lays her first egg, incubation begins, just as is true for boobies. Because there are 1 or 2 days between egg layings in a three-egg clutch, the young hatch out asynchronously, with the firstborn getting a head start in growth. As a result, this chick is much larger than the third-born chick, which helps ensure that the senior chick monopolizes the small fish its parents bring to the nest. The senior chick is not only bigger but also more aggressive because, at least in the cattle egret, the first-laid eggs contain relatively large quantities of androgens—a hormonal aggression facilitator given to the first-hatched offspring by its mother. The unequal feeding rates that result further exaggerate the size differences in the siblings, creating a runt of the litter that often dies from the combined effects of starvation and assault.[75,91]

Discussion Questions

9.16   Female cattle egrets adjust the amount of androgen they supply to their eggs in relation to the egg's position in the laying sequence. Female canaries add male sex hormones to the eggs fertilized by male partners with attractive songs.[25] Female blue tits provide less food for the offspring of partners whose crown feathers have been manipulated so that they reflect less ultraviolet light.[39] Why can these different decisions all be considered examples of parental favoritism, and what do the three examples have in common with respect to how maternal decisions advance the fitness of the mother?

9.17   Birds are not the only animals in which intense, and sometimes fatal, sibling conflicts are known to occur. For example, spotted hyena females (see page 68) often give birth to twin pups, which compete aggressively, even lethally, for their mother's milk. Develop one or more adaptationist hypotheses about siblicide, and then make use of the following four findings: (1) the total amount of milk given to pairs of offspring in which siblicide eventually occurs is lower than from mothers with surviving twins, (2) females do not reduce the amount of milk they provide after siblicide has occurred, (3) siblicide is more common when females have to travel great distances in search of prey, and (4) females sometimes separate fighting twins and may preferentially nurse the subordinate cub.[35,110]

Parent egrets not only tolerate siblicide, they actually promote it. Why? Perhaps because parental interests are served by having the chicks themselves eliminate those members of the brood that are unlikely to survive to reproduce. Although in good years parents can supply a large brood with enough for all to eat, in most years food will be moderately scarce, making it impossible for the adults to rear all three offspring. When there is not enough food to go around, a reduction in the brood, accomplished by siblicide, saves the parents the time and energy that would otherwise be wasted in feeding offspring with little or no chance of reaching adulthood even if not attacked by their siblings.

One way to test this hypothesis is to create unnaturally synchronous broods of cattle egrets,[73] which was done by taking chicks that had hatched in several nests on the same day and putting them in the same nest. Normal asynchronous broods were assembled as well, by bringing chicks together that differed in age by the typical 1.5-day interval. A category of exaggeratedly asynchronous broods was also created by putting chicks that had hatched 3 days apart into the same nests. If the normal hatching interval is optimal in promoting

TABLE 9.1 The effect of hatching asynchrony on parental efficiency in cattle egrets

	Mean survivors per nest	Food brought to nest per day (ml)	Parental efficiency[a]
Synchronous brood	1.9	68.3	2.8
Normal asynchronous brood	2.3	53.1	4.4
Exaggerated asynchronous brood	2.3	65.1	3.5

Source: Mock and Ploger[73]

[a]The number of surviving chicks divided by the volume of food brought to the nest per day × 100

efficient brood reduction, then the number of offspring fledged per unit of parental effort should be highest for the normal asynchronous broods. This prediction was confirmed. Members of synchronous broods not only fought more and survived less well but also required more food per day than normal broods, resulting in low parental efficiency (Table 9.1). The same result occurred in similar experiments with the blue-footed booby.[80]

Cattle egret parents and others like them seem to know (unconsciously) what they are doing when they manipulate the hormone content of their eggs and incubate them in ways that lead to differences in size and fighting ability among their chicks. Sibling rivalry and siblicide help parents deliver their care only to offspring that have a good chance of eventually reproducing, while keeping their food delivery costs to a minimum. Although cases of this sort represent extreme examples of parental favoritism, even those nesting birds that bias their allocation of food resources toward vigorous offspring are really practicing infanticide, increasing the likelihood of death for those progeny unlikely to reproduce even if well fed.

Discussion Question

9.18   In species like boobies and egrets, parents' decisions about incubation and hormone allocation put their second or third offspring at great risk. If the second or third egg laid is destined to produce a chick that will die within a few days, why don't parents save the energy that goes into the superfluous egg, which would also enable parents to give more to their favored offspring? One possibility goes under the label of the insurance hypothesis: adults invest in a backup egg in order to have a replacement for a favored first-laid egg should something happen to that egg or to the nestling itself after the egg hatches.[11] How would you test this hypothesis experimentally?

Parental Behavior in Relation to Offspring Value

Individuals have a limited amount of time, energy, and risk-taking that they can devote to any given offspring. (Some behavioral biologists prefer to use the label **parental effort** for these activities rather than *parental investment*, because of the problems associated with measuring the future cost of parental behaviors in terms of reduced reproductive output; parental effort requires no such measurements for its analysis.) If parents have limited food resources at their disposal, they may use their offspring's behavior to decide how to allocate these goods most effectively. For example, a parent may judge the physiological state of their juvenile offspring by their appearance and by their begging behavior, the better to give more to youngsters more likely to

FIGURE 9.32   An honest signal of condition? The red mouth gape of nestling lark buntings is exposed when the birds beg for food from their parents. The brightness of a baby bird's gape may reveal something about the strength of its immune system or its general health. Photograph by Bruce Lyon.

FIGURE 9.33   The color of the mouth gape affects the amount of food that nestling barn swallows are given by their parents. After experimenters colored the gapes of some nestlings with two drops of red food coloring, they received more food. In contrast, nestlings that received two drops of yellow food coloring or water were not fed more. After Saino et al.[87]

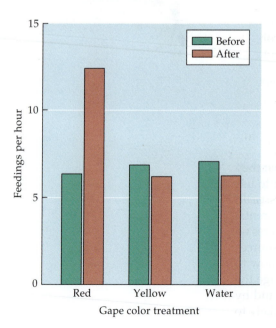

become reproductively successful adults in the fullness of time (see page 84). One informative aspect of a nestling's appearance could be the bright red lining of its mouth, which is conspicuously displayed by many nestling songbirds as they stretch up to solicit food from a returning parent (Figure 9.32). Because the red color of the mouth lining is generated by carotenoid pigments in the blood, and because carotenoids are believed to contribute to immune function, a bright red gape could signal a healthy nestling capable of making the most of any food it receives. (Alternatively perhaps only healthy nestlings can move any carotenoids they receive to the tissues in their mouths.) In any case, barn swallow chicks with redder mouths weighed more 6 days after hatching and had greater feather growth at 12 days of age than chicks with paler gapes.[21]

Parents that preferentially fed those members of their brood that had bright red mouths would be investing in nestlings of high **reproductive value**, that is, those youngsters that were healthy and on the road to becoming successful breeders. If parents do indeed make adaptive parental decisions of this sort, then nestlings made ill by injection of a foreign material should have paler mouth linings than their healthier nestmates. Furthermore, parents should feed offspring with artificially reddened gapes more than they feed offspring with unaltered mouth coloration. When Nicola Saino's research team tested both predictions in the barn swallow, the results were positive (Figure 9.33).[87]

On the other hand, among the alternative explanations for bright gape coloration[12] is the possibility that young birds gain by having colored gapes simply because these make the begging bird's maw more visible to its parents, especially when the nest is placed in a dark tree hollow or other cavity.[33] Philipp Heeb and his coworkers found that great tit nestlings with yellow-painted gapes were fed more often in relatively dark nest boxes than red-painted birds, whose mouths were less visible under low-light conditions. In nest boxes with clear Plexiglas windows on top, however, the red-mouthed nestlings suffered no begging handicap, as shown by their ability to achieve the same weight as their yellow-mouthed companions.[33]

However, the long, orange-tipped plumes on the heads and throats of baby coots can hardly help food-bearing adults stuff edibles into the gapes of the young birds. Therefore, Bruce Lyon and his colleagues decided to test the hypothesis that these feathers might provide cues used by the parents to determine which individuals to feed and which to ignore. Coots produce large clutches of eggs, but soon after the young begin to hatch, the adult birds often begin what to humans seems an especially unpleasant process of brood reduction. As some babies swim up to beg for food from a parent, the parent may not only refuse to provide something to eat but also aggressively peck the head of its youngster. Eventually, these chicks stop begging, and expire face-down in the water.

To test the link between chick ornaments and parental care, Lyon and company trimmed the thin orange tips from these special feathers on half of the chicks in a brood, while leaving the other

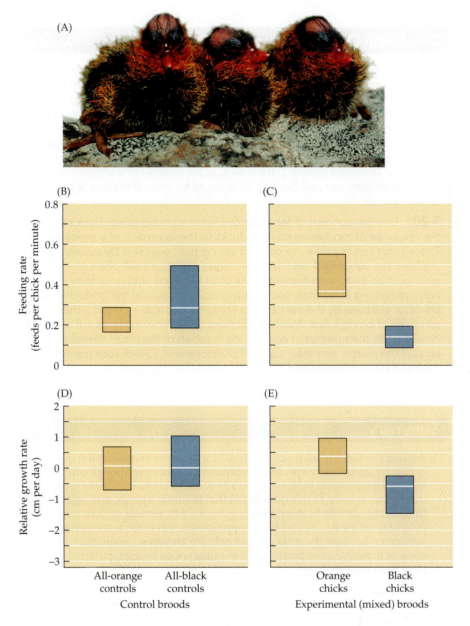

FIGURE 9.34 The effect of orange feather ornaments of baby coots on parental care. (A) Baby coots have unusually colorful feathers near the head. (B) Control groups composed entirely of either unaltered (orange) chicks or chicks that had had the orange tips trimmed from their ornamental feathers (black) were fed at the same mean rate (mean rates for each group are shown as white lines in the orange and blue rectangles). Although some black chicks were fed at a higher rate, the difference between the mean values for the two groups was not statistically signficant. (C) In experimental broods in which half the chicks were orange and half were black, the ornamented individuals received more frequent feedings from parent birds. (D) The relative growth rates of chicks in both control groups were the same, but (E) ornamented chicks grew faster in mixed broods compared with the experimentally altered chicks. A, photograph by Bruce Lyon; B–E, after Lyon et al.[59]

members of the brood untouched. The unaltered orange-plumed chicks were fed more frequently by their parents, and they grew more rapidly as well (Figure 9.34).[59] In control broods in which all of the chicks had had their orange feathers trimmed, the youngsters were fed as often as control broods consisting only of untouched orange-feathered chicks. This result shows that the parents of the experimental mixed broods discriminated against the chicks without orange feathers because they were not as well-ornamented as their feather-intact brood mates, not because the parents failed to recognize them as their offspring.[59] The fundamental message provided by coots and so many other animals is that parents do not necessarily treat each offspring the same. Cases in which parents help some of their brood survive at the expense of others remind us that selection acts not on variation in the number of offspring produced but on the number that survive to reproduce and pass on the parents' genes.

Discussion Questions

9.19   In fish with paternal care of egg clutches, it is not uncommon for the brooding male to consume a portion of the eggs he has received from his mate(s). Hope Klug found that in the flagfish, for example, brooding males sometimes consumed every egg they were brooding; other males occasionally ate only a part of the clutch under their control.[43] When males devoured the entire clutch, the eggs they ate tended to be of higher energy content than the eggs eaten by males that consumed a part of the clutch they were brooding. There are several Darwinian puzzles here; what are they and what hypotheses can you develop to account for male decisions about egg consumption?

9.20   Use the concept of reproductive value to make predictions about the response of incubating mallard ducks to the approach of a predator. These birds can improve the odds of saving their own skin by quickly flushing from the concealed nest, but their noisy departure will often give the nest location away, with the likely loss of all the eggs within. Or they can sit tight, remaining as inconspicuous as possible, improving the odds that the predator will pass by the nest and its contents but also increasing their personal risk of being killed. Make predictions about how mallards will respond in relation to the number of eggs being incubated, the mean size of the eggs in the clutch, and their stage of development. Check your predictions against data in Albrecht and Klvaňa.[1]

Summary

1. Among the Darwinian puzzles offered by parental animals is the question, why do so many species lack biparental care even though offspring that receive parental help are generally more likely to survive to reproduce? One answer to this question is that the time, energy, and resources that parents devote to their offspring have costs, including reduced fecundity in the future and fewer opportunities to mate in the present. Moreover, because the costs of paternal care are often greater than the costs of maternal care, particularly in polygynous species, paternal care is rarer than maternal care.

2. Exceptions to the rule are instructive. Thus, uniparental care by male fishes may have evolved because males caring for eggs laid in the nests they guard can be more attractive to potential mates than if they lacked eggs to protect. In contrast, the costs of maternal care to female fish may include large reductions in growth rate and consequent losses of fecundity. Fecundity losses for females may also be involved in the evolution of exclusive paternal care by some water bugs.

3. An evolutionary approach to parental care yields the expectation that when the risk of investing in genetic strangers is high, parents will be able to identify their own offspring. As predicted, offspring recognition is widespread, particularly in colonial species in which adults have many opportunities to misdirect their care to foreign offspring. However, adults do sometimes adopt nongenetic offspring, including those of specialist brood parasites, with consequent losses of fitness. Multiple hypotheses exist to account for these puzzling cases, including the possibility that highly discriminating host adults could lose fitness by sometimes erroneously rejecting their own offspring.

4. Another Darwinian puzzle is the indifference shown by some animals to lethal aggression among their young offspring. Cases like these may be explained as part of a parental strategy to let the offspring themselves identify which individuals are most likely to provide an eventual genetic payoff to the parents that care for them. The more general principle is that selection rarely favors completely even treatment of offspring, because some youngsters are more likely than others to survive to reproduce.

Suggested Reading

Robert Trivers wrote a now classic article on parental investment.[107] Good books on parental care include *The Evolution of Parental Care* by Timothy Clutton-Brock[13] and *Mother Nature* by Sarah Hrdy.[38] A vast literature exists on brood parasites and their interactions with their hosts; for a superb review, see *Cuckoos, Cowbirds and Other Cheats* by Nick Davies.[19] Doug Mock and his coworkers have written an excellent article about siblicide;[74] Mock has also produced a book on the subject for a general audience.[76]

10 Proximate and Ultimate Causes of Behavior

 The focus of this book to this point has been on how biologists use Darwinian natural selection theory and its modern amendments, primarily kin selection theory, to study behavioral adaptations. From time to time, we have also looked at how Darwin's other evolutionary theory, the theory of descent with modification, can help us outline the historical sequence of events underlying the evolution of interesting behaviors. The two Darwinian "levels of analysis" are clearly complementary, but one element focuses on the process by which evolution occurs and the other on the chain of events produced by selection over time. Together these evolutionary, or **ultimate**, **causes** of behavior provide us with a perspective on the adaptive value and the history of a trait of interest.

But there is much more to behavioral research than studies of adaptation and history. As the great behavioral biologist Niko Tinbergen pointed out long ago, biologists also need to study how a behavior develops over the lifetime of an individual and how an animal's physiological systems make behaving possible.[90] Because the mechanisms that underlie development and neurophysiology act within the life span of individuals (unlike evolutionary causes, which act across generations), they are considered **proximate** (immediate) **causes** of behavior. Every single behavioral trait is caused by both proximate and ultimate factors; the self-sacrificing behavior of worker termites, the helping-at-the-nest behavior of Seychelles warblers, the hyena's greeting behavior, the territorial behavior of a poplar aphid, you name it, they all are driven by immediate underlying mechanisms acting within the bodies of warblers and aphids and every other animal. To paint a complete picture of how and why an animal behaves, you need to explore and integrate both the proximate and the ultimate causes of the behavior (Figure 10.1).

There are both immediate physiological and long-term evolutionary causes for why this male blue-cheeked bee-eater produces vocalizations when communicating with other bee-eaters.

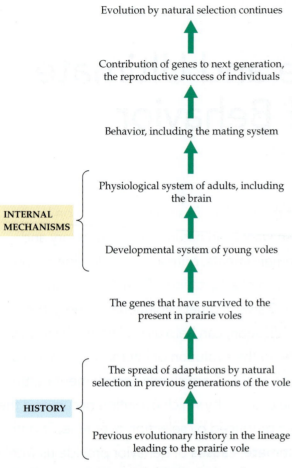

FIGURE 10.1 Evolution by natural selection shapes the mechanisms of behavior, as illustrated by the prairie vole's mating behavior. Evolutionary processes determine which genes survive over time. The genes that animals possess influence the proximate development of the neural mechanisms that make behavior possible. Behavior affects the genetic success of individuals in the current generation. Evolutionary change is ongoing.

The next several chapters will examine both the developmental bases of behavior and the neural and hormonal factors that influence animal behavior. As is true throughout this text, I have selected only a tiny fraction of the available studies in order to illustrate certain issues that seem important to me. The first of these points is that proximate mechanisms evolve, which ties proximate causes to evolutionary ones.

Connecting the Four Levels of Analysis

The four levels of analysis (Table 10.1) are linked to one another because behaviors that have spread through a species due to their positive effects on fitness must have underlying proximate mechanisms that can be inherited. For example, as we outlined in Chapter 1, a digger bee male can find emerging females efficiently because this skill is adaptive, which is to say that it evolved and is now maintained by natural selection. The ability to smell hidden females (a proximate ability) enabled males to reproduce more successfully than rivals without the same capacity. But this adaptationist explanation also means that there is a history to the behavior, with adaptive changes, such as extreme sensitivity to odors linked to buried females, layered on traits that already had evolved previously in the history of the digger bee species.

The ultimate causes of behavior helped determine which genes survived to the present and are in the bodies of living digger bees today. These genes have proximate effects when they, in conjunction with the cellular environment in which they operate, cause the bee to develop in a particular manner. The proximate developmental mechanisms within the egg, larva, prepupa, and pupa of digger bees have follow-on consequences as they influence the production of the sensory and motor mechanisms with which adult male digger bees are endowed. The scent-detecting sensory cells in the bee's antennae in concert with olfactory neural networks in the bee's nervous system make it possible for males to find pre-emergent females. As a result, males are ready to pounce on sexually receptive females when they come out of their emergence tunnels. Because the immediate mechanisms of behavior have an evolutionary basis, the proximate and ultimate causes of behavior are related, and both are required for a full explanation of any behavioral trait.[1]

TABLE 10.1 Levels of analysis in the study of animal behavior	
Proximate	**Ultimate**
How **genetic–developmental mechanisms** influence the assembly of an animal and its internal components, including its nervous and endocrine systems	The **evolutionary history** of a behavioral trait as affected by descent with modification from ancestral species
How **neuronal–hormonal mechanisms**, which develop within an animal during its lifetime, control what an animal can do behaviorally	The **adaptive value** of a behavioral trait as affected by the process of evolution by natural selection

Sources: Holekamp and Sherman;[40] Tinbergen[90]

Discussion Questions

10.1   When Stephen Jay Gould wrote his article on the pseudopenis of the spotted hyena (see Chapter 4),[36] he concluded that it was unnecessary to ask the question, what is the pseudopenis for? In fact, according to Gould, it was unwise in that the question distracted attention from the more important question: How is it built? What do you think?

10.2   When a honey bee swarm is moving to a new hive, scouts fly out from the swarm in search of new locations. When they return to the swarm, the scouts' pattern of dancing differs markedly from the pattern exhibited by foragers returning to a hive (Figure 10.2). Whereas foragers that have found a good food patch continue to recruit over many trips back and forth, individual scouts stop recruiting for a potential new hive site after only a relatively few trips. These differences affect how the population of foragers distributes themselves among different pollen- and nectar-producing flower patches and how a swarm eventually decides to move to a single site to set up housekeeping there. Some have said that we do not need evolutionary explanations for why the bees are able to make collective decisions about where to forage and where to move as a swarm. Instead, according to these researchers, colony-wide decisions are the inevitable consequence of mindless behavioral rules used by individual members of the group. Indeed, groups are said to possess self-organizing properties that stem from the simple behaviors of their members, and knowledge of the "emergent properties" of groups negates the need for other kinds of explanations of their social activities. Does the proximate–ultimate distinction help you evaluate this claim?

(A)

(B)

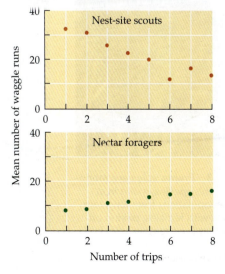

FIGURE 10.2   Recruitment patterns in the honey bee. Scouts travel out from (A) a swarm in search of a new hive site. (B) When they return with information about a new site, they behave differently than foragers that recruit others to a patch of flowers. The more times a hive-hunting scout returns to the swarm, the fewer the number of dances it performs there. A, photograph by the author; B, after Beering[11] and Visscher.[92]

The Proximate and Ultimate Causes of Monogamy in Prairie Voles

In order to see how the proximate and ultimate causes of behavior are intertwined, it is helpful to look at a single species that has been studied from every angle. Unfortunately, few animals have been subject to behavioral research that demonstrates the complementarity and utility of all four levels of analysis. But there are some suitable species, one of which is the prairie vole (*Microtus ochrogaster*), a mouselike mammal that lives in burrows in grasslands in the central parts of the United States and southern Canada. In almost every respect, the drab little prairie vole is nothing to write home about. But one feature of its behavior does stand out: male prairie voles are often monogamous (Figure 10.3),[33] although there is geographic variation in the degree to which monogamy occurs in this species.[88] As you know from Chapter 8, monogamy offers a Darwinian puzzle because it would appear that by remaining with a single sexual partner for a breeding season or even longer, monogamous males are losing valuable additional opportunities to mate and pass on their genes.

You may recall that one selectionist hypothesis for male monogamy (see page 218) is that by remaining with one female, the male increases the odds that he will father all that female's offspring. A research team led by Jerry Wolff tested the mate-guarding hypothesis for male monogamy in the prairie vole. As Wolff and his team predicted, when they experimentally prevented male prairie voles from staying with their mates in a laboratory setting, 55 percent of the females copulated with other males. (The females in this experiment could choose among three males that had been tethered so that they could not interfere with one another, nor could they prevent the females from leaving them and moving on to other males.) In nature, males that left their mates might well lose them to other males, which would translate into reduced paternity for the less monogamous males.[95]

So, at least under some conditions, the costs of monogamy could be outweighed by the fitness benefits that come from preventing a mate from accepting sperm from other males. This adaptationist explanation is complemented by another ultimate explanation, one that traces the sequence of events that took place during evolution as monogamy originated and spread among some vole lineages. There was surely a time when the prairie vole, or a distant ancestor, was not monogamous. As noted in Chapter 7, in most mammals, males attempt to mate with more than one female, which is also true for some voles, like the red-backed vole, a species that retains the presumptive ancestral pattern of polygyny today (Figure 10.4). Judging from the data presented in Figure 10.4, the shift toward monogamy may have occurred in a now-extinct species that gave rise to species in two modern genera, *Lemmiscus* and *Microtus*. A number of these species combine elements of male parental care and monogamy. Within this group is the prairie vole, whose monogamy was almost certainly the result of modifications of a polygynous mating system.[95]

So now we have two complementary evolutionary or ultimate explanations for the unusual mating system of the prairie vole. Neither of these explanations, one on the adaptive value of the trait, the other on its long-term history, deals with the proximate factors that cause prairie vole males to bond with their mates. This gap, however, has been filled by other research teams, including one led by Larry Young. As a result, we know now that cells in certain parts of the brains of male prairie voles[98] contain many protein receptors that bind chemically with a hormone called vasopressin. Vasopressin is produced and released into the bloodstream by other brain cells when a vole copulates several times with a given female. Molecules of vasopressin are carried to the

FIGURE 10.3 Prairie voles are monogamous. In this species, males in at least some populations form long-term relationships with females, pairing off as couples that live together, with both parents caring for offspring. Photograph by Lowell Getz.

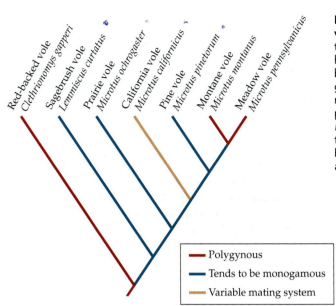

FIGURE 10.4   The evolutionary relationships of the prairie vole and six of its relatives. The diagram suggests that a tendency toward monogamy may have originated early in the history of the genus *Microtus* in a now-extinct ancestor and has been retained in the monogamous extant species within this group, which includes prairie and pine voles. Polygyny as the male mating system was almost certainly the ancestral state that preceded the evolution of paternal care and mate guarding in *Microtus*, but a few modern species in the genus (e.g., the montane and meadow voles) have evidently shifted from monogamy back to polygyny. Data from Conroy and Cook[24] and McGuire and Bemis.[61]

ventral pallidum, the label given a particular structure near the brain's base in mammals and other vertebrates that helps provide <u>rewarding sensations</u> upon completion of certain behaviors. When receptor proteins in the ventral pallidum, called V1a receptors, are stimulated by vasopressin, they trigger activity in the receptor-rich cells (Figure 10.5). This activity in turn affects neural pathways in the brain that provide the vole with positive feedback for its mating behavior.

Young and his coworkers believe that these rewards encourage the male vole to remain in the company of his mate, forming a long-term social bond with her. In contrast, the reward system in the brains of males of polygynous vole species is different, in part because V1a receptors are less numerous in the ventral pallidum in these species. As a result, when these males copulate, their brains do not provide the same kind of feedback that leads to the formation of a durable social attachment between a male and his mate. Polygynous males move on after copulating with a female, instead of staying put.

In addition to explaining male monogamy in terms of brain physiology, Young and his colleagues have also looked at a possible genetic basis for the monogamous mating system.[77] Young's group knew that the V1a receptor protein, which is so important in the prairie vole's vasopressin-based system of social bonding, is encoded by a specific gene, the *avpr1a* gene. The prairie vole's *avpr1a* gene has a specific chunk of DNA that is lacking in the polygy-

FIGURE 10.5   The brain of the prairie vole, like that of all mammals, is a complex, evolved mechanism with special features whose operation helps explain vole behavior. (A) A cross section of the brain with just a few of its anatomically distinct regions labeled. The ventral pallidum contains many cells with receptor proteins that bind to the hormone vasopressin. (B) A brain section that has been treated in such a way that regions with large numbers of vasopressin receptors appear black. The ventral pallidum occurs in both the left and right halves of the brain; the left-hand portion of the ventral pallidum is outlined in black (arrow). After Lim et al.[51]

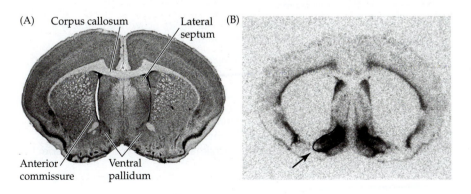

nous montane vole's version of the same gene. The extra DNA of the prairie vole's gene might have developmental effects that result in an increase in the abundance of vasopressin receptors in the ventral pallidum.

There is, however, considerable debate about the existence and significance of differences in this part of the gene for the occurrence of monogamy in prairie voles. Although some researchers have reached a negative conclusion about the gene's role in the development of monogamous behavior,[30,54] others have not.[38] A recent finding of note is that female prairie voles in the laboratory apparently prefer partners with a particular form of the *avpr1a* gene. Males with this form of the gene are more monogamous and paternal than males with other alleles, providing an adaptive basis for the female preference as well as suggesting that specific genetic differences among voles within and between species do indeed lead to differences in their mating and parental behavior.[21] Doubtless, other factors influence male mating behavior in this species as well.

Leaving aside the differences of opinion on this matter, Larry Young and his team reasoned that if the *avpr1a* gene had something to do with brain development in the male prairie vole (and thus the mating system of this species), and if they could transfer extra copies of the prairie vole's form of the *avpr1a* gene into the right cells of the correct part of the brain of male prairie voles, then they should be able to make these voles bond even more eagerly with a female than they would naturally. Using a safe viral vector, the researchers inserted extra copies of the gene in question directly into cells in the ventral pallidum in their prairie vole subjects. Once in place, the "extra" genes enabled these particular cells to make even more V1a receptor proteins than they would have otherwise. The genetically modified, receptor-rich males did indeed form especially strong social bonds with female companions, even if they had not mated with them, when given a choice between a familiar female and one with which they had not previously interacted (Figure 10.6). The research team concluded, therefore, that *avpr1a* contributes in some way to the development of monogamous behavior of male prairie voles.[98]

Imagine that you could induce males of the polygynous meadow vole to express copies of their *avpr1a* gene in their ventral pallidum at unusually high levels. The experiment has been done[52] and the degree to which the genetically altered males associated closely with their sexual partners was compared with the behavior of control males. What did the experimenters expect to find?

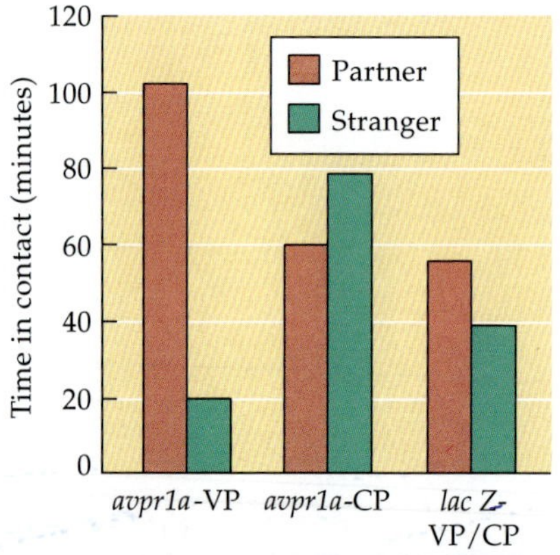

FIGURE 10.6 A gene that affects male pairing behavior in the prairie vole. Males of the monogamous prairie vole that received added copies of the *avpr1a* gene in the ventral pallidum (VP) spent significantly more time with a familiar female versus an unfamiliar one, when given a choice over a 3-hour trial. No such effect occurred when males were given extra copies of the gene in the caudate putamen (CP) of the brain. Likewise, males given a different gene (*lacZ*) in the ventral pallidum or the caudate putamen showed no special preference for a familiar female. After Pitkow et al.[77]

Discussion Question

10.3   The four main questions for behavioral researchers according to Niko Tinbergen[90] can be paraphrased as follows:

1. How does the behavior promote an animal's ability to survive and reproduce?
2. How does an animal use its sensory and motor abilities to activate and modify its behavior patterns?
3. How does an animal's behavior change during its growth, especially in response to the experiences that it has while maturing?
4. How does an animal's behavior compare with that of other closely related species, and what does this tell us about the origins of its behavior and the changes that have occurred during the history of the species?

Place these questions within the four-levels-of-analysis framework, and then assign each to the proximate or ultimate category. If you heard that because evolutionary questions are "ultimate" ones, they are therefore more important than questions about proximate causes, you would respectfully disagree. Why?

Discussion Question

10.4 When a female baboon copulates, she vocalizes loudly, but her cries are longer and louder if her partner happens to be a high-ranking, "alpha" male.[81] A primate researcher has suggested that females cry out more vigorously when copulating with top males because this warns low-ranking baboons to stay clear. (Subordinate males sometimes harass mating pairs to such an extent that the copulation ends prematurely, but if that happens, they may be attacked by the dominant male whose mating they so rudely interrupted.) The same researcher also says, however, that the more vigorous cries may simply reflect the fact that females are more strongly stimulated by the larger, more energetic, alpha males. Use the proximate–ultimate distinction to establish which view is correct.

The Proximate Causes of Bird Song

Having used vole behavior to illustrate how the different levels of analysis have been studied by researchers, I have selected learning of song dialects by songbirds as a second example. There are other possibilities, but as is true for vole monogamy, avian song learning has been explored extensively from both proximate and ultimate perspectives, and birdsong is full of intriguing mysteries that make it an ideal topic to examine here.

We begin with Peter Marler's work on why males in different populations of some bird species sing distinctive variants of their species' song just as people in different parts of America and Great Britain speak different dialects of English. Marler became aware of bird dialects at a time when he was traveling from one British lake to another as part of a limnological study. As he casually listened to the local songs of chaffinches, a common European songbird, he realized that each lake's population sang a somewhat different version of this species' standard song, which has been described as a rattling series of chips terminated by a descending flourish.

When Marler moved to the University of California in the 1960s, he and his students investigated the same phenomenon in white-crowned sparrows, the males of which sing a complex whistled vocalization during the breeding season. They found that white-crowns living in Marin, north of San Francisco Bay, sing a song type that is easily distinguished from that produced by white-crowns that live in Berkeley, and the same is true for white-crowns in Sunset Beach to the south of Berkeley (Figure 10.7).[58] Although white-crowned sparrow dialects sometimes change gradually over time,[68] in at least some populations, the local dialect has persisted for decades with only modest changes,[39] showing that this bird's dialects can be relatively stable (just like human ones).

So what developmental factors are responsible for avian dialects? Marler knew that one proximate explanation for the dialect differences was that Marin white-crowns might differ genetically from Berkeley birds in ways that affected the construction of their nervous systems, with the result that birds in the two populations came to sing different songs. One way to test the genetic differences hypothesis is to check the prediction that groups of birds singing different dialects will be genetically distinct from one another. In one such study, however, researchers found little genetic differentiation among six different dialect groups of white-crowned sparrows.[86]

So let us consider an alternative hypothesis for dialect differences, namely that these differences might not be hereditary but rather might be caused by differences in the birds' social environments. Perhaps young males in Marin learn to sing the dialect of that region by listening to what adult Marin males are sing-

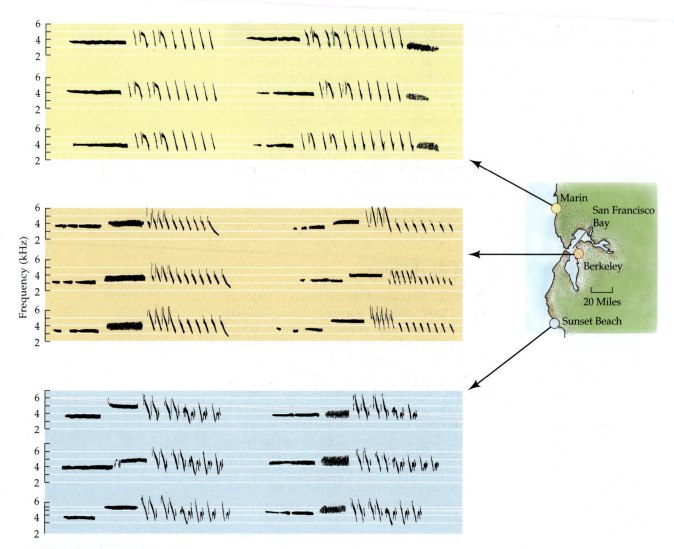

FIGURE 10.7 Song dialects in white-crowned sparrows. Males from Marin, Berkeley, and Sunset Beach, California, have their own distinctive song dialects, as revealed in these sonograms of the songs of six birds from each location provided by Peter Marler. Sonograms shown in the same color are of songs with the same dialect.

ing, while farther south in Berkeley, young male white-crowns have different formative experiences as a result of hearing males singing the Berkeley dialect. After all, a young person growing up in Mobile, Alabama, acquires a dialect different from someone in Bangor, Maine, simply because the Alabaman child hears a different brand of English than the youngster reared in downeast Maine.

Marler and his colleagues explored the development of white-crown singing behavior by taking eggs from the nests of white-crowned sparrows and hatching them in the laboratory, where the baby birds were hand-reared. Even when these young birds were kept isolated from the sounds made by singing birds, they still started singing when they were about 150 days old, but the best they could ever do was a twittering vocalization that never took on the rich character of the full song of a wild male white-crowned sparrow from Marin or Berkeley or anywhere else.[59]

This result suggested that something critical was missing from the hand-reared birds' environment, perhaps the opportunity to hear the songs of adult male white-crowned sparrows. If this was the key factor, then a young male isolated in a soundproof chamber but exposed to tapes of white-crowned sparrow song ought to be able to sing a complete white-crown song in due course. And that is exactly what happened when 10- to 50-day-old white-crowns were allowed to listen to tapes of white-crowned sparrow song. These birds also started singing on schedule when they were about 150 days old. At first their songs were incomplete, but by the age of 200 days, the isolated birds not only sang the species-typical form of their song, they closely mimicked the exact version that they had heard on tape. Play a Berkeley song to an isolated young male white-crown, and that male will come to sing the Berkeley dialect. Play a Marin song to another male, and he will eventually sing the Marin dialect.

These results offer powerful support for the hypothesis that the experience of hearing neighboring male white-crowns sing affects the development of white-crown dialects in young males. Birds that grow up in the vicinity of Marin hear only the Marin dialect as sung by older males in their neighborhood. They evidently store the acoustical information they acquire from their tutors and later match their own initially incomplete song versions against their memories of tutor song, gradually coming to duplicate a particular dialect. Along these lines, if a young hand-reared white-crown is unable to hear itself sing (as a result of being deafened after hearing others sing but before beginning to vocalize itself), then it never produces anything like a normal song, let alone a duplicate copy of the one it heard earlier in life.[44] Indeed, the ability to hear oneself sing appears to be critical for the development of a complete song in a host of songbirds (Figure 10.8).

Marler and others did many more experiments designed to determine just how song development takes place in white-crowns. For example, they wondered whether young males were more easily influenced by the stimuli provided by singing adults of their own species than by those of other species. In fact, young, isolated, hand-reared white-crowns that hear only songs of another species almost never come to sing that kind of song (although they may incorporate notes from the other species' song into their vocalizations). If 10- to 50-day-old birds listen only to tapes of song sparrows instead of white-crowns, they develop aberrant songs similar to the "songs" produced by males that never hear any bird song at all. But if an experimental bird has the chance to listen to tapes of the white-crowned sparrow along with songs of another sparrow species, then by 200 days of age it will sing the white-crowned sparrow dialect that it heard earlier.[45] The young bird's developmental system is such that listening to the other sparrow's song has little effect on its later singing behavior.

Zebra finch

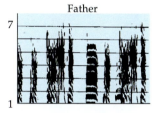

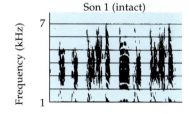

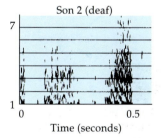

FIGURE 10.8 Hearing and song learning. In white-crowned sparrows and many other songbirds, young males have to hear themselves singing in order to produce an accurate copy of their species' song. A sonogram of a male zebra finch's song is shown with those of two of his male offspring. The first son's hearing was intact, and he was able to copy his father's song. The second son was experimentally deafened early in life, and as a consequence, he never sang a typical zebra finch song, let alone one that resembled his father's song. Photograph courtesy of Atsuko Takahashi; sonograms from Wilbrecht et al.[94]

Discussion Questions

10.5   Male white-crowned sparrows must learn to sing a particular dialect of the full song of their species. But this fact does not mean that genetic information present in the cells of white-crowned sparrows is irrelevant for the development of the bird's singing behavior. Why not? In this regard, remember that white-crown males can learn their species' song far more easily than the song of, say, the white-throated sparrow. What about the finding that white-crown males that hear white-crown song only during a 40-day period early in life can nevertheless generate a complete song, although they do not start singing themselves for several months after their early exposure to a tutor's song?

10.6   If you were to say that the *scientific conclusion* of Marler's research on young captive white-crowns was that the birds could pick up a dialect by listening to tapes of white-crown song in the laboratory, I would say that you were mistaken. What is the real point of this and all other scientific experiments?

FIGURE 10.9 Song learning hypothesis based on laboratory experiments with white-crowned sparrows. According to this hypothesis, young white-crowns have a special sensitive period 10 to 50 days after hatching, when their neural systems can acquire information from listening to white-crown song, but not to any other species' song. Later in life, the bird matches his own plastic song with his memory of the tutor's song and eventually imitates it perfectly—but does not sing elements of the song sparrow's song that he heard during his development. Based on a diagram by Peter Marler.

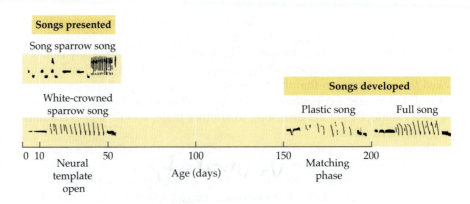

Social Experience and Song Development

The experiments with isolated white-crowns exposed to taped songs in laboratory cages led Marler to summarize the path of song development in this species, as shown in Figure 10.9. At a very early age, the white-crown's still immature brain is able to selectively store information about the sounds made by singing white-crowns while ignoring other species' songs. At this stage of life, it is as if the brain possesses a restricted computer file capable of recording only one kind of sound input. Then, months later, when the bird begins to sing, it accesses the file. By listening to its own plastic songs (incomplete versions of the more complex full song that it will eventually sing) and comparing those sounds against its memories of the full song it has heard, the maturing bird is able to shape its own songs to match its memory of the song it has on file. When it gets a good match, it then repeatedly practices this "right" song, and in so doing crystallizes a full song of its own, which it can then sing for the rest of its life.

The ability of male white-crowned sparrows to learn the songs of other male white-crowns in their birthplace just by listening to them sing provides a plausible proximate explanation for how males come to sing a particular dialect of their species' full song. However, very occasionally, observers have heard wild white-crowns singing songs like those of other species, including the song sparrow (Figure 10.10). These rare exceptions led Luis Baptista to

FIGURE 10.10 Social experience influences song development. A white-crowned sparrow that has been caged next to a strawberry finch will learn the song of its social tutor but will not learn the song of a nearby, but unseen, white-crowned sparrow. (A) The song of a tutor strawberry finch. (B) The song of a white-crowned sparrow caged nearby. The letters beneath the sonograms label the syllables of the finch song and their counterparts in the song learned by the sparrow. Sonograms courtesy of Luis Baptista.

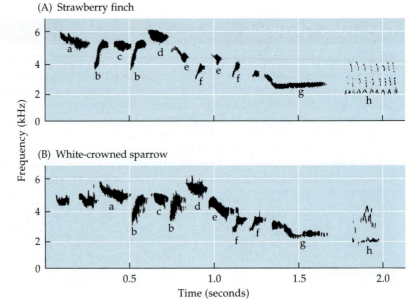

FIGURE 10.11   Social effects on song learning. Kuro the starling learned to include words in his vocalizations because he had a close relationship with the family of Keigo Iizuka. Photograph by Birgitte Nielsen, courtesy of Keigo Iizuka.

wonder whether some other factor, in addition to acoustical experience, might influence song development in white-crowns. One such factor might be social interactions, a variable excluded from Marler's famous experiments with isolated, hand-reared birds whose environments offered acoustical stimuli but not the opportunity to interact with living, breathing companions.

To test whether social stimuli can influence song learning in white-crowns, Baptista and his colleague Lewis Petrinovich placed young hand-reared white-crowns in cages where they could see and hear living adult song sparrows or strawberry finches.[5] Under these circumstances, white-crowns learned their social tutor's song, even when they could hear, but not see, adult male white-crowned sparrows (see Figure 10.10). Clearly, social acoustical experience can override purely acoustical stimulation during the development of white-crowned sparrow singing behavior, nor is the white-crown unique in this regard (Figure 10.11).

Discussion Question

10.7   A natural experiment occurred in Australian woodlands when galahs (a species of parrot) laid eggs in tree hole nests that were stolen from them by pink cockatoos (another species of parrot), which then became unwitting foster parents for baby galahs. The young cockatoo-reared galahs produced begging calls and alarm calls that were identical to those produced by galahs cared for by their genetic parents. However, the adopted galahs eventually produced contact calls (used to promote flock cohesion) very much like those of their adoptive cockatoo parents, as you can see from Figure 10.12.[79] If you heard that galah begging and alarm calls are genetically determined, whereas contact calls are environmentally determined, you would know better. Why? But it does make sense to claim that the alarm calls of the adopted galahs and their cockatoo foster parents are different as a result of genetic differences between them. Why? (You may want to read ahead in the next chapter on behavioral development.) What other behavioral differences are the result of differences between the social environment of the adopted galahs and that of certain other individuals?

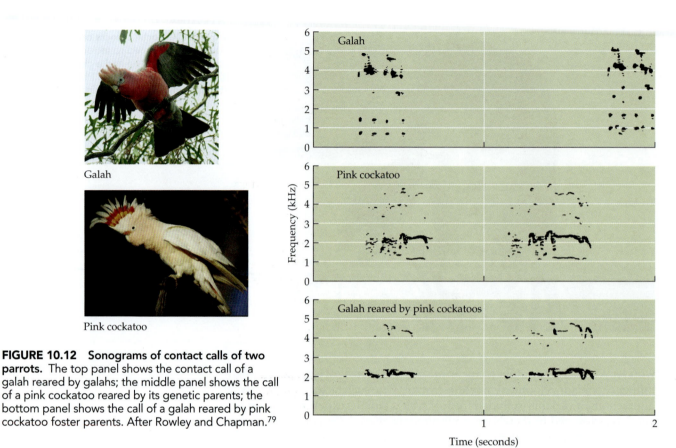

FIGURE 10.12 Sonograms of contact calls of two parrots. The top panel shows the contact call of a galah reared by galahs; the middle panel shows the call of a pink cockatoo reared by its genetic parents; the bottom panel shows the call of a galah reared by pink cockatoo foster parents. After Rowley and Chapman.[79]

Bird Brains and Bird Songs

If we want to understand more about the ability of white-crowned sparrow males to learn a dialect of their species' song, we have to go beyond identifying those elements of the young bird's social acoustical environment that affect the development of its behavior at a later date. The process of development in a songbird results in the production of a brain. Understanding how a bird's brain works requires another level of analysis, the neurophysiological angle, which we must investigate if we are to figure out such things as where in a month-old white-crown's brain are memories stored of the songs it has listened to. And what part of the brain controls the sounds that the bird will produce when it is 5 months old? And how does the young male come to match his song memories with his own initially simpler songs? These questions require us to consider the internal devices that the young male possesses that are capable of using social and acoustical inputs to steer his singing behavior along a particular developmental pathway.

When a young bird is bombarded with sounds produced by singing adults of its own species, these sounds must activate special sensory signals that get relayed to particular parts of the brain. In response to these distinctive inputs, some cells in these locations must then alter their biochemistry in order to change the bird's behavior. If we trace back the biochemical changes taking place in the brain cells of a sparrow, we will eventually find that these are linked to changes in gene expression (in which the information encoded in a gene is actually used to produce a product, such as an enzyme). So, for example, when a young white-crown hears its species' song, certain patterns of sensory signals generated by acoustical receptors in the bird's ears are relayed to song control centers in the brain where learning occurs. These sensory inputs are believed to alter the activity of certain genes in the set of responding cells, leading to new patterns of protein production and follow-on changes in cell biochemistry that reshape those cells. Once the cells have been altered, the modified song control system can do things that it could not do before the bird was exposed to the song of other males of its species.

One of the genes that contributes to these changes is called *ZENK*. The protein that it codes for (ZENK) is expressed in particular parts of the brain after a bird, like a young zebra finch or canary, hears the songs of its species.[62] For birds that only listen to these songs but do not sing in response, the protein appears only in certain brain structures associated with auditory processing. For those birds that respond after hearing their species' song, the ZENK protein is strongly produced in the nuclei that control song production. The ZENK protein is a transcription factor, which is to say that it is part of the regulatory apparatus that determines whether information in one or more genes is expressed and to what extent.[64] What these other genes do in the case of bird song appears to affect which proteins appear in the connections (synapses) between certain brain cells. Therefore, ZENK and the genes the protein regulates influence how one nerve cell communicates with another; changes in these connections can alter a bird's behavior.

Studies of the relationship between genes and neural changes illustrate the intimate connection between the developmental and physiological levels of analysis of bird song. Changes in genetic activity in response to key environmental stimuli translate into changes in neurophysiological mechanisms that control the learning process. An understanding of both developmental and neurophysiological systems is therefore necessary to give us a full account of the proximate causes of behavior.

Discussion Question

10.8 Female starlings possess the same *ZENK* gene males do. In the female brain, the ventral caudomedial neostriatum, or NCMv, responds to signals sent to it from auditory neurons that fire when the bird is exposed to sounds, such as those made by singing male starlings. When captive female starlings are given a choice between perching next to a nest box where they can hear a long song versus perching next to another nest box where a shorter song is played, they spend more time at the long song site (Figure 10.13).[32] What proximate hypothesis could account for the song preferences of female starlings? What prediction can you make about the activity of *ZENK* in the cells that make up the NCMv of female starlings exposed to long versus short songs? How might you check your prediction? What would be the scientific point of collecting the data necessary to evaluate your prediction? For an ultimate explanation for the female preference for long songs see Farrell et al.[28]

How the Avian Song Control System Works

Having identified just a few of the many factors involved in the operation of the avian song-learning mechanism, let's now consider the parts of the brain that are essential for learning a song dialect. White-crowned sparrows, and other songbirds, have a brain with many anatomically distinct clusters of neurons, or **nuclei**, such as the just-mentioned NCMv, as well as neural connections that link one nucleus to another. The various components of the brain are made up of cells (neurons) that communicate with one another via bioelectric messages (action

European starling

FIGURE 10.13 The song preferences of female starlings. In this experiment, song preference was measured by the willingness of females to perch either near a nest box from which a long starling song (lasting about 50 seconds) was played or near one from which a shorter song (lasting about half as long) was played. After Gentner and Hulse.[32]

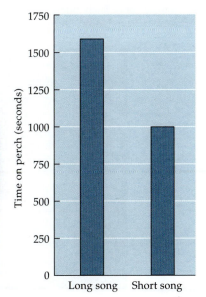

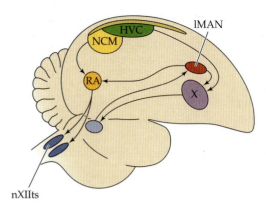

FIGURE 10.14 The song system of a typical songbird. The major components, or nuclei, involved in song production include the robust nucleus of the arcopallium (RA), the HVC, the lateral portion of the magnocellular nucleus of the anterior nidopallium (lMAN), the caudomedial neostriatum (NCM), and area X (X). Neural pathways carry signals from the HVC to the tracheosyringeal portion of the hypoglossal nucleus (nXIIts) to the muscles of the song-producing syrinx. Other pathways connect the nuclei, such as lMAN and area X, that are involved in song learning rather than song production. After Brenowitz et al.[16]

FIGURE 10.15 Difference in the size of one nucleus of the song system, the robust nucleus of the arcopallium (RA), in (A) the male and (B) the female zebra finch. Photographs courtesy of Art Arnold; from Nottebohm and Arnold.[71]

potentials) that travel from one neuron to another via elongate extensions of the neurons (axons) (see Figure 12.8). Some components of the brain are deeply involved in the memory of songs, while others are necessary for the imitative production of memorized song patterns.[35] Figuring out which anatomical unit does what is a task for neurophysiologists more interested in the operational mechanics of the nervous system than in its development.

These researchers have long focused on a region of the brains of white-crowned sparrows and other songbirds called the HVC. This dense collection of neurons connects to the robust nucleus of the arcopallium (mercifully shortened to RA by anatomists), which in turn is linked with the tracheosyringeal portion of the hypoglossal nucleus (whose less successful acronym is nXIIts). This bit of brain anatomy sends messages to the syrinx, the sound-producing structure of birds that is analogous to the larynx in humans. The fact that the HVC and RA can communicate with the nXIIts, which connects to the syrinx, immediately suggests that these brain elements exert control over singing behavior (Figure 10.14).

This hypothesis about the neural control of bird song has been tested. For example, if neural messages from the RA cause songs to be produced, then the experimental destruction of this center, or surgical cuts through the neural pathway leading from the RA to the nXIIts, should have devastating effects on a bird's ability to sing, which they do.[22] If the RA plays an important role in controlling bird song, then in bird species like the white-crowned sparrow, in which males sing and females do not, the RA should be larger in male brains than in female brains, and it is (Figure 10.15).[2,66,71]

Moreover, the RA can respond to the social environment of the bird. In starlings, for example, the RA is the only song control nucleus that grows substantially in males that are exposed to high-quality (longer) songs of other males for a week (Figure 10.16).[85] The suggestion here is that when some males are singing songs that are especially attractive to females, other eavesdropping males in the neighborhood benefit from modulating the motor region controlling their song production in order to sing more attractively themselves.

Different brain nuclei appear to be essential for song learning, rather than for the production of vocal signals. If, as suspected, the lateral mag-

(A)
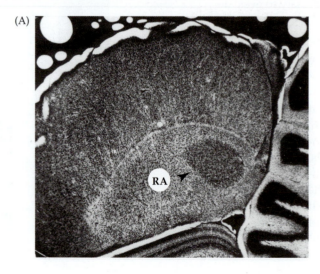

(B)
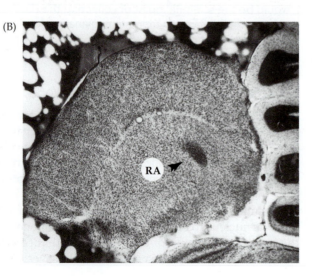

Male starling

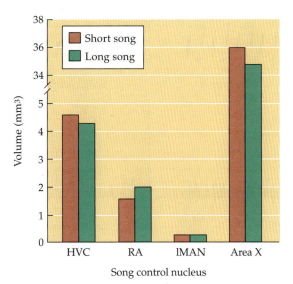

FIGURE 10.16   Song competition in the starling changes the size of the robust nucleus of the arcopallium (RA). The RA of males exposed to high-quality (long) songs of other starlings when they are adults increases in size, unlike some other song control nuclei. After Sockman et al.[85]

nocellular nucleus of the anterior nidopallium (lMAN) is a key element needed for song learning, then the lMAN should be much reduced or absent in bird species that sing but do not learn their songs (of which there are many—see below). Sure enough, some birds of this type lack the well-defined lMANs (and certain other forebrain nuclei) present in white-crowned sparrows and other vocal learners.[22]

But what about the assumption that large amounts of neural tissue are required if a male sparrow or other songbird is to learn a complex song or songs. One could argue instead that the experience of singing a complex song or acquiring a large repertoire causes the HVC to expand in response to stimulation of this region of the brain. If true, then isolated males should have smaller HVCs than those males that learn songs. This experiment has been done with sedge warblers; isolated males had brains that were in no way different from those of males that had learned their songs by listening to the songs of other males.[50] Prior to this work, similar results came from a study in which some male marsh wrens were given a chance to learn a mere handful of songs while another group listened to and learned up to 45 songs.[15] Thus, in both the warbler and the wren, the male brain develops largely independently of the learning experiences of its owner, suggesting that the production of a large HVC is required for learning, rather than the other way around.

Although the neurophysiology of song learning has been explored at the level of entire brain nuclei, neuroscientists could in theory find out how a given neuron contributes to communication between birds. In fact, this kind of research has been done by Richard Mooney and his coworkers in their work with swamp sparrows.[63] Males of this species sing two to five song types, each type consisting of a "syllable" of sound that is repeated over and over in a trill that lasts for a couple of seconds. If a young male swamp sparrow is to learn a set of song types, he must be able to discriminate between the types being sung by males around him, and then later he must be able to tell the difference between his own song types as he listens to himself sing.

One mechanism that could help a young male control his song type output would be a set of specialized neurons in the HVC that respond selectively to a specific song type. Activity in these cells could contribute to his ability to monitor what he is singing so that he could adjust his repertoire in a strategic manner (by, for example, selecting a song type that would be particularly effective in communicating with a neighboring male, as we will see below).

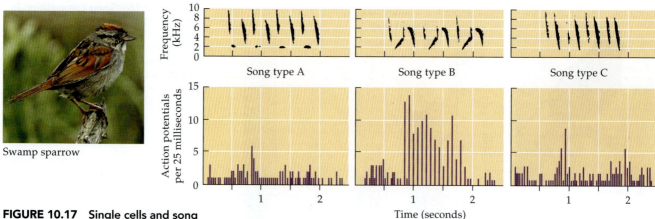

FIGURE 10.17 Single cells and song learning in the swamp sparrow. (Top) Sonograms of three song types: A, B, and C. (Bottom) When the three songs are presented to the sparrow, one of its HVC relay neurons reacts substantially only to song type B. Other cells not shown here respond strongly only to song type A, while still others fire rapidly only when song type C is the stimulus. After Mooney et al.[63]

The technology exists to permit researchers to record the responses of single cells in a swamp sparrow's HVC to playbacks of that bird's own songs. In so doing, Mooney and his associates discovered a number of HVC relay neurons that generated intense volleys of action potentials when receiving neural signals from other cells upon exposure to one song type only.[63] Thus, one relay neuron, whose responses to three different song types are shown in Figure 10.17, produces large numbers of action potentials in a short period when the song stimulus is song type B. This same cell, however, is relatively unresponsive when the stimulus is song type A or C. So here we have a special kind of cell that could (in conjunction with many other neurons) help the sparrow identify which song type it is hearing, the better to select the best response to that signal.

The Ultimate Causes of Bird Songs

Although a great deal has been learned about how the song control system of songbirds develops and operates, we still have much to learn about the underlying proximate mechanisms of singing behavior. But even if we had this information in hand, our understanding of singing by white-crowned sparrows would still be incomplete until we dealt with the ultimate causes of the behavior. Because the complex and elaborate proximate mechanisms underlying bird song are unlikely to have materialized out of thin air, we can ask questions such as, when in the distant past did an ancestral bird species start learning its species-specific song, thereby setting in motion the changes that led eventually to dialect-learning abilities in birds like the white-crowned sparrow?

Evidence relevant to this historical question includes the finding that song learning occurs in members of just 3 of the 23 avian orders: the parrots, the hummingbirds, and the "songbirds," which belong to that portion of the Passeriformes that includes the sparrows and warblers, among other species (Figure 10.18).[14] Members of the remaining 20 orders of birds produce complex vocalizations, but they apparently do not have to learn how to do so, as shown in some cases by experiments in which young birds that were never permitted to hear a song tutor, or were deafened early in life prior to the onset of song practice, nevertheless came to sing normally.[46]

One question about the historical sequence underlying the evolution of song learning is, did the song-learning ability exhibited by the three groups of birds evolve independently or not? There has been considerable debate about phylogenetic relationships between the three orders of song-learners. If we were to accept the phylogeny shown in Figure 10.18A, this evolutionary tree

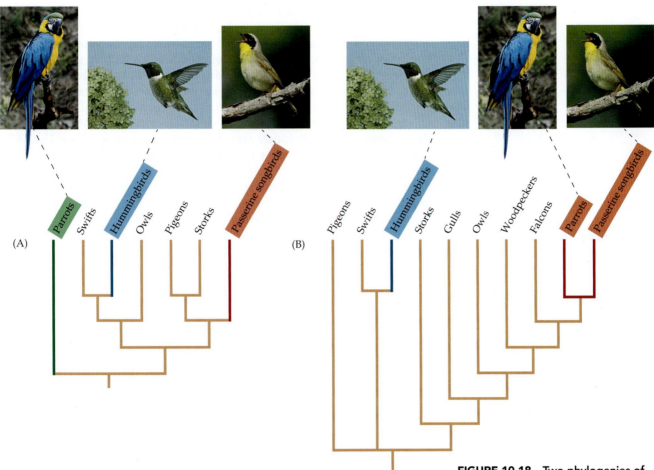

FIGURE 10.18 Two phylogenies of song learning in birds. (A) The bird phylogeny available in 1991 suggests that song learning evolved independently in three different lineages of modern birds (the passerine songbirds, the hummingbirds, and the parrots). (B) A newer phylogeny published in 2008 indicates that song learning evolved just twice, in the ancestor of the songbirds and parrots, which are close relatives, and independently in the hummingbirds. New evidence yields new conclusions. A, after Brenowitz;[14] B, after Hackett et al.[37]

means that the nearest living relatives of each of the three song-learning orders do not learn their songs. This conclusion suggests one of two things: either song learning originated three different times in the approximately 65 million years since the first truly modern bird appeared, or song learning originated in the ancestor of all the lineages shown in Figure 10.18A but then was lost three times: (1) at the base of the evolutionary line leading to the swifts, (2) at the base of the owl lineage, and (3) at the base of the lines constituting the pigeons and storks. In other words, either there have been three independent origins of song learning or one origin followed by three separate losses of the trait (see Figure 10.18A).[42]

Recently, however, a new phylogeny has been assembled by two different research teams on the basis of new information that indicates that, surprisingly enough, parrots are the closest relatives of songbirds.[89,93] If correct, then there need be only two, not three, independent evolutionary origins of song learning (or one followed by two losses of the trait). If the two-independent-origins scenario applies, we would expect to find major differences in the song control systems of hummingbirds versus parrots and songbirds. To check this prediction, we need to identify the brain structures that endow these birds with their singing abilities. One way to do so is to capture birds that have just been singing (or listening to others sing) and kill them immediately in order to examine their brains for regions in which a product of the *ZENK* gene can be found.[42] As noted above, this gene's activity is stimulated in particular parts of the brain when birds produce songs and in different parts of the brain when birds hear songs. Other stimuli and activities do not turn the gene on, so its product is absent or very scarce when the birds are not communicating. By comparing how much ZENK protein appears in different parts of the brains of birds that were engaged in singing or listening to songs immediately before their deaths, one can effectively map those brain regions that are associated with bird song.

The *ZENK* activity maps of the brains of selected parrots, hummingbirds, and songbirds reveal strong similarities in the number and organization of discrete centers devoted to song production and processing. In all three groups, for example, cells that form the caudomedial neostriatum (NCM) activate their *ZENK* genes when individuals are exposed to the songs of others. The NCM is located in roughly the same part of the brain in all three groups, where it constitutes part of a larger aggregation of anatomically distinct elements that contribute to the processing of song stimuli. When parrots, hummingbirds, and songbirds vocalize, other brain centers, many in the anterior part of the forebrain, respond with heightened *ZENK* gene activity. Once again, there is considerable (but not perfect) correspondence in the locations of these distinctive song production centers in parrots, hummingbirds, and sparrows (Figure 10.19). The many similarities in brain anatomy among the three groups of vocal learners argue against the hypothesis that song-learning abilities

FIGURE 10.19 The song control systems of parrots, hummingbirds, and oscine songbirds are distributed throughout the brain in remarkably similar patterns. On the left is a diagram of the evolutionary relationships among some major groups of birds, including the three orders of vocal learners. On the right are diagrams of the brains of these groups, with the various equivalent components of the song control systems labeled (e.g., the NCM). After Jarvis et al.[42]

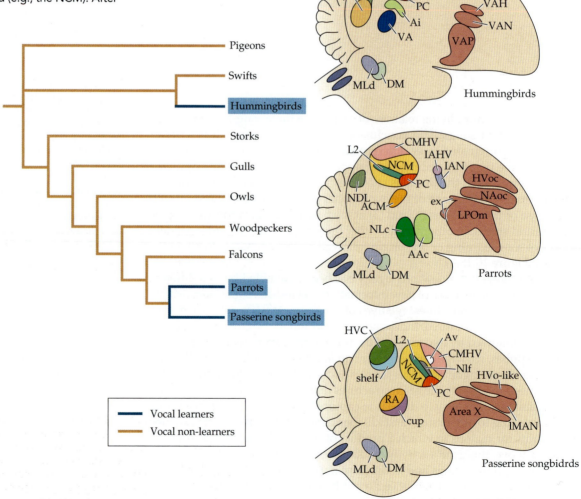

evolved independently in the three groups of birds. Thus, we have to take seriously the possibility that the ancestor of all those birds in the orders sandwiched between the hummingbirds and the parrots-songbirds was a song learner and that the mechanism for vocal learning possessed by that bird was then retained in some lineages while dropping out early on in others—a hypothesis that continues to generate debate.[29,42]

The Adaptive Value of Song Learning

Whatever the historical sequence of events underlying the evolutionary history of song learning in birds, we can still ask, why did the ability to learn spread through one or more ancestors of today's song learners? What are the fitness benefits of song learning for individual parrots, hummingbirds, and songbirds? To answer this question, let us focus on white-crowned sparrows and their learned dialects. The puzzle here is that song dialect learning has some obvious reproductive disadvantages including the time, energy, and special neural mechanisms that this behavior requires. What are the compensatory benefits associated with song learning that could make a white-crowned male communicate more effectively with male rivals and/or potential mates?

One possible advantage of song learning might be the ability of a young male to fine-tune his song as he acquires a dialect that can be transmitted unusually effectively in the particular habitat where he lives. A young male that learns his song from his older neighbors might then generate songs that travel farther and with less degradation than if he sang another dialect better suited to a different acoustical environment.[22] And in fact, male white-crowned sparrows in densely vegetated habitats sing songs with slower trills and lower frequencies than birds in places with lower vegetation density.[25] This same pattern has been observed for other birds, like the satin bowerbird and the great tit, which employ lower frequencies in dense forests than in more open areas (Figure 10.20).[41,69] Here we have an example of how the environment either influences the evolution of communication signals or favors individuals with the flexibility to adjust their songs to different conditions, topics with real significance for the adaptationist,[65,91] and not just with respect to sound communication.[27,57]

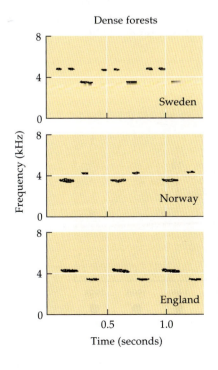

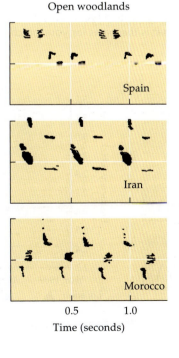

Great tit

FIGURE 10.20 Songs match habitats. Great tits from dense forests produce pure whistles of relatively low frequency, whereas males of the same species that live in more open woodlands use more and higher sound frequencies in their more complex songs. After Hunter and Krebs.[41]

A second hypothesis on the benefits of song learning centers on the advantages of matching songs to the singer's social environment.[17] The idea is that males able to learn the local version of their species-specific song can communicate better with rivals that will also be singing that particular learned song variant. If this is so, then we can predict that young males should learn directly from their territorial neighbors, using those individuals as social tutors. By adopting the song of a neighbor, a new boy on the block could signal his recognition of that male as an individual and demonstrate that he was not a novice competitor but old enough and experienced enough to have utilized the information provided by his neighbor's song. If true, young males should alter their song selection to match that of older territorial neighbors, and these older males should take males that match their songs more seriously as rivals than young novice challengers with unmatched song types, and both these predictions are correct.[6,78]

FIGURE 10.21   The dialects of white-crowned sparrows in three parts of San Francisco. (A) The three dialects recorded over several decades of increasing urbanization at the three locations, each represented by a different symbol. (B) The minimum frequency contained in the songs in the three locations. The numbers above the vertical bars represent the number of birds recorded. After Luther and Baptista.[53]

White-crowned sparrow

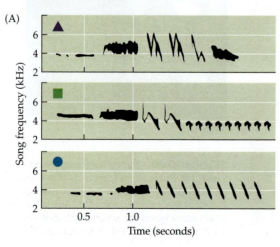

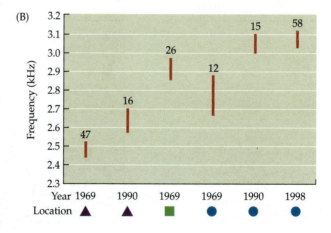

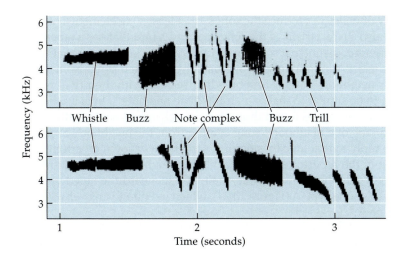

FIGURE 10.22 **Two white-crowned sparrow songs from different dialect populations.** The component parts of the songs are labeled, including the "note complex" and the "trill" at the end of the song. After Nelson et al.[68]

Discussion Question

10.11   At the proximate level, the ability of some yearling white-crowned sparrows to match the dialect of neighboring males can be explained by the *late acquisition hypothesis*. This proposal states that yearlings have a developmental window similar to that of very young birds, which enables them to listen to and learn directly from their immediate neighbors, overriding (if need be) any dialect learning that took place earlier in life. In contrast, the *selective attrition hypothesis* argues that fledglings can memorize a number of dialect versions of their species' song early in life; then, after settling next to some older males, these birds gradually discard certain of their learned variants until they are left with the one that best matches the dialect of their neighbors. What predictions follow from these two hypotheses? Figure 10.23 should help you evaluate these predictions if you know that when the two yearlings (FCN33 and PPM) settled in next to their neighbors, each initially sang two somewhat different songs but later dropped the song less like that of its territorial neighbors.

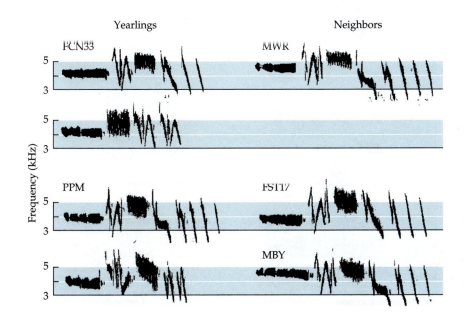

FIGURE 10.23 **Dialect selection by male white-crowned sparrows.** After arriving on their first territories, yearling males FCN33 and PPM initially each sang two different dialects, but after a time they each sang only one song type (the upper of the two song types shown for each male in the left side of the figure). After Nelson.[67]

The ability of males to produce the songs, or at least song elements, of rival neighbors is also characteristic of the song sparrow, a bird that has a repertoire of several different, distinctive song types rather than a single vocalization of a particular dialect. Young song sparrows usually learn their songs from tutors that are their neighbors in their first breeding season, with their final repertoire tending to be similar to that of an immediate neighbor.[70] In fact, Michael Beecher and his colleagues found through playback experiments that when a male heard a tape of a neighbor's song coming from the neighbor's territory, he tended to reply to that tape by singing a song from his own repertoire that matched one in the repertoire of that particular neighbor. This kind of type matching occurred when, for example, male BGMG heard the taped version of male MBGB's song type A (Figure 10.24), which led him to answer with his own song type A.[9]

Song sparrow

FIGURE 10.24 Song type matching in the song sparrow. Males BGMG and MBGB occupy neighboring territories and share three song types (A, B, and C: the top three rows of sonograms); six unshared song types (D, E, F, G, H, and I) appear on the bottom three rows. After Beecher et al.[7]

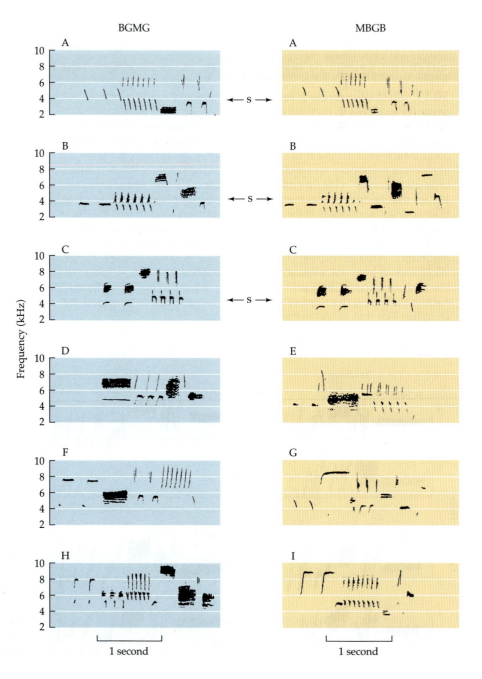

Adaptive Repertoire Matching

Another kind of response involves repertoire matching, in which a bird exposed to a song type from a neighbor responds not with the exact same type but with a song type drawn from their shared repertoire. For example, repertoire matching would occur if male BGMG answered back with song type B or C when he heard song type A from male MBGB's territory (see Figure 10.24).

A third option for a male song sparrow is to reply to a song from a neighbor with an unshared song. This kind of mismatched response would be illustrated by male BGMG singing song type D, F, or H (see Figure 10.24) upon hearing male MGBG's song type A.

The fact that song sparrows type-match and repertoire-match so often indicates that they recognize their neighbors and know what songs they sing and that they use this information to shape their replies.[9] But how do male sparrows benefit from their selection of a song type to reply to a neighbor? Perhaps their choice enables them to send graded threat signals to their neighbors, with type matches telling the targeted receiver that the singer is highly aggressive, whereas replying with an unshared song could signal a desire to back off, while a repertoire match might signal an intermediate level of aggressiveness. If so, then playback experiments should reveal differences in the responses of territorial birds to taped songs that contain a type match, a repertoire match, or neither. When Michael Beecher and colleagues checked this prediction, they found that a tape containing a type match did indeed elicit the most aggressive response from a listening neighbor, while a repertoire match song generated an intermediate reaction, and nonshared song types were treated less aggressively still (Figure 10.25).[10,19] This result supports the hypothesis that a male song sparrow that is able to learn songs from a neighbor can convey information about just how strongly it is prepared to challenge that individual.

If the ability to modulate song challenges is truly adaptive, then we can predict that male territorial success, as measured by the duration of territorial tenure, should be a function of the number of song types that a male shares with his neighbors. Song sparrows can hold onto their territories for up to 8 years, which tends to create a relatively stable community of neighbors. And, in fact, the number of years that an individual male song sparrow holds his territory increases about fourfold as the number of song types shared with his neighbors goes from fewer than 5 to more than 20.[8] This finding suggests strongly that the potential for neighbors to form long-term associations in song sparrow communities selects for males able to advertise their aggressive intentions to particular individuals. If so, then in the fluid social settings of certain other species in which territorial males come and go quickly, song learning and sharing of the sort exhibited by song sparrows should not occur.

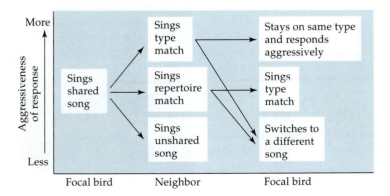

FIGURE 10.25 Song matching and communication of aggressive intent in the song sparrow. Song sparrow males can control the level of conflict with a neighbor by their selection of songs. When a focal male sings a shared song at a rival, the neighbor has three options: one that will escalate the contest, one that keeps it at the same level, and another that de-escalates the interaction. Likewise, the initiator of the contest can use his ability to select a matching song type, a repertoire match, or an unshared song type. The three different kinds of songs convey information about the readiness of the singer to escalate or defuse an aggressive encounter. After Beecher and Campbell.[10]

Instead, under these different ecological circumstances, males should acquire songs characteristic of their species as a whole to facilitate communication with any and all conspecifics, rather than developing a dialect or a set of shared song types characteristic of a small, stable population.

Donald Kroodsma and his colleagues have tested this proposition by taking advantage of the existence of two very different populations of the sedge wren. In the Great Plains of North America, sedge wren males are highly nomadic, moving from breeding site to breeding site throughout the summer. In this population, song matching and dialects are absent; instead, the birds improvise variations on songs heard early in life as well as inventing entirely novel songs of their own, albeit employing the general pattern characteristic of their species.[47] In contrast, sedge wren males living in Costa Rica and Brazil remain on their territories year-round. In these areas, dialects and song matching are predicted to occur, and these learning-proficient males do have dialects and do sing like their neighbors.[48] The differences in song learning that occur within this one species indicate that learning and dialects do evolve when males gain by communicating specifically with other males that are their long-standing neighbors.

Female Preferences and Song Learning

Still other ultimate hypotheses on song learning take aim at the social environment provided by females, which are often the intended recipients of bird songs. So we need to expand our focus on the role of song in male-male interactions to include a look at male-female communication, as illustrated by the fact that when a female Cassin's finch disappears, leaving her partner without a mate, the number of songs he sings and the time he spends singing increase dramatically, either to lure his partner back or to attract a replacement. The evidence reviewed in the caption to Figure 10.26 suggests that the first hypothesis is correct.[84]

But could the fact that male Cassin's finches learn their songs contribute to a male's ability to attract females? Consider a species that is divided into stable subpopulations. In such a species, males in each of these groups are likely to have genes that have been passed down for generations by those males' successful ancestors. Therefore, by learning to sing the dialect associated with their place of birth, males could announce their possession of traits (and underlying genes) well adapted for that particular area. Females hatched in that area might gain by having a preference for males that sang the local dialect, because they would endow their offspring with genetic information that would promote the development of locally adapted characteristics.[3]

This hypothesis matches the discovery that male white-crowned sparrows that sang the local dialect around Tioga Pass, California, were less infected

FIGURE 10.26 Male Cassin's finches sing to attract females. When a male is paired off with a female on day 1, he sings relatively little. But when the female is removed on day 2, the male invests considerable time in singing, perhaps to call her back. If the male is caged without a female on day 1, he allocates little time to singing, and he sings even less when a female is introduced into his cage, indicating that it is the loss of a potential mate that stimulates the male to sing. After Sockman et al.[84]

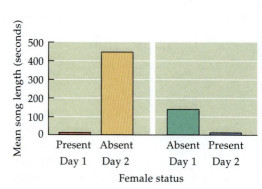

Cassin's finch

with a blood parasite than non-dialect singers. Therefore, female white-crowns in this population could potentially use song information to secure healthier mates. And, in fact, males with songs not matched to the local dialect fathered fewer offspring than local dialect singers, a result consistent with the hypothesis that females prefer to mate with males that sing the local dialect.[56]

On the other hand, if female preferences are designed to get them to mate with males produced in their natal area, then female white-crowns should prefer males with the dialect that they heard while they were nestlings—namely, their father's dialect—and they do not, at least in one Canadian population.[23] Furthermore, young male white-crowns are not always locked into their natal dialects but may be able to change them,[26] if they happen to move from one dialect zone to another; therefore, female white-crowns cannot rely on a male's dialect to identify his birthplace with complete certainty. All of this casts doubt on the proposition that female song preferences enable them to endow their offspring with locally adapted gene complexes from locally hatched partners, as does the lack of genetic differentiation found among dialect groups, which was noted earlier.[86]

Another version of the female choice hypothesis for male song learning is based on the possibility that females look to the learned songs of potential mates for information about their developmental history. If a female could tell just by listening to a male that he was unusually healthy, she could acquire a mate whose genes had worked well during his development and so should be worth passing on to her offspring. In song sparrows, males with larger song repertoires have a larger-than-average HVC and are also in better condition as judged by their relatively large fat reserves and by their apparently more robust immune systems.[76] Females that preferred males of this sort might be able to endow their male offspring with genes that could give them a competitive edge (assuming that there is a hereditary component to HVC size and body condition).

Alternatively, a healthier partner might be able to provide his offspring with above-average parental care. In some European warblers, females prefer to mate with males with large song repertoires.[73] Katherine Buchanan and Clive Catchpole have shown that song-rich males of the sedge warbler bring more food to their offspring, which grow bigger—a result that almost certainly raises the reproductive success of females that find large song repertoires sexually appealing, which in turn selects for males that are able to sing in the favored manner.[18]

BIRD SONG AS AN INDICATOR OF MALE DEVELOPMENTAL HISTORY   But the question still remains, why might female choice favor males that learn their complex songs, rather than simply manufacturing an innately complex song with no learned component? One hypothesis is that the quality of vocal learning could give female songbirds a valuable clue about the quality of the singers as potential mates. The key point here is that song learning occurs when males are very young and growing rapidly. If rapid growth is difficult to sustain, then young males that are handicapped by genetic defects or nutritional stress should be unable to keep up, resulting in suboptimal brain development. According to this hypothesis, individuals with even slightly deficient brains may be less able to meet the demands of learning a complex species-specific song.[72]

These predictions have been tested in experiments with swamp sparrows. By bringing very young nestling males into the laboratory, Steve Nowicki and his coworkers were able to control what they were fed. One group of nine males (the controls) received as much food as they could eat, while another group of seven (the experimentals) received 70 percent of the food volume

(A)

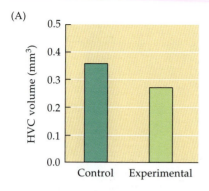

(B)

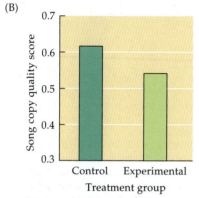

Treatment group

FIGURE 10.27 Nutritional stress early in life has large effects on song learning by male swamp sparrows. (A) Stress and brain development, as expressed in the volume of the HVC, and (B) stress and song learning, as measured by the match between learned songs and tutor tapes. Nestling swamp sparrows were hand reared and exposed to tutor tapes. The control group was fed all they could eat, whereas the experimental group received only 70 percent of that amount for about 2 weeks. After Nowicki et al.[74]

consumed by the control males. During the first 2 weeks or so, when the sparrows were totally dependent on their handlers for their meals, the controls came to weigh about a third more than the experimentals, a difference that was then gradually eliminated over the next 2 weeks as the sparrows came to feed themselves on abundant seeds and mealworms. Even though the period of nutritional stress was brief, this handicap had large effects on both brain development and song learning (Figure 10.27). Components of the song system in the food-deprived experimental group were significantly smaller than the equivalent regions in the control sparrows.[74]

Likewise, in the related song sparrow, nestlings that were subject to food shortages soon after hatching developed smaller HVCs than those fed as much as they wanted. The effect manifested itself by the time of fledging, even before the young birds had begun to learn their songs.[55] In the case of swamp sparrows, the experimentally deprived birds came to sing poorer copies, compared with the controls, of the taped song that both groups listened to during their early weeks of captivity.[74] These results, and similar ones from other studies,[87] have not been replicated by some other researchers,[34] which is why we should be cautious about accepting the relationship between nutritional shortfalls, stress, development of the song nuclei, and song learning. As is true for many issues in science, researchers will have to continue to test the hypothesis that adult female songbirds learn something about the developmental history, and thus the quality, of potential mates just by listening to their learned songs.

Poor nutrition during a bird's early development could affect more than a male's capacity to learn its species song. When young zebra finches were divided into two groups and fed either a high-protein diet or a low-protein alternative, the birds with the better diet grew larger and "solved" a food-related trial-and-error task faster than their less fortunate companions.[13] This result suggests that a superior diet translates into birds with superior learning abilities, which could affect the ability of a male to forage efficiently, a factor that might come into play for paternal males that provision their offspring. If so, females that gained information about a male's developmental history could choose superior mates by preferring those that ate well during their early development.

But do female birds actually pay attention to the information about mate quality that appears to be encoded in male songs? Nowicki's team collected songs of song sparrows that had been copied from tape tutors with varying degrees of accuracy and played them to female sparrows in the laboratory.[75] Female sparrows can be hormonally primed to respond with a tail-up precopulatory display to male songs they find sexually stimulating. The more accurately copied songs elicited significantly more precopulatory displays from the females than the songs that were copied less perfectly (Figure 10.28). Somewhat similar results have been obtained by researchers working with zebra finches who have found that stressed young males come to produce shorter and less complex songs than less handicapped males.[87]

Female songbirds may prefer songs that are relatively difficult to produce.[96] For example, the faster a male swamp sparrow sings its trilled song (see Figure 10.17), the harder it is to sing, especially if the syllables of the trill cover a relatively wide range of sound frequencies. Females are attracted to males that sing right up to the limits of performance.[4] Likewise, female serins (a species of small finch) prefer males that sing at relatively high frequencies, another physiological challenge for males.[20] By basing their mate preferences on male performance, females are in effect favoring males likely to be healthy and in good condition, attributes that could make the male a superior donor of genes or a superior caregiver to offspring.

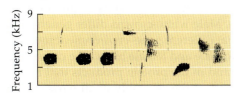

(A) Tape tutor song

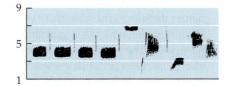

(B) Good copy of (A)

(C) Tape tutor song

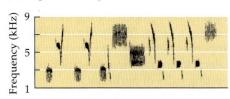

(D) Poor copy of (C)

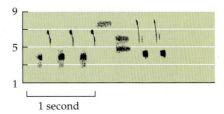

1 second

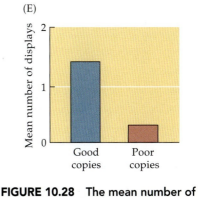

(E)

FIGURE 10.28 **The mean number of precopulatory displays in response to differences in song quality.** Female song sparrows were tested with playback of the songs of males that had been able to copy their tutor songs very accurately versus males with lower copying accuracy. The upper panel shows (A) a tape tutor song and (B) a copy sung by a male able to copy a high proportion of the notes in the taped song it listened to; the bottom panel shows (C) a tape tutor song that was (D) poorly copied by another individual. (E) The females' response to good and poor song copies. After Nowicki et al.[75]

Of course, these social interactions depend on the ability of both male and female birds to develop "normally." Sadly, a variety of industrial pollutants, such as the break-down components of the pesticide DDT, alter the development of the brains of both sexes in some birds and thus change the mate choice and mate attraction behaviors of males and females,[82] yet another way in which humans are altering the lives of animals around us.

Discussion Questions

10.12   William Searcy and a team of researchers played taped songs to captive female song sparrows that had been given hormone implants shortly after being taken to the laboratory.[80] The recorded songs came from male song sparrows that lived in the population from which the females had been taken, as well as from males living various distances (18, 34, 68, 135, and 540 kilometers) from the female subjects. Songs from males living 34 or more kilometers from the populations from which the females came were not nearly as effective in eliciting the precopulatory display as songs from local males; in contrast, songs from males living only 18 kilometers away were about as sexually stimulating as local songs. These data have relevance for more than one ultimate hypothesis on song learning by male sparrows. What are the hypotheses, and what importance do these findings have for them?

10.13   Parasites are often microscopic in size but have large negative effects on their hosts. If this is true for the parasites of songbirds, what predictions follow about their effects on male song performance, and how should females respond to the song of infected males as opposed to uninfected individuals? Check your predictions by reading Garamszegi.[31]

Proximate and Ultimate Causes Are Complementary

We have reviewed just a small portion of what is known about the proximate and ultimate causes of learning and dialects in bird song. The wealth of competing hypotheses within any one level of analysis can be a bit overwhelming. But a crucial point to which I return here is that the proximate and ultimate causes of behavior are complementary. On the proximate side, if we want to understand why male white-crowns from Marin and Berkeley sing somewhat different songs, we must understand how the song control mechanisms develop and how they operate. Male white-crowns differ genetically from females, a fact that guarantees that different gene–environment interactions will take place within embryonic male and female birds. The cascading effects of differences in gene activity and in the protein products that result from gene–environment interactions affect the assembly of all parts of the bird's body, including its brain and nervous system. The way in which a male white-crown's brain, with its special subsystems, can respond to experience is a function of the way in which the male has developed. By the time a male sparrow has become an adult, he possesses a large, highly organized song control system whose physiological properties permit him to produce a very particular kind of song. The learned dialect he sings is a manifestation of both his developmental history and the operating rules of his brain.

Ultimately, the proximate differences among white-crowns that generate differences in their dialects have an evolutionary basis. In the past, in the lineage leading to today's white-crowns, males surely differed in their song control systems and thus in their ability to learn to sing a dialect accurately. Some of these differences were doubtless hereditary. If the genetic differences among males translated into differences in acquiring a dialect and thereby their desirability as mates or their capacity to cope with rival males, natural selection would have resulted in the spread of those hereditary attributes associated with greater reproductive success. The males that made the greatest genetic contributions to the next generation would have supplied more of the genes available to participate in interactive developmental processes within members of that generation. The hereditary attributes of those animals would in turn be measured against one another in terms of their ability to promote genetic success in a new round of selection. This selective process links the proximate bases of behavior with their ultimate causes in a never-ending spiral through time (see Figure 10.1).

Discussion Questions

10.14   Both young males and young females of the sac-winged bat appear to learn and eventually reproduce the territorial songs of harem-controlling males.[43] Why might males imitate a song tutor of this sort? And why might females do the same?

10.15   What features of language learning in humans are similar to song learning in birds? What do these similarities suggest about the genetic and developmental bases of human language learning? Do comparisons with birds also suggest some interesting hypotheses on the adaptive value of learned language for members of our species? After you have attempted to answer these questions, go to the Web of Knowledge if it is available at your college or university library, and locate references on the shared proximate and ultimate causes of vocal learning by humans and birds. Some of those who have written knowledgeably on this subject include Peter Marler, Fernando Nottebohm, Steven Pinker, and Erich Jarvis.

Summary

1. Behavioral traits have both ultimate (evolutionary) and proximate (immediate) causes that are complementary, not mutually exclusive.

2. Questions about ultimate causes are those that focus on the possible adaptive value of a characteristic as well as those that ask how an ancestral trait became modified over time, leading to a modern characteristic of interest. Questions about proximate causes can be categorized as those concerned with the genetic–developmental bases for behavior as well as those that deal with how the physiological (neural and hormonal) systems provide the basis for behavior.

3. The complementary nature of research into the ultimate and proximate causes of behavior can be illustrated by asking what causes male prairie voles to be monogamous and what causes some male birds to sing different versions (dialects) of their species' standard territorial songs.

4. The prairie vole's monogamy and the vocalizations of white-crowned sparrows, as well as all other behavioral traits, have a history influenced by natural selection. As a result of many generations of selection acting on the members of a given species, some alleles will have been passed on while others will have gradually disappeared from populations. The genes present in any one generation interact with the environment to control the development of the members of that generation. Developmental processes lead to multicelled organisms with nerve cells and nervous systems, and with hormones and endocrine systems, the proximate physiological bases for behavioral abilities. If these abilities are the product of past selection, they ultimately will help individuals reproduce successfully, passing on the genes associated with above-average individual fitness.

Suggested Reading

The levels of analysis are the subject of two classic papers, one by Niko Tinbergen[90] and the other by Ernst Mayr.[60] For a recent review of the causes of monogamy in the prairie vole, see the work of Young et al.[97] The proximate causes of bird song have been studied by a host of researchers including Moorman et al.[64] and Zeigler and Marler.[99] The ultimate causes of this form of communication are examined in *Bird Song: Biological Themes and Variations*.[22] Eliot Brenowitz and Mike Beecher have written a concise review that focuses on the differences in what birds learn and how they learn their songs.[17] Donald Kroodsma's *The Singing Life of Birds* has been written for the general reader, who will learn how exciting it can be to listen to and understand bird song.[49] You can hear the complex vocalizations of many North American birds in *Bird Songs* by Les Beletsky.[12]

The Development of Behavior

The preceding chapter introduced the concept of the proximate (immediate) causes of behavior. There I made the key point that one can categorize proximate causes into two complementary (actually overlapping) levels of analysis, the developmental component and the physiological component. This chapter will expand on what is known about the developmental basis of behavior, especially the role that genes play as proximate factors underlying animal behavior.

Genes are involved in the development of a white-crowned sparrow's song (see Chapter 10). The developmental process is dependent on not only the genetic information the bird possesses but also a host of environmental influences, ranging from the nongenetic materials in the egg yolk, to the hormones that certain of the bird's cells manufacture and transport to other cells, to the sensory signals generated when a baby sparrow hears its species' song, to say nothing of the neural activity that occurs when a young adult male interacts with a neighboring territorial male. The sparrow's song therefore highlights a key idea, namely, that development is an interactive process in which genetic information interacts with changing internal and external environments in ways that assemble an organism with special properties and abilities. The process occurs because some of the genes in the organism's cell nuclei can be turned on or off by the appropriate signals that come from the developing animal's external environment. As genetic activity changes within an organism, the chemical reactions within its cells change, building (or modifying) the proximate mechanisms that underlie an organism's characteristics and capacities.

The role that genes play in this process means that genetic differences among individuals can lead to differences in the way

Both genes and environment contribute to the development of foraging behavior in the honey bee.

proximate mechanisms develop, which in turn can cause individuals to differ behaviorally. Behavioral variation among individuals can affect their ability to pass on their genes. In other words, as the previous chapter also stressed, there are strong links between evolutionary processes and the proximate causes of behavior, a point that this chapter will re-emphasize. But first, we need to deal with the mistaken view that some traits are "genetically determined" as well as the equally erroneous claim that some traits are "environmentally determined."

The Nature or Nurture Misconception

Evolutionary analyses of behavioral development cannot proceed if we believe that some behaviors are "genetic" or "genetically determined" and so could evolve, while other behaviors are learned and so are environmentally determined and, by extension, are immune to natural selection. This major misconception, and that is what it is, is applied with particular fervor to humans, whose behavior is of course highly influenced by learning. Because so much of human behavior is learned, many have concluded that nature (our genetic heritage) is trumped by nurture (growing up in a cultural environment where learning is of paramount importance). But neither we nor white-crowned sparrows could learn a thing from our environments without our brains, which are composed of cells, each of which contains a nucleus, which encapsulates our DNA, our genes. The activity of those genes in conjunction with environmentally supplied chemicals endowed our cells with special properties. As a result, our brain cells and those of white-crowned sparrows help us respond to our cultural and our acoustical environments, respectively, by learning certain things. So the idea of a purely environmentally determined behavior (one that would develop even in the absence of genetic information) is as nonsensical as the idea of a purely genetically determined behavior (one that would develop even in the absence of environmental inputs) (see Chapter 2 in Dawkins;[21] also Robinson[84]).

White-crowned sparrow males can learn a particular song dialect of their species' basic song because of an interaction between nature, inherited (genetic) information, and nurture, including the social and acoustical environment to which the young bird is exposed. Change the genes, and the genetic information available to respond to environmental inputs could change, leading to an alteration in the development of the bird's brain, which could affect its capacity to learn its song from adult male white-crowns singing nearby. Change the environment, for example, by playing the young male white-crown tapes of a dialect he would otherwise never hear, and the brain–acoustical environment interaction changes, affecting the bird's brain development and eventually its singing behavior.

The Interactive Theory of Development

We shall further illustrate the interactive theory of development by considering the changes that take place in the roles played by a worker bee over the course of her lifetime. After a worker bee metamorphoses into a young adult, her first job within the hive is a humble one, the cleaning of comb cells. She then becomes a nurse bee that feeds honey to larvae in the brood comb before making the transition to being a distributor of food to her fellow workers. The last phase of her life, which begins when she is about 3 weeks old, is spent foraging for pollen and nectar outside the hive (Figure 11.1).[58]

So what causes a worker to go through these different stages? If behavioral development requires both genetic information and environmental inputs, the

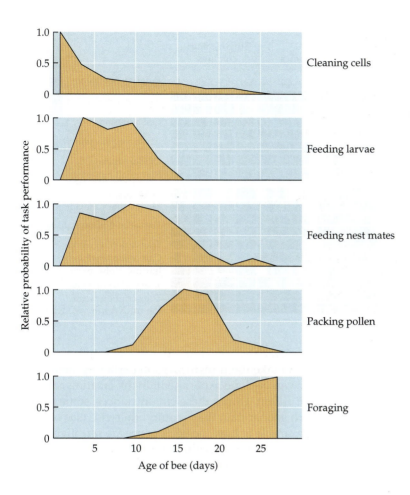

FIGURE 11.1 **Development of worker behavior in honey bees.** The tasks, such as foraging for pollen (seen here), adopted by worker bees are linked to their age, as demonstrated by following marked individuals over their lifetimes. After Seeley.[85]

information in some of the bee's many thousands of genes (the bee's **geno-type**) must respond to the environment in ways that influence the development of her measurable extra-genetic characteristics (the color of her body, the length of her antennae, the quantity of juvenile hormone in her hemolymph, all examples of the bee's **phenotype**). In other words, if the interactionist theory of development is correct, as a bee's behavioral phenotype changes from nurse to forager during her life, there must be changes in the interplay between genes and some aspect of the individual's environment.

The search for changes in the genetic component of these interactions has been facilitated by the invention of microarray technology, a procedure that makes it possible to analyze the activity of a large set of genes in particular tissues, like the brain, by detecting certain products (messenger RNAs) made when the genetic information is being used to produce biochemical products. In order to see which gene products are abundant and which are not, biologists scan a sheet or glass slide on which minute amounts of tissue have been spread before being subjected to a fluorescent dye that reacts with nucleic acids. When Charles Whitfield and his coworkers ran brain extracts from nurses and foragers through a variety of microarrays, they were able to compare the activity of about 5500 genes (of the roughly 14,000 in the bee genome) for these two kinds of individuals.[103] Some of the genes that were turned on in nurse brains differed substantially and consistently from those active in foragers' brains, and vice versa. Indeed, about 40 percent of the surveyed genes showed different levels of product output in the two kinds of bees. These changes in genetic activity are correlated with the typical age-related transformation of a nurse into a forager (Figure 11.2). They also, however, occur

FIGURE 11.2 Gene activity varies in the brains of nurse bees and foragers. Shown here are individual records (each bar represents one bee's gene expression record for a particular gene) of bees from typical and manipulated colonies. The 17 selected genes surveyed here were the ones that showed the largest difference in activity between the brains of nurses and of foragers. In addition, these 17 genes are highly similar to genes found in fruit flies (*Drosophila*); the function of the corresponding gene in fruit flies has been established and is noted here. The activity level of a gene in forager, relative to nurse, bee brains is color coded from high (>2) to low (<0.5); see the reference code at the upper right. The left-hand portion of the figure shows gene activity for young nurses (YN) and old foragers (OF) from unmanipulated colonies; the right-hand portion shows those for young foragers (YF) paired with nurses of the same age and old nurses (ON) paired with foragers of the same age taken from colonies manipulated to contain either all young or all old workers. The expression ratio (F/N) is calculated by dividing the activity score of a given gene in foragers by that for nurses. After Whitfield et al.[103]

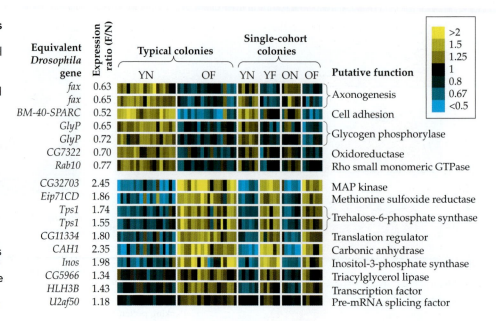

when nurses are induced to make the transformation earlier in their adult lives than is normal, which can be done experimentally by creating colonies in which older foragers are absent. Under these circumstances, some nurses adopt the forager role much sooner than they would otherwise (see below). This study provides a key example of how the bee's genes are sensitive to the environment in which the developing individual finds itself, providing bees with the flexibility to switch between tasks in a way that meets the needs of the queen and her brood.

The microarray approach has been expanded to examine gene activity in the brain cells of hundreds of honey bees that were collected as they engaged in different activities in the wild. This work identified large numbers of genes whose expression differed depending on the behavioral state of the bee; the set of activated genes depended, for example, on whether the bees had been collected while defending the hive or were sacrificed as they were foraging. Large numbers of genes were involved in the network, including, importantly, those that code for various transcription factors, proteins that regulate the activity of a number of target genes. The different behaviors exhibited by the honey bees in the study could be linked to modules that generate a specific set of transcription factors, which in turn promote or suppress the activity of particular groups of target genes in the bee's brain.[17,30] Other transcription factors are involved in the control of neural genes with more general effects on the bee's behavior.

A demonstration that the environment affects gene activity comes from a study designed to see whether bees that differ in their perception of how far they have flown exhibit differences in which of the genes in their brains are active. When foragers return to the hive after drinking sugar water at a feeding station, they monitor the distance they have traveled by observing the flow of environmental images on the retina. They can be tricked into "thinking" that they have flown a longer distance than they have actually, by having them fly through a tunnel with vertical-striped walls.[86] Using this technique, a research team created two categories of bees, both of which had flown the same distance, but one group had flown through a tunnel with vertical stripes while the other had made its way through a tunnel without that pattern. The genes expressed in the two groups of bees differed strongly, demonstrating that the different perceptions of distance flown were based in part on a difference in genetic activity in particular parts of the bee's brain that differed as a result of flight experience.

Note that studies of this sort do *not* demonstrate that the environment is irrelevant to behavioral development and brain functioning. Far from it. Indeed, environmental factors are critical for every element of gene expression within organisms, in part because the environment supplies the molecular building blocks that are essential if the information in DNA is to be used to make messenger RNAs, some of which are involved in gene regulation via transcription factors, others of which carry the information about making specific proteins from the nucleus of the cell to the protein assembly sites in the cytoplasm. In addition, the cellular environment must contain the precursors of these constituents of living things if they are to be produced. These chemicals ultimately come from substances consumed by the queen prior to making her eggs, as well as from the honey and pollen eaten by the larvae and adults that develop from those eggs. Some of the resultant developmental products may play a special role in changing the activity of one or more key genes in an individual, initiating a cascading series of biochemical changes that eventually alters the development of the brain and, thus, the behavior of the bee.

One particularly potent developmental product appears to be a substance called juvenile hormone, which is found in low concentrations in the blood of young nurse workers but in much higher concentrations in older foragers. As one might predict, if young bees are treated with juvenile hormone, they become precocious foragers,[83] but if one removes a bee's corpora allata (the glands that produce juvenile hormone), the bee delays its transition to foraging. Moreover, bees without corpora allata that receive hormone treatment regain the normal timing of the switch to foraging.[97]

So it appears that changes in juvenile hormone production have fundamental developmental effects in honey bees, a widespread feature of hormones in many other animals as well, with, for example, the sex hormones (such as testosterone and estrogen) often intimately involved in the construction of neural mechanisms that are responsible for the differences between male and female mammals in their sexual behavior.[57] In honey bees, hormone production responds to the social environment of individuals. When researchers have created experimental honey bee colonies with a worker force consisting of uniformly young bees of the same age, a division of labor still manifested itself, with some individuals remaining nurses much longer than usual, while others began foraging as much as 2 weeks sooner than average.

What enabled the bees to make these developmental adjustments? Perhaps a deficit in social encounters with older foragers may have stimulated an early developmental transition from nurse to forager behavior. This possibility has been tested by adding groups of older foragers to experimental colonies made up of only young workers. The higher the proportion of added older bees, the lower the proportion of young nurse bees that undergo an early transformation into foragers (Figure 11.3).[44] As it turns out, the behavioral interactions between the young residents and the older transplants inhibit the development of foraging behavior, because transplants of young bees have no such effect on young resident bees. The inhibiting agent has been traced to a fatty acid compound called ethyl oleate, which only foragers manufacture and store in a chamber (the crop) off the digestive tract.[56] When returning foragers pass nectar contained in the crop to nurses back

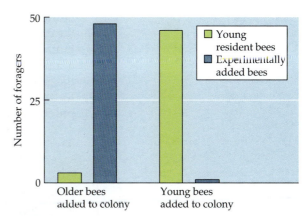

FIGURE 11.3 Social environment and task specialization by worker honey bees. In experimental colonies composed exclusively of young workers (residents), the young bees do not forage if older forager bees are added to their hive. But if young bees are added instead, the young residents develop into foragers very rapidly. After Huang and Robinson.[44]

at the hive, ethyl oleate is probably transferred as well. The more foragers in a hive, the more likely nurses are to receive quantities of this chemical, which slows their transition to foraging status.

Discussion Questions

11.1   Why do queen bees behave very differently from their workers even though a queen has essentially the same genome as her worker sisters and daughters? Develop at least one proximate hypothesis on why the two categories of bees behave so differently after you find out about how worker bee and queen larvae are reared. The Internet will be helpful in this regard.

11.2   Honey bees possess a gene (labeled *for*) that contains information for the production of a particular enzyme called PKG. If this genetic information is important (when expressed in brain cells) for foraging activity in worker honey bees, what prediction follows about the levels of the messenger RNA needed to produce the PKG enzyme present in foragers versus nonforagers taken from a typical bee colony? Figure 11.4 presents the relevant data, which you can use to evaluate your prediction. But since foragers are older than nurses in typical colonies, perhaps the greater activity of *for* is simply an age-related change that has nothing to do with foraging. What additional prediction and experiment are required to reach a solid conclusion about the causal role of *for* in regulating foraging in the honey bee? Suggestion: Take advantage of the ability to create experimental colonies of same-age workers. Given that the *for* gene contributes to the onset of foraging by worker bees,[5] how could this switch also be influenced by the workers' social environment? Produce a hypothesis that integrates the genetic and environmental contributions to the age-related change in bee behavior.

The honey bee example clearly shows why it would be an utter error to say that some behavioral phenotypes are genetic or more genetic than others. Worker foraging behavior cannot be purely "genetically determined" because the behavior is the product of literally thousands of chemical interactions between the bee's genes and its environment, all of which are required to construct the bee's brain and the rest of its body. The information in a gene is expressed only when the gene is in the appropriate environment. As Gene Robinson puts it, "DNA is both inherited and environmentally responsive."[84] Environmental signals, such as those provided by juvenile hormone and ethyl oleate, influence gene activity. When a gene is turned on or off by changes in the environment, the resulting changes in protein production can directly or indirectly alter the activity of other genes within affected cells. A multitude of precisely timed and well-integrated biochemical changes involving both the genes and the cellular environments that surround them are responsible for the construction of all traits, every last one. Therefore, no trait can be purely "genetic."

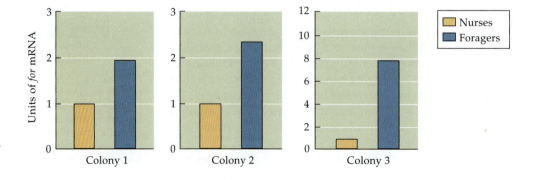

FIGURE 11.4   Levels of the messenger RNA produced when the *for* gene is expressed in the brains of nurses and of foragers in three typical honey bee colonies. After Ben-Shahar et al.[5]

By the same token, it is flat wrong to say that a given phenotype is "environmentally determined." The development of every attribute in every living thing requires the information contained in large numbers of genes. Neither genotype nor environment can be said to be more important than the other, just as no one would say that a chocolate cake owed more to the recipe used by the cook than to the ingredients that actually went into the finished product.

Discussion Question

11.3 The nature–nurture controversy involves those who believe that our nature (essentially our genes) dominates our behavioral development and others who argue just as forcefully that our nurture (especially our upbringing as children) is what shapes our personalities. Some have dismissed the controversy by saying that the two sides might as well be fighting about whether a rectangle's area is primarily a matter of its height or mostly a function of its width. What's the point of the rectangle analogy? Does the analogy have any weaknesses?

Learning Requires Both Genes and Environment

The development of the ability to learn depends on specialized features of the brain, which in turn arise developmentally through the interplay between information-rich genes and key elements of the animal's environment. A classic example of the dependence of learning on genes is provided by **imprinting**, in which a young animal's early social interactions, usually with its parents, lead to its learning such things as what constitutes an appropriate sexual partner. Thus, a group of greylag goslings, having imprinted on behavioral biologist Konrad Lorenz (Figure 11.5) rather than a mother goose, formed both a learned attachment to Lorenz[60] and, in the case of the male greylags when they reached adulthood, a sexual preference for humans as mates. The experience of following a particular individual early in life must have somehow altered those regions of the male goose's nervous system responsible for sexual recognition and courtship in the adult bird. The special effects of imprinting could not have occurred without a "prepared" brain, one whose genetically influenced development enabled it to respond in special ways to particular kinds of information available from its social environment.

FIGURE 11.5 Imprinting in greylag geese. These goslings have imprinted on the behavioral biologist Konrad Lorenz and are following him wherever he goes.

(A)

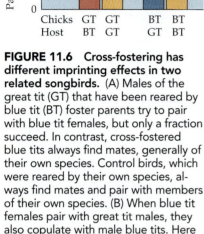

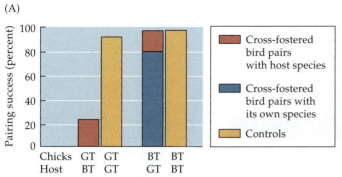

Legend:
- Cross-fostered bird pairs with host species
- Cross-fostered bird pairs with its own species
- Controls

X-axis labels:
| Chicks | GT | GT | BT | BT |
| Host | BT | GT | GT | BT |

Y-axis: Pairing success (percent), 0 to 100

FIGURE 11.6 Cross-fostering has different imprinting effects in two related songbirds. (A) Males of the great tit (GT) that have been reared by blue tit (BT) foster parents try to pair with blue tit females, but only a fraction succeed. In contrast, cross-fostered blue tits always find mates, generally of their own species. Control birds, which were reared by their own species, always find mates and pair with members of their own species. (B) When blue tit females pair with great tit males, they also copulate with male blue tits. Here a female blue tit (far left) paired with a male great tit (upper left) rear a brood together that consists entirely of blue tit nestlings. After Slagsvold;[92] photograph by Tore Slagsvold.

(B)

The fact that different species exhibit different imprinting tendencies provides further circumstantial evidence for the genetic contribution to learning. A group of Norwegian researchers provided this kind of evidence when they switched broods from blue tit nests into the nests of breeding great tits, and vice versa. Some of the cross-fostered youngsters grew up and survived to court and form pair bonds with members of the opposite sex (Figure 11.6A). Of the surviving fostered great tits, only 3 of 11 found mates—all of which were blue tit females that had been fostered by great tits. Of the surviving fostered blue tits, all 17 found mates, although 3 of these were females that socially mated with cross-fostered male great tits.[94]

However, although some individuals of both species became imprinted on another species as a result of their foster care experiences, the degree to which individuals imprinted on their foster parents differed between the two species of songbirds. Note that none of the cross-fostered great tits mated with a member of its own species, whereas most of the cross-fostered blue tits did. Moreover, each blue tit female that had a great tit as a social partner must have mated with a blue tit male on the side, because all 33 offspring produced by those females were blue tits, not hybrids (Figure 11.6B). Thus, although misimprinting occurred in both species, the developmental effect of being reared by members of another species was far greater for great tits than for blue tits, an indication that the hereditary basis of the imprinting mechanism was very different for these two related species.

Discussion Question

11.4 When great tits were experimentally reared in blue tit nests, they survived quite well but failed to mate with their own species, as just discussed. The reproductive success of these individuals was consequently low.[93] Discuss both the negative and positive effects that imprinting of this sort could have had on the evolution of *interspecific* brood parasitism.

In addition to imprinting, bird species possess other specialized learning abilities, including the ability to remember where they have hidden food. The black-capped chickadee is especially good at this task. This bird's spatial memory enables it to relocate large numbers of seeds or small insects that it has hidden in bark crevices or patches of moss scattered throughout its woodland environment, an ability that David Sherry investigated by providing captive chickadees with a chance to store food in holes drilled in small

(A)

(B)

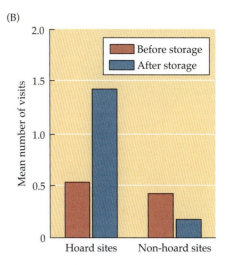

Black-capped chickadee

FIGURE 11.7 Spatial learning by chickadees. (A) Black-capped chickadees spent much more time at sites in an aviary where they had stored food 24 hours previously (hoard sites) than they had spent during their initial exposure to those sites, even though experimenters had removed the stored food. (B) The chickadees also made many more visits to hoard sites than to other sites, evidently because they remembered having stored food there. After Sherry.[87]

trees placed in an aviary. After the chickadees had placed sunflower seeds in 4 or 5 of 72 possible storage sites, they were shooed into a holding cage for 24 hours. During this time, Sherry removed the seeds and closed each of the 72 storage sites with a Velcro cover. When the birds were released back into the aviary, they spent much more time inspecting and pulling at the covers at their hoard sites than at sites where they had not stored food 24 hours before (Figure 11.7). Because the storage sites were all empty and covered, there were no olfactory or visual cues provided by stored food to guide the birds in their search; they had to rely solely on their memories of where they had hidden seeds.[87] In nature, these birds store only one food item per hiding spot and never use the same location twice, yet they can relocate their caches as much as 28 days later.[40]

Clark's nutcrackers may have an even more impressive memory, for they scatter as many as 33,000 seeds in up to 5000 caches that may be as many as 25 kilometers from the harvest site (Figure 11.8). The bird digs a little hole in the earth for each store of seeds, then completely covers the cache. A nutcracker does this work in the fall and then relies on its stores through the winter and

FIGURE 11.8 A Clark's nutcracker holding a seed in its bill that the bird is about to cache underground. Photograph by Russ Balda.

into the spring, recovering an estimated two-thirds of the hidden stores, often months after hiding them.[3,67]

It could be that nutcrackers do not really remember where each and every seed cache is but instead rely on a simple rule of thumb, such as "look near little tufts of grass." Or they might remember only the general location where food was stored and, once there, look around until they see disturbed soil or some other indicator of a cache. But experiments similar to those performed with chickadees show that the birds do remember exactly where they hid their food. In one such test, a nutcracker was given a chance to store seeds in a large outdoor aviary, after which it was moved to another cage. The observer, Russ Balda, mapped the location of each cache and then removed the buried seeds and swept the cage floor, removing any signs of cache making. No visual or olfactory cues were available to the bird when it was permitted to go back to the aviary a week later and hunt for the food. Balda mapped the locations where the nutcracker probed with its bill, searching for the nonexistent caches. The bird's spatial memory served it well, for it dug into as many as 80 percent of its ex-cache sites, while only very rarely digging in other places.[3] Other long-term experiments on nutcracker memory have demonstrated that nutcrackers can remember where they have hidden food for at least 6, and perhaps as long as 9, months.[4] Indeed, when Balda tested one of his graduate students as if he were a food-storing bird, the student did only about half as well as a typical nutcracker when tested a month after making his caches.[67] The birds can even remember the size of the seeds they have hidden, as demonstrated by their tendency to spread their bills farther apart when probing the earth for large cached seeds as opposed to smaller ones. Because nutcrackers retrieve one seed at a time from their underground stores, they can secure and process them more efficiently by opening their beaks just the right distance to grasp and pluck a seed of a given size out of a cache.[69]

The extraordinary ability of nutcrackers and chickadees to store spatial information in their brains is surely related to the ability of certain brain mechanisms to change biochemically and structurally in response to the kinds of sensory stimulation associated with hiding food. These changes could not occur without the genes needed to construct the learning system and the genes that are responsive to key sensory stimuli relevant to the learning task. The general point is that even learned behaviors, which are obviously environment-dependent, are gene-dependent as well.

Discussion Question

11.5 Return to the chocolate cake analogy (see page 329), and use it to illustrate how a change in either genes or environment could lead to developmental differences between individuals.

Environmental Differences Can Cause Behavioral Differences

Although no behavior is either purely genetic or purely environmental, *differences between individuals* can arise as a result of developmental differences stemming from differences in *either* their genes *or* their environments. You remember from Chapter 10 that the dialect differences among male white-crowned sparrows are a product of differences in the birds' acoustic and social environments, which affect what nestling sparrows learn as they listen to males singing nearby. Environmental differences are important whenever members of a species differ in a learned behavior.

For example, the nonbreeding workers on a *Polistes* paper wasp nest typically react calmly toward other female workers that have been reared on that nest and have stayed on to become helpers there. But should another worker from another nest try to join the colony, the workers react very differently, by attacking these "foreigners." Workers appear to learn to recognize one another as nest mates in large part because they have acquired the special odor of the nest (see Figure 2.17) in which they were reared by other adult members of the colony.[32]

Of course, the ability to record information about the odor or appearance of one's nest mates requires genetic information, which is needed for the construction of a nervous system with the capacity for this kind of learning. This point can be made with reference to a study of Belding's ground squirrels in which the newborn offspring of captive females were switched around at birth, creating four classes of individuals: (1) siblings reared apart, (2) siblings reared together, (3) nonsiblings reared apart, and (4) nonsiblings reared together. After having been reared and weaned, the juvenile ground squirrels were placed in an arena in pairs and given a chance to interact. In most cases, animals that were reared together, whether actual siblings or not, tolerated each other, whereas animals that had been reared apart tended to react aggressively to each other.[41] Here, the young squirrels learned something about their nest mates thanks to a nervous system primed to record certain information about the olfactory cues provided by their companions.

But in addition, biological sisters reared apart engaged in fewer aggressive interactions than nonsiblings reared apart (Figure 11.9). In other words, the squirrels had some way of recognizing their sibs that was not dependent on living with them as youngsters.[41,43] Instead, a different kind of learning was probably involved, one that goes informally by the indelicate label of the "armpit effect."[63] That is, if individuals can learn what they themselves smell like, then they can use this information as a reference against which to compare the odors of other individuals.[42]

This hypothesis has been examined by Jill Mateo,[64] who noted that Belding's ground squirrels possess several scent-producing glands,[65] including one around the mouth and another on the animal's back. Moreover, these

Belding's ground squirrel

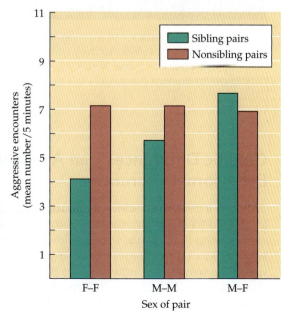

FIGURE 11.9 Kin discrimination in Belding's ground squirrels. Sisters reared apart display significantly less aggression toward each other than other combinations of siblings reared apart, which are as aggressive to each other when they meet in an experimental arena as nonsiblings reared apart. After Holmes and Sherman.[41]

FIGURE 11.10 Belding's ground squirrels learn their own odor. Juvenile squirrels were first given three trials during which they could investigate their own odors applied to plastic cubes. Note the squirrels' decline in responsiveness to their own dorsal gland scents over these initial trials. Then the squirrels were provided with plastic cubes daubed with dorsal gland odors from four categories of individuals. Cubes with scents from close relatives received less attention than those with scents from distant relatives or nonkin. Numbers in parentheses show the degree of genetic relatedness (see page 22) of the individual to the tested ground squirrel. After Mateo.[64]

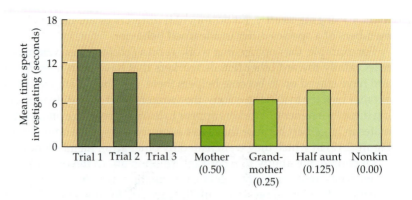

squirrels regularly sniff the oral glands of other individuals, as if they were acquiring odor information that could conceivably be compared with the sniffers' own scents. By capturing pregnant ground squirrels and moving them to laboratory enclosures, Mateo was able to observe their juvenile offspring investigating objects (plastic cubes) that had been rubbed on the dorsal glands of other squirrels of varying degrees of relatedness to the youngsters. Because the captive squirrels had been separated from some of their relatives, they had never met them and so had no prior experience with their odors. If, however, the test animals had learned what they themselves smelled like, and if close relatives produced scents more similar to their own than do distant relatives, an inexperienced youngster could in theory discriminate between unfamiliar relatives and nonrelatives on the basis of odor cues alone. As it turns out, the longer a Belding's ground squirrel sniffs an object, the greater its interest in that object. Cubes that have been rubbed on a fairly close relative, genetically speaking, receive only a cursory inspection. Items smeared with odors of a more distant relative are given a significantly longer sniff, and the inspection time increases again for cubes daubed with a nonrelative's odor. Belding's ground squirrels are odor analyzers, spending less time with scents similar to their own and increasingly more time with odors less like their own (Figure 11.10). They therefore have the capacity to treat individuals differently on the basis of this highly specific learned label of relatedness.[64]

Discussion Question

11.6 A good predictor of a young person's vocabulary is the amount of time parents spent talking to their child when he or she was very young. Some have concluded that family environment is therefore the essential factor in determining a person's language skills. What's a logical problem with this conclusion?

Genetic Differences Can Also Cause Behavioral Differences

Although a great many differences in behavioral phenotypes have been traced to differences in the environment, others have been linked to genetic differences among individuals, which follows from the interactive theory of development. An example comes from a study of blackcap warblers, some of which spend the winter in southern Great Britain while others go south to Africa (Figure 11.11).[7] If the differences between the two groups of blackcaps occurred because of genetic differences between them, the offspring of winter-in-Britain

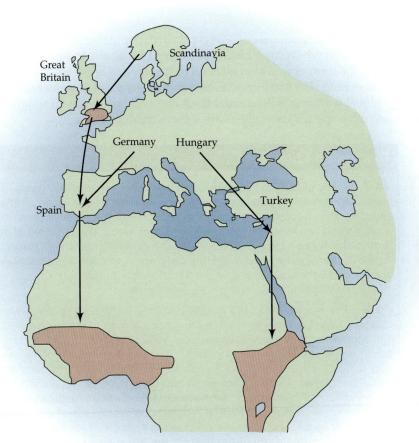

Blackcap warbler

FIGURE 11.11   Migratory routes taken by blackcap warblers in the fall. Blackcaps living in southern Germany and Scandinavia first go southwest to Spain before turning south to western Africa. Blackcaps living in eastern Europe migrate southeast before turning south to fly to eastern Africa. Where do the birds that winter in Great Britain come from? (It turns out they do not come from Scandinavia.)

birds should differ in their migratory behavior from the offspring of blackcaps that winter elsewhere. Therefore, Peter Berthold and his colleagues captured some wild blackcaps in Britain during the winter and took them to a laboratory in Germany, where the birds were kept indoors. Then, with the advent of spring, pairs of warblers were released into outdoor aviaries, where they bred, providing Berthold with a crop of youngsters that had never migrated.[8]

Once the young birds were several months old, Berthold's team placed some in special cages that had been electronically wired to record the number of times a bird hopped from one perch to another. The electronic data revealed that when fall arrived, the young warblers became increasingly restless at night, exhibiting the kind of heightened activity characteristic of songbirds preparing to migrate. The immature blackcaps' parents also became nocturnally restless when placed in the same kind of cages in the fall. The floors of these funnel-shaped cages were lined with paper that acquires marks when touched (see Figure 6.23). Whenever a bird leapt up from the base of the funnel in an attempt to take off, it landed on the paper and left scratch marks, which indicated the direction in which the bird was trying to go. Berthold's subjects, experienced adults and young novices alike, oriented due west, jumping up in that direction over and over, judging from the footmarks left on the paper. These data showed that the adults, which had been captured in wintery Britain, must have traveled there by flying west from Belgium or from central Germany, a point eventually confirmed by the discovery of some blackcaps in Britain that had been banded earlier in Germany. Young birds whose parents oriented in a southerly direction did the same in fall, convincing Berthold that he had a case in which behavioral differences in migratory orientation were caused by genetic differences between the two populations of individuals.[7] The rapidity with which some blackcaps evolved the ability to fly to a new migratory destination as well as evidence that this species can cease to be migratory in response to selection by researchers provides a ray of hope for birds affected by global warming.[80] Behavioral flexibility will be required if songbirds and most other species are to cope with a fast-warming planet.

Discussion Questions

11.7   A few blackcaps live year-round in southern France, although 75 percent of the breeding population migrates from this area in winter. Perhaps the difference between the two behavioral phenotypes is environmentally induced and not hereditary. Make a prediction about the outcome of an **artificial selection** experiment in which the experimenter tries to select for both nonmigratory and migratory behavior in this species. Describe the procedure and present your predicted results graphically. Check your predictions against the actual results (see Berthold[7]).

11.8   The black redstart is a bird species that migrates a relatively short distance from Germany to the Mediterranean region of Europe, whereas the common redstart travels as much as 5000 kilometers from Germany to central Africa. The scale in Figure 11.12 shows the duration of migratory restlessness in captive black redstarts, common redstarts, and hybrids between the two species, all of which had been hand-raised under identical conditions. Why do black redstarts exhibit migratory restlessness at night for fewer days than common redstarts? What does the behavior of the hybrids tell us about the genetic differences hypothesis for the difference in the duration of migratory restlessness in the two parental species?

FIGURE 11.12   Differences in the migratory behavior of two closely related birds, the black redstart and the common redstart. The scale at the bottom shows the periods after hatching (day 0) when the young birds exhibit migratory restlessness at night. After Berthold and Querner.[9]

Black redstart

Common redstart

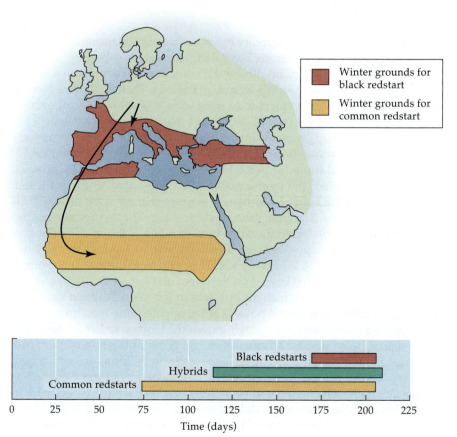

Another example of a case in which the differences among individuals of the same species have been traced to differences in the genes possessed by these individuals involves the garter snake *Thamnophis elegans*. This species occupies much of dry inland western North America as well as foggy coastal California.[2] The diets of snakes in the two areas, referred to hereafter as inland and coastal snakes, differ markedly. Whereas inland snakes feed primarily on the fish and frogs found in lakes and streams in the arid West, coastal snakes regularly eat the banana slugs that thrive in the wet forests of coastal California (Figure 11.13). You can watch a brief video of a snake swallowing a slug at http://www.birdsamore.com/videos/snake-eatingslug.htm. I can only marvel at the ability of the snakes to consume these prey. When I once made the mistake of picking up a banana slug, the creature promptly covered my hand with massive amounts of a repulsive mucus that greatly reduced my desire to touch these animals ever again.

If the preference for banana slugs exhibited by coastal garter snakes has a hereditary basis, then these snakes should differ genetically from inland snakes. To check this prediction, Steve Arnold took pregnant female snakes from the two populations into the laboratory, where they were held under identical conditions. When the females gave birth to a litter (garter snakes produce live young rather than laying eggs), each baby snake was placed in a separate cage, away from its litter mates and its mother, to remove these possible environmental influences on its behavior. Some days later, Arnold offered each baby snake a chance to eat a small chunk of freshly thawed banana slug by placing it on the floor of the young snake's cage. Most naive young coastal snakes ate all the slug hors d'oeuvres they received; most of the inland snakes did not (Figure 11.14). In both populations, slug-refusing snakes ignored the slug cube completely.

FIGURE 11.13 **A coastal Californian garter snake about to consume a banana slug,** a favorite food of snakes in this region. Photograph by Steve Arnold.

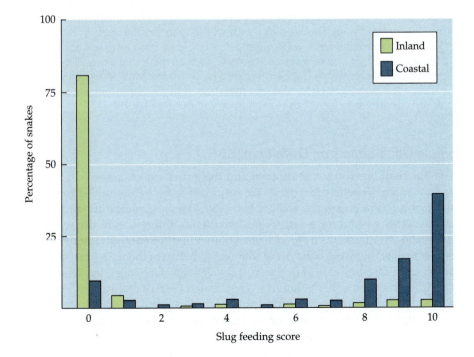

FIGURE 11.14 **Response of newborn, naive garter snakes to slug cubes.** Young snakes from coastal populations tended to have high slug feeding scores (e.g., a score of 10 indicates that the snake ate a slug cube on each of the 10 days of the experiment). Inland garter snakes were much less likely than coastal snakes to eat even one slug cube (which would yield a score of 1). After Arnold.[2]

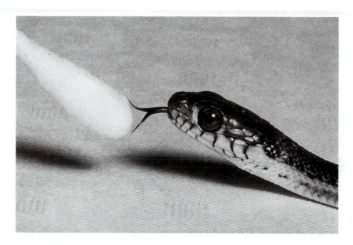

FIGURE 11.15   A tongue-flicking newborn garter snake senses odors from a cotton swab that has been dipped in slug extract. Photograph by Steve Arnold.

Arnold took another group of isolated newborn snakes that had never fed on anything and offered them a chance to respond to the odors of different prey items. He took advantage of the readiness of newborn snakes to flick their tongues at, and even attack, cotton swabs that have been dipped in fluids from prey animals (Figure 11.15). Chemical scents are carried by the tongue to the vomeronasal organ in the roof of the snake's mouth, where the odor molecules are analyzed as part of the process of detecting prey. By counting the number of tongue flicks that hit the swab during a 1-minute trial, Arnold measured the relative responsiveness of inexperienced baby snakes to different odors.

Populations of inland and coastal snakes reacted about the same to swabs dipped in toad tadpole solution (a prey of both groups), but they behaved very differently toward swabs daubed with slug scent. Almost all inland snakes ignored the slug odor, whereas almost all coastal snakes rapidly flicked their tongues at it. Because all the young snakes had been reared in the same environment, the differences in their willingness to eat slugs and to tongue-flick in reaction to slug odor appear to have been caused by their genetic differences.

If the feeding differences between the two populations arise because most coastal snakes have a different allele or alleles than most inland snakes, then crossing adults from the two populations should generate a great deal of genetic and phenotypic variation in the resulting hybrid group. Arnold conducted the appropriate experiment and found the expected result, confirming again that the behavioral differences between populations have a strong genetic component.[2]

Discussion Question

11.9   Debi Fadool at Florida State University headed a research team that studied a strain of genetically modified mice that lacked the ability to make a protein called Kv1.3.[27] In unaltered mice, this protein is found in regions of the brain that process olfactory information, leading to the prediction that the two kinds of mice should differ in their ability to smell things. In fact, the genetically modified mice were able to smell scents, such as those associated with food, at much lower concentrations than mice that possessed the protein. What Darwinian puzzle is created by these findings? What ultimate explanation do you have for the fact that mice with Kv1.3 protein are actually less sensitive to food odors than mice without that protein?

Single Gene Effects on Development

The blackcap breeding experiments and the garter snake crosses do not tell us how many genetic differences are responsible for behavioral differences among blackcaps and garter snakes. In theory, a single genetic difference could be the starting point for a series of downstream differences in the way that genes and environments influence one another in different individuals, which may translate into large behavioral differences between them. In Chapter 5 we discussed a case of this sort in the fruit fly *Drosophila melanogaster*, whose larvae differ in how far they roam in a yeast-coated petri dish during a 5-minute period.[22] When adults reared from rover larvae are bred with adults reared from sitter larvae, these pairs of flies produce larval offspring (the F_1 generation) that are all rovers. When these larvae mature and interbreed, they

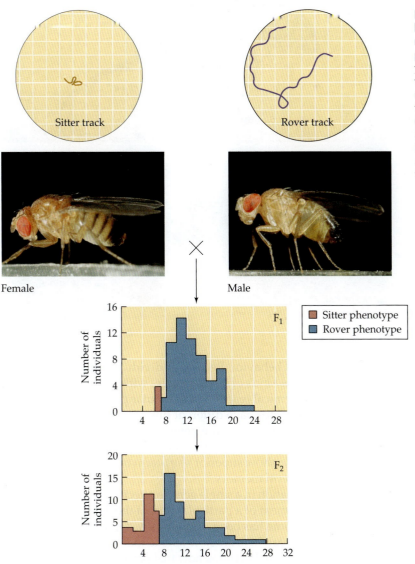

FIGURE 11.16 Genetic differences cause behavioral differences in fruit fly larvae. Representative tracks made by sitter and rover phenotypes feeding in a petri dish appear at the top of the figure. When adult female flies of the sitter strain mate with adult male flies of the rover strain, their larval offspring (the F_1 generation) almost all exhibit the rover phenotype (that is, they move more than 7.6 centimeters in 5 minutes). When flies from the F_1 generation interbreed, their offspring (the F_2 generation) are composed of rovers (blue) and sitters (red) in the ratio of 3:1. After de Belle and Sokolowski.[22]

produce an F_2 generation with three times as many rovers as sitters (Figure 11.16). Persons familiar with Mendelian genetics will recognize that rovers are likely to have at least one copy of the dominant allele of a gene affecting larval foraging behavior, whereas sitters are likely to have two copies of the recessive allele. If this analysis is correct, and if one could transfer the dominant allele associated with rover behavior to an individual of the sitter genotype, then the genetically altered larva should exhibit rover behavior. This experiment has been done, with positive results.[72] Thus, the difference in foraging behavior between rovers and sitters has its origins in a difference in the information contained in a single gene—just one of the 13,061 genes[74] located on the four chromosomes of *Drosophila melanogaster*.[23]

The techniques now available to biologists have been used to identify the gene in question, which has been labeled *for*. The gene codes for a particular protein, which goes by the label cGMP-dependent kinase, more of which is produced by larvae carrying the rover form of the allele (*forR*) than by those endowed with the sitter allele (*fors*).[72] This enzyme is produced in certain cells of the larval fruit fly's brain, where it presumably affects neuronal activity and

FIGURE 11.17 **A single genetic difference has a large effect on maternal behavior.** Wild-type female mice gather their pups together and crouch over them (above), but females with inactivated *fosB* genes (below) do not exhibit these behaviors (the pups can be seen scattered in the foreground). Photographs courtesy of Michael Greenberg, from Brown et al.[11]

thereby shapes the larva's behavior. Because individuals that differ in their *for* alleles produce different kinds of a key protein, with different efficacies in promoting a particular chemical reaction in particular parts of the fly's brain, larvae with different genotypes have different physiologies and they behave differently as a result.[18,50] Moreover, the gene affects not only the foraging behavior of larval fruit flies but also their associative learning abilities; individuals with the rover allele show superior *short*-term memory whereas sitter larvae exhibit better *long*-term memory.[66]

The power of the information in a single gene to affect a complex network of genes involved in development of a particular part of the brain at a particular period in an animal's lifetime has been dramatically demonstrated via gene knockout experiments. Researchers are now able to inactivate a given gene in an animal's genome in order to determine how that particular gene contributes to development in a particular environment. Consider the developmental effect of scrambling the genetic code of the *fosB* gene in laboratory mice. Females with the experimental "mutation" are normal in most respects but are totally indifferent to their newborn pups, which they fail to retrieve should they wriggle away from the nest. In contrast, normal females with two copies of the active *fosB* gene invariably gather displaced pups together and crouch over them, keeping them warm and permitting them to nurse (Figure 11.17).[11]

Discussion Question

11.10 Use the *fosB* knockout experiment to illustrate the difference between claiming that maternal behavior is genetically determined and claiming that certain differences among individuals in their maternal behavior phenotypes are genetically determined. How is the idea that genes are responsive to particular kinds of environmental inputs illustrated by the fact that when a female mouse inspects her pups after birth, she receives olfactory stimulation, which affects the mouse's brain and triggers *fosB* gene activity in a mouse with the typical genotype? How might this gene's activity initiate additional changes in other genes, leading to a specific pattern of biochemical events?

(A)

(B)

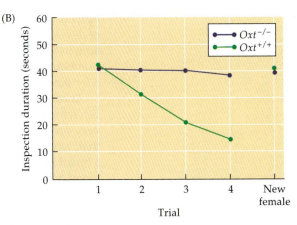

FIGURE 11.18 **Social amnesia is related to the loss of a single gene.** (A) Male mouse inspecting a female. (B) A knockout male mouse that lacks a functional *Oxt* gene carefully inspects the same female every time she is reintroduced into his cage, whereas a male with the typical genotype shows less and less interest in a female that he has inspected previously. A, photograph by Larry J. Young; B, after Ferguson et al.[28]

Other knockout mutations also have highly specific developmental consequences for mice. Males whose *Oxt* gene has been knocked out cannot produce oxytocin, an important brain hormone,[28] with the correlated effect that these males cannot remember females with whom they have recently interacted. Each time a given female is removed from and then returned to the cage she shares with an *Oxt* mutant male, the male gives her a thorough and lengthy sniffing that is no different from his response the very first time they met (Figure 11.18). In contrast, if a female is placed in the cage of a normal male with a functional *Oxt* gene, he remembers what she smells like, so if she is taken from his cage but then later returned to it, he spends less time sniffing her on the second occasion than he did the first time they met.

Evolution and Behavioral Development

As we showed in the preceding chapter, proximate mechanisms have ultimate causes, and this ties the different levels of analysis together. This point applies of course to developmental mechanisms, which have an evolutionary history that can sometimes be described as a series of modifications of an ancestral pattern and its reconfiguration into a modern attribute. This kind of research is at the heart of what has been called the field of evolutionary development, or "evo-devo."[16,99] Much of this work has focused on the history behind the development of structural features rather than behavioral traits per se, but the approach applies to both morphological and behavioral development. With respect to the often dramatic physical differences among species, even closely related ones, Darwin's theory of descent with modification has been used to identify the underlying genetic–developmental changes responsible for these differences. Thus, a gene shared by honey bees and fruit flies by virtue of their common ancestor contributes to the development of castes in bees by promoting the development of the relatively large size of queen bees as well as their large, fully functional ovaries.[47] The same gene in fruit flies codes for the same protein (royalactin) that can also promote the development of large body size in these insects.

A spectacular product of this approach has been the discovery that creatures as different as fruit flies and humans share a set of *homeobox* (or *Hox*) genes whose operation is critical for the developmental organization of their bodies. These genes, which must have originated in a very distant common ancestor, have been retained in flies, humans, and many other organisms as a result of their usefulness during the development of functional body structures. The base sequence of the genes has, of course, been altered somewhat from species to species, and the way in which their products influence the process of development differ markedly, leading to major differences in the structural products of development. Nonetheless, the imprint of history on the process can still be seen in the information contained within the *Hox* genes, the "toolkit" that vast numbers of descendants from a very ancient ancestor use as they develop from a fertilized egg to a multi-celled organism with a body and limbs.[16]

Another example of the phenomenon that relates specifically to animal behavior involves the *for* gene in *Drosophila* fruit flies, a gene that also occurs in very similar form in the honey bee.[99] The fruit fly *for* gene codes for a protein that, when produced, affects the operation of many other genes,[51] leading to a broad range of chemical changes affecting the operation of the brain in both larval and adult fruit flies. These developmental changes control the larval flies' tendency to move about in a petri dish. In adults, the gene has a variety of additional metabolic and behavioral effects. The honey bee has inherited this same gene from a common ancestor of flies and bees. But over evolutionary time, the now modified gene has taken on a different but allied function in the honey bee by helping to regulate the transition of a bee from a sedentary young adult that stays within the hive to a long-distance forager worker that collects food for the colony outside the hive. This transition is linked to an increase in the expression of the gene in the brains of the older workers.[5]

Another example of the conservative nature of behavioral development comes from studies of vasopressin, the neuropeptide that may play an important role in the mate-guarding monogamy of the prairie vole. As it turns out, all mammals, including our species, carry the gene for the production of this protein, which varies hardly at all from one mammalian species to another. In fact, molecules very similar to vasopressin have been found in everything from birds to insects to worms. Yet despite the similarities between the kinds of "vasopressin" produced in different animals, the chemical is involved in many different developmental processes.[25] Thus, in the prairie vole, vasopressin seems to play a part in the formation of a social attachment between a male and a female; however, the montane vole, which also has the gene for vasopressin and which also produces the chemical, does not form monogamous relationships.[104] If vasopressin is involved in the regulation of many other genes, whose expression and products can differ from species to species, many different developmental and behavioral outcomes could result.

Indeed, perhaps the major conclusion of evo-devo is that although changes in the base sequences of enzyme-coding genes are clearly important in the evolution of certain differences among species, another category of changes may often be even more important. I speak of changes in the mechanisms regulating the expression of other genes, namely mutations in those genes whose transcriptional products exert their effect by influencing the activity of other genes, often many other genes.[10,15] Altering the regulatory functions of the genome will affect how development proceeds because of the multiplier effect that comes from a mutation in one gene that changes the way in which the regulated genes work. As Sean Carroll has said, the process is analogous to teaching old genes new tricks rather than having each species evolve its own particular novel major genes for any given developmental task.[15] In fact, in a laboratory experiment in which the researchers artificially selected for speed of development of fruit flies from egg to hatching, they found that a great many genes were involved to some extent, not one or two mutant genes that swept through the population via natural selection, leading to the elimination of competing alleles over time.[12]

Adaptive Developmental Homeostasis

In addition to the historical level of analysis, evolutionary biologists can examine all traits in terms of their possible adaptive value. One of the features of developmental mechanisms that can be approached in this way is the ability of organisms to develop more or less normally even when endowed with genetic mutations that could be injurious. Despite the potential for genetic–developmental problems caused by the widespread distribution of novel mutations, most animals look and behave reasonably normally, a helpful outcome in terms of individual reproductive success. In fact, although gene knockout experiments sometimes do have dramatic phenotypic effects, in many cases, blocking the activity of a particular gene has little or no developmental effect. These findings have led some geneticists to conclude that genomes exhibit considerable information redundancy, which would explain why the loss of one protein linked to a particular gene is not fatal to the acquisition of certain traits of importance to the individual.[52,78]

We also know that many animals overcome what you might think would constitute great environmental obstacles to normal development. For example, some young birds lack the opportunity to interact with their parents and so cannot acquire the information that in other species is essential for normal social and sexual development via imprinting. When chicks of the Australian brush turkey hatch from eggs placed deep within the immense compost heap of a typical nest, they dig their way out and walk away, often without ever seeing a parent or sibling. So how do they manage to recognize other members of their species? Ann Göth and Christopher Evans studied captive young brush turkeys in an aviary in which they were exposed to feathered robots that looked like other youngsters. All that was required to elicit an approach from a naive youngster was a peck or two at the ground by the robot. Thus, young brush turkeys do not require extensive social experience in order to develop their basic social behaviors.[37] As adults, the birds are completely capable of normal sexual behavior despite having lived primarily by themselves beforehand.

Other experimenters have created genuinely weird rearing environments, only to find that various forms of deprivation have little or no effect on the development of normal behavior. Bring up baby Belding's ground squirrels without their mothers, and they still stop what they are doing to look around when they hear a tape of the alarm call of their species.[62] Male crickets that live in complete isolation sing a normal species-specific song despite their severely restricted social and acoustical environment.[6] Captive hand-reared female cowbirds that have never heard a male cowbird sing nevertheless adopt the appropriate precopulatory pose when they hear cowbird song for the first time, if they have mature eggs to be fertilized.[53]

The ability of many animals to develop more or less normally, despite defective genes and deficient environments, has been attributed to a process called **developmental homeostasis**. This property of developmental systems reduces the variation around a mean value for a phenotype and reflects the ability of developmental processes to produce an adaptive phenotype quite reliably. A clear demonstration of this ability comes from a classic experiment on the development of social behavior in young rhesus monkeys deprived of contact with others of their species by Margaret and Harry Harlow.[38,39] (The Harlows' experiments were conducted nearly 4 decades ago, when animal rights were not an issue; do you think the Harlows' harsh treatment of infant monkeys can be justified?) In one such study, the Harlows separated a young rhesus from its mother shortly after birth. The baby was placed in a cage with an artificial surrogate mother (Figure 11.19), which might be a wire cylinder or a terry cloth figure with a nursing bottle. The baby rhesus gained weight normally and developed physically in the same way that non-isolated rhesus

FIGURE 11.19 Surrogate mothers used in social deprivation experiments. This isolated rhesus infant was reared with wire cylinder and terry cloth dummies as substitutes for its mother.

FIGURE 11.20 **Socially isolated rhesus infants** that are permitted to interact with one another for short periods each day at first cling to each other during the contact period.

infants do. However, it soon began to spend its days crouched in a corner, rocking back and forth, biting itself. If confronted with a strange object or another monkey, the isolated baby withdrew in apparent terror.

The isolation experiment demonstrated that a young rhesus needs social experience to develop normal social behavior. But what kind of social experience—and how much—is necessary? Interactions with a mother are insufficient since infants reared only with their mothers fail to develop truly normal sexual, play, and aggressive behavior. Perhaps standard social development in rhesus monkeys requires the young animals to interact with one another. To test this hypothesis, the Harlows isolated some infants from their mothers but gave these infants a chance to be with three other such infants for just 15 minutes each day.[39] At first, the young rhesus monkeys simply clung to one another (Figure 11.20), but later they began to play. In their natural habitat, rhesus babies start to play when they are about 1 month old, and by 6 months they spend practically every waking moment in the company of their peers. Even so, the 15-minute-play group developed nearly normal social behavior. As adolescents and adults, they were capable of interacting sexually and socially with other rhesus monkeys without exhibiting the intense aggression or withdrawal of individuals that had been completely isolated from other youngsters.

Naturally one wonders about the relevance of these studies for another primate species, *Homo sapiens*, whose intellectual development is often said to be dependent on the early experiences that children have with their parents and peers. But is this true? We cannot, of course, do social isolation experiments with human babies, but we can examine evidence of another sort regarding the resilience of intellectual development in the face of nutritional deprivation. Consider, for example, the results of a study of young Dutch men who were born or conceived during the Nazi transport embargo during the winter of 1944–1945, which caused many deaths from starvation in the larger Dutch cities.[98] For most of the winter, the average caloric intake of city people was about 750 calories per day. As a result, urban women living under famine conditions produced babies of very low birth weights. In contrast, rural women were less dependent on food transported to them, and the babies they had that were conceived at the same time were born at more or less normal birth weights.

One would think that full brain development depends on adequate nutrition during pregnancy, when much of brain growth occurs. However, Dutch boys who were born in urban famine areas did not exhibit a higher incidence of mental retardation at age 19 than rural boys whose early nutrition was far superior (Figure 11.21A). Nor did those boys born to food-deprived mothers score more poorly than their relatively well-nourished rural counterparts when they took the Dutch intelligence test administered to draft-aged men (Figure 11.21B).[98] These results are buttressed by the discovery that Finnish adults who experienced severe nutritional shortfalls in utero (during a nineteenth-century famine) lived just as long on average as those who were born after the famine was over.[48]

No one believes that pregnant women or young children should be deprived of food,[70] and some continue to argue that the nutritional state of the fetus is critical for a person's health later in life (see review in Rasmussen[81]). But if the

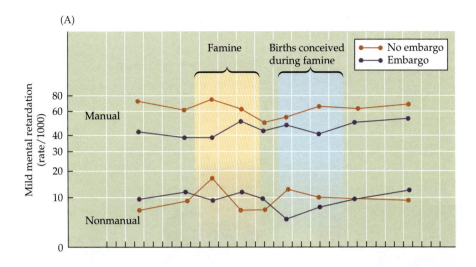

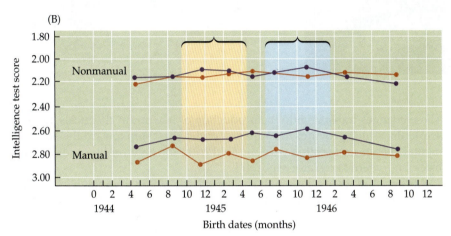

FIGURE 11.21 Developmental homeostasis in humans. Maternal starvation has surprisingly few effects on intellectual development in humans, judging from (A) rates of mild mental retardation and (B) intelligence test scores among 19-year-old Dutch men whose mothers lived under Nazi occupation while pregnant. (In this case, the lower the intelligence test score, the greater the intelligence of the subject.) The subjects were grouped according to the occupations of their fathers (manual or nonmanual) and whether their mothers lived or gave birth to them in a city subjected to food embargo by the Nazis (embargo) or in a rural area unaffected by the embargo (no embargo). Those who were conceived or born under famine conditions exhibited the same rates of mental retardation and the same levels of intelligence test scores as men conceived by or born to unstarved rural women. After Stein et al.[95]

survival of a fetus, our early intellectual development, and our later health are not necessarily harmed even by highly adverse early-life conditions,[20] perhaps selection in ancestral environments, where episodes of nutritional deprivation were not uncommon, has endowed us with resilient developmental systems.

Adaptive Developmental Switch Mechanisms

A selectionist approach to development has led to an appreciation of the adaptive value of developmental homeostasis, which helps individuals in diverse environments acquire traits that are essential for individual reproductive success. But there are many species in which two or three quite distinct alternative phenotypes coexist comfortably, with the differences arising proximately as a result of environmental differences among the individuals in question (Figure 11.22).[90,101] A challenge associated with such **polyphenisms** is to identify the proximate environmental cues that activate the mechanisms underlying the development of either one or another distinct phenotype, not an intermediate form of some sort. For example, we now know that the solitary and gregarious forms of the migratory locust, a major consumer of crops in Africa, develop into the swarming gregarious form when the nymphs see and smell other locusts and when grasshopper densities are high enough so that the insects' hindlegs are touched repeatedly.[89] Under these conditions, mechanoreceptors on the hindlegs relay their tactile messages to the central nervous system, where the neural messages trigger genetic changes in target cells.

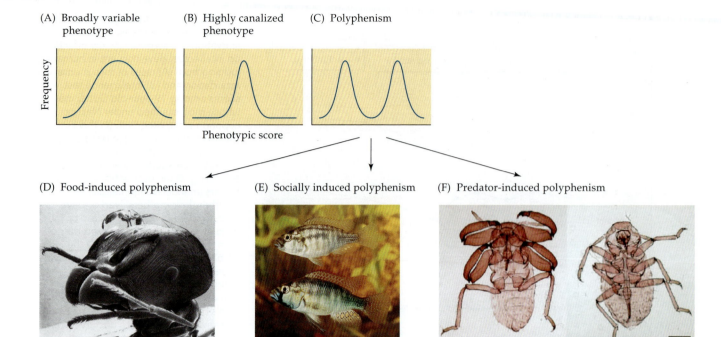

(A) Broadly variable
     phenotype

(B) Highly canalized
     phenotype

(C) Polyphenism

Frequency

Phenotypic score

(D) Food-induced polyphenism

(E) Socially induced polyphenism

(F) Predator-induced polyphenism

FIGURE 11.22 Developmental switch mechanisms can produce polyphenisms within the same species. Different phenotypes can arise when developmental switch mechanisms are activated in response to critical environmental cues. Top panel: Phenotypic variation within a species can range from (A) continuous, broad variation about a single mean value to (B) continuous but narrow variation about a single mean value to (C) discontinuous variation that generates several distinct peaks, each representing a different phenotype. Bottom panel: (D) In some cases, the amount or nature of the food eaten contributes to the production of certain polyphenisms, as in the castes of ants and other social insects. (E) In others, social interactions play a key role in switching phenotypes, as in the territorial and nonterritorial forms of the cichlid *Astatotilapia burtoni*. (F) In still other instances, the presence or activity of predators contributes to the development of an antipredator phenotype, as in the soldier caste (left) of some aphids, which possess more powerful grasping legs and a larger, stabbing proboscis than nonsoldier forms (right). D, photograph by Mark Moffett; E, photograph by Russ Fernald; F, photograph by Takema Fukatsu; from Ijichi et al.[45]

These changes translate into the production of new brain chemicals, notably serotonin, which provides the apparent critical first step toward the developmental changes associated with the shift from the green solitary form to the black-and-yellow migratory morph.[1]

At the ultimate level of analysis, we can ask what advantage individual locusts gain from having the developmental flexibility to be migratory or sedentary. The migratory morph is the one that does so much damage when the grasshoppers gather together in areas where they eat everything in sight before flying away to new locations where there may still be plants to consume. You can imagine how well assorted sedentary and semi-migratory types would do in competition with a truly migratory form when there are millions of grasshoppers competing for the palatable plants that they must eat to survive and reproduce.

Discussion Question

11.11   Serotonin has been shown to be sufficient to trigger the development of the gregarious form of the migratory locust. What predictions must have been checked to arrive at this conclusion? You should know that you can inject serotonin as well as drugs that disable serotonin directly into nymphal locusts.

There are also two very different tiger salamander phenotypes: (1) a typical aquatic immature form, which eats small pond invertebrates such as dragonfly nymphs, and (2) a cannibal morph, which grows much larger, has more powerful teeth, and feeds on other tiger salamander larvae unfortunate enough to live in its pond (Figure 11.23). The development of the cannibal type, with its distinctive form and behavior, depends proximately on the salamanders' social environment. For example, cannibals develop only when many salamander larvae live together.[19] Moreover, they appear more often when the larvae in a pond (or aquarium) differ greatly in size, with the

FIGURE 11.23 Tiger salamanders occur in two forms. The typical form (being eaten here) feeds on small invertebrates and grows more slowly than the cannibal form (which is doing the eating). Cannibals have broader heads and larger teeth than their insect-eating victims. Photograph by David Pfennig, courtesy of James Collins.

largest individual much more likely to become a cannibal than its smaller companions.[61] In addition, the cannibal form is more likely to develop when the population consists largely of unrelated individuals than when many siblings live together.[75] If a larger-than-average salamander larva occupies a pond with many other young salamanders that do not smell like its close relatives,[76] its development may well be switched from the typical track to the one that produces a giant, fierce-toothed cannibal.

What selective advantages do tiger salamanders derive from having two possible developmental pathways and a switch mechanism that enables them to "choose" how to grow and behave? Individuals with some developmental flexibility may be able to exploit a particular resource niche better than individuals stuck with a one-size-fits-all phenotype. Larval salamanders can access two different sources of potential nutrients, insect prey and their fellow salamanders. If numerous salamander larvae occupy a pond, and if most are smaller than the individual that becomes a cannibal, then shifting to the cannibal phenotype gives that individual an abundant food source that is not being exploited by its fellows, so it can grow quickly. But a relatively small individual that was locked into becoming a big-jawed cannibal form would surely starve to death in a pond that lacked numerous potential victims of appropriate size. Because salamanders have no way of knowing in advance which of two different food sources will be more available in the place where they happen to be developing, selection appears to have favored individuals with the ability to develop in either one of two ways depending on the properties of their environments.

More generally, when a species faces discrete ecological problems that require different developmental solutions, the stage may be set for the evolution by natural selection of sophisticated developmental switch mechanisms that enable individuals to develop the phenotype best suited for their particular circumstances. The existence of two nonoverlapping categories of food (large versus small) or two levels of risk (predators present versus predators absent)[77] may select for the kind of developmental mechanism that can produce a limited number of very different specialist phenotypes, rather than a mechanism that generates a full range of intermediate forms.

So, for example, the fact that males of the cichlid fish *Astatotilapia burtoni* are either competitively superior or socially inferior to others helps explain why they have the capacity to shift between two different phenotypes. In this fish,

FIGURE 11.24 Activity of the gene that codes for gonadotropin-releasing hormone in the cichlid fish *Astatotilapia burtoni*. After males switch from nonterritorial to territorial status, the *GnRH* gene becomes increasingly active over time in certain brain cells. Conversely, those males that switch from territorial to nonterritorial status show reduced activity in the *GnRH* gene. After White et al.[102]

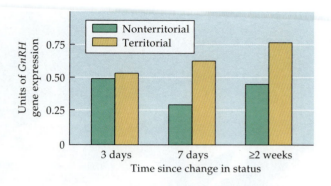

males compete for a mate-attracting territory (a location they defend), with winners holding sites until ousted by stronger intruders. In such an either-or social environment, it pays to be either aggressively territorial (and to signal that state with bright colors) or nonaggressive (and to signal that state with dull colors).[29,31] Fish that behave in some intermediate fashion will almost certainly fail to hold a territory against motivated rivals, but they will also fail to conserve their energy, which they can do only by dropping out (at least temporarily) from the competition to secure a territory. To this end, the fish respond to changes in their social status with changes in gene activity (Figure 11.24) within specific brain cells;[102] indeed, when a subordinate male is experimentally given a chance to become dominant (through the removal of a rival), gonadotropin-releasing (GnRH1) nerve cells in the anterior parvocellular preoptic nucleus quickly begin to ramp up the activity of a gene (*egr-1*) that codes for a protein (GnRH) that regulates another gene. In socially ascendant males, *egr-1* is expressed twice as much in the target cells as in males that have been stable subordinates or longtime dominant territorial individuals (Figure 11.25). By a week later, the males had been transformed not only in terms of their appearance and behavior but also in the size of their GnRH1 neurons and the size of their testes. The initial rapid rise in *egr-1* expression in the brains of previously subordinate males in response to the chance to become dominant appears to act as a trigger for a whole series of genetic and developmental changes. These changes enable an ascendant subordinate to take advantage of his good fortune and become reproductively active while suppressing reproduction in the other males in his neighborhood.[13]

Although polyphenisms are common, they are far from universal, perhaps because many environmental features vary continuously rather than discontinuously. Under these conditions, individuals may not benefit from developmental systems that produce a particular phenotype targeted at a narrow part of the entire range of environmental variation. Instead, selection may favor the ability to shift the phenotype by degrees in such a way as to generate a broad

FIGURE 11.25 Subordinate males of the fish *Astatotilapia burtoni* react very quickly to the absence of a dominant rival. (A) Within minutes of the removal of the dominant male, a subordinate may begin to behave more aggressively than before. (B) This change in behavior is correlated with a surge in the activity of a specific gene in the preoptic region of the fish's brain. This gene may initiate a sequence of other genetic changes that provide the male with the physiological foundation for dominance behaviors. Note that the gene *egr-1* ramps up its activity during the transition from subordinate to dominant status but then falls back once the male has become truly dominant. A, photograph courtesy of Russell Fernald; B, after Burmeister et al.[13]

(A)

(B)

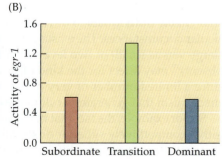

FIGURE 11.26 Developmental flexibility in redback spiders. (A) An adult male redback. (B–D) Immature males that grow to adulthood in the presence of females (red bars) develop more rapidly and so reach a smaller adult size in poorer physical condition than immatures that develop in the absence of females (orange bars). Adult males that develop in places without females are likely to encounter male competitors when they reach a web with a female somewhere else. A, photograph by Ken Jones and copyright by M. C. B. Andrade; B–D, after Kasumovic and Andrade.[49]

(A)

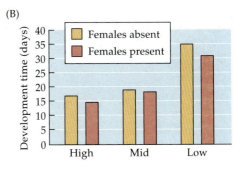

(B)

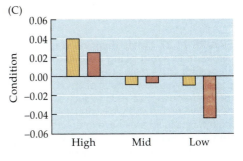

(C)

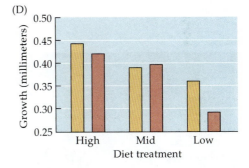

(D)

distribution of phenotypes, each one representing an adaptive response to one or more variable factors. So, for example, the final body size of a male redback spider varies considerably in response to variation in available food and the cues associated with virgin female redbacks (potential mates) and male redbacks (rivals for females).[49] When males are reared in the presence of virgin females, they develop more quickly and achieve adulthood at a smaller size compared with males reared on the same diet but in the absence of the odor cues associated with virgin females (Figure 11.26). The developmental effects of growing up in the presence of males are exactly the opposite, a fine example of how evolutionary pressures can favor developmental plasticity that produces a wide range of phenotypes within one species.

Discussion Questions

11.12 Identify the probable adaptive basis for the flexible development of body size in the redback spider. Predict what effect large body size must have on female choice in this species versus the effect of large body size on the ability of male redbacks to compete physically with rival males. Check your answer with Kasumovic and Andrade.[49]

11.13 Some marine fishes exhibit a spectacular polyphenism, in that individuals can, under special circumstances, change their sex from female to male (in other species, the switch goes from male to female). This developmental change involves reproductive organs, hormones, and mating behavior.[100] In some species, the removal of a dominant, breeding male from a cluster of females triggers a sex change in the largest female present. Identify the apparent developmental restrictions imposed on this system, such as the requirement that a female be transformed into a male rather than some sort of intermediate sex. Speculate on the benefits associated with each restriction.

The Adaptive Value of Learning

We continue our adaptationist analysis of development by considering the adaptive value of the neural mechanisms that make **learning** possible. Animals that learn modify their behavior based on certain experiences that they have had. As such, learning can be considered a polyphenism of sorts because it too confers a highly focused behavioral flexibility arising from developmental modifications in the nervous system. Learning does not produce behavioral change just for the sake of change. Instead, selection favors investment in the mechanisms underlying learning only when there is environmental unpredictability that has reproductive relevance for individuals. What we have here is another cost–benefit argument, which presupposes that any proximate mechanisms that enable individuals to learn come with a price tag. We can check this assumption by predicting, for example, that the brains of male long-billed marsh wrens living in the western United States should be larger than those of their East Coast counterparts because young West Coast wrens learn nearly 100 songs by listening to others, whereas East Coast wrens have much smaller learned repertoires of about 40 songs.[55] When the birds' brains were examined, the song control systems of West Coast wrens weighed on average 25 percent more than the equivalent nuclei of East Coast wrens.

If learning mechanisms are costly, then we can expect learning to evolve only when there is some major counterbalancing benefit. Male thynnine wasps do indeed derive a considerable benefit by learning to avoid sites where they have encountered sexually deceptive orchids. You will recall that some orchids have flowers with a lip petal that smells and looks vaguely like a female thynnine wasp. A male can be fooled into rushing to one of these flowers and attempting to mate with the petal (see Figure 4.28);[96] indeed, in some cases, males are so stimulated by the experience that they ejaculate upon grasping the orchid's decoy petal.[35] But having once been deceived by a particular flower, male wasps generally learn to avoid the spot where that flower occurs. But if the orchid is then moved to a new spot, males quickly approach and sometimes grab the lip petal; the number of males fooled in this way then decreases very sharply over time (Figure 11.27).

Male thynnine wasps evidently store information about the locations of pseudofemale orchids and learn to avoid alluring scents coming from those sites.[73] The reproductive benefits of the male wasp's behavioral flexibility are clear. Male wasps cannot be programmed in advance to know where female wasps and deceptive orchid flowers are on any given day. By using experience to learn where particular orchids are (in order to avoid them) while remaining responsive to novel sources of sex pheromone, the male wasp saves time and energy and improves his chance of encountering a receptive female that has begun to release sex pheromone.

That spatial learning evolves in response to particular ecological pressures can also be seen by comparing the learning abilities of four bird species, all members of the crow family (Corvidae), that vary in their predisposition to store food—a task that puts a premium on spatial memory. As we have seen, Clark's nutcracker is a food-storing specialist, and it has a large pouch for the transport of pine seeds to storage sites. The pinyon jay also has a special anatomical feature, an expandable esophagus, for carrying large quantities of seeds to hiding places. In contrast, the scrub jay and Mexican jay lack special seed transport devices and appear to hide substantially less food than their relatives.

Individuals from the four species were tested on two different learning tasks in which they had to peck a computer screen to receive rewards. One task required the birds to remember the color of a circle on the screen (a nonspatial learning task), and the other required them to remember the location of a circle on the screen (a spatial task). When it came to the nonspatial learning test, pinyon jays and Mexican jays did substantially better than scrub jays and nutcrackers. But in the spatial learning experiment, the nutcracker went to the head of the class, followed by the pinyon jay, then

FIGURE 11.27   Male thynnine wasps can learn to avoid being deceived by an orchid. (A) A male attempting to copulate with a flower petal of an orchid. (B) The frequency of visits to such a deceptive orchid soon falls after the male wasps in an area have interacted with it and learned that an unrewarding source of sex pheromone is associated with that particular location. B, after Peakall.[73]

(A)

(B)

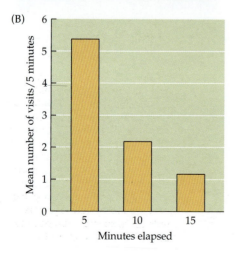

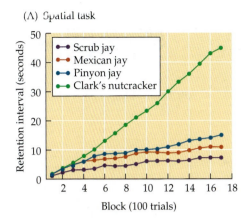

(A) Spatial task

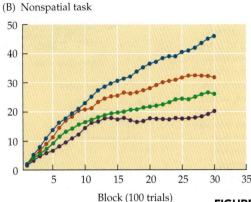

(B) Nonspatial task

FIGURE 11.28 **Spatial learning abilities differ among members of the crow family.** (A) Captive Clark's nutcrackers performed much better than three other corvid species in experiments that required the birds to retain information about the location of a circle. (B) But when the birds' ability to remember the color of a circle was tested, the nutcrackers did not excel in this nonspatial learning task. After Olson et al.[71]

the Mexican jay, and finally the scrub jay (Figure 11.28) (but see de Kort and Clayton[24] and Pravosudov and de Kort[79] on examples that do not fit this pattern and why they might occur). To the extent that nutcrackers and pinyon jays really are especially good at spatial learning only, we can conclude that birds have not evolved all-purpose learning abilities but rather their learning skills are designed to promote success in solving ecologically relevant problems.[71]

The logic of an evolutionary approach to learning leads us to expect that if males and females of the same species differ in the benefits derived from a particular learned task, then a sex difference in learning skills should evolve. The pinyon jay provides a case in point. As just noted, the jay hides large numbers of pinyon seeds, when they are available; it retrieves them up to 5 months later when food is scarce. But males are more likely than females to have to relocate old caches because they provide their mates and young with recovered food while their female partners spend their time instead at the nest, incubating their eggs and young. As predicted, males appear to have evolved better long-term memory than females. When captive birds of both sexes were tested during what was the nesting season, males made fewer errors than females (Figure 11.29).[26] The poorer performance of females suggests that the ability to learn has costs that require special benefits if the trait is to evolve.

The sex differences hypothesis has also been tested by Steve Gaulin and Randall FitzGerald in their studies of spatial learning in three species of voles, all members of the genus *Microtus*. Males of the polygynous and wide-ranging meadow vole do better on maze-learning tests than females of this species. In

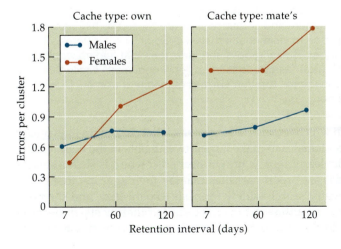

Pinyon jay

FIGURE 11.29 **Male pinyon jays make fewer errors than females do when retrieving seeds** from caches they have made, especially after intervals of 2 to 4 months. This result accords with expectation, because females are the incubators of eggs and youngsters while the males provide the female and offspring with seeds relocated in caches made up to several months previously. After Dunlap et al.[26]

FIGURE 11.30   Sex differences in spatial learning ability are linked to home range size. Spatial learning in voles was tested by giving individuals opportunities to travel through seven different mazes of increasing complexity in the laboratory and then letting them run through each maze again. (A) Polygynous male meadow voles, which roam over wide areas in nature, consistently made fewer errors (wrong turns) on average than the more sedentary females of their species. (B) In contrast, females matched male performance in the monogamous prairie vole, a species in which males and females live together on the same territory. After Gaulin and FitzGerald.[36]

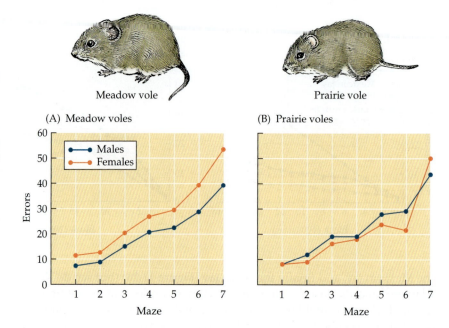

the monogamous prairie vole, whose males and females share the same living space, the two sexes did equally well on the same set of maze tests given to the meadow vole (Figure 11.30).[36]

Discussion Questions

11.14   In a study in which men and women were asked to sit at a computer and navigate through a virtual maze (Figure 11.31), the men were able to complete the task more quickly and with fewer errors over five trials than the women.[68] See also Jones and Healy.[46] (Note, however, that in other tests, involving language skills, women score higher on average than men.) What possible proximate developmental mechanisms might be responsible for this sex difference in navigational ability? Use the evolutionary explanation for sex differences in spatial learning ability by voles to make a prediction about the nature of human mating systems over evolutionary time.

11.15   Another sex difference in spatial skill exhibited by our species is the slightly greater ability of men compared with women when it comes to visualizing what a three-dimensional object would look like if rotated in space. This difference has been linked in part to differences in the parietal lobe of the two sexes.[54] One of the several authors on this report commented that it remains to be seen whether the differences in brain structure and cognitive skills are caused by nature or nurture. He went on to claim that if there were significant differences in the parietal lobes of young boys and girls, then that finding would support a "biological" as opposed to an "environmental" cause for the differences in mental rotation abilities of men and women. Do you agree? (I hope not.)

In a species in which females face greater spatial challenges than males, we would expect females to make larger investments in the neural foundations of spatial learning. The brown-headed cowbird is such a species because cowbirds are brood parasites that lay their eggs in other birds' nests. A female must search widely for nests to parasitize, and she must remember where potential victims have started their nests, in order to return to them up to several days later when the time is ripe for her to

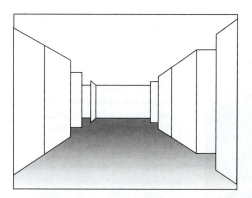

FIGURE 11.31   A virtual maze used for computer-based studies of navigational skills. After Moffat et al.[68]

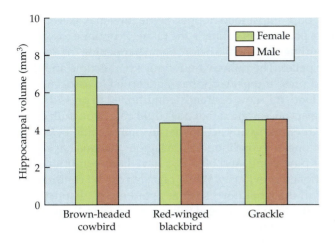

FIGURE 11.32 Sex differences in the hippocampus. Female brown-headed cowbirds have a larger hippocampus than males, as would be expected if this brain structure promotes spatial learning and if selection for spatial learning ability is greater on female than on male cowbirds. Red-winged blackbirds and common grackles do not exhibit this sex difference. After Sherry et al.[88]

add one of her eggs to those already laid by the host bird. In contrast, male cowbirds do not confront such difficult spatial problems. As predicted, the hippocampus (but no other brain structure) is considerably larger in female brown-headed cowbirds than in males. The hippocampus is thought to play an important role in spatial learning; in the nonparasitic relatives of the cowbird, the hippocampus is the same size in both sexes (Figure 11.32).[88]

Moreover, it is not just spatial learning that bears the clear imprint of natural selection. Consider **operant conditioning**, in which an animal learns to associate a voluntary action with the consequences that follow from that action.[91] Operant conditioning (or trial-and-error learning) does occur outside psychology laboratories, but it has been studied extensively in Skinner boxes, named after the psychologist B. F. Skinner. After a white rat has been introduced into a Skinner box, it may accidentally press a bar on the wall of the cage (Figure 11.33), perhaps as it reaches up to look for a way out. When the bar is pressed

FIGURE 11.33 Operant conditioning exhibited by a rat in a Skinner box. (A) The rat approaches the bar and (B) presses it. (C) The animal awaits the arrival of a pellet of rat chow and (D) consumes it, so the bar-pressing behavior is reinforced. Photographs by Larry Stein.

(A)

(B)

(C)

(D)

down, a rat chow pellet pops into a food hopper. Some time may pass before the rat happens upon the pellet. After eating it, the rat may continue to explore its rather limited surroundings for a while before again happening to press the bar. Out comes another pellet. The rat may find it quickly this time and then turn back to the bar and press it repeatedly, having learned to associate this particular activity with food. It is now operantly conditioned to press the bar.

Skinnerian psychologists once argued that one could condition with equal ease almost any operant (defined as any action that an animal could perform). Indeed, the successes of operant conditioning are legion; pigeons have been taught to play a kind of ping-pong, and blue jays have been conditioned to use computers, as mentioned previously. White rats also can be conditioned to do all sorts of things in the laboratory, such as avoid novel, distinctively flavored foods or fluids after they are exposed to nausea-inducing X-ray radiation. However, John Garcia and his colleagues found that the ability of these animals to learn to avoid certain punishing foods or liquids had some restrictions.[33,34] The degree to which an irradiated rat rejects a food or fluid is proportional to (1) the intensity of the resulting illness, (2) the intensity of the taste of the substance, (3) the novelty of the substance, and (4) the shortness of the interval between consumption and illness.[34] But even if there is a long delay (up to 7 hours) between eating a distinctively flavored food and exposure to radiation and consequent illness, the rat still links the two events and uses the information to modify its behavior.

In contrast, white rats never learn that a distinctive sound (a click) is a signal that always precedes an event associated with nausea. Nor can rats easily make an association between a particular taste and shock punishment (Figure 11.34). If, after drinking a sweet-tasting fluid, the rat receives a shock on its feet, it often remains as fond of the fluid as it was before, as measured by the amount drunk per unit of time, no matter how often it is shocked after drinking this liquid. These failures surely relate to the fact that in nature, particular sounds are never associated with illness-inducing meals, any more than the consumption of certain fluids causes a rat's feet to hurt.

Understanding the natural environment of the ancestor of white rats, the Norway rat, also helps explain why white rats are so adept at learning to avoid novel foods with distinctive tastes that are associated with illness, even hours after ingesting the food. Under natural conditions, a Norway rat becomes

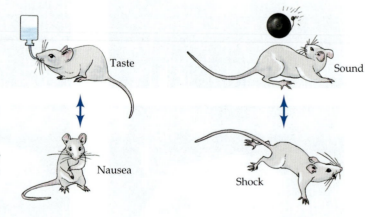

FIGURE 11.34 Biases in taste aversion learning. Although white rats can easily learn that certain taste cues will be followed by sensations of nausea and that certain sounds will be followed by skin pain caused by shock, they have great difficulty forming learned associations between taste and consequent skin pain or between sound and subsequent nausea. After Garcia et al.[34]

completely familiar with the area around its burrow, foraging within that area for a wide variety of foods, plant and animal.[59] Some of these foods are edible and nutritious; others are toxic and potentially lethal. A rat cannot clear its digestive system of toxic foods by vomiting. Therefore, the animal takes only a small bite of anything new. If it gets sick later, it avoids this food or liquid in the future, as it should because eating large amounts might kill it.[34] This case suggests that even what appears to be a general, all-purpose form of learning is actually a specialized response to particular kinds of biologically significant associations that occur in nature.

If this argument is correct, then other mammals that are dietary generalists, which also run the risk of consuming dangerous, toxic items, should behave like the Norway rat, which is to say that they should also quickly form taste aversions to bad-tasting, illness-inducing items. And they do. Three bat species that feed on a range of foods behaved in the predicted manner: they rapidly formed taste aversions when fed a meal laced with an unfamiliar flavor, cinnamon or citric acid, before being injected with a chemical that made them vomit. When later offered a choice between food with and without cinnamon or citric acid, these three generalist consumers avoided foods spiced with the novel additives.[82]

In contrast, dietary specialists, which concentrate exclusively on one or a very few safe foods, should be unable to acquire taste aversions in this manner. The vampire bat, a blood-feeding specialist, is in fact quite incapable of learning that consumption of an unusual-tasting fluid will lead to gastrointestinal distress (Figure 11.35).[82] The difference between the specialist vampire bat and its generalist relatives supports the hypothesis that taste aversion learning is an evolved response to the risk of food or fluid poisoning. Just as is true for all aspects of behavioral development, the changes associated with learned behavior are worth the cost only if they confer a net fitness benefit on individuals capable of modifying their behavior in a particular way.

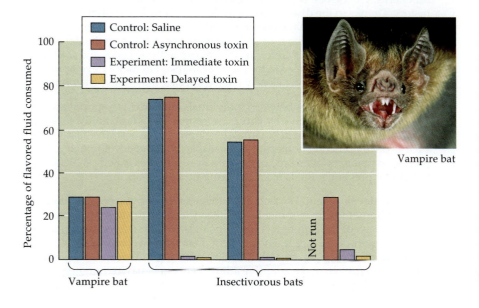

Vampire bat

FIGURE 11.35 Vampire bats could not form learned taste aversions. Instead they continued to consume a flavored fluid even if, immediately after accepting this novel substance, they were injected with a toxin that caused gastrointestinal distress. In contrast, three insect-eating bat species completely rejected the novel dietary item when it was combined with injection of the toxin, no matter whether this was done immediately after feeding or after a delay. Two control groups were also used in the experiment, one in which the consumption of the novel food was paired with a harmless injection of saline solution, and another in which the toxin was injected but not in conjunction with feeding on the fluid. After Ratcliffe et al.[82]

Summary

1. The development of any trait, including a behavioral ability, is the result of an interaction between the genotype of a developing organism and its environment, which consists of not only the food it receives and the metabolic products produced by its cells (the material environment) but also its sensory experiences (the experiential environment). Genes can respond to signals from the environment by altering their activity, leading to changes in the gene products available to the developing organism.

2. Despite the nature–nurture misconception, no measurable product of development (a phenotype) can be genetically determined. The statement "in garter snakes, there is a gene for eating banana slugs" really means "a particular allele in a garter snake's genotype codes for a distinctive protein; if the protein is actually made, which requires an interaction between the gene and its environment, the protein may influence the development or operation of specific physiological mechanisms underlying the snake's ability to recognize slugs as food."

3. By the same token, the interactive nature of development means that no phenotype can be purely environmentally determined. The statement "the white-crowned sparrow's dialect is environmentally determined" really means "acoustic experiences early in the sparrow's life led to chemical changes in the bird's brain, which altered the pattern of genetic activity in some parts of its nervous system. These changes set in motion subsequent genetic and neural changes in the physiological systems that an adult sparrow uses when singing a version of its species' song."

4. Because development is interactive, changes in either the genetic information or the environmental inputs available to an individual can potentially alter the course of its development by changing the interplay between genes and environment within that individual. Therefore, the behavioral *differences* between two individuals can be genetically determined or environmentally determined or both.

5. The genetic mechanisms that affect development can be analyzed evolutionarily. On the one hand, these mechanisms have an evolutionary history. Thanks to the discipline of evo-devo we now know of many examples of genes that have been retained in modern species from a very distant ancestor. These genes often have regulatory effects on the development of important traits. Modest modifications in these genes and/or the target genes whose expression they affect can result in major developmental differences, even among closely related species, that evolve over time.

6. Because some differences between individuals are hereditary, populations have the potential to evolve by natural selection. The adaptationist level of analysis has revealed that behavioral development has adaptive features, such as developmental homeostasis, the capacity of the developmental process to ignore or overcome certain environmental or genetic shortfalls that might prevent animals from acquiring fitness-enhancing traits.

7. Likewise, developmental switch mechanisms are adaptive because they control alternative developmental pathways leading to alternative phenotypes. Each of the different traits helps individuals succeed within a particular part of the larger environment. Learning mechanisms provide another form of developmental flexibility, enabling individuals to use their experiences to make adaptive adjustments in behavior that help individuals cope with a variable environment.

Suggested Reading

The nature–nurture debate has been nicely dissected by Gene Robinson.[84] The relatively new field of evo-devo is reviewed by Sean Carroll in his delightful book *Endless Forms Most Wonderful*, which explains the interplay between descent with modification and the genetic contribution to development.[14] Mary Jane West-Eberhard has also weighed in on the evolutionary significance of the developmental flexibility caused by the interaction of genes with a variable environment in the production of alternative phenotypes.[101] Scott Simpson and his colleagues have written a review article that integrates the proximate and ultimate causes of an assortment of behavioral polyphenisms.[90]

12 Evolution, Nervous Systems, and Behavior

 In the preceding chapter, our focus was primarily on the developmental basis for animal behavior, one of the two major proximate levels of analysis that define the research of behavioral biologists. As multicelled animals develop, they acquire a nervous system, whose properties are crucially important to the proximate control of their behavior. Because nervous systems are so significant for an understanding of behavior, biologists have been eager to learn how these systems work, as we have already seen in reviewing the connection between bird song and bird brains. Although our current understanding of these systems has been greatly helped by the application of sophisticated technologies, even simple observations of animals in action can sometimes provide considerable information about the proximate properties of their neural mechanisms. So, the fact that males of the bee *Centris pallida* are able to dig up dead females of their species that were buried by an experimenter indicates that they have olfactory systems that are exquisitely sensitive to odors coming from female bees. The relationship between a sensory ability and reproductive success is obvious in this case, but what evolutionary explanation exists for why males of *C. pallida* will try to copulate with a person's thumb (Figure 12.1)?

I learned that males of the digger bee could be turned on by my thumb after I had unsympathetically pulled a male bee from his sexual partner, the better to measure his head width with a pair of calipers. In the midst of this exercise, I found more or less by accident that if I perched the male on my upturned thumb, he would grasp it firmly and stroke it with his legs and antennae as if he were holding a female of his species. (Incidentally, male bees don't sting,

(A)

(B)

FIGURE 12.1 A complex response to simple stimuli. (A) A male of *Centris pallida* copulating with a female of his species. (B) A male of this species attempting (unsuccessfully) to copulate with the author's thumb. Photographs by the author.

so my actions were neither courageous nor stupid.) Despite the fact that my thumb has only the vaguest similarity to a female *C. pallida*, it is evidently close enough for males of this species.

Even though I am not a neurophysiologist, I could see that the bee's nervous system had some special operating rules. Apparently, when a sexually motivated male *C. pallida* grasps an object approximately the size of a female of its species, the sensory signals generated by its touch receptors travel to other parts of its nervous system, where messages are produced that eventually translate into a complex series of muscle commands. The behavioral result is the sequence of movements that passes for courtship in *C. pallida*. That these activities can be stimulated by my thumb instead of a female bee indicates that the nervous system of a male *C. pallida* is not terribly discriminating. Nor is "my" bee at all unusual in this respect, given that males of the ivy bee, *Colletes hederae*, will attempt to mate, not with a thumb, but with a mass of tiny blister beetle larvae (Figure 12.2)[82] (see also Saul-Gershenz and Millar[65]).

Complex Responses to Simple Stimuli

Observations of bees trying to mate with objects other than females of their species are puzzling. How can the underlying neural mechanisms be adaptive if they generate such clearly maladaptive responses? These and other

FIGURE 12.2 A male of the solitary bee *Colletes hederae* attracted to a cluster of larval blister beetles (some of which are indicated with the arrow). After attempting to mate with the mass of beetles, the bee becomes covered in the tiny larvae, which climb aboard for transport to a female bee, should the male succeed in finding a mate of the appropriate species. Photograph by Nicolas Vereecken.

puzzles have been solved in part by studying how complex responses to simple stimuli are triggered by animal nervous systems. Consider Niko Tinbergen's classic work on gull chick begging behavior.[77] Recently hatched baby gulls will peck at the tip of the beak of their parent (Figure 12.3), an action that induces the adult gull to regurgitate a half-digested fragment of fish or other delicacy, which the chick enthusiastically consumes.

You might think that a baby gull would find the beak of a living three-dimensional gull more stimulating than painted sticks and two-dimensional cardboard cutouts, but sticks and cutouts work fine in eliciting the pecking response from very young chicks (Figure 12.4). Experiments with these and other models have revealed that recently hatched herring gull chicks ignore almost everything except the shape of the "bill" and the red dot at the end of the beak. Tinbergen proposed that when a young gull sees a pointed object with a contrasting dot at the tip, the resulting sensory signals reach the brain, where other **neurons** eventually generate the motor commands that cause the chick to peck at the stimulus—whether it is located on its mother's bill or a piece of cardboard or the end of a stick.

Tinbergen and his friend Konrad Lorenz collaborated on another famous experiment that also identified a simple stimulus capable of triggering a complex behavior. They found that if they removed an egg from under an incubating greylag goose and put it a half meter away, the goose would retrieve the egg by stretching its neck forward, tucking the egg under its lower bill, and rolling the egg carefully back into its nest. If they replaced the goose egg with almost any roughly egg-shaped object, the goose would invariably run through its egg retrieval routine. And if the researchers removed the object as it was being retrieved, the bird would continue pulling its head back just as if an egg were still balanced against the underside of its bill.[78] From these results, Tinbergen and Lorenz concluded that the goose must have a special perceptual mechanism that is highly sensitive to visual cues provided by egg-shaped objects. Moreover, that sensory mechanism must send its information to neurons in the brain that automatically activate a more or less invariant motor program for egg retrieval.

The gull chick's pecking response and the greylag goose's retrieval behavior are only two of many instincts that Tinbergen and Lorenz studied. These founders of **ethology**, the first discipline dedicated to the study of both the proximate and ultimate causes of animal behavior, were especially interested in the instincts exhibited by wild animals living under natural conditions.[6,44] An **instinct** can be

FIGURE 12.3 Begging behavior by a gull chick. A silver gull chick is being fed regurgitated food by its parent after pecking at the adult's bill. Photograph by the author.

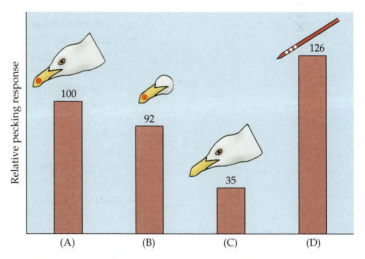

FIGURE 12.4 Effectiveness of different visual stimuli in triggering the begging behavior of young herring gull chicks. A two-dimensional cardboard cutout of the head of a gull with a red dot on its bill (A) is not much more effective in eliciting begging behavior in a gull chick than is a model of the bill alone (B), provided the red dot is present. Moreover, a model of a gull head without the red dot (C) is a far less effective stimulus than is an unrealistically long "bill" with contrasting bars at the end (D). After Tinbergen and Perdeck.[77]

FIGURE 12.5 The architects of instinct theory.
(A) Niko Tinbergen (and Konrad Lorenz—see Figure 11.4) proposed that simple stimuli, such as (B) the red dot on a parent gull's bill, can activate or release complex behaviors, such as a gull chick's begging behavior. This effect is achieved, according to these ethologists, because certain sensory messages from the releaser are processed by innate releasing mechanisms (neuronal clusters) higher in the nervous system, leading to motor commands that control a fixed action pattern, a preprogrammed series of movements that constitute an adaptive reaction to the releasing stimulus. Photograph by B. Tschanz.

(A)

Niko Tinbergen

(B)

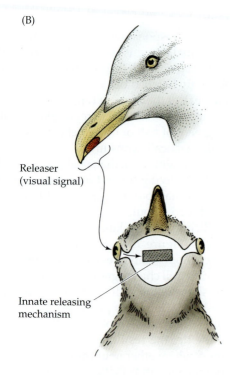

Releaser
(visual signal)

Innate releasing
mechanism

defined as a behavior pattern that appears in fully functional form the first time it is performed, even though the animal may have had no previous experience with the cues that elicit the behavior. Note that instincts are not "genetically determined," because these behaviors, like learned ones as well, are dependent on the gene–environment interactions that took place during development. In the case of a herring gull chick, these interactions led to the construction of a nervous system that contains a network that enables the young gull to respond to the red dot at the end of an adult gull's bill. The neural network responsible for detecting the simple cue (the **sign stimulus** or **releaser**) and activating the instinct, or **fixed action pattern** (FAP), was given the label of **innate releasing mechanism** by Tinbergen and Lorenz (Figure 12.5).[78]

The simple relationship between an innate releasing mechanism, sign stimulus, and FAP is highlighted by the ability of some species to exploit the FAPs of other species, a tactic sometimes referred to as code breaking.[86] In Chapter 4, we met orchids whose flower petals provide the visual, tactile, and olfactory releasers that trigger attempted copulation by certain male wasps (see Figure 4.28).[73] Likewise, the blister beetle larvae mentioned at the start of this chapter use scents that mimic those released by receptive female ivy bees. When a male ivy bee pounces on a ball of beetle larvae, he gets covered with little parasites that may later be transferred to the female bee, if the male is fortunate enough to find a real sexual partner. After moving onto a female, the larvae will eventually be transported to the underground nest of the bee, where they can drop off and make their way into the food-containing brood cells. There they consume the provisions the mother bee collected and stored for her own offspring.[82] We have also discussed deception by brood parasites, such as the European cuckoo and North American cowbird, the nestlings of which look and behave in ways that stimulate their adult hosts to feed them (see Figure 9.16).[18]

The exploitation of the proximate neural mechanisms of certain animals has even enabled some families to make a living collecting earthworms for sale to fishermen. People in Florida have learned that earthworms will come rushing to the surface to crawl away when a stake is driven into the earth and vibrated, making the collection of the worms easy for "worm grunters," as these folks are called. The mystery of why worms behave this way has been solved by the discovery that worm-eating American moles also generate vibrations in the soil as they dig underground.[14] By using soil vibrations to activate escape behavior, worms generally leave places before they are captured by moles. Humans have come to take advantage of this innate response of earthworms since the 1960s, indicating that the current maladaptive behavior of the worms can be explained in terms of novel environment theory (see page 89).

Discussion Questions

12.1   Suggest how a modern behavioral biologist might explore the effect of a releaser, such as a red dot on a moving gull bill, in terms of changes in gene expression and neural activity in selected portions of the brain of the gull chick.

12.2   Males of certain Australian beetles have been seen trying to copulate with everything from beer bottles to large orange signs (Figure 12.6). Apply ethological terminology to these cases by identifying the releaser, the fixed action pattern, and the innate releasing mechanism. Then develop an ultimate hypothesis to account for what clearly is maladaptive behavior on the part of these obtuse beetles (which sometimes die rather than leave the inanimate and unresponsive copulatory partners that they have chosen).

How Moths Avoid Bats

Natural selection has endowed digger bees and herring gull chicks with neural circuitry that usually enables these creatures to make wise choices about how to mate and how to get fed, respectively. The role of selection in shaping individual nerve cells so that animals behave adaptively was the subject of classic research by Kenneth Roeder on how night-flying moths manage to escape from night-hunting bats. Roeder told his readers that they too could watch moths fleeing from bats by spending some time outdoors on summer nights, armed with "a minimum amount of illumination, perhaps a 100-watt

(A)

(B)

FIGURE 12.6   Male beetles trying to mate with objects other than female beetles. This large Australian beetle will attempt to mate with any object with approximately the same color as a female of his species, such as (A) a beer bottle or (B) a telecommunication sign. A, photograph courtesy of Darryl Gwynne; B, photograph by the author.

bulb with a reflector, and a fair amount of patience and mosquito repellent."[60] Were you to follow Roeder's suggestion, you might see a moth come to the light only to turn abruptly away just before a bat swooped into view. Or you might observe a moth dive straight down just as a bat appears. You might also see moths turn away or dive down when you jangled a set of keys, suggesting that an acoustical cue provides the trigger for certain moth behaviors, just as simple visual cues are sufficient to activate begging behavior in a baby gull.

As it turns out, the hypothesis that acoustical stimuli trigger the turning or diving behavior of moths is correct, but the sounds that the moth hears when a bat approaches or keys are rattled together are not the sounds that you and I hear. Instead, the moth detects the very-high-frequency sounds produced by bats or clashing keys, which makes sense when you consider that most night-hunting bats vocalize ultrasonically using sound frequencies between 20 and 80 kilohertz (kHz)—well above the hearing range of humans, but not moths.

Bats use ultrasonic calls to navigate at night—something that was not suspected until the 1930s, when researchers with ultrasound detectors were able to eavesdrop on bats. At that time, Donald Griffin suggested that night-flying bats use high-frequency cries in order to listen for weak ultrasonic echoes reflected back from objects in their flight paths.[31] Skeptics of the echolocation hypothesis came round after reading about Griffin's experiments with little brown bats, a common North American species. When Griffin placed captive bats in a dark room filled with fruit flies and wires strung from ceiling to floor, his subjects had no trouble catching the insects while negotiating the obstacle course—until Griffin turned on a machine that filled the room with high-frequency sound. As soon as the machine-produced ultrasound bombarded the bats, they began to collide with the wires and crash to the floor, where they remained until Griffin turned off the jamming device. In contrast, loud sounds with frequencies between 1 and 15 kHz (which humans can easily hear) had no effect on the ultrasound-using bats because these stimuli did not mask the high-frequency echoes bouncing back from objects in the room. Griffin rightly concluded that the little brown bat employs a sonar system to avoid obstacles and detect prey at night.

As Roeder watched moths evading echolocating bats, he felt sure that the insects were able to hear pulses of bat ultrasound. He began to test his idea by finding the ears of a noctuid moth, one on each side of the thorax (Figure 12.7). Each ear consists of a thin, flexible sheet of cuticle—the tympanic membrane, or tympanum—lying over a chamber on the side of the thorax. Attached to the tympanum are two neurons, the A1 and A2 auditory receptors. These receptor cells are deformed when the tympanum vibrates, which it does when intense sound pressure waves sweep over the moth's body.

FIGURE 12.7 Noctuid moth ears. (A) The location of the ear. (B) The design of the ear, which features two auditory receptors (A1, A2) linked to a tympanum that vibrates when exposed to sounds. After Roeder.[60]

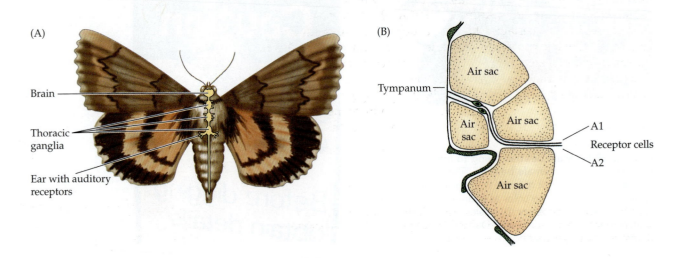

FIGURE 12.8 **Neurons and their operation.** This diagram illustrates the structure of a generalized neuron with its dendrites, cell body, axon, and synapses. Electrical activity in a neuron originates with the effects of certain stimuli on the dendrites. Electrical changes in a dendrite's cell membrane can, if sufficiently great, trigger an action potential, which begins near the cell body and travels along the axon toward the next cell in the network.

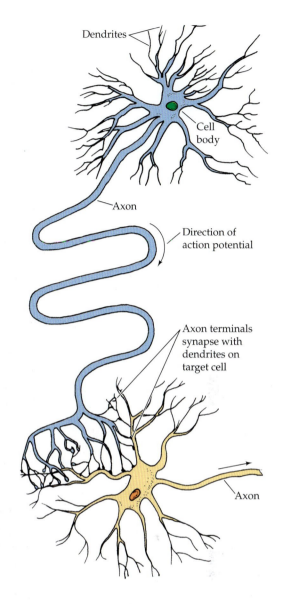

The moth's A1 and A2 receptor cells work much like other neurons: they respond to the energy contained in selected stimuli by changing the permeability of their cell membranes to positively charged ions. The effective stimuli for a moth auditory receptor appear to be provided by the movement of the tympanum, which mechanically stimulates the receptor cell, opening stretch-sensitive channels in the cell membrane. As positively charged ions flow in, they cause the normally negative inside of the cell to become more positive. If the inward movement of ions is sufficiently great, a substantial, abrupt, local change in the electrical charge difference across the membrane may occur and spread to neighboring portions of the membrane, sweeping around the cell body and down the axon—the "transmission line" of the cell (Figure 12.8). This brief, all-or-nothing change in electrical charge, called an **action potential**, is the signal that one neuron uses to communicate with another.

When an action potential arrives at the end of an axon, it may cause the release of a neurotransmitter at this point. This chemical signal diffuses across the narrow gap, or **synapse**, separating the axon tip of one cell from the surface of the next cell in the network. Neurotransmitters can affect the membrane permeability of the next cell in a chain of cells in ways that increase (or decrease) the probability that this second neuron will produce its own action potential(s). If a neuron generates an action potential in response to stimulation provided by the preceding cell in the network, the message may be relayed on to the next cell, and on and on. Volleys of action potentials initiated by distant receptors may have excitatory (or inhibitory) effects that reach deep into the nervous system, eventually resulting in action potential outputs that reach the animal's muscles and cause them to contract.

In the case of the noctuid moth studied by Roeder, the A1 and A2 receptor cells are linked to relay cells called **interneurons**, whose action potentials can change the activity of other cells in one or more of the insect's thoracic ganglia (a ganglion is a neural structure composed of a highly organized mass of neurons) that relay messages on to the moth's brain (Figure 12.9). As messages flow through these parts of the nervous system, certain patterns of action potentials produced by cells in the thoracic ganglia trigger other interneurons, whose action potentials in turn reach motor neurons that are connected to the wing muscles of the moth. When a motor neuron fires, the neurotransmitter it releases at the synapse with a muscle fiber changes the membrane permeability of the muscle cell. These changes initiate the contraction or relaxation of muscles, and this drives the wings and thereby affects the moth's movements.

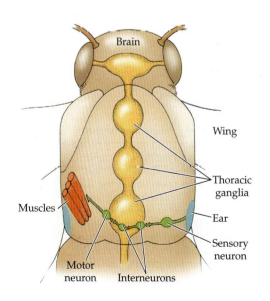

FIGURE 12.9 **Neural network of a moth.** Receptors in the ear relay information to interneurons in the thoracic ganglia, which communicate with motor neurons that control the wing muscles.

Thus, the moth's behavior, like that of any animal, is the product of an integrated series of chemical and biophysical changes in a network of cells. Because these changes occur with remarkable rapidity, a moth can react to certain acoustical stimuli in fractions of a second, which helps the moth avoid bats zooming in for the kill.

Although the neurons of many animals are very much alike at a fundamental level, they differ greatly in their functions. Thus, the auditory receptors of noctuid moths are highly specialized for the detection of ultrasonic stimuli. Roeder demonstrated this point by attaching recording electrodes to the A1 and A2 receptors of living, but restrained, moths.[60] When he projected a variety of sounds at the moths, the electrical responses of the receptors were relayed to an oscilloscope, which produced a visible record. These recordings revealed the following features (Figure 12.10):

1. The A1 receptor is sensitive to ultrasounds of low to moderate intensity, whereas the A2 receptor begins to produce action potentials only when an ultrasound is relatively loud.

2. As a sound increases in intensity, the A1 receptor fires more often and with a shorter delay between the arrival of the stimulus at the tympanum and the onset of the first action potential.

3. The A1 receptor fires much more frequently in response to pulses of sound than to steady, uninterrupted sounds.

4. Neither receptor responds differently to sounds of different frequencies over a broad ultrasonic range. Thus, a burst of 30 kHz sound elicits much the same pattern of firing as an equally intense sound of 50 kHz.

5. The receptor cells do not respond at all to low-frequency sounds, which means that moths are deaf to stimuli that we can easily hear,

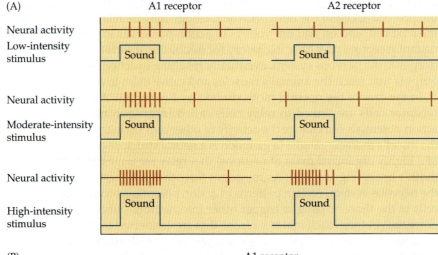

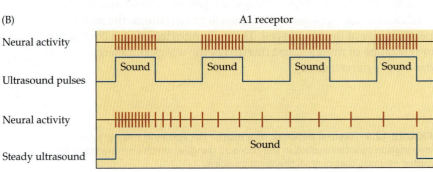

FIGURE 12.10 Properties of the ultrasound-detecting auditory receptors of a noctuid moth. (A) Sounds of low or moderate intensity do not generate action potentials in the A2 receptor. The A1 receptor fires sooner and more often as sound intensity increases. (B) The A1 receptor initially reacts strongly to pulses of ultrasound but then reduces its rate of firing if the stimulus is a constant sound. After Roeder.[60]

such as the chirping calls and trills of night-singing crickets. Of course, we are deaf to sounds that nocturnal moths have no trouble hearing.

Although the moth's ears have just two receptors each, they could potentially provide an impressive amount of information to the moth's nervous system about echolocating bats. The key property of the A1 receptor is its great sensitivity to pulses of ultrasound, which enables it to begin generating action potentials in response to the faint cries of a little brown bat 30 meters away, long before the bat can detect the moth. In addition, because the rate of firing in the A1 cell is proportional to the loudness of the sound, the insect has a system for determining whether a bat is getting closer or going farther away.

The moth's ears also gather information that could be used to locate the bat in space. If a hunting bat is on the moth's left, for example, the A1 receptor in its left ear will be stimulated a fraction of a second sooner and somewhat more strongly than the A1 receptor in its right ear, which is shielded from the sound by the moth's body. As a result, the left receptor will fire sooner and more often than the right receptor (Figure 12.11A). The moth's nervous system could also detect whether a bat is above it or below it. If the predator is higher than the moth, then with every up-and-down movement of the insect's wings, there will be a corresponding fluctuation in the firing rate of the A1 receptors as they are exposed to bat cries and then shielded from them by the wings (Figure 12.11B). If the bat is directly behind the moth, there will be no such fluctuation in neural activity (Figure 12.11C).

As neural signals initiated by the receptors race through the moth's nervous system, they may ultimately generate motor messages that cause the moth to turn and fly directly away from the source of ultrasonic stimuli.[59] When a moth is moving away from a bat, it exposes less echo-reflecting area than if it were presenting the full surface of its wings at right angles to the bat's vocalizations. If a bat receives no insect-related echoes from its calls, it cannot detect a meal. Bats rarely fly in a straight line for long, and therefore the odds are good that a moth will remain undetected if it can stay out of range for a few seconds. By then the bat will have found something else within its 3-meter moth detection range and will have veered off to pursue it.

In order to employ its antidetection response, a moth must orient itself so as to synchronize the activity of the two A1 receptors. Differences in the rate of firing by the A1 receptors in the two ears are probably monitored by the brain, which relays neural messages to the wing muscles via the thoracic ganglia and allied motor neurons. The resulting changes in muscular action steer the moth away from the side of its body with the ear that is more strongly stimulated. (You can imagine what would happen if the moth

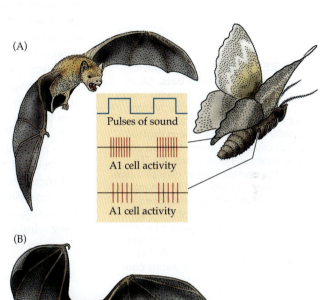

(A)

Pulses of sound

A1 cell activity

A1 cell activity

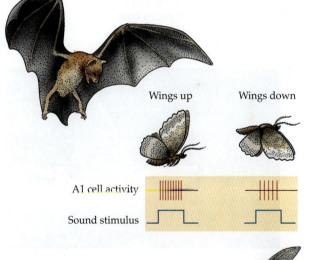

(B)

Wings up   Wings down

A1 cell activity

Sound stimulus

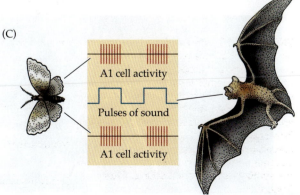

(C)

A1 cell activity

Pulses of sound

A1 cell activity

FIGURE 12.11 How moths might locate bats in space. (A) When a bat is to one side of the moth, the A1 receptor on the side closer to the predator fires sooner and more often than the shielded A1 receptor in the other ear. (B) When a bat is above the moth, activity in the A1 receptors fluctuates in synchrony with the moth's wingbeats. (C) When a bat is directly behind the moth, both A1 receptors fire at the same rate and time. (Figures not drawn to scale.)

turned itself toward the side of its body with the more strongly stimulated ear!) As the moth turns away from the relatively intense ultrasound reaching one side of its body, it will reach a point at which both A1 cells are equally active; at this moment, it will be facing in the opposite direction from the bat and will be heading away from danger (see Figure 12.11C).

Although this reaction is effective if the moth has not been detected, it is useless if a speedy bat has come within 3 meters of the moth. At this point, a moth has at most a second, and probably less, before it and the predator will collide.[40] Therefore, moths in this situation do not try to outrun their enemies but instead employ drastic evasive maneuvers, such as wild loops and power dives, that make it harder for bats to intercept them. A moth that executes a successful power dive and reaches a bush or grassy spot is safe from further attack because echoes from the leaves or grass at the moth's crash-landing site mask those coming from the moth itself.[59] Other nocturnal insects have independently evolved the capacity to sense ultrasound, and they also take evasive action when bats approach them (Figure 12.12).[49,88,89]

Roeder speculated that the physiological basis for this erratic escape flight lies in the neural circuitry leading from the A2 receptors to the brain and then back to the thoracic ganglia.[61] When a bat is about to catch a moth, the intensity of the sound waves reaching the insect's ears is high. It is under these conditions that the A2 cells fire. Roeder believed that the A2 signals, once relayed to the brain, might shut down the central steering mechanism that regulates the activity of the flight motor neurons. If the steering mechanism was inhibited, the moth's wings would begin beating out of synchrony, irregularly, or not at all. As a result, the insect might not know where it was going—but neither would the pursuing bat, whose inability to plot the path of its prey could permit the insect to escape.

Although Roeder's hypotheses about the functions of the A1 and A2 cells were plausible and supported by considerable evidence, especially with respect

FIGURE 12.12   Evasive behavior of insects attacked by bats. (A) In normal flight, a praying mantis holds its forelegs close its the body (top), but when it detects ultrasound, it rapidly extends its forelegs (bottom), which causes the insect to loop and dive erratically downward. (B) The anti-interception power dive of a lacewing when approached by a hunting bat. The numbers superimposed on this multiple-exposure photograph show the relative positions of a lacewing and a bat over time. (The lacewing survived that attack.) A, from Yager and May;[89] photographs courtesy of D. D. Yager and M. L. May; B, photograph by Lee Miller.

(A)

(B)

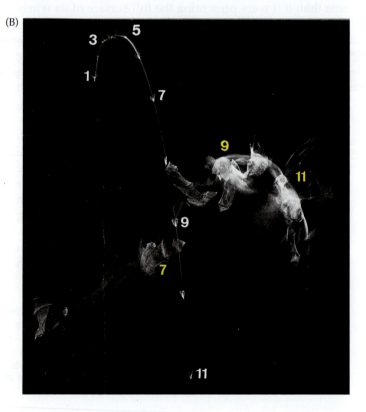

to the A1 receptor, other persons continued to look at how these cells control the response of moths to bats. As a result, we now know that notodontid moths, even though they have just one auditory receptor per ear, still appear to exhibit a two-part response to approaching bats: the turning away from distant hunters and then the last-second erratic flight pattern when death is at hand. Thus, two cells may not be necessary for the double-barreled response of moths to their hunters.[75] Even in moths with two receptors per ear, the A1 cell's activity changes greatly as a bat comes sailing in toward it, because the bat's ultrasonic cries speed up and become much more intense (Figure 12.13).[26] Presumably, higher-order neurons up the chain of command could analyze the correlated

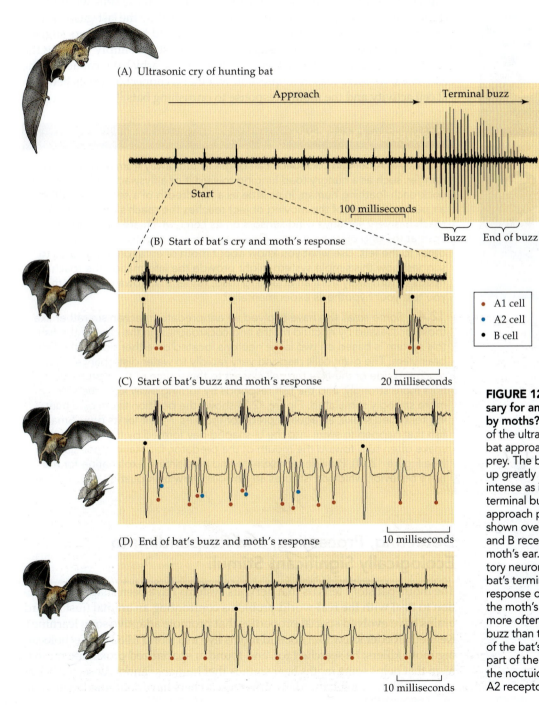

(A) Ultrasonic cry of hunting bat

Approach   Terminal buzz

Start

100 milliseconds

Buzz   End of buzz

(B) Start of bat's cry and moth's response

- ● A1 cell
- ● A2 cell
- ● B cell

(C) Start of bat's buzz and moth's response   20 milliseconds

(D) End of bat's buzz and moth's response   10 milliseconds

10 milliseconds

FIGURE 12.13   Is the A2 cell necessary for anti-interception behavior by moths? (A) A visual representation of the ultrasonic pulses produced by a bat approaching and then attacking a prey. The bat's ultrasonic cries speed up greatly and become much more intense as it closes in on the prey (the terminal buzz phase). (B) Part of the approach portion of the bat's cry is shown over the response of the A1 and B receptor cells in the noctuid moth's ear. (The B cell is a nonauditory neuron.) (C) The initial part of the bat's terminal buzz is shown over the response of the A1, A2, and B cells in the moth's ear. The A1 receptor fires more often in response to the terminal buzz than to the approach component of the bat's cry. (D) During the latter part of the bat's terminal buzz, only the noctuid's A1 cell is active, not the A2 receptor. After Fullard et al.[26]

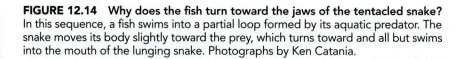

FIGURE 12.14   Why does the fish turn toward the jaws of the tentacled snake? In this sequence, a fish swims into a partial loop formed by its aquatic predator. The snake moves its body slightly toward the prey, which turns toward and all but swims into the mouth of the lunging snake. Photographs by Ken Catania.

changes in the activity of the A1 receptor alone and make adaptive adjustments accordingly without involvement of the A2 receptor.

More doubts that the A2 cell is necessary to trigger erratic evasive behavior come from the finding that in some noctuid moths, both the A1 and A2 cells may more or less stop firing during the terminal buzz phase—the last 150 milliseconds—of a bat attack. One would think that these cells would keep signaling if either one was truly important in controlling the last-gasp evasive maneuvers of moths under attack. James Fullard and his coworkers suggest that perhaps these cells fail to signal at this late stage simply because the extremely loud and rapid attack vocalizations of a nearby bat incapacitate the cells. Thus, we have reason to question whether a connection exists between A2 cell activity and the moth's response to onrushing bats.[26]

Discussion Questions

12.3   An American cockroach can begin to turn away from approaching danger, such as a hungry toad lunging toward it or a flyswatter wielded by a cockroach-loathing human, in as little as a hundredth of a second after the air pushed in front of the moving object reaches the roach's body. A cockroach has very sensitive wind sensors on its cerci, which are small appendages at the end of its abdomen. One cercus points slightly to the right, the other to the left. Use what you know about moth orientation to bat cries to suggest how this simple system might provide the information the roach needs to turn away from an attacking toad, rather than toward it. How might you test your hypothesis experimentally?

12.4   Some small fish have evolved an antipredator mechanism rather like those possessed by noctuid moths and cockroaches. In the case of the fish, this system is based on two large neurons located on either side of the fish's brainstem. The giant cells respond very rapidly to water disturbances that reach the side of the fish; the cell closest to the source of the disturbance fires more strongly, and the fish responds by turning away from the side more strongly stimulated. Figure 12.14 shows the position taken by a hunting tentacled snake, an aquatic predator that lies in wait for its prey. Small fish that happen to swim into the U-shaped bay formed by the body of the snake often turn *toward* the mouth of the predator, facilitating their capture (see Figure 12.14). Provide a proximate and an ultimate explanation for the fish's behavior, which constitutes an obvious Darwinian puzzle.

Detecting, Processing, and Responding to Ecologically Significant Stimuli

The moth–bat story tells us that nervous systems have been shaped by selection to deal with the real-world problems confronting an animal (just as bird brains have evolved special properties that promote adaptive song learning). Kenneth Roeder knew that noctuid moths live in a world filled with echolocating moth killers; he searched for, and found, a specialized proximate mechanism that helps some moths hear sounds that humans cannot sense, the better to deal with hunting bats. Many other researchers have continued to explore

(A)

(B)

FIGURE 12.15 **Ultraviolet-reflecting patterns are detected by many animals.** In both sets of photographs, the image on the left shows the organism as it appears to humans, while the image on the right shows the organism's UV-reflecting (pale) surfaces. (A) The ultraviolet pattern on this daisy advertises the central location of food for insect pollinators. (B) Only males (top specimens) of this sulfur butterfly species have UV-reflecting patches on their wings, which helps signal their sex to other individuals of their species. A, photographs by Tom Eisner; B, photographs by Randi Papke and Ron Rutowski.

the proximate and ultimate aspects of moth–bat interactions and, in so doing, have produced a vast literature on how these creatures have influenced each other's evolution.[16] Among the many fascinating discoveries made in the post-Roeder era is that some species of edible moths respond to bat ultrasound by generating ultrasonic clicks of their own. At first glance, this reaction would seem to be suicidal, as the moths appear to be offering signals that their predators could use to track them down. However, the moth clicks are in fact loud jamming signals, which interfere with a bat's ability to zero in on a prey, causing the predator to misjudge the location of the moth.[17]

Both the noctuid moths studied by Roeder and the tiger moths that jam bat sonar use their evolved acoustical mechanisms to detect sounds outside the range of human hearing. Many other creatures have evolved specialized sensory skills in other modalities. Take ultraviolet radiation, which comes from many objects in nature and yet we humans are blind to it. Many bees see these stimuli, which enable some species to use the ultraviolet-reflecting bee guides on flowers to locate the nectar source more readily (Figure 12.15). Ultraviolet-reflecting patches on the wings of certain male butterflies are seen by and are sexually attractive to females of these butterflies.[43,63] The same is true for female sticklebacks, which respond positively to the carotenoid pigments on male bodies (Figure 12.16).[4] When we look at carotenoids, we see reds, oranges, and yellows but not the ultraviolet signals that stickleback

FIGURE 12.16 Ultraviolet reflectance and mate choice in the stickleback. (A) The experimental setup. A female could view males in adjacent chambers that were separated from her compartment by a screen and filters, which could either permit the passage of UV (UV+) or block UV (UV–). (B) When given a choice of which male to spend time close to, the females preferred the male whose UV signals could reach them via the UV+ filter. After Boulcott et al.[4]

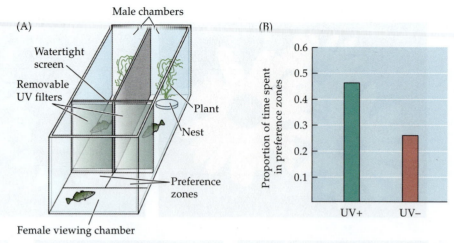

females apparently use to determine the quantity of carotenoids incorporated in a male's skin.[56] Female preferences based on these signals help them choose males capable of surviving the breeding season, during which time the males will provide parental care for their choosy mates' offspring.[55]

Signals containing UV wavelengths play a role in certain aggressive, as well as sexual, interactions. For example, adult male collared lizards aggressively display to one another by opening their mouths wide, and when they do so, whitish UV-reflecting patches at the corners of their mouths become visible to other lizards (see Figure 4.20) [45]. In another lizard species, males with experimentally reduced UV-reflecting throat patches were subject to more attacks than control males, an indication that the size of the UV patch is a signal of male fighting ability in this lizard as well.[72]

Finally, the ability to detect UV radiation plays a role in the extraordinary navigational abilities of migratory monarch butterflies. Migrant monarchs fly south from Canada toward a select group of forest groves high in the mountains of central Mexico, a trip of as much as 3600 kilometers (Figure 12.17). Once in Mexico, the butterflies congregate by the millions in stands of Oyamel firs, where they mostly sit and wait, day after day, through the cool winter season until spring comes. Then the monarchs rouse themselves and begin a return trip to the Gulf Coast, arriving in the southern United States in time to lay their eggs on milkweed plants growing there, which their caterpillar progeny will consume.

Migrating monarchs fly during the daytime, which suggests that they might use the sun as a compass to guide them in a southwesterly direction during the fall migration (and in a northeasterly direction during the spring return flight). Sunlight is composed of many different wavelengths of light, and so the question arises, which wavelengths are critical for monarch navigation? To test the hypothesis that monarchs depend on ultraviolet radiation, experimenters captured migrating butterflies in the fall and then permitted tethered individuals to start flying in the flight cage where they could orient in a direction of their choosing. The team then covered the flight cage with a UV-interference filter, which screened out this component of sunlight. The monarchs quickly became confused, and many stopped flying altogether. Most individuals (11 out of 13) resumed flight, however, as soon as the filter was removed, evidence that the sun's UV radiation is indeed essential for monarch navigation.[25]

Ultraviolet light may help monarchs get started flying in the right direction. But once airborne, the butterflies need to maintain their compass orientation, and as it turns out, they achieve this end with assistance from the cues provided by the polarized skylight pattern. Polarized light, which you and I cannot see, is produced when sunlight enters the Earth's atmosphere

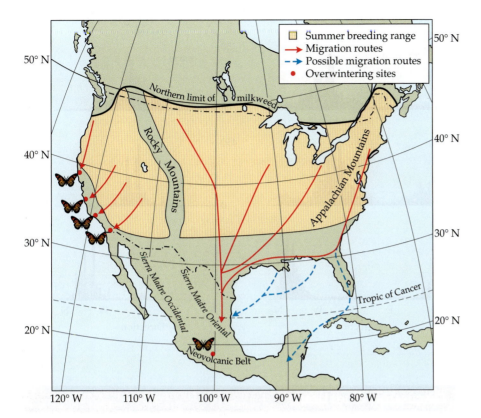

FIGURE 12.17 The fall migration routes of monarch butterflies. Monarchs that reach southern Canada then turn around to travel to Mexico, where they cluster in a few forest patches high in the central Mexican mountains. Monarchs in the western United States migrate to coastal wintering sites. After Brower.[5]

and the light scatters so that some light waves are vibrating perpendicular to the direction of actual sun rays. The three-dimensional pattern of polarized light in the sky created in this fashion depends upon the position of the sun relative to the Earth, and so this pattern changes as the sun moves across the sky. Therefore, if an animal on the ground is able to perceive the pattern of polarized light in the sky, this stimulus can indicate the position of the sun at any given time. As a result, creatures with a **biological clock** and the capacity to see polarized light can use the information in skylight as a compass in the same way that the sun's position in the sky can be used by humans as a compass. Being able to make use of the directional information in polarized light has real advantages for monarchs and some other animals, such as migrating salmon,[34] because this cue is available even when the sun is hidden behind clouds or mountains.

A team of researchers established that monarchs can orient to polarized light information by tethering butterflies captured on their fall migration in a small walled arena where the insects could not see the sun but could look at the sky overhead. Under these conditions, when the butterflies flew, they were able to orient consistently to the southwest. Flying tethered monarchs retained this ability when a light filter was placed over the apparatus and aligned so as to permit the entry of linear polarized light waves in the same plane as occurs in the sky at the zenith (the highest point in the sky). However, when the filter was turned 90 degrees from its original orientation, changing the angle of entry of the polarized light visible to the monarchs at the zenith, they altered their flight orientation by 90 degrees as well, demonstrating their reliance on a polarized-light compass (Figure 12.18).[57]

The point is that many species have visual systems very different from our own, and they put their extraordinary (from our perspective) abilities to use in activities that promote their survival and reproduction.

(A) Skylight   Treatment
         e-vector

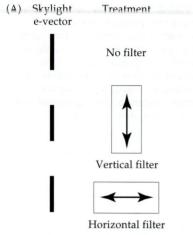

No filter

Vertical filter

Horizontal filter

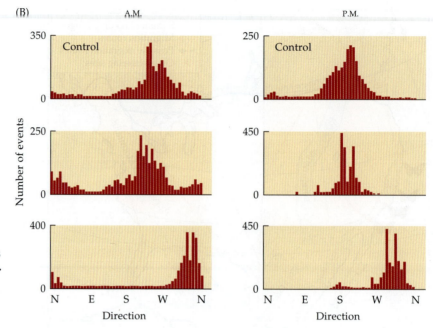

FIGURE 12.18 Polarized light affects the orientation of monarch butterflies. (A) Butterflies tethered in a flight cage received three treatments with respect to the angle of polarized light reaching them from the sky (the skylight e-vector): (1) no filter, which meant that they could see the natural pattern of polarized light in the sky; (2) vertical filter, which did not interfere with the pattern of polarized light visible from the flight cage; and (3) horizontal filter, which shifted the angle of polarized light arriving from the sky by 90 degrees. (B) Orientation choices made by two flying monarchs in the flight cage, one tested in the morning (A.M.) and the other in the afternoon (P.M.). During the period of flight, a computer automatically recorded the orientation of the butterfly every 200 milliseconds, yielding a record composed of many "events" during the minutes when the monarch was flying. The data show that when the insects could observe polarized light in the natural skylight pattern, they tended to fly to the southwest, but when the filter shifted the angle of polarized light, altering the skylight pattern by 90 degrees, the monarchs shifted their flight orientation accordingly. After Reppert et al.[57]

Discussion Questions

12.5 The blue tit is a songbird with a blue patch of feathers that reflect ultraviolet radiation. Female blue tits apparently prefer to pair with males that have relatively bright UV-reflecting feathers on their crowns.[39] But at least in one population, male blue tits whose crowns reflected less UV produced more offspring than males with more UV-ornamented crowns. A research team suggested that perhaps the males with low-UV crowns were better able to sneak onto neighboring territories and sire "extra-pair" offspring (see page 225) with their neighbors' mates than were males with high-UV crowns. How would you test this hypothesis experimentally? List your predictions and then check the results presented in Delhey et al.[21]

12.6 Homing pigeons are very good at returning home after being released in a strange location (Figure 12.19). Internal biological clocks (see Chapter 13) play a role in this ability.[83] You can reset a pigeon's biological clock by placing the bird in a closed room with artificial lighting that comes on and goes off out of phase with sunrise and sunset in the real world. For example, if sunrise is at 6:00 A.M. and sunset at 6:00 P.M., let's set the lights in the room to go on at noon (6 hours later than the actual sunrise) and off at midnight (6 hours later than actual sunset). A pigeon exposed to this routine for several days will become clock-shifted 6 hours out of phase with the natural day. If taken from the room and released *at noon* 30 kilometers due east of its loft, the bird will behave as if the sun had just come up in the east. Why then does the bird fly north instead of west?

Selective Relaying of Sensory Inputs

In monarchs, sticklebacks, and many other species, sensory receptors collect information about ultraviolet radiation. To act on that information, however, the animal's receptors must forward messages to those parts of its central nervous system that can process selected sensory inputs and order appropriate responses based on that information. We can illustrate how neural mechanisms contribute to the selective transfer of data by returning to an insect that

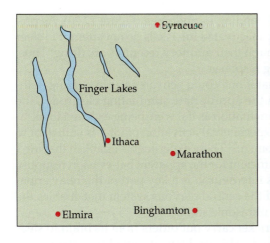

FIGURE 12.19 **Navigation by homing pigeons.** The birds were released at Marathon, New York. If their biological clocks were not manipulated, they flew west for about 30 kilometers until they reached Ithaca, where their home loft was located. After Keeton.[41]

Homing pigeon

hears ultrasound, the cricket *Teleogryllus oceanicus*.[50] As in noctuid moths, the cricket uses this ability to avoid hunting bats. The process begins with the firing of certain ultrasound-sensitive auditory receptors in its ears, which are found on the cricket's forelegs. Sensory messages from these receptors travel to other cells in the cricket's central nervous system. Among the receivers of these messages is a pair of sensory interneurons called int-1, also known as AN2, one of which is located on each side of the insect's body. Ron Hoy and his coworkers established that int-1 plays a key role in the perception of ultrasound by playing sounds of different frequencies to a cricket and recording the resulting neural activity. They found that these cells became highly excited when the cricket's ears were bathed in ultrasound. The more intense a sound in the 40 to 50 kHz range, the more action potentials the cells produced and the shorter the latency between stimulus and response—two properties that match those of the A1 receptor in noctuid moths.[50]

The int-1 cell seems to be part of a neural circuit that helps the cricket respond to ultrasound. If this is true, then it follows that if one could experimentally inactivate int-1, ultrasonic stimulation should not generate the typical reaction of a tethered cricket suspended in midair, which is to turn away

FIGURE 12.20 **Avoidance of and attraction to different sound frequencies by crickets.** (A) In the absence of sound, a flying tethered cricket holds its abdomen straight. (B) If it hears low-frequency sound, the cricket turns toward the source of the sound. (C) If it hears high-frequency sound, it turns away. Males of this species produce sounds in the 5 kHz range; some predatory bats produce high-frequency calls of about 40 kHz. (D) The tuning curve of the cricket's int-1 interneuron. Sounds in the 5 to 6 kHz range need only be about 55 decibels (dB) loud in order to trigger firing. The intensity threshold also dips in the neighborhood of 40 kHz, which is in the ultrasonic range commonly produced by bats. After Moiseff et al.[50]

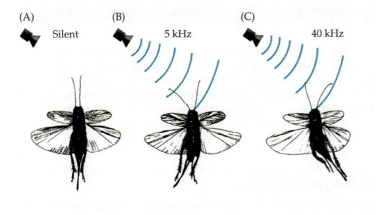

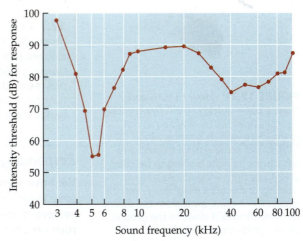

from the source of the sound by bending its abdomen (Figure 12.20). As predicted, crickets with temporarily inactivated int-1 cells do not attempt to steer away from ultrasound, even though their auditory receptors are firing. Thus, int-1 is necessary for the steering response.

The corollary prediction is that if one could activate int-1 in a flying, tethered cricket (and one can, with the appropriate stimulating electrode), the cricket should change its body orientation as if it were being exposed to ultrasound, even when it was not. Experimental activation of int-1 is sufficient to cause the cricket to bend its abdomen.[52] These experiments establish that int-1 activity is both necessary and sufficient for the apparent bat-evasion response of flying crickets; therefore, this interneuron is a key part of the relay apparatus between the receptors and the central nervous system that enables the cricket to react adaptively to an ultrasonic stimulus.

And just how does a flying cricket carry out orders from its brain to steer away from a source of ultrasound? This problem attracted the attention of Mike May, who began his study not by conducting a carefully designed experiment but instead by "toying with the ultrasound stimulus and watching the responses of a tethered cricket."[47] As May zapped the cricket with bursts of ultrasound, he noticed that the beating of one hindwing seemed to slow down with each application of the stimulus. Crickets have four wings, but only the two hindwings are directly involved in flight. If the hindwing opposite the source of ultrasound really did slow down, that would reduce power or thrust on one side of the cricket's body, with a corresponding turning (or yawing) of the cricket away from the stimulus.

On the basis of his informal observations, May proposed that the flight path of a cricket was controlled by the position of the insect's hindleg, which when lifted into a hindwing, altered the beat of that wing and thereby changed the cricket's position in space (Figure 12.21). May went on to take a number of high-speed photographs of crickets with and without hindlegs. Without the appropriate hindleg to act as a brake, both hindwings continued beating unimpeded when the cricket was exposed to ultrasound. As a result, crickets without hindlegs required about 140 milliseconds to begin to turn, whereas intact crickets started their turns in about 100 milliseconds. These findings led May to assert that neurons in the ultrasound detection network order the appropriate motor neurons to induce muscle contractions in the opposite-side hindleg of the cricket. As these muscles contract, they lift the leg into the wing, interfering with its beating movement, thereby causing the cricket to veer rapidly away from an ultrasound-producing bat.[47]

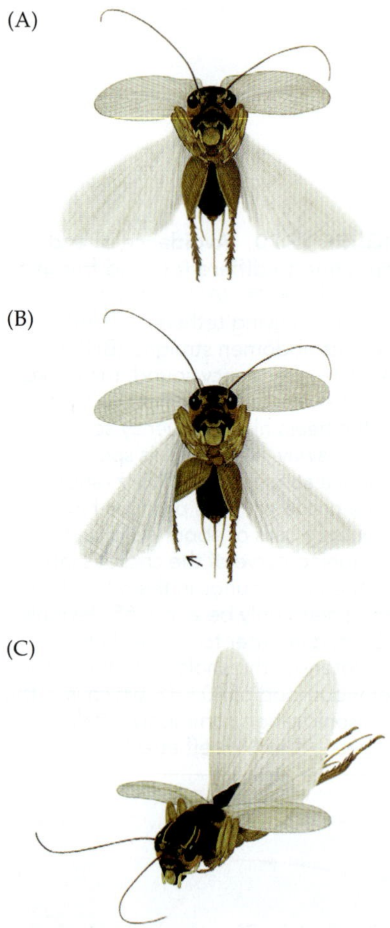

FIGURE 12.21 How to turn away from a bat—quickly. (A) A flying cricket typically holds its hindlegs so as not to interfere with its beating wings. (B) Ultrasound coming from the cricket's left causes its right hindleg to be lifted up into its right wing. (C) As a result, the beating of the right wing slows, and the cricket turns right, away from the source of ultrasound as it dives to the ground. Drawings by Virge Kask, from May.[47]

Discussion Question

12.7 Outline Mike May's research in terms of the question that provoked his study and his hypothesis, prediction(s), evidence, and scientific conclusion. In addition, what contribution to this research could come from learning that locusts, a group of insects not closely related to crickets, also possess a special mechanism for very rapidly altering leg positions and wingbeat patterns in reaction to ultrasound, such that a flying individual banks sharply downward away from the side stimulated by the stimulus.[19]

Responding to Relayed Messages

We have seen how sensory receptors detect key stimuli before relaying sensory signals via interneurons to other neurons within the central nervous system. When these central cells respond, they can generate signals that turn on a set of motor neurons. The escape dives of moths and the evasive

FIGURE 12.22 Escape behavior by a sea slug. The slug in this photograph has just begun to swim away from its deadly enemy, a predatory sea star. The slug's dorsal muscles are maximally contracted, drawing the slug's head and tail together. Soon the ventral muscles will contract and the slug will begin to thrash away to safety. Photograph by William Frost.

swerves of flying crickets pursued by bats are effective one-step responses triggered by simple releasing stimuli in these animals' environments. Most behaviors, however, involve a coordinated series of muscular responses, which cannot result from a single command from a neuron or neural network. Consider, for example, the escape behavior of the sea slug *Tritonia diomedea*, which is activated when the slug comes in contact with a predatory sea star (Figure 12.22). Stimuli associated with this event cause the slug to swim in the ungainly fashion of sea slugs, by bending its body up and down.[87] If all goes well, the slug will move far enough away from the sea star to live another day.

How does *Tritonia* manage its multistep swimming response, which requires from 2 to 20 alternating bends, each involving the contraction of a sheet of muscles on the slug's back followed by a contraction of the muscles on its belly? As it turns out, the dorsal and ventral muscles are under the control of a small number of motor neurons. The dorsal flexion neurons (DFN) are active when the animal is being bent into a U, and the ventral flexion neuron (VFN) produces a pulse of action potentials that turn the slug into an inverted U (Figure 12.23). But what controls the alternating pattern of DFN and VFN activity?

The escape reaction begins when sensory receptor cells (S) in the skin of *Tritonia* detect certain chemicals on the tube feet of its sea star enemy (Figure 12.24). The receptors then relay messages to interneurons, among them the dorsal ramp interneurons (DRI), which, upon receipt of sufficiently strong

Dorsal flexion

Ventral flexion

DFN

VFN

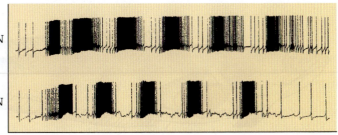

FIGURE 12.23 Neural control of escape behavior in *Tritonia*. The dorsal and ventral muscles of the sea slug are under the control of two dorsal flexion neurons (DFN) and a ventral flexion neuron (VFN). The alternating pattern of activity in these two categories of motor neurons translates into alternating bouts of dorsal and ventral bending—the movements that cause this animal to swim. After Willows.[87]

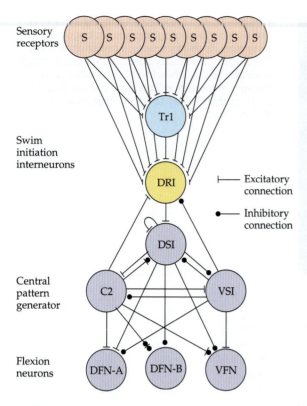

Sensory receptors

Swim initiation interneurons

├── Excitatory connection

●── Inhibitory connection

Central pattern generator

Flexion neurons

FIGURE 12.24 The central pattern generator of *Tritonia* in relation to the dorsal ramp interneurons (DRI) that maintain activity in the cells that generate the sequence of signals necessary for the sea slug to swim to safety. These interneurons receive excitatory input from receptor cells (S) and from another interneuron (Tr1). DRI cells in turn interact with three other categories of neurons (DSI, C2, and VSI), which are the cells that send messages to the flexion neurons. There are two kinds of dorsal flexion neurons (DFN-A and DFN-B) with somewhat different properties and one kind of ventral flexion neuron (VFN). After Frost et al.[24]

stimulation, begin to fire steadily. This category of interneurons sends a stream of excitatory signals to several interneurons (the dorsal swim interneurons, or DSI), which in turn are part of an assembly of interconnected cells, among them the ventral swim interneurons (VSI) and cerebral neuron 2 (C2), as well as the flexion neurons mentioned already.[27,28] A web of excitatory and inhibitory relations exists within this cluster of interneurons such that, for example, activity in DSI turns on C2, which leads to excitation of the DFN and the contraction of the dorsal flexion muscles. After a short period of excitation, however, C2 begins to block the DFN while sending excitatory messages to the VSI, leading to activation of the VFN and contraction of the ventral flexion muscles. The situation then reverses. Alternating bouts of activity in the interneurons regulating the DFN and VFN lead to alternating bouts of DFN and VFN firing and, thus, alternation of dorsal and ventral bending.[24]

The capacity of the simple neural network headed by the DRI to impose order on the activity of the motor neurons that control the dorsal and ventral flexion muscles means that this mechanism qualifies as a **central pattern generator**. Systems of this sort have been particularly well studied in invertebrates, especially with regard to locomotion, because the small number of neurons involved and their relatively large size facilitate their investigation.[15] The neural clusters labeled central pattern generators play a preprogrammed set of messages—a motor tape, if you will—that helps organize the motor output underlying movements of the sort that Tinbergen would have labeled a fixed action pattern.

Central pattern generators are also found in vertebrates, as we can illustrate by looking at the plainfin midshipman, a fish that "sings" by contracting and relaxing certain muscles in a highly coordinated fashion.[3] Only the large males of this rather grotesque fish sing "humming" songs, which last more than a minute, and they do so only at night during spring and summer while guarding certain rocks. Their songs are so loud that they can annoy houseboat owners in the Pacific Northwest. The male fish sings to attract females of his species; the fish spawn at the defended rocks, and the male guards the eggs his mates lay in his territory.

How do the male fish produce their songs? When Andrew Bass and his coworkers inspected the anatomy of the fish's abdomen, they found a large, air-filled swim bladder sandwiched between layers of muscles (Figure 12.25). The bladder serves as a drum; rhythmic contractions of the muscles "beat" the drum, generating vibrations that other fish can hear. Muscle contractions require signals from motor neurons, which Bass found connected to the sonic muscles. He applied a cellular dye called biocytin to the cut ends of these

FIGURE 12.25 Song-producing apparatus of the male plainfin midshipman fish. The sonic muscles control the movement of the swim bladder, thereby controlling the fish's ability to sing. After an illustration by Margaret Nelson, in Bass.[3]

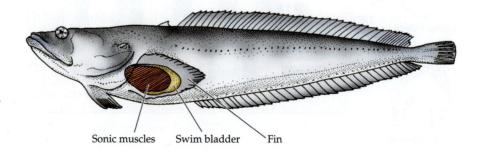

Sonic muscles   Swim bladder   Fin

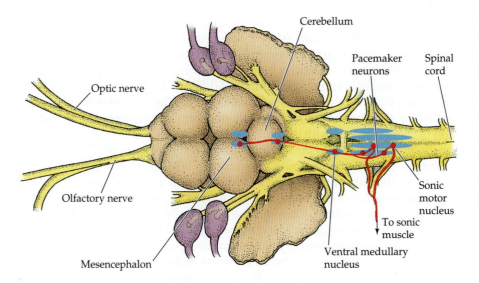

FIGURE 12.26   Neural control of the sonic muscles in the plainfin midshipman fish. Signals from the central region of the brain (the mesencephalon) travel by way of the cerebellum and ventral medullary nuclei to the sonic motor nuclei in the upper part of the spinal cord. The firing of the pacemaker neurons regulates the frequency of firing by the neurons in the sonic motor nuclei; these signals, in turn, set the rate of contraction of the sonic muscles and thus the frequency of the sounds produced by the fish. After an illustration by Margaret Nelson, in Bass.[3]

motor neurons, which absorbed the material, staining the cells brown. And the stain kept moving along, crossing the synapses between the first cells to receive it and the next ones in the circuit, and so on, through the whole network of cells connected to the sonic muscles. By cutting the brain into fine sections and searching for cells stained brown by biocytin, Bass and his colleagues mapped the fish's sonic control system. In so doing, they discovered two discrete collections of interrelated neurons that generate the signals controlling the coordinated muscle contractions required for midshipman humming. These two clusters, called the sonic motor nuclei, consist of about 2000 neurons each and are located in the upper part of the spinal cord near the base of the brain (Figure 12.26). The long axons of their neurons travel out from the brain, fusing together to form nerves that reach the sonic muscles.

In addition to these components, two other anatomically distinct elements of the nervous system ally themselves with the sonic motor nuclei. First, a sheet of pacemaker neurons—a multicellular central pattern generator—lies next to each nucleus and adjusts the activity of the sonic motor neurons so that the sequence of muscle contractions will yield a proper song. Second, there are some special neurons in front of the pair of sonic motor nuclei that appear to connect the two nuclei, probably to coordinate the firing patterns coming from the left and the right nucleus so that the muscle contractions and relaxations will be synchronous, the better to produce a humming sound.[3]

The Proximate Basis of Stimulus Filtering

We have now looked at the attributes of individual neurons and neural clusters that are involved in the detection of certain kinds of sensory information, the relaying of messages to other cells in the nervous system, and the control of motor commands that are sent to muscles. The effective performance of these basic functions is promoted by **stimulus filtering**, the ability of neurons and neural networks to ignore—to filter out—vast amounts of potential information in order to focus on biologically relevant elements within the diverse stimuli bombarding an animal.

The noctuid moth's auditory system offers an object lesson on the operation and utility of stimulus filtering. First, the A1 receptors are activated only by acoustical stimuli, not by other forms of stimulation. Moreover, as noted, these cells completely ignore sounds of relatively low frequencies, which means that moths are not sensitive to the stimuli produced by chirping crickets or croaking frogs—sounds they can safely ignore. Finally, even when the A1 receptors do fire in response to ultrasound, they do little to discriminate between different ultrasonic frequencies (whereas human auditory receptors produce distinct messages in response to sounds of different frequencies, which is why we can tell the difference between C and C-sharp). The noctuid moth's sensory apparatus appears to have just one task of paramount importance: the detection of cues associated with its echolocating predators. To this end, its auditory capabilities are tuned to pulsed ultrasound at the expense of all else. Upon detection of these critical inputs, the moth can take effective action.

The relationship between stimulus filtering and a species' special route to reproductive success is evident in every animal whose sensory systems have been carefully examined. Consider the male midshipman fish that listen to the underwater grunts, growls, and hums produced by others of their species. These signals are dominated by sounds in the 60 to 120 hertz (Hz) range. The auditory receptor cells in the hearing organs of these fish are most sensitive to sounds in exactly that range.[70] In summer, however, when the humming "songs" of territorial males incorporate higher-frequency sounds, the female's auditory system changes, providing females in search of spawning partners with sensitivity to sounds that range up to 400 Hz.[69]

Female midshipman fish employ stimulus filtering on a seasonal basis as they listen to sounds in their underwater world. The screening of acoustical stimuli also occurs in certain parasitoid flies, which use their hearing to locate singing male crickets, the better to place their larvae on these insects. The little maggots burrow into the unlucky crickets and proceed to devour them from the inside out. Larvae-laden female flies of the fly *Ormia ochracea* can find food for their offspring because they have ears tuned to cricket calls, as researchers discovered when they found *Ormia* coming to loudspeakers that were playing tapes of cricket song at night.

The unique ears of the female fly consist of two air-filled structures with tympanic membranes and associated auditory receptors on the front of the thorax. Vibration of the fly's tympanic "eardrums" activates the receptors, just as in noctuid moth ears, and thus provides the fly with information about sound in its environment—but not every sound. As predicted by a trio of evolutionary biologists, Daniel Robert, John Amoroso, and Ronald Hoy, the female fly's auditory system is tuned to (i.e., most sensitive to) the dominant frequencies in cricket songs (Figure 12.27). That is, the female fly can hear

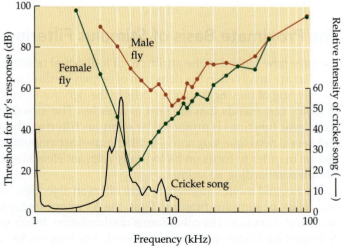

FIGURE 12.27   Tuning curves of a parasitoid fly. Females, but not males, of the fly *Ormia ochracea* find their victims by listening for the calls of male crickets, which produce sound with a frequency–intensity spectrum that peaks between 4 and 5 kHz. The female fly, unlike the male fly, is maximally sensitive to sounds around 5 kHz. After Robert et al.[58]

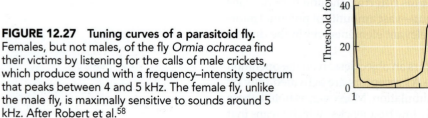

sounds of 4 to 5 kHz (the sort produced by crickets) more easily than sounds of 7 to 10 kHz, which have to be much louder if they are to generate any response.[58] In contrast, male *Ormia* are not especially sensitive to sounds of 4 to 5 kHz, which makes ecological sense since males do not track down singing male crickets.

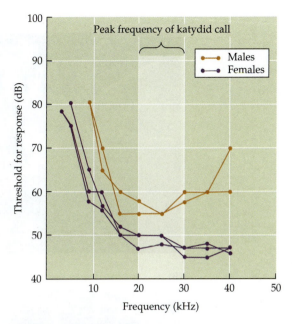

FIGURE 12.28 **Tuning curves of a katydid killer.** Females of the fly *Therobia leonidei* parasitize male katydids, whose stridulatory calls contain most of their energy in the range of 20 to 30 kHz. After Stumpner and Lakes-Harlan.[74]

Discussion Question

12.8 Females of another parasitoid fly related to *Ormia* search for singing male katydids (*Poecilimon veluchianus*), whose ultrasonic mate-attracting calls fall largely in the 20 to 30 kHz range.[74] What sound frequencies should elicit maximal response in the ears of this katydid-hunting parasitoid if stimulus filtering enables the animal to achieve biologically relevant goals? What conclusion comes from the data in Figure 12.28?

Stimulus Filtering via Cortical Magnification

Stimulus filtering takes place at the level of the sensory receptors and sensory interneurons—and at the level of the brain as well, as we will see from the following sketch of the operating rules of the star-nosed mole's brain. This weird mammal lives in wet, marshy soil, where it burrows about in search of earthworms and other prey. In its dark tunnels, earthworms cannot be seen, and indeed, the mole's eyes are greatly reduced in size, so it largely ignores visual information even when light is available. Instead, the mole relies heavily on touch to find its food, using its wonderfully strange nose to sweep the tunnel walls as it moves forward. Its two nostrils are ringed by 22 fleshy appendages, 11 on each side of the nose (Figure 12.29). These

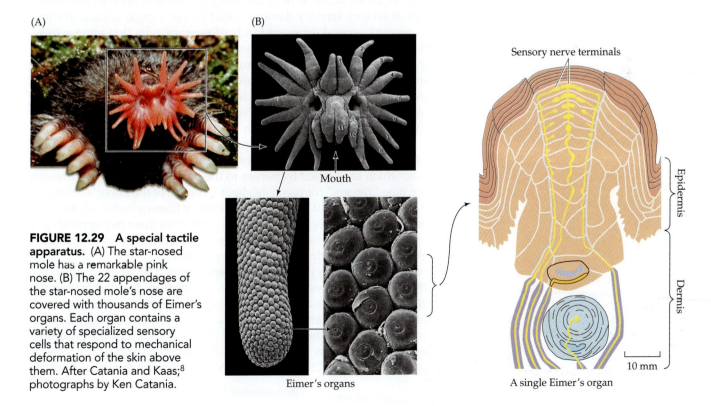

FIGURE 12.29 **A special tactile apparatus.** (A) The star-nosed mole has a remarkable pink nose. (B) The 22 appendages of the star-nosed mole's nose are covered with thousands of Eimer's organs. Each organ contains a variety of specialized sensory cells that respond to mechanical deformation of the skin above them. After Catania and Kaas;[8] photographs by Ken Catania.

(A)

(B)

(C)

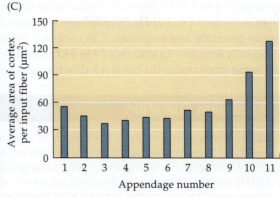

FIGURE 12.30 The cortical sensory map of the star-nosed mole. The map gives disproportionate weight to appendage 11 of the mole's nose. (A) The nose of the mole, with each appendage numbered on one side. (B) A section through the area of the somatosensory cortex that is responsible for analyzing sensory inputs from the nose. The cortical areas that receive information from each appendage are numbered. (C) The amount of somatosensory cortex devoted to the analysis of information from each nerve fiber carrying sensory signals from the different nose appendages. After Catania and Kaas;[9] photographs by Ken Catania.

appendages cannot grasp or hold anything but instead are covered with a thousand or so tiny sensory devices called Eimer's organs. Each of these organs contains several different kinds of sensory cells that appear to be dedicated to the detection of objects touching the nose.[46] With these mechanoreceptors, the animal can collect extremely complex patterns of information about the things it encounters underground, enabling it to identify prey items in the darkness with great rapidity.[8,12]

Whenever the mole brushes an earthworm with, say, appendage 5, it instantly sweeps its nose over the prey so that the two projections closest to the mouth, labeled appendage 11 (Figure 12.30A), come in contact with the object of interest. The tactile receptors on each appendage 11 generate a volley of signals, which are carried by nerves to the brain of the animal. Although these two nose "fingers" contain only about 7 percent of the Eimer's organs on the star nose, more than 10 percent of all the nerve fibers relaying information from the nose's touch receptors to the brain come from these two appendages. In other words, the mole uses relatively more neurons to relay information from appendage 11 than from any other appendage.

Not only is the relay system biased toward inputs from appendage 11, but the animal's brain also gives extra weight to signals from this part of the nose. The information from the nose travels through nerves to the somatosensory cortex, the part of the brain that receives and decodes sensory signals from touch receptors all over the animal's body (Figure 12.30B). Of the portion of the somatosensory cortex that is dedicated to decoding inputs from the 22 nose appendages, about 25 percent deals exclusively with messages from the two appendage 11s (Figure 12.30C).[9] This discovery was made by Kenneth Catania and Jon Kaas when they recorded the responses of cortical neurons as they touched different parts of the anesthetized mole's nose. Perhaps the mole's brain is "more interested in" information from appendage 11 because of its location right above the mouth; should signals from this appendage activate a cortical order to capture a worm and consume it, the animal is in the right position to carry out the action immediately.[8]

In the star-nosed mole, the disproportionate investment in brain tissue to decode tactile signals from one part of the nose is mirrored on a larger scale by the biases evident in the somatosensory cortex as a whole, which focuses on signals from the mole's hands and nose at the expense of other parts of the body. This pattern of cortical magnification makes adaptive sense for this species because of the biological importance of the mole's hands for burrowing and the mole's nose for locating prey (Figure 12.31A).

If this argument is correct, then we can predict that the allocation of cortical tissue to somatosensory inputs will differ from species to species in ways that make sense given the special environmental problems each species has to solve. And it is true that other insectivores, even though they are related to the star-nosed mole, exhibit their own adaptively different patterns of cortical magnification (Figure 12.31B–D).

FIGURE 12.31 Sensory analysis in four insectivores. In each case, the smaller drawing shows the actual anatomical proportions of the animal; the larger drawing shows how the body is proportionally represented in the somatosensory cortex of the animal's brain. (A) The star-nosed mole devotes much more somatosensory cortex to processing inputs from its nose and forelimbs than it does to processing those coming from receptors in other parts of its body. (B) Cortical magnification in the eastern mole focuses on sensory inputs from the feet, nose, and sensory hairs, or vibrissae, around the nose. (C) Cortical magnification in the masked shrew also reveals the importance of the vibrissae. (D) Sensory signals from the vibrissae are magnified cortically to a lesser degree in the African hedgehog. After Catania and Kaas[8] and Catania.[10]

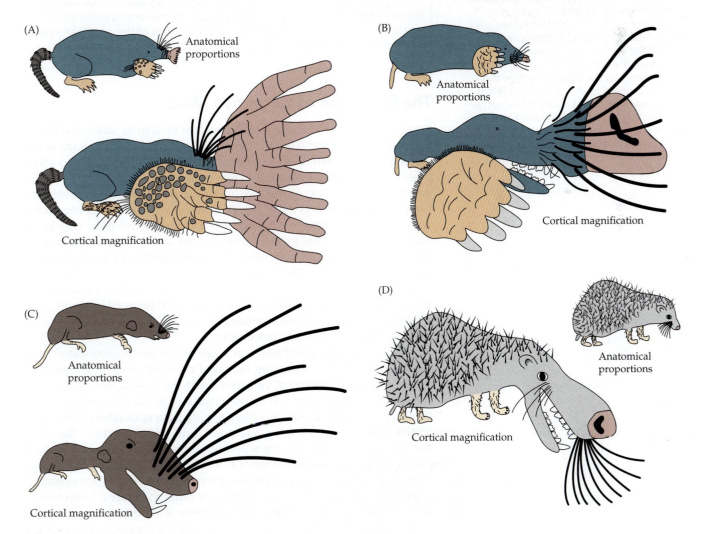

(A) Anatomical proportions

Cortical magnification

(B) Anatomical proportions

Cortical magnification

(C) Anatomical proportions

Cortical magnification

(D) Anatomical proportions

Cortical magnification

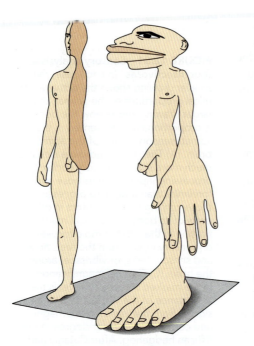

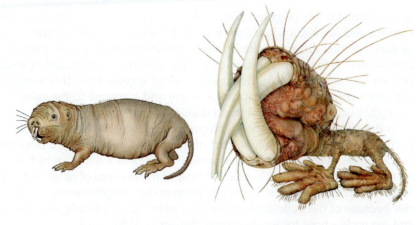

FIGURE 12.32 Sensory analysis in humans and naked mole rats. Brains evolve in response to selection pressures associated with particular physical and social environments. For each species, the drawing on the left shows the actual anatomical proportions; the drawing on the right shows how the body is proportionally represented in the somatosensory cortex. Cortical map of human male is based on data from Kell et al.[42] Cortical map of naked mole rat drawn by Lana Finch, from Catania and Remple.[11]

Discussion Question

12.9 Cortical magnification occurs in all mammals. The drawings in Figure 12.32 are based on the amount of brain tissue devoted to tactile signals coming from different parts of the bodies of human beings and of the naked mole rat (see Figure 3.22), which you may remember lives in groups in underground tunnels that the animals dig. In what ways do these two maps support the argument that animal brains exhibit adaptive sensory biases? For an additional comparison of the cortical maps of laboratory rats and the naked mole rat, see Catania and Henry.[13]

The Evolution of Cognitive Skills

Because of our considerable intelligence, we tend to be interested in the evolution of this characteristic even if it is difficult to define (as is the allied term *cognition*). If, as is generally assumed, our capacity to solve problems, to employ rational thought, to think logically is related to our very large cerebral cortex, then we have a Darwinian puzzle to solve. The human brain is an expensive organ, especially for a developing infant and the infant's mother.[62] What benefits could we derive from a big brain that might compensate for the metabolic and developmental costs of all those neurons? After all, very small brains are capable of remarkable feats of learning, as witness the honey bee's ability to learn where and when flowers will be available to provide the bee and its recruits with pollen and nectar. Honey bees have only about 960,000 neurons in their grass-seed-sized brains,[48] whereas humans have 10,000 times as many neurons packed in a much, much larger brain.

True, our cognitive skills appear more varied and sophisticated than the learning abilities of honey bees and creatures with much smaller brains. On the other hand, consider what the New Caledonian crow can do with a brain only a fraction of the size of the one nestled in our skulls. This bird routinely modifies a wide variety of objects, like pandanus palm leaves, sticks, grass stems, and the like, which the crow then uses to extract beetle grubs and other food items from their hiding places (Figure 12.33). In the lab, captive birds have proven to be ingenious as they solve novel problems of food extraction that (for example) require the crow to use one short stick in order to secure a longer stick, which it then uses to secure some meat that is otherwise out of reach; four of seven birds tested made use of the two tools successfully on the

FIGURE 12.33 A New Caledonian crow using a tool. This bird is unusual in its ability to modify objects in its environments in ways that improve their usefulness as tools for the extraction of beetle larvae out of reach inside their burrows.

first attempt.[76] Just why this crow, an inhabitant of a remote Pacific island nation, should have evolved such impressive problem-solving abilities is a bit of a mystery. It is true, however, that there are no woodpeckers on New Caledonia, creating an open niche for a tool-using woodpecker substitute. Moreover, related species, such as the common raven, are widely recognized to be highly intelligent animals.[35]

Discussion Question

12.10 Cowbirds are another bird with a special cognitive skill.[85] If a cowbird female is to successfully lay an egg into a nest of one of its many host species, she had better do so after the host parent has finished laying her clutch (usually three to six eggs laid one at a time every day or so) but before incubation of the eggs in the nest has begun. If, during this brief window of time, a female cowbird can make her way to the victim's nest, toss out one of the host's eggs, and lay one of her own, then her egg is more likely to be accepted. Since cowbird eggs develop very rapidly, the cowbird chick will be first out of the egg and well on the road to monopolizing the food provided by its hosts. How might a female cowbird tell when a host's clutch of several eggs is complete and incubation is about to start? Why would a kind of mathematical skill come in handy for the parasite?

The rather asocial New Caledonian crow[37] does not really fit the mold for a widely discussed hypothesis for the evolution of cognitive abilities, namely, the social brain or Machiavellian intelligence hypothesis.[20] This hypothesis proposes that advanced problem solving and the like evolved in the context of dealing with the obstacles to reproductive success posed by members of one's own species.[23] (For a variant on this hypothesis as it applies particularly to humans, see Chapter 14.) The logic of this explanation is that an increase in intelligence in some individuals could favor an increase in (social) intelligence in others, setting up a positive feedback loop leading to ever greater species-wide cognitive abilities up to a point.

Although the New Caledonian crow is not particularly social, it is a bird with long-lasting pair bonds and biparental care, social features of birds known to have relatively large brains.[68] Humans also form pair bonds, of course, and they and many other species live in groups in which one's reproductive success may be linked to one's social standing and ability to persuade, coerce, and manipulate others to engage in tasks that require cooperation (such as defending the group's territory against intruders from other bands).

FIGURE 12.34 **An experimental demonstration of hyena cooperation.** Two spotted hyenas have spontaneously figured out that they both have to pull ropes at the same time in order to secure food that is out of reach. Photograph by Christine Drea.

If the social brain hypothesis is correct, we would predict that social mammals would have particularly large brains for animals of their size, and they do.[54] A case in point is the spotted hyena, which as you will recall from Chapter 4, lives in highly stratified clans whose members cooperatively hunt prey and work together to battle other clans and other predators over control of food and good hunting areas. This species does, as predicted from the social brain hypothesis, have a larger brain relative to body volume than other closely related but solitary hyena species.[64]

Christine Drea and Allisa Carter used knowledge about the socio-ecological factors affecting spotted hyenas to predict that this animal would use its large brain to cooperatively solve ecologically relevant problems. To check on this possibility, they devised a task for captive hyenas that required the coordinated activity of two individuals, each of which had to pull on a separate rope at the same time to open a trap door and spill food onto the floor of the cage (Figure 12.34). Pairs of hyenas spontaneously and quickly solved the problem, coordinating their rope pulling so as to gain access to the food[22]—using the capacity to work together to get food, an ability that helps spotted hyenas capture big game animals.

The kind of intelligence exhibited by spotted hyenas suggests that selection arising from social interactions in an animal's real world can generate special problem-solving skills. Another test of this proposition comes from studies of how the "intelligence" of wolves differs from that of dogs. Dogs are descended from wolves but are domesticated and have therefore interacted and evolved with humans for thousands of years. Brian Hare and others predicted therefore that dogs should be much more capable than wolves of detecting and responding to cues about the

---

BOX 12.1 Controversy about whether the cognitive skills of dogs have been evolved via domestication

We have repeatedly noted in passing that a particular finding or conclusion by one researcher or research team has been challenged by other biologists. Published results are not always accepted universally. In fact, skepticism by researchers is a fundamental and very productive part of science, as mentioned in Chapter 1, because repeated tests of a hypothesis can force a re-evaluation or an improvement in its analysis.

An example of an ongoing scientific dispute revolves around the conclusion that dogs have evolved a special sensitivity to the cues provided by humans because of selection associated with domestication. This conclusion arises in part from tests of the prediction that dogs, which have been domesticated for thousands of years, will be more alert to signals from humans, such as pointing to hidden food, than are wolves, which are not the products of domestication. Data collected

by Brian Hare and his associates and outlined here provide support for this prediction, thus confirming the domestication hypothesis in their minds.[32]

However, Monique Udell and her colleagues tested this prediction again, and they claimed that their results showed that wolves were actually more attentive to signals provided by human trainers than were dogs tested outdoors.[80] Moreover, shelter dogs, as opposed to pet dogs, failed to match the responses of wolves to pointing humans in the experiments reported by Udell.

You can decide for yourself which research team is more persuasive by reading the work of Hare et al.[32] and Udell et al.,[80] followed by a rebuttal of Udell's work by Hare et al.[33] Then you can see if Udell and her coauthors have backed down by reading a recent review by this group.[81] Might some part of the disagreement between the teams stem from mixing proximate and ultimate hypotheses?

(A)

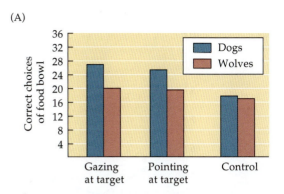

(B)

FIGURE 12.35   Dogs have evolved the ability to pay attention to humans. (A) Dogs do better than wolves reared in the same way when it comes to using human-provided cues, such as gazing at a food bowl or pointing at it, to go to the correct bowl with food. (B) This dog is staring intently at its master, the better to attend to the behavior of its human companion. Photograph by Evan MacLean.

location of food provided by humans.[32] Hare and some other researchers have found that indeed dogs, even as puppies, are far more attentive to the actions of humans than are wolves, even those that have been raised by people (see Box 12.1 on the controversy surrounding this conclusion). For example, when dogs are presented with two dishes, one covering a food item and the other not, they respond to a person pointing to or even gazing at the dish with food (Figure 12.35) whereas wolves are significantly less likely to do so.

The eagerness with which dogs attend to human handlers is reflected in the results of another experiment in which a researcher hid a toy a number of times at a particular location, after which the researcher repeated the experiment after having hidden the toy in another place. The dogs typically continued to look for the toy at the first hiding place, just as human infants do; socialized wolves were less likely to keep looking in the original spot.[79] These findings suggest that dogs are like 10-month-old babies in that they pay more attention to the instructions they receive from a human companion than to their own observations. According to some, dogs and humans are creatures with evolved brains that reflect the effect of a social environment associated with domestication.

Studies of this sort demonstrate that the kinds of problem-solving behaviors exhibited by animal species differ in ways related to the environments in which those species evolved. A final demonstration of this point comes from studies of *Polistes* paper wasps in which Elizabeth Tibbetts and her coworkers have shown that *P. fuscatus* does an excellent job of recognizing individual wasps by their facial color patterns.[66,67] These wasps demonstrate they can tell the difference between faces when they are put into a T-maze with one non-electrified arm and another where a wandering wasp will receive a mild shock. The two arms of the maze are associated with images of two different wasp faces. Members of *P. fuscatus* colonies quickly learn to avoid the image of a wasp face linked with an electric shock. In contrast, another member of the genus, *P. metricus*, performs essentially at random in tests of this sort (Figure 12.36).

FIGURE 12.36 Face recognition learning occurs in some species of paper wasps, but not others. (A) Females of *Polistes fuscatus* learn to associate a particular face of a female of their species with a safe refuge in a T-maze. (B) Females of *P. metricus* perform no better when required to discriminate between faces of their own species than they do with faces of *P. fuscatus*. (C) The faces of female *P. fuscatus* differ among individuals whereas (D) the faces of *P. metricus* are similar, indicating that selection has acted to facilitate face recognition in the one species, but not the other. After Sheehan and Tibbetts.[66,67]

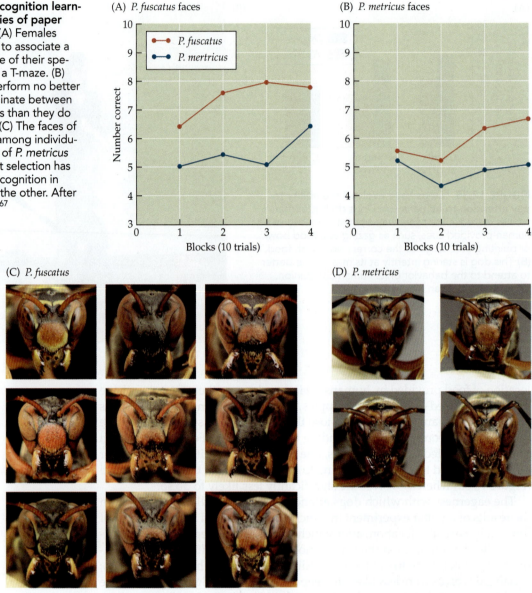

(A) *P. fuscatus* faces

(B) *P. metricus* faces

(C) *P. fuscatus*

(D) *P. metricus*

Why the difference in face learning between the two species? Females of *P. fuscatus* regularly join others in founding colonies where individuals compete with their companions for dominance and, thus, the right to reproduce. Knowing the identity of others in the dominance hierarchy helps them know when to compete for higher status and when to accept a subordinate position to particular rivals. In contrast, females of *P. metricus* are solitary creatures that found nests by themselves. They gain no advantage by recognizing other wasps as individuals, and indeed, the faces of females of this species are far more uniform in appearance than the faces of *P. fuscatus*.[67]

What paper wasps, dogs, and our own species appear to tell us is that cognitive skills evolve in response to particular selection pressures associated with particular environments, of which the social environment is especially significant.

Summary

1. The operating rules of neural mechanisms constitute proximate causes of behavior. Receptor cells acquire sensory information from the environment, interneurons relay and process that information, and other nerve cells in the central nervous system order appropriate motor responses to the events an animal can detect. Different species have different neural mechanisms and therefore perform these tasks differently, providing proximate reasons for why species differ in their behavior.

2. At the ultimate level, animals differ in their neurophysiology because their proximate mechanisms have been shaped by different selection pressures, as can be seen in the highly specialized sensory cells possessed by different species. For example, because of their auditory receptors, moths and other ultrasound-detecting species can hear sounds humans cannot sense, while certain bees, birds, fish, and lizards, among others, easily detect ultraviolet radiation that is invisible to us.

3. In addition to adaptively specialized sensory receptors, stimulus filtering (the selective processing of potential stimuli) is apparent in all the components of nervous systems. Sensory receptors ignore some stimuli in favor of others, while interneurons relay some, but not all, of the messages they receive from receptor cells. Within the central nervous system, many cells and circuits are devoted to the analysis of certain categories of information, although this means that other inputs are discarded. As a result, animals focus on the biologically relevant stimuli in their environments, increasing the odds of a prompt and effective reaction to the items that really matter.

4. The cognitive abilities of humans and other animals are also based on the evolved properties of nerve cells and nervous systems. Species that live with others in a competitive social environment possess special problem-solving skills associated with this environment, another demonstration of the power of natural selection in shaping the proximate neural mechanisms of behavior.

Suggested Reading

Hans Kruuk describes Niko Tinbergen's life and science dispassionately and well,[44] while Richard Burkhardt's *Patterns of Behavior* tells us how Tinbergen and Konrad Lorenz founded the field of ethology.[6] Kenneth Roeder's *Nerve Cells and Insect Behavior*[60] is a classic guide to research on the physiology and adaptive value of behavior. *Mapping the Mind* by Rita Carter covers the operation of the human brain.[7] *Evolving Brains* by John Allman focuses on the diversity of brains produced by descent with modification.[1]

A major finding of neurobiologists is that different species have evolved specialized sensory abilities, including the electric sense of certain fishes (see Hopkins[38] and Babineau et al.[2]), infrasound-detecting elephants,[53,71] and infrared perception in some snakes[29,51] and vampire bats.[30] See also the adaptive differences among vertebrates in their vomeronasal organs (see Michael Meredith's website, http://www.neuro.fsu.edu/faculty/meredith/vomer/). The different mechanisms underlying the feats of animal navigation are fascinating (see Wehner et al.[84]); these neural systems include those that provide songbirds with the capacity to detect the Earth's magnetic field.[36] Sea turtles are extremely good long-distance navigators; to learn about the mechanisms underlying this ability, go to Ken Lohmann's website at http://www.unc.edu/depts/oceanweb/turtles/.

13

How Neurons and Hormones Organize Behavior

 In this chapter, we shall continue to examine the relationship between the proximate mechanisms of behavior and their adaptive outcomes. A very widespread problem facing the members of animal species is how to select which of several options to perform when competing responses are possible. Natural selection appears to have favored individuals whose proximate systems not only activate useful responses to key stimuli but also organize an animal's repertoire of behaviors in ways that establish an adaptive set of priorities among these options. For example, the neurons in a noctuid moth help the insect sense and then respond to an approaching source of ultrasound (generated by a hunting bat). But the moth can also detect and track down a distant source of sex pheromone (produced by a female of its species). As a male noctuid moth flies toward that distant female, he may encounter a hunting bat. What should he do? Fortunately for the moth, his nervous system "knows" what to do. If the moth hears very loud ultrasonic pulses, he generally aborts his scent-tracking activity and dives for cover, an adaptive decision that enables the male to live to copulate on another day.[1,84]

Although noctuid moths and most other animals have the capacity to do many different things in response to many different stimuli, they rarely try to do two things at once—for good reason. One way in which nervous systems might be organized to avoid maladaptive conflicts involves **command centers** that communicate with one another. These centers need not be separate units in the brain but might instead consist of an assortment of cells capable of unified decision making, such as the innate releasing mechanisms, central pattern generators, song control systems, and the like that we met

The nervous system of the male praying mantis makes it possible for him to copulate even if his head has been removed and eaten by a hungry mate.

in preceding chapters. If each command center were primarily responsible for activating a particular response but could inhibit or be inhibited by signals from other centers, an individual might avoid incapacitating or dangerous indecision or conflict.

Neural Command and Control

Kenneth Roeder used the command center hypothesis to examine decision making in the praying mantis.[77] A mantis can do many things: search for mates, sunbathe, copulate, fly, dive away from bats, and so on. Most of the time, however, the typical mantis remains motionless on a leaf until an unsuspecting bug wanders within striking distance. When the mantis's visual receptors alert it to the presence of the prey, the mantis makes very rapid, accurate, and powerful grasping movements with its front pair of legs.

Roeder proposed that the mantis's nervous system sorts out its options thanks to inhibitory relationships between an assortment of command centers. The design of the mantis's nervous system (Figure 13.1) suggested that the control of muscles in each of the insect's segments is the responsibility of that segment's ganglion. Roeder tested this possibility by cutting one segmental ganglion's connections with the rest of the nervous system. Not surprisingly, the muscles within the neurally isolated segment subsequently failed to react when the mantis's nervous system became active elsewhere. However, when the isolated ganglion was stimulated electrically, the muscles and any limbs in that segment made vigorous, complete movements.

If the segmental ganglia are indeed responsible for telling the muscles in particular segments to carry out given movements, what is the mantis's brain doing? Roeder suspected that certain brain cells are responsible for inhibiting (blocking) neural activity in the segmental ganglia, keeping cells in a ganglion quiet until they are specifically ordered into action by an excitatory command center in the brain. If so, then cutting the connection between the inhibitory brain cells and the segmental ganglia should have the effect of removing this

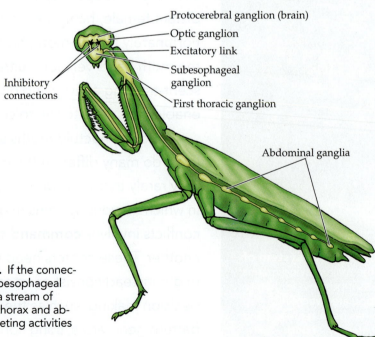

Protocerebral ganglion (brain)
Optic ganglion
Excitatory link
Subesophageal ganglion
First thoracic ganglion
Inhibitory connections
Abdominal ganglia

FIGURE 13.1 Nervous system of a praying mantis. If the connections between the protocerebral ganglion and the subesophageal ganglion are cut, the subesophageal ganglion sends a stream of excitatory messages to the segmental ganglia in the thorax and abdomen; the mantis then attempts to do several competing activities simultaneously.

inhibition and inducing inappropriate, conflicting responses. When Roeder severed the connections between the protocerebral ganglion (the mantis's brain) and the rest of its nervous system, he produced an insect that walked and grasped simultaneously, not an adaptive thing to do. The protocerebral ganglion apparently makes certain that an intact mantis either walks or grasps but does not do both at the same time.

When Roeder removed the entire head, however—a procedure that eliminates the subesophageal ganglion as well as the protocerebral ganglion—the mantis became immobile. Roeder could induce single, irrelevant movements by poking the creature sharply, but nothing more. These results suggest that the protocerebral ganglion of an intact mantis typically sends out a stream of inhibitory messages to the subesophageal ganglion, preventing these neurons from communicating with the other ganglia. When certain sensory signals reach the protocerebral ganglion, however, neurons there stop inhibiting certain modules in the subesophageal ganglion. Freed from suppression, these subesophageal neurons send excitatory messages to various segmental ganglia, where new signals are generated that order muscles to take specific actions. Depending on what sections of the subesophageal ganglion are no longer inhibited, the mantis walks forward, or strikes out with its forelegs, or flies, or does something else.

Interestingly, mature male mantises do not always obey the "rule" that complete removal of the head eliminates behavior. Instead, a headless adult male performs a series of rotary movements that swing its body sideways in a circle. While this is happening, the mantis's abdomen is twisting around and down, movements that are normally blocked by signals coming from the protocerebral ganglion. This odd response to decapitation begins to make sense when you consider that a male mantis sometimes literally loses his head over a female, when she grabs him and consumes him, head first. Even under these difficult circumstances, the male can still copulate with his cannibalistic partner (Figure 13.2), thanks to the nature of the control system regulating his mating behavior. Headless, his legs carry what is left of him in a circular path

FIGURE 13.2 A no-brainer. This male praying mantis, unlike the one shown in the chapter opener, lost his head to his predatory mate. Even so, the headless (brainless) male mounted on the back of the intact female was able to copulate with his lethal partner. Photograph by Mike Maxwell.

until his body touches the female's, at which point he climbs onto her back and twists his abdomen down to copulate competently.[77]

Discussion Questions

13.1   Figure 13.3 shows a record of the activity of a particular neuron in a cricket's brain that appears to participate in a command center controlling the male's chirping call, which is produced when the male rubs his wings together. About 3 seconds into the recording, the cricket was subjected to a sharp puff of air that struck its cerci, a pair of sensory appendages that project out from the rear of the abdomen. What is the relevance of this figure to the topic that we have just discussed? What adaptive value do you attach to the proximate mechanisms that help male crickets make behavioral decisions like the one shown here?

13.2   Inhibitory neural messages often play a key role in organizing the behavior of an animal, as the mantis demonstrates. Mature female crickets (*Gryllus bimaculatus*) approach chirping males. About one hour after mating, during which the male transfers a spermatophore to his partner, the female stops tracking the calls of males of her species. If you found that emptying the sperm storage organ caused the female to resume responding to calling males, you could speculate on how the female cricket's nervous system controlled this aspect of her behavior. How might inhibitory messages be involved? What is the adaptive significance of this proximate mechanism? See Loher et al.[58]

(A)

(B)

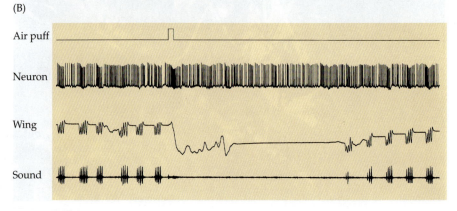

FIGURE 13.3   Record of neural and behavioral activity of a calling cricket. (A) Male crickets call by rubbing their forewings together. (B) The record of a command neuron in the brain of a calling male that was subjected to a brief tactile stimulus, a puff of air on the tip of the abdomen. The cricket's wing movements and the sounds of its calls are shown on separate lines. A, photograph by Edward S. Ross; B, from Hedwig.[38]

Daily Changes in Behavioral Priorities

The ability of neural command units to communicate with and inhibit one another helps set an animal's behavioral priorities. But the relationship between different components of the brain can change adaptively over the course of a day (or longer period). Female *Teleogryllus* crickets, for example, usually hide in burrows or under leaf litter during the day and move about only after dusk, when it is relatively safe to search for mates.[57] Not surprisingly, male crickets wait for the evening to start calling for mates.[56] These observations suggest that the inhibitory relationships between the calling center in a male cricket's brain and other neural elements responsible for other behaviors must change cyclically over a 24-hour period. A clock mechanism could be useful for the control of singing behavior, just as the monarch butterfly's biological clock enables it to adjust its sun compass as the hours pass during the daytime (Figure 13.4).[103]

Two competing theories have been developed to explain how animals can change their priorities over time. First, individuals might manage their priorities in response to a biological clock, a timing mechanism with a built-in schedule that acts independently of any cues from the animal's surroundings. The idea that at least partially environment-independent timing mechanisms might exist should be plausible to anyone who has flown across several time zones and then tried to adjust immediately to local conditions. The second theory suggests that animals alter the relationships between command centers in their nervous systems strictly on the basis of feedback information gathered by mechanisms that monitor the surrounding environment. Such

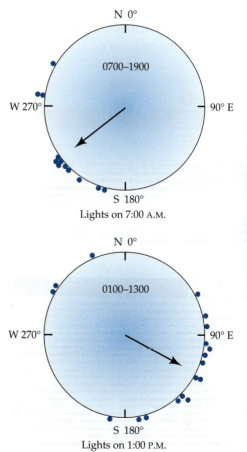

Lights on 7:00 A.M.

Lights on 1:00 P.M.

Monarch butterfly

FIGURE 13.4 The monarch butterfly possesses a biological clock. Experimental manipulation of the clock changes the orientation of migrating monarchs. Individuals were tested in an outdoor flight cage after one group had been held indoors under an artificial light–dark cycle with lights on at 7:00 A.M. and lights off at 7:00 P.M. This group tried to fly in a southwesterly direction. A second group of butterflies that had been held with lights on at 1:00 A.M. and lights off at 1:00 P.M. flew in a southeasterly direction, evidence of the importance of a clock and sun compass in helping monarchs stay on course. After Froy et al.[20]

devices would enable individuals to modulate their behavior in response to certain changes in the world around them, such as a decrease in light intensity as evening comes on.

Let's consider these two possibilities in the context of the calling cycle of male *Teleogryllus* crickets. Each day's calling bout could begin at much the same time because the crickets possess an internal timer that measures how long it has been since the last bout began; they could use this environment-independent system to activate the onset of a new round of chirping each evening at dusk. Alternatively, the insect's neural mechanisms might be designed to initiate calling when light intensity falls below a particular level. If this second hypothesis is correct, then crickets held under constant bright light should not call. But in fact, laboratory crickets held in rooms in which the temperature stays the same and the lights are on 24 hours a day still continue to call regularly for several hours each day. Under conditions of constant light, calling starts about 25 to 26 hours after it did the previous day (Figure 13.5). A cycle of activity that is not matched to environmental cues is called a **free-running cycle**. Because the length, or period, of the free-running cycle of cricket calling deviates from the many 24-hour environmental cycles caused by the Earth's daily rotation about its axis, we can conclude that the cyclical pattern of cricket calling is caused in part by an environment-independent internal **circadian rhythm** (circadian means "about a day").

Now let's place our crickets in a regime of 12 hours of light and 12 hours of darkness. The switch from light to darkness offers an external environmental cue that the crickets may use to adjust their timing mechanism. In fact, they do, just as monarch butterflies and pigeons can reset their clocks when moved into a laboratory with artificial lighting. After a few days, the males start calling about 2 hours before the lights go off, accurately anticipating "nightfall," and they continue until about 2.5 hours before the lights go on again in the "morning" (see Figure 13.5). This cycle of calling matches the natural one, which is synchronized with dusk; unlike the free-running cycle, it does not

FIGURE 13.5 Circadian rhythms in cricket calling behavior. (A) A Female *Teleogryllus* cricket (right) has tracked down a male (left) by his calls. (B) Calling occurs on a daily schedule. Each horizontal line on the grid represents one day; each vertical line represents a half hour on a 24-hour timescale. Dark marks indicate periods of activity—in this case, calling. The bars at the top and middle of the figure represent the lighting conditions; thus, for the first 12 days of this experiment, male crickets are kept in constant light (LL), and for the remainder, they are subjected to 12-hour cycles of light and dark (LD). Male crickets held under constant light exhibit a daily cycle of calling and noncalling, but the calling starts later each day. The onset of "nightfall" on day 13 acts as a cue that resets the calling rhythm, which soon stops shifting and eventually begins an hour or two before the lights are turned off each day. A, photograph by Leigh Simmons; B, after Loher.[56]

(A)

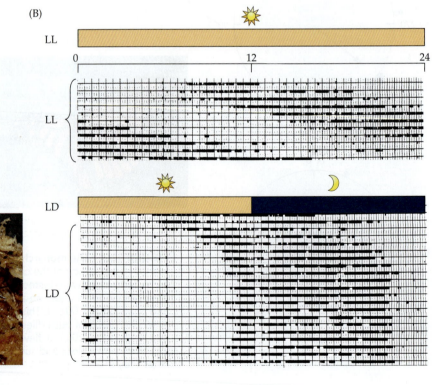

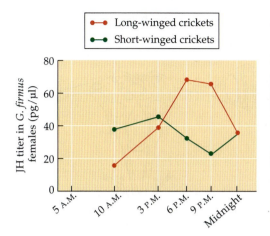

FIGURE 13.6 In the early part of the night, the long-winged, flight-capable form of the cricket *Gryllus firmus* (on the right in the photo) has higher concentrations of juvenile hormone (JH) in its blood than the short-winged form of the cricket (on the left). After Zera et al;[102] photograph by Derek Roff.

drift out of phase with the 24-hour day but is reset, or **entrained**, each day so that it begins at the same time in relation to lights-out.[56] From these results, we can conclude that the complete control system for cricket calling has both an environment-independent timer, or biological clock, set on a cycle that is not exactly 24 hours long, and an environment-activated entrainment device that synchronizes the clock with local light conditions.

What about a species that exhibits a polyphenism, such as the sand cricket *Gryllus firmus*, a species that comes in different forms: a long-winged, flight-capable, nocturnally active morph and a short-winged, flightless morph that is more likely to be active during the day. The long-winged form flies primarily at night, a safe time for an insect to leave its burrow and fly off to find mates; the short-winged form cannot fly and so stays in the safety of concealing vegetation on the ground.[78] To compare the circadian rhythms of the two types of sand crickets, Anthony Zera and colleagues brought recently collected crickets to the laboratory and then took blood samples at intervals over 24-hour periods. When they analyzed the juvenile hormone (JH) concentrations in the blood of short-winged crickets, they found no significant changes in relation to time of day. In contrast, however, the blood of flight-capable crickets revealed a very different story, with JH concentrations rising sharply in the late afternoon or evening from baseline levels comparable to those in the flightless crickets (Figure 13.6).[102] The different daily patterns of circulating juvenile hormone in the two kinds of crickets are correlated with the different circadian rhythms of the two forms. It seems likely that the nicely timed surge in JH in fully winged crickets in some way helps prepare these individuals for a round of nocturnal flight.

How Do Circadian Mechanisms Work?

Researchers interested in the proximate causes of behavior have worked out how circadian mechanisms work in some cases. For example, the daily schedule of behavioral activity in long-winged *Gryllus firmus* appears to be linked to the cyclical release of JH in these crickets. Other aspects of the circadian rhythm of these insects involve the optic lobes of their brain. If one cuts the nerves carrying sensory information from the eyes of a male cricket to the optic lobes (Figure 13.7), the insect enters a free-running cycle. Visual signals of some sort are evidently needed to entrain the daily rhythm to local conditions, but a rhythm persists in the absence of this information. If, however, one cuts the connections between both optic lobes and the rest of the brain, the calling cycle breaks down completely; all hours are now equally probable times for cricket calling. These results are consistent with the hypothesis that

FIGURE 13.7   The cricket nervous system. Visual information from the eyes is relayed to the optic lobes of the cricket's brain. If the optic lobes are surgically disconnected from the rest of the brain, the cricket loses its capacity to maintain a circadian rhythm. Based on diagrams by F. Huber and W. F. Shurmann.

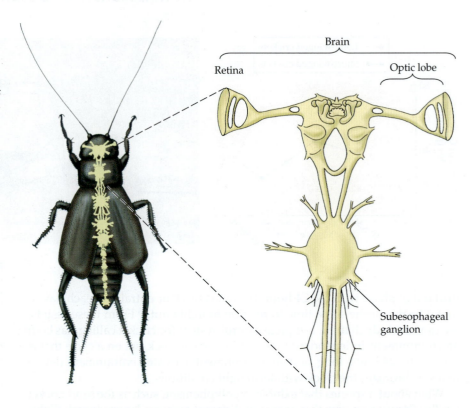

a master clock mechanism (Figure 13.8) resides within the optic lobes, sending messages to other regions of the nervous system[42,69] as well as receiving hormonal signals generated by the animal's endocrine system.

Biologists interested in the control of circadian rhythms in mammals and other vertebrates have focused on the hypothalamus of the brain, with special emphasis on the suprachiasmatic nucleus (SCN), a pair of hypothalamic neural clusters that receive inputs from nerves originating in the retina. The SCN is therefore a likely element of the mechanism that secures information about day and night length, information that could be used to adjust a master biological clock.

If the SCN contains a master clock or pacemaker that is critical for maintaining circadian rhythms, then neurons in the putative clock should change their activity in a regular fashion over the course of a 24-hour period. They do, with some groups of cells in the SCN that are electrically silent at night but which become active at dawn and continue to fire during the day.[8] Another prediction is that damage to the SCN should interfere with circadian rhythms.

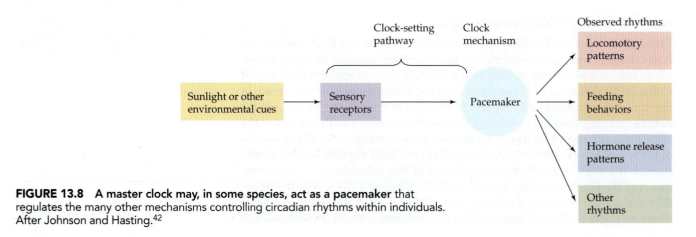

FIGURE 13.8   A master clock may, in some species, act as a pacemaker that regulates the many other mechanisms controlling circadian rhythms within individuals. After Johnson and Hasting.[42]

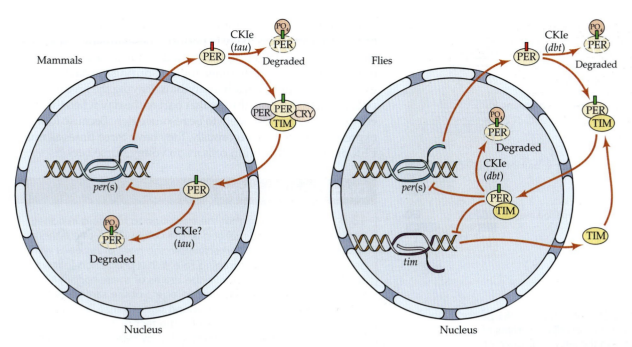

FIGURE 13.9   The genetics of biological clocks in mammals and fruit flies. In both groups, a set of three key genes produces proteins that interact with one another to regulate the activity of certain other genes on a cycle lasting approximately 24 hours. One of the genes (*per*) codes for a protein (PER) that gradually builds up inside and outside the cell nucleus over time. Another key gene, called *tau* in mammals and *dbt* in flies, codes for an enzyme, CKIe, that helps break down PER, slowing its rate of accumulation in the cell. But during peak periods of production of PER, more PER is available to bind with another protein (TIM), coded for by a third gene (*tim*). When the PER protein is bound in complexes with TIM (and another protein, CRY, in the case of mammals), it cannot be broken down as quickly by CKIe. Therefore, more intact PER is carried back into the nucleus, where it blocks the activity of the very gene that produces it, though only temporarily. Then a new cycle of *per* gene activity and PER protein production begins. After Young.[101]

In experiments in which SCN neurons have been destroyed in the brains of hamsters and white rats, the animals have subsequently exhibited arrhythmic patterns of hormone secretion, locomotion, and feeding.[104] If arrhythmic hamsters receive transplants of SCN tissue from fetal hamsters, they sometimes regain their circadian rhythms, but not if the tissue transplants come from other parts of the fetal hamster brain.[14] Moreover, if an arrhythmic hamster gets an SCN transplant from a mutant hamster with a circadian period that is much shorter than the standard one of approximately 24 hours, the experimental subject adopts the circadian rhythm of the donor hamster, further evidence in support of the hypothesis that the SCN controls this aspect of hamster behavior.[74]

Perhaps the SCN clock of hamsters, rats, and other mammals operates via rhythmic changes in gene activity. A key candidate gene in this regard appears to be the *per* gene, which codes for a protein (PER) whose production varies over a 24-hour schedule in concert with that of the product of another mammalian gene, called *tau* (Figure 13.9). The product of *tau* is an enzyme whose production is turned on when PER is at peak abundance in the cell. The enzyme degrades PER, contributing to a 24-hour cycle in which PER first increases in abundance and then falls.[101]

A striking feature of this system, whose complexity is daunting, is that the key clock genes regulating cellular circadian rhythms in mammals are also present in insects, a legacy of a very ancient shared ancestor. *Drosophila* fruit flies and honey bees also have the *per* gene, a chain of DNA composed of somewhat over 3500 base pairs, which provides the information needed to produce the PER protein chain of nearly 1200 amino acids. Alterations in the base sequence involving as little as a single base substitution can result in dramatically different circadian rhythms in fruit flies (Figure 13.10), as well as in humans (carriers of one *per* mutation typically fall asleep around 7:30 in the evening and arise at about 4:30 in the morning[89]). These results strongly suggest that the gene's information plays a critical role in enabling circadian rhythms.

If expression of *per* does have this effect, then animals in which the *per* gene is relatively inactive should behave in an arrhythmic fashion. Very little PER protein is manufactured in young honey bees, which generally remain

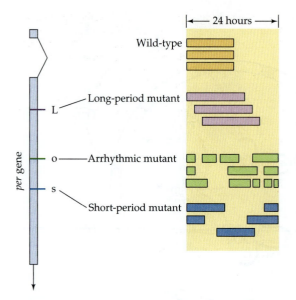

FIGURE 13.10 **Mutations of the** *per* **gene affect the circadian rhythms of fruit flies.** On the left is a diagram of the DNA sequence that constitutes the *per* gene. The locations of the base substitutions present in three mutant alleles found in fruit flies are indicated on the diagram. The activity patterns of wild-type flies, and those associated with each mutation, are shown on the right. After Baylies et al.[4]

within the hive to care for eggs and larvae. And young honey bees are in fact just as likely to perform these nursing tasks at any time during the day or night over a 24-hour period. In contrast, older honey bees, which forage for food during the daytime only, exhibit well-defined circadian rhythms, leaving the hive to collect pollen and nectar only during that part of the day when the flowers they seek are most likely to be resource rich. Because foragers express their *per* gene vigorously, they have three times as much PER protein in their brain cells as do young nurse bees.[91]

Discussion Question

13.3   You may recall that the transition to foraging in honey bees depends on the makeup of the colony, such that if there is a shortage of nurse workers within the hive, older bees will delay their shift to the foraging role. What prediction follows about *per* gene expression in the brains of these socially delayed older nurses relative to foragers of the same age from other colonies with numerous young nurse bees? Provide proximate and ultimate hypotheses for the fact that social interactions can alter circadian rhythms in honey bees—and even fruit flies,[52] which do not live in highly organized societies.

Because the fruit fly, the hamster, the honey bee, and you and I all have the same gene serving much the same clock function, evolutionary biologists believe that we inherited this gene, as well as some others involved in the regulation of activity patterns, from a creature that lived perhaps 550 million years ago.[101] In mammals and some other vertebrates, the *per* gene and certain others are expressed in the neurons of the SCN, which is usually viewed as the home of a master clock or circadian pacemaker that regulates many other tissues, thereby keeping many different behaviors on a daily schedule. However, other parts of the brain may also have their own biological clocks. For example, the olfactory bulb in mice exhibits environment-independent cyclical changes in genetic activity, with the result that mice are more sensitive to odors at night than during the day, an adaptive effect of a timing mechanism for a nocturnally active creature.[25]

Nonetheless, the SCN clearly contains a major biological clock, which must send chemical signals to the target systems it controls. If so, we can make three predictions: (1) the molecule that relays the clock's information ought to be secreted by the SCN in a manner regulated by the clock's genes, (2) there should be receptor proteins for that chemical messenger in cells of the target tissues, and (3) experimental administration of the chemical messenger should disrupt the normal timing of an animal's behavior.

The classic candidate chemical of this sort is melatonin, which has been the subject of much research demonstrating its involvement in the regulation of animal circadian rhythms.[93] But melatonin is not the only such chemical. More recently, researchers have demonstrated that a protein, prokineticin 2 (PK2), has the key properties expected of a clock messenger.[7] So, in normal mice living under a cyclical regime of 12 hours of light and 12 hours of darkness, the PK2 protein is produced in a strongly circadian pattern by the SCN (Figure 13.11). Moreover, mice with certain mutations in key clock genes lack the circadian rhythm of PK2 production. Furthermore, as also predicted, only certain structures within the brain of the mouse produce a receptor protein that binds with PK2. These regions are linked to the SCN by neural pathways and are believed to contain command centers that control various behavioral activities

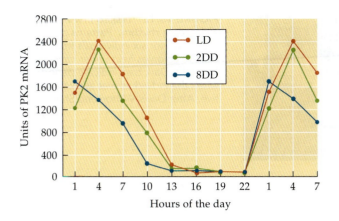

FIGURE 13.11 Expression of the gene that codes for PK2 in the SCN. Mice were exposed to a standard light–dark cycle of 12 hours of light and 12 hours of darkness (LD) prior to the start of the experiment. The graph shows hourly changes in the production of the messenger RNA encoded by the *PK2* gene. The gene was expressed in a circadian rhythm whether the animals were held under the standard light–dark cycle, kept in complete darkness for 2 days (2DD), or held in darkness for 8 days (8DD). After Cheng et al.[7]

in a circadian fashion. Finally, if one injects PK2 into the brains of white rats during the night, when the animals are normally active, the behavior of these animals changes dramatically. Instead of running in their running wheel, they sleep, shifting to daytime activity instead (Figure 13.12).[7]

All of these lines of evidence point to PK2 as the chemical messenger that the mammalian SCN uses to communicate with and regulate target centers in the brain. The SCN, in turn, receives information from the retina about the light–dark cycle of the animal's environment so that it can fine-tune its autoregulated pattern of gene expression. The adaptive value of the environment-independent component of the clock system may be that a clock of this sort enables individuals to alter the timing of their behavioral and physiological cycles without having to constantly check the environment to see what time it is. The presence of an environment-dependent element, however, permits individuals to adjust their cycles in keeping with local conditions. As a result, a typical nocturnal mammal automatically becomes active at about the right time each night, while retaining the capacity to shift its activity period gradually to accommodate the changes in day length that occur as spring becomes summer, or summer becomes fall.

One way to test this ultimate hypothesis about the adaptive value of circadian rhythms would be to predict that if there were animals for which the day–night cycle was biologically irrelevant, then these species should lack a circadian pattern of activity. Nurse honey bees spend their days in the darkness of the hive where they feed and care for their larval sisters 24/7; their brains exhibit no circadian rhythmicity in the expression of key genes associated with the honey bee's circadian clock. Foragers, on the other hand, go outside the hive to collect pollen and nectar during the day; their brains exhibit regular oscillations in the activity of many genes believed to be linked to the biological clock system of the bee.[76]

The naked mole rat is a similar species in that these animals live underground in total darkness and almost never come to the surface. As predicted, naked mole rats lack a circadian rhythm. Instead, individuals engage in generally brief episodes of activity within longer periods of inactivity, with the pattern changing irregularly from day to day (Figure 13.13).[12]

Seasonal and Annual Cycles of Behavior

Because of their unusual lifestyle, naked mole rats do not have to deal with cyclically changing environments, and they have apparently lost their circadian rhythm as a result. But almost all other creatures confront not only daily changes in food availability or risk of predation but also changes that cover periods longer than 24 hours, especially the seasonal changes that occur

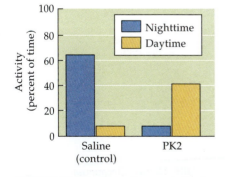

FIGURE 13.12 Circadian control of wheel running by white rats changes when the brains of rats are injected with PK2. These rats are active primarily during the daylight hours, whereas control rats injected with saline exhibit the standard preference for nighttime activity. After Cheng et al.[7]

Time

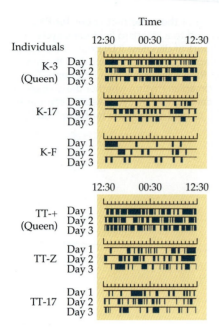

FIGURE 13.13   Naked mole rats lack a circadian rhythm. Patterns of activity are shown for six individuals from two captive colonies held under constant low light. Dark bars indicate periods when the individual was awake and active. After Davis-Walton and Sherman.[12]

Naked mole rat

in many parts of the world. If circadian rhythms enable animals to prepare physiologically and behaviorally for certain predictable daily changes in the environment, might not some animals possess a **circannual rhythm** that runs on an approximately 365-day cycle?[28] A circannual clock mechanism could be similar to the circadian master clock, with an environment-independent timer capable of generating a circannual rhythm in conjunction with a mechanism that keeps the clock entrained to local conditions.

Testing the proximate hypothesis that an animal has a circannual rhythm is technically difficult because individuals must be maintained under constant conditions for at least 2 years after being removed from their natural environments. One successful study of this sort involved the golden-mantled ground squirrel[70] of North America, which in nature spends the late fall and frigid winter hibernating in an underground chamber. Five members of this species were born in captivity, then blinded and held thereafter in constant darkness and at a constant temperature while supplied with an abundance of food. Year after year, these individuals entered hibernation at about the same time as other golden-mantled ground squirrels living in the wild (Figure 13.14).

In another study, several nestling stonechats were taken from Kenya to Germany to be reared in laboratory chambers in which the temperature and **photoperiod** (the number of hours of light in a 24-hour period) were always the same. Needless to say, these birds, and their offspring, never had a chance to encounter the spring rainy season in Kenya, which heralds a period of insect abundance and is the time when Kenyan stonechats must reproduce if they are to find sufficient food for their nestlings. Thus, wild stonechats

FIGURE 13.14   Circannual rhythm of the golden-mantled ground squirrel. Animals held in constant darkness and at a constant temperature nevertheless entered hibernation (green bars) at certain times year after year. After Pengelley and Asmundson.[70]

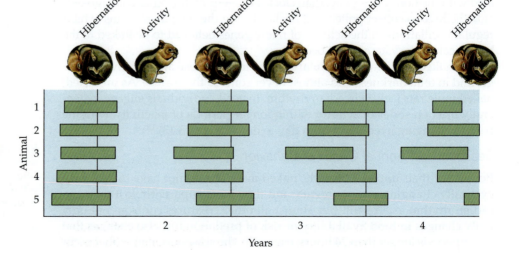

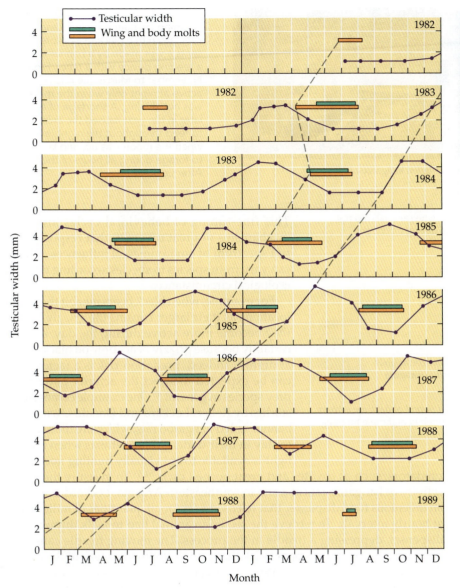

Stonechat

FIGURE 13.15 Circannual rhythm in a stonechat. When transferred from Kenya to Germany and held under constant conditions, this male stonechat still underwent a regular long-term cycle of testicular growth and decline (purple lines), as well as regular feather molts (the two bars, which refer to wing and body molts). The cycle was not exactly 12 months long, however, so the timing of molting and testicular growth gradually shifted over the years (see the dashed lines that angle downward from right to left). After Gwinner and Dittami.[27]

exhibit an annual cycle of reproductive physiology and behavior. The transplanted stonechats, despite their constant environment, also exhibited an annual reproductive cycle, but one that shifted out of phase with that of their Kenyan compatriots over time (Figure 13.15). One male, for example, went through nine cycles of testicular growth and decline during the 7.5 years of the experiment. Evidently, the stonechat's circannual rhythm is generated in part by an internal, environment-independent mechanism, just as is true for the golden-mantled ground squirrel.[27] Moreover, different populations of stonechats living in different places have evolved different circannual mechanisms that predispose the birds to molt and reproduce at the appropriate season for their geographic location.[39]

CUES THAT ENTRAIN SEASONAL AND ANNUAL CYCLES OF BEHAVIOR In nature, environmental cues reset circadian and circannual clocks and thus produce behavioral rhythms that match the particular features of the animal's environment, such as the times of sunrise and sunset, or the onset of the rainy season in a given year, or the increasing day lengths associated with spring.

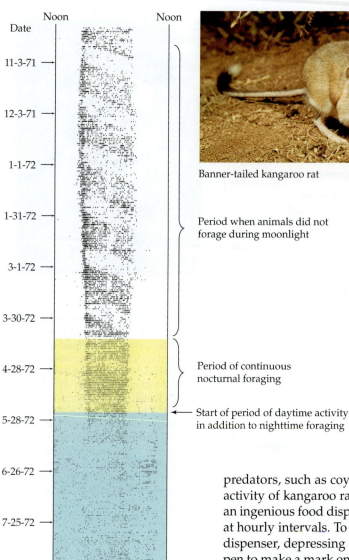

Noon   Noon

Date

11-3-71

12-3-71

1-1-72

1-31-72

3-1-72

3-30-72

4-28-72

5-28-72

6-26-72

7-25-72

8-24-72

Banner-tailed kangaroo rat

Period when animals did not
forage during moonlight

Period of continuous
nocturnal foraging

Start of period of daytime activity
in addition to nighttime foraging

FIGURE 13.16   The lunar cycle and foraging by banner-tailed kangaroo rats. Each thin black mark represents a visit made by a banner-tail to a feeding device with a timer. From November to March, the rats were active at night only when the moon was not shining. A shortage of seeds later in the year caused the animals to feed throughout the night, even when the moon was up, and later still, to forage during all hours of the day. After Lockard.[55]

This fine-tuning of behavioral cycles involves mechanisms of great diversity that respond to a full spectrum of environmental influences, which can vary from species to species according to their ecological circumstances.

For example, when black bears inhabit the same locations as their larger cousins the grizzly bears, they become more active during the day, almost certainly because grizzly bears prey on black bears at night.[82] Indeed, predation risk is a factor that can be expected to affect the activity pattern of prey species in general.[48] One example comes from studies by Robert Lockard and Donald Owings on banner-tailed kangaroo rats, which are more likely to stay in their underground retreats when moonlight is available to aid visually hunting nocturnal predators, such as coyotes and great horned owls.[54,55] To measure the daily activity of kangaroo rats at a site in southeastern Arizona, Lockard invented an ingenious food dispenser that released very small quantities of millet seed at hourly intervals. To retrieve the seeds, an animal had to walk through the dispenser, depressing a treadle in the process. The moving treadle caused a pen to make a mark on a paper disk that turned slowly throughout the night, driven by a clock mechanism. When the paper disk was collected in the morning, it carried a temporal record of all nocturnal visits to the dispenser.

Lockard's data showed that in the fall when the kangaroo rats had accumulated a large cache of seeds, they were selective about the timing of foraging, usually coming out of their underground burrows only at night when the moon was not shining (Figure 13.16). Because the predators of kangaroo rats (coyotes and owls) can see their prey more easily in moonlight, banner-tails are probably safer when they forage in complete darkness. For this reason, the kangaroo rats apparently possess a mechanism that enables them to shift their foraging schedule in keeping with nightly moonlight conditions.

Of course, if more kangaroo rats are out and about when it is completely dark or nearly so, perhaps their predators also respond to the lunar cycle similarly. If so, the resulting overlap in their activity periods in relation to nighttime darkness would be an example of an arms race between predator and prey. Although it is not known how coyotes deal with moonlight, Craig Packer and his colleagues have established that African lions secure more food on dark nights near the new moon, when it is difficult for their prey to detect them (Figure 13.17). On the other hand, when the moon is full, lions are more likely to hunt or scavenge during the day, a pattern indicative of a shift in foraging behavior linked to the lunar cycle. Because lions prefer to

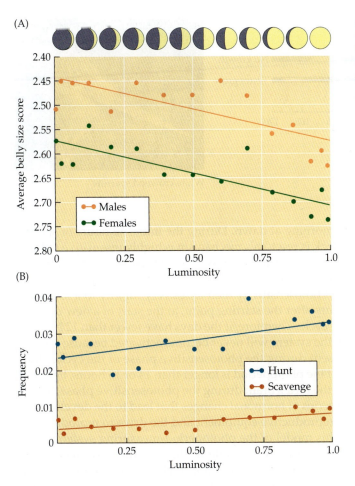

(A)

(B)

FIGURE 13.17 **The lunar cycle affects lion hunting success.** (A) As nighttime luminosity (brightness) increases, the bellies of males and females decrease in fullness, indicating a greater difficulty in making kills under brighter conditions. (B) As the moon becomes full, lions are more likely to have to hunt during the day and to scavenge from others. After Packer et al.[68]

hunt when it is dark, people as well as antelopes and the like are more likely to fall prey to lions at this time.[68] Given that *Homo sapiens* evolved in areas with lions and other large nocturnal predators, perhaps we can understand why we tend to be inactive at night, a decision fueled by the fear of darkness, a common feature of human behavior.

Whereas desert kangaroo rats (and lions) forage at night in accordance with the lunar cycle and Kenyan stonechats employ a circannual rhythm to breed at the right time in their tropical environment, temperate-zone birds such as white-crowned sparrows have their own behavioral control system that is well suited for coping with the dramatic seasonal changes that occur where they live. In the spring, males fly from their wintering grounds in Mexico and the southwestern United States to their distant summer headquarters in the northern United States, Canada, or Alaska. There they establish breeding territories, fight with rivals, and court sexually receptive females. In concert with these striking behavioral changes, the gonads of the birds grow with dramatic rapidity, regaining all the weight lost during the fall, when they shrink to 1 percent of their breeding-season weight. In order to restore their gonads in time for the onset of reproduction, the birds must somehow anticipate the spring breeding season. How do they manage this feat?

The sparrows' ability to change their physiology and behavior depends on their capacity to detect changes in the photoperiod, which grows longer as spring advances in temperate North America.[18] One proximate hypothesis on how such a system might work proposes that the clock mechanism of white-crowns exhibits a daily change in sensitivity to light, with a cycle that is reset

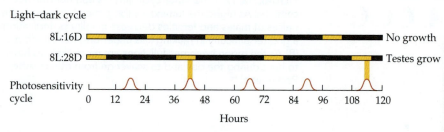

White-crowned sparrow

FIGURE 13.18 A cycle of photosensitivity. An experiment with white-crowned sparrows tested the hypothesis that these birds possess a clock mechanism that is especially sensitive to light between hours 17 and 19 of each day. The lower line represents these hypothetical periods of photosensitivity. The yellow and black sections of the two upper horizontal bars show the light and dark periods of two different light–dark regimes. Only sparrows under the 8L:28D experimental regime were exposed to light during the supposed photosensitive phase of the cycle, and only they responded with testicular growth. After Farner.[17]

FIGURE 13.19 A hormonal response to light. Different groups of white-crowned sparrows were kept in complete darkness for different lengths of time before being exposed to a single 8-hour period of light. The upper diagram illustrates when the 8-hour light exposures occurred. The lower graph shows the change from the start of the experiment in luteinizing hormone (LH) concentrations in the blood in each group of birds. LH is a hormone known to stimulate the growth of the gonads. After Follett et al.[19]

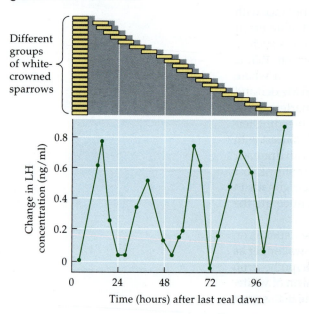

each morning at dawn. During the initial 12 hours or so after the clock is reset, this mechanism is highly insensitive to light, an insensitivity that gives way to increasing sensitivity, which reaches a peak 16 to 20 hours after the starting point in the cycle. Photosensitivity then fades very rapidly to a low point 24 hours after the starting point, at the start of a new day and a new cycle. Therefore, if the days have fewer than 12 hours of light, the system will never become activated because no light is present during the photosensitive phase of the cycle. However, if the photoperiod is longer than 14 or 15 hours, light will reach the bird's brain during the photosensitive phase, initiating a series of hormonal changes that lead to the development of its reproductive equipment and the drive to reproduce.

If this model of the photoperiod-measuring system is correct, it should be possible to deceive the system. William Hamner, working with house finches,[35] and Donald Farner, in similar studies with white-crowned sparrows,[17] stimulated testicular growth by exposing captive birds to light during the hypothesized photosensitive phase of their circadian rhythms. In Farner's experiment, birds that had been on a regular schedule of 8 hours of light and 16 hours of darkness (8L:16D) were shifted to an 8L:28D schedule. Because the light periods were now out of phase with a 24-hour cycle, these birds sometimes experienced light during the time when their brains were predicted to be highly photosensitive. The male birds' testes grew under these conditions, even though there was a lower ratio of light to dark hours than under the 8L:16D cycle, which did not stimulate testicular growth (Figure 13.18).[17]

Discussion Question

13.4 In another experiment with white-crowned sparrows, groups of males that had been held on an 8L:16D schedule were housed in complete darkness for variable lengths of time (anywhere from 2 to 100 hours) before being exposed to an 8-hour period of light. A few hours later, the researchers measured the concentrations in the birds' blood of luteinizing hormone (a hormone released by the anterior pituitary and carried to the testes, where it stimulates the growth of these tissues). Do the collected data shown in Figure 13.19 provide a test for the photosensitivity hypothesis just described?

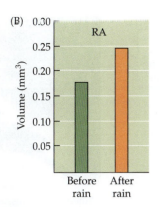

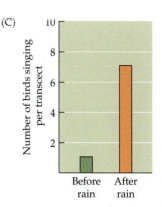

(A)

(B)

(C)

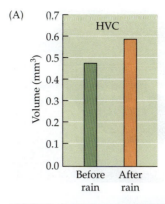

Rufous-winged sparrow

FIGURE 13.20 **Seasonal changes in the song control regions of the rufous-winged sparrow's brain.** Summer rainfall triggers these changes, which lead to an increase in singing behavior. The sizes of (A) the HVC (formerly called the higher vocal center) and (B) the robust nucleus of the arcopallium (RA) in the sparrow increase following monsoon thunderstorms in southern Arizona. (C) The neural changes are linked to an increase in singing after the monsoon has begun. Green bars represent data collected before the first thunderstorms. Orange bars represent data collected after the start of the monsoon. Photograph of a rufous-winged sparrow by Pierre Deviche; A–C after Strand et al.[87]

Photoperiod changes are useful guides for breeding activity in birds that live in seasonal environments with predictable changes in food resources. But what about the rufous-winged sparrow, a bird that lives in the drought-prone Sonoran Desert, where the arrival of spring does not guarantee abundant supplies of the insects needed to rear a batch of offspring? Instead, food for this sparrow's brood may not become available until summer rains have fallen, in which case they wait to reproduce until after the onset of the monsoon, which can begin anywhere from early July to mid August, when the photoperiod is already declining.[85] Summer rainfall after a dry spring appears to stimulate the production of luteinizing hormone, which in turn may cause the testes to produce testosterone, priming the male to establish a territory, start singing, and if all goes well, reproduce (Figure 13.20).[87]

Rufous-winged sparrows have evolved proximate mechanisms that respond to summer rainfall because in their desert environment, rainfall is a more reliable indicator of future food for offspring than photoperiod changes are. For two seed-eating finch species, the white-winged and red crossbills, it is food intake itself that acts as the primary cue for breeding.[5] Craig Benkman has found that in years when conifer seed production is so high that the birds can find ample food for themselves and a brood in almost any month, crossbills take advantage of the good times to breed. This pattern can be linked to a distinctive feature of crossbill hormonal physiology, which is that birds that have been on long photoperiod days (20L:4D) do not completely shut down their gonadotropin-releasing neurons, as do other, related birds such as redpolls and pine siskins.[71] Instead, they appear to retain the capacity to stimulate the release of reproduction-regulating hormones should conditions for reproduction become especially favorable.

This flexibility, however, does not mean that crossbills ignore all environmental cues except seed abundance.[32] In studies of crossbills in the wild, Thomas Hahn noticed a break in crossbill breeding in December and January (Figure 13.21), even in environments where food was plentiful. Therefore,

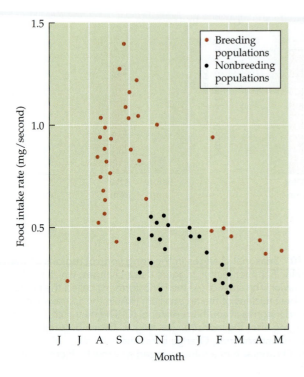

White-winged crossbill

FIGURE 13.21 **Food intake and reproductive timing in the white-winged crossbill.** Breeding populations usually occur in areas with relatively high food availability. Nonbreeding populations generally occur in areas where the birds have low food intake. Note, however, the absence of breeding populations in all locations in December and January. After Benkman.[5]

although these birds are more flexible and opportunistic than the average songbird, Hahn wondered if crossbills still have an underlying reproductive cycle dependent on photoperiod. When he held red crossbills at a constant temperature with unlimited access to food while letting the birds experience natural photoperiod changes, he found that male testis length fluctuated in a cyclical fashion (Figure 13.22), becoming smaller during October through December of each year, even when the birds had all the seeds a crossbill could hope to eat. In addition, free-living red crossbills also always undergo declines in gonad size and in the concentrations of sex hormones in their blood at this time of year.[31] Therefore, the reproductive opportunism of these birds is not absolute, but rather superimposed on the photoperiod-driven timing mechanism inherited from a common ancestor of temperate-zone songbirds, a point that applies to some other opportunistically breeding birds as well.[33]

The partial persistence of the standard timing system in the flexibly breeding crossbills could be explained at the evolutionary level in at least two ways: (1) the photoperiod-driven mechanism might be a nonadaptive holdover from a distant ancestor with the standard pattern, or (2) crossbills might derive

FIGURE 13.22 **Photoperiod affects testis size in the red crossbill.** Six captive birds were held under natural photoperiods, which changed over the seasons, but temperature and food supply were held constant. The data are the average testis length among these birds at different times of the year. After Hahn.[29]

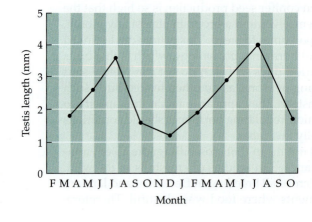

Red crossbill

reproductive benefits from the retention of a physiological system that reduces the likelihood that they will attempt to reproduce at times when other costly, competing activities take priority (such as the need to molt and replace their feathers in the fall).[15] On the other hand, by maintaining their gonadal equipment in at least partial readiness for much of the year, crossbills and other birds that encounter unpredictable variation in their environments are able to react more quickly to a chance to reproduce.[90]

Social Conditions and Changing Priorities

As we have just seen, different features of the physical environment, such as moonlight, day length, rainfall, or food supply, are used by different species to set their priorities and thereby forage instead of hiding, or reproduce instead of waiting some more. In addition, animals can use changes in their social environments to make adaptive adjustments in their physiology and behavior. Thus, for example, when Hahn and several coworkers performed an experiment on crossbills in which some captive males were caged with their mates while others were forced into bachelorhood but were kept within sight and sound of the paired crossbills in a neighboring aviary, the bachelor males experienced a slower return to reproductive condition after the winter break than did the paired males.[30] Even more dramatically, when free-living female song sparrows were given estrogen implants, which greatly prolonged their sexual receptivity, their hormonally untreated mates also remained in reproductive mode for months longer than males paired with control females at this location. The sexual and territorial responses of male song sparrows in this experiment demonstrate the sensitivity of animals to their social environment.[79]

Much the same thing occurs in house mice. When female mice are given a chance to show a preference for a potential mate, they spend time sniffing the male of their choice when placed in the center cubicle with males in two compartments at either end of the cage. Females that have had prior experience only with a subordinate male do not exhibit a preference when given a choice between sniffing a dominant and a subordinate male. But those that have had prior experience with the odor of a dominant male spend much more time near the dominant individual (Figure 13.23). Exposure to the scent of the dominant male promotes the addition of neurons in two regions of the female brain, effectively rewiring her neural machinery such that she can identify a dominant male mouse should her social environment provide her with such a partner.[60]

Dominant male house mice are also programmed to permit social experience to alter their behavioral decisions. For example, after a male house mouse mounts a female and ejaculates, he immediately becomes highly aggressive toward mouse pups, killing any he finds. For almost 3 weeks after mating, he is likely to commit infanticide, but after that time, he becomes more and more likely to protect any young pups he encounters. When about 7 weeks have passed since ejaculation, he becomes infanticidal once again.[72]

FIGURE 13.23   Dominant male odors change female mate preferences in the house mouse. (A) The three-compartment cage in which a female (center chamber) can approach and smell males in either of two end compartments. (B) Females that have experienced only subordinate male pheromones do not exhibit a preference for either subordinate or dominant males; females that have been exposed to dominant male mouse odors spend much more time sniffing the more dominant of two males in the experimental cage. A, photograph courtesy of Gloria Mak; B, after Mak et al.[60]

(A)

(B)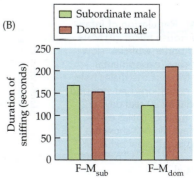

This remarkable cycle has clear adaptive value. After a male transfers sperm to a partner, 3 weeks pass before she gives birth. Attacks on pups during these 3 weeks will invariably be directed against a rival male's offspring, with all the benefits attendant on their elimination (see page 10). After 3 weeks, a male that switches to paternal behavior will almost always care for his own neonatal offspring. After 7 weeks, his weaned pups will have dispersed, so once again he can practice infanticide advantageously.

At the proximate level, what kind of mechanism could enable a male to switch from infanticidal Mr. Hyde to paternal Dr. Jekyll 3 weeks after a mating? One possible explanation involves an internal timing device that records the number of days since the male last copulated. If such a sexually activated timing mechanism exists, then an experimental manipulation that either increases or decreases the length of a "day," as perceived by the mouse, ought to have an effect on the absolute amount of time that passes before the male makes the transition from killer to caregiver.

Glenn Perrigo and his coworkers manipulated day length by placing groups of mice under two different laboratory conditions, one with "fast days," in which 11 hours of light were followed by 11 hours of darkness (11L:11D) to make a 22-hour "day," and another with "slow days" (13.5L:13.5D) that lasted for 27 hours. As predicted, the total number of light–dark cycles, not the number of 24-hour periods, controlled the infanticidal tendencies of males (Figure 13.24). Thus, when male mice in the fast-day group were exposed to mouse pups 20 real days after mating, only a small minority committed infanticide, because these males had experienced 22 light–dark cycles during this period. In contrast, more than 50 percent of the males in the slow-day group attacked newborn pups at 20 real days after mating, because these males had experienced only 18 light–dark cycles during this time. These results demonstrate that a timing device registers the number of light–dark cycles that have occurred since mating and that this information provides the proximate basis for the control of the infanticidal response.[72]

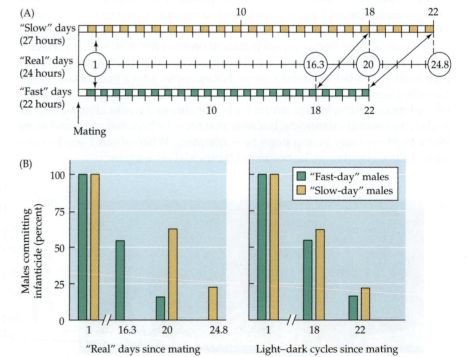

FIGURE 13.24 Regulation of infanticide by male house mice. (A) Male mice were held under artificial "slow-day" and "fast-day" experimental conditions. (B) Most of the males held under fast-day conditions had stopped being infanticidal by 20 real days (22 fast days) after mating; males experiencing slow days did not show the same decline in infanticidal behavior until nearly 25 real days had passed. After Perrigo et al.[72]

Hormonal Modulation of Behavior

The proximate mechanism that controls the male mouse's treatment of mouse pups probably influences the hormonal state of the male, which in turn alters the animal's behavioral tendencies. If true, then during their infanticidal phase, male mice might have high concentrations of testosterone in their blood, given the well-established relationship between testosterone and male aggression. An alternative hypothesis is that extreme male aggression toward mouse pups occurs when the males are under the influence of progesterone, because this hormone is known to suppress parental behavior in female rodents.

To test these alternatives, Jon Levine and his coworkers used genetic knockout techniques to create a population of laboratory mice that lacked progesterone receptors, which meant that males of this line could not detect the progesterone in their bodies. When the knockout males were exposed to a pup, they never attacked it, whereas over half of the males from a genetically unmodified strain of laboratory mice assaulted a test infant (Figure 13.25). (This strain is an unusually aggressive one in which untreated males may kill their own pups.) There were no significant differences in testosterone or progesterone levels between the two groups of mice, only a difference in the ability of their brain cells to detect progesterone This work suggests that when progesterone is present in certain concentrations within an intact male, the mouse is primed to be an infanticidal killer.[81] As progesterone levels slowly fall (or sensitivity to the hormone declines in brain cells) after mating, the paternal capacity of the male may slowly increase in time, which has the adaptive effect of keeping him from harming his own brood.

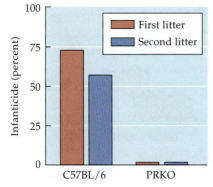

FIGURE 13.25 A hormonal effect on infanticidal behavior in laboratory mice. Males of the strain C57BL/6 are highly likely to attack their own first or second litters, rather than caring for their young. If, however, the progesterone receptor gene is removed from the genome, the resulting progesterone knockout (PRKO) mice cannot detect the progesterone in their bodies, and they do not exhibit infanticidal behavior toward their offspring. After Schneider et al.[81]

Discussion Question

13.5 In the California mouse, *Peromyscus californicus*, males are highly paternal. Explanations for this behavior include the reduced progesterone hypothesis that we have just explored for laboratory mice. Alternatively, it is possible that decreased levels of testosterone are responsible for male parental behavior. Make some predictions derived from these two proximate hypotheses. Then examine Figure 13.26 and evaluate your hypotheses in light of these data.

Hormones have many regulatory effects that enable individuals to modify their response to key social stimuli in adaptive ways. A dramatic example is provided by male Japanese quail, which behave differently after mating with a female. Prior to copulating, a male housed next to a female in a two-compartment cage will spend relatively little time peering through a window between the compartments in order to look at her. But after mating, a male appears positively fascinated by his partner, to the extent that he will stare

FIGURE 13.26 Does testosterone or progesterone modulate male parental behavior in California mice? (A) The testosterone differences between inexperienced males (with neither mates nor offspring) and fathers (with a mate and offspring) are not statistically significant. (B) In contrast, progesterone occurs in far higher amounts in the blood of inexperienced males than in males with offspring. A and B, after Trainor et al.;[92] photograph by Brian Trainor.

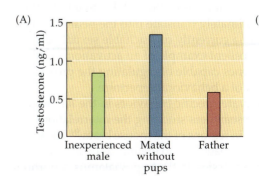

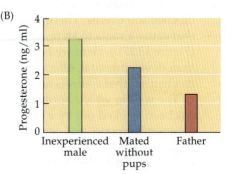

California mouse

(A)

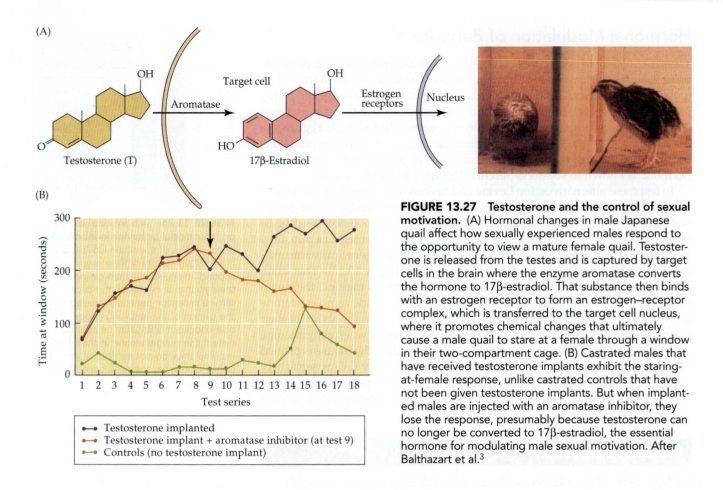

FIGURE 13.27 Testosterone and the control of sexual motivation. (A) Hormonal changes in male Japanese quail affect how sexually experienced males respond to the opportunity to view a mature female quail. Testosterone is released from the testes and is captured by target cells in the brain where the enzyme aromatase converts the hormone to 17β-estradiol. That substance then binds with an estrogen receptor to form an estrogen–receptor complex, which is transferred to the target cell nucleus, where it promotes chemical changes that ultimately cause a male quail to stare at a female through a window in their two-compartment cage. (B) Castrated males that have received testosterone implants exhibit the staring-at-female response, unlike castrated controls that have not been given testosterone implants. But when implanted males are injected with an aromatase inhibitor, they lose the response, presumably because testosterone can no longer be converted to 17β-estradiol, the essential hormone for modulating male sexual motivation. After Balthazart et al.[3]

at her for hours on end (Figure 13.27A).[3] This behavior, which presumably encourages unconfined males to get together with a receptive female for yet another copulation, is heavily influenced by the presence of testosterone in particular regions of the male's brain. This conclusion is based in part on the finding that removal of the male's testes eliminates all sexual behavior—unless the male receives an implant containing testosterone, in which case he regains his motivation to remain close to a sexual partner.

Interestingly, testosterone itself is not the signal that turns a male Japanese quail into an apparently lovelorn individual when separated from his mate. Instead, testosterone is converted into 17β-estradiol, an estrogen, in target cells within the preoptic area of the brain. The conversion requires an enzyme, aromatase, which is coded for by a gene that becomes much more active after a male quail has mated. The estrogen produced in the presence of aromatase then binds with an estrogen receptor protein, and the resulting estrogen–receptor complex relays a signal to the nucleus of the cell, leading to further biochemical events. These events ultimately translate into the neural signals that cause a male to stare intently at a currently unreachable sexual partner (Figure 13.27B). Here we have an excellent example of an activational, rather than developmental, effect of hormones capable of regulating the proximate mechanism that controls an element of a behavioral repertoire.

The way in which hormones activate specific behaviors in the lab mouse has been worked out by a team of researchers studying the interrelationship between hormones and genes in certain regions of the mouse brain.[99] The team began by using microarray analysis (see page 325) to identify a set of genes that differed between the sexes in their expression in the hypothalamus, a portion of the brain that plays a major role in controlling sexual behavior in adult mice. Having identified a number of genes that had the potential to influence the reproductive behavior of mice, they then compared the behavior of individuals with and without the typical (wild-type) alleles of these genes. Males without wild-type alleles of the gene *Sytl4* were much slower to copulate with a receptive female than were nonmutant males, although the sexual behavior of the mutants was otherwise largely normal. In contrast, females with the mutant *Sytl4* genotype behaved normally, but those without the wild-type allele of a different gene (*Irs4*) were much slower in retrieving displaced pups and much less likely to attack intruders than wild-type mothers (Figure 13.28). Results of this sort indicate that male and female sex hormones affect the expression of certain genes in particular brain cells in ways that affect the performance of specific male and female behaviors. A hormonally induced change in the activity of even a single gene can apparently alter the way in which neural circuits in the brain operate, changing the response of males to potential mates or how quickly mother mice react to pups that have wandered away.

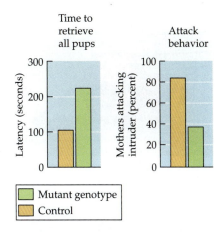

FIGURE 13.28 One gene's effect in activating two maternal behaviors in mice. Females with a mutant genotype with respect to the gene *Irs4* take longer to retrieve a group of pups moved away from them. They also are less likely to attack an intruder that approaches their pups. After Xu et al.[99]

Discussion Question

13.6 In addition to vasopressin, the hormone oxytocin contributes to the development of attachment behavior in monogamous prairie voles.[34] This hormone plays a role in rewarding a male for mating and staying with a partner. Oxytocin also affects the social behavior of humans; the hormone appears to make us feel more warmly toward others, more trusting and generous than we might be otherwise.[88] Some researchers have suggested that application of oxytocin via nasal sprays would provide therapeutic benefits by boosting our willingness to cooperate with others. With these points in mind, how did Carsten De Dreu and his coworkers use evolutionary theory to interpret the results shown in Figure 13.29?[13]

Hormones and Reproductive Behavior

Unlike Japanese quail and prairie voles, mice and many other animals do not form long-lasting sexual attachments. Indeed, soon after a fruit fly female mates, she becomes completely unreceptive, rejecting not only her previous partner but all other males, should they attempt to court her. The female fly then spends her time laying eggs, an adaptive shift in behavior given that she can fertilize all her mature eggs with stored sperm received from a single mating. As it turns out, the female's dramatic switch from sexual receptivity to sexual refusal is regulated by a hormone—not one that she manufactures herself, but one that she receives from the seminal fluid the male transfers to her during copulation.

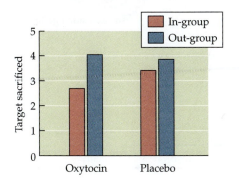

FIGURE 13.29 Oxytocin and human behavior. In these experiments done in the Netherlands, some Dutch citizens self-administered a nasal spray with oxytocin while others applied a nasal spray that did not contain the hormone. The Dutch subjects were then given a moral choice dilemma task that required that they choose whether they would sacrifice a target individual by, for example, switching a trolley so that it would be diverted onto a track where that one person was standing versus continuing on a track where five persons were in mortal danger. The target individual was given either a Dutch name (as a member of the in-group) or a Muslim or German name (indicating that the target was a member of an out-group, a non-Dutch group). There were no significant differences in the choices made by individuals who had the placebo spray, but those who had given themselves a dose of oxytocin were significantly more likely to sacrifice an individual with a Muslim or German name as opposed to one with a Dutch name. After De Dreu et al.[13]

(A)

(B)

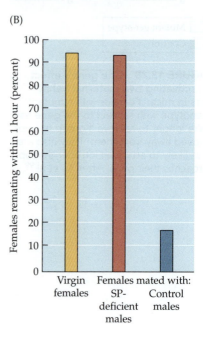

FIGURE 13.30   What controls female receptivity in a fruit fly? A sex peptide (SP) transferred to females in the male seminal donation appears to be a critical factor. Female fruit flies mated to males unable to supply sex peptide are as likely to copulate with a new male 48 hours after mating with an SP-deficient male as are virgin females to mate on their first sexual encounter. A, photograph of mating *Drosophila* by Brian Valentine, www.flickr.com/photos/lordv; B, after Chapman et al.[6]

The male-donated hormone is called sex peptide (SP). Using a technique called RNA interference, it is possible to block the specific gene in male fruit flies that codes for the hormone SP. The blocked males are normal except for the fact that they cannot make SP and, therefore, when they copulate with females, they are unable to donate SP to their mates. If SP is indeed the critical signal that stops females from responding to courting males, then females mated with SP-deficient males should remain receptive following copulation, despite having received sperm and seminal fluid. This prediction is correct (Figure 13.30).[6]

As noted earlier, a chemical messenger acts by binding with receptor molecules on the surface of a target cell. The fruit fly hormone SP should therefore bind with a specific receptor protein, and accordingly, the appropriate molecule, unimaginatively called sex peptide receptor (SPR), has been found. Here too it is possible to block the single gene that codes for SPR, creating "mutant" females whose cells are incapable of binding with the male-donated hormone. When this is done experimentally, SPR-deficient females that have mated once copulate again when given a chance, just as if they were still virgins.[100]

Hormonal signals that regulate animal sexual behavior often lead to increases in both gamete production and sexual activity (Figure 13.31). This **associated reproductive pattern** is, for example, evident in the seasonal reproductive behavior of British red deer. Stags that have been living peacefully with one another all summer become aggressive as September approaches prior to the mating season. At this time, their testes generate sperm and testosterone. Adult males that have been castrated prior to the fall rutting season show little aggression and do not try to mate with sexually receptive females. If the behavioral differences between castrated and intact males stem from an absence of circulating testosterone in the castrated stags, then testosterone implants should restore their aggressive and sexual behavior during the rut, and they do.[53]

Likewise, in the green anole when males first become active after a winter dormant phase, their circulating concentrations of the hormone testosterone are very low. As this hormone begins to be produced in greater amounts, however, the males' testes grow in size, and mature sperm are made. At this time

FIGURE 13.31   Gonadal hormones and two different reproductive patterns. In species with an associated reproductive pattern, environmental cues trigger internal hormonal changes, which then activate behavioral responses relatively quickly (as illustrated by the green anole). In a dissociated reproductive pattern, mating may be dissociated in time from hormonal activity (as in the red-sided garter snake). After Crews.[10]

some males begin to defend territories and court females.[41] These individuals tend to be larger and more powerful biters, to have much more testosterone in their blood than smaller male anoles, and to have larger dewlaps with which to court females and threaten rivals.[40]

These effects of testosterone are in part dependent upon the season of the year, as Jennifer Neal and Juli Wade showed by dividing captive male green anoles with implanted testosterone capsules into two groups: one that was exposed to conditions (warm temperatures and long day lengths) that mimicked the normal summer breeding season environment, the other kept under cooler conditions and shorter day lengths. The males in the "breeding season" (BS) group were much more likely to display to females and to copulate with them than were males in the "nonbreeding season" (NBS) group. The brains of BS males were also different from those of NBS males in that neurons in their amygdalae were larger, suggesting that here, too, specific regions of the brain are under hormonal regulation. Thus the associated reproductive pattern of the green anole is dependent on the season of the year.[66]

Discussion Question

13.7   In humans, the menstrual cycle involves hormonally mediated physiological changes that lead to the production of a mature egg. Historically, physiologists have not looked for an associated pattern of sexual activity in women.[22] But evolutionary biologists have recently proposed that there should be a relationship between the menstrual cycle and sexual desire. Why would they predict a difference in this relationship for married women versus those without a steady partner? Why might ovulating women find males with masculinized facial features especially attractive? (See Figure 14.5 for an example of variation in male facial features.) Check your responses against Gangestad et al.,[21] Macrae et al.,[59] and Pillsworth et al.[73]

TESTOSTERONE AND REPRODUCTIVE BEHAVIOR   Many animal species, whether closely related or not, have much the same set of hormones, a result of the conservative nature of evolution in which ancestral molecules are often modified only slightly during evolutionary time as they take on very different roles in different species. Testosterone is a prime example. Although white-crowned sparrows, green anoles, and red deer all produce similar forms of the hormone, the sparrows are *not* testosterone-dependent when it comes to mating. Even without his testes, a male white-crown will mount females that solicit copulations, provided that he has been exposed to long photoperiods.[64] In addition, when male white-crowns mate with females to produce a second or third clutch of fertilized eggs in the summer, they do so without much testosterone in their blood (Figure 13.32), further evidence that high levels of circulating testosterone are not essential for male sexual behavior in this species.

Discussion Question

13.8   To study the hormonal control of behavior, researchers often remove an animal's ovaries or testes and then inject the creature with assorted hormones to see what behavioral effects they have. What advantage does this technique have over another approach, which is simply to measure the concentrations of specific hormones in the blood of animal subjects from time to time? The far less invasive direct measurement approach would show, for example, whether testosterone or estrogen concentrations were elevated when mating was occurring.

FIGURE 13.32 Hormonal and behavioral cycles in single-brood and multiple-brood populations of white-crowned sparrows. Testosterone concentrations in the blood of male white-crowns peak shortly before the time when the males mate with females (M) in their first breeding cycle of the season. In populations that breed twice in one season, however, copulation also occurs during the second breeding cycle, at a time when testosterone concentrations are declining. After Wingfield and Moore.[95]

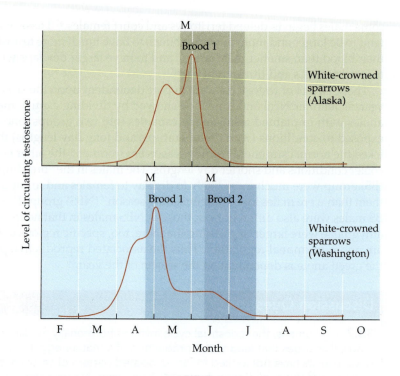

FIGURE 13.33 Testosterone and female aggression in the dunnock. Testosterone concentrations were higher in females competing for males in polygynous groups than in females in monogamous relationships. After Langmore et al.[49]

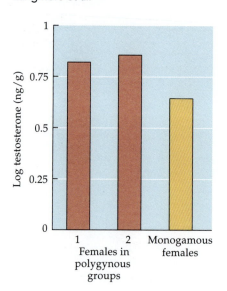

The male white-crowned sparrow is not the only bird in which courtship and mating do not require high levels of circulating testosterone.[37] Perhaps in species of this sort testosterone facilitates aggression rather than sexual behavior. If this hypothesis is correct, then we can predict that in seasonally territorial birds, testosterone concentrations should be especially high early in the breeding season, when males are aggressively defending a territory against rivals—as is the case for white-crowned sparrows (see Figure 13.32) and some other birds.[95] Even territorial spotted antbirds, which live in the tropics and do not exhibit elevated testosterone concentrations in any particular seasonal pattern, respond to recorded songs of their species with a rapid buildup of circulating testosterone.[36]

The hypothesis that testosterone is linked to aggression leads to another prediction, which is that when competition among females is a regular feature of a bird's life history, then aggressive females, which occur in the dunnock, should possess relatively high testosterone levels. A number of female dunnocks often live in the same general area where they share the sexual favors of one or more resident males. Because a female's reproductive success depends on how much assistance she can secure from her male partner(s), which depends on how often they mate with her, females try to keep other females away from "their" mates. When the testosterone levels of aggressive females in multi-female groups were compared with those of unchallenged females that were each paired with a single male, the results indicated that testosterone does make females more aggressive (Figure 13.33),[49] although it is possible that fighting causes testosterone levels to rise, rather than the other way around.

The Costs of Hormonal Regulation

As you know, adaptationists think about the benefits *and* the costs of the traits that interest them, and testosterone, for example, has many disadvantages to consider. So although male gray-headed flying foxes with higher levels of testosterone are more likely to reclaim their complete harems of females after being experimentally removed for a few days,[45] the hormone almost certainly exacts a price from the males in question. For example, testosterone may interfere with the immune response,[26,105] which may explain why males of so many vertebrates are more likely than females to be infected by viruses, bacteria, and parasites.[44,96] Thus, in a population of chimpanzees, the higher-ranking males with higher testosterone levels supported a greater diversity of helminth parasites, a group that includes the familiar tapeworms, as revealed by the undoubtedly unpleasant inspection of chimp feces.[65]

In addition, the direct behavioral effects of testosterone can be costly, leading individuals under the influence of the hormone to expend energy at a much higher rate than otherwise (see Figure 6.3).[61] For example, when male barn swallows had their chest feathers painted a darker red, they became a target of other aggressive males. As a result, their testosterone concentrations went up—and their body weight went down over time, surely because the hormone induced the birds to become more physically active.[80] Testosterone-saturated males can also become so focused on mating with females or fighting with rivals that they become sitting ducks for predators, which may be why in some species males with higher testosterone concentrations are less likely to survive than those with more modest amounts of the hormone in their blood (Figure 13.34).[83] Even if a testosterone-soaked male overcomes these handicaps, he may spend more time fighting with other males than caring for his offspring. Thus, although male dark-eyed juncos that have received testosterone implants do not experience higher mortality than control birds, the implanted males do feed their young less often,[43] to the detriment of these dependent individuals.[75] Perhaps the trade-off between aggression and parental care is one reason why in humans, testosterone levels decline significantly after a man becomes a father.[23]

The paternal costs of testosterone can occur even when the increases in the hormone are only brief and transitory. A team of behavioral ecologists led by Ellen Ketterson injected gonadotropin-releasing hormone (GnRH) into captured male juncos. This treatment induced a brief and variable increase in testosterone concentrations in the birds, which were then released back to their nesting territories. Those males whose testosterone concentrations were relatively high after the GnRH challenge behaved aggressively toward a simulated intruder (a caged male junco placed in the center of the resident's territory). But the males that increased their testosterone concentrations the most, in response to the GnRH challenge, fed their nestlings at the lowest rate.[62] Even a short period of elevated testosterone concentrations apparently has the potential to make a male a less helpful parent.

FIGURE 13.34 A survival cost of testosterone. (A) The three different forms of the common side-blotched lizard have different throat colors and different amounts of testosterone circulating through their blood on average. (B) The form with the highest testosterone concentrations also has a very high annual mortality rate.
A and B, after Sinervo et al.;[83] photograph by the author.

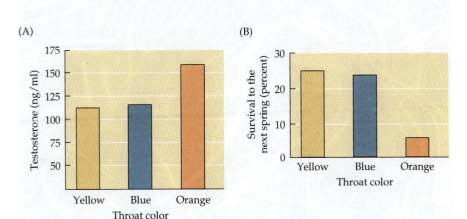

Discussion Question

13.9   Some bird species live short but highly reproductive lives, mostly in north temperate zones; other species, generally tropical ones, live longer but invest less each year in egg production and the rearing of offspring. What prediction can you make about the levels of circulating testosterone in the two categories of birds? Relate the results of testosterone measurements for temperate-zone and tropical birds (in Hau et al.[37]) to the work of Ghalambor and Martin[24] that we discussed in Chapter 9 (see page 258).

The various disadvantages of testosterone help explain why after becoming aggressive (and often sexually motivated) early in the breeding season, males of many bird species reduce their circulating levels of testosterone dramatically over time.[97] In this way, the males may reduce any of several disadvantages of prolonged exposure to the hormone, such as a reduction in their parental behavior.

Although male red-sided garter snakes do not care for their young in any way, they too provide evidence that testosterone is an expensive tool to use in regulating reproduction and aggression.[86] This snake lives as far north as southern Canada, where it spends much of the year dormant, sheltered underground. On warm days in the late spring, the snakes begin to stir, and they soon emerge, sometimes by the thousands, from their hibernacula (Figure 13.35). Before a group of snakes leaves their retreat, they engage in an orgy of sexual activity, with males slithering after females and attempting to copulate with them to the exclusion of all else. Later in the season, when the odds of finding receptive females are very low, male snakes are more likely to forage for food, a nice example of how proximate physiological mechanisms enable animals to resolve competition among their behavioral options.[67]

During the mating frenzy, males compete for females by trying to contact receptive partners before other males do, but they do not fight with one another for the privilege of copulation. Examination of the sex hormone concentrations in their blood reveals that these nonaggressive snakes have almost no circulating testosterone. Yet they have no trouble mating, so red-sided garter snakes are animals with a **dissociated reproductive pattern** (see

FIGURE 13.35 Spring mating aggregation of red-sided garter snakes. The males search for and avidly court females emerging from hibernation. Photograph by Nic Bishop.

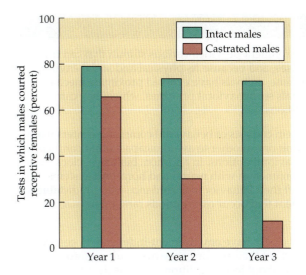

FIGURE 13.36 Testosterone and the long-term maintenance of mating behavior. Male garter snakes whose testes were removed shortly before the breeding season in year 1 remained sexually active during that breeding season, despite the absence of testosterone. But in years 2 and 3, these males became less and less likely to court receptive females, compared with males that still possessed their testes. After Crews.[11]

Figure 13.31).[9] However, there is some evidence that an enzyme responsible for converting testosterone into estrogen is present at relatively high levels in certain parts of the snake's brain in those males that are actively seeking females in the spring,[47] which suggests that there are elements of an associated reproductive pattern at work in the red-sided garter snake.

Moreover, testosterone production, which normally begins in the male snake's second or third year, appears to be necessary for the full development and maintenance of the neural mechanisms that control the snake's sexual behavior (Figure 13.36).[11] The fact that the enzyme that converts testosterone into estrogen also becomes more common in particular regions of the snake's brain in the fall is consistent with the hypothesis that seasonal changes in estrogen (derived from testosterone) prepare the key neural circuits to promote sexual activity in the spring. So once again we have an example of a proximate mechanism, in this case an endocrine–neuronal system, that has surely been retained in part from a distant ancestor but is used in a distinctive manner to achieve adaptive goals for a modern species.

Discussion Question

13.10   The endangered Amargosa River pupfish lives in Death Valley, where different populations of the species live in total isolation from one another in tiny permanent pools and short stream segments.[50] Although these populations have been separated from one another for only 400 to 4000 years, males in some places aggressively defend territories and court females drawn to them, while males in other populations are not aggressive toward one another and do not defend territories. How could these changes occur so rapidly? How might you establish experimentally that the hormone arginine vasotocin (AVT) decreases aggressive behavior in pupfish? If the hormone does lower aggression in this species, what predictions could you make about AVT or about AVT receptor protein differences between territorial and nonterritorial males of the same species?

Summary

1. Because an animal's environment provides various stimuli that can trigger contradictory responses, and because its physical and social environment often change over time, animals have evolved proximate mechanisms that help individuals set behavioral priorities, which they can adjust to changing conditions.

2. One such proximate system consists of behavioral command centers that have the capacity to inhibit one another to avoid incapacitating conflicts between an animal's behavioral options. In addition, biological clock mechanisms, acting in conjunction with neural and hormonal systems, enable individuals to shift their behavior in accordance with predictable changes in the environment that occur over periods ranging from a day to a year.

3. Circadian and circannual clocks have environment-independent components, but they can also adjust their performance by acquiring information from the environment about local conditions, such as the time of sunrise or sunset. This information can entrain (reset) the clock so that changing behavioral priorities will not get out of synchrony with the physical environment.

4. The endocrine system also contributes to the operation of neurophysiological mechanisms that respond to changes in the physical environment (such as seasonal changes in photoperiod) as well as to changes in the social environment (such as the presence of potential mates). Hormones produced by endocrine organs often set in motion a cascading series of physiological changes that make reproductive activity the top priority at times when the production of surviving offspring is most likely to occur.

5. The fact that so many species employ essentially the same hormonal tool kit is a reflection of the conservative consequences of evolution by descent with modification (from ancestral species).

6. Hormonal control, like all proximate mechanisms, comes with costs as well as potential benefits. The many damaging effects of testosterone provide a case in point. The cost–benefit approach helps explain why different animal species use the same hormone, such as testosterone, for very different functions.

Suggested Reading

Kenneth Roeder's *Nerve Cells and Insect Behavior* discusses how animals avoid conflicting responses while structuring their behavior over the short haul.[77] Circadian rhythms are reviewed by Isaac Edery.[16] The hormonal modulation of behavior is the subject of a book by Elizabeth Adkins-Regan.[2]

The dissociated reproductive pattern of the red-sided garter snake has attracted a great deal of attention, which has resulted in many interesting papers on some factors influencing male and female behavior, including those by Krohmer,[46] Lemaster and Mason,[51] and Mendonca et al.[63] The snake's proximate mechanisms for organizing sexual behavior can be contrasted with those of the green anole,[94] about which there has been some debate, as is so often true in science (e.g., see Jenssen et al.[41] and then Winkler and Wade[98] on the importance of testosterone for male behavior in the green anole).

14 The Evolution of Human Behavior

 Humans are an animal species. Yes, we are unusual and quite wonderful in our own way, but so too are fruit flies, kittiwakes, and rhinoceroses. Having applied the four-levels-of-analysis approach to other animals throughout this textbook, we will do the same for *Homo sapiens* in the final chapter. After all, our behavior owes its existence to our genetic–developmental systems, which make it possible for each of us to grow from a single cell into a creature with a brain, spinal cord, and endocrine system. These neural–hormonal mechanisms in turn make behaving possible for us. What we do affects how many copies of our genes we transmit to the next generation. There have surely been hereditary differences among people living in preceding generations that influenced their inclusive fitness, which determined the frequency of competing alleles in succeeding generations. Thus, natural selection has shaped our evolutionary history, as it has for all other organisms. Moreover, our species is derived from other, still older species in an evolutionary lineage that ultimately goes back to a primordial single-celled protobacterium of some sort. Understanding even a part of this still longer history will tell us when changes were layered on older traits and what these changes were. We are a species with an evolutionary history, and so we have proximate mechanisms inherited from ancestral species but modified in ways that help us behave in an adaptive fashion.

Of course controversies abound with respect to any sort of research on humans, but here I intend simply to illustrate how biologists and psychologists have gone about studying the behavior of our species from a four-levels-of-analysis perspective. We begin with what some researchers think we now know about the ability of people to speak to one another.

Although the behavior of the young African men in the Gerewol ceremony probably would seem very strange to most readers of this textbook, evolutionary analyses might persuade you that the cultural differences between us and them conceal some very important ultimate similarities.

Language and the Four Levels of Analysis

Although organisms of all sorts, even bacteria, communicate (see Chapter 4), our species is the only one that uses exceedingly complex languages to do the job. When we speak, whether in Urdu or English, we can produce a huge array of meaningful sentences and be understood, because both signaler and receiver have acquired unconsciously at an early age the operating rules, the grammar, of the language of their culture. This ability is revealed in many ways, such as in the errors made by young English-speaking children who say things like, "I speaked to my daddy on the phone." This mistake shows that the young person has learned the rule of adding *ed* to a regular verb to create the past tense. The rule doesn't work for irregular verbs, something that the child will eventually learn with or without formal instruction from others.

So one evolutionary question about our stunning language ability is, where did this skill come from? Answering it is a challenging task given that some features of language are often considered unique to our species. But were there antecedents of modern speech present in now extinct species in the lineage that gave rise to modern *Homo sapiens*? One way in which some scientists have explored this possibility is to see if modern chimpanzees, our closest living relatives, have retained some elements of these protolinguistic abilities to this day. The search for these abilities has led some researchers to try to rear chimpanzees with humans, the better to teach language to chimps, which after all are very intelligent and very social creatures. In one such early experiment, Winthrop and Luella Kellogg raised a female chimpanzee named Gua along with their own young son. The study helped demonstrate that the vocal tracts of chimpanzees are physiologically unsuited to producing speech; Gua never produced a recognizable spoken word.[95]

Trying to get chimpanzees to use language without speaking has been more successful. The young chimpanzees that have taken part in nonverbal language experiments have been able to learn to associate a host of visual symbols with objects and to respond to spoken commands and requests by, for example, touching certain symbols on a computer screen or keyboard. Other chimpanzees have been taught elements of sign language (Figure 14.1),[151] a task that takes advantage of the fact that chimps use a modest

FIGURE 14.1 Chimpanzee trained to use "language." A young chimpanzee being taught sign language by its caretaker.

number of communicative gestures in their natural environments.[83] Some observers, but not all, believe that these experiments reveal that our close relatives, chimpanzees and bonobos (also known as pygmy chimpanzees), have evolved abilities that enable them to employ rudimentary elements of human speech or language.[180] But there are many problems and uncertainties associated with this claim.[164]

If, however, we agree that captive chimpanzees are able to acquire some rudiments of language, then it seems likely that similar abilities were present in an ancestral species in a lineage that eventually produced both today's chimpanzees and today's human beings. Nonetheless, it is instructive that chimps have to be trained and trained and trained some more in order to achieve the capacity to communicate (after a fashion) via computer screens or with American Sign Language. In contrast, babies pick up the language spoken (or signed) around them on their own without dedicated instructors trying consciously to impart these language skills to their charges. Moreover, chimps lack the vocal apparatus that makes speech possible for people. In addition, there is considerable debate on whether chimpanzees can acquire any grammatical rules on how to string words together to make even simple short sentences.[142,151] There is no doubt that there are big differences between our close relatives and us on the language front, which is not too surprising given that chimpanzees and humans have been separated evolutionarily for roughly 6 million years or so,[194] during which time humans have clearly evolved under very different selection pressures than our anthropoid ape relatives.

Discussion Question

14.1   Another way to explore the origins of language involves a comparison of the number of phonemes (the simplest speech sounds that are used together to make up word sounds) in modern languages. As it turns out, the languages with the greatest number of phonemes are African; languages spoken by peoples whose dispersal routes took them farthest from Africa have the fewest phonemes. This result parallels the finding that the greatest amount of genetic diversity is found in African populations, with the least in peoples whose ancestors moved the farthest from Africa. The genetic result is attributed to the fact that as humans moved farther and farther from Africa in stages, each new colonizing group was small and so did not possess all the genetic alleles found in the population from which it came.[15] If we apply this kind of argument to language diversity, where did the first language-using members of *Homo sapiens* live? Since language evolution is culturally controlled,[126] does it make sense to apply techniques developed for genetic evolution to this case?

The History of Human Speech

Although attempts to detect language learning abilities in nonhuman animals have yielded cloudy results, there is another line of evidence that hints at a connection between our species and ancestral ones that may reveal something about the history of human language. I speak of the discovery of a gene, *FOXP2*, that has been linked to language ability because persons in one British family with a mutant allele of this gene have severe speech deficits, including difficulty in speaking words and probably also in using words grammatically.[85] The fact that *FOXP2* codes for a transcription factor means that when the gene is expressed, the resulting protein controls the operation of additional genes, whose protein products affect the development of neural circuitry in the brain that is necessary for adept speech.

There are almost certainly other regulatory genes involved in the development of "speech and language centers" in the human brain, but *FOXP2* clearly has something to do with our capacity for language.[160] What is striking about this gene is its occurrence in chimpanzees and gorillas, as well as in frogs, fish, rodents, and especially certain songbirds that learn their species' vocalizations. There is no question therefore that *FOXP2* originated in the distant past in an ancestor of many modern species. This gene may have been independently modified in ways that enable birds like the white-crowned sparrow to learn their communication signals by listening to others of their species vocalize. Humans and their close relatives have retained the gene in nearly the same form from a more recent mammalian ancestor; in humans, the modifications of *FOXP2* presumably affect how the human gene's transcription regulates other genes involved in the development of the neural substrate for the acquisition of learned speech.[160] Thus, as is so often the case, modern species carry genes that appeared in the distant past and that have taken on new adaptive functions as mutations occurred that affected what these genes do,[40] as is the case for *FOXP2*.

Remember, however, that language is not genetically determined. The parts of the brain that are essential for language acquisition, production, and comprehension develop as a result of an interaction between *FOXP2* and many other genes expressed in brain cells and the chemical and experiential environment in and around those cells. For example, the cultural environment that a baby experiences obviously has a huge effect on the interactive events that take place as the physiological foundation for language use is built. A child born and reared by Muslim parents in Pakistan will probably speak Urdu; a child born and reared in Mexico City will probably speak Spanish. But this does not mean that either child's language is environmentally determined, only that the *differences* between them are caused by an environmental difference that affects the gene–environment interactions that underlie their speech.

Discussion Questions

14.2   In 1981, the first vocational school for deaf children was opened in Nicaragua. Although the children at the school had never been taught a sign language, they invented one of their own that became increasingly complex.[162] This unique language has many of the fundamental properties of all other languages, including the breakdown of information into discrete units and a grammatical presentation of words (gestures in this case). What is the significance of the children's behavior for discussions about the four-levels-of-analysis approach to language-learning ability in human beings?

14.3   Robert Plomin and his colleagues have compared the cognitive abilities of children with those of their genetic and adoptive parents.[144] What significance do you attach to these data (Figure 14.2) in the context of determining whether genetic or environmental differences are responsible for the differences between humans in their spatial and verbal abilities? If environmental differences are the key to understanding differences in these human phenotypes, what is the predicted relationship between the number of years a child has spent in an adoptive home and the degree to which the child's spatial and verbal attributes differ from those of his or her genetic parents?

The Neurophysiology of Speech

The gene *FOXP2* exerts its developmental effect on our ability to talk by regulating a battery of other genes. The products of these genes contribute in some way to the development of particular parts of the brain that are essential for language learning or speech itself. It has long been known that damage to

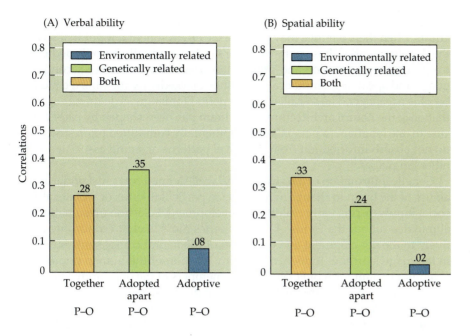

(A) Verbal ability

(B) Spatial ability

FIGURE 14.2 **Does heredity play a role in the development of verbal and spatial intelligence?** Correlations in (A) the verbal ability scores and (B) the spatial ability scores of parents (P) and their genetic or adoptive offspring (O). Note the relatively high correlations in the scores for parents and their children who were reared in adoptive homes (adopted apart). In contrast, the correlation of the two scores between children and the adults who adopted them was very low. After Plomin et al.[144]

what is called Broca's area in the cerebral cortex is associated with the loss of the ability to speak a language, while a different anatomical brain region, Wernicke's area, is often said to be responsible for the ability of adults to comprehend spoken language (Figure 14.3). Neither of these regions is independent of other anatomical components of the brain; these circuits and structures certainly work together with Broca's and Wernicke's areas in the production and comprehension of speech.[111] Therefore, if we wish to understand the neurophysiological mechanisms of language and speech, we have to discover how various structural and connective elements in the brain operate in an integrated fashion, a task that is still far from complete. However, the fact that in almost all adults the speech mechanisms are located on one side of the brain and include well-defined parts of the cerebral cortex offers evidence that human brain development has evolved to make language and

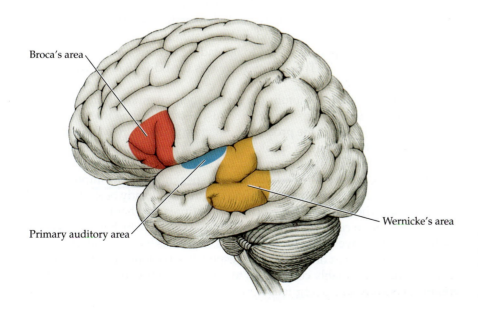

Broca's area

Primary auditory area

Wernicke's area

FIGURE 14.3 **Regions of the brain implicated in language abilities.** Broca's and Wernicke's areas are believed to play important roles in the ability of people to understand and produce speech.

speech possible. In infants, exposure to speech during social interactions with adults, often their parents, may provide the environmental input that shapes the development of those parts of the brain that gradually become dedicated to speech comprehension and production. The ability to learn a second language and speak it grammatically declines markedly by the time a child has become a teenager. The so-called critical period for language learning suggests that once the neural architecture of the brain has been shaped by early exposure to one language, a person's developmentally committed brain structures do not absorb information from the novel patterns of sound associated with other languages.[101]

Genetic information of a restricted and well-defined sort must underlie the ability of the brain to change in response to particular kinds of acoustical and social inputs during early childhood development. Selection presumably has eliminated all but the forms of the key genes most likely to contribute positively to the development and operation of the neural components involved in language learning. Indeed, only a tiny fraction of the human population exhibits a variant form of *FOXP2*. Those members of the one family known to possess this rare allele are unable to control the muscles of the face, lips, and mouth in the manner necessary for speech. Functional magnetic resonance images taken of these individuals have also revealed abnormalities in Broca's area and other components of the neural circuitry related to speech, further evidence that particular parts of the brain are essential for this very important human activity. Work with these individuals and studies of the form of the gene found in other mammals have revealed that it is likely that *FOXP2* is usually expressed in cells in several parts of the brain. Without the normal form of the gene, the cells in the speech circuits cannot carry out tasks that are standard operating procedures for the rest of us.[181]

Many other elements of the human brain provide skills that contribute to our facility with language. So, for example, activity in our visual cortex complements our acoustical analysis of language in a most interesting way, a point discovered when researchers created videos in which a person was filmed saying one thing but the audio track played a different sound. For example, if the film showed a person saying "ga" but the audio track played "ba," an observer both watching and listening would hear "da."[125] This means that when someone sees a video of a person mouthing the nonsense sentence "my gag kok me koo grive" while an audiotape synchronously plays the nonsense sentence "my bab pop me poo brive," the "test subject" will hear quite clearly "my dad taught me to drive." This result demonstrates that our brains have circuits that integrate the visual and auditory stimuli associated with speech, using the visual component linked to lipreading to adjust our perception of the auditory channel.[123] This ability increases the odds that we will understand what others say to us.

Indeed, babies about 6 months old switch their attention from the eyes of speaking adults to their lips. They do so just about the time that the infants begin babbling as part of the process of integrating the sounds they are hearing with the movements of a speaker's lips.[107] When a baby begins to speak his or her first words, usually around the first birthday, the child switches back to looking at the eyes of adults that are speaking. But the importance of lipreading for verbal communication has apparently already been established in the toddler. The neurophysiological basis of lipreading has been traced to a region of the brain called the superior temporal sulcus, which becomes especially active when we view moving mouths, as well as hands and eyes (Figure 14.4).[6] Cells in this part of the brain are responsive to subtle movements of the lips, and this presumably contributes to our great ability to comprehend the complex vocalizations of our fellow man.

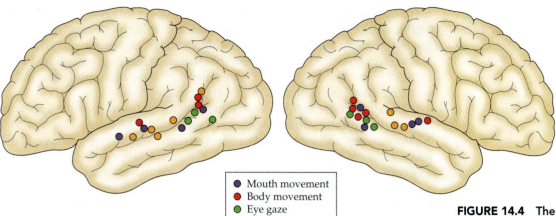

- Mouth movement
- Body movement
- Eye gaze
- Hand movement

FIGURE 14.4   The function of the superior temporal sulcus of the brain. Socially relevant movements of the eyes, mouth, hands, and body activate neurons in different parts of this part of the cerebral cortex. The left and right hemispheres are shown on the left and right, respectively. Each circle represents findings by a particular research team. After Allison et al.[7]

Discussion Questions

14.4   Considerable disagreement exists about the extent to which our brains are composed of well-defined modules, each shaped by selection to carry out specialized tasks, as opposed to a brain composed of generalized networks with much functional plasticity. In light of this argument, what do you make of the following findings? Humans are extremely good at recognizing familiar faces, thanks to various elements of the brain.[69] Persons with certain kinds of damage to the fusiform gyrus, which is on the underside of the cerebral cortex, lose the ability to recognize familiar faces, a deficit much more likely to occur after damage to the right hemisphere of the brain.[55] (See *The Man Who Mistook His Wife for a Hat* by Oliver Sacks.[157]) Functional magnetic resonance imaging reveals that neurons in a small part of the posterior fusiform gyrus, the facial fusiform area, fire only when a person looks at a face. This part of the brain does not respond to pictures of inanimate objects, although another nearby region of the brain does.[124]

14.5   In the past, aspiring London taxicab drivers underwent a rigorous education that required them to learn the location of about 25,000 streets in the city. The average posterior hippocampus of London cabbies, as revealed by magnetic resonance imaging, was larger than that in a comparable group of London bus operators, who follow a fixed route and so do not have to learn a detailed map of the city.[121] In addition, the more years of taxi driving, the larger the posterior hippocampus.[122] What does this research tell us about the interplay of the environment and genetics in the development of adaptive navigational skills in human beings?

The Adaptive Value of Speech

Having sketched a few aspects of the proximate causes of speech and language, let us return to an ultimate cause, namely the evolved function of language, a level of analysis that we need to consider if we are to determine why natural selection led to the spread of alleles that contributed to the development of a brain capable of speech and language comprehension. It is not difficult to imagine why those of our ancestors with a protolanguage experienced higher fitness than those with less speaking facility. Indeed so many benefits to individuals currently flow from being able to speak that it may seem hardly worth thinking about the adaptive value of the trait. But we must do so because the underlying neural mechanisms of speech are extremely costly thanks to the fact that brain tissue requires about 16 times as much energy to maintain as an equivalent weight of muscle tissue. Indeed, the human brain's metabolic demands may have had all manner of evolutionary consequences, including the need for a nutritionally dense diet high in energy, as well as the requirement that infants be provided with large amounts of fat to fuel brain development.[106] To the extent that language learning contributed to an increase in our brain's volume, the benefits must have been substantial if they were to outweigh the fitness costs of an expensive brain.

As I say, there are many potential benefits for an individual that appear to stem from the ability to speak a language, notably the transmission of useful information about one's culture and environment to one's offspring and relatives as well as to nonkin who will return the favor or help with mutually beneficial tasks.[143] People also appear particularly interested in talking about the behavior of other people.[52] While some of the information received may be accurate and benefit the listener, let me add the important caveat that language also facilitates the transmission of *mis*information to signal receivers from manipulative persons whose speech benefits them at the expense of those they deceive,[178] another cost of language learning.

Rather than review the more familiar possible adaptive features of language, let's briefly examine two less frequently discussed ideas. The first has to do with the role that language, and especially language dialects, might play in the ability of humans to use the way they talk to signal group membership, an advantageous use of communication as a foundation for cooperative endeavors. Just as chaffinches and white-crowned sparrows differ in the vocalizations produced in different regions, so do humans and for the same reason—the learning mechanisms underlying speech promote the learned acquisition of slight variations from group to group as speakers in different locations happen to differ in their speech. Given enough time and isolation, these differences can accumulate, and at some point, dialects can become separate languages. But even in the early stages of differentiation, the subtle or not-so-subtle differences between groups can serve as badges of group or tribal membership. The dialect that one speaks can convey information about the identity of an individual, where the person lives or was raised, his or her social status, and the person's desire to affiliate with others.[29]

Indeed, the drive to be part of a group is generally so strong that people will unconsciously mimic the speech of people around them, no matter if they normally speak the same dialect or not. In fact, we can mimic a dialect when the only cues available come from lipreading (an ability we are good at, as noted already). Dialect mimicry occurred in a study in which the human subjects were exposed to a silent video showing a person saying words like *tennis* or *cabbage*. The participants, who had had no experience lipreading, were then asked to say the word they had seen being spoken in the silent video. When they said out loud "tennis" or "cabbage" or whatever word they had lipread, they spoke this word closer to the way in which the silent model had spoken it than when asked previously to read the word in their own customary voice.[130] People evidently copy the dialect that they lipread and hear, even if this requires a change in how they usually speak, perhaps as a way to announce unconsciously to others their desire to be part of the group composed of speakers of another dialect.

The "affiliation function" of speech, if it exists, provides an example of a benefit that arises from the role that learning plays in language acquisition and production, even among adults. Another possible adaptive value of language learning has been championed by Geoffrey Miller, a strong advocate for the role of sexual selection in the evolution of speech and many other features of human behavior. Miller's argument is that skill in language use comes into play during what he calls "verbal courtship."[127] According to Miller, men compete with one another to display their verbal competence to women who evaluate potential mates at least partly on the basis of their capacity to use language creatively, amusingly, flatteringly, and so on (Figure 14.5). In other words, speech has become a focus of sexual selection, with female choice favoring males with exceptional language skills, just as selection by choosy females is responsible for the evolution of the plumes of birds of paradise, which are far more extravagant than needed for flight or any other more prosaic function. On average, adult humans have a vocabulary of about 60,000 words; our fellow primates produce at most a few dozen different vocalizations.[127] Deaf humans who use American Sign Language have 9000-plus signs at their disposal; chimpanzees have a repertoire of less than 100 gestures,[83] their analog of sign language. The point is that most people know and use far more words, many of which are redundant, than are actually needed to convey basic information to their listeners. If the adaptive significance of speech were simply to transmit functional information from one person to another, natural selection would presumably have acted against the extravagances associated with human language.

In keeping with Miller's proposal on the sexually selected nature of language, when young women were given the opportunity to watch videos showing young men reading out loud and talking (among other things), performers that had previously scored well on a verbal intelligence test were judged more sexually attractive.[146] The women independently judged male verbal intelligence accurately on the basis of what they observed from the videos. Thus it would make sense that men attempt to be verbally proficient when interacting with women, while the opposite sex uses their equally large vocabularies and speech capabilities to judge the verbal proficiency of potential partners, perhaps as a proxy for intelligence.

The sexual selection hypothesis also led to the finding that, although both men and women say that they value a sense of humor in the opposite sex, men think that a woman's sense of humor can be judged on how she responds to their jokes whereas women think that a man's sense of humor is a function of whether he says things that strike them as funny.[23] In other words, perhaps men are not interested in female comedians but instead are looking for appreciative females—unlike women, who do like amusing men. If so, men rated relatively highly as humorists should have relatively high levels of sexual success. As predicted, the men who were considered funnier (by six student judges, based on captions that the subjects wrote for a set of untitled *New Yorker* cartoons) reported having more sexual partners than those who scored lower on the humor test.[75]

The follow-on argument here is that the language skills and capacity for humor are linked to the developmental history of men. Given that early deficits in nutrition can have long-lasting negative effects on survival and reproductive success in our species,[118] individuals that have been well fed as youngsters (or blessed with genes that buffer their development against disturbances) should be able to afford well-developed brains, which you will remember are very expensive organs to produce and maintain. By choosing developmentally advantaged men, females could gain good genes for their offspring or good parents for their children, or both (as noted in the discussion of female mate choice in birds in Chapter 10).

"Let's go someplace where I can talk."

FIGURE 14.5 This cartoon captures the sexual selectionist interpretation of language use by men.

Discussion Question

14.6   Humans are cultural animals. As such, we have major differences between cultures in the behavior of people, as illustrated by the 5000 or so languages spoken within historical times in the world. Exposure to a particular language leads children to adopt that language. A popular view in some circles is that the behavioral differences between men and women likewise stem largely from exposure to a culture's view of how males and females should behave. In our culture, young boys typically receive guns and miniature trucks to play with while young girls get dolls and baby carriages. Cultural stereotyping of this sort is said to push boys into "masculine roles" while guiding girls into culturally approved "feminine roles." Two researchers gave young male and female monkeys both types of toys and measured the amount of time they spent with each kind of toy. Why did they do this? How could the evidence they gathered help them test the cultural stereotyping hypothesis? Check your answer against Alexander and Hines.[2]

The Evolutionary Analysis of Mate Choice

With these issues behind us, we now return to the evolutionary analysis of mate choice in our species. Tackling this component of our behavior is a real challenge because of the diversity of cultural rules and regulations surrounding human sexual behavior. There are monogamous, polygynous, and polyandrous societies, some in which you are not allowed to marry an unrelated person who belongs to your clan, others in which adult men can marry prepubescent girls, some in which males and their relatives provide payment for a bride, and others in which women must bring a costly dowry with them to their wedding.

Despite all this cultural variety and other factors that make humans difficult to study (such as the uncertainties associated with self-reported data), certain biological facts of life still apply to our species. For one thing, women are typical female mammals in that they retain control of reproduction by virtue of their physiological investments in producing eggs, nurturing embryos, and providing infants with breast milk after they are born. Although men are also able and often willing to make large parental investments in their offspring, their reproductive decisions nevertheless take place in a setting defined by female physiology and psychology.[46,67] Not every woman has been able to exercise complete control over her reproduction, but at least under some conditions some women have been able to choose from a number of potential partners (or their relatives have done so on their behalf).

Mate Choice by Women

From an evolutionary perspective, we would expect women to be attracted to men whose physical attributes are indicative of high genetic quality or parental ability, both of which vary among males and have the potential to affect the fitness of a woman and her children. Some researchers have found that women do indeed prefer men with "masculine" facial features, namely, a prominent chin and strong cheekbones (Figure 14.6).[89] In addition, facial symmetry has been identified as a plus (and not just in Western societies[115]), and so is a muscular upper body,[67] above-average height,[133] and a deep voice.[56] In Slovakia, a European country, physically attractive men are more likely to marry, and of those who are in the married cohort, the more attractive individuals (as judged by a group of young women on the basis of photographs taken of the men in their 20s) have more children.[145]

(A)

(B)

FIGURE 14.6 **Mate choice based on facial appearance.** Computer-modified facial images provide a way to test the preferences of subjects for (A) masculinized faces versus (B) feminized ones. Photographs by Ben Jones and Lisa De Bruine of the Face Research Laboratory, University of Aberdeen, from Jones et al.[89]

The greater reproductive success of physically attractive men might be linked to high testosterone levels, good current health, and perhaps most importantly, good health during juvenile development. The development of males is at special risk because of the potentially damaging side effects of testosterone. Therefore, the ability of a man to develop normally, despite high levels of circulating testosterone, is a possible indicator of a strong immune system capable of overcoming the handicap imposed by the sex hormone.[60] If strong, healthy men can pass on defense against disease to their offspring, their children are more likely to survive and their mates will benefit as a result.

In addition, healthy men probably can compete effectively with rivals for dominance within their group. In this context, it is relevant that the men rated as having the most dominant, masculine, and attractive faces are those with the greatest handgrip strength, which correlates well with overall physical strength.[58] Moreover, men with high handgrip strength do develop sexual relationships sooner and have more (self-reported) sexual partners than individuals with less powerful grips.[62] Over the course of human evolution, dominant, powerful men probably have been able to protect their partners, as well as supply them and their offspring with the resources that usually are possessed by males of high social and political status.[38,73]

Discussion Question

14.7 How might the following findings be understood in terms of the adaptive value of female mate preferences? Deep-voiced men have more children in a traditional hunter-gatherer culture, the Hazda of Tanzania.[14] Taller men are more likely to be chosen in speed-dating competitions than their shorter rivals[18] and are more likely to be chosen as sperm donors by women in California.[54] Images of men with slightly bloodshot eyes are judged less attractive than photographs of the same men in which the whites of the eyes are clear.[147]

The importance for women of having a good provider as a mate has been established in studies showing that females in cultures without birth control who secure relatively rich husbands do tend to have higher fitness than females whose partners cannot offer many material benefits. Among the Aché of Paraguay, for example, the children of men who were good hunters were in fact more likely to survive to reproductive age than the children of less skillful hunters.[92] Likewise, several studies of traditional societies in Africa and Iran have revealed a positive correlation between a woman's reproductive success and her husband's wealth, as measured by land owned or number of domestic animals in the husband's herds.[21,87,120] Even in modern societies, household income is correlated with children's health, with the effect growing larger as children get older. Chronic illnesses in childhood can reduce the earning power of children who reach adulthood, thus perpetuating poverty across generations,[41] with all of the negative reproductive consequences this has for human beings.

Evidence of this sort has convinced some evolutionary psychologists to predict that females will usually put wealth, social status, and political influence ahead of good looks when it comes to selecting mates. This evolutionary prediction has been supported by questionnaire and interview studies by David Buss[31] and others. However, even when researchers have found clear differences between the sexes in the value they attach to "good financial prospects" versus "good looks," the absolute measures of importance given to these attributes have not necessarily been especially high for either sex. But the men and women in these studies typically have not had to specify which items among a list of attributes are absolutely essential to their choice of mates versus which would be nice to have in a partner but are not crucial. Therefore, a team of social psychologists led by Norm Li attempted to put constraints on the choices made by adults they interviewed in the United States by giving them a limited budget to expend on designing a hypothetical ideal mate.[108] Subjects were given a list of traits and a supply of "mate dollars" and told to decide how many of their mate dollars to spend on traits such as physical attractiveness, creativity, and yearly income. Subjects could spend up to 10 mate dollars to get the highest level of traits they particularly valued, or nothing at all on traits of no importance to them. When the persons interviewed had only 20 mate dollars to work with, their investments differed greatly according to their sex. Men devoted 21 percent of the total budget to the acquisition of a physically attractive partner; women spent 10 percent of the same total amount to the same end. On the other hand, women on this tight budget devoted 17 percent of their money to boost the yearly income of an ideal mate, whereas men invested just 3 percent of their mate dollars on this attribute.

After spending their first 20 mate dollars, the participants were given two additional 20-dollar increments. By the time they reached the third 20 mate dollars, the sexes did not differ markedly with respect to the attributes they were buying. Having already purchased what they really valued, they could and did spend on other attributes. Confirmation of these results comes from a similar study in a Singaporean population, suggesting that this evolutionary analysis has cross-cultural validity.[109] These experiments tell us that the criteria for mate choice are not the same for men and women, as predicted by an evolutionary approach.[108]

Personal ads, whose cost limits the number of words used by the advertisers, also provide relevant evidence on what people consider fundamentally important in a real mating market. So, for example, women seeking partners through newspapers are far more likely than men to specify that they are looking for someone who is relatively rich.[184] In keeping with this goal, women advertisers in both Arizonan and Indian newspapers also often specify an interest in someone older than they are;[97] older men usually have

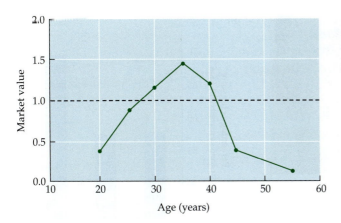

FIGURE 14.7 Age and the market value of men. Market value is calculated by taking the number of newspaper "personal" advertisements by women requesting men of certain age classes and dividing that by the number of men of those age classes announcing their availability. After Pawłowski and Dunbar.[138]

larger incomes than younger men.[31] If women really are highly interested in a partner's wealth and capacity to provide for offspring, then men in their 30s should be most desirable because men of this age have relatively high incomes and are likely to live long enough to invest large amounts in their children over many years. One can calculate the "market value" of men of different ages by using samples of personal ads and dividing the number of women requesting a particular age class of partner in their advertisements by the number of men in that age class who are advertising their availability; this measure thus combines both demand and supply. Men in their late 30s do indeed have the highest market value (Figure 14.7).[138]

Discussion Question

14.8 Use an adaptationist approach to predict what the market value curve for women should look like by adding your predicted data points to the graph shown in Figure 14.7. Check your prediction against data in Pawłowski and Dunbar.[138]

It could be, however, that women's interest in the earning power of potential mates is a purely rational response to the fact that males in almost every culture control their society's economy, making it difficult for a woman to achieve material well-being on her own. If this non-evolutionary hypothesis explains why females favor wealthy males, then women who are themselves well off and not dependent on a partner's resources should place much less importance on male earning power. Contrary to this prediction, however, several surveys have shown that women with relatively high expected incomes actually put more emphasis, not less, on the financial status of prospective mates.[177,187] For example, relatively wealthy female undergraduates seek wealth and high status in potential long-term mates.[39]

The general rule is that a woman's own social standing affects her mate choice criteria. Thus, women who rate themselves as "highly attractive" show a stronger preference for both relatively masculine and relatively symmetrical faces than women who believe that they are of average or low attractiveness (Figure 14.8).[113] David Buss and Todd Shackelford report that not only are physically attractive women desirous of a highly attractive mate, they also have higher standards when it comes to partner wealth, commitment, and parental abilities,[35] a finding that matches the results of studies based on dating advertisements.[138] Finally, Peter Buston and Steve Emlen found that both men and women who consider themselves high-ranking prospects for a long-

(A)

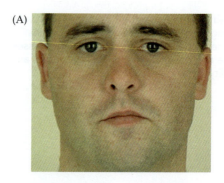
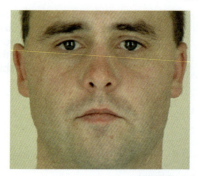

(B)

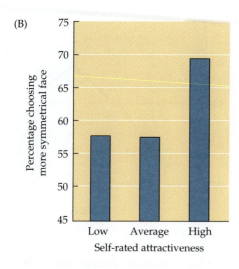

FIGURE 14.8 Women who think they are highly attractive prefer more attractive men. (A) When women were given a choice between two digitally altered photographs of men, one of which was more symmetrical than the other (right versus left), they differed in the degree to which they said they preferred the more symmetrical face. (B) Women who rated themselves highly attractive chose the more symmetrical face nearly 70 percent of the time, whereas women of lower self-rated attractiveness did so less than 60 percent of the time. After Little et al.[113]

term relationship expressed a preference for an equally high-ranking partner, whereas those individuals with a lower self-perception of market value were less demanding (Figure 14.9).[39] The less-demanding requirements of women who are unlikely to be competitive in securing and maintaining long-term relationships may help these individuals avoid the costs of searching unrealistically for men they are unlikely to attract, as well as reducing the odds of being deserted, should they enter into a relationship with a high-ranking male.[113]

Interestingly, similar results have been achieved in experimental studies of zebra finch females, some of which were reared in small broods while others grew up in large broods. These conditions affected the size and metabolic efficiency of the birds, with "privileged" small-brood females becoming high-quality adult finches whereas the large-brood, low-quality females were handicapped as a result of the increased competition for food that they had experienced when young. The two kinds of females were given an opportunity to listen to the songs of either high-quality males or low-quality ones, which they could turn on by pecking a key; high-quality females preferred the songs of high-quality males, while low-quality females chose the songs of low-quality males. In so doing, as has been suggested for humans,[113] low-quality females may benefit by avoiding a competition for high-quality mates that they could not win or by ignoring high-quality males likely to reject them sooner or later.[84]

FIGURE 14.9 Self-perception of attractiveness affects mate preferences in both sexes. The degree to which women and men consider themselves attractive is correlated with their mate preferences. Less attractive individuals are willing to settle for less in a partner. The values for mate preferences are mean scores of the subjects' answers to questions about the importance of ten attributes in their dating decisions such as, on a scale of 1 to 9 how important is physical attractiveness to you? The values for self-perception are the mean scores of the respondents' own evaluations of how they would score on these ten attributes. After Buston and Emlen.[39]

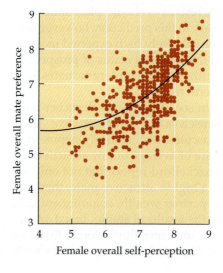
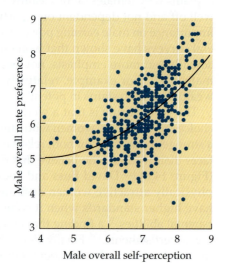

Sexual preferences in zebra finches and humans are one thing, but mating behavior is another. Are women actually more likely to copulate with men who possess the attributes they prefer than with men who lack them? Among the Aché of eastern Paraguay, good hunters with high social status are more likely to have extramarital affairs and produce illegitimate children than poor hunters, suggesting that females in this society find skillful providers sexually attractive.[92] Likewise, in Renaissance Portugal, noblemen were more likely to marry more than once, and more likely to produce illegitimate children, than men of lower social rank. These results are consistent with the prediction that women use possession of resources as a cue when selecting a father for their children.[20]

If we bear the imprint of past evolution on our psyches, then women in today's Western societies should also use resource control and its correlates, such as high social status, when deciding which men to accept as sexual partners. In order to study the relationship between male income and copulatory success in modern Quebec, Daniel Perusse secured data from a large sample of respondents on how often they copulated with each of their sexual partners in the preceding year. With this information, Perusse was able to estimate the number of potential conceptions (NPC) a male would have been responsible for, had he and his partner(s) abstained from birth control. Male mating success, as measured by NPC, was highly correlated with male income, especially for unmarried men (Figure 14.10). Perusse concluded that single Canadian men attempt to mate often with more than one woman, but their ability to do so is affected by their wealth and social standing.[141] These findings have been replicated in great detail with a much larger random sample of men living in the United States.[91] Thus, male striving for high income and status may be the product of past selection by choosy females, which occurred in environments in which potential conceptions were very likely to be actual ones.[141] In the past, those men who secured wealth and high social standing almost certainly had higher fitness on average than men who cared little about these attributes.

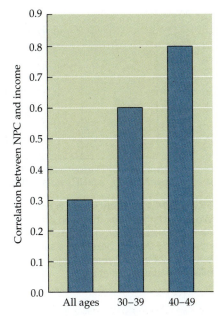

FIGURE 14.10 **Higher income increases male copulatory success.** Income is positively correlated with the number of potential conceptions (NPC) in the preceding year for unmarried Canadian men of various age groups, but especially for older men. After Perusse.[141]

Discussion Question

14.9 Although women seem to prefer wealthy men, in most modern cultures, high family income is not positively correlated with the number of children produced (Figure 14.11). Indeed, poor couples often have more surviving children than do rich ones. Does this finding invalidate an evolutionary analysis of human behavior, as some believe?[183] You might want to contrast aspects of the current human environment with our ancestors' environment. Can you make use of the finding that in preindustrial Finland, for women of high fecundity, the number of *surviving* offspring was less in resource-poor landless families than for women in landowning families?[68] Also fit the following finding into your analysis: in a survey of modern data from 145 countries, human fertility was negatively linked to population density.[119]

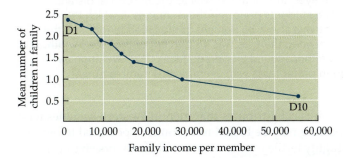

FIGURE 14.11 **Fertility declines as family income increases.** Shown here is the average number of children per family in relation to the amount of income per family member. Data are for families in the United States in 1994 as grouped by income decile, so the bottom 10 percent are shown at D1, and the top 10 percent are shown at D10. After a graph drawn by the Institute for Social Research at the University of Michigan.

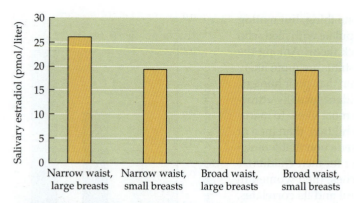

FIGURE 14.12 Body shape is correlated with fertility in women. Women with features that men often claim they prefer—namely, narrow waists and large breasts—are more fertile than women with other body shapes, based on a sample of healthy Polish women between 24 and 37 years of age who were not taking birth control pills. After Jasiénska et al.[88]

Mate Choice by Men

You will recall the prediction that men will place more value on "good looks" than will women, who should tend to be more interested in the earning potential or resources controlled by a would-be mate. From an evolutionary perspective, a good-looking woman is predicted to be a fertile one, given that fertility varies considerably from woman to woman as a result of her age, health, and body weight, among other things. Preadolescent and postmenopausal women obviously cannot become pregnant. Women in their 20s are more likely to become pregnant than women in their 40s. Healthy women are more fertile than sick ones. Women who are substantially overweight or underweight are less likely to become pregnant than women of average weight.[150]

Men in Western cultures favor feminine features that are associated with developmental homeostasis, a strong immune system, good health, high estrogen levels, and most importantly, youthfulness.[67,174] In particular, men favor full lips, small noses, large breasts, an intermediate weight rather than extreme thinness or obesity,[176] and a waist circumference that is substantially smaller than hip circumference.[165] The level of circulating estrogen in the large-breasted, narrow-waisted participants in a sample of healthy Polish women meant that these women were about three times more likely to conceive a child than the other women in the study (Figure 14.12).[88] The fact that in different cultures[24] men also prefer the waist-to-hip ratio associated with the hourglass figure, as do congenitally blind individuals who assess this ratio by touch, suggests that the Western visual media is not responsible for this aspect of male choice.[93]

Discussion Questions

14.10 We have just offered the fertility hypothesis for the male preference for women with an hourglass figure. Produce a different hypothesis for this sexual preference, based on the following facts: the body fat stored in a pregnant woman's lower body is of a type that promotes the growth of the fetal brain, whereas upper-body (abdominal) fat differs in its composition and is not used for the development of the embryo's brain. Once you have come up with your hypothesis, use it to produce at least one testable prediction. Compare your explanation and prediction with that of William Lassek and Steve Gaulin.[104]

14.11 University students were asked to judge the attractiveness of different versions of faces that had been digitally altered to change the vertical distance between the eyes and mouth as well as the horizontal distance between the eyes. The students preferred those images in which the relevant vertical distance was about 36 percent of the face's length and the horizontal distance between the eyes was 46 percent of the face's width. These proportions are those found in average faces.[136] In evolutionary terms, why might men prefer faces with average structural features?

Most or all of the physical attributes that men in Western societies tend to find sexually appealing may be linked to a woman's potential to become pregnant. But even highly fertile women are most likely to conceive only during the few days each month when they are ovulating. During this time, mating with high-quality men could provide genetic benefits for a woman's offspring, and if so, we can predict that women will be especially interested in competing for and securing high-quality sexual partners.

Ovulation is linked to a spike in circulating estrogen levels; the higher these levels, the more likely a woman is to become pregnant should she copulate at this time.[112] In other nonhuman primates, the fertile window is often widely and conspicuously advertised, such as in the bright red sexual swelling in baboons. Although it was once thought that neither men nor women could detect when a woman was fertile, there is growing evidence that both sexes can detect the changes associated with ovulation. For example, women with relatively high estrogen levels are considered more attractive by others.[53] Ovulating women provide cues of this event with changes in facial appearance (Figure 14.13),[105,152] voice

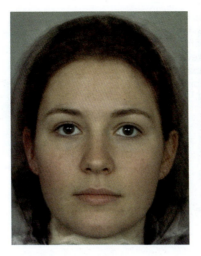

FIGURE 14.13 Facial appearance provides a cue of circulating estrogen levels. The faces shown are composite images generated by overlaying ten photographs of women with the highest urinary estrogen levels and ten photographs of women with the lowest urinary estrogen levels. You should be able to guess which composite image was derived from photographs of women at peak fertility. From Law-Smith et al.[105]

pitch[28] (but see Fischer et al.[59]), the way they walk,[148] and their body odor. Devendra Singh and Matthew Bronstad showed that men find the smell of a T-shirt worn by an ovulating woman to be "more pleasant and sexy" than the scent of a T-shirt worn by the same woman when she is not fertile.[166] The odor of women in their fertile phase also induces men to boost the levels of circulating testosterone in their bodies.[131]

Geoffrey Miller and his colleagues used a novel method to demonstrate that men are quite capable of assessing (unconsciously) female fertility. They recruited a number of lap dancers working in Albuquerque, some of whom were on the pill and others who were not. These women were willing to cooperate with the researchers by telling them the amount they received as tips after their erotic performances. Before reading on, you may want to identify what predictions the adaptationist Miller and his coworkers must have had in mind.

Because the research team considered it likely that men could detect cues associated with female fertility and because men ought to find these cues attractive, the team predicted that lap dancers who were not on the pill and were ovulating would receive more tip money from their customers than would those women who were not ovulating, either because they were taking birth control pills or because they were in the nonovulatory phase of the menstrual cycle. The actual results were that the lap dancers took in about twice as much in tips when they were fertile compared with when they were menstruating or were on the pill.[128]

Discussion Question

14.12 In our species, males typically prefer to mate with youthful women. But in our closest relative, the chimpanzee, older females in estrus are approached more often by sexually motivated males, and they copulate more frequently with dominant males (Figure 14.14).[134] What prediction can you make therefore about the relation between age and fertility of female chimpanzees? Check your prediction against the data in Thompson et al.[171]

FIGURE 14.14   Mating preferences of dominant chimpanzee males. These individuals prefer to copulate with old parous females (those that have had infants previously), like the one in the photograph. Nulliparous females (those that have yet to give birth) are the least preferred. After Muller et al.[134]

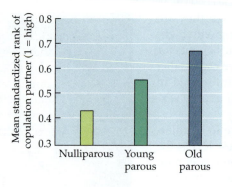

Old chimp mother with youngster

Discussion Question

14.13   Some persons believe that men in Western societies generally prefer youthful sexual partners because they have been conditioned through advertisements and the like that they should find females *younger than they are* more appealing. What prediction follows from this hypothesis with respect to the dating preferences of teenage males for females of different ages? What conclusion follows from the data in Figure 14.15?[99]

Contraceptive pills not only affect how women are perceived by men but may also influence female mate choice as well by, for example, reducing the attractiveness of genetically dissimilar partners under some conditions.[11,153] More generally, the sexual preferences of women, as well as men, change in relation to the menstrual cycle.[63,89] For example, when asked to evaluate potential partners for a brief sexual encounter, fertile women tended to favor the men with more masculine faces[140] and bodies[114] (Figure 14.16), and they found men with symmetrical faces more attractive at this time.[115] Indeed, women at their most fertile stage also more strongly prefer the smell of symmetrical men.[173] At the proximate level, changes in neural responses may underlie these changes in preferences; the right medial orbitofrontal cortex of ovulating women is activated more strongly when these women are shown images of masculinized male faces as opposed to feminized ones.[156]

Hormones are surely also involved proximately in changes in the perception of male features that occur over the lifetime of a female. Before puberty and after menopause, women have low levels of estrogen. In keeping with this hormonal reality, preferences for masculine male faces are relatively low before puberty and after menopause.[116] In this way, changes in hormone levels (and in the analysis of men's faces) across the lifespan of a female mirror the changes that occur during the menstrual cycle of a postpubescent woman.

These changes in brain activity and mate preferences might occur because ovulating women with social partners of average or low quality are unconsciously attempting to secure "good genes" for their offspring, which could make their sons especially attractive or dominant. Alternatively, selective women might be more likely to secure superior "complementary genes" from an extra-pair mate, which could generate better offspring genotypes, perhaps particularly with respect to immune system development (see also Chapter

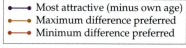

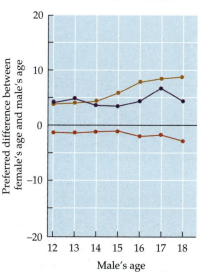

FIGURE 14.15   Preferred dating partners by age for male teenagers. The age of preferred dates is shown in relation to the age of the male subjects. After Kenrick et al.[99]

8, page 234). This second hypothesis is supported by the finding that women usually prefer the smell of men with dissimilar major histocompatibility complex (MHC) genes.[110,185] (The MHC complex plays a critical role in the immune response to pathogens.[25]) The offspring of MHC-dissimilar couples will be more heterozygous and so should have better immune systems and greater resistance to infection. Interestingly, women in relationships with men whose MHC genes are similar to their own report lower satisfaction with their mates and a greater number of extra-pair partners than do those women in relationships with men who are genetically dissimilar with respect to these genes.[64]

Discussion Question

14.14   The research we have reviewed tells us what people want in an ideal mate, but for most of us, ideal mates are in short supply. Not every man is paired off with an exceedingly fertile, gorgeous model, nor is every woman married to an extremely wealthy, highly parental, loving individual with outstanding genes.[35] The mate choices of men and women often diverge considerably from what would seem to be best for their genes. Why? Perhaps because trade-offs are involved in any pairing. Men that marry extremely attractive women may lose paternity to other men who are also attracted to their partners. Women that marry extremely strong, powerful men may lose resources to other women attracted to their partners. In light of the trade-off hypothesis, evaluate the data presented in Figure 14.17.[169]

The Evolutionary Analysis of Sexual Conflict

Because the mate preferences and genetic interests of men and women are not the same, we expect to find sexual conflict in our species, as we do in most other animals (see Chapter 7). One significant source of sexual conflict

FIGURE 14.16   The menstrual cycle affects female mate choice. Women change their evaluation of two male body images, one relatively feminized and the other masculinized, depending on the phase of their menstrual cycles. Fertile (ovulating) females (blue bars) who are asked to judge the two images for a prospective long-term relationship favor the feminized image, whereas they pick the masculinized image when asked about a short-term relationship preference. After Little et al.[114]

(A)
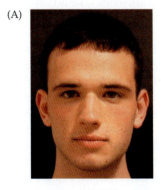

(B)
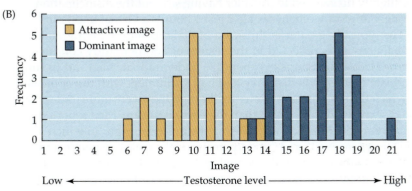

FIGURE 14.17   Females differ with respect to facial features they associate with dominant men versus attractive men. (A) In this study, digital images of the same male's face were altered to reflect the developmental effects of testosterone, from low to high levels (left to right). (B) When young women were asked to judge these photographs for physical attractiveness, they tended to pick images like the one in the middle, whereas when they were asked to rate the images in terms of social dominance, they tended to pick those similar to the right-hand image. After Swaddle and Reierson.[169]

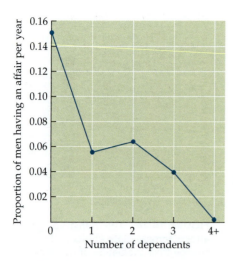

FIGURE 14.18 Adaptive regulation of extra-pair affairs. Men in a traditional society are less likely to engage in extramarital affairs in a given year if they have a relatively large number of dependents to care for with their wives. After Winking et al.[193]

arises if men are, on average, more interested in acquiring multiple sexual partners than women are. Although monogamous men may advance their reproductive success by helping one mate, polygynous men with several wives are likely to have more children than their monogamous counterparts. Polygyny has almost certainly been an option for men with substantial resources throughout our species' history, judging from the fact that the acquisition of several wives was culturally sanctioned in 83 percent of all preindustrial societies.[135]

Extramarital affairs also have the potential to increase the fitness of a male substantially, especially if his illegitimate children are cared for by his extramarital partners and their husbands. However, extramarital activity has potential costs as well as benefits for males if a diversion of resources from existing children results from their pursuit of an additional partner (or two). Jeffrey Winking and his colleagues attempted to test whether this cost would moderate the polygynous tendencies of men in a traditional society, the Tsimane of Bolivia.[193] They predicted that if extramarital activity reduces the chances that existing offspring will achieve their maximum reproductive potential, due to loss of paternal investment, then the frequency with which men have sexual affairs will decline as a man and his primary mate have more children. The predicted pattern does occur (Figure 14.18).

Despite the costs of extramarital activity for men, many are still motivated, at least under some conditions, to seek out sexual variety, which expresses itself in the willingness of some men to patronize female prostitutes; women, on the other hand, almost never pay men to copulate with them. Moreover, males, not females, also support a huge pornography industry in Western societies because men, not women, are willing to pay just to look at nude women. Note that in modern societies these particular aspects of male sexual behavior are surely maladaptive; prostitutes almost universally employ effective birth control or undergo abortions when pregnant, and payment for pornography is unlikely to boost a man's reproductive success. The prostitution and pornography industries take advantage of the male psyche, which evolved prior to modern birth control and the publication of *Playboy*.[170]

David Buss and David Schmitt illuminated these differences between the sexes simply by asking a sample of college undergraduates how many sexual partners they would like to have over different periods of time. The men in their study wanted many more mates than the women did (Figure 14.19A). Moreover, when Buss and Schmitt asked their subjects to evaluate the likelihood that they would be willing to have sex with a desirable potential mate after having known this person for periods ranging from 1 hour to 5 years, the differences between men and women were also dramatic (Figure 14.19B): "After knowing a potential mate for just 1 hour, men are slightly disinclined to consider having sex, but the disinclination is not strong. For most women, sex after just 1 hour is a virtual impossibility."[33]

The typically greater enthusiasm of males compared with females for sexual activity is reflected in the results of another study conducted by Martie Haselton.[81] She asked about 100 undergraduate men and 100 undergraduate women whether they had had encounters with the opposite sex in which the other person evidently thought they were more (or less) interested sexually in this person than they actually were. Men reported about equal numbers of encounters during the preceding year in which women had "overperceived" and "underperceived" the males' romantic intentions. Women, on the other hand, claimed that men were far more likely to think that they were sexually interested in them, when in fact the women were

(A)

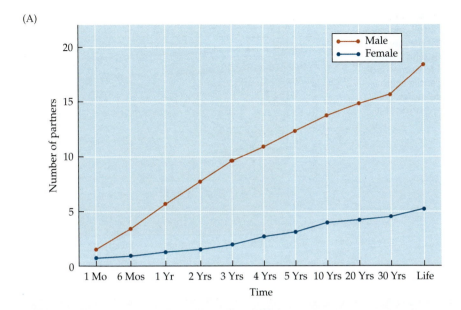

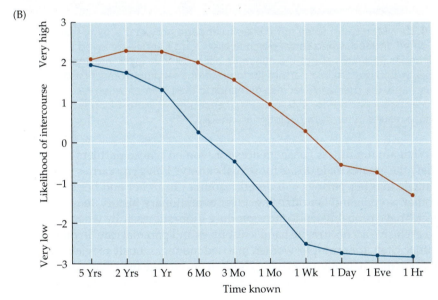

(B)

FIGURE 14.19 **Sex differences in the desire for sexual variety.** (A) Men and women differ in the number of sexual partners that they say they would ideally like to have over different periods of time. (B) Men and women also differ in their estimates of the likelihood that they would agree to have sexual intercourse with an attractive member of the opposite sex after having known that individual for varying lengths of time. After Buss and Schmitt.[33]

not, than to make the opposite mistake of underperception of sexual intent (Figure 14.20).

This kind of bias was documented in another way by two social psychologists who sent confederates, an attractive young man and an attractive young woman, on the following mission: they were to approach strangers of the opposite sex on a college campus, asking some of them, "Would you go to bed with me tonight?" Not one woman agreed to the proposition, but 75 percent of the men said yes. Remember that the male subjects had known the woman in question for about a minute.[43] A study of French men and women produced similar results.[77]

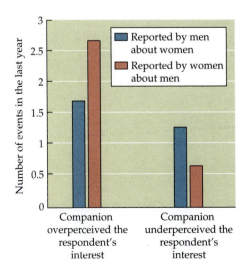

FIGURE 14.20 **A male bias toward sexual overperception.** Men tend to think that women have romantic inclinations toward them even when they do not. After Haselton.[81]

FIGURE 14.21   Sex differences in mate selectivity. College men differ from college women in the minimum intelligence that they say they would require in a casual sexual partner. However, men and women have similar standards with respect to the minimum intelligence they say is essential for a marriage partner. "Intelligence" was scored on an IQ percentile scale such that a score of 50 meant that the acceptable individual had an IQ higher than half the population. After Kenrick et al.[96]

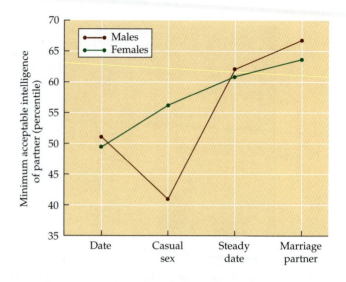

Discussion Question

14.15   Men seem to be more keen to engage in sexual activity than women. Indeed, when social psychologist Doug Kenrick surveyed a group of undergraduates about the minimum acceptable level of intelligence they would require in a partner for interactions ranging from a first date to marriage, the men were far less demanding than women when it came to accepting a partner for a casual sexual encounter (Figure 14.21).[96] Evaluate these results in terms of evolutionary theory.

What are the fitness consequences for women whose partners are able to satisfy the greater male desire for sexual variety? Polygyny carries the risk for a female that her partner will divert resources from her and her children to another woman and her offspring. Over evolutionary history, females of our species have probably tended to reproduce more successfully when they have had exclusive access to a husband's resources and parental care. In nineteenth-century Utah, for example, women married monogamously to relatively poor Mormon men had more surviving children on average (6.9) than women married to rich polygynous Mormons (5.5).[82] In fact, for every woman added to a polygynist's collection of wives, the reproductive success of the previous wives fell by about one child on average, further evidence that polygyny was probably not reproductively advantageous to Mormon women in the past.[132]

The potential benefits to men of polygyny and extramarital affairs increase the likelihood of conflict between husbands and wives. On the other side of the coin, some women are receptive to extramarital affairs, which may enable them to acquire additional material goods, or more protection, or better genes for their offspring from their extra-pair partners. If a wealthy or powerful extra-pair partner becomes a woman's primary partner, she may be able to exchange a low-ranking husband for a socially superior one, with all the positive effects on her fitness that trading up affords.[161]

Because of the risk of lost paternity due to unfaithful wives, men are well known to pay special attention to the resemblance between themselves and their putative children. Wives are well aware of this interest, and they are often quick to suggest that their newborns look very much like their husbands (even when impartial judges detect a greater similarity between a baby's appearance and that of the mother).[8] Moreover, a father's similarity to his child (as judged by himself or by others) clearly influences his investment in the child.[9,13]

Another result of the reproductive conflicts between the sexes may be a capacity for sexual jealousy, an emotional state that helps individuals prevent the fitness damage caused by an unfaithful mate. But the nature of sexual jealousy should differ between the sexes because adulterous partners harm men and women in different ways. A wife whose husband acquires another partner loses some or all of her access to her husband's wealth, which she needs to support herself and her children. Therefore, from an ultimate perspective, a woman's sexual jealousy should be focused on the loss of a provider, which is more likely to occur when a man becomes emotionally involved with another woman. On the other hand, a husband whose wife mates with another man may eventually care unknowingly for offspring fathered by that man. A man's sexual jealousy should therefore revolve around the potential loss of paternity and parental investment arising from a wife's extramarital sexual activity rather than the potential loss of resources and emotional commitment.[45]

If this view is correct, then if men and women were asked to imagine their responses to two scenarios—one in which a partner develops a deep friendship with another individual, and one in which a partner engages in sexual intercourse with another individual—women should find the first scenario more disturbing than the second, whereas men should be more upset at the thought of a mate copulating with another man. Data from several cultures confirm these predictions.[32,45] For example, in a study involving Swedish university students, who live in a fairly sexually permissive culture, 63 percent of the women found the prospect of emotional infidelity more troubling, whereas almost exactly the same percentage of the men deemed sexual infidelity more upsetting.[188]

Everywhere, men aspire to monopolize or restrict sexual access to their mates, although they do not always succeed. Marriage institutionalizes these ambitions. Although one sometimes hears of societies in which complete sexual freedom is the norm, the notion that such cultures actually exist appears to have been a (wishful?) misinterpretation on the part of outside observers. Sexual infidelity by a woman carries with it the risk of lost paternity (and fitness) for her mate; this risk appears to lie behind much sexual violence committed by men against their partners.[30,37,48] In some societies, a woman known to have cuckolded her husband may be legally killed by her aggrieved mate.[45]

Discussion Question

14.18   Mate guarding is an evolved response to sperm competition (see Chapter 7). Chimpanzee females regularly mate with several males in the same estrous cycle, whereas gorilla females almost never do, since they typically live in bands, each controlled by a single, powerful male. How large (as a proportion of body size) should the testes of chimpanzee males be relative to gorilla testes? If the testes of men are more similar to those of chimpanzees, what would this tell us about the intensity of sperm competition during our evolutionary past? If, on the other hand, human testes resemble those of gorillas, what conclusion is justified? Check your predictions against Harcourt et al.[78]

Coercive Sex

The fact that the murder of women suspected of adultery is still condoned in some parts of the world is one of the least attractive manifestations of sexual conflict in our species. Another is the regular occurrence of forced copulation, a phenomenon that is not limited to human beings, by the way (Figure 14.22). Despite the fact that human rapists are often severely punished, rape occurs in every culture studied to date.[175]

Previous analyses of rape include the highly influential book by Susan Brownmiller, *Against Our Will*.[26] In her view, rapists act on behalf of all men to instill fear in all women, the better to intimidate and control them, thus keeping them "in their place." Brownmiller's intimidation hypothesis implies that some males are willing to take the substantial punishment risks associated with rape in order to provide a benefit for many other men. This argument suffers from all the logical problems inherent in "for the good of the group" hypotheses (with the added difficulty that groups composed of only one sex cannot be the focus of any realistic sort of group selection), but we can test it anyway. If the evolved function of the trait is to subjugate all women, then the rapist element in male society can be predicted to target older, dominant women (or young women who aspire to positions of power) to demonstrate the penalty that comes from stepping outside the traditional subordinate role. This prediction is not supported: most rape victims are young, poor women.[172]

An alternative evolutionary hypothesis proposed by Randy and Nancy Thornhill is that rape is an adaptive tactic controlled by a conditional sexual strategy. According to the Thornhills, sexual selection has favored males with the capacity to commit rape under some conditions as a means of fertilizing

FIGURE 14.22 Rape occurs in animals other than humans. (A) In the beetle *Tegrodera aloga*, a male (right) can court a female (left) decorously by repeatedly drawing her antennae into grooves on his head; copulation ensues only if the female permits it. (B) Alternatively, a male (below) can force a female (above) to mate by running to her, grasping her, throwing her on her side, and inserting his everted genitalia as the female struggles to break free. (See also Figure 7.41). Photographs by the author.

(A)

(B)

eggs and leaving descendants.[172] As such, rape of strangers by men is analogous to forced copulation in *Panorpa* scorpionflies (see page 187), in which males unable to offer nuptial gifts use the low-gain, last-chance tactic of trying to force females to copulate with them.

Discussion Question

14.19 The proposition that rape might serve an adaptive sexual function has angered many people, including Brownmiller, who wrote, "It is reductive and reactionary to isolate rape from other kinds of violent antisocial behavior and to dignify it with adaptive significance."[27] And the philosopher and critic David Buller said that the view that rape could be a consequence of evolved psychological mechanisms was a "get-out-of-jail-free card" for rapists (quoted in Sharon Begley[17]). How might the Thornhills reply to these criticisms?

The explanatory hypothesis that human rape is an evolved reproductive tactic controlled by a conditional strategy generates the prediction that some raped women will become pregnant, which they do, even in modern societies in which many women take birth control pills.[175] In fact, copulatory rape apparently is more likely to result in pregnancy than consensual sex.[70] In the past, in the absence of reliable birth control technology and abortion procedures, rapists would have had a still higher probability of fathering children through forced copulation. Furthermore, if rape really is a product of an evolved reproductive mechanism, then rapists should more often target women of high fertility, just as bank swallows and other birds identify fertile (egg-laying) females and try to force those individuals to copulate with them.[16,168] In contrast, Brownmiller's view is that rape has nothing to do with reproduction but is merely another form of violent antisocial behavior driven by a desire of men to dominate women. Note that this hypothesis focuses on the proximate cause of rape, not its ultimate reproductive consequences; it is possible for rapists to be motivated purely by a desire to hurt women and yet to have children as a result of their aggression. The notion, however, that rape has no proximate sexual component at all leads to the prediction that the age distribution of rape victims should be the same as that of women robbed[57] or murdered[172] by male assailants. Crime data are at odds with this prediction (Figure 14.23).

Although these findings suggest that rape may increase the fitness of some men, it is also entirely possible that rape is not adaptive per se but is instead a maladaptive by-product of the male sexual psyche, which causes quick sexual arousal, a desire for variety in sexual partners, and an interest in impersonal sex, all attributes that generate many adaptive (i.e., fitness-enhancing) conse-

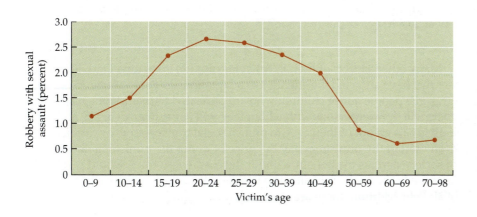

FIGURE 14.23 Testing alternative hypotheses for sexual assault. If rape were motivated purely by the intent to attack women (a proximate hypothesis), we would expect that the percentage of robbery victims who were sexually assaulted would not differ across age classes of the victims. Instead, women who were robbed *and* sexually assaulted are especially likely to be between 15 and 29 years of age, a result consistent with ultimate hypotheses proposing that coercive sex is linked to male sexual drive. Crime statistics come from a large database of 44,237 cases in which a lone male robbed a lone female. After Felson and Cundiff.[57]

quences while also incidentally leading some men to rape some women.[175] After all, men engage in many decidedly nonreproductive sexual activities, including masturbation, homosexual sex and rape of postmenopausal women and prepubertal girls, just as males of many nonhuman species also exhibit sexual activity that cannot possibly result in offspring, such as the copulatory mounting of weaned pups by male elephant seals.[155] Moreover, one attempt to estimate the reproductive consequences of rape for men in a traditional society yielded the conclusion that the fitness cost to a rapist exceeded the benefit by a factor of about ten,[167] a result that strongly suggests rape is not an adaptation, at least not in this society.

Even if coercive copulation usually reduces the fitness of its practitioners, the rape as by-product hypothesis would be tenable if the systems motivating male sexual behavior had a net positive effect on fitness. One prediction unique to the by-product hypothesis is that rapists will have unusually high levels of sexual activity with consenting as well as nonconsenting partners. Some evidence supports this hypothesis,[103,137] but as is true of many issues in human sexual behavior, more data are required. Nevertheless, in this case as in so many others, the adaptationist approach has generated novel hypotheses that are entirely testable in principle and practice. The evolutionary angle on rape is now available for skeptical scrutiny, and as a result, we may eventually gain a better understanding of the ultimate causes of the behavior. When we do, we will not in any way be obliged to be more understanding of the illegal and immoral activity of rapists.[175]

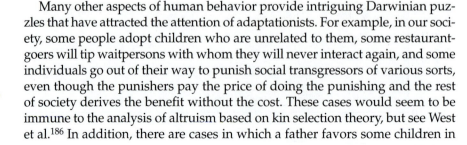

Discussion Questions

14.20   Discussions of rape are invariably emotionally charged. From an evolutionary perspective, why might women have an especially strong visceral response to the topic and an intense desire to punish rapists? Would an understanding of evolutionary theory have led to a revision in a legal ruling by the U.S. Supreme Court that contained the following claim: "Rape is without doubt deserving of serious punishment; but in terms of moral depravity and of the injury to the person and to the public, it does not compare with murder. ... [Rape] does not include ... even the serious injury to another person" (quoted in Jones[90])?

14.21   Natalie Angier states that married men have the same probability of fertilizing an egg per copulation with their wives as rapists do when forcing copulation on a victim.[12] In the past, the survival probability for an offspring of a married man was almost certainly much higher than that for a rapist's child, because married men often assist their children whereas rapists do not. Is Angier correct, therefore, in claiming that rape cannot be an adaptive tactic? (Remember that *adaptive* means "reproductively useful.") What do you make of the fact that low-status men are more likely to rape women unknown to them whereas high-status men dominate the category of acquaintance or partner rape?[182]

Many other aspects of human behavior provide intriguing Darwinian puzzles that have attracted the attention of adaptationists. For example, in our society, some people adopt children who are unrelated to them, some restaurant-goers will tip waitpersons with whom they will never interact again, and some individuals go out of their way to punish social transgressors of various sorts, even though the punishers pay the price of doing the punishing and the rest of society derives the benefit without the cost. These cases would seem to be immune to the analysis of altruism based on kin selection theory, but see West et al.[186] In addition, there are cases in which a father favors some children in his family more than others; given that fathers share 50 percent of their genes with their genetic offspring, favoritism would seem to be maladaptive, but it is not.[10] And then there are the cultural differences in human behavior, such as the requirement in some places that a bride bring a dowry to her husband or his family upon the bride's marriage, while in other groups the husband to be must ante up a bridewealth payment to the family of his bride to be. The view that these differences are the arbitary effects of cultural mores has been debunked.[65,79] Finally, there are all sorts of examples of irrational or erroneous thinking on the part of our fellow humans, such as the gambler's "hot hand" fallacy,[74] spider phobias,[50] and assorted "false memories,"[61] despite the fact that we pride ourselves on our intelligence and rationality. Although these may seem a hodgepodge of cognitive errors, they have been explained in terms of the adaptive properties of the human mind.[129,159]

Discussion Question

14.22   Use an evolutionary approach to explain why persons will give substantial tips to waiters and waitresses whom they do not know and whom they will never meet again.

Practical Applications of Evolutionary Theory

Although the ability of evolutionary theory to contribute to a purely academic understanding of our behavior pleases many biologists and psychologists, there is also a hardheaded, applied side to this approach. Perhaps we will be able to link the male drive for high status to an environmental good by making it possible for young men to conspicuously consume expensive "green" products, as an advertisement to women of their income and social consciousness.[76] A less hypothetical demonstration of the practicality of evolutionary thinking comes from the work of two evolutionary psychologists, Martin Daly and Margo Wilson, who were the first to show that child abuse was associated more strongly with stepparents than with genetic parents.[47] Criminal investigators working on cases of child abuse have surely benefited from an awareness of the statistical relationship between this kind of crime and the presence of a stepparent in the family. Likewise, some evolutionary biologists believe that if young women recognized that coercive sex, including rape, is not practiced solely by classic criminal sociopaths, they might be more on guard against more ordinary men, whose intense sex drive enables them to feel sexual desire even when their "partners" do not.

Moreover, everyone should know that our perfectly natural, highly adaptive drive to have children is directly linked to an exploding human population that has passed 7 billion on its way to 9 or 10 billion, much to the detriment of us all. Our efforts at getting things under control might be advanced if we all knew why we think fat, healthy babies are wonderful and why we believe nothing is too good for our children and grandchildren (Figure 14.24), even though the effect of having lots of plump babies and affording them a middle-class upbringing has pushed the life support system of our globe ever closer to the edge.[139]

Not only do we want large quantities of goods to provision our offspring adequately, many of us want to acquire wealth and use it ostentatiously, a key component of the reproductive strategy of men, or to marry someone who is rich, a common wish of unmarried women.[129] These factors and others have led the average resident of the United States to need all the resources that can

FIGURE 14.24   **The author with his two grandchildren,** Abby and Jake Alcock, both much admired additions to an already overcrowded world. Photograph by Nick Alcock.

be "harvested" from about 10 hectares of land. Since there are only about 12 billion hectares of productive land on the planet, and more than 300 million citizens of the USA, our single nation accounts for one-fourth of that entire amount. If all the 7 billion people of the world were to achieve a standard of living equal to that of the average American, we would need 70 billion hectares of productive land to do the job. Not going to happen. Even now we are eating our planet alive in an unsustainable manner (Figure 14.25), destroying the capital that our descendants will need.[149] Sadly, the current inhabitants of the Earth show little interest in taking the steps needed to reduce world population or avoid the destruction of our planet—but they should.[94]

Part of the reason why we are so reluctant to think of the long-range consequences of overpopulation and overconsumption is that people often discount the future when it comes to resource use, which is to say that most of us tend to spend what we have now at the expense of setting something aside for the long term.[192] Some persons have linked this aspect of human psychology to the high mortality rates that have characterized most of human evolution until very recently. Given the reality that the odds were against a long life, it would have been generally adaptive for people during our history as a species to consume resources as quickly as possible and to produce children promptly.

This argument can be tested by comparing the behavior of people in modern communities where life expectancy varies. Margo Wilson and Martin Daly have done the test, taking advantage of the fact that during a period around 1990, male life expectancy in different Chicago neighborhoods varied by about 25 years. If men possess a conditional strategy whose operating

FIGURE 14.25   **Unsustainable exploitation of the Earth's resources.** Huge numbers of poor Peruvian men literally dig up entire tropical rivers in their country in the search for small amounts of gold.

rules are affected by the likelihood of early death, then we would predict that individuals in high-mortality neighborhoods should be prepared to take greater risks than those living in places where death, from natural causes, is likely to be postponed. One dramatic manifestation of risk taking by men is their willingness to engage in extreme violence, which in turn affects homicide rates. As predicted, the rate at which homicides occurred in a neighborhood was tightly linked to male life expectancy (with the values for this figure adjusted by removal of the effects of homicide itself) (Figure 14.26).[191] In other words, when men perceive that their lives may well be short, they are more likely to take up a life of crime and violence—a high-risk tactic, but one with a potentially high immediate payoff for the successful risk taker.

The broader point is that not one product of natural selection, whether it be our capacity for risk taking or our ability to secure material wealth or our love of grandchildren, is guaranteed to produce socially or morally desirable results, either for us or for our species as a whole. Take our great capacity for cooperation. On the one hand, humans do often band together for goals that seem admirable: making a computer program work better, trying to eliminate polio, putting white-collar criminals in jail, staging a play or opera—the list is all but endless. But the ability to identify with a group and to form strong bonds with its members can also be employed in profoundly aggressive ways against other groups formed of people with similarly powerful cooperative abilities.

Indeed, Richard Alexander has argued that competition among human groups for resources led to the evolution of the extraordinary cooperation and the intensely group-centered sense of morality that people exhibit today.[3] The adaptively unpleasant nature of human morality can be illustrated by examining what the injunction "thou shalt not kill" has meant to people who proclaim it loudly. This commandment has almost always been interpreted to mean "thou shalt not kill members of one's own tribe or religious group or community or nation—but the destruction of others living outside the band, city, or country will be accepted, even encouraged, if these others pose a threat to the communal welfare or if the resources they control can be taken from them with only modest risk of personal injury."[80] Highly selective morality of this sort can be seen in the historically frequent and widespread practice of genocide.[51]

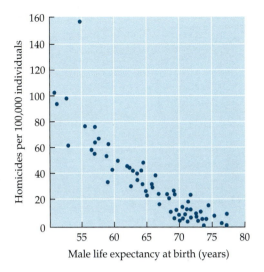

FIGURE 14.26 Homicide rates are highly correlated with male life expectancy. These data came from a number of neighborhoods in Chicago in which the average male age at death ranged from less than 55 to more than 75 years (after the effects of homicides themselves on life expectancy were removed). After Wilson and Daly.[191]

Discussion Question

14.23 In writing about genocide, Stephen Jay Gould reviewed the adaptationist hypothesis that the capacity for large-scale murder evolved as a result of intense competition between small bands during our evolutionary history.[72] Gould dismissed this hypothesis, saying the following: "An evolutionary speculation can only help if it teaches us something we don't know already—if, for example, we learned that genocide was biologically enjoined by certain genes, or even that a positive propensity, rather than a mere capacity, regulated our murderous potentiality. But the observational facts of human history speak against determination and only for potentiality. Each case of genocide can be matched with numerous incidents of social benevolence; each murderous clan can be paired with a pacific clan." Evaluate Gould's argument critically in the light of what you know about (1) the proximate–ultimate distinction, (2) conditional strategies, and (3) how to test adaptationist hypotheses.

The Triumph of an Evolutionary Analysis of Human Behavior

In the preceding pages, the ideas of many adaptationists have been reviewed with respect to the adaptive value of language and other intriguing features of human behavior. However, evolutionary explanations for elements of human behavior have not always been well received, especially by social scientists. The hostility to an evolutionary analysis of human behavior by a component of the biological community reached a peak soon after the 1975 publication of E. O. Wilson's *Sociobiology: The New Synthesis*.[190] The controversy surrounding **sociobiology**, now generally known as behavioral ecology or **evolutionary psychology**, has largely faded, but certainly not entirely.[36] Indeed, one still occasionally hears the very same criticisms given by Wilson's opponents in 1975 (e.g., see Begley[17]). Here are rebuttals for some of the major assertions of those who would dismiss the adaptationist approach to human behavior, an approach whose value has been demonstrated many times since 1975.

"WE HUMANS DON'T DO THINGS JUST BECAUSE WE WANT TO RAISE OUR INCLUSIVE FITNESS." Some persons have pointed out that, although humans have a great many desires, the wish to maximize our inclusive fitness rarely, if ever, is at the top of our list.[154] True, but if a baby cuckoo could talk, it would not tell you that it rolled its host's eggs out of the nest "because I want to propagate as many copies of my genes as possible." Neither cuckoos nor humans need be aware of the ultimate reasons for their activities in order to behave adaptively. It is enough that proximate mechanisms, like a well-developed sex drive, motivate individuals to do things, like copulate, that are correlated with the production of surviving offspring. On the proximate level, we want to have sex, we enjoy sweet foods, and we derive satisfaction from our ability to be funny because we possess physiological mechanisms that generate these emotions. Because honey tastes good, we want to eat it, and when we do, we acquire useful calories that may contribute to our survival and reproductive success even if we are completely unaware of the evolved function of our fondness for sweets. You do not have to be conscious of the ultimate reasons for a preference for sugar, in order to pass on the genes that contribute to the development of a sweet tooth.

"BUT NOT ALL HUMAN BEHAVIOR IS BIOLOGICALLY ADAPTIVE!" Over the years, critics have claimed that an array of cultural practices, such as circumcision, prohibitions against eating perfectly edible foods, and the use of birth control devices to limit fertility, seem unlikely to advance individual fitness. If some humans do things that reduce their fitness, these persons argue, then an adaptationist approach to human behavior must be invalid. This claim is based on the belief that natural selection theory requires that every aspect of every organism must be currently adaptive.[71] This assumption is incorrect, as noted in Chapter 5 (see Table 5.1), in part because some current environments of humans are different in many respects from the environments in which we evolved. Behavioral ecologists know this. Their goal is not to assert that a behavior of interest is an adaptation but to identify evolutionary puzzles, produce plausible hypotheses, and test alternative explanations. There is no guarantee that any selectionist hypothesis will withstand testing. This is as it should be. T. H. Huxley, the great defender of Darwinian theory, wrote, "There is a wonderful truth in [the] saying [that] next to being right in this world, the best of all things is to be clearly and definitely wrong, because you will come out somewhere."[86] If evolutionary hypotheses about the adaptive value of language are incorrect, effective tests will tell us so, which enables us to remove some ideas from the table.

"*EVOLUTIONARY APPROACHES TO HUMAN BEHAVIOR ARE BASED ON A POLITICALLY REACTIONARY DOCTRINE THAT SUPPORTS SOCIAL INJUSTICE AND INEQUALITY.*" The original critics of sociobiology denounced it as politically dangerous by providing scientific cover for immoral social policies of the sort advanced by racist and fascist demagogues in the past.[4,5] According to this view, the claim that such and such a trait is adaptive implies that it is both genetically fixed and good and therefore cannot and should not be changed. The implication of the critics was that sociobiologists were right-wing types who let their conservative politics affect their "science" (see Tybur et al. for an actual test of this claim[179]). Even if sociobiologists are as left-wing in their politics as most academics, which they are, critics might still assert that they are worried that evolutionary research might be used by malevolent politicians and their ilk to promote racist or sexist policies. This fear, however, could be put to rest if it were widely acknowledged that evolutionary researchers are trying to explain why humans behave the way they do, rather than attempting to justify any particular trait. This distinction is easily understood in cases involving other organisms. Biologists who study infanticide in Hanuman langurs or how a small marine copepod feeds on the eyes of Greenland sharks are never accused of approving of infanticide or the blinding of sharks. To say that something is biologically or evolutionarily adaptive means only that it tends to elevate the inclusive fitness of individuals with the trait—nothing more. It most certainly does not mean that an evolved behavior is moral or ethical. Think of infanticide if you wish.

Moreover, a hypothesis that a behavioral ability is adaptive does not mean that the characteristic is developmentally fixed. All biologists now understand that development is an interactive process involving both genes and environment. Change the environment, and you will change the gene–environment interactions underlying a behavioral phenotype, with the result that the phenotype may change. If, for example, you change the cultural environment that a baby is exposed to so that, say, a child of Spanish-speaking parents is raised by adults who speak Urdu, you know what will happen. But if infants are to become language users, they need very special genes, especially those genes that code for proteins that promote the development and maintenance of the brain elements that underlie language learning. Natural selection has surely had a role to play in the evolution of these mechanisms. This is an explanatory hypothesis, not a claim about developmental inflexibility, not a statement about the moral desirability of "natural" phenomena.

Discussion Questions

14.24   Marshall Sahlins argued that sociobiology is contradicted because people in most cultures do not even have words to express fractions. Without fractions, a person cannot possibly calculate coefficients of relatedness, and without this information (Sahlins claims), people cannot determine how to behave in order to maximize their indirect fitness.[158] Did Sahlins deliver a knockout blow to sociobiological theory?

14.25   Philip Kitcher states that "socially relevant science," such as sociobiology, demands "higher standards of evidence" because if a mistake is made (a hypothesis presented as confirmed when it is false), the societal consequences may be especially severe. For example, a hypothesis that men are more disposed to seek political power and high status in business and science than women is dangerous because it "threaten(s) to stifle the aspirations of millions."[100] Is Kitcher right?

One thing should be clear to everyone, whether an opponent or proponent of an adaptationist aproach to human behavior, namely that evolutionary analyses have demonstrated beyond doubt that just because something is "adaptive" or "natural" or "evolved" does not mean that it is "good" or "moral" or "desirable." Once you know that our genes have the capacity to make us work on their behalf without regard for our welfare or that of most other people, then it should be easier to fight against our evolved impulses.[189] Knowledge of evolutionary outcomes could, to name just one thing, make us less susceptible to exploitation by unscrupulous individuals.[42,44] Informed persons could also better avoid the moral certainty and self-righteousness that enables us to demonize and dehumanize our opponents prior to attacking and killing them. When Pogo declared, "We have met the enemy and he is us," he was correct, and those of us who understand the role of natural selection in shaping our evolutionary history know why he was right. Perhaps some of you can use that understanding to help make the human species less an enemy to itself.

Summary

1. Human beings are an animal species. Therefore, all four levels of analysis can be used to identify the causes of our behavior, including how and why we use language to communicate. All the neural and endocrine mechanisms in our bodies develop as a result of the continuous interplay between our genes and the environments in which they operate. The way in which our brains work provides the proximate foundation for our behavior. Some genetic differences lead to neurophysiological differences among people, with reproductive consequences for individuals. As a result of behavioral differences among people, some genes are passed on to the next generation more than others, leading to the evolution of our species and our behavior.

2. The adaptationist approach is the primary focus of sociobiology, aka behavioral ecology, evolutionary psychology, or evolutionary anthropology. Researchers in these disciplines use natural selection theory to generate testable hypotheses about the possible adaptive value of our species' behavior. These hypotheses are designed to *explain* why people do what they do (*not* to justify these actions as moral or desirable).

3. The utility of the sociobiological or adaptationist approach for an understanding of the ultimate causes of our behavior is especially apparent in the studies that have been done on mate choice by human beings. Sexual selection theory leads us to suspect that men and women will differ in how they evaluate potential mates. Women may focus on male traits ranging from skill in verbal courtship to indicators of wealth to evidence of parental potential. Men instead prefer the characteristics associated with high fertility in women, such as a youthful appearance, a low waist-to-hip ratio, and an intermediate body weight.

4. The hypothesis that our evolved psychological mechanisms increased the fitness of individuals living in the pre-contraceptive past is open to test. These tests have led some researchers to conclude that we possess adaptive mechanisms that lead to behavior almost universally considered immoral and undesirable, such as violent conflict between the sexes, including coercive sex.

5. When a claim is made that something is adaptive or natural, this does not mean that it is desirable, moral, or necessary, only that it tends to propagate our genes. Widespread acceptance of this point might enable us to understand that trusting our impulses is more likely to help us pass on our genes than to produce wise decisions that maximize either our personal happiness or the general good.

Suggested Reading

Sara Shettleworth has written a comprehensive and insightful review of the various attempts to determine whether chimpanzees (and other animals) have some capacity to learn a language.[164] The role of the *FOXP2* gene in the development of human language is the subject of a paper by Constance Scharff and Jana Petri.[160] The possible effect of sexual selection on the evolution of a capacity for language is spelled out in *The Mating Mind* by Geoffrey Miller.[127]

The misconceptions about E. O. Wilson's *Sociobiology*[190] have been examined in my book *The Triumph of Sociobiology*[1] as well as in articles by many other persons, including Doug Kenrick[98] and Owen Jones.[90] Martin Daly and Margo Wilson critically examine some recent critiques of the evolutionary explanation for child abuse by stepparents, one of the best examples of sociobiological research.[49]

Books that deal with the adaptive value of human mate choice (in addition to *The Mating Mind*) include those by David Geary[67] and David Buss.[34] The evolutionary significance of facial appearance is reviewed by Anthony Little and his colleagues.[117] Many classic papers on the evolutionary analysis of human behavior can be found, along with updates and critiques, in *Human Nature*, edited by Laura Betzig.[19] Robert Boyd and Joan Silk have written a wide-ranging textbook entitled *How Humans Evolved*.[22] David Buss[38] and Steve Gaulin and Donald McBurney[66] have authored textbooks on evolutionary psychology per se. Readers can also find many interesting articles on human behavior in the journal *Evolution and Human Behavior*.

Glossary

A

Action potential The neural signal; a self-regenerating change in membrane electrical charge that travels the length of a nerve cell.

Adaptation A characteristic that confers higher inclusive fitness to individuals than any other existing alternative exhibited by other individuals within the population; a trait that has spread or is spreading or is being maintained in a population as a result of natural selection or indirect (kin) selection.

Adaptationist A behavioral biologist who develops and tests hypotheses on the possible adaptive value of a particular trait. Persons using an adaptationist approach test whether a given trait enables individuals to propagate their special genes more effectively than if they had an alternative trait.

Adaptive altruism See *Altruism*.

Adaptive value The contribution that a trait or gene makes to inclusive fitness.

Allele A form of a gene. Different alleles typically code for distinctive variants of the same enzyme.

Altruism Helpful behavior that raises the recipient's direct fitness while lowering the donor's direct fitness.

> **Adaptive altruism** Altruism that increases the helper's inclusive fitness, in contrast to maladaptive altruism, which decreases the helper's inclusive fitness.
>
> **Facultative altruism** Altruism that the helper can employ at its discretion, in contrast to **obligate altruism**, in which helpers are locked into providing assistance to others.
>
> **Reciprocal altruism** See *Reciprocity*.

Arms race The result of selection acting on two parties that are in opposition to one another, as in the increasing sophistication of defensive mechanisms in a species that is preyed upon by an increasingly sophisticated predator.

Artificial selection See *Selection*.

Associated reproductive pattern A seasonal change in reproductive behavior that is tightly correlated with changes in the gonads and hormones, in contrast to a dissociated reproductive pattern, in which the onset of reproductive behavior is apparently not triggered by a sharp change in circulating hormones.

B

Behavioral strategy See *Strategy*.

Biological clock An internal physiological mechanism that enables organisms to time any of a wide assortment of biological processes and activities.

Brood parasite An animal that exploits the parental care of individuals other than its parents.

By-product hypothesis An explanation for a maladaptive or nonadaptive attribute that is said to occur as a by-product of a proximate mechanism that has some other adaptive consequence for individuals.

C

Central pattern generator A group of cells in the central nervous system that produces a particular pattern of signals necessary for a functional behavioral response.

Chase-away selection See *Selection*.

Circadian rhythm A roughly 24-hour cycle of behavior that expresses itself independent of environmental changes.

Circannual rhythm An annual cycle of behavior that expresses itself independent of environmental changes.

Coefficient of relatedness The probability that an allele present in one individual will be present in a close relative; the proportion of the total genotype of one individual present in the other, as a result of shared ancestry.

Command center A neural cluster or an integrated set of clusters that has primary responsibility for the control of a particular behavioral activity.

Comparative method A procedure for testing evolutionary hypotheses based on disciplined comparisons among species of known evolutionary relationships.

Conditional strategy   See *Strategy*.

Convergent evolution   The independent acquisition over time through natural selection of similar characteristics in two or more unrelated species.

Cooperation   A mutually helpful action that may have immediate benefits for both parties or postponed benefits for one of the cooperators.

Cost–benefit approach   A method for studying the adaptive value of alternative traits based on the recognition that phenotypes come with fitness costs and fitness benefits. An adaptation has a better cost–benefit ratio than alternative versions of that trait.

Cryptic female choice   The ability of a female in receipt of sperm from more than one male to choose whose sperm get to fertilize her eggs.

D

Darwinian puzzle   A trait that appears to reduce the fitness of individuals that possess it. Traits of this sort attract the attention of evolutionary biologists.

Developmental homeostasis   The capacity of developmental mechanisms within individuals to produce adaptive traits, despite potentially disruptive effects of mutant genes and suboptimal environmental conditions.

Dilution effect   Safety in numbers that comes from swamping the ability of local predators to consume prey.

Diploid   Having two copies of each gene in one's genotype.

Direct fitness   See *Fitness*.

Direct selection   See *Selection*.

Display   A stereotyped action used as a communication signal by individuals.

Dissociated reproductive pattern   See *Associated reproductive pattern*.

Divergent evolution   The evolution by natural selection of differences between closely related species that live in different environments and are therefore subject to different selection pressures.

Dominance hierarchy   A social ranking within a group, in which some individuals give way to others, often conceding useful resources to others without a fight.

E

Eavesdropping   The detection of signals from a legitimate signaler by an illegitimate receiver to the detriment of the signaler and the benefit of the receiver.

Entrainment   The process of resetting a biological clock so that an organism's activities are scheduled in keeping with local conditions.

Ethology   The study of the proximate mechanisms and adaptive value of animal behavior.

Eusocial   Refers to species in which colonies contain specialized nonreproducing castes that work for the reproductive members of the group.

Evolutionary history   The sequence of changes that have occurred over time as an ancestral trait becomes modified and takes on a new form (and sometimes, a new function).

Evolutionarily stable strategy   See *Strategy*.

Evolutionary psychology   The study of the adaptive value of psychological mechanisms, especially of human beings; a key component of sociobiology.

Explosive breeding assemblage   The temporary formation of large groups of mating individuals.

Extra-pair copulation   A mating by a male or female with someone other than his or her primary partner in a seemingly monogamous species.

F

Facultative altruism   See *Altruism*.

Female defense polygyny   See *Polygyny*.

Female-enforced monogamy   See *Monogamy*.

Female preference hypothesis   An explanation for the formation of leks in which females prefer to choose mates from a group rather than by inspecting potential mates one by one.

Fertility insurance hypothesis   An explanation for why females might mate with more than one male per breeding cycle, with the benefit being an increase in egg fertilization rate.

Fitness   A measure of the genes contributed to the next generation by an individual, often stated in terms of the number of surviving offspring produced by the individual.

   Direct fitness   The genes contributed by an individual via personal reproduction to the bodies of surviving offspring.

   Inclusive fitness   The sum of an individual's direct and indirect fitness.

   Indirect fitness   The genes contributed by an individual indirectly by helping nondescendant kin, in effect creating relatives that would not have existed without the help of the individual.

Fitness benefit   That aspect of a trait that tends to raise the inclusive fitness of individuals.

Fitness cost   That aspect of a trait that tends to reduce the inclusive fitness of individuals.

Fixed action pattern   An innate, highly stereotyped response that is triggered by a well-defined, simple stimulus. Once the pattern is activated, the response is performed in its entirety.

Free-running cycle   The cycle of activity of an individual that is expressed in a constant environment.

Frequency-dependent selection   See *Selection*.

G

Game theory   An evolutionary approach to study of adaptive value in which the payoffs to individuals associated with one behavioral tactic are dependent on what the other members in the group are doing.

Gene   A segment of DNA, typically one that encodes information about the sequence of amino acids that makes up a protein.

Genetic compatibility hypothesis   An explanation for why females may exhibit mate choice designed to acquire genes present in the sperm of particular males because those genes complement the genes present in their eggs, resulting in an increased likelihood of the development of superior offspring.

Genetic–developmental mechanism   An internal system based on genetic information that affects the way in which an individual develops during its lifetime.

Genotype   The genetic constitution of an individual; refers to either the alleles of one gene possessed by the individual or to its complete set of genes.

Good genes theory   The argument that mate choice advances individual fitness because it provides the offspring of choosy individuals with genes that promote reproductive success by advancing the offspring's chances of survival or reproductive success.

Good parent theory   An explanation for female preferences for males whose appearance or behavior signals that these potential mates are likely to provide above-average parental care for their offspring.

Group selection   See *Selection*.

H

Hamilton's rule   The argument made by W. D. Hamilton that altruism can spread through a population where $rB > C$ (with r being the coefficient of relatedness between the altruist and the individual helped, B being the fitness benefit received by the helped individual, and C being the cost of altruism in terms of the direct fitness lost by the altruist due to its actions).

Haploid   Having only one copy of each gene in the genotype, as for example in the sperm and eggs of diploid organisms.

Healthy mates theory   An explanation for preferences by females for males whose appearance or behavior signals that these potential mates are unlikely to transmit communicable diseases or parasites to them.

Home range   An area that an animal occupies but does not defend, in contrast to a territory, which is defended.

Honest signal   A signal that conveys accurate information about some aspect of the signaler's quality, such as its fighting ability or value as a potential mate.

Hotshot   A male whose attributes are especially appealing to sexually receptive females.

Hotshot hypothesis   A tentative explanation for why males cluster their display territories in a lek. The clustering is said to occur because subordinate males are attracted to dominant individuals that are especially appealing to females, in order perhaps to gain sexual access to some of these females.

Hotspot   A location whose properties attract sexually receptive females to the male able to hold the site against rival males.

Hotspot hypothesis   A tentative explanation for why males cluster their display territories in a lek. The clustering is said to occur because males compete to display near a place that tends to attract passing females.

I

Ideal free distribution theory   The general theory that explains why individuals distribute themselves spatially in certain way in terms of the fitness gains generated by the decisions of individuals that are free to go wherever they choose.

Illegitimate receiver   An individual that listens to the signals of others, thereby gaining information that it uses to reduce the fitness of the signaler.

Illegitimate signaler   An individual that produces signals that may deceive others into responding in ways that reduce the fitness of the signal receiver.

Imprinting   A form of learning in which individuals exposed to certain key stimuli, usually early in life, form an association with an object (or individual) and may later attempt to mate with similar objects.

Inclusive fitness   See *Fitness*.

Indirect fitness   See *Fitness*.

Indirect reciprocity   See *Reciprocity*.

Indirect selection   See *Selection*.

Innate releasing mechanism   A hypothetical neural mechanism thought to control an innate response to a sign stimulus.

Instinct   A behavior pattern that reliably develops in most individuals, promoting a functional response to a releaser stimulus the first time the behavior is performed.

Intelligent design theory   The argument that biological features are too complex to have evolved by natural processes, such as natural selection, and that therefore an intelligent designer, such as a god, is required to account for the existence of these complex features.

Interneuron   A nerve cell that relays messages either from receptor neurons to the central nervous system (a sensory interneuron) or from the central nervous system to neurons commanding muscle cells (a motor interneuron).

K

Kin selection   See *Selection*.

L

Learning   A durable and usually adaptive change in an animal's behavior traceable to a specific experience in the individual's life.

Lek   A traditional display site that females visit to select a mate from among the males displaying at their small resource-free territories.

Lek polygyny   See *Polygyny*.

M

Mate-assistance hypothesis   A possible explanation for why in monogamous species males might forgo attempts to be polygynous. By assisting a single partner, a male might achieve higher fitness than if he were to attempt to be polygynous.

Mate-assistance monogamy   See *Monogamy*.

Mate guarding   Actions taken by males (usually) to prevent sexual partners from acquiring sperm from other males.

Mate-guarding hypothesis   A possible explanation for why in monogamous species males might forgo attempts to be polygynous. By remaining with a single partner in order to guard her against other males, the monogamous male might fertilize more of his mate's eggs than otherwise, compensating him for a reduction in his interactions with other females.

Mate-guarding monogamy   See *Monogamy*.

Material benefits   The food or other key resources that males supply their mates, which can make it advantageous for females of some species to mate with several males per breeding cycle.

Migration   The regular movement back and forth between two relatively distant locations by animals that use resources concentrated in these different sites.

Monogamy   A mating system in which one male mates with just one female, and one female mates with just one male, in a breeding season.

> **Female-enforced monogamy**   A mating system in which females prevent their mates from copulating with more than one individual, resulting in partnerships of one male and one female.

> **Mate-assistance monogamy**   Monogamy that arises because a male gains more fitness by offering parental care for the offspring of his mate than by seeking out additional sexual partners.

> **Mate-guarding monogamy**   The mating system that occurs when one or the other member of a pair guards its partner in ways that prevent that partner from acquiring an additional mate.

N

Natural selection   See *Selection*.

Neuron   A nerve cell.

Neuronal–hormonal mechanism   A system of nervous cells and/or endocrine cells whose effects underlie the ability of an individual to carry out a physiological or behavioral function.

Nondescendant kin   Relatives other than offspring.

Nuclei   Dense clusters of cell bodies of neurons within nervous systems.

Nuptial gift   A food item transferred by a male to a female just prior to or during copulation.

O

Obligate altruism   See *Altruism*.

Operant conditioning   A kind of learning based on trial and error, in which an action, or operant, becomes more frequently performed if it is rewarded.

Operational sex ratio   The ratio of receptive males to receptive females over a given period.

Optimality theory   An evolutionary theory based on the assumption that the attributes of organisms are optimal, that is, better than others in terms of the ratio of fitness benefits to costs. The theory is used to generate hypotheses about the possible adaptive value of traits in terms of the net fitness gained by individuals that exhibit these attributes.

P

Parent–offspring conflict   The clash of interests that occurs when parents can gain fitness by withholding parental care or resources from some offspring in order to invest in others now or later, even though the deprived offspring would gain more fitness by receipt of parental care or resources.

Parental effort   Time, energy, or risks taken by an adult to produce or care for its offspring.

Parental investment   Costly parental activities that increase the likelihood of survival for some existing offspring but that reduce the parent's chances of producing offspring in the future.

Phenotype   Any measurable aspect of an individual that arises from an interaction of the individual's genes with its environment.

Pheromone   A volatile chemical released by an individual as a scent signal for another.

Photoperiod   The number of hours of light in 24 hours.

Phylogeny   An evolutionary genealogy of the relationships between a number of species or clusters of species that can be used to develop hypotheses on the evolutionary history of a given trait.

Pleiotropy   The capacity of a gene to have multiple developmental effects on individuals.

Polyandry   A mating system in which a female has several partners in a breeding season.

Polygyny   A mating system in which a male fertilizes the eggs of several partners in a breeding season.

> **Female defense polygyny**   Polygyny in which males directly defend several mates.

> **Lek polygyny**   Polygyny in which males attract several mates to a display territory.

> **Resource defense polygyny**   Polygyny in which males acquire several mates attracted to resources under the males' control.

> **Scramble competition polygyny**   Polygyny in which males acquire several widely scattered mates by finding them first.

Polygyny threshold hypothesis   An explanation for polygyny based on the premise that females will gain fitness by mating with an already paired male if the resources controlled by that male greatly exceed those under the control of unmated males.

Polyphenism   The occurrence within a species of two or more alternative phenotypes whose differences are induced by key differences in the environments experienced by individual members of this species.

Prisoner's dilemma   A game theory construct in which the fitness payoffs to individuals are set such that mutual cooperation between the players generates a lower return than defection, which occurs when one individual accepts assistance from the other but does not return the favor.

Proximate cause   An immediate, underlying cause based on the operation of internal mechanisms possessed by an individual.

R

Reciprocity   Also known as **reciprocal altruism** in which a helpful action is repaid at a later date by the recipient of assistance.

> **Indirect reciprocity**   A form of reciprocity in which a helpful action is repaid at a later date by individuals other than the recipient of assistance.

Releaser   A sign stimulus given by an individual as a social signal to another.

Reproductive success   The number of surviving offspring produced by an individual; direct fitness.

Reproductive value   A measure of the probability that a given offspring will reach the age of reproduction, or the potential of an individual to leave surviving descendants in the future.

Resource defense polygyny   See *Polygyny.*

Resource-holding power   The inherent capacity of an individual to defeat others when competing for useful resources.

Runaway selection   See *Selection.*

S

Satellite male   A male that waits near another male to intercept females drawn to the signals produced by the other male or attracted by the resources defended by the other male.

Scramble competition polygyny   See *Polygyny.*

Selection   The effects of differences between individuals in their ability to transmit copies of their genes to the next generation.

> **Artificial selection**   A process that is identical to natural selection, except that humans control the reproductive success of alternative types within the selected population.

> **Chase-away selection**   The reciprocal, spiraling effects of males attempting to exploit female mate choice

mechanisms while females are evolving resistance to these attempts.

> **Direct selection**   Natural selection; a term that refers to selection that results from hereditary differences between individuals in the production of surviving offspring.

> **Frequency-dependent selection**   A form of natural selection in which those individuals that happen to belong to the less common of two types in the population are the ones that are more fit because of their lower frequency in the population.

> **Group selection**   The process that occurs when groups differ in their collective attributes and the differences affect the survival chances of the groups.

> **Kin selection (indirect selection)**   The process that occurs when individuals differ in ways that affect their parental care or helping behavior, and thus the survival of their own offspring or the survival of nondescendant kin.

> **Natural selection (direct selection)**   The process that occurs when individuals differ in their traits and the differences are correlated with differences in reproductive success. Natural selection can produce evolutionary change when these differences are inherited.

> **Runaway selection**   A form of sexual selection that occurs when female mating preferences for certain male attributes create a positive feedback loop favoring both males with these attributes and females that prefer them.

> **Sexual selection**   A form of natural selection that occurs when individuals vary in their ability to compete with others for mates or to attract members of the opposite sex. When the variation between individuals is correlated with genetic differences, sexual selection leads to genetic changes in the population.

Selfish herd   A group of individuals whose members use others as living shields against predators.

Sensory exploitation   The evolution of signals that happen to activate established sensory systems of signal receivers in ways that elicit responses favorable to the signal sender.

Sex role reversal   A change in the typical behavior patterns of males and females, as when, for example, females compete for access to males or when males choose selectively among potential mates.

Sexual selection   See *Selection.*

Siblicide   The killing of a sibling by a brother or sister.

Sign stimulus   The effective component of an action or object that triggers a fixed action pattern in an animal.

Sociobiology   A discipline that uses evolutionary theory as the foundation for the study of social behavior; often used to refer to studies of this sort involving human beings.

Sperm competition   The competition between males that determines whose sperm will fertilize a female's eggs when both males' sperm have been accepted by that female.

Stimulus filtering   The capacity of nerve cells and neural networks to ignore stimuli that could potentially elicit a response from them.

Strategy   A genetically distinctive set of rules for traits exhibited by individuals.

Behavioral strategy   A genetically distinctive set of behavioral decision-making rules exhibited by individuals.

Conditional strategy   A set of rules that enables individuals to use different tactics under different environmental conditions; the inherited behavioral capacity to be flexible in response to certain cues or situations.

Evolutionarily stable strategy   That set of rules of behavior that when adopted by a certain proportion of the population cannot be replaced by any alternative strategy.

Synapse   The point of near contact between one nerve cell and another.

T

Tactic   A behavior pattern that is enabled by an underlying hereditary mechanism of some sort. Tactics are often referred to when someone wishes to distinguish between an option available to an individual, as opposed to a strategy, which specifies a fixed response to a particular situation.

Territorial   Exhibiting a readiness to defend an area against intruders.

Theory of descent with modification   Darwin's argument that over evolutionary history, changes accumulate gradually in ancestral species, altering them more and more as these ancestral species evolve into more recent forms derived from their predecessors.

U

Ultimate cause   The evolutionary, historical reason why something is the way it is.

Bibliography

Chapter 1

1. Alcock, J., Jones, C. E., and Buchmann, S. L. 1976. Location before emergence of female bee, *Centris pallida*, by its male (Hymenoptera: Anthophoridae). *Journal of Zoology* 179: 189–199.

2. Bartoš, L., Bartošová, J., Pluháček, J., and Sindelářová, J. 2011. Promiscuous behaviour disrupts pregnancy block in domestic horse mares. *Behavioral Ecology and Sociobiology* 65: 1567–1572.

3. Beehner, J. C. and Bergman, T. J. 2008. Infant mortality following male takeovers in wild geladas. *American Journal of Primatology* 70: 1152–1159.

4. Birkhead, T. R. 2010. Undiminished passion. In T. Székely, M. C. Moore, and J. Komdeur (Eds.), *Social Behaviour: Genes, Ecology and Evolution*. Cambridge University Press, Cambridge, England.

5. Borries, C., Launhardt, K., Epplen, C., Epplen, J. T., and Winkler, P. 1999. DNA analyses support the hypothesis that infanticide is adaptive in langur monkeys. *Proceedings of the Royal Society of London B* 266: 901–904.

6. Costa, J. T. 2009. *The Annotated Origin: A Facsimile of the First Edition of* On the Origin of Species. Harvard University Press, Cambridge, MA.

7. Curtin, R. and Dolhinow, P. 1978. Primate social behavior in a changing world. *American Scientist* 66: 468–475.

8. Darwin, C. 1859. *On the Origin of Species*. Murray, London, England.

9. Dawkins, R. 1986. *The Blind Watchmaker*. W. W. Norton, New York, NY.

10. Dawkins, R. 1989. *The Selfish Gene* (2nd ed.). Oxford University Press, Oxford, England.

11. Dobzhansky, T. 1973. Nothing makes sense except in the light of evolution. *American Biology Teacher* 35: 125–129.

12. Evans, H. E. 1966. *Life on a Little Known Planet*. Dell, New York, NY.

13. Evans, H. E. 1973. *Wasp Farm*. Anchor Press, Garden City, NY.

14. Griggs, D. J. and Kestin, T. S. 2011. Bridging the gap between climate scientists and decision makers. *Climate Research* 47: 139–144.

15. Heaney, V. and Monaghan, P. 1995. A within-clutch trade-off between egg-production and rearing in birds. *Proceedings of the Royal Society of London B* 261: 361–365.

16. Heinrich, B. 1989. *Ravens in Winter*. Summit Books, New York, NY.

17. Heinrich, B. 2004. *The Geese of Beaver Bog*. HarperCollins, New York, NY.

18. Hölldobler, B. and Wilson, E. O. 1994. *Journey to the Ants*. Harvard University Press, Cambridge, MA.

19. Hrdy, S. B. 1977. Infanticide as a primate reproductive strategy. *American Scientist* 65: 40–49.

20. Knörnschild, M., Ueberschaer, K., Helbig, M., and Kalko, E. K. V. 2011. Sexually selected infanticide in a polygynous bat. PLoS ONE 6(9): e25001.doi:10.1371/journal.pone.0025001.

21. Le Page, M. 2008, April 16. Evolution: 24 myths and misconceptions. *New Scientist* (http://www.newscientist.com/article/dn13620-evolution-24-myths-and-misconceptions.html).

22. Lorenz, K. Z. 1952. *King Solomon's Ring*. Crowell, New York, NY.

23. Lyon, J. E., Pandit, S. A., van Schalk, C. P., and Pradhan, G. R. 2011. Mating strategies in primates: A game theoretical approach to infanticide. *Journal of Theoretical Biology* 274: 103–108.

24. Monaghan, P. and Nager, R. G. 1997. Why don't birds lay more eggs? *Trends in Ecology & Evolution* 12: 270–274.

25. Packer, C. 1994. *Into Africa*. University of Chicago Press, Chicago, IL.

26. Schaller, G. B. 1964. *The Year of the Gorilla*. University of Chicago Press, Chicago, IL.

27. Schaller, G. B. 1972. *The Serengeti Lion*. University of Chicago Press, Chicago, IL.

28. Sommer, V. 1994. Infanticide among free-ranging langurs (*Presbytis entellus*) of Jodhpur (Rajasthan/India): Recent observations and a reconsideration of hypotheses. *Primates* 28: 163–197.

29. Tinbergen, N. 1958. *Curious Naturalists*. Doubleday, Garden City, NY.

30. Vanderwerf, E. 1992. Lack's clutch size hypothesis: An examination of the evidence using meta-analysis. *Ecology* 73: 1699–1705.

31. Williams, G. C. 1966. *Adaptation and Natural Selection*. Princeton University Press, Princeton, NJ.

32. Woodward, J. and Goodstein, D. 1996. Conduct, misconduct and the structure of science. *American Scientist* 84: 479–490.

Chapter 2

1. Bernasconi, G. and Strassmann, J. E. 1999. Cooperation among unrelated individuals: The ant foundress case. *Trends in Ecology and Evolution* 14: 477–482.

2. Biesmeijer, J. C. and Seeley, T. D. 2005. The use of waggle dance information by honey bees throughout

their foraging careers. *Behavioral Ecology and Sociobiology* 59: 133–142.

3. Boomsma, J. J. 2009. Lifetime monogamy and the evolution of eusociality. *Philosophical Transactions of the Royal Society B* 364: 3191–3207.

4. Bourke, A. F. G. 2008. Social evolution: Daily self-sacrifice by worker ants. *Current Biology* 18: R1100–R1101.

5. Bourke, A. F. G. 2011. *Principles of Social Evolution*. Oxford University Press, Oxford, England.

6. Bourke, A. F. G. 2011. The validity and value of inclusive fitness theory. *Proceedings of the Royal Society B* 278: 3313–3320.

7. Brown, J. L. 1987. *Helping and Communal Breeding in Birds: Ecology and Evolution*. Princeton University Press, Princeton, NJ.

8. Cardinal, S. and Danforth, B. N. 2011. The antiquity and evolutionary history of social behavior in bees. *PLoS ONE* 6(6): e21086.doi:10.1371/journal.pone.0021086.

9. Dawkins, R. 1986. *The Blind Watchmaker*. W. W. Norton, New York, NY.

10. Dawkins, R. 1989. *The Selfish Gene* (2nd ed.) Oxford University Press, Oxford, England.

11. Dornhaus, A. and Chittka, L. 2001. Food alert in bumblebees (*Bombus terrestris*): Possible mechanisms and evolutionary implications. *Behavioral Ecology and Sociobiology* 50: 570–576.

12. Eldakar, O. T. and Wilson, D. S. 2011. Eight criticisms not to make about group selection. *Evolution* 65: 1523–1526.

13. Flowers, J. M., Li, S. I., Stathos, A., Saxer, G., Ostrowski, E. A., Queller, D. C., Strassmann, J. E., and Purugganan, M. D. 2010. Variation, sex, and social cooperation: Molecular population genetics of the social amoeba *Dictyostelium discoideum*. *PLoS Genetics* 6(7): e1001013.doi: 10.1371/journal.pgen.1001013.

14. Foster, K. R., Wenseleers, T., and Ratnieks, F. L. W. 2006. Kin selection is the key to altruism. *Trends in Ecology and Evolution* 21: 57–60.

15. Fox, M., Sear, R., Beise, J., Ragsdale, G., Voland, E., and Knapp, L. A. 2010. Grandma plays favourites: X-chromosome relatedness and sex-specific childhood mortality. *Proceedings of the Royal Society B* 277: 567–573.

16. Gardner, A., Alpedrinha, J., and West, S. A. 2012. Haplodiploidy and the evolution of eusocialty: Split sex ratios. *American Naturalist* 179: 240–256.

17. Gitschier, J. 2008. Taken to school: An interview with the Honorable Judge John E. Jones III. *PLoS Genetics* 4(12): e1001096.doi:10.1371/journal.pgen.1000297.

18. Hamilton, W. D. 1964. The genetical theory of social behaviour, I, II. *Journal of Theoretical Biology* 7: 1–52.

19. Hamilton, W. D. 1996. Foreword. In S. Turillazzi and M. J. West-Eberhard (Eds.), *Natural History and the Evolution of Paper Wasps* (pp. v–vi). Oxford University Press, Oxford, England.

20. Heinze, J., Hölldobler, B., and Peeters, C. 1994. Conflict and cooperation in ant societies. *Naturwissenschaften* 81: 489–497.

21. Hughes, W. O. H., Oldroyd, B. P., Beekman, M., and Ratnieks, F. L. W. 2008. Ancestral monogamy shows kin selection is key to evolution of sociality. *Science* 320: 1213–1216.

22. Johns, P. M., Howard, K. J., Breisch, N. L., Rivera, A., and Thorne, B. L. 2009. Nonrelatives inherit colony resources in a primitive termite. *Proceedings of the National Academy of Sciences* 106: 17452–17456.

23. Kirchner, W. H. and Grasser, A. 1998. The significance of odor cues and dance language information for the food search behavior of honeybees (Hymenoptera: Apidae). *Journal of Insect Behavior* 11: 169–178.

24. Liebig, J., Peeters, C., and Hölldobler, B. 1999. Worker policing limits the number of reproductives in a ponerine ant. *Proceedings of the Royal Society of London B* 266: 1865–1870.

25. Lindauer, M. 1961. *Communication among Social Bees*. Harvard University Press, Cambridge, MA.

26. Maguire, G. S. 2006. Territory quality, survival and reproductive success in southern emu-wrens *Stipiturus malachurus*. *Journal of Avian Biology* 37: 579–593.

27. Marshall, J. A. R. 2011. Group selection and kin selection: Formally equivalent approaches. *Trends in Ecology & Evolution* 26: 325–332.

28. Maschwitz, U. and Maschwitz, E. 1974. Platzende Arbeiterinnen: Eine neue Art der Feindabwehr bei sozialen Hautflüglern. *Oecologia* 14: 289–294.

29. Meunier, J., West, S. A., and Chapuisat, M. 2008. Split sex ratios in the social Hymenoptera: A meta-analysis. *Behavioral Ecology* 19: 382–390.

30. Monnin, T., Ratnieks, F. L. W., Jones, G. R., and Beard, R. 2002. Pretender punishment induced by chemical signalling in a queenless ant. *Nature* 419: 61–65.

31. Nieh, J. C. 2004. Recruitment communication in stingless bees (Hymenoptera, Apidae, Meliponini). *Apidologie* 35: 159–182.

32. Nowak, M. A., Tarnita, C. E., and Wilson, E. O. 2011. The evolution of eusociality. *Nature* 466: 1057–1062.

33. Ostrowski, E. A., Katoh, M., Shaulsky, G., Queller, D. C., and Strassmann, J. E. 2008. Kin discrimination increases with genetic distance in a social amoeba. *PLoS Biology* 6: 2376–2382.

34. Page, R. E., Robinson, G. E., Fondrk, M. K., and Nasr, M. E. 1995. Effects of worker genotypic diversity on honey-bee colony development and behavior (*Apis mellifera* L.). *Behavioral Ecology and Sociobiology* 36: 387–396.

35. Penny, D. 2011. Darwin's theory of descent with modification, versus the biblical tree of life. *PLoS Biology* 9(7): e1001096.doi:10.1371/journal.pbio.1001096.

36. Queller, D. C. 1996. The measurement and meaning of inclusive fitness. *Animal Behaviour* 51: 229–232.

37. Ratnieks, F. L. W. and Helanterä, H. 2009. The evolution of extreme altruism and inequality in insect societies. *Philosophical Transactions of the Royal Society B* 364: 3169–3179.

38. Ratnieks, F. L. W., Foster, K. R., and Wenseleers, T. 2011. Darwin's special difficulty: The evolution of "neuter insects" and current theory. *Behavioral Ecology and Sociobiology* 65: 481–492.

39. Reeve, H. K. 2000. Review of *Unto Others: The Evolution and Psychology of Unselfish Behavior*. *Evolution and Human Behavior* 21: 65–72.

40. Richard, F.-J., Tarpy, D., and Grozinger, C. 2007. Effects of insemination quantity on honey bee queen physiology. *PLoS ONE* 2(10): e980.doi: 10:1371/journal.pone.0000980.

41. Strassmann, J. E. and Queller, D. C. 2007. Insect societies as divided organisms: The complexities of purpose and cross-purpose. *Proceedings of the National Academy of Sciences* 104: 8619–8626.

42. Strassmann, J. E., Page, R. E., Robinson, G. E., and Seeley, T. D. 2011. Kin selection and eusociality. *Nature* 471: E5–E6.

43. Strassmann, J. E., Queller, D. C., Avise, J. C., and Ayala, F. J. 2011. In the light of evolution V: Cooperation and conflict. *Proceedings of the National Academy of Sciences* 108: 10787–10791.

44. Tarpy, D. R. 2003. Genetic diversity within honeybee colonies prevents severe infections and promotes colony growth. *Proceedings of the Royal Society of London B* 270: 99–103.

45. Tofilski, A., Couvillon, M. J., Evison, S. E. F., Helanterä, H., Robinson, E. J. H., and Ratnieks, F. L. W. 2008. Preemptive defensive self-sacrifice by ant workers. *American Naturalist* 172: E239–E243.

46. Trivers, R. L. and Hare, H. 1976. Haplodiploidy and the evolution of the social insects. *Science* 191: 249–263.

47. von Frisch, K. 1967. *The Dance Language and Orientation of Bees*. Harvard University Press, Cambridge, MA.

48. Wenseleers, T. and Ratnieks, F. L. W. 2004. Tragedy of the commons in *Melipona* bees. *Proceedings of the Royal Society of London B* 271: S310–S312.

49. Wenseleers, T. and Ratnieks, F. L. W. 2006. Comparative analysis of worker reproduction and policing in eusocial Hymenoptera supports relatedness theory. *American Naturalist* 168: E163–E179.

50. Wenseleers, T. and Ratnieks, F. L. W. 2006. Enforced altruism in insect societies. *Nature* 444: 50.

51. West, S. A., El Mouden, C., and Gardner, A. 2011. Sixteen common misconceptions about the evolution of cooperation in humans. *Evolution and Human Behavior* 32: 231–262.

52. Williams, G. C. 1966. *Adaptation and Natural Selection*. Princeton University Press, Princeton, NJ.

53. Wilson, D. S. 1975. Theory of group selection. *Proceedings of the National Academy of Sciences* 72: 143–146.

54. Wilson, D. S. and Wilson, E. O. 2007. Rethinking the theoretical foundation of sociobiology. *Quarterly Review of Biology* 82: 327–348.

55. Wilson, D. S. and Wilson, E. O. 2008. Evolution "for the good of the group." *American Scientist* 96: 380–389.

56. Wilson, E. O. 1971. *The Insect Societies*. Harvard University Press, Cambridge, MA.

57. Wilson, E. O. 1975. *Sociobiology, the New Synthesis*. Harvard University Press, Cambridge, MA.

58. Wynne-Edwards, V. C. 1962. *Animal Dispersion in Relation to Social Behaviour*. Oliver & Boyd, Edinburgh, Scotland.

Chapter 3

1. Alexander, R. D. 1974. The evolution of social behavior. *Annual Review of Ecology and Systematics* 5: 325–383.

2. Axelrod, R. and Hamilton, W. D. 1981. The evolution of cooperation. *Science* 211: 1390–1396.

3. Baglione, V., Marcos, J. M., Canestrari, D., and Ekman, J. 2002. Direct fitness benefits of group living in a complex cooperative society of carrion crows, *Corvus corone corone*. *Animal Behaviour* 64: 887–893.

4. Bilde, T., Coates, K. S., Birkhofer, K., Bird, T., Maklakov, A. A., Lubin, Y., and Aviles, L. 2007. Survival benefits select for group living in a social spider despite reproductive costs. *Journal of Evolutionary Biology* 20: 2412–2426.

5. Bourke, A. F. G. 2011. *Principles of Social Evolution*. Oxford University Press, Oxford, England.

6. Braude, S. 2000. Dispersal and new colony formation in wild naked mole-rats: Evidence against inbreeding as the system of mating. *Behavioral Ecology* 11: 7–12.

7. Brown, C. R. and Brown, M. B. 1986. Ecto-parasitism as a cost of coloniality in cliff swallows (*Hirundo pyrrhonota*). *Ecology* 67: 1206–1218.

8. Brown, C. R. 1988. Social foraging in cliff swallows: Local enhancement, risk sensitivity, competition and the avoidance of predators. *Animal Behaviour* 36: 780–792.

9. Brown, C. R. and Brown, M. B. 2004. Empirical measurement of parasite transmission between groups in a colonial bird. *Ecology* 85: 1619–1626.

10. Brown, J. L. 1987. *Helping and Communal Breeding in Birds: Ecology and Evolution*. Princeton University Press, Princeton, NJ.

11. Brown, J. L. and Vleck, C. M. 1998. Prolactin and helping in birds: Has natural selection strengthened helping behavior? *Behavioral Ecology* 9: 541–545.

12. Burda, H., Honeycutt, R. L., Begall, S., Locker-Grütjen, O., and Scharff, A. 2000. Are naked and common mole-rats eusocial and if so, why? *Behavioral Ecology and Sociobiology* 47: 293–303.

13. Canestrari, D., Marcos, J. M., and Baglione, V. 2008. Reproductive success increases with group size in cooperative carrion crows, *Corvus corone corone*. *Animal Behaviour* 75: 403–416.

14. Clutton-Brock, T. 2009. Structure and function in mammalian societies. *Philosophical Transactions of the Royal Society B* 364: 3229–3242.

15. Cockburn, A. 1998. Evolution of helping behavior in cooperatively breeding birds. *Annual Review of Ecology and Systematics* 29: 141–177.

16. Cockburn, A. 2006. Prevalence of different modes of parental care in birds. *Proceedings of the Royal Society B* 273: 1375–1383.

17. Cockburn, A., Sims, R. A., Osmond, H. L., Green, D. J., Double, M. C., and Mulder, R. A. 2008. Can we measure the benefits of help in cooperatively breeding birds: The case of superb fairy-wrens *Malurus cyaneus*? *Journal of Animal Ecology* 77: 430–438.

18. Connor, R. C. 2010. Cooperation beyond the dyad: On simple models and a complex society. *Philosophical Transactions of the Royal Society B* 365: 2687–2697.

19. Cornwallis, C. K., West, S. A., Davis, K. E., and Griffin, A. S. 2010. Promiscuity and the evolutionary transition to complex societies. *Nature* 466: 969–972.

20. Danforth, B. N., Conway, L., and Ji, S. Q. 2003. Phylogeny of eusocial *Lasioglossum* reveals multiple losses of eusociality within a primitively eusocial clade of bees (Hymenoptera: Halictidae). *Systematic Biology* 52: 23–36.

21. Davis, A. R., Corl, A., Surget-Groba, Y., and Sinervo, B. 2011. Convergent evolution of kin-based sociality in a lizard. *Proceedings of the Royal Society B* 278: 1507–1514.

22. East, M. L. and Hofer, H. 2010. Social environments, social tactics and their fitness consequences in complex mammalian societies. In T. Székely, M. C. Moore, and J. Komdeur (Eds.), *Social Behaviour: Genes, Ecology and Evolution*. Cambridge University Press, Cambridge, England.

23. Eikenaar, C., Richardson, D. S., Brouwer, L., and Komdeur, J. 2007. Parent presence, delayed dispersal, and territory acquisition in the Seychelles warbler. *Behavioral Ecology* 18: 874–879.

24. Emlen, S. T. 1978. Cooperative breeding. In J. R. Krebs and N. B. Davies (Eds.), *Behavioural Ecology: An Evolutionary Approach*. Blackwell, Oxford, England.

25. Emlen, S. T., Wrege, P. H., and Demong, N. J. 1995. Making decisions in the family: An evolutionary perspective. *American Scientist* 83: 148–157.

26. Foster, M. S. 1977. Odd couples in manakins: A study of social organization and cooperative breeding in *Chiroxiphia linearis*. *American Naturalist* 111: 845–853.

27. Greene, E., Lyon, B. E., Muehter, V. R., Ratcliffe, L., Oliver, S. J., and Boag, P. T. 2000. Disruptive sexual selection for plumage colouration in a passerine bird. *Nature* 407: 1000–1003.

28. Griesser, M. and Ekman, J. 2005. Nepotistic mobbing behaviour in the Siberian jay, *Perisoreus infaustus*. *Animal Behaviour* 69: 345–352.

29. Griffin, A. S. and West, S. A. 2003. Kin discrimination and the benefit of helping in cooperatively breeding vertebrates. *Science* 302: 634–636.

30. Hamilton, W. D. 1964. The genetical theory of social behaviour, I, II. *Journal of Theoretical Biology* 7: 1–52.

31. Hatchwell, B. J. 2009. The evolution of cooperative breeding in birds: Kinship, dispersal and life history. *Philosophical Transactions of the Royal Society B* 364: 3217–3227.

32. Haydock, J. and Koenig, W. D. 2002. Reproductive skew in the polygynandrous acorn woodpecker. *Proceedings of the National Academy of Sciences* 99: 7178–7183.

33. Heg, D., Jutzeler, E., Mitchell, J. S., and Hamilton, I. M. 2009. Helpful female subordinate cichlids are more likely to reproduce. *PLoS ONE* 4(5): e5458. doi:10.1371/journal.pone.0005458.

34. Jamieson, I. G. 1991. The unselected hypothesis for the evolution of helping behavior: Too much or too little emphasis on natural selection? *American Naturalist* 138: 271–282.

35. Jarvis, J. U. M. and Bennett, N. C. 1993. Eusociality has evolved independently in two genera of bathyergid mole-rats—but occurs in no other subterranean mammal. *Behavioral Ecology and Sociobiology* 33: 253–260.

36. Jennions, M. D. and Macdonald, D. W. 1994. Cooperative breeding in mammals. *Trends in Ecology and Evolution* 9: 89–93.

37. Kingma, S. A., Hall, M. L., Arriero, E., and Peters, A. 2010. Multiple benefits of cooperative breeding in purple-crowned fairy-wrens: A consequence of fidelity? *Journal of Animal Ecology* 79: 757–768.

38. Kingma, S. A., Hall, M. L., and Peters, A. 2011. Multiple benefits drive helping behavior in a cooperatively breeding bird: An integrated analysis. *American Naturalist* 177: 486–495.

39. Kleiman, D. G. 2011. Canid mating systems, social behavior, parental care and ontogeny: Are they flexible? *Behavior Genetics* 41: 803–809.

40. Koenig, W. D. and Mumme, R. L. 1987. *Population Ecology of the Cooperatively Breeding Acorn Woodpecker*. Princeton University Press, Princeton, NJ.

41. Koenig, W. D., Mumme, R. L., Stanback, M. T., and Pitelka, F. A. 1995. Patterns and consequences of egg destruction among joint-nesting acorn woodpeckers. *Animal Behaviour* 50: 607–621.

42. Komdeur, J. 1992. Importance of habitat saturation and territory quality for evolution of cooperative breeding in the Seychelles warbler. *Nature* 358: 493–495.

43. Komdeur, J., Huffstadt, A., Prast, W., Castle, G., Mileto, R., and Wattel, J. 1995. Transfer experiments of Seychelles warblers to new islands: Changes in dispersal and helping behavior. *Animal Behaviour* 49: 695–708.

44. Komdeur, J. 2003. Daughters on request: About helpers and egg sexes in the Seychelles warbler. *Proceedings of the Royal Society of London B* 270: 3–11.

45. Krams, I., Krama, T., Igaune, K., and Mand, R. 2008. Experimental evidence of reciprocal altruism in the pied flycatcher. *Behavioral Ecology and Sociobiology* 62: 599–605.

46. Lacey, E. A. and Sherman, P. W. 1991. Social organization of naked mole-rat colonies: Evidence for divisions of labor. In P. W. Sherman, J. U. M. Jarvis, and A. D. Alexander (Eds.), *The Biology of the Naked Mole-Rat*. Princeton University Press, Princeton, NJ.

47. Leadbeater, E., Carruthers, J. M., Green, J. P., Rosser, N. S., and Field, J. 2011. Nest inheritance is the missing source of direct fitness in a primitively eusocial insect. *Science* 333: 874–876.

48. Loiselle, B. A., Ryder, T. B., Duraes, R., Tori, W., Blake, J. G., and Parker, P. G. 2007. Kin selection does not explain male aggregation at leks of 4 manakin species. *Behavioral Ecology* 18: 287–291.

49. Machado, G. 2002. Maternal care, defensive behavior, and sociality in neotropical *Goniosoma* harvestmen (Arachnida, Opiliones). *Insectes Sociaux* 49: 388–393.

50. McDonald, D. B. and Potts, W. K. 1994. Cooperative display and relatedness among males in a lek-mating bird. *Science* 266: 1030–1032.

51. McDonald, D. B. 2010. A spatial dance to the music of time in the leks of long-tailed manakins. *Advances in the Study of Behavior* 42: 55–81.

52. Mosser, A. and Packer, C. 2009. Group territoriality and the benefits of sociality in the African lion, *Panthera leo. Animal Behaviour* 78: 359–370.

53. Mumme, R. L., Koenig, W. D., and Ratnieks, F. L. W. 1989. Helping behaviour, reproductive value, and the future component of indirect fitness. *Animal Behaviour* 38: 331–343.

54. Mumme, R. L. 1992. Do helpers increase reproductive success? An experimental analysis in the Florida scrub jay. *Behavioral Ecology and Sociobiology* 31: 319–328.

55. Ostrowski, E. A., Katoh, M., Shaulsky, G., Queller, D. C., and Strassmann, J. E. 2008. Kin discrimination increases with genetic distance in a social amoeba. *PLoS Biology* 6(11): e287. doi:10.1371/journal.pbio.0060287.

56. Packer, C., Scheel, D., and Pusey, A. E. 1990. Why lions form groups: Food is not enough. *American Naturalist* 136: 1–19.

57. Pike, N., Whitfield, J. A., and Foster, W. A. 2007. Ecological correlates of sociality in *Pemphigus* aphids, with a partial phylogeny of the genus. *BMC Evolutionary Biology* 7: 185.

58. Raihani, N. J. and Bshary, R. 2011. Resolving the iterated prisoner's dilemma: Theory and reality. *Journal of Evolutionary Biology* 24: 1628–1639.

59. Reyer, H.-U. 1984. Investment and relatedness: A cost/benefit analysis of breeding and helping in the pied kingfisher. *Animal Behaviour* 32: 1163–1178.

60. Riehl, C. 2011. Living with strangers: Direct benefits favour non-kin cooperation in a communally nesting bird. *Proceedings of the Royal Society B* 278: 1728–1735.

61. Russell, A. F., Langmore, N. E., Cockburn, A., Astheimer, L. B., and Kilner, R. M. 2007. Reduced egg investment can conceal helper effects in cooperatively breeding birds. *Science* 317: 941–944.

62. Russell, A. F., Young, A. J., Spong, G., Jordan, N. R., and Clutton-Brock, T. H. 2007. Helpers increase the reproductive potential of offspring in cooperative meerkats. *Proceedings of the Royal Society B* 274: 513–520.

63. Ryder, T. B., Blake, J. G., Parker, P. G., and Loiselle, B. A. 2011. The composition, stability, and kinship of reproductive coalitions in a lekking bird. *Behavioral Ecology* 22: 282–290.

64. Schino, G. and Aureli, F. 2010. The relative roles of kinship and reciprocity in explaining primate altruism. *Ecology Letters* 13: 45–50.

65. Schoech, S. J. 1998. Physiology of helping in Florida scrub-jays. *American Scientist* 86: 70–77.

66. Sherman, P. W. 1977. Nepotism and the evolution of alarm calls. *Science* 197: 1246–1253.

67. Sherman, P. W., Jarvis, J. U. M., and Alexander, R. D. (Eds.). 1991. *The Biology of the Naked Mole-Rat*. Princeton University Press, Princeton, NJ.

68. St-Pierre, A., Larose, K., and Dubois, F. 2009. Long-term social bonds promote cooperation in the iterated Prisoner's Dilemma. *Proceedings of the Royal Society B* 276: 4223–4228.

69. Stern, D. L. and Foster, W. A. 1996. The evolution of soldiers in aphids. *Biological Reviews* 71: 27–80.

70. Stern, D. L. and Foster, W. A. 1997. The evolution of sociality in aphids: A clone's-eye-view. In J. Choe and B. Crespi (Eds.), *Social Competition and Cooperation in Insects and Arachnids: II Evolution of Sociality*. Princeton University Press, Princeton, NJ.

71. Székely, T., Moore, M. C., and Komdeur, J. 2010. *Social Behaviour: Genes, Ecology and Evolution*. Cambridge University Press, Cambridge, England.

72. Taborsky, M. 1994. Sneakers, satellites, and helpers: Parasitic and cooperative behavior in fish reproduction. *Advances in the Study of Behavior* 23: 1–100.

73. Trivers, R. L. 1971. The evolution of reciprocal altruism. *Quarterly Review of Biology* 46: 35–57.

74. Turner, V. L. G., Lynch, S. M., Paterson, L., León-Cortés, J. L., and Thorpe, J. P. 2003. Aggression as a function of genetic relatedness in the sea anemone *Actinia equina* (Anthozoa: Actiniaria). *Marine Ecology Progress Series* 247: 85–92.

75. Vleck, C. M. and Brown, J. L. 1999. Testosterone and social and reproductive behaviour in *Aphelocoma* jays. *Animal Behaviour* 58: 943–951.

76. Webster, M. S., Varian, C. W., and Karubian, J. 2008. Plumage color and reproduction in the red-backed fairy-wren: Why be a dull breeder? *Behavioral Ecology* 19: 517–524.

77. Wedekind, C. and Milinski, M. 1996. Human cooperation in the simultaneous and the alternating Prisoner's Dilemma: Pavlov versus Generous Tit-for-Tat. *Proceedings of the National Academy of Sciences* 93: 2686–2689.

78. West-Eberhard, M. J. 1975. The evolution of social behavior by kin selection. *Quarterly Review of Biology* 50: 1–33.

79. West, S. A., Griffin, A. S., Gardner, A., and Diggle, S. P. 2006. Social evolution theory for microorganisms. *Nature Reviews Microbiology* 4: 597–607.

80. West, S. A., El Mouden, C., and Gardner, A. 2011. Sixteen common misconceptions about the evolution of cooperation in humans. *Evolution and Human Behavior* 32: 231–262.

81. Wilkinson, G. S. 1984. Reciprocal food sharing in the vampire bat. *Nature* 308: 181–184.

82. Wilson, E. O. 1975. *Sociobiology: The New Synthesis*. Harvard University Press, Cambridge, MA.

83. Wolff, J. O. and Sherman, P. W. (Eds.). 2007. *Rodent Societies: An Ecological and Evolutionary Perspective*. University of Chicago Press, Chicago, IL.

84. Woolfenden, G. E. and Fitzpatrick, J. W. 1984. *The Florida Scrub Jay: Demography of a Cooperative-Breeding Bird*. Princeton University Press, Princeton, NJ.

85. Yosef, R. and Yosef, N. 2009. Cooperative hunting in brown-necked raven (*Corvus rufficollis*) on Egyptian mastigure (*Uromastyx aegyptius*). *Journal of Ethology* 28: 385–388.

86. Young, A. J., Carlson, A. A., Monfort, S. L., Russell, A. F., Bennett, N. C., and Clutton-Brock, T. H. 2006. Stress and the suppression of subordinate reproduction in cooperatively breeding meerkats. *Proceedings of the National Academy of Sciences* 103: 12005–12010.

Chapter 4

1. Alcock, J. and Bailey, W. J. 1995. Acoustical communication and the mating system of the Australian whistling moth *Hecatesia exultans* (Noctuidae: Agaristinae). *Journal of Zoology* 237: 337–352.

2. Alcock, J. 2000. Interactions between the sexually deceptive orchid *Spiculaea ciliata* and its wasp pollinator *Thynnoturneria* sp. (Hymenoptera: Thynninae). *Journal of Natural History* 34: 629–636.

3. Barbero, F., Bonelli, S., Thomas, J. A., Balletto, E., and Schonrogge, K. 2009. Acoustical mimicry in a predatory social parasite of ants. *Journal of Experimental Biology* 212: 4084–4090.

4. Basolo, A. L. 1990. Female preference predates the evolution of the sword in swordtail fish. *Science* 250: 808–810.

5. Basolo, A. L. 1995. Phylogenetic evidence for the role of a pre-existing bias in sexual selection. *Proceedings of the Royal Society of London B* 259: 307–311.

6. Baugh, A. T. and Ryan, M. J. 2011. The relative value of call embellishment in túngara frogs. *Behavioral Ecology and Sociobiology* 65: 359–367.

7. Bernal, X. E., Page, R. A., Rand, A. S., and Ryan, M. J. 2007. Cues for eavesdroppers: Do frog calls indicate prey density and quality? *American Naturalist* 169: 409–415.

8. Borgia, G. 2006. Preexisting traits are important in the evolution of elaborated male sexual display. *Advances in the Study of Behavior* 36: 249–303.

9. Bradbury, J. W. and Vehrencamp, S. L. 2011. *Principles of Animal Communication* (2nd ed.). Sinauer Associates, Sunderland, MA.

10. Brandt, Y. 2003. Lizard threat display handicaps endurance. *Proceedings of the Royal Society of London B* 270: 1061–1068.

11. Briefer, E., Vannoni, E., and McElligott, A. G. 2010. Quality prevails over identity in the sexually selected vocalisations of an ageing mammal. *BMC Biology* 8. doi:10.1186/1741-7007-8-35.

12. Bro-Jørgensen, J. and Pangle, W. M. 2010. Male topi antelopes alarm snort deceptively to retain females for mating. *American Naturalist* 176: E33–E39.

13. Burgener, N., Dehnhard, M., Hofer, H., and East, M. L. 2009. Does anal gland scent signal identity in the spotted hyaena? *Animal Behaviour* 77: 707–715.

14. Burley, N. T. and Symanski, R. 1998. "A taste for the beautiful": Latent aesthetic mate preferences for white crests in two species of Australian grassfinches. *American Naturalist* 152: 792–802.

15. Calleia, F. D. O., Rohe, F., and Gordo, M. 2009. Hunting strategy of the margay (*Leopardus wiedii*) to attract the wild tamarin (*Saguinus bicolor*). *Neotropical Primates* 16: 31–34.

16. Carlsen, S. M., Jacobsen, G., and Romundstad, P. 2006. Maternal testosterone levels during pregnancy are associated with offspring size at birth. *European Journal of Endocrinology* 155: 365–370.

17. Christy, J. H. 1995. Mimicry, mate choice, and the sensory trap hypothesis. *American Naturalist* 146: 171–181.

18. Chuang, C.-Y., Yang, E.-C., and Tso, I.-M. 2007. Diurnal and nocturnal prey luring of a colorful predator. *Journal of Experimental Biology* 210: 3830–3837.

19. Chuang, C.-Y., Yang, E.-C., and Tso, I.-M. 2008. Deceptive color signaling in the night: A nocturnal predator attracts prey with visual lures. *Behavioral Ecology* 19: 237–244.

20. Conley, A. J., Corbin, C. J., Browne, P., Mapes, S. M., Place, N. J., Hughes, A. L., and Glickman, S. E. 2006. Placental expression and molecular characterization of aromatase cytochrome P450 in the spotted hyena (*Crocuta crocuta*). *Placenta* 28: 668–675.

21. Crespi, B. J. 2000. The evolution of maladaptation. *Heredity* 84: 623–629.

22. Crews, D. and Moore, M. C. 1986. Evolution of mechanisms controlling mating behavior. *Science* 231: 121–125.

23. Cummings, M. E., Rosenthal, G. G., and Ryan, M. J. 2003. A private ultraviolet channel in visual communication. *Proceedings of the Royal Society of London B* 270: 897–904.

24. Darwin, C. 1859. *On the Origin of Species*. Murray, London, England.

25. Darwin, C. 1892. *The Various Contrivances by which Orchids Are Fertilised by Insects*. D. Appleton, New York, NY.

26. Darwin, C. 1896. *Insectivorous Plants*. D. Appleton, New York, NY.

27. Davies, N. B. and Halliday, T. R. 1978. Deep croaks and fighting assessment in toads *Bufo bufo*. *Nature* 275: 683–685.

28. Dawkins, R. and Krebs, J. 1978. Animal signals: Information or manipulation? In J. R. Krebs and N. B. Davies (Eds.), *Behavioural Ecology: An Evolutionary Approach*. Blackwell, Oxford, England.

29. Dawkins, R. 1982. *The Extended Phenotype*. Freeman, San Francisco, CA.

30. Dloniak, S. M., French, J. A., Place, N. J., Weldele, M. L., Glickman, S. E., and Holekamp, K. E. 2004. Non-invasive

monitoring of fecal androgens in spotted hyenas (*Crocuta crocuta*). *General and Comparative Endocrinology* 135: 51–61.

31. Dodson, G. N. 1997. Resource defense mating system in antlered flies, *Phytalmia* spp. (Diptera: Tephritidae). *Annals of the Entomological Society of America* 90: 496–504.

32. Drea, C. M., Weldele, M. L., Forger, N. G., Coscia, E. M., Frank, L. G., Licht, P., and Glickman, S. E. 1998. Androgens and masculinization of genitalia in the spotted hyaena (*Crocuta crocuta*). 2. Effects of prenatal anti-androgens. *Journal of Reproduction and Fertility* 113: 117–127.

33. Drea, C. M., Place, N. J., Weldele, M. L., Coscia, E. M., Licht, P., and Glickman, S. E. 2002. Exposure to naturally circulating androgens during foetal life incurs direct reproductive costs in female spotted hyenas, but is prerequisite for male mating. *Proceedings of the Royal Society of London B* 269: 1981–1987.

34. Drea, C. M. 2009. Endocrine mediators of masculinization in female mammals. *Current Directions in Psychological Science* 18: 221–226.

35. Drea, C. M. 2011. Endocrine correlates of pregnancy in the ring-tailed lemur (*Lemur catta*): Implications for the masculinization of daughters. *Hormones and Behavior* 59: 417–427.

36. East, M. L., Hofer, H., and Wickler, W. 1993. The erect "penis" is a flag of submission in a female-dominated society: Greetings in Serengeti spotted hyenas. *Behavioral Ecology and Sociobiology* 33: 355–370.

37. East, M. L., Burke, T., Wilhelm, K., Greig, C., and Hofer, H. 2003. Sexual conflicts in spotted hyenas: Male and female mating tactics and their reproductive outcome with respect to age, social status and tenure. *Proceedings of the Royal Society of London B* 270: 1247–1254.

38. East, M. L., Höner, O. P., Wachter, B., Wilhelm, K., Burke, T., and Hofer, H. 2009. Maternal effects on offspring social status in spotted hyenas. *Behavioral Ecology* 20: 478–483.

39. East, M. L. and Hofer, H. 2012. The penile clitoris of the spotted hyena: A review of its evolution, fitness costs and benefits, and use in greeting ceremonies. Unpublished.

40. Egger, B., Klaefiger, Y., Theis, A., and Salzburger, W. 2011. A sensory bias has triggered the evolution of egg-spots in cichlid fishes. *PLoS ONE* 6(10): e25601. doi:10.1371/journal.pone.0025601.

41. Estes, R. D. 1991. *The Behavior Guide to African Mammals*. University of California Press, Berkeley, CA.

42. Ewer, R. F. 1973. *The Carnivores*. Cornell University Press, Ithaca, NY.

43. Frank, L. G., Holekamp, H. E., and Smale, L. 1995. Dominance, demographics and reproductive success in female spotted hyenas: A long-term study. In A. R. E. Sinclair and P. Arcese (Eds.), *Serengeti II: Research, Management, and Conservation of an Ecosystem*. University of Chicago Press, Chicago, IL.

44. Frank, L. G., Weldele, M. L., and Glickman, S. E. 1995. Masculinization costs in hyaenas. *Nature* 377: 584–585.

45. Fullard, J. H. 1997. The sensory coevolution of moths and bats. In R. R. Hoy, A. N. Popper, and R. R. Fay (Eds.), *Comparative Hearing: Insects*. Springer, New York, NY.

46. Furlow, F. B. 1997. Human neonatal cry quality as an honest signal of fitness. *Evolution and Human Behavior* 18: 175–194.

47. Gaskett, A. C. and Herberstein, M. E. 2008. Orchid sexual deceit provokes pollinator ejaculation. *American Naturalist* 171: E206–E212.

48. Glickman, S. E., Frank, L. G., Licht, P., Yalckinkaya, T., Siiteri, P. K., and Davidson, J. 1993. Sexual differentiation of the female spotted hyena: One of nature's experiments. *Annals of the New York Academy of Sciences* 662: 135–159.

49. Glickman, S. E., Cunha, G. R., Drea, C. M., Conley, A. J., and Place, N. J. 2006. Mammalian sexual differentiation:

Lessons from the spotted hyena. *Trends in Endocrinology and Metabolism* 17: 349–356.

50. Golla, W., Hofer, H., and East, M. L. 1999. Within-litter sibling aggression in spotted hyaenas: Effect of maternal nursing, sex and age. *Animal Behaviour* 58: 715–726.

51. Gould, S. J. 1981. Hyena myths and realities. *Natural History* 90: 16–24.

52. Gould, S. J. 1986. Evolution and the triumph of homology, or why history matters. *American Scientist* 74: 60–69.

53. Goymann, W., East, M. L., and Hofer, H. 2001. Androgens and the role of female "hyperaggressiveness" in spotted hyenas (*Crocuta crocuta*). *Hormones and Behavior* 39: 83–92.

54. Gray, D. A. and Cade, W. H. 1999. Sex, death and genetic variation: Natural and sexual selection on cricket song. *Proceedings of the Royal Society of London B* 266: 707–709.

55. Grether, G. F. 2000. Carotenoid limitation and mate preference evolution: A test of the indicator hypothesis in guppies (*Poecilia reticulata*). *Evolution* 54: 1712–1714.

56. Grether, G. F. 2010. The evolution of mate preferences, sensory biases, and indicator traits. *Advances in the Study of Behavior* 41: 35–76.

57. Haskell, D. G. 1999. The effect of predation on begging-call evolution in nestling wood warblers. *Animal Behaviour* 57: 893–901.

58. Heinrich, B. 1988. Winter foraging at carcasses by three sympatric corvids, with emphasis on recruitment by the raven, *Corvus corax*. *Behavioral Ecology and Sociobiology* 23: 141–156.

59. Heinrich, B. 1989. *Ravens in Winter*. Summit, New York, NY.

60. Hibbitts, T. J., Whiting, M. J., and Stuart-Fox, D. M. 2007. Shouting the odds: Vocalization signals status in a lizard. *Behavioral Ecology and Sociobiology* 61: 1169–1176.

61. Hofer, H. and East, M. L. 2000. Conflict management in female-dominated spotted hyenas. In F. Aureli and F. M. B. de Waal (Eds.), *Natural Conflict Resolution*. University of California Press, Berkeley, CA.

62. Hofer, H. and East, M. L. 2003. Behavioral processes and costs of co-existence in female spotted hyenas: A life history perspective. *Evolutionary Ecology* 17: 315–331.

63. Höner, O. P., Wachter, B., Hofer, H., Wilhelm, K., Thierer, D., Trillmich, F., Burke, T., and East, M. L. 2010. The fitness of dispersing spotted hyaena sons is influenced by maternal social status. *Nature Communications* 1: 7.

64. Huffard, C. L., Saarman, N., Hamilton, H., and Simison, W. B. 2010. The evolution of conspicuous facultative mimicry in octopuses: An example of secondary adaptation? *Biological Journal of the Linnean Society* 101: 68–77.

65. Jones, I. L. and Hunter, F. M. 1998. Heterospecific mating preferences for a feather ornament in least auklets. *Behavioral Ecology* 9: 187–192.

66. Kessel, E. L. 1955. Mating activities of balloon flies. *Systematic Zoology* 4: 97–104.

67. Klump, G. M., Kretzschmar, E., and Curio, E. 1986. The hearing of an avian predator and its avian prey. *Behavioral Ecology and Sociobiology* 18: 317–324.

68. Kocher, S. D. and Grozinger, C. M. 2011. Cooperation, conflict, and the evolution of queen pheromones. *Journal of Chemical Ecology* 37: 1263–1275.

69. Kruuk, H. 1972. *The Spotted Hyena*. University of Chicago Press, Chicago, IL.

70. LaBas, N. and Hockman, L. R. 2005. An invasion of cheats: The evolution of worthless nuptial gifts. *Current Biology* 15: 64–67.

71. Lailvaux, S. P., Reaney, L. T., and Backwell, P. R. Y. 2009. Dishonest signalling of fighting ability and

multiple performance traits in the fiddler crab *Uca mjoebergi*. *Functional Ecology* 23: 359–366.

72. Lappin, A. K., Brandt, Y., Husak, J. F., Macedonia, J. M., and Kemp, D. J. 2006. Gaping displays reveal and amplify a mechanically based index of weapon performance. *American Naturalist* 168: 100–113.

73. Leech, S. M. and Leonard, M. L. 1997. Begging and the risk of predation in nestling birds. *Behavioral Ecology* 8: 644–646.

74. Lewis, S. M. and Cratsley, C. K. 2008. Flash signal evolution, mate choice, and predation in fireflies. *Annual Review of Entomology* 53: 293–321.

75. Lloyd, J. E. 1965. Aggressive mimicry in *Photuris*: Firefly femmes fatales. *Science* 149: 653–654.

76. Lloyd, J. E. 1966. Studies on the flash communication systems of *Photinus* fireflies. *University of Michigan* 130: 1–95.

77. Lloyd, J. E. 1975. Aggressive mimicry in *Photuris* fireflies: Signal repertoires by femmes fatales. *Science* 197: 452–453.

78. Logue, D. M., Abiola, I. O., Rains, D., Bailey, N. W., Zuk, M., and Cade, W. H. 2010. Does signalling mitigate the cost of agonistic interactions? A test in a cricket that has lost its song. *Proceedings of the Royal Society B* 277: 2571–2575.

79. Lummaa, V., Vuorisalo, T., Barr, R. G., and Lehtonen, L. 1998. Why cry? Adaptive significance of intensive crying in human infants. *Evolution and Human Behavior* 19: 193–202.

80. Marler, P. 1955. Characteristics of some animal calls. *Nature* 176: 6–8.

81. Marshall, D. C. and Hill, K. B. R. 2009. Versatile aggressive mimicry of cicadas by an Australian predatory katydid. *PLoS ONE* 4(1): e4185. doi:10.1371/journal.pone.0004185.

82. Matthews, L. H. 1939. Reproduction in the spotted hyena *Crocuta crocuta* (Erxleben). *Philosophical Transactions of the Royal Society of London B* 230: 1–78.

83. Maynard Smith, J. 1974. The theory of games and the evolution of animal conflicts. *Journal of Theoretical Biology* 47: 209–221.

84. Meyer, A., Morrisey, J. M., and Schartl, M. 1994. Recurrent origin of a sexually selected trait in *Xiphophorus* fishes inferred from a molecular phylogeny. *Nature* 368: 539–542.

85. Mills, M. G. L. 1990. *Kalahari Hyaenas: Comparative Behavioural Ecology of Two Species*. Unwin Hyman, London, England.

86. Money, J. and Ehrhardt, A. A. 1972. *Man and Woman, Boy and Girl*. Johns Hopkins University Press, Baltimore, MD.

87. Muller, M. N. and Wrangham, R. 2002. Sexual mimicry in hyenas. *Quarterly Review of Biology* 77: 3–16.

88. Nakano, R., Takanashi, T., Skals, N., Surlykke, A., and Ishikawa, Y. 2010. To females of a noctuid moth, male courtship songs are nothing more than bat echolocation calls. *Biology Letters* 6: 582–584.

89. Östlund-Nilsson, S. and Holmlund, M. 2003. The artistic three-spined stickleback (*Gasterosteus aculeatus*). *Behavioral Ecology and Sociobiology* 53: 214–220.

90. Ostner, J. and Heistermann, M. 2003. Intersexual dominance, masculinized genitals and prenatal steroids: Comparative data from lemurid primates. *Naturwissenschaften* 90: 141–144.

91. Otte, D. 1974. Effects and functions in the evolution of signaling systems. *Annual Review of Ecology, Evolution and Systematics* 5: 385–471.

92. Owens, D. D. and Owens, M. J. 1996. Social dominance and reproductive patterns in brown hyenas, *Hyaena brunnea*, of the central Kalahari desert. *Animal Behaviour* 51: 535–551.

93. Place, N. J. and Glickman, S. E. 2004. Masculinization of female mammals: Lessons from nature. In L. Baskin (Ed.), *Hypospadias and Genital Development*. Kluwer Academic/Plenum Publishers, New York, NY.

94. Place, N. J., Coscia, E. M., Dahl, N. J., Drea, C. M., Holekamp, K. E., Roser, J. F., Sisk, C. L., Weldele, M. L., and Glickman, S. E. 2011. The anti-androgen combination, flutamide plus finasteride, paradoxically suppressed LH and androgen concentrations in pregnant spotted hyenas, but not in males. *General and Comparative Endocrinology* 170: 455–459.

95. Proctor, H. C. 1991. Courtship in the water mite *Neumania papillator*: Males capitalize on female adaptations for predation. *Animal Behaviour* 42: 589–598.

96. Proctor, H. C. 1992. Sensory exploitation and the evolution of male mating behaviour: A cladistic test. *Animal Behaviour* 44: 745–752.

97. Quinn, V. S. and Hews, D. K. 2000. Signals and behavioural responses are not coupled in males: Aggression affected by replacement of an evolutionarily lost colour signal. *Proceedings of the Royal Society of London B* 267: 755–758.

98. Racey, P. A. and Skinner, J. D. 1979. Endocrine aspects of sexual mimicry in spotted hyenas *Crocuta crocuta*. *Journal of Zoology* 187: 315–326.

99. Reichert, M. S. and Gerhardt, H. C. 2011. The role of body size on the outcome, escalation and duration of contests in the grey treefrog, *Hyla versicolor*. *Animal Behaviour* 82: 1357–1366.

100. Ridley, A. R., Child, M. F., and Bell, M. B. V. 2007. Interspecific audience effects on the alarm-calling behaviour of a kleptoparasitic bird. *Biology Letters* 3: 589–591.

101. Ridsdill-Smith, T. J. 1970. The biology of *Hemithynnus hyalinatus* (Hymenoptera, Tiphiidae), a parasite of scarabaeid larvae. *Journal of the Australian Entomological Society* 9: 183–195.

102. Rodd, F. H., Hughes, K. A., Grether, G. F., and Baril, C. T. 2002. A possible non-sexual origin of mate preference: Are male guppies mimicking fruit? *Proceedings of the Royal Society of London B* 269: 475–481.

103. Rodríguez, R. L. and Snedden, W. A. 2004. On the functional design of mate preferences and receiver biases. *Animal Behaviour* 68: 427–432.

104. Ryan, M. J., Tuttle, M. D., and Taft, L. K. 1981. The costs and benefits of frog chorusing behavior. *Behavioral Ecology and Sociobiology* 8: 273–278.

105. Ryan, M. J. 1985. *The Túngara Frog*. University of Chicago Press, Chicago, IL.

106. Ryan, M. J. and Wagner, W. E., Jr. 1987. Asymmetries in mating behavior between species: Female swordtails prefer heterospecific males. *Science* 236: 595–597.

107. Ryan, M. J., Fox, J. H., Wilczynski, W., and Rand, A. S. 1990. Sexual selection for sensory exploitation in the frog *Physalaemus pustulosus*. *Nature* 343: 66–67.

108. Ryan, M. J., Bernal, X. E., and Rand, A. S. 2010. Female mate choice and the potential for ornament evolution in túngara frogs *Physalaemus pustulosus*. *Current Zoology* 56: 343–357.

109. Schiestl, F. P. 2010. Pollination: Sexual mimicry abounds. *Current Biology* 20: R1020–R1022.

110. Schiestl, F. P. and Dötteri, S. 2012. The evolution of floral scent and olfactory preferences in pollinators: Coevolution or pre-existing bias? *Evolution*. doi:10.1111/j.1558-5646.2012.01593.x.

111. Smith, J. E., Van Horn, R. C., Powning, K. S., Cole, A. R., Graham, K. E., Memenis, S. K., and Holekamp, K. E. 2010. Evolutionary forces favoring intragroup coalitions among spotted hyenas and other animals. *Behavioral Ecology* 21: 284–303.

112. Smith, J. E., Powning, K. S., Dawes, S. E., Estrada, J. R., Hopper, A. L., Piotrowski, S. L., and Holekamp, K. E. 2011. Greetings promote cooperation and reinforce social bonds among spotted hyaenas. *Animal Behaviour* 81: 401–415.

113. Suzuki, T. N. 2011. Parental alarm calls warn nestlings about different predatory threats. *Current Biology* 21: R15–R16.

114. Tibbetts, E. A. and Izzo, A. 2010. Social punishment of dishonest signalers caused by mismatch between signal and behavior. *Current Biology* 20: 1637–1640.

115. Vanpé, C., Gaillard, J. M., Kjellander, P., Mysterud, A., Magnien, P., Delorme, D., Van Laere, G., Klein, F., Liberg, O., and Hewison, A. J. M. 2007. Antler size provides an honest signal of male phenotypic quality in roe deer. *American Naturalist* 169: 481–493.

116. Watts, H. E. and Holekamp, K. E. 2007. Hyena societies. *Current Biology* 17: R657–R660.

117. Wells, J. C. K. 2003. Parent-offspring conflict theory, signaling of need, and weight gain in early life. *Quarterly Review of Biology* 78: 169–202.

118. West-Eberhard, M. J. 1979. Sexual selection, social competition, and evolution. *Proceedings of the American Philosophical Society* 123: 222–234.

119. Wickler, W. 1968. *Mimicry in Plants and Animals*. World University Library, London, England.

120. Wiens, J. J. 2001. Widespread loss of sexually selected traits: How the peacock lost its spots. *Trends in Ecology and Evolution* 19: 517–523.

121. Wignall, A. E. and Taylor, P. W. 2011. Assassin bug uses aggressive mimicry to lure spider prey. *Proceedings of the Royal Society B* 278: 1427–1433.

122. Wilkinson, G. S. and Dodson, G. N. 1997. Function and evolution of antlers and eye stalks in flies. In J. C. Choe and B. J. Crespi (Eds.), *The Evolution of Mating Systems in Insects and Arachnids*. Cambridge University Press, Cambridge, England.

123. Wilson, R. S., Angelitta, M. J., Jr, James, R. S., Navas, C., and Seebacher, F. 2007. Dishonest signals of strength in male slender crayfish (*Cherax dispar*) during agonistic encounters. *American Naturalist* 170: 284–291.

124. Wyatt, T. D. 2003. *Pheromones and Animal Behaviour*. Cambridge University Press, Cambridge, England.

125. Wyatt, T. D. 2009. Pheromones and other chemical communication in animals. *Encyclopedia of Neuroscience* 7: 611–616.

126. Zahavi, A. 1975. Mate selection: A selection for a handicap. *Journal of Theoretical Biology* 53: 205–214.

127. Zuk, M., Rotenberry, J. T., and Tinghitella, R. M. 2006. Silent night: Adaptive disappearance of a sexual signal in a parasitized population of field crickets. *Biology Letters* 2: 521–524.

Chapter 5

1. Baum, D. A. and Larson, A. 1991. Adaptation reviewed: A phylogenetic methodology for studying character macroevolution. *Systematic Zoology* 40: 1–18.

2. Beckmann, C. and Shine, R. 2011. Toad's tongue for breakfast: Exploitation of a novel prey type, the invasive cane toad, by scavenging raptors in tropical Australia. *Biological Invasions* 13: 1447–1455.

3. Blackledge, T. A. and Wenzel, J. W. 1999. Do stabilimenta in orb webs attract prey or defend spiders? *Behavioral Ecology* 10: 372–376.

4. Blamires, S. J., Hochuli, D. F., and Thompson, M. B. 2008. Why cross the web: Decoration spectral properties and prey capture in an orb spider (*Argiope keyserlingi*) web. *Biological Journal of the Linnean Society* 94: 221–229.

5. Brakefield, P. M. and Liebert, T. G. 2000. Evolutionary dynamics of declining melanism in the peppered moth in The Netherlands. *Proceedings of the Royal Society of London B* 267: 1953–1957.

6. Brooks, D. R. and McLennan, D. A. 1991. *Phylogeny, Ecology, and Behavior*. University of Chicago Press, Chicago, IL.

7. Brower, J. V. Z. 1958. Experimental studies of mimicry in some North American butterflies. 1. The monarch, *Danaus plexippus*, and viceroy, *Limenitis archippus*. *Evolution* 12: 3–47.

8. Brower, L. P. and Calvert, W. H. 1984. Chemical defence in butterflies. In: R. I. Vane-Wright and P. R. Ackery (Eds.), *The Biology of Butterflies*. Academic Press, London, England.

9. Bugoni, L., Krause, L., and Petry, M. V. 2001. Marine debris and human impacts on sea turtles in southern Brazil. *Marine Pollution Bulletin* 42: 1330–1334.

10. Bura, V. L., Rohwer, V. G., Martin, P. R., and Yack, J. E. 2011. Whistling in caterpillars (*Amorpha juglandis*, Bombycoidea): Sound-producing mechanism and function. *Journal of Experimental Biology* 214: 30–37.

11. Burger, J. and Gochfeld, M. 2001. Smooth-billed ani (*Crotophaga ani*) predation on butterflies in Mato Grosso, Brazil: Risk decreases with increased group size. *Behavioral Ecology and Sociobiology* 49: 482–492.

12. Caro, T. M. 1986. The functions of stotting: A review of the hypotheses. *Animal Behaviour* 34: 649–662.

13. Caro, T. M. 1986. The functions of stotting in Thomson's gazelles: Some tests of the predictions. *Animal Behaviour* 34: 663–684.

14. Caro, T. M. 2005. *Antipredator Defenses in Birds and Mammals*. University of Chicago Press, Chicago, IL.

15. Christianson, D. and Creel, S. 2010. A nutritionally mediated risk effect of wolves on elk. *Ecology* 91: 1184–1191.

16. Clutton-Brock, T. H. and Harvey, P. H. 1984. Comparative approaches to investigating adaptation. In J. R. Krebs and N. B. Davies (Eds.), *Behavioural Ecology: An Evolutionary Approach*. Blackwell, Oxford, England.

17. Cook, L. M. 2003. The rise and fall of the *carbonaria* form of the peppered moth. *Quarterly Review of Biology* 78: 399–417.

18. Cook, L. M., Grant, B. S., Saccheri, I. J., and Mallet, J. 2012. Selective bird predation on the peppered moth: The last experiment of Michael Majerus. *Biology Letters* doi:10.1098/rsbl.2011.1136.

19. Coss, R. G. and Goldthwaite, R. O. 1995. The persistence of old designs for perception. *Perspectives in Ethology* 11: 83–148.

20. Coyne, J. 1998. Not black and white. *Nature* 396: 35–36.

21. Creel, S., Winnie, J., Jr, Maxwell, B., Hamlin, K., and Creel, M. 2005. Elk alter habitat selection as an antipredator response to wolves. *Ecology* 86: 3387–3397.

22. Creel, S. and Christianson, D. 2008. Relationships between direct predation and risk effects. *Trends in Ecology and Evolution* 23: 194–201.

23. Crespi, B. J. 2000. The evolution of maladaptation. *Heredity* 84: 623–629.

24. Cresswell, W. and Quinn, J. L. 2011. Predicting the optimal prey group size from predator hunting behaviour. *Journal of Animal Ecology* 80: 310–319.

25. Cristol, D. A. and Switzer, P. V. 1999. Avian prey-dropping behavior. II. American crows and walnuts. *Behavioral Ecology* 10: 220–226.

26. Cullen, E. 1957. Adaptations in the kittiwake to cliff nesting. *Ibis* 99: 275–302.

27. Dawkins, R. 1980. Good strategy or evolutionarily stable strategy? In G. W. Barlow and J. Silverberg (Eds.), *Sociobiology: Beyond Nature/Nurture?* Westview Press, Boulder, CO.

28. Dawkins, R. 1989. *The Selfish Gene* (3rd ed.). Oxford University Press, Oxford, England.

29. de Belle, J. S. and Sokolowski, M. B. 1987. Heredity of rover/sitter: Alternative foraging strategies of *Drosophila melanogaster* larvae. *Heredity* 59: 73–83.

30. Eberhard, W. G. 1982. Beetle horn dimorphism: Making the best of a bad lot. *American Naturalist* 119: 420–426.

31. Eberhard, W. G. 2006. Stabilimenta of *Philoponella vicina* (Araneae: Uloboridae) and *Gasteracantha cancriformis*

(Araneae: Araneidae). Evidence against a prey attraction function. *Biotropica* 39: 216–220.

32. Endler, J. A. 1991. Interactions between predators and prey. In J. R. Krebs and N. B. Davies (Eds.), *Behavioural Ecology: An Evolutionary Approach*. Blackwell, Oxford, England.

33. Farley, C. T. and Taylor, C. R. 1991. A mechanical trigger for the trot-gallop transition in horses. *Science* 253: 306–308.

34. Fitzpatrick, M. J., Feder, E., Rowe, L., and Sokolowski, M. B. 2007. Maintaining a behaviour polymorphism by frequency-dependent selection on a single gene. *Nature* 447: 210–212.

35. Ford, E. B. 1955. *Moths*. Collins, London, England.

36. Fullard, J. H. 2000. Day-flying butterflies remain day-flying in a Polynesian, bat-free habitat. *Proceedings of the Royal Society of London B* 267: 2295–2300.

37. Fullard, J. H., Otero, L. D., Orellana, A., and Surlykke, A. 2000. Auditory sensitivity and diel flight activity in Neotropical Lepidoptera. *Annals of the Entomological Society of America* 93: 956–965.

38. Fullard, J. H., Dawson, J. W., and Jacobs, D. S. 2003. Auditory encoding during the last moment of a moth's life. *Journal of Experimental Biology* 206: 281–294.

39. Fullard, J. H., Ratcliffe, J. M., and Soutar, A. R. 2004. Extinction of the acoustic startle response in moths endemic to a bat-free habitat. *Journal of Evolutionary Biology* 17: 856–861.

40. Gill, F. B. and Wolf, L. L. 1975. Economics of feeding territoriality in the golden-winged sunbird. *Ecology* 56: 333–345.

41. Gill, F. B. and Wolf, L. L. 1975. Foraging strategies and energetics of East African sunbirds at mistletoe flowers. *American Naturalist* 109: 491–510.

42. Goldbogen, J. A., Calambokidis, J., Oleson, E., Potvin, J., Pyenson, N. D., Schorr, G., and Shadwick, R. E. 2011. Mechanics, hydrodynamics and energetics of blue whale lunge feeding: Efficiency dependence on krill density. *Journal of Experimental Biology* 214: 131–146.

43. Gould, S. J. and Lewontin, R. C. 1979. The spandrels of San Marco and the Panglossian paradigm: A critique of the adaptationist programme. *Proceedings of the Royal Society of London B* 205: 581–598.

44. Grant, B. S., Owen, D. F., and Clarke, C. A. 1996. Parallel rise and fall of melanic peppered moths in America and Britain. *Journal of Heredity* 87: 351–357.

45. Grant, B. S. 1999. Fine tuning the peppered moth paradigm. *Evolution* 53: 980–984.

46. Grant, B. S. and Wiseman, L. L. 2002. Recent history of melanism in American peppered moths. *Journal of Heredity* 93: 86–90.

47. Greene, E., Orsak, L. T., and Whitman, D. W. 1987. A tephritid fly mimics the territorial displays of its jumping spider predators. *Science* 236: 310–312.

48. Grémillet, D., Pichegru, L., Kuntz, G., Woakes, A. G., Wilkinson, S., Crawford, R. J. M., and Ryan, P. G. 2008. A junk-food hypothesis for gannets feeding on fishery waste. *Proceedings of the Royal Society B* 275: 1149–1156.

49. Gross, M. R. and MacMillan, A. M. 1981. Predation and the evolution of colonial nesting in bluegill sunfish (*Lepomis macrochirus*). *Behavioral Ecology and Sociobiology* 8: 163–174.

50. Gross, M. R. 1996. Alternative reproductive strategies and tactics: Diversity within species. *Trends in Ecology and Evolution* 11: 92–98.

51. Gwynne, D. T. 1983. Beetles on the bottle. *Journal of the Australian Entomological Society* 23: 79.

52. Hamilton, W. D. 1971. Geometry for the selfish herd. *Journal of Theoretical Biology* 31: 295–311.

53. Harvey, P. H. and Pagel, M. D. 1991. *The Comparative Method in Evolutionary Biology*. Oxford University Press, London, England.

54. Hayes, L. D. 2000. To nest communally or not to nest communally: A review of rodent communal nesting and nursing. *Animal Behaviour* 59: 677–688.

55. Herberstein, M. E., Craig, C. L., Coddington, J. A., and Elgar, M. A. 2000. The functional significance of silk decorations of orb-web spiders: A critical review of the empirical evidence. *Biological Reviews* 75: 649–669.

56. Hoogland, J. L. and Sherman, P. W. 1976. Advantages and disadvantages of bank swallow (*Riparia riparia*) coloniality. *Ecological Monographs* 46: 33–58.

57. Hori, M. 1993. Frequency-dependent natural selection in the handedness of scale-eating cichlid fish. *Science* 260: 216–219.

58. Howlett, R. J. and Majerus, M. E. N. 1987. The understanding of industrial melanism in the peppered moth (*Biston betularia*) (Lepidoptera: Geometridae). *Biological Journal of the Linnean Society* 30: 31–44.

59. Ioannou, C. C., Bartumeus, F., Krause, J., and Ruxton, G. D. 2011. Unified effects of aggregation reveal larger prey groups take longer to find. *Proceedings of the Royal Society B* 278: 2985–2990.

60. Johansson, J., Turesson, H., and Persson, A. 2004. Active selection for large guppies, *Poecilia reticulata*, by the pike cichlid, *Crenicichla saxatilis*. *Oikos* 105: 595–605.

61. Johnsson, J. I. and Sundström, F. 2007. Social transfer of predation risk information reduces food locating ability in European minnows (*Phoxinus phoxinus*). *Ethology* 113: 166–173.

62. Kettlewell, H. B. D. 1955. Selection experiments on industrial melanism in the Lepidoptera. *Heredity* 9: 323–343.

63. Krebs, J. R. and Kacelnik, A. 1991. Decision-making. In J. R. Krebs and N. B. Davies (Eds.), *Behavioural Ecology: An Evolutionary Approach*. Blackwell, Oxford, England.

64. Kruuk, H. 1964. Predators and anti-predator behaviour of the black-headed gull *Larus ridibundus*. *Behaviour Supplements* 11: 1–129.

65. Laist, D. W. 1987. Overview of the biological effects of lost and discarded plastic debris in the marine environment. *Marine Pollution Bulletin* 18: 319–326.

66. Lauder, G. V., Leroi, A. M., and Rose, M. R. 1993. Adaptations and history. *Trends in Ecology and Evolution* 8: 294–297.

67. Leal, M. 1999. Honest signalling during prey-predator interactions in the lizard *Anolis cristatellus*. *Animal Behaviour* 58: 521–526.

68. Lemon, W. C. and Barth, R. H. 1992. The effects of feeding rate on reproductive success in the zebra finch, *Taeniopyga guttata*. *Animal Behaviour* 44: 851–857.

69. Majerus, M. E. N. 2008. Non-morph specific predation of peppered moths (*Biston betularia*) by bats. *Ecological Entomology* 33: 679–683.

70. Massaro, M., Chardine, J. W., and Jones, I. L. 2001. Relationships between black-legged kittiwake nest site characteristics and susceptibility to predation by large gulls. *Condor* 103: 793–801.

71. Mather, M. H. and Roitberg, B. D. 1987. A sheep in wolf's clothing: Tephritid flies mimic spider predators. *Science* 236: 308–310.

72. Meehan, C. J., Olson, E. J., Reudink, M. W., Kyser, T. K., and Curry, R. L. 2009. Herbivory in a spider through exploitation of an ant-plant mutualism. *Current Biology* 19: R892–R893.

73. Meire, P. M. and Ervynck, A. 1986. Are oystercatchers (*Haematopus ostralegus*) selecting the most profitable mussels (*Mytilus edulis*)? *Animal Behaviour* 34: 1427–1435.

74. Molleman, F. 2011. Puddling: From natural history to understanding how it affects fitness. *Entomologia Experimentalis et Applicata* 134: 107–113.

75. Morrell, L. J., Ruxton, G. D., and James, R. 2011. Spatial positioning in the selfish herd. *Behavioral Ecology* 22: 16–22.

76. Morrison, C. D. and Berthoud H.-R. 2007. Neurobiology of nutrition and obesity. *Nutrition Reviews* 65: 517–534.

77. Nesse, R. M. 2005. Maladaptation and natural selection. *Quarterly Review of Biology* 80: 62–70.

78. Olofsson, M., Vallin, A., Jakobsson, S., and Wiklund, C. 2011. Winter predation on two species of hibernating butterflies: Monitoring rodent attacks with infrared cameras. *Animal Behaviour* 81: 529–534.

79. Oosthuizen, J. H. and Davies, R. W. 1994. The biology and adaptations of the hippopotamus leech *Placobdelloides jaegerskioeldi* (Glossiphoniidae) to its host. *Canadian Journal of Zoology* 72: 418–422.

80. Owings, D. H. and Coss, R. G. 1977. Snake mobbing by California ground squirrels: Adaptive variation and ontogeny. *Behaviour* 62: 50–69.

81. Palmer, A. R. 2010. Scale-eating cichlids: From hand(ed) to mouth. *Journal of Biology* 9: 11.

82. Phillips, R. A., Furness, R. W., and Stewart, F. M. 1998. The influence of territory density on the vulnerability of Arctic skuas *Stercorarius parasiticus* to predation. *Biological Conservation* 86: 21–31.

83. Pierce, G. J. and Ollason, J. G. 1987. Eight reasons why optimal foraging theory is a complete waste of time. *Oikos* 49: 111–118.

84. Pietrewicz, A. T. and Kamil, A. C. 1977. Visual detection of cryptic prey by blue jays (*Cyanocitta cristata*). *Science* 195: 580–582.

85. Pitcher, T. 1979. He who hesitates lives: Is stotting antiambush behavior? *American Naturalist* 113: 453–456.

86. Plowright, R. C., Fuller, G. A., and Paloheimo, J. E. 1989. Shell dropping by northwestern crows: A reexamination of an optimal foraging study. *Canadian Journal of Zoology* 67: 770–771.

87. Preston-Mafham, R. and Preston-Mafham, K. 1993. *The Encyclopedia of Land Invertebrate Behaviour*. MIT Press, Cambridge, MA.

88. Prum, R. O. and Brush, A. H. 2002. The evolutionary origin and diversification of feathers. *Quarterly Review of Biology* 77: 261–295.

89. Queller, D. C. 1995. The spaniels of St. Marx and the Panglossian paradox: A critique of a rhetorical programme. *Quarterly Review of Biology* 70: 485–490.

90. Quinn, J. L. and Creswell, W. 2006. Testing domains of danger in the selfish herd: Sparrowhawks target widely spaced redshanks in flocks. *Proceedings of the Royal Society B* 273: 2521–2526.

91. Reeve, H. K. and Sherman, P. W. 1993. Adaptation and the goals of evolutionary research. *Quarterly Review of Biology* 68: 1–32.

92. Reeve, H. K. and Sherman, P. W. 2001. Optimality and phylogeny. In S. H. Orzack and E. Sober (Eds.), *Adaptationism and Optimality*. Cambridge University Press, Cambridge, England.

93. Rudge, D. W. 2006. Myths about moths: A study in contrasts. *Endeavour* 30: 19–23.

94. Rundus, A. S., Owings, D. S., Joshi, S. S., Chinn, E., and Giannini, N. 2007. Ground squirrels use an infrared signal to deter rattlesnake predation. *Proceedings of the National Academy of Sciences* 104: 14372–14374.

95. Ruxton, G. D., Sherratt, T. N., and Speed, M. P. 2004. *Avoiding Attack: The Evolutionary Ecology of Crypsis, Warning Signals, and Mimicry*. Oxford University Press, Oxford, England.

96. Rydale, J., Roininen, H., and Philip, K. W. 2000. Persistence of bat defence reactions in high Arctic moths (Lepidoptera). *Proceedings of the Royal Society of London B* 267: 553–557.

97. Sargent, T. D. 1976. *Legion of Night: The Underwing Moths*. University of Massachusetts Press, Amherst, MA.

98. Schlaepfer, M. A., Runge, M. C., and Sherman, P. W. 2002. Ecological and evolutionary traps. *Trends in Ecology and Evolution* 17: 474–480.

99. Sinervo, B. and Calsbeek, R. 2006. The physiological, behavioral, and genetical causes and consequences of frequency-dependent selection in the wild. *Annual Review of Ecology, Evolution and Systematics* 37: 581–610.

100. Smith, M. D. and Conway, C. J. 2007. Use of mammalian manure by nesting burrowing owls: A test of four functional hypotheses. *Animal Behaviour* 73: 65–73.

101. Smith, M. D. and Conway, C. J. 2011. Collection of mammal manure and other debris by nesting burrowing owls. *Journal of Raptor Research* 45: 220–228.

102. Snowberg, L. K. and Benkman, C. W. 2009. Mate choice based on a key ecological performance trait. *Journal of Evolutionary Biology* 22: 762–769.

103. Sordahl, T. A. 2004. Field evidence of predator discrimination abilities in American Avocets and Black-necked Stilts. *Journal of Field Ornithology* 75: 376–386.

104. Starks, P. T. 2002. The adaptive significance of stabilimenta in orb-webs: A hierarchical approach. *Annales Zoologici Fennici* 39: 307–315.

105. Stuart-Fox, D. M., Moussalli, A., Marshall, N. J., and Owens, I. P. F. 2003. Conspicuous males suffer higher predation risk: Visual modelling and experimental evidence from lizards. *Animal Behaviour* 66: 541–550.

106. Sweeney, B. W. and Vannote, R. L. 1982. Population synchrony in mayflies: A predator satiation hypothesis. *Evolution* 36: 810–821.

107. Tinbergen, N. 1959. Comparative studies of the behaviour of gulls (Laridae): A progress report. *Behaviour* 15: 1–70.

108. Townsend, S. W., Zottl, M., and Manser, M. B. 2011. All clear? Meerkats attend to contextual information in close calls to coordinate vigilance. *Behavioral Ecology and Sociobiology* 65: 1927–1934.

109. Turner, G. F. and Pitcher, T. J. 1986. Attack abatement: A model for group protection by combined avoidance and dilution. *American Naturalist* 128: 228–240.

110. Urban, M. C. 2007. Risky prey behavior evolves in risky habitats. *Proceedings of the National Academy of Sciences* 104: 14377–14382.

111. Walter, A. and Elgar, M. A. 2011. Signals for damage control: Web decorations in *Argiope keyserlingi* (Araneae: Araneidae). *Behavioral Ecology and Sociobiology* 65: 1909–1915.

112. Walther, B. A. and Gosler, A. G. 2001. The effects of food availability and distance to protective cover on the winter foraging behaviour of tits (Aves: *Parus*). *Oecologia* 129: 312–320.

113. Watt, P. J. and Chapman, R. 1998. Whirligig beetle aggregations: What are the costs and the benefits? *Behavioral Ecology and Sociobiology* 42: 179–184.

114. Whitfield, D. P. 1990. Individual feeding specializations of wintering turnstone *Arenaria interpres*. *Journal of Animal Ecology* 59: 193–211.

115. Wiersma, P. and Verhulst, S. 2005. Effects of intake rate on energy expenditure, somatic repair and reproduction of zebra finches. *Journal of Experimental Biology* 208: 4091–4098.

116. Williams, C. K., Lutz, R. S., and Applegate, R. D. 2003. Optimal group size and northern bobwhite coveys. *Animal Behaviour* 66: 377–387.

117. Wirsing, A. J., Heithaus, M. R., and Dill, L. M. 2007. Can you dig it? Use of excavation, a risky foraging tactic, by dugongs is sensitive to predation danger. *Animal Behaviour* 74: 1085–1091.

118. Yack, J. E. and Fullard, J. H. 2000. Ultrasonic hearing in nocturnal butterflies. *Nature* 403: 265–266.

119. Zach, R. 1979. Shell-dropping: Decision-making and optimal foraging in northwestern crows. *Behaviour* 68: 106–117.

Chapter 6

1. Alcock, J. and Bailey, W. J. 1997. Success in territorial defence by male tarantula hawk wasps *Hemipepsis ustulata*: The role of residency. *Ecological Entomology* 22: 377–383.

2. Altizer, S., Bartel, R., and Han, B. A. 2011. Animal migration and infectious disease risk. *Science* 331: 296–302.

3. Anderson, J. B and Brower, L. P. 1996. Freeze-protection of overwintering monarch butterflies in Mexico: Critical role of the forest as a blanket and an umbrella. *Ecological Entomology* 21: 107–116.

4. Anderson, R. C. 2009. Do dragonflies migrate across the western Indian Ocean? *Journal of Tropical Ecology* 25: 347–358.

5. Baird, T. A. and Curtis, J. L. 2010. Context-dependent acquisition of territories by male collared lizards: The role of mortality. *Behavioral Ecology* 21: 753–758.

6. Baker, A. J., Gonzalez, P. M., Piersma, T., Niles, L. J., do Nascimento, I. D. S., Atkinson, P. W., Clark, N. A., Minton, C. D. T., Peck, M. K., and Aarts, G. 2004. Rapid population decline in red knots: Fitness consequences of decreased refuelling rates and late arrival in Delaware Bay. *Proceedings of the Royal Society of London B* 275: 875–882.

7. Beletsky, L. D. and Orians, G. H. 1989. Territoriality among male red-winged blackbirds. III. Testing hypotheses of territorial dominance. *Behavioral Ecology and Sociobiology* 24: 333–339.

8. Bell, C. P. 1997. Leap-frog migration in the fox sparrow: Minimizing the cost of spring migration. *Condor* 99: 470–477.

9. Bell, C. P. 2000. Process in the evolution of bird migration and pattern in avian ecogeography. *Journal of Avian Biology* 31: 258–265.

10. Bergman, M., Gotthard, K., Berger, D., Olofsson, M., Kemp, D. J., and Wiklund, C. 2007. Mating success of resident versus non-resident males in a territorial butterfly. *Proceedings of the Royal Society B* 274: 1659–1665.

11. Berthold, P., Helbig, A. J., Mohr, G., and Querner, U. 1992. Rapid microevolution of migratory behaviour in a wild bird species. *Nature* 360: 668–670.

12. Boyle, W. A., Guglielmo, C. G., Hobson, K. A., and Norris, D. R. 2011. Lekking birds in a tropical forest forego sex for migration. *Biology Letters* 7: 661–663.

13. Bridge, E. S., Thorup, K., Bowlin, M. S., Chilson, P. B., Diehl, R. H., Fleron, R. W., Hartl, P., Kays, R., Kelly, J. F., Robinson, W. D., and Wikelski, M. 2011. Technology on the move: Recent and forthcoming innovations for tracking migratory birds. *BioScience* 61: 689–698.

14. Brower, L. P. 1996. Monarch butterfly orientation: Missing pieces of a magnificent puzzle. *Journal of Experimental Biology* 199: 93–103.

15. Brower, L. P., Fink, L. S., and Walford, P. 2006. Fueling the fall migration of the monarch butterfly. *Integrative and Comparative Biology* 46: 1123–1142.

16. Brown, J. L. 1969. Territorial behavior and population regulation in birds. *Wilson Bulletin* 81: 293–329.

17. Brown, J. L. 1975. *The Evolution of Behavior*. W. W. Norton, New York, NY.

18. Bull, C. M. and Freake, M. J. 1999. Home-range fidelity in the Australian sleepy lizard, *Tiliqua rugosa*. *Australian Journal of Zoology* 47: 125–132.

19. Buston, P. M. 2004. Territory inheritance in clownfish. *Proceedings of the Royal Society of London B* 271: S252–S254.

20. Calvert, W. H. and Brower, L. P. 1986. The location of monarch butterfly (*Danaus plexippus* L.) overwintering colonies in Mexico in relation to topography and climate. *Journal of the Lepidopterists' Society* 40: 164–187.

21. Caro, T. (Ed.) 1998. *Behavioral Ecology and Conservation Biology*. Oxford University Press, New York, NY.

22. Caro, T. and Sherman, P. W. 2011. Endangered species and a threatened discipline: Behavioural ecology. *Trends in Ecology and Evolution* 26: 111–118.

23. Carpenter, S. J., Erickson, J. M., and Holland, F. D. 2003. Migration of a Late Cretaceous fish. *Nature* 423: 70–74.

24. Clapham, J. 2001. Why do baleen whales migrate? A response to Corkeron and Connor. *Marine Mammal Science* 17: 432–436.

25. Coombs, W. P., Jr. 1990. Behavior patterns of dinosaurs. In D. B. Weishampel, P. Dodson, and H. Osmólska (Eds.), *The Dinosauria*. University of California Press, Berkeley, CA.

26. Corkeron, P. J. and Connor, R. C. 1999. Why do baleen whales migrate? *Marine Mammal Science* 15: 1228–1245.

27. Cox, G. W. 1985. The evolution of avian migration systems between temperate and tropical regions of the New World. *American Naturalist* 126: 452–474.

28. Davies, N. B. 1978. Territorial defence in the speckled wood butterfly (*Pararge aegeria*): The resident always wins. *Animal Behaviour* 26: 138–147.

29. Davies, N. B. and Houston, A. I. 1984. Territory economics. In J. R. Krebs and N. B. Davies (Eds.), *Behavioural Ecology: An Evolutionary Approach*. Blackwell, Oxford, England.

30. Dawkins, R. 1986. *The Blind Watchmaker*. W. W. Norton, New York, NY.

31. Dawkins, R. 1989. *The Selfish Gene* (2nd ed.). Oxford University Press, Oxford, England.

32. Dhondt, A. A. and Schillemans, J. 1983. Reproductive success of the great tit in relation to its territorial status. *Animal Behaviour* 31: 902–912.

33. Dugatkin, L. A. and Reeve, H. K. 1998. *Game Theory and Animal Behavior*. Oxford University Press, New York, NY.

34. Fischer, K., Perlick, J., and Galetz, T. 2008. Residual reproductive value and male mating success: Older males do better. *Proceedings of the Royal Society B* 275: 1517–1524.

35. Fisher, J. 1954. Evolution and bird sociality. In J. Huxley, A. C. Hardy, and E. B. Ford (Eds.), *Evolution as a Process*. Allen & Unwin, London, England.

36. Fretwell, S. D. and Lucas, H. K., Jr. 1969. On territorial behavior and other factors influencing habitat distribution in birds. I. Theoretical development. *Acta Biotheoretica* 19: 16–36.

37. Fricke, H. C., Hencecroth, J., and Hoerner, M. E. 2011. Lowland-upland migration of sauropod dinosaurs during the Late Jurassic epoch. *Nature* 480: 513–515.

38. Greenwood, P. J. 1980. Mating systems, philopatry, and dispersal in birds and mammals. *Animal Behaviour* 28: 1140–1162.

39. Gross, M. R. 1996. Alternative reproductive strategies and tactics: Diversity within species. *Trends in Ecology and Evolution* 11: 92–98.

40. Hanby, J. P. and Bygott, J. D. 1987. Emigration of subadult lions. *Animal Behaviour* 35: 161–169.

41. Hasselquist, D. 1998. Polygyny in great reed warblers: A long-term study of factors contributing to fitness. *Ecology* 79: 2376–2390.

42. Hedenström, A. 2010. Extreme endurance migration: What is the limit to non-stop flight? *PLoS Biology* 8(5): e1000362. doi:10.1371/journal.pbio.1000362.

43. Hobson, K. A. 2011. Isotopic ornithology: A perspective. *Journal of Ornithology* 152: 49–66.

44. Holekamp, K. E. 1984. Natal dispersal in Belding's ground squirrels (*Spermophilus beldingi*). *Behavioral Ecology and Sociobiology* 16: 21–30.

45. Höner, O. P., Wachter, B., East, M. L., Streich, W. J., Wilhelm, K., Burke, T., and Hofer, H. 2007. Female mate-choice drives the evolution of male-biased dispersal in a social mammal. *Nature* 448: 798–801.

46. Hyman, J., Hughes, M., Searcy, W. A., and Nowicki, S. 2004. Individual variation in the strength of territory defense in male song sparrows: Correlates of age, territory tenure, and neighbor aggressiveness. *Behaviour* 141: 15–27.

47. Jiménez, J. A., Hughes, K. A., Alaks, G., Graham, L., and Lacy, R. C. 1994. An experimental study of inbreeding depression in a natural habitat. *Science* 266: 271–273.

48. Kemp, D. J. 2002. Sexual selection constrained by life history in a butterfly. *Proceedings of the Royal Society of London B* 269: 1341–1345.

49. Kemp, D. J. and Wiklund, C. 2003. Residency effects in animal contests. *Proceedings of the Royal Society of London B* 271: 1707–1711.

50. Krebs, J. R. 1982. Territorial defence in the great tit (*Parus major*): Do residents always win? *Behavioral Ecology and Sociobiology* 11: 185–194.

51. Lacey, E. A. and Wieczorek, J. R. 2001. Territoriality and male reproductive success in arctic ground squirrels. *Behavioral Ecology* 12: 626–631.

52. Latta, S. C. and Brown, C. 1999. Autumn stopover ecology of the blackpoll warbler (*Dendroica striata*) in thorn scrub forest of the Dominican Republic. *Canadian Journal of Zoology* 77: 1147–1156.

53. Levey, D. J. and Stiles, F. G. 1992. Evolutionary precursors of long-distance migration: Resource availability and movement patterns in Neotropical landbirds. *American Naturalist* 140: 447–476.

54. Lucia, K. E. and Keane, B. 2012. A field test of the effects of familiarity and relatedness on social associations and reproduction in prairie voles. *Behavioral Ecology and Sociobiology* 66: 13–27.

55. Lundberg, P. 1985. Dominance behavior, body-weight and fat variations, and partial migration in European blackbirds *Turdus merula*. *Behavioral Ecology and Sociobiology* 17: 185–189.

56. Lundberg, P. 1988. The evolution of partial migration in birds. *Trends in Ecology and Evolution* 3: 172–176.

57. Malcolm, J. R., Liu, C. R., Neilson, R. P., Hansen, L., and Hannah, L. 2006. Global warming and extinctions of endemic species from biodiversity hotspots. *Conservation Biology* 20: 538–548.

58. Marden, J. H. and Waage, J. K. 1990. Escalated damselfly territorial contests and energetic wars of attrition. *Animal Behaviour* 39: 954–959.

59. Marden, J. H. and Rollins, R. A. 1994. Assessment of energy reserves by damselflies engaged in aerial contests for mating territories. *Animal Behaviour* 48: 1023–1030.

60. Margulis, S. W. and Altmann, J. 1997. Behavioural risk factors in the reproduction of inbred and outbred oldfield mice. *Animal Behaviour* 54: 397–408.

61. Marler, C. A. and Moore, M. C. 1989. Time and energy costs of aggression in testosterone-implanted free-living male mountain spiny lizards (*Sceloporus jarrovi*). *Physiological Zoology* 62: 1334–1350.

62. Marler, C. A. and Moore, M. C. 1991. Supplementary feeding compensates for testosterone-induced costs of aggression in male mountain spiny lizards, *Sceloporus jarrovi*. *Animal Behaviour* 42: 209–219.

63. Marra, P. P. and Holmes, R. T. 2001. Consequences of dominance-mediated habitat segregation in American redstarts during the nonbreeding season. *Auk* 118: 92–104.

64. Marvin, G. A. 2001. Age, growth, and long-term site fidelity in the terrestrial plethodontid salamander *Plethodon kentucki*. *Copeia* 2001: 108–117.

65. Maynard Smith, J. and Parker, G. A. 1976. The logic of asymmetric contests. *Animal Behaviour* 24: 159–175.

66. McGowan, C. P., Smith, D. R., Sweka, J. A., Martin, J., Nichols, J. D., Wong, R., Lyons, J. E., Niles, L. J., Kalasz, K., and Brust, J. 2011. Multispecies modeling for adaptive management of horseshoe crabs and red knots in the Delaware Bay. *Natural Resource Modeling* 24: 117–156.

67. McNair, D. B., Massiah, E. B., and Frost, M. D. 2002. Ground-based autumn migration of blackpoll warblers at Harrison Point, Barbados. *Caribbean Journal of Science* 38: 239–248.

68. Mitani, J. C., Watts, D. P., and Amsler, S. J. 2010. Lethal intergroup aggression leads to territorial expansion in wild chimpanzees. *Current Biology* 20: R507–R508.

69. Moore, J. and Ali, R. 1984. Are dispersal and inbreeding avoidance related? *Animal Behaviour* 32: 94–112.

70. Müller, C. A. and Manser, M. B. 2007. "Nasty neighbours" rather than "dear enemies" in a social carnivore. *Proceedings of the Royal Society B* 274: 959–965.

71. Musiega, D. E., Kazadi, S. N., and Fukuyama, K. 2006. A framework for predicting and visualizing the East African wildebeest migration-route patterns in variable climatic conditions using geographic information system and remote sensing. *Ecological Research* 21: 530–543.

72. Niles, L. J., Bart, J., Sitters, H. P., Dey, A. D., Clark, K. E., Atkinson, P. W., Baker, A. J., Bennett, K. A., Kalasz, K. S., Clark, N. A., Clark, J., Gillings, S., Gates, A. S., González, P. M., Hernandez, D. E., Minton, C. D. T., Morrison, R. I. G., Porter, R. R., Ross, R. K., and Veitch, C. R. 2009. Effects of horseshoe crab harvest in Delaware Bay on red knots: Are harvest restrictions working? *BioScience* 59: 153–164.

73. Norris, D. R., Marra, P. P., Kyser, T. K., Sherry, T. W., and Ratcliffe, L. M. 2004. Tropical winter habitat limits reproductive success on the temperate breeding grounds in a migratory bird. *Proceedings of the Royal Society of London B* 271: 59–64.

74. O'Neill, K. M. 1983. Territoriality, body size, and spacing in males of the bee wolf *Philanthus basilaris* (Hymenoptera: Sphecidae). *Behaviour* 86: 295–321.

75. Orians, G. H. 1969. On the evolution of mating systems in birds and mammals. *American Naturalist* 103: 589–603.

76. Outlaw, D. C., Voelker, G., Mila, B., and Girman, D. J. 2003. Evolution of long-distance migration in and historical biogeography of *Catharus* thrushes: A molecular phylogenetic approach. *Auk* 120: 299–310.

77. Papaj, D. R. and Messing, R. H. 1998. Asymmetries in physiological state as a possible cause of resident advantage in contests. *Behaviour* 135: 1013–1030.

78. Pimm, S. L. and Askins, R. A. 1995. Forest losses predict bird extinctions in eastern North America. *Proceedings of the National Academy of Sciences* 92: 9343–9347.

79. Piper, W. H. 2011. Making habitat selection more "familiar": A review. *Behavioral Ecology and Sociobiology* 65: 1329–1351.

80. Plaistow, S. and Siva-Jothy, M. T. 1996. Energetic constraints and male mate securing tactics in the damselfly *Calopteryx splendens xanthosoma* (Charpentier). *Proceedings of the Royal Society of London B* 263: 1233–1238.

81. Powell, G. V. N. and Bjork, R. D. 2004. Habitat linkages and the conservation of tropical biodiversity as indicated by seasonal migrations of three-wattled bellbirds. *Conservation Biology* 18: 500–509.

82. Pryke, S. R. and Andersson, S. 2003. Carotenoid-based epaulettes reveal male competitive ability: Experiments with resident and floater red-shouldered widowbirds. *Animal Behaviour* 66: 217–224.

83. Pusey, A. E. and Wolf, M. 1996. Inbreeding avoidance in animals. *Trends in Ecology and Evolution* 11: 201–206.

84. Quaintenne, G., van Gils, J. A., Bocher, P., Dekinga, A., and Piersma, T. 2011. Scaling up ideals to freedom: Are densities of red knots across western Europe consistent with ideal free distribution? *Proceedings of the Royal Society B* 278: 2728–2736.

85. Rachlow, J. L., Berkeley, E. V., and Berger, J. 1998. Correlates of male mating strategies in white rhinos (*Ceratotherium simum*). *Journal of Mammalogy* 79: 1317–1324.

86. Ralls, K., Brugger, K., and Ballou, J. 1979. Inbreeding and juvenile mortality in small populations of ungulates. *Science* 206: 1101–1103.

87. Ramírez, M. I., Azcárate, J. G., and Luna, L. 2003. Effects of human activities on monarch butterfly habitat in protected mountain forests, Mexico. *Forestry Chronicle* 79: 242–246.

88. Rodenhouse, N. L., Sillett, T. S., Doran, P. J., and Holmes, R. T. 2003. Multiple density-dependence mechanisms regulate a migratory bird population during the breeding season. *Proceedings of the Royal Society of London B* 270: 2105–2110.

89. Rubenstein, D. R., Chamberlain, C. P., Holmes, R. T., Ayres, M. P., Waldbauer, J. R., Graves, G. R., and Tuross, N. C. 2002. Linking breeding and wintering ranges of a migratory songbird using stable isotopes. *Science* 295: 1062–1065.

90. Rubenstein, D. R. and Hobson, K. A. 2004. From birds to butterflies: Animal movement patterns and stable isotopes. *Trends in Ecology and Evolution* 19: 256–263.

91. Ruegg, K. C. and Smith, T. B. 2002. Not as the crow flies: A historical explanation for circuitous migration in Swainson's thrush (*Catharus ustulatus*). *Proceedings of the Royal Society of London B* 269: 1375–1381.

92. Ruegg, K. C., Hijmans, R. J., and Moritz, C. 2006. Climate change and the origin of migratory pathways in the Swainson's thrush, *Catharus ustulatus*. *Journal of Biogeography* 33: 1172–1182.

93. Sandberg, R. and Moore, F. R. 1996. Migratory orientation of red-eyed vireos, *Vireo olivaceus*, in relation to energetic condition and ecological context. *Behavioral Ecology and Sociobiology* 39: 1–10.

94. Sandercock, B. K. and Jaramillo, A. 2002. Annual survival rates of wintering sparrows: Assessing demographic consequences of migration. *Auk* 119: 149–165.

95. Schwabl, H. 1983. Ausprägung und Bedeutung des Teilzugverhaltnes einer sudwestdeutschen Population der Amsel *Turdus merula*. *Journal für Ornithologie* 124: 101–116.

96. Shaffer, S. A., Tremblay, Y., Weimerskirch, H., Scott, D., Thompson, D. R., Sagar, P. M., Moller, H., Taylor, G. A., Foley, D. G., Block, B. A., and Costa, D. P. 2006. Migratory shearwaters integrate oceanic resources across the Pacific Ocean in an endless summer. *Proceedings of the National Academy of Sciences* 103: 12799–12802.

97. Shier, D. M. and Swaisgood, R. R. 2011. Fitness costs of neighborhood disruption in translocations of a solitary mammal. *Conservation Biology* 26: 116–123.

98. Shuster, S. M. and Wade, M. J. 2003. *Mating Systems and Strategies*. Princeton University Press, Princeton, NJ.

99. Smith, S. M. 1978. The "underworld" in a territorial species: Adaptive strategy for floaters. *American Naturalist* 112: 571–582.

100. Stutchbury, B. J. M., Tarof, S. A., Done, T., Gow, E., Kramer, P. M., Tautin, J., Fox, J. W., and Afanasyev, V. 2009. Tracking long-distance songbird migration by using geolocators. *Science* 323: 896.

101. Tobias, J. 1997. Asymmetric territorial contests in the European robin: The role of settlement costs. *Animal Behaviour* 54: 9–21.

102. Urquhart, F. A. 1960. *The Monarch Butterfly.* University of Toronto Press, Toronto, Canada.

103. Weber, J. M. 2009. The physiology of long-distance migration: Extending the limits of endurance metabolism. *Journal of Experimental Biology* 212: 593–597.

104. Weidinger, K. 2000. The breeding performance of blackcap *Sylvia atricapilla* in two types of forest habitat. *Ardea* 88: 225–233.

105. Weimerskirch, H., Martin, J., Clerquin, Y., Alexandre, P., and Jiraskova, S. 2001. Energy saving in flight formation. *Nature* 413: 697–698.

106. Welty, J. 1982. *The Life of Birds* (3rd ed.). Saunders College Publishing, Philadelphia, PA.

107. West-Eberhard, M. J. 2003. *Developmental Plasticity and Evolution*. Oxford University Press, New York, NY.

108. Whitham, T. G. 1979. Territorial defense in a gall aphid. *Nature* 279: 324–325.

109. Whitham, T. G. 1979. Habitat selection by *Pemphigus* aphids in response to resource limitation and competition. *Ecology* 59: 1164–1176.

110. Whitham, T. G. 1980. The theory of habitat selection examined and extended using *Pemphigus* aphids. *American Naturalist* 115: 449–466.

111. Whiting, M. J. 1999. When to be neighbourly: Differential agonistic responses in the lizard *Platysaurus broadleyi*. *Behavioral Ecology and Sociobiology* 46: 210–214.

112. Williams, T. C. and Williams, J. M. 1978. An oceanic mass migration of land birds. *Scientific American* 239 (Oct.): 166–176.

113. Winger, B. M., Lovette, I. J., and Winkler, D. W. 2012. Ancestry and evolution of seasonal migration in the Parulidae. *Proceedings of the Royal Society B* 279: 610–618.

114. Winker, K. and Pruett, C. L. 2006. Seasonal migration, speciation, and morphological convergence in the genus *Catharus* (Turdidae). *Auk* 123: 1052–1068.

115. Wolanski, E., Gereta, E., Borner, M., and Mduma, S. 1999. Water, migration and the Serengeti ecosystem. *American Scientist* 87: 526–533.

116. Yoder, J. M., Marschall, E. A., and Swanson, D. A. 2004. The cost of dispersal: Predation as a function of movement and site familiarity in ruffed grouse. *Behavioral Ecology* 15: 469–476.

117. Zedrosser, A., Støenm, O.-G., Saebø, S., and Swenson, J. R. 2007. Should I stay or should I go? Natal dispersal in the brown bear. *Animal Behaviour* 74: 369–376.

118. Zeh, D. W. and Zeh, J. A. 1992. Dispersal-generated sexual selection in a beetle-riding pseudoscorpion. *Behavioral Ecology and Sociobiology* 30: 135–142.

119. Zeh, J. A. 1997. Polyandry and enhanced reproductive success in the harlequin beetle-riding pseudoscorpion. *Behavioral Ecology and Sociobiology* 40: 111–118.

Chapter 7

1. Aisenberg, A. and Eberhard, W. G. 2009. Female cooperation in plug formation in a spider: Effects of male copulatory courtship. *Behavioral Ecology* 20: 1236–1241.

2. Aisenberg, A. and Barrantes, G. 2011. Sexual behavior, cannibalism, and mating plugs as sticky traps in the orb weaver spider *Leucauge argyra* (Tetragnathidae). *Naturwissenschaften* 98: 605–613.

3. Akçay, C., Searcy, W. A., Campbell, S. E., Reed, V. A., Templeton, C. N., Hardwick, K. M., and Beecher, M. D. 2012. Who initiates extrapair mating in song sparrows? *Behavioral Ecology* 23: 44–50.

4. Alberts, S. C., Watts, H. E., and Altmann, J. 2003. Queuing and queue-jumping: Long-term patterns of reproductive skew in male savannah baboons, *Papio cynocephalus*. *Animal Behaviour* 65: 821–840.

5. Alberts, S. C., Buchan, J. C., and Altmann, J. 2006. Sexual selection in wild baboons: From mating opportunities to paternity success. *Animal Behaviour* 72: 1177–1196.

6. Alcock, J., Jones, C. E., and Buchmann, S. L. 1977. Male mating strategies in the bee *Centris pallida* Fox (Hymenoptera: Anthophoridae). *American Naturalist* 111: 145–155.

7. Ancona, S., Drummond, H., and Zaldívar-Rae, J. 2010. Male whiptail lizards adjust energetically costly mate guarding to male-male competition and female reproductive value. *Animal Behaviour* 79: 75–82.

8. Andersson, M. 1982. Female choice selects for extreme tail length in a widowbird. *Nature* 299: 818–820.

9. Andersson, M. 1994. *Sexual Selection*. Princeton University Press, Princeton, NJ.

10. Andrade, M. C. B. 1996. Sexual selection for male sacrifice in the Australian redback spider. *Science* 271: 70–72.

11. Andrade, M. C. B. 2003. Risky mate search and male self-sacrifice in redback spiders. *Behavioral Ecology* 14: 531–538.

12. Arnold, S. J. 1983. Sexual selection: The interface of theory and empiricism. In P. P. G. Bateson (Ed.), *Mate Choice*. Cambridge University Press, Cambridge, England.

13. Arnqvist, G. and Kirkpatrick, M. 2005. The evolution of infidelity in socially monogamous passerines: The strength of direct and indirect selection on extrapair copulation behavior in females. *American Naturalist* 165: S26–S37.

14. Bakst, M. R. 1998. Structure of the avian oviduct with emphasis on sperm storage in poultry. *Journal of Experimental Zoology* 282: 618–626.

15. Barske, J., Schlinger, B. A., Wikelski, M., and Fusani, L. 2011. Female choice for male motor skills. *Proceedings of the Royal Society B* 278: 3523–3528.

16. Berglund, A., Rosenqvist, G., and Svensson, I. 1986. Mate choice, fecundity and sexual dimorphism in two pipefish species (Syngnathidae). *Behavioral Ecology and Sociobiology* 19: 301–307.

17. Berglund, A., Rosenqvist, G., and Robinson-Wolrath, S. 2006. Food or sex—males and females in a sex role reversed pipefish have different interests. *Behavioral Ecology and Sociobiology* 60: 281–287.

18. Bērziņš, A., Krama, T., Krams, I., Freeberg, T. M., Kivleniece, I., Kullberg, C., and Rantala, M. J. 2010. Mobbing as a trade-off between safety and reproduction in a songbird. *Behavioral Ecology* 21: 1054–1060.

19. Birkhead, T. R. and Møller, A. P. 1992. *Sperm Competition in Birds: Evolutionary Causes and Consequences*. Academic Press, London, England.

20. Birkhead, T. R. 2002. *Promiscuity: An Evolutionary History of Sperm Competition*. Harvard University Press, Cambridge, MA.

21. Bjork, A., Dallai, I., and Pitnick, S. 2007. Adaptive modulation of sperm production rate in *Drosophila bifurca*, a species with giant sperm. *Biology Letters* 3: 517–519.

22. Blanckenhorn, W. U. 2005. Behavioral causes and consequences of sexual size dimorphism. *Ethology* 11: 977–1016.

23. Blount, J. D., Metcalfe, N. B., Birkhead, T. R., and Surai, P. F. 2003. Carotenoid modulation of immune function and sexual attractiveness in zebra finches. *Science* 300: 125–127.

24. Borgia, G. 1985. Bower quality, number of decorations and mating success of male satin bowerbirds (*Ptilonorhynchus violaceus*). *Animal Behaviour* 33: 266–271.

25. Borgia, G. 1986. Sexual selection in bowerbirds. *Scientific American* 254 (June): 92–100.

26. Borgia, G., Egeth, M., Uy, J. A. C., and Patricelli, G. L. 2004. Juvenile infection and male display: Testing the bright male hypothesis across individual life histories. *Behavioral Ecology* 15: 722–728.

27. Borgia, G. 2006. Preexisting traits are important in the evolution of elaborated male sexual display. *Advances in the Study of Behavior* 36: 249–303.

28. Bro-Jørgensen, J. 2007. Reversed sexual conflict in a promiscuous antelope. *Current Biology* 17: 2157–2161.

29. Bro-Jørgensen, J., Johnstone, R. A., and Evans, M. R. 2007. Uninformative exaggeration of male sexual ornaments in barn swallows. *Current Biology* 17: 850–855.

30. Brockmann, H. J. and Penn, D. 1992. Male mating tactics in the horseshoe crab, *Limulus polyphemus*. *Animal Behaviour* 44: 653–665.

31. Brockmann, H. J., Colson, T., and Potts, W. 1994. Sperm competition in horseshoe crabs (*Limulus polyphemus*). *Behavioral Ecology and Sociobiology* 35: 153–160.

32. Brockmann, H. J. 2002. An experimental approach to alternative mating tactics in male horseshoe crabs (*Limulus polyphemus*). *Behavioral Ecology* 13: 232–238.

33. Brownmiller, S. 1975. *Against Our Will*. Simon and Schuster, New York, NY.

34. Bruning, B., Phillips, B. L., and Shine, R. 2010. Turgid female toads give males the slip: A new mechanism of female mate choice in the Anura. *Biology Letters* 6: 322–324.

35. Buskirk, R. E., Frolich, C., and Ross, K. G. 1984. The natural selection of sexual cannibalism. *American Naturalist* 123: 612–625.

36. Buzatto, B. A. and Machado, G. 2008. Resource defense polygyny shifts to female defense polygyny over the course of the reproductive season of a Neotropical harvestman. *Behavioral Ecology and Sociobiology* 63: 85–94.

37. Byers, J., Hebets, E., and Podos, J. 2010. Female mate choice based upon male motor performance. *Animal Behaviour* 79: 771–778.

38. Casselman, S. J. and Montgomerie, R. 2004. Sperm traits in relation to male quality in colonial spawning bluegill. *Journal of Fish Biology* 64: 1700–1711.

39. Clutton-Brock, T. 2007. Sexual selection in males and females. *Science* 318: 1882–1885.

40. Clutton-Brock, T. H., Hodge, S. J., Spong, G., Russell, A. F., Jordan, N. R., Bennett, N. C., Sharpe, L. L., and Manser, M. B. 2006. Intrasexual competition and sexual selection in cooperative mammals. *Nature* 444: 1065–1068.

41. Cordero, C. and Eberhard, W. G. 2003. Female choice of sexually antagonistic male adaptations: A critical review of some current research. *Journal of Evolutionary Biology* 16: 1–6.

42. Córdoba-Aguilar, A., Uhía, E., and Rivera, A. C. 2003. Sperm competition in Odonata (Insecta): The evolution of female sperm storage and rivals' sperm displacement. *Journal of Zoology* 261: 381–398.

43. Cornwallis, C. K., and Birkhead, T. R. 2007. Changes in sperm quality and numbers in response to experimental manipulation of male social status and female attractiveness. *American Naturalist* 170: 758–770.

44. Cornwallis, C. K. and O'Connor, E. A. 2009. Sperm: Seminal fluid interactions and the adjustment of sperm quality in relation to female attractiveness. *Proceedings of the Royal Society B* 276: 3467–3475.

45. Cowlishaw, G. and Dunbar, R. I. M. 1991. Dominance rank and mating success in male primates. *Animal Behaviour* 41: 1045–1056.

46. Crockford, C., Wittig, R. M., Seyfarth, R. M., and Cheney, D. L. 2007. Baboons eavesdrop to deduce mating opportunities. *Animal Behaviour* 73: 885–890.

47. Cunningham, E. J. A. and Russell, A. F. 2000. Egg investment is influenced by male attractiveness in the mallard. *Nature* 404: 74–77.

48. Dakin, R. and Montgomerie, R. 2011. Peahens prefer peacocks displaying more eyespots, but rarely. *Animal Behaviour* 82: 21–28.

49. Daly, M. and Wilson, M. 1983. *Sex, Evolution and Behavior* (2nd ed.). Willard Grant Press, Boston, MA.

50. Darwin, C. 1871. *The Descent of Man and Selection in Relation to Sex.* Murray, London, England.

51. Davies, N. B. 1983. Polyandry, cloaca-pecking and sperm competition in dunnocks. *Nature* 302: 334–336.

52. Dawkins, R. 1980. Good strategy or evolutionarily stable strategy? In G. W. Barlow and J. Silverberg (Eds.), *Sociobiology: Beyond Nature/Nurture?* Westview Press, Boulder, CO.

53. Dawkins, R. 1986. *The Blind Watchmaker.* W. W. Norton, New York, NY.

54. Day, L. B., Westcott, D. A., and Olster, D. H. 2005. Evolution of bower complexity and cerebellum size in bowerbirds. *Brain Behavior and Evolution* 66: 62–72.

55. Dean, R., Nakagawa, S., and Pizzari, T. 2011. The risk and intensity of sperm ejection in female birds. *American Naturalist* 178: 343–354.

56. DeWoody, J. A., Fletcher, D. E., Mackiewicz, M., Wilkins, S. D., and Avise, J. C. 2000. The genetic mating system of spotted sunfish (*Lepomis punctatus*): Mate numbers and the influence of male reproductive parasites. *Molecular Ecology* 9: 2119–2128.

57. Dickinson, J. L. and Rutowski, R. L. 1989. The function of the mating plug in the chalcedon checkerspot butterfly. *Animal Behaviour* 38: 154–162.

58. Dickinson, J. L. 1995. Trade-offs between postcopulatory riding and mate location in the blue milkweed beetle. *Behavioral Ecology* 6: 280–286.

59. Dodson, G. N. and Beck, M. W. 1993. Pre-copulatory guarding of penultimate females by male crab spiders, *Misumenoides formosipes. Animal Behaviour* 46: 951–959.

60. Doucet, S. M. and Montgomerie, R. 2003. Multiple sexual ornaments in satin bowerbirds: Ultraviolet plumage and bowers signal different aspects of male quality. *Behavioral Ecology* 14: 503–509.

61. Drăgăniou, T. I., Nagle, L., and Kreutzer, M. 2002. Directional female preference for an exaggerated male trait in canary (*Serinus canaria*) song. *Proceedings of the Royal Society of London B* 269: 2525–2531.

62. Drews, C. 1996. Contests and patterns of injuries in free-ranging male baboons (*Papio cynocephalus*). *Behaviour* 133: 443–474.

63. Eberhard, W. G. 1982. Beetle horn dimorphism: Making the best of a bad lot. *American Naturalist* 119: 420–426.

64. Eberhard, W. G. 1990. Animal genitalia and female choice. *American Scientist* 78: 134–141.

65. Eberhard, W. G. 1996. *Female Control: Sexual Selection by Cryptic Female Choice.* Princeton University Press, Princeton, NJ.

66. Eberhard, W. G. 2005. Evolutionary conflicts of interest: Are female conflicts of interest different? *American Naturalist* 165: S19–S25.

67. Edward, D. A. and Chapman, T. 2011. The evolution and significance of male mate choice. *Trends in Ecology and Evolution* 26: 647–654.

68. Emlen, D. J. 2008. The evolution of animal weapons. *Annual Review of Ecology, Evolution and Systematics* 39: 387–413.

69. Emlen, S. T. and Oring, L. W. 1977. Ecology, sexual selection and the evolution of mating systems. *Science* 197: 215–223.

70. Field, S. A. and Keller, M. A. 1993. Alternative mating tactics and female mimicry as post-copulatory mate-guarding behaviour in the parasitic wasp *Cotesia rubecula. Animal Behaviour* 46: 1183–1189.

71. Foellmer, M. W. and Fairbairn, D. J. 2003. Spontaneous male death during copulation in an orb-weaving spider. *Proceedings of the Royal Society of London B* 270: S183–S185.

72. Forsgren, E., Amundsen, T., Borg, A. A., and Bjelvenmark, J. 2004. Unusually dynamic sex roles in a fish. *Nature* 429: 551–554.

73. Forster, L. M. 1992. The stereotyped behaviour of sexual cannibalism in *Latrodectus hasselti* Thorell (Araneae: Theridiidae), the Australian redback spider. *Australian Journal of Zoology* 40: 1–11.

74. Fox, E. A. 2002. Female tactics to reduce sexual harassment in the Sumatran orangutan (*Pongo pygmaeus abelii*). *Behavioral Ecology and Sociobiology* 52: 93–101.

75. Friberg, U. and Arnqvist, G. 2003. Fitness effects of female mate choice: Preferred males are detrimental for *Drosophila melanogaster* females. *Journal of Evolutionary Biology* 16: 797–811.

76. Fu, P., Neff, B. D., and Gross, M. R. 2001. Tactic-specific success in sperm competition. *Proceedings of the Royal Society of London B* 268: 1105–1112.

77. Funk, D. H. and Tallamy, D. W. 2000. Courtship role reversal and deceptive signals in the long-tailed dance fly, *Rhamphomyia longicauda. Animal Behaviour* 59: 411–421.

78. García-Navas, V., Ferrer, E. S., and Sanz, J. J. 2012. Plumage yellowness predicts foraging ability in the blue tit *Cyanistes caeruleus. Biological Journal of the Linnean Society* 106: 418–429.

79. Gesquiere, L. R., Wango, E. O., Alberts, S. C., and Altmann, J. 2007. Mechanisms of sexual selection: Sexual swellings and estrogen concentrations as fertility indicators and cues for male consort decisions in wild baboons. *Hormones and Behavior* 51: 114–125.

80. Gil, D., Leboucher, G., Lacroix, A., Cue, R., and Kreutzer, M. 2004. Female canaries produce eggs with greater amounts of testosterone when exposed to preferred male song. *Hormones and Behavior* 45: 64–70.

81. Gilbert, L., Williamson, K. A., and Graves, J. A. 2012. Male attractiveness regulates daughter fecundity non-genetically via maternal investment. *Proceedings of the Royal Society B* 279: 523–528.

82. Gomes, C. M. and Boesch, C. 2009. Wild chimpanzees exchange meat for sex on a long-term basis. *PLoS ONE* 4(4): e5116. doi:10.1371/journal.pone.0005116.

83. Grether, G. F. 2000. Carotenoid limitation and mate preference evolution: A test of the indicator hypothesis in guppies (*Poecilia reticulata*). *Evolution* 54: 1712–1714.

84. Griggio, M., Biard, C., Penn, D. J., and Hoi, H. 2011. Female house sparrows "count on" male genes: Experimental evidence for MHC-dependent mate preference in birds. *BMC Evolutionary Biology* 11(44). doi:10.1186/1471-2148-11-44.

85. Gross, M. R. 1982. Sneakers, satellites, and parentals: Polymorphic mating strategies in North American sunfishes. *Zeitschrift für Tierpsychologie* 60: 1–26.

86. Gross, M. R. 1996. Alternative reproductive strategies and tactics: Diversity within species. *Trends in Ecology and Evolution* 11: 92–98.

87. Gwynne, D. T. 1981. Sexual difference theory: Mormon crickets show role reversal in mate choice. *Science* 213: 779–780.

88. Gwynne, D. T. and Simmons, L. W. 1990. Experimental reversal of courtship roles in an insect. *Nature* 346: 172–174.

89. Gwynne, D. T., Bussiere, L. F., and Ivy, T. M. 2007. Female ornaments hinder escape from spider webs in a role-reversed swarming dance fly. *Animal Behaviour* 73: 1077–1082.

90. Hamilton, W. D. and Zuk, M. 1982. Heritable true fitness and bright birds: A role for parasites? *Science* 218: 384–387.

91. Hausfater, G. 1975. Dominance and reproduction in baboons (*Papio cynocephalus*): A quantitative analysis. *Contributions in Primatology* 7: 1–150.

92. Healy, S. D. and Rowe, C. 2007. A critique of comparative studies of brain size. *Proceedings of the Royal Society of London B* 274: 453–464.

93. Hidalgo-García, S. 2006. The carotenoid-based plumage coloration of adult Blue Tits *Cyanistes caeruleus* correlates with the health status of their brood. *Ibis* 148: 727–734.

94. Hill, G. E. and Montgomerie, R. 1994. Plumage color signals nutritional condition in the house finch. *Proceedings of the Royal Society of London B* 258: 47–52.

95. Himuro, C. and Fujisaki, K. 2008. Males of the seed bug *Togo hemipterus* (Heteroptera: Lygaeidae) use accessory gland substances to inhibit remating by females. *Journal of Insect Physiology* 54: 1538–1542.

96. Holland, B. and Rice, W. R. 1998. Chase-away sexual selection: Antagonistic seduction versus resistance. *Evolution* 52: 1–7.

97. Holland, B. and Rice, W. R. 1999. Experimental removal of sexual selection reverses intersexual antagonistic coevolution and removes a reproductive load. *Proceedings of the National Academy of Sciences* 96: 5083–5088.

98. Hosken, D. J. and Stockley, P. 2004. Sexual selection and genital evolution. *Trends in Ecology and Evolution* 19: 87–93.

99. Hoving, H. J. T., Lipinski, M. R., Videler, J. J., and Bolstad, K. S. R. 2010. Sperm storage and mating in the deep-sea squid *Taningia danae* Joubin, 1931 (Oegopsida: Octopoteuthidae). *Marine Biology* 157: 393–400.

100. Hunt, J. and Simmons, L. W. 2002. Confidence of paternity and paternal care: Covariation revealed through the experimental manipulation of the mating system in the beetle *Onthophagus taurus*. *Journal of Evolutionary Biology* 15: 784–795.

101. Johns, J. L., Roberts, J. A., Clark, D. L., and Uetz, G. W. 2009. Love bites: Male fang use during coercive mating in wolf spiders. *Behavioral Ecology and Sociobiology* 64: 13–18.

102. Johnsen, A., Pärn, H., Fossøy, F., Kleven, O., Laskemoen, T., and Lifjeld, T. J. 2008. Is female promiscuity constrained by the presence of her social mate? An experiment with bluethroats *Luscinia svecica*. *Behavioral Ecology and Sociobiology* 62: 1761–1767.

103. Johnson, J. C., Trubl, P., Blackmore, V., and Miles, L. 2011. Male black widows court well-fed females more than starved females: Silken cues indicate sexual cannibalism risk. *Animal Behaviour* 82: 383–390.

104. Jukema, J. and Piersma, T. 2005. Permanent female mimics in a lekking shorebird. *Biology Letters* 2: 161–164.

105. Keagy, J., Savard, J. F., and Borgia, G. 2011. Complex relationship between multiple measures of cognitive ability and male mating success in satin bowerbirds, *Ptilonorhynchus violaceus*. *Animal Behaviour* 81: 1063–1070.

106. Kim, Y. J., Bartalska, K., Audsley, N., Yamanaka, N., Yapici, N., Lee, J. Y., Kim, Y. C., Markovic, M., Isaac, E., Tanaka, Y., and Dickson, B. J. 2009. MIPs are ancestral ligands for the sex peptide receptor. *Proceedings of the National Academy of Sciences* 107: 6520–6525.

107. Kimball, R. T., Braun, E. L., Ligon, J. D., Lucchini, V., and Randi, E. 2001. A molecular phylogeny of the peacock-pheasants (Galliformes : *Polyplectron* spp.) indicates loss and reduction of ornamental traits and display behaviours. *Biological Journal of the Linnean Society* 73: 187–198.

108. Kirkpatrick, M. 1982. Sexual selection and the evolution of female choice. *Evolution* 36: 1–12.

109. Kodric-Brown, A. and Brown, J. H. 1984. Truth in advertising: The kinds of traits favored by sexual selection. *American Naturalist* 124: 309–323.

110. Kodric-Brown, A. 1993. Female choice of multiple male criteria in guppies: Interacting effects of dominance, coloration and courtship. *Behavioral Ecology and Sociobiology* 32: 415–420.

111. Kokko, H., Jennions, M. D., and Brooks, D. R. 2006. Unifying and testing models of sexual selection. *Annual Review of Ecology, Evolution and Systematics* 37: 43–46.

112. Komdeur, J., Kraaijeveld-Smit, F., Kraaijeveld, K., and Edelaar, P. 1999. Explicit experimental evidence for the role of mate guarding in minimizing loss of paternity in the Seychelles warbler. *Proceedings of the Royal Society of London B* 266: 2075–2081.

113. Komdeur, J., Burke, T., and Richardson, D. S. 2007. Explicit experimental evidence for the effectiveness of proximity as mate-guarding behaviour in reducing extra-pair fertilization in the Seychelles warbler. *Molecular Ecology* 16: 3679–3688.

114. Lack, D. 1968. *Ecological Adaptations for Breeding in Birds*. Methuen, London, England.

115. Lande, R. 1981. Models of speciation by sexual selection of polygenic traits. *Proceedings of the National Academy of Sciences* 78: 3721–3725.

116. Lank, D. B., Smith, C. M., Hanotte, O., Burke, T., and Cooke, F. 1995. Genetic polymorphism for alternative mating behaviour in lekking male ruff *Philomachus pugnax*. *Nature* 378: 59–62.

117. Lelito, J. P. and Brown, W. D. 2008. Mate attraction by females in a sexually cannibalistic praying mantis. *Behavioral Ecology and Sociobiology* 63: 313–320.

118. Lewis, S. M. and Cratsley, C. K. 2008. Flash signal evolution, mate choice, and predation in fireflies. *Annual Review of Entomology* 53: 293–321.

119. Li, D., Oh, J., Kralj-Fisher, S., and Kuntner, M. 2012. Remote copulation: Male adaptation to female cannibalism. *Biology Letters* doi:10.1098/rsbl.2011.1202.

120. Low, M. 2005. Female resistance and male force: Context and patterns of copulation in the New Zealand stitchbird *Notiomystis cincta*. *Journal of Avian Biology* 36: 436–448.

121. Low, M. 2006. The energetic cost of mate guarding is correlated with territorial intrusions in the New Zealand stitchbird. *Behavioral Ecology* 17: 270–276.

122. Loyau, A., Saint Jalme, M., Cagniant, C., and Sorci, G. 2005. Multiple sexual advertisements honestly reflect health status in peacocks (*Pavo cristatus*). *Behavioral Ecology and Sociobiology* 58: 552–557.

123. Loyau, A., Saint Jalme, M., Mauget, R., and Sorci, G. 2007. Male sexual attractiveness affects the investment of maternal resources into the eggs in peafowl (*Pavo cristatus*). *Behavioral Ecology and Sociobiology* 61: 1043–1052.

124. Lung, O., Tram, U., Finnerty, C. M., Eipper-Mains, M. A., Kalb, J. M., and Wolfner, M. F. 2002. The *Drosophila melanogaster* seminal fluid protein Acp62F is a protease inhibitor that is toxic upon ectopic expression. *Genetics* 160: 211–224.

125. Lyon, B. E. and Montgomerie, R. 2012. Sexual selection is a form of social selection. *Philosophical Transactions of the Royal Society B* 367: 2266–2273.

126. Madden, J. R. 2001. Sex, bowers and brains. *Proceedings of the Royal Society of London B* 268: 833–838.

127. Madden, J. R. 2003. Male spotted bowerbirds preferentially choose, arrange and proffer objects that are good predictors of mating success. *Behavioral Ecology and Sociobiology* 53: 263–268.

128. Madden, J. R. 2003. Bower decorations are good predictors of mating success in the spotted bowerbird. *Behavioral Ecology and Sociobiology* 53: 269–277.

129. Marshall, R. C., Buchanan, K. L., and Catchpole, C. K. 2003. Sexual selection and individual genetic diversity in a songbird. *Proceedings of the Royal Society of London B* 270: S248–S250.

130. Matsumoto-Oda, A., Hamai, M., Hayaki, H., Hosaka, K., Hunt, K. D., Kasuya, E., Kawanaka, K., Mitani, J. C., Takasaki, H., and Takahata, Y. 2007. Estrus cycle asynchrony in wild female chimpanzees, *Pan troglodytes schweinfurthii*. *Behavioral Ecology and Sociobiology* 61: 661–668.

131. Mattle, B. and Wilson, A. B. 2009. Body size preferences in the pot-bellied seahorse *Hippocampus abdominalis*: Choosy males and indiscriminate females. *Behavioral Ecology and Sociobiology* 63: 1403–1410.

132. Maxwell, M. R., Gallego, K. M., and Barry, K. L. 2010. Effects of female feeding regime in a sexually cannibalistic mantid: Fecundity, cannibalism, and male response in *Stagmomantis limbata* (Mantodea). *Ecological Entomology* 35: 775–787.

133. McGraw, K. J. and Hill, G. E. 2000. Differential effects of endoparasitism on the expression of carotenoid- and melanin-based ornamental coloration. *Proceedings of the Royal Society of London B* 267: 1525–1531.

134. McGraw, K. J. and Ardia, D. R. 2003. Carotenoids, immunocompetence, and the information content of sexual colors: An experimental test. *American Naturalist* 162: 704–712.

135. Michl, G., Török, J., Griffith, S. C., and Sheldon, B. C. 2002. Experimental analysis of sperm competition mechanisms in a wild bird population. *Proceedings of the National Academy of Sciences* 99: 5466–5470.

136. Miller, J. A. 2007. Repeated evolution of male sacrifice behavior in spiders correlated with genital mutilation. *Evolution* 61: 1301–1315.

137. Møller, A. P. 1988. Female choice selects for male sexual tail ornaments in the monogamous swallow. *Nature* 332: 640–642.

138. Moran, N. A. and Dunbar, H. E. 2006. Sexual acquisition of beneficial symbionts in aphids. *Proceedings of the National Academy of Sciences* 103: 12803–12806.

139. Muller, M. N., Kahlenberg, S. M., Thompson, M. E., and Wrangham, R. W. 2007. Male coercion and the costs of promiscuous mating for female chimpanzees. *Proceedings of the Royal Society B* 274: 1009–1014.

140. Noë, R. and Sluijter, A. A. 1990. Reproductive tactics of male savanna baboons. *Behaviour* 113: 117–170.

141. Obara, Y., Fukano, Y., Watanabe, K., Ozawa, G., and Sasaki, K. 2011. Serotonin-induced mate rejection in the female cabbage butterfly, *Pieris rapae crucivora*. *Naturwissenschaften* 98: 989–993.

142. Östlund, S. and Ahnesjö, I. 1998. Female fifteen-spined sticklebacks prefer better fathers. *Animal Behaviour* 56: 1177–1183.

143. Palombit, R. A., Seyfarth, R. M., and Cheney, D. L. 1997. The adaptive value of "friendships" to female baboons: Experimental and observational evidence. *Animal Behaviour* 54: 599–614.

144. Parker, G. A. 1970. Sperm competition and its evolutionary consequences in the insects. *Biological Reviews* 45: 526–567.

145. Parker, G. A. 2006. Sexual conflict over mating and fertilization: An overview. *Philosophical Transactions of the Royal Society B* 361: 235–259.

146. Pasch, B., George, A. S., Campbell, P., and Phelps, S. M. 2011. Androgen-dependent male vocal performance influences female preference in Neotropical singing mice. *Animal Behaviour* 82: 177–183.

147. Patricelli, G. L., Uy, J. A. C., Walsh, G., and Borgia, G. 2002. Male displays adjusted to female's response. *Nature* 415: 279–280.

148. Patricelli, G. L., Uy, J. A. C., and Borgia, G. 2003. Multiple male traits interact: Attractive bower decorations facilitate attractive behavioural displays in satin bowerbirds. *Proceedings of the Royal Society of London B* 270: 2389–2395.

149. Patricelli, G. L., Uy, J. A. C., and Borgia, G. 2004. Female signals enhance the efficiency of mate assessment in satin bowerbirds (*Ptilonorhynchus violaceus*). *Behavioral Ecology* 15: 297–304.

150. Petrie, M. 1992. Peacocks with low mating success are more likely to suffer predation. *Animal Behaviour* 44: 585–586.

151. Petrie, M. 1994. Improved growth and survival of offspring of peacocks with more elaborate trains. *Nature* 371: 585–586.

152. Pike, T. W., Blount, J. D., Lindstrom, J., and Metcalfe, N. B. 2007. Dietary carotenoid availability influences a male's ability to provide parental care. *Behavioral Ecology* 18: 1100–1105.

153. Pischedda, A. and Chippindale, A. K. 2006. Intralocus sexual conflict diminishes the benefits of sexual selection. *PLoS Biology* 4(11): e356. doi:10.1371/journal.pbio.0040356.

154. Pitnick, S. and García-González, F. 2002. Harm to females increases with male body size in *Drosophila melanogaster*. *Proceedings of the Royal Society of London B* 269: 1821–1828.

155. Pitnick, S., Jones, K. E., and Wilkinson, G. S. 2006. Mating system and brain size in bats. *Proceedings of the Royal Society B* 273: 719–724.

156. Pizzari, T. and Birkhead, T. R. 2000. Female feral fowl eject sperm of subdominant males. *Nature* 405: 787–789.

157. Polak, M., Wolf, L. L., Starmer, W. T., and Barker, J. S. F. 2001. Function of the mating plug in *Drosophila hibisci* Bock. *Behavioral Ecology and Sociobiology* 49: 196–205.

158. Pratt, D. M. and Anderson, V. H. 1985. Giraffe social behavior. *Journal of Natural History* 19: 771–781.

159. Preston, B. T., Saint Jalme, M., Hingrat, Y., Lacroix, F., and Sorci, G. 2011. Sexually extravagant males age more rapidly. *Ecology Letters* 14: 1017–1024.

160. Prete, F. R. 1995. Designing behavior: A case study. *Perspectives in Ethology* 11: 255–277.

161. Pruett-Jones, S. and Pruett-Jones, M. 1994. Sexual competition and courtship disruptions: Why do male bowerbirds destroy each other's bowers? *Animal Behaviour* 47: 607–620.

162. Prum, R. O. 2010. The Lande-Kirkpatrick mechanism is the null model of evolution by intersexual selection: Implications for meaning, honesty and design in intersexual signals. *Evolution* 64: 3085–3100.

163. Queller, D. C. 1997. Why do females care more than males? *Proceedings of the Royal Society of London B* 264: 1555–1557.

164. Reinhardt, K., Naylor, R., and Siva-Jothy, M. T. 2003. Reducing a cost of traumatic insemination: Female bedbugs evolve a unique organ. *Proceedings of the Royal Society of London B* 270: 2371–2375.

165. Reinhardt, K., Naylor, R. A., and Siva-Jothy, M. T. 2009. Ejaculate components delay reproductive senescence while elevating female reproductive rate in an insect. *Proceedings of the National Academy of Sciences* 106: 21743–21747.

166. Reynolds, J. D. and Gross, M. R. 1990. Costs and benefits of female mate choice: Is there a lek paradox? *American Naturalist* 136: 230–243.

167. Reynolds, S. M., Dryer, K., Bollback, J., Uy, J. A. C., Patricelli, G. L., Robson, T., Borgia, G., and Braun, M. J. 2007. Behavioral paternity predicts genetic paternity in Satin Bowerbirds (*Ptilonorhynchus violaceus*), a species with a non-resource-based mating system. *Auk* 124: 857–867.

168. Rezac, M. 2009. The spider *Harpactea sadistica*: Co-evolution of traumatic insemination and complex female genital morphology in spiders. *Proceedings of the Royal Society B* 276: 2697–2701.

169. Rintamäki, P. T., Lundberg, A., Alatalo, R. V., and Höglund, J. 1998. Assortative mating and female clutch investment in black grouse. *Animal Behaviour* 56: 1399–1403.

170. Rosenqvist, G. 1990. Male mate choice and female-female competition for mates in the pipefish *Nerophis ophidion*. *Animal Behaviour* 39: 1110–1116.

171. Rowland, W. J. 1994. Proximate determinants of stickleback behavior: An evolutionary perspective. In M. Bell and S. Foster (Eds.), *The Evolutionary Biology of the Threespine Stickleback*. Oxford University Press, Oxford, England.

172. Ryan, M. J., Fox, J. H., Wilczynski, W., and Rand, A. S. 1990. Sexual selection for sensory exploitation in the frog *Physalaemus pustulosus*. *Nature* 343: 66–67.

173. Sato, H. 1998. Male participation in nest building in the dung beetle *Scarabaeus catenatus* (Coleoptera: Scarabaeidae): Mating effort versus paternal effort. *Journal of Insect Behavior* 11: 833–843.

174. Schärer, L., Rowe, L., and Arnqvist, G. 2012. Anisogamy, chance and the evolution of sex roles. *Trends in Ecology and Evolution* 27: 260–264.

175. Schwabl, H., Mock, D. W., and Gieg, J. A. 1997. A hormonal mechanism for parental favouritism. *Nature* 386: 231.

176. Senar, J. C., Figuerola, J., and Pascual, J. 2002. Brighter yellow blue tits make better parents. *Proceedings of the Royal Society of London B* 269: 257–261.

177. Shuster, S. M. 1989. Male alternative reproductive strategies in a marine isopod crustacean (*Paracerceis sculpta*): The use of genetic markers to measure differences in the fertilization success among alpha, beta, and gamma-males. *Evolution* 43: 1683–1689.

178. Shuster, S. M. and Wade, M. J. 1991. Equal mating success among male reproductive strategies in a marine isopod. *Nature* 350: 608–610.

179. Shuster, S. M. 1992. The reproductive behaviour of alpha, beta, and gamma morphs in *Paracerceis sculpta*: A marine isopod crustacean. *Behaviour* 121: 231–258.

180. Shuster, S. M. and Wade, M. J. 2003. *Mating Systems and Strategies*. Princeton University Press, Princeton, NJ.

181. Simmons, L. W. 2001. *Sperm Competition and Its Evolutionary Consequences in the Insects*. Princeton University Press, Princeton, NJ.

182. Simmons, L. W. and Emlen, D. J. 2006. Evolutionary trade-off between weapons and testes. *Proceedings of the National Academy of Sciences* 103: 16346–16351.

183. Simmons, R. E. and Scheepers, L. 1996. Winning by a neck: Sexual selection in the evolution of giraffe. *American Naturalist* 148: 771–786.

184. Simpson, S. J., Sword, G. A., Lorch, P. D., and Couzin, I. D. 2006. Cannibal crickets on a forced march for protein and salt. *Proceedings of the National Academy of Sciences* 103: 4152–4156.

185. Smith, H. G. and Montgomerie, R. D. 1991. Sexual selection and the tail ornaments of North American barn swallows. *Behavioral Ecology and Sociobiology* 28: 195–201.

186. Stoltz, J. A., Elias, D. O., and Andrade, M. C. B. 2008. Females reward courtship by competing males in a cannibalistic spider. *Behavioral Ecology and Sociobiology* 62: 689–697.

187. Strum, S. C. 1987. *Almost Human*. W. W. Norton, New York, NY.

188. Stutt, A. D. and Siva-Jothy, M. T. 2001. Traumatic insemination and sexual conflict in the bed bug *Cimex lectularius*. *Proceedings of the National Academy of Sciences* 98: 5683–5687.

189. Svensson, B. G. 1997. Swarming behavior, sexual dimorphism, and female reproductive status in the sex role-reversed dance fly species *Rhamphomyia marginata*. *Journal of Insect Behavior* 10: 783–804.

190. Takahashi, M., Arita, H., Hiraiwa-Hasegawa, M., and Hasegawa, T. 2008. Peahens do not prefer peacocks with more elaborate trains. *Animal Behaviour* 75: 1209–1219.

191. Thornhill, R. 1976. Sexual selection and nuptial feeding behavior in *Bittacus apicalis* (Insecta: Mecoptera). *American Naturalist* 119: 529–548.

192. Thornhill, R. 1981. *Panorpa* (Mecoptera: Panorpidae) scorpionflies: Systems for understanding resource-defense polygyny and alternative male reproductive efforts. *Annual Review of Ecology and Systematics* 12: 355–386.

193. Tomkins, J. L., Lebas, N. R., Witton, M. P., Martill, D. M., and Humphries, S. 2010. Positive allometry and the prehistory of sexual selection. *American Naturalist* 176: 141–148.

194. Tregenza, T., Simmons, L. W., Wedell, N., and Zuk, M. 2006. Female preference for male courtship song and its role as a signal of immune function and condition. *Animal Behaviour* 72: 809–818.

195. Trivers, R. L. 1972. Parental investment and sexual selection. In B. Campbell (Ed.), *Sexual Selection and the Descent of Man*. Aldine, Chicago, IL.

196. Tsai, M. L. and Dai, C. F. 2003. Cannibalism within mating pairs of the parasitic isopod, *Ichthyoxenus fushanensis*. *Journal of Crustacean Biology* 23: 662–668.

197. Tsubaki, Y., Samejima, Y., and Siva-Jothy, M. T. 2010. Damselfly females prefer hot males: Higher courtship success in males in sunspots. *Behavioral Ecology and Sociobiology* 64: 1547–1554.

198. Tuttle, E. M., Pruett-Jones, S., and Webster, M. S. 1996. Cloacal protuberances and extreme sperm production in Australian fairy-wrens. *Proceedings of the Royal Society of London B* 263: 1359–1364.

199. Uhl, G., Nessler, S., and Schneider, J. 2011. Securing paternity in spiders? A review on occurrence and effects of mating plugs and male genital mutilation. *Genetica* 138: 75–104.

200. Uy, J. A. C., Patricelli, G. L., and Borgia, G. 2001. Complex mate searching in the satin bowerbird *Ptilonorhynchus violaceus*. *American Naturalist* 158: 530–542.

201. Vahed, K. 1998. The function of nuptial feeding in insects: Review of empirical studies. *Biological Reviews* 73: 43–78.

202. Vallet, E., Beme, I., and Kreutzer, M. 1998. Two-note syllables in canary songs elicit high levels of sexual display. *Animal Behaviour* 55: 291–297.

203. Vieites, D. R., Nieto-Román, S., Barluenga, M., Palanca, A., Vences, M., and Meyer, A. 2004. Post-mating clutch piracy in an amphibian. *Nature* 431: 305–308.

204. Waage, J. K. 1979. Dual function of the damselfly penis: Sperm removal and transfer. *Science* 203: 916–918.

205. Waage, J. K. 1997. Parental investment—minding the kids or keeping control? In P. A. Gowaty (Ed.), *Feminism and Evolutionary Biology: Boundaries, Interactions, and Frontiers*. Chapman and Hall, New York, NY.

206. Wenninger, E. J. and Averill, A. L. 2006. Influence of body and genital morphology on relative male fertilization success in oriental beetle. *Behavioral Ecology* 17: 656–663.

207. West-Eberhard, M. J. 1979. Sexual selection, social competition, and evolution. *Proceedings of the American Philosophical Society* 123: 222–234.

208. Whiting, M. J., Webb, J. K., and Keogh, J. S. 2009. Flat lizard female mimics use sexual deception in visual but not chemical signals. *Proceedings of the Royal Society B* 276: 1585–1591.

209. Widdig, A., Bercovitch, F. B., Streich, W. J., Sauermann, U., Nürnberg, P., and Krawczak, M. 2004. A longitudinal analysis of reproductive skew in male rhesus macaques. *Proceedings of the Royal Society of London B* 271: 819–826.

210. Wigby, S. and Chapman, T. 2004. Female resistance to male harm evolves in response to manipulation of sexual conflict. *Evolution* 58: 1028–1037.

211. Wikelski, M. and Baurle, S. 1996. Pre-copulatory ejaculation solves time constraints during copulations in marine iguanas. *Proceedings of the Royal Society of London B* 263: 439–444.

212. Wojcieszek, J. M., Nicholls, J. A., and Goldizen, A. W. 2007. Stealing behavior and the maintenance of a visual display in the satin bowerbird. *Behavioral Ecology* 18: 689–695.

213. Wojcieszek, J. M. and Simmons, L. W. 2011. Male genital morphology influences paternity success in the millipede *Antichiropus variabilis*. *Behavioral Ecology and Sociobiology* 65: 1843–1856.

214. Wolff, J. O., Mech, S. G., Dunlap, A. S., and Hodges, K. E. 2002. Multi-male mating by paired and unpaired female prairie voles (*Microtus ochrogaster*). *Behaviour* 139: 1147–1160.

215. Zahavi, A. 1975. Mate selection: A selection for a handicap. *Journal of Theoretical Biology* 53: 205–214.

216. Zeh, J. A. and Zeh, D. W. 2003. Toward a new sexual selection paradigm: Polyandry, conflict and incompatibility. *Ethology* 109: 929–950.

Chapter 8

1. Alcock, J., Eickwort, G. C., and Eickwort, K. R. 1977. The reproductive behavior of *Anthidium maculosum* and the evolutionary significance of multiple copulations by females. *Behavioral Ecology and Sociobiology* 2: 385–396.

2. Andersson, M. 1994. *Sexual Selection*. Princeton University Press, Princeton, NJ.

3. Andrade, M. C. B. 1996. Sexual selection for male sacrifice in the Australian redback spider. *Science* 271: 70–72.

4. Andrade, M. C. B. 2003. Risky mate search and male self-sacrifice in redback spiders. *Behavioral Ecology* 14: 531–538.

5. Apollonio, M., Festa-Bianchet, M., Mari, F., Bruno, E., and Locati, M. 1998. Habitat manipulation modifies lek use in fallow deer. *Ethology* 104: 603–612.

6. Arnold, T. W. 1999. What limits clutch size in waders? *Journal of Avian Biology* 30: 216–220.

7. Arnqvist, G. and Kirkpatrick, M. 2005. The evolution of infidelity in socially monogamous passerines: The strength of direct and indirect selection on extrapair copulation behavior in females. *American Naturalist* 165: S26–S37.

8. Arnqvist, G. and Kirkpatrick, M. 2007. The evolution of infidelity in socially monogamous passerines revisited: A reply to Griffith. *American Naturalist* 169: 282–283.

9. Avise, J. C. 2004. *Molecular Markers, Natural History and Evolution*. Sinauer Associates, Sunderland, MA.

10. Balmford, A., Deutsch, J. C., Nefdt, R. J. C., and Clutton-Brock, T. 1993. Testing hotspot models of lek evolution: Data from three species of ungulates. *Behavioral Ecology and Sociobiology* 33: 57–65.

11. Bateman, A. J. 1948. Intra-sexual selection in *Drosophila*. *Heredity* 2: 349–368.

12. Birkhead, T. R. and Møller, A. P. (Eds.). 1998. *Sperm Competition and Sexual Selection*. Academic Press, San Diego, CA.

13. Bissoondath, C. J. and Wiklund, C. 1995. Protein content of spermatophores in relation to monandry/polyandry in butterflies. *Behavioral Ecology and Sociobiology* 37: 365–372.

14. Borgia, G. 1985. Bower quality, number of decorations and mating success of male satin bowerbirds (*Ptilonorhynchus violaceus*). *Animal Behaviour* 33: 266–271.

15. Borries, C., Launhardt, K., Epplen, C., Epplen, J. T., and Winkler, P. 1999. Males as infant protectors in Hanuman langurs (*Presbytis entellus*) living in multimale groups—defence pattern, paternity and sexual behavior. *Behavioral Ecology and Sociobiology* 46: 350–356.

16. Borries, C., Savini, T., and Koenig, A. 2011. Social monogamy and the threat of infanticide in larger mammals. *Behavioral Ecology and Sociobiology* 65: 685–693.

17. Bradbury, J. W. 1977. Lek mating behavior in the hammer-headed bat. *Zeitschrift für Tierpsychologie* 45: 225–255.

18. Bradbury, J. W. 1981. The evolution of leks. In R. D. Alexander and D. W. Tinkle (Eds.), *Natural Selection and Social Behavior*. Chiron Press, New York, NY.

19. Bradbury, J. W. and Gibson, R. M. 1983. Leks and mate choice. In P. Bateson (Ed.), *Mate Choice*. Cambridge University Press, Cambridge, England.

20. Bradbury, J. W., Vehrencamp, S. L., and Gibson, R. M. 1989. Dispersion of displaying male sage grouse. I. Patterns of temporal variation. *Behavioral Ecology and Sociobiology* 24: 1–14.

21. Breven, K. A. 1981. Mate choice in the wood frog, *Rana sylvatica*. *Evolution* 35: 707–722.

22. Bro-Jørgensen, J. and Durant, S. M. 2003. Mating strategies of topi bulls: Getting in the centre of attention. *Animal Behaviour* 65: 585–594.

23. Brockmann, H. J. and Penn, D. 1992. Male mating tactics in the horseshoe crab, *Limulus polyphemus*. *Animal Behaviour* 44: 653–665.

24. Brotherton, P. N. M. and Manser, M. B. 1997. Female dispersion and the evolution of monogamy in the dik-dik. *Animal Behaviour* 54: 1413–1424.

25. Brown, J. L., Morales, V., and Summers, K. 2008. Divergence in parental care, habitat selection and larval life history between two species of Peruvian poison frogs: An experimental analysis. *Journal of Evolutionary Biology* 21: 1534–1543.

26. Burke, T. and Bruford, M. W. 1987. DNA fingerprinting in birds. *Nature* 327: 149–152.

27. Burke, T. 1989. DNA fingerprinting and other methods for the study of mating success. *Trends in Ecology and Evolution* 4: 139–144.

28. Byrne, P. G. and Whiting, M. J. 2011. Effects of simultaneous polyandry on offspring fitness in an African tree frog. *Behavioral Ecology* 22: 385–391.

29. Cantoni, D. and Brown, R. 1997. Paternal investment and reproductive success in the California mouse, *Peromyscus californicus*. *Animal Behaviour* 54: 377–386.

30. Carey, M. and Nolan, V., Jr. 1975. Polygyny in indigo buntings: A hypothesis tested. *Science* 190: 1296–1297.

31. Cézilly, F. and Nager, R. G. 1995. Comparative evidence for a positive association between divorce and extra-pair paternity in birds. *Proceedings of the Royal Society of London B* 262: 7–12.

32. Chaine, A. S. and Lyon, B. E. 2008. Adaptive plasticity in female mate choice dampens sexual selection on male ornaments in the lark bunting. *Science* 319: 459–462.

33. Concannon, M. R., Stein, A. C., and Uy, A. L. 2012. Kin selection may contribute to lek evolution and trait introgression across an avian hybrid zone. *Molecular Ecology* 21:1477–1486.

34. Cox, R. M., Duryea, M. C., Najarro, M., and Calsbeek, R. 2011. Paternal condition drives progeny sex-ratio bias in a lizard that lacks parental care. *Evolution* 65: 220–230.

35. Cunningham, E. J. A. 2003. Female mate preferences and subsequent resistance to copulation in the mallard. *Behavioral Ecology* 14: 326–333.

36. Dale, J., Montgomerie, R., Michaud, D., and Boag, P. 1999. Frequency and timing of extrapair fertilisation in the polyandrous red phalarope (*Phalaropus fulicarius*). *Behavioral Ecology and Sociobiology* 46: 50–56.

37. Dale, S., Rinden, H., and Slagsvold, T. 1992. Competition for a mate restricts mate search of female pied flycatchers. *Behavioral Ecology and Sociobiology* 30: 165–176.

38. Davies, N. B. and Lundberg, A. 1984. Food distribution and a variable mating system in the dunnock, *Prunella modularis*. *Journal of Animal Ecology* 53: 895–912.

39. Davies, N. B. 1985. Cooperation and conflict among dunnocks, *Prunella modularis*, in a variable mating system. *Animal Behaviour* 33: 628–648.

40. Davies, N. B. 1992. *Dunnock Behaviour and Social Evolution.* Oxford University Press, Oxford, England.

41. Davies, N. B., Hartley, I. R., Hatchwell, B. J., and Langmore, N. E. 1996. Female control of copulations to maximize male help: A comparison of polygynandrous alpine accentors, *Prunella collaris*, and dunnocks, *P. modularis*. *Animal Behaviour* 51: 27–47.

42. Deutsch, J. C. 1994. Uganda kob mating success does not increase on larger leks. *Behavioral Ecology and Sociobiology* 34: 451–459.

43. Dibattista, J. D., Feldheim, K. A., Gruber, S. H., and Hendry, A. P. 2008. Are indirect genetic benefits associated with polyandry? Testing predictions in a natural population of lemon sharks. *Molecular Ecology* 17: 783–795.

44. Dickinson, J. L. 2001. Extrapair copulations in western bluebirds (*Sialia mexicana*): Female receptivity favors older males. *Behavioral Ecology and Sociobiology* 50: 423–429.

45. Dixson, A. F., Bossi, T., and Wickings, E. J. 1993. Male-dominance and genetically-determined reproductive success in the mandrill (*Mandrillus sphinx*). *Primates* 34: 525–532.

46. Dodson, G. N. 1997. Resource defense mating system in antlered flies, *Phytalmia* spp. (Diptera: Tephritidae). *Annals of the Entomological Society of America* 90: 496–504.

47. Double, M. C. and Cockburn, A. 2003. Subordinate superb fairy-wrens (*Malurus cyaneus*) parasitize the reproductive success of attractive dominant males. *Proceedings of the Royal Society of London B* 270: 379–384.

48. Edvardsson, M. 2007. Female *Callosobruchus maculatus* mate when they are thirsty: Resource-rich ejaculates as mating effort in a beetle. *Animal Behaviour* 74: 183–188.

49. Eggert, A.-K. and Sakaluk, S. K. 1995. Female-coerced monogamy in burying beetles. *Behavioral Ecology and Sociobiology* 37: 147–154.

50. Eliassen, S. and Kokko, H. 2008. Current analyses do not resolve whether extra-pair paternity is male or female driven. *Behavioral Ecology and Sociobiology* 62: 1795–1804.

51. Emlen, S. T. and Oring, L. W. 1977. Ecology, sexual selection and the evolution of mating systems. *Science* 197: 215–223.

52. Emlen, S. T., Wrege, P. H., and Webster, M. S. 1998. Cuckoldry as a cost of polyandry in the sex-role-reversed wattled jacana, *Jacana jacana*. *Proceedings of the Royal Society of London B* 265: 2359–2364.

53. Faaborg, J., Parker, P. G., DeLay, L., de Vries, T. J., Bednarz, J. C., Paz, S. M., Naranjo, J., and Waite, T. A. 1995. Confirmation of cooperative polyandry in the Galapagos hawk (*Buteo galapagoensis*). *Behavioral Ecology and Sociobiology* 36: 83–90.

54. Fiske, P., Rintamäki, P. T., and Karvonen, E. 1998. Mating success in lekking males: A meta-analysis. *Behavioral Ecology* 9: 328–338.

55. Foltz, D. W. and Schwagmeyer, P. L. 1989. Sperm competition in the thirteen-lined ground squirrel: Differential fertilization success under field conditions. *American Naturalist* 133: 257–265.

56. Fossøy, F., Johnsen, A., and Lifjeld, J. T. 2006. Evidence of obligate female promiscuity in a socially monogamous passerine. *Behavioral Ecology and Sociobiology* 60: 255–259.

57. Fossøy, F., Johnsen, A., and Lifjeld, J. T. 2008. Multiple genetic benefits of female promiscuity in a socially monogamous passerine. *Evolution* 62: 145–156.

58. Francis, C. M., Elp, A., Brunton, J. A., and Kunz, T. H. 1994. Lactation in male fruit bats. *Nature* 367: 691–692.

59. Frith, C. B. and Frith, D. W. 2001. Nesting biology of the spotted catbird, *Ailuroedus melanotis*, a monogamous bowerbird (Ptilonorhynchidae), in Australian Wet Tropics upland rainforests. *Australian Journal of Zoology* 49: 279–310.

60. Gerlach, N. M., McGlothlin, J. W., Parker, P. G., and Ketterson, E. D. 2012. Promiscuous mating produces offspring with higher lifetime fitness. *Proceedings of the Royal Society B* 279: 860–866.

61. Getz, L. L. and Carter, C. S. 1996. Prairie-vole partnerships. *American Scientist* 84: 56–62.

62. Gibson, R. M. 1996. A re-evaluation of hotspot settlement in lekking sage grouse. *Animal Behaviour* 52: 993–1005.

63. Gibson, R. M., Aspbury, A. S., and McDaniel, L. L. 2002. Active formation of mixed-species grouse leks: A role for predation in lek evolution? *Proceedings of the Royal Society of London B* 269: 2503–2507.

64. Gray, E. M. 1997. Do red-winged blackbirds benefit genetically from seeking copulations with extra-pair males? *Animal Behaviour* 53: 605–623.

65. Gray, E. M. 1997. Female red-winged blackbirds accrue material benefits from copulating with extra-pair males. *Animal Behaviour* 53: 625–639.

66. Griffith, S. C., Owens, I. P. F., and Thuman, K. A. 2002. Extra pair paternity in birds: A review of interspecific variation and adaptive function. *Molecular Ecology* 11: 2195–2212.

67. Griffith, S. C. 2007. The evolution of infidelity in socially monogamous passerines: Neglected components of direct and indirect selection. *American Naturalist* 169: 274–281.

68. Gubernick, D. J. and Teferi, T. 2000. Adaptive significance of male parental care in a monogamous mammal. *Proceedings of the Royal Society of London B* 267: 147–150.

69. Harcourt, A. H., Harvey, P. H., Larson, S. G., and Short, R. V. 1981. Testis weight, body weight and breeding system in primates. *Nature* 293: 55–57.

70. Hartke, T. R. and Baer, B. 2011. The mating biology of termites: A comparative review. *Animal Behaviour* 82: 927–936.

71. Heinze, J., Hölldobler, B., and Yamauchi, K. 1998. Male competition in *Cardiocondyla* ants. *Behavioral Ecology and Sociobiology* 42: 239–246.

72. Heistermann, M., Ziegler, T., van Schaik, C. P., Launhardt, K., Winkler, P., and Hodges, J. K. 2001. Loss of oestrus, concealed ovulation and paternity confusion in free-ranging Hanuman langurs. *Proceedings of the Royal Society of London B* 268: 2445–2451.

73. Himuro, C. and Fujisaki, K. 2008. Males of the seed bug *Togo hemipterus* (Heteroptera: Lygaeidae) use accessory gland substances to inhibit remating by females. *Journal of Insect Physiology* 54: 1538–1542.

74. Höglund, J. and Lundberg, A. 1987. Sexual selection in a monomorphic lek-breeding bird: Correlates of male mating success in the great snipe *Gallinago media*. *Behavioral Ecology and Sociobiology* 21: 211–216.

75. Hohoff, C., Franzen, K., and Sachser, N. 2003. Female choice in a promiscuous wild guinea pig, the yellow-toothed cavy (*Galea musteloides*). *Behavioral Ecology and Sociobiology* 53: 341–349.

76. Hoogland, J. L. 1998. Why do female Gunnison's prairie dogs copulate with more than one male? *Animal Behaviour* 55: 351–359.

77. Huk, T. and Winkel, W. 2006. Polygyny and its fitness consequences for primary and secondary female pied flycatchers. *Proceedings of the Royal Society B* 273: 1681–1688.

78. Jamieson, I. G. 1997. Testing reproductive skew models in a communally breeding bird, the pukeko, *Porphyrio porphyrio*. *Proceedings of the Royal Society of London B* 264: 335–340.

79. Jeffreys, A. J., Wilson, V., and Thein, S. L. 1985. Hypervariable "minisatellite" regions in human DNA. *Nature* 314: 67–73.

80. Jennions, M. D. and Petrie, M. 2000. Why do females mate multiply? A review of the genetic benefits. *Biological Reviews* 75: 21–64.

81. Jiguet, F. and Bretagnolle, V. 2006. Manipulating lek size and composition using decoys: An experimental investigation of lek evolution models. *American Naturalist* 168: 758–768.

82. Johnsen, A., Andersson, S., Ornberg, J., and Lifjeld, J. T. 1998. Ultraviolet plumage ornamentation affects social mate choice and sperm competition in bluethroats (Aves: *Luscinia s. svecica*): A field experiment. *Proceedings of the Royal Society of London B* 265: 1313–1318.

83. Johnsen, A., Andersen,V., Sunding, C., and Lifjeld, J. T. 2000. Female bluethroats enhance offspring immunocompetence through extra-pair copulations. *Nature* 406: 296–299.

84. Jones, J. S. and Wynne-Edwards, K. E. 2000. Paternal hamsters mechanically assist the delivery, consume amniotic fluid and placenta, remove fetal membranes, and provide parental care during the birth process. *Hormones and Behavior* 37: 116–125.

85. Just, J. 1988. Siphonoecetinae (Corophiidae) 6: A survey of phylogeny, distribution, and biology. *Crustaceana* 13(Suppl.): 193–208.

86. Kempenaers, B., Verheyen, G. R., van der Broeck, M., Burke, T., van Broeckhoven, C., and Dhondt, A. A. 1992. Extra-pair paternity results from female preference for high-quality males in the blue tit. *Nature* 357: 494–496.

87. Kempenaers, B., Verheyen, G. R., and Dhondt, A. A. 1997. Extrapair paternity in the blue tit (*Parus caeruleus*): Female choice, male characteristics, and offspring quality. *Behavioral Ecology* 8: 481–492.

88. Komers, P. E and Brotherton, P. N. M. 1997. Female space use is the best predictor of monogamy in mammals. *Proceedings of the Royal Society of London B* 264: 1261–1270.

89. Lack, D. 1968. *Ecological Adaptations for Breeding in Birds*. Methuen, London, England.

90. Lank, D. B., Oring, L. W., and Maxson, S. J. 1985. Mate and nutrient limitation of egg-laying in a polyandrous shorebird. *Ecology* 66: 1513–1524.

91. Lill, A. 1974. Sexual behavior of the lek-forming white-bearded manakin, *Manacus manacus trinitatis* Hartert. *Zeitschrift für Tierpsychologie* 36: 1–36.

92. Lloyd, J. E. 1980. Insect behavioral ecology: Coming of age in bionomics or compleat biologists have revolutions too. *Florida Entomologist* 63: 1–4.

93. Loyau, A., Saint Jalme, M., and Sorci, G. 2007. Non-defendable resources affect peafowl lek organization: A male removal experiment. *Behavioural Processes* 74: 64–70.

94. Lyon, B. E., Montgomerie, R. D., and Hamilton, L. D. 1987. Male parental care and monogamy in snow buntings. *Behavioral Ecology and Sociobiology* 20: 377–382.

95. Mattila, H. R. and Seeley, T. D. 2007. Genetic diversity in honey bee colonies enhances productivity and fitness. *Science* 317: 362–364.

96. McCracken, G. F. and Bradbury, J. W. 1981. Social organization and kinship in the polygynous bat *Phyllostomus hastatus*. *Behavioral Ecology and Sociobiology* 8: 11–34.

97. Miller, J. A. 2007. Repeated evolution of male sacrifice behavior in spiders correlated with genital mutilation. *Evolution* 61: 1301–1315.

98. Moreno, J., Veiga, J. P., Cordero, P. J., and Mínguez, E. 1999. Effects of paternal care on reproductive success in the polygynous spotless starling *Sturnus unicolor*. *Behavioral Ecology and Sociobiology* 47: 47–53.

99. Murphy, C. G. 2003. The cause of correlations between nightly numbers of male and female barking treefrogs (*Hyla gratiosa*) attending choruses. *Behavioral Ecology* 14: 274–281.

100. Nazareth, T. M. and Machado, G. 2010. Mating system and exclusive postzygotic paternal care in a Neotropical harvestman (Arachnida: Opiliones). *Animal Behaviour* 79: 547–554.

101. Newcomer, S. D., Zeh, J. A., and Zeh, D. W. 1999. Genetic benefits enhance the reproductive success of polyandrous females. *Proceedings of the National Academy of Sciences* 96: 10236–10241.

102. Nunn, C. L., Gittleman, J. L., and Antonovics, J. 2000. Promiscuity and the primate immune system. *Science* 290: 1168–1170.

103. Oldroyd, B. P. and Fewell, J. H. 2007. Genetic diversity promotes homeostasis in insect colonies. *Trends in Ecology and Evolution* 22: 408–413.

104. Orians, G. H. 1969. On the evolution of mating systems in birds and mammals. *American Naturalist* 103: 589–603.

105. Oring, L. W. and Knudson, M. L 1973. Monogamy and polyandry in the spotted sandpiper. *The Living Bird* 11: 59–73.

106. Oring, L. W. 1985. Avian polyandry. *Current Ornithology* 3: 309–351.

107. Oring, L. W., Colwell, M. A., and Reed, J. M. 1991. Lifetime reproductive success in the spotted sandpiper (*Actitis macularia*): Sex differences and variance components. *Behavioral Ecology and Sociobiology* 28: 425–432.

108. Oring, L. W., Fleischer, R. C., Reed, J. M., and Marsden, K. E. 1992. Cuckoldry through stored sperm in the sequentially polyandrous spotted sandpiper. *Nature* 359: 631–633.

109. Packer, C., Scheel, D., and Pusey, A. E. 1990. Why lions form groups: Food is not enough. *American Naturalist* 136: 1–19.

110. Partecke, J., von Haeseler, A., and Wikelski, M. 2002. Territory establishment in lekking marine iguanas, *Amblyrhynchus cristatus*: Support for the hotshot mechanism. *Behavioral Ecology and Sociobiology* 51: 579–587.

111. Piper, W. H., Evers, D. C., Meyer, M. W., Tischler, K. B., Kaplan, J. D., and Fleischer, R. C. 1997. Genetic monogamy in the common loon (*Gavia immer*). *Behavioral Ecology and Sociobiology* 41: 25–32.

112. Pribil, S. and Searcy, W. A. 2001. Experimental confirmation of the polygyny threshold model for red-winged blackbirds. *Proceedings of the Royal Society of London B* 268: 1643–1646.

113. Queller, D. C., Strassmann, J. E., and Hughes, C. R. 1993. Microsatellites and kinship. *Trends in Ecology and Evolution* 8: 285–288.

114. Quinn, J. S., Woolfenden, G. E., Fitzpatrick, J. W., and White, B. N. 1999. Multi-locus DNA fingerprinting supports genetic monogamy in Florida scrub-jays. *Behavioral Ecology and Sociobiology* 45: 1–10.

115. Reid, J. M., Monaghan, P., and Ruxton, G. D. 2002. Males matter: The occurrence and consequences of male incubation in starlings (*Sturnus vulgaris*). *Behavioral Ecology and Sociobiology* 51: 255–261.

116. Rintamäki, P. T., Alatalo, R. V., Höglund, J., and Lundberg, A. 1995. Male territoriality and female choice on black grouse leks. *Animal Behaviour* 49: 759–767.

117. Rintamäki, P. T., Höglund, J., Alatalo, R. V., and Lundberg, A. 2001. Correlates of male mating success on black grouse (*Tetrao tetrix* L.) leks. *Annales Zoologici Fennici* 38: 99–109.

118. Rittenhouse, T. A. G., Semlitsch, R. D., and Thompson, F. R. 2009. Survival costs associated with wood frog breeding migrations: Effects of timber harvest and drought. *Ecology* 90: 1620–1630.

119. Rodríguez-Muñoz, R., Bretman, A., and Tregenza, T. 2011. Guarding males protect females from predation in a wild insect. *Current Biology* 21: 1716–1719.

120. Rubenstein, D. R. 2007. Female extrapair mate choice in a cooperative breeder: Trading sex for help and increasing offspring heterozygosity. *Proceedings of the Royal Society B* 274: 1895–1903.

121. Runcie, M. J. 2000. Biparental care and obligate monogamy in the rock-haunting possum, *Petropseudes dahli*, from tropical Australia. *Animal Behaviour* 59: 1001–1008.

122. Ryder, T. B., Blake, J. G., and Loiselle, B. R. A. 2006. A test of the environmental hotspot hypothesis for lek placement in three species of Manakins (Pipridae) in Ecuador. *Auk* 123: 247–258.

123. Sardell, R. J., Arcese, P., Keller, L. F., and Reid, J. M. 2011. Sex-specific differential survival of extra-pair and within-pair offspring in song sparrows, *Melospiza melodia*. *Proceedings of the Royal Society B* 278: 3251–3259.

124. Sato, T. 1994. Active accumulation of spawning substrate: A determinant of extreme polygyny in a shell-brooding cichlid fish. *Animal Behaviour* 48: 669–678.

125. Schamel, D., Tracy, D. M., and Lank, D. B. 2004. Male mate choice, male availability and egg production as limitations on polyandry in the red-necked phalarope. *Animal Behaviour* 67: 847–853.

126. Schwagmeyer, P. L. 1988. Scramble-competition polygyny in an asocial mammal: Male mobility and mating success. *American Naturalist* 131: 885–892.

127. Schwagmeyer, P. L. 1994. Competitive mate searching in the 13-lined ground squirrel: Potential roles of spatial memory? *Ethology* 98: 265–276.

128. Schwagmeyer, P. L. 1995. Searching today for tomorrow's mates. *Animal Behaviour* 50: 759–767.

129. Schwensow, N., Eberle, M., and Sommer, S. 2008. Compatibility counts: MHC-associated mate choice in a wild promiscuous primate. *Proceedings of the Royal Society B* 275: 555–564.

130. Seeley, T. D. and Tarpy, D. R. 2007. Queen promiscuity lowers disease within honeybee colonies. *Proceedings of the Royal Society B* 274: 67–72.

131. Setchell, J. M., Charpentier, M. J. E., Abbott, K. M., Wickings, E. J., and Knapp, L. A. 2010. Opposites attract: MHC-associated mate choice in a polygynous primate. *Journal of Evolutionary Biology* 23: 136–148.

132. Sheldon, B. C. 1993. Sexually-transmitted disease in birds: Occurrence and evolutionary significance. *Philosophical Transactions of the Royal Society of London B* 339: 491–497.

133. Shorey, L. 2002. Mating success on white-bearded manakin (*Manacus manacus*) leks: Male characteristics and relatedness. *Behavioral Ecology and Sociobiology* 52: 451–457.

134. Shuster, S. M. and Wade, M. J. 2003. *Mating Systems and Strategies*. Princeton University Press, Princeton, NJ.

135. Snow, D. W. 1956. Courtship ritual: The dance of the manakins. *Animal Kingdom* 59: 86–91.

136. Sogabe, A., Matsumoto, K., and Yanagisawa, Y. 2007. Mate change reduces the reproductive rate of males in a monogamous pipefish *Corythoichthys haematopterus*: The benefit of long-term pair bonding. *Ethology* 113: 764–771.

137. Soukup, S. S. and Thompson, C. F. 1998. Social mating system and reproductive success in house wrens. *Behavioral Ecology* 9: 43–48.

138. Stenmark, G., Slagsvold, T., and Lifjeld, J. T. 1988. Polygyny in the pied flycatcher, *Ficedula hypoleuca*: A test of the deception hypothesis. *Animal Behaviour* 36: 1646–1657.

139. Strassmann, J. E. 2001. The rarity of multiple mating by females in the social Hymenoptera. *Insectes Sociaux* 48: 1–13.

140. Tarpy, D. R. and Nielsen, D. I. 2002. Sampling error, effective paternity, and estimating the genetic structure of honey bee colonies (Hymenoptera : Apidae). *Annals of the Entomological Society of America* 95: 513–528.

141. Taylor, M. L., Wedell, N., and Hosken, D. J. 2007. The heritability of attractiveness. *Current Biology* 17: R959–R960.

142. Temrin, H. and Arak, A. 1989. Polyterritoriality and deception in passerine birds. *Trends in Ecology and Evolution* 4: 106–108.

143. Thornhill, R. and Alcock, J. 1983. *The Evolution of Insect Mating Systems*. Harvard University Press, Cambridge, MA.

144. Townsend, A. K., Clark, A. B., and McGowan, K. J. 2010. Direct benefits and genetic costs of extrapair paternity for female American crows (*Corvus brachyrhynchos*). *American Naturalist* 175: E1–E9.

145. Tregenza, T. and Wedell, N. 2000. Genetic compatibility, mate choice and patterns of parentage: Invited review. *Molecular Ecology* 9: 1013–1027.

146. Tschirren, B., Postma, E., Rutstein, A. N., and Griffith, S. C. 2012. When mothers make sons sexy: Maternal effects contribute to the increased sexual attractiveness of extra-pair offspring. *Proceedings of the Royal Society B* 279: 1233–1240.

147. Vincent, A. C. J. and Sadler, L. M. 1995. Faithful pair bonds in wild seahorses, *Hippocampus whitei*. *Animal Behaviour* 50: 1557–1569.

148. Vincent, A. C. J., Marsden, A. D., Evans, K. L., and Sadler, L. M. 2004. Temporal and spatial opportunities for polygamy in a monogamous seahorse, *Hippocampus whitei*. *Behaviour* 141: 141–156.

149. Waage, J. K. 1973. Reproductive behavior and its relation to territoriality in *Calopteryx maculata* (Beauvois) (Odonata: Calopterygidae). *Behaviour* 47: 240–256.

150. Walker, R. S., Flinn, M. V., and Hill, K. R. 2010. Evolutionary history of partible paternity in lowland South America. *Proceedings of the National Academy of Sciences* 107: 19195–19200.

151. Weatherhead, P. J. and Robertson, R. J. 1979. Offspring quality and the polygyny threshold: The "sexy son" hypothesis. *American Naturalist* 113: 201–208.

152. Weathers, W. W. and Sullivan, K. A. 1989. Juvenile foraging proficiency, parental effort, and avian reproductive success. *Ecological Monographs* 59: 223–246.

153. Webster, M. S. 1994. Female-defence polygyny in a Neotropical bird, the Montezuma oropendula. *Animal Behaviour* 48: 779–794.

154. Webster, M. S. and Robinson, S. K. 1999. Courtship disruptions and male mating strategies: Examples from female-defense mating systems. *American Naturalist* 154: 717–729.

155. Webster, M. S. and Reichart, L. 2005. Use of microsatellites for parentage and kinship analyses in animals. *Molecular Evolution* 395: 222–238.

156. Webster, M. S., Tarvin, K. A., Tuttle, E. M., and Pruett-Jones, S. 2007. Promiscuity drives sexual selection in a socially monogamous bird. *Evolution* 61: 2205–2211.

157. Wedell, N. and Tregenza, T. 1999. Successful fathers sire successful sons. *Evolution* 53: 620–625.

158. Westneat, D. F., Sherman, P. W., and Morton, M. L. 1990. The ecology and evolution of extra-pair copulations in birds. *Current Ornithology* 7: 330–369.

159. Wickler, W. and Seibt, U. 1981. Monogamy in Crustacea and man. *Zeitschrift für Tierpsychologie* 57: 215–234.

160. Wiklund, C., Karlsson, B., and Leimar, O. 2001. Sexual conflict and cooperation in butterfly reproduction: A comparative study of polyandry and female fitness. *Proceedings of the Royal Society of London B* 268: 1661–1667.

161. Wilson, A. B. and Martin-Smith, K. M. 2007. Genetic monogamy despite social promiscuity in the pot-bellied seahorse (*Hippocampus abdominalis*). *Molecular Ecology* 16: 2345–2352.

162. Wong, M. Y. L., Munday, P. L., Buston, P. M, and Jones, G. P. 2008. Monogamy when there is potential for polygyny: Tests of multiple hypotheses in a group-living fish. *Behavioral Ecology* 19: 353–361.

163. Woodroffe, R. and Vincent, A. 1994. Mother's little helpers: Patterns of male care in mammals. *Trends in Ecology and Evolution* 9: 294–297.

164. Woyciechowski, M., Kabat, L., and Król, E. 1994. The function of the mating sign in honey bees, *Apis mellifera* L.: New evidence. *Animal Behaviour* 47: 733–735.

165. Zeh, J. A. and Zeh, D. W. 1996. The evolution of polyandry I: Intragenomic conflict and genetic incompatibility. *Proceedings of the Royal Society of London B* 263: 1711–1717.

166. Zeh, J. A. 1997. Polyandry and enhanced reproductive success in the harlequin beetle-riding pseudoscorpion. *Behavioral Ecology and Sociobiology* 40: 111–118.

167. Zeh, J. A. and Zeh, D. W. 1997. The evolution of polyandry II: Post-copulatory defenses against genetic incompatibility. *Proceedings of the Royal Society of London B* 264: 69–75.

168. Zeh, J. A., Newcomer, S. D., and Zeh, D. W. 1998. Polyandrous females discriminate against previous mates. *Proceedings of the National Academy of Sciences* 95: 13273–13736.

Chapter 9

1. Albrecht, T. and Klvaňa, P. 2004. Nest crypsis, reproductive value of a clutch and escape decisions in incubating female mallards *Anas platyrhynchos*. *Ethology* 110: 603–614.

2. Allman, J., Rosin, A., Kumar, R., and Hasenstaub, A. 1998. Parenting and survival in anthropoid primates: Caretakers live longer. *Proceedings of the National Academy of Sciences* 95: 6866–6869.

3. Aragón, S., Møller, A. P., Soler, J. J., and Soler, M. 1999. Molecular phylogeny of cuckoos supports a polyphyletic origin of brood parasitism. *Journal of Evolutionary Biology* 12: 495–506.

4. Balcombe, J. P. 1990. Vocal recognition of pups by mother Mexican free-tailed bats, *Tadarida brasiliensis mexicana*. *Animal Behaviour* 39: 960–966.

5. Balshine-Earn, S. 1995. The costs of parental care in Galilee St Peter's fish, *Sarotherodon galilaeus*. *Animal Behaviour* 50: 1–7.

6. Beecher, M. D., Beecher, I. M., and Hahn, S. 1981. Parent-offspring recognition in bank swallows, *Riparia riparia*: II. Development and acoustic basis. *Animal Behaviour* 29: 95–101.

7. Brown, K. M. 1998. Proximate and ultimate causes of adoption in ring-billed gulls. *Animal Behaviour* 56: 1529–1543.

8. Buchan, J. C., Alberts, S. C., Silk, J. B., and Altmann, J. 2003. True paternal care in a multi-male primate society. *Nature* 425: 179–181.

9. Buzatto, B. A., Requena, G. S., Martins, E. G., and Machado, G. 2007. Effects of maternal care on the lifetime reproductive success of females in a neotropical harvestman. *Journal of Animal Ecology* 76: 937–945.

10. Charrier, I., Mathevon, N., and Jouventin, P. 2003. Vocal signature recognition of mothers by fur seal pups. *Animal Behaviour* 65: 543–550.

11. Clifford, L. D. and Anderson, D. J. 2001. Experimental demonstration of the insurance value of extra eggs in an obligately siblicidal seabird. *Behavioral Ecology* 12: 340–347.

12. Clotfelter, E. D., Schubert, K. A., Nolan, V., and Ketterson, E. D. 2003. Mouth color signals thermal state of nestling dark-eyed juncos (*Junco hyemalis*). *Ethology* 109: 171–182.

13. Clutton-Brock, T. H. 1991. *The Evolution of Parental Care*. Princeton University Press, Princeton, NJ.

14. Clutton-Brock, T. H. and Parker, G. A. 1992. Potential reproductive rates and the operation of sexual selection. *Quarterly Review of Biology* 67: 437–456.

15. Creighton, J. C., Heflin, N. D., and Belk, M. C. 2009. Cost of reproduction, resource quality, and terminal investment in a burying beetle. *American Naturalist* 174: 673–684.

16. Davies, N. B. and de L. Brooke, M. 1988. Cuckoos versus reed warblers: Adaptations and counteradaptations. *Animal Behaviour* 36: 262–284.

17. Davies, N. B., de L. Brooke, M., and Kacelnik, A. 1996. Recognition errors and probability of parasitism determine whether reed warblers should accept or reject mimetic cuckoo eggs. *Proceedings of the Royal Society of London B* 263: 925–931.

18. Davies, N. B., Kilner, R. M., and Noble, D. G. 1998. Nestling cuckoos *Cuculus canorus* exploit hosts with begging calls that mimic a brood. *Proceedings of the Royal Society of London B* 265: 673–678.

19. Davies, N. B. 2000. *Cuckoos, Cowbirds and Other Cheats*. T & AD Poyser, London, England.

20. Dawkins, R. and Carlisle, T. R. 1976. Parental investment, mate desertion and a fallacy. *Nature* 262: 131–133.

21. de Ayala, R. M., Saino, N., Møller, A. P., and Anselmi, C. 2007. Mouth coloration of nestlings covaries with offspring quality and influences parental feeding behavior. *Behavioral Ecology* 18: 526–534.

22. Del-Claro, K. and Tizo-Pedroso, E. 2009. Ecological and evolutionary pathways of social behavior in pseudoscorpions (Arachnida: Pseudoscorpiones). *Acta Ethologica* 12: 13–22.

23. Eadie, J. M. and Fryxell, J. M. 1992. Density dependence, frequency dependence, and alternative nesting strategies in goldeneyes. *American Naturalist* 140: 621–641.

24. Ghalambor, C. K. and Martin, T. E. 2001. Fecundity-survival trade-offs and parental risk-taking in birds. *Science* 292: 494–497.

25. Gil, D., Leboucher, G., Lacroix, A., Cue, R., and Kreutzer, M. 2004. Female canaries produce eggs with greater amounts of testosterone when exposed to preferred male song. *Hormones and Behavior* 45: 64–70.

26. Gomendio, M., García-González, F., Reguera, P., and Rivero, A. 2008. Male egg carrying in *Phyllomorpha laciniata* is favoured by natural not sexual selection. *Animal Behaviour* 75: 763–770.

27. Graves, J. A. and Whiten, A. 1980. Adoption of strange chicks by herring gulls, *Larus argentatus*. *Zeitschrift für Tierpsychologie* 54: 267–278.

28. Grim, T. 2007. Experimental evidence for chick discrimination without recognition in a brood parasite host. *Proceedings of the Royal Society B* 274: 373–381.

29. Grim, T. 2011. Ejecting chick cheats: A changing paradigm? *Frontiers in Zoology* 8. doi:10.1186/1742–9994–8–14.

30. Gross, M. R. and Sargent, R. C. 1985. The evolution of male and female parental care in fishes. *American Zoologist* 25: 807–822.

31. Hartung, J. 1982. Polygyny and inheritance of wealth. *Current Anthropology* 23: 1–12.

32. Hauber, M. E. and Kilner, R. M. 2007. Coevolution, communication, and host chick mimicry in parasitic finches: Who mimics whom? *Behavioral Ecology and Sociobiology* 61: 497–503.

33. Heeb, P., Schwander, T., and Faoro, S. 2003. Nestling detectability affects parental feeding preferences in a cavity-nesting bird. *Animal Behaviour* 66: 637–642.

34. Heinsohn, R., Langmore, N. E., Cockburn, A., and Kokko, H. 2011. Adaptive secondary sex ratio adjustments via sex-specific infanticide in a bird. *Current Biology* 21: 1744–1747.

35. Hofer, H. and East, M. L. 2008. Siblicide in Serengeti spotted hyenas: A long-term study of maternal input and cub survival. *Behavioral Ecology and Sociobiology* 62: 341–351.

36. Holley, A. J. F. 1984. Adoption, parent-chick recognition, and maladaptation in the herring gull *Larus argentatus. Zeitschrift für Tierpsychologie* 64: 9–14.

37. Hoover, J. P. and Robinson, S. K. 2007. Retaliatory mafia behavior by a parasitic cowbird favors host acceptance of parasitic eggs. *Proceedings of the National Academy of Sciences* 104: 4479–4483.

38. Hrdy, S. B. 1999. *Mother Nature: A History of Mothers, Infants, and Natural Selection.* Pantheon, New York, NY.

39. Hunt, S., Cuthill, I. C., Bennett, A. T. D., and Griffiths, R. 1999. Preferences for ultraviolet partners in the blue tit. *Animal Behaviour* 58: 809–815.

40. Jesseau, S. A., Holmes, W. G., and Lee, T. M. 2008. Mother-offspring recognition in communally nesting degus, *Octodon degus. Animal Behaviour* 75: 573–582.

41. Kemp, D. J. 2002. Sexual selection constrained by life history in a butterfly. *Proceedings of the Royal Society of London B* 269: 1341–1345.

42. Kilner, R. M. and Langmore, N. E. 2011. Cuckoos versus hosts in insects and birds: Adaptations, counter-adaptations and outcomes. *Biological Reviews* 86: 836–852.

43. Klug, H. 2009. Relationship between filial cannibalism, egg energetic content and parental condition in the flagfish. *Animal Behaviour* 77: 1313–1319.

44. Knudsen, B. and Evans, R. M. 1986. Parent-young recognition in herring gulls (*Larus argentatus*). *Animal Behaviour* 34: 77–80.

45. Kölliker, M. 2007. Benefits and costs of earwig (*Forficula auricularia*) family life. *Behavioral Ecology and Sociobiology* 61: 1489–1497.

46. Krüger, O. and Davies, N. B. 2002. The evolution of cuckoo parasitism: A comparative analysis. *Proceedings of the Royal Society of London B* 269: 375–381.

47. Lack, D. 1968. *Ecological Adaptations for Breeding in Birds.* Methuen, London, England.

48. Langmore, N. E., Hunt, S., and Kilner, R. M. 2003. Escalation of a coevolutionary arms race through host rejection of brood parasitic young. *Nature* 422: 157–160.

49. Langmore, N. E. and Kilner, R. M. 2007. Breeding site and host selection by Horsfield's bronze-cuckoos, *Chalcites basalis. Animal Behaviour* 74: 995–1004.

50. Langmore, N. E., Cockburn, A., Russell, A. F., and Kilner, R. M. 2009. Flexible cuckoo chick-rejection rules in the superb fairy-wren. *Behavioral Ecology* 20: 978–984.

51. Lanyon, S. M. 1992. Interspecific brood parasitism in blackbirds (Icterinae): A phylogenetic perspective. *Science* 255: 77–79.

52. Lin, C. P., Danforth, B. N., and Wood, T. K. 2004. Molecular phylogenetics and evolution of maternal care in Membracine treehoppers. *Systematic Biology* 53: 400–421.

53. Lotem, A. 1993. Learning to recognize nestlings is maladaptive for cuckoo *Cuculus canorus* hosts. *Nature* 362: 743–745.

54. Lotem, A., Nakamura, H., and Zahavi, A. 1995. Constraints on egg discrimination and cuckoo-host co-evolution. *Animal Behaviour* 49: 1185–1209.

55. Lougheed, L. W. and Anderson, D. J. 1999. Parent blue-footed boobies suppress siblicidal behavior of offspring. *Behavioral Ecology and Sociobiology* 45: 11–18.

56. Lycett, J. E. and Dunbar, R. I. M. 1999. Abortion rates reflect the optimization of parental investment strategies. *Proceedings of the Royal Society of London B* 266: 2355–2358.

57. Lyon, B. E. and Eadie, J. M. 1991. Mode of development and interspecific avian brood parasitism. *Behavioral Ecology* 2: 309–318.

58. Lyon, B. E. 1993. Conspecific brood parasitism as a flexible female reproductive tactic in American coots. *Animal Behaviour* 46: 911–928.

59. Lyon, B. E., Eadie, J. M., and Hamilton, L. D. 1994. Parental choice selects for ornamental plumage in American coot chicks. *Nature* 371: 240–243.

60. Lyon, B. E. 2003. Ecological and social constraints on conspecific brood parasitism by nesting female American coots (*Fulica americana*). *Journal of Animal Ecology* 72: 47–60.

61. Lyon, B. E. 2007. Mechanism of egg recognition in defenses against conspecific brood parasitism: American coots (*Fulica americana*) know their own eggs. *Behavioral Ecology and Sociobiology* 61: 455–463.

62. Lyon, B. E. and Eadie, J. M. 2008. Conspecific brood parasitism in birds: A life-history perspective. *Annual Review of Ecology, Evolution, and Systematics* 39: 343–363.

63. Machado, G., Requena, G. S., Buzatto, B. A., Osses, F., and Rossetto, L. M. 2004. Five new cases of paternal care in harvestmen (Arachnida: Opiliones): Implications for the evolution of male guarding in the neotropical family Gonyleptidae. *Sociobiology* 44: 577–598.

64. Martin, T. E. and Schwabl, H. 2008. Variation in maternal effects and embryonic development rates among passerine species. *Philosophical Transactions of the Royal Society B* 363: 1663–1674.

65. Mas, F., Haynes, K. F., and Kölliker, M. 2009. A chemical signal of offspring quality affects maternal care in a social insect. *Proceedings of the Royal Society B* 276: 2847–2853.

66. Massaro, M., Starling-Windhof, A., Briskie, J. V., and Martin, T. E. 2008. Introduced mammalian predators induce behavioural changes in parental care in an endemic New Zealand bird. *PLoS ONE* 3(6): e2331. doi:10.1371/journal.pone.0002331.

67. McCracken, G. F. 1984. Communal nursing in Mexican free-tailed bat maternity colonies. *Science* 223: 1090–1091.

68. McCracken, G. F. and Gustin, M. K. 1991. Nursing behavior in Mexican free-tailed bat maternity colonies. *Ethology* 89: 305–321.

69. Medvin, M. B. and Beecher, M. D. 1986. Parent-offspring recognition in the barn swallow (*Hirundo rustica*). *Animal Behaviour* 34: 1627–1639.

70. Medvin, M. B., Stoddard, P. K., and Beecher, M. D. 1993. Signals for parent-offspring recognition: A comparative analysis of the begging calls of cliff swallows and barn swallows. *Animal Behaviour* 45: 841–850.

71. Miller, D. E. and Emlen, J. T., Jr. 1975. Individual chick recognition and family integrity in the ring-billed gull. *Behaviour* 52: 124–144.

72. Mock, D. W. 1984. Siblicidal aggression and resource monopolization in birds. *Science* 225: 731–733.

73. Mock, D. W. and Ploger, B. J. 1987. Parental manipulation of optimal hatch asynchrony in cattle egrets: An experimental study. *Animal Behaviour* 35: 150–160.

74. Mock, D. W., Drummond, H., and Stinson, C. H. 1990. Avian siblicide. *American Scientist* 78: 438–449.

75. Mock, D. W. and Parker, G. A. 1997. *The Evolution of Sibling Rivalry.* Oxford University Press, Oxford, England.

76. Mock, D. W. 2004. *More Than Kin and Less Than Kind: The Evolution of Family Conflict.* Harvard University Press, Cambridge, MA.

77. Nazareth, T. M. and Machado, G. 2010. Mating system and exclusive postzygotic paternal care in a Neotropical harvestman (Arachnida: Opiliones). *Animal Behaviour* 79: 547–554.

78. Neff, B. D. 2003. Decisions about parental care in response to perceived paternity. *Nature* 422: 716–719.

79. Neff, B. D., Fu, P., and Gross, M. R. 2003. Sperm investment and alternative mating tactics in bluegill sunfish (*Lepomis macrochirus*). *Behavioral Ecology* 14: 634–641.

80. Osorno, J. L. and Drummond, H. 1995. The function of hatching asynchrony in the blue-footed booby. *Behavioral Ecology and Sociobiology* 37: 265–274.

81. Petit, L. J. 1991. Adaptive tolerance of cowbird parasitism by prothonotary warblers: A consequence of site limitation? *Animal Behaviour* 41: 425–432.

82. Pierotti, R. and Murphy, E. C. 1987. Intergenerational conflicts in gulls. *Animal Behaviour* 35: 435–444.

83. Pollet, T. V., Fawcett, T. W., Buunk, A. P., and Nettle, D. 2009. Sex-ratio biasing towards daughters among lower-ranking co-wives in Rwanda. *Biology Letters* 5: 765–768.

84. Queller, D. C. 1997. Why do females care more than males? *Proceedings of the Royal Society of London B* 264: 1555–1557.

85. Requena, G. S., Buzatto, B. A., Munguía-Steyer, R., and Machado, G. 2009. Efficiency of uniparental male and female care against egg predators in two closely related syntopic harvestmen. *Animal Behaviour* 78: 1169–1176.

86. Rohwer, S. and Spaw, C. D. 1988. Evolutionary lag versus bill-size constraints: A comparative study of the acceptance of cowbird eggs by old hosts. *Evolutionary Ecology* 2: 27–36.

87. Saino, N., Ninni, P., Calza, S., Martinelli, R., de Bernardi, F., and Møller, A. P. 2000. Better red than dead: Carotenoid-based mouth coloration reveals infection in barn swallow nestlings. *Proceedings of the Royal Society of London B* 267: 57–61.

88. Salomon, M., Schneider, J., and Lubin, Y. 2005. Maternal investment in a spider with suicidal maternal care, *Stegodyphus lineatus* (Araneae, Eresidae). *Oikos* 109: 614–622.

89. Sargent, R. C. 1989. Allopaternal care in the fathead minnow, *Pimephales promelas*: Stepfathers discriminate against their adopted eggs. *Behavioral Ecology and Sociobiology* 25: 379–386.

90. Schielzeth, H. and Bolund, E. 2010. Patterns of conspecific brood parasitism in zebra finches. *Animal Behaviour* 79: 1329–1337.

91. Schwabl, H., Mock, D. W., and Gieg, J. A. 1997. A hormonal mechanism for parental favouritism. *Nature* 386: 231.

92. Sealy, S. G. 1995. Burial of cowbird eggs by parasitized yellow warblers: An empirical and experimental study. *Animal Behaviour* 49: 877–889.

93. Seidelmann, K. 2006. Open-cell parasitism shapes maternal investment patterns in the Red Mason bee *Osmia rufa*. *Behavioral Ecology* 17: 839–846.

94. Seidelmann, K., Ulbrich, K., and Mielenz, N. 2010. Conditional sex allocation in the Red Mason bee, *Osmia rufa*. *Behavioral Ecology and Sociobiology* 64: 337–347.

95. Semel, B. and Sherman, P. W. 2001. Intraspecific parasitism and nest-site competition in wood ducks. *Animal Behaviour* 61: 787–803.

96. Shizuka, D. and Lyon, B. E. 2011. Hosts improve the reliability of chick recognition by delaying the hatching of brood parasitic eggs. *Current Biology* 21: 515–519.

97. Slagsvold, T. 1998. On the origin and rarity of interspecific nest parasitism in birds. *American Naturalist* 152: 264–272.

98. Smiseth, P. T., Lennox, L., and Moore, A. J. 2007. Interaction between parental care and sibling competition: Parents enhance offspring growth and exacerbate sibling competition. *Evolution* 61: 2331–2339.

99. Smiseth, P. T., Ward, R. J. S., and Moore, A. J. 2007. Parents influence asymmetric sibling competition: Experimental evidence with partially dependent young. *Ecology* 88: 3174–3182.

100. Smith, M. S., Kish, B. J., and Crawford, C. B. 1987. Inheritance of wealth as human kin investment. *Ethology and Sociobiology* 8: 171–182.

101. Smith, R. L. and Larsen, E. 1993. Egg attendance and brooding by males of the giant water bug *Lethocerus medius* (Guerin) in the field (Heteroptera, Belostomatidae). *Journal of Insect Behavior* 6: 93–106.

102. Smith, R. L. 1997. Evolution of paternal care in giant water bugs (Heteroptera: Belostomatidae). In J. C. Choe and B. J. Crespi (Eds.), *The Evolution of Social Behaviour in Insects and Arachnids*. Cambridge University Press, Cambridge, England.

103. Soler, M., Soler, J. J., Martínez, J. G., and Møller, A. P. 1995. Magpie host manipulation by great spotted cuckoos: Evidence for an avian Mafia? *Evolution* 49: 770–775.

104. Sorenson, M. D. and Payne, R. B. 2001. A single ancient origin of brood parasitism in African finches: Implications for host-parasite coevolution. *Evolution* 55: 2550–2567.

105. Tallamy, D. W. 2001. Evolution of exclusive paternal care in arthropods. *Annual Review of Entomology* 46: 139–165.

106. Tokue, K. and Ueda, K. 2010. Mangrove gerygones *Gerygone laevigaster* eject little bronze-cuckoo *Chalcites minutillus* hatchlings from parasitized nests. *Ibis* 152: 835–839.

107. Trivers, R. L. 1972. Parental investment and sexual selection. In B. Campbell (Ed.), *Sexual Selection and the Descent of Man*. Aldine, Chicago, IL.

108. Trivers, R. L. and Willard, D. E. 1973. Natural selection of parental ability to vary the sex ratio of offspring. *Science* 179: 90–92.

109. Trivers, R. L. 1974. Parent-offspring conflict. *American Zoologist* 14: 249–264.

110. White, P. A. 2008. Maternal response to neonatal sibling conflict in the spotted hyena, *Crocuta crocuta*. *Behavioral Ecology and Sociobiology* 62: 353–361.

111. Wilkinson, M., Kupfer, A., Marques-Porto, R., Jeffkins, H., Antoniazzi, M. M., and Jared, C. 2008. One hundred million years of skin feeding? Extended parental care in a Neotropical caecilian (Amphibia: Gymnophiona). *Biology Letters* 4: 358–361.

112. Winfree, R. 1999. Cuckoos, cowbirds and the persistence of brood parasitism. *Trends in Ecology and Evolution* 14: 338–343.

113. Yom-Tov, Y. and Geffen, E. 2006. On the origin of brood parasitism in altricial birds. *Behavioral Ecology* 17: 196–205.

Chapter 10

1. Alcock, J. and Sherman, P. W. 1994. On the utility of the proximate-ultimate dichotomy in biology. *Ethology* 96: 58–62.

2. Baker, M. C., Bottjer, S. W., and Arnold, A. P. 1984. Sexual dimorphism and lack of seasonal changes in vocal control regions of the white-crowned sparrow brain. *Brain Research* 295: 85–89.

3. Baker, M. C. and Cunningham, M. A. 1985. The biology of bird-song dialects. *Behavioral and Brain Sciences* 8: 85–133.

4. Ballentine, B., Hyman, J., and Nowicki, S. 2004. Vocal performance influences female response to male bird song: An experimental test. *Behavioral Ecology* 15: 163–168.

5. Baptista, L. F. and Petrinovich, L. 1986. Song development in the white-crowned sparrow: Social factors and sex differences. *Animal Behaviour* 34: 1359–1371.

6. Baptista, L. F. and Morton, M. L. 1988. Song learning in montane white-crowned sparrows: From whom and when. *Animal Behaviour* 36: 1753–1764.

7. Beecher, M. D., Stoddard, P. K., Campbell, S. E., and Horning, C. L. 1996. Repertoire matching between neighbouring song sparrows. *Animal Behaviour* 51: 917–923.

8. Beecher, M. D., Campbell, E., and Nordby, J. C. 2000. Territory tenure in song sparrows is related to song sharing with neighbours, but not to repertoire size. *Animal Behaviour* 59: 29–37.

9. Beecher, M. D., Campbell, S. E., Burt, J. M., Hill, C. E., and Nordby, J. C. 2000. Song-type matching between neighbouring song sparrows. *Animal Behaviour* 59: 21–27.

10. Beecher, M. D. and Campbell, E. 2005. The role of unshared song in the singing interactions between neighbouring song sparrows. *Animal Behaviour* 70: 1297–1304.

11. Beering, M. 2001. A comparison of the patterns of dance language behavior in house-hunting and nectar-foraging honey bees. PhD thesis. University of California, Riverside.

12. Beletsky, L. D. 2006. *Bird Songs.* Chronicle Books, San Francisco, CA.

13. Bonaparte, K. M., Riffle-Yokoi, C., and Burley, N. T. 2011. Getting a head start: Diet, sub-adult growth, and associative learning in a seed-eating passerine. *PLoS ONE* 6(9): e23775. doi:10.1371/journal.pone.0023775.

14. Brenowitz, E. A. 1991. Evolution of the vocal control system in the avian brain. *Seminars in the Neurosciences* 3: 399–407.

15. Brenowitz, E. A., Lent, K., and Kroodsma, D. E. 1995. Brain space for learned song in birds develops independently of song learning. *Journal of Neuroscience* 15: 6281–6286.

16. Brenowitz, E. A., Margoliash, D., and Nordeen, K. W. 1997. An introduction to birdsong and the avian song system. *Journal of Neurobiology* 33: 495–500.

17. Brenowitz, E. A. and Beecher, M. D. 2005. Song learning in birds: Diversity and plasticity, opportunities and challenges. *Trends in Ecology and Evolution* 28: 127–132.

18. Buchanan, K. L. and Catchpole, C. K. 2000. Song as an indicator of male parental effort in the sedge warbler. *Proceedings of the Royal Society of London B* 267: 321–326.

19. Burt, J. M., Campbell, S. E., and Beecher, M. D. 2001. Song type matching as threat: A test using interactive playback. *Animal Behaviour* 62: 1163–1170.

20. Cardoso, G. C., Mota, P. G., and Depraz, V. 2007. Female and male serins (*Serinus serinus*) respond differently to derived song traits. *Behavioral Ecology and Sociobiology* 61: 1425–1436.

21. Castelli, F. R., Kelley, R. A., Keane, B., and Solomon, N. G. 2011. Female prairie voles show social and sexual preferences for males with longer *avpr1a* microsatellite alleles. *Animal Behaviour* 82: 1117–1126.

22. Catchpole, C. K. and Slater, P. J. B. 2008. *Bird Song: Biological Themes and Variations* (2nd ed.). Cambridge University Press, Cambridge, England.

23. Chilton, G., Lein, M. R., and Baptista, L. F. 1990. Mate choice by female white-crowned sparrows in a mixed-dialect population. *Behavioral Ecology and Sociobiology* 27: 223–227.

24. Conroy, C. J. and Cook, J. A. 2000. Molecular systematics of a holarctic rodent (*Microtus*: Muridae). *Journal of Mammalogy* 81: 344–359.

25. Derryberry, E. P. 2007. Evolution of bird song affects signal efficacy: An experimental test using historical and current signals. *Evolution* 61: 1938–1945.

26. DeWolfe, B. B., Baptista, L. F., and Petrinovich, L. 1989. Song development and territory establishment in Nuttall's white-crowned sparrow. *Condor* 91: 297–407.

27. Endler, J. A. and Day, L. B. 2006. Ornament colour selection, visual contrast and the shape of colour preference functions in great bowerbirds, *Chlamydera nuchalis*. *Animal Behaviour* 72: 1405–1416.

28. Farrell, T. M., Weaver, K., An, Y. S., and MacDougall-Shackleton, S. A. 2012. Song bout length is indicative of spatial learning in European starlings. *Behavioral Ecology* 23: 101–111.

29. Farries, M. A. 2001. The oscine song system considered in the context of the avian brain: Lessons learned from comparative neurobiology. *Brain Behavior and Evolution* 58: 80–100.

30. Fink, S., Excoffier, L., and Heckel, G. 2006. Mammalian monogamy is not controlled by a single gene. *Proceedings of the National Academy of Sciences* 103: 10956–10960.

31. Garamszegi, L. Z. 2005. Bird songs and parasites. *Behavioral Ecology and Sociobiology* 59: 169–180.

32. Gentner, T.Q. and Hulse, S. H. 2000. European starling preference and choice for variation in conspecific male song. *Animal Behaviour* 59: 443–458.

33. Getz, L. L., and Carter, C. S. 1996. Prairie-vole partnerships. *American Scientist* 84: 56–62.

34. Gil, D., Naguid, M. K. R., Rutstein, A., and Gahr, M. 2006. Early condition, song learning, and the volume of song brain nuclei in the zebra finch (*Taeniopygia guttata*). *Journal of Neurobiology* 66: 1602–1612.

35. Gobes, S. M. H. and Bolhuis, J. J. 2007. Birdsong memory: A neural dissociation between song recognition and production. *Current Biology* 17: 789–793.

36. Gould, S. J. 1981. Hyena myths and realities. *Natural History* 90: 16–24.

37. Hackett, S. J., Kimball, R. T., Reddy, S., Bowie, R. C. K., Braun, E. L., Braun, M. J., Chojnowski, J. L., Cox, W. A., Han, K. L., Harshman, J., Huddleston, C. J., Marks, B. D., Miglia, K. J., Moore, W. S., Sheldon, F. H., Steadman, D. W., Witt, C. C., and Yuri, T. 2008. A phylogenomic study of birds reveals their evolutionary history. *Science* 320: 1763–1768.

38. Hammock, E. A. D. and Young, L. J. 2006. Oxytocin, vasopressin and pair bonding: Implications for autism. *Philosophical Transactions of the Royal Society B* 361: 2187–2198.

39. Harbison, H., Nelson, D. A., and Hahn, T. P. 1999. Long-term persistence of song dialects in the mountain white-crowned sparrow. *Condor* 101: 133–148.

40. Holekamp, K. E. and Sherman, P. W. 1989. Why male ground squirrels disperse. *American Scientist* 77: 232–239.

41. Hunter, M. L. and Krebs, J. R. 1979. Geographic variation in the song of the great tit (*Parus major*) in relation to ecological factors. *Journal of Animal Ecology* 48: 759–785.

42. Jarvis, E. D., Ribeiro, S., da Silva, M. L., Ventura, D., Vielliard, J., and Mello, C. V. 2000. Behaviourally driven gene expression reveals song nuclei in hummingbird brain. *Nature* 406: 628–632.

43. Knörnschild, M., Nagy, M., Metz, M., Mayer, F., and von Helversen, O. 2010. Complex vocal imitation during ontogeny in a bat. *Biology Letters* 6: 156–159.

44. Konishi, M. 1965. The role of auditory feedback in the control of vocalization in the white-crowned sparrow. *Zeitschrift für Tierpsychologie* 22: 770–783.

45. Konishi, M. 1985. Birdsong: From behavior to neurons. *Annual Review of Neuroscience* 8: 125–170.

46. Kroodsma, D. E. and Konishi, M. 1991. A suboscine bird (Eastern Phoebe, *Sayornis phoebe*) develops normal song without auditory feedback. *Animal Behaviour* 42: 477–487.

47. Kroodsma, D. E., Liu, W. C., Goodwin, E., and Bedell, P. A. 1999. The ecology of song improvisation as illustrated by North American sedge wrens. *Auk* 116: 373–386.

48. Kroodsma, D. E., Sánchez, J., Stemple, D. W., Goodwin, E., da Silva, M. L., and Vielliard, J. M. E. 1999. Sedentary life style of Neotropical sedge wrens promotes song imitation. *Animal Behaviour* 57: 855–863.

49. Kroodsma, D. E. 2005. *The Singing Life of Birds: The Art and Science of Listening to Birdsong.* Houghton Mifflin, New York, NY.

50. Leitner, S., Nicholson, J., Leisler, B., DeVoogd, T. J., and Catchpole, C. K. 2002. Song and the song control pathway in the brain can develop independently of exposure to song in the sedge warbler. *Proceedings of the Royal Society of London B* 269: 2519–2524.

51. Lim, M. M., Murphy, A. Z., and Young, L. J. 2004. Ventral striatopallidal oxytocin and vasopressin V1a receptors in the monogamous prairie vole (*Microtus ochrogaster*). *Journal of Comparative Neurology* 468: 555–570.

52. Lim, M. M., Wang, X., Olazábal, D. E., Ren, X., Terwilliger, E. F., and Young, L. J. 2004. Enhanced partner preference in a promiscuous species by manipulating the expression of a single gene. *Nature* 429: 754–757.

53. Luther, D. and Baptista, L. 2010. Urban noise and the cultural evolution of bird songs. *Proceedings of the Royal Society B* 277: 469–473.

54. Mabry, K. E., Streatfeild, C. A., Keane, B., and Solomon, N. G. 2010. *avpr1a* length polymorphism is not associated with either social or genetic monogamy in free-living prairie voles. *Animal Behaviour* 81: 11–18.

55. MacDonald, I. F., Kempster, B., Zanette, L., and MacDougall-Shackleton, S. A. 2006. Early nutritional stress impairs development of a song-control brain region in both male and female juvenile song sparrows (*Melospiza melodia*) at the onset of song learning. *Proceedings of the Royal Society B* 273: 2559–2564.

56. MacDougall-Shackleton, E. A., Derryberry, E. P., and Hahn, T. P. 2002. Nonlocal male mountain white-crowned sparrows have lower paternity and higher parasite loads than males singing local dialect. *Behavioral Ecology* 13: 682–689.

57. Marchetti, K. 1993. Dark habitats and bright birds illustrate the role of the environment in species divergence. *Nature* 362: 149–152.

58. Marler, P. and Tamura, M. 1964. Culturally transmitted patterns of vocal behavior in sparrows. *Science* 146: 1483–1486.

59. Marler, P. 1970. Birdsong and speech development: Could there be parallels? *American Scientist* 58: 669–673.

60. Mayr, E. 1961. Cause and effect in biology. *Science* 134: 1501–1506.

61. McGuire, B. and Bemis, W. E. 2007. Parental care. In J. O. Wolff and P. W. Sherman (Eds.), *Rodent Societies: An Ecological and Evolutionary Perspective.* University of Chicago Press, Chicago, IL.

62. Mello, C. V. and Ribeiro, S. 1998. ZENK protein regulation by song in the brain of songbirds. *Journal of Comparative Neurology* 383: 426–438.

63. Mooney, R., Hoese, W., and Nowicki, S. 2001. Auditory representation of the vocal repertoire in a songbird with multiple song types. *Proceedings of the National Academy of Sciences* 98: 12778–12783.

64. Moorman, S., Mello, C. V., and Bolhuis, J. J. 2011. From songs to synapses: Molecular mechanisms of birdsong memory. *Bioessays* 33: 377–385.

65. Morton, E. S. 1975. Ecological sources of selection on avian sounds. *American Naturalist* 109: 17–34.

66. Nealen, P. M. and Perkel, D. J. 2000. Sexual dimorphism in the song system of the Carolina wren *Thryothorus ludovicianus. Journal of Comparative Neurology* 418: 346–360.

67. Nelson, D. A. 2000. Song overproduction, selective attrition and song dialects in the white-crowned sparrow. *Animal Behaviour* 60: 887–898.

68. Nelson, D. A., Hallberg, K. I., and Soha, J. A. 2004. Cultural evolution of Puget Sound white-crowned sparrow song dialects. *Ethology* 110: 879–908.

69. Nicholls, J. A and Goldizen, A. W. 2006. Habitat type and density influence vocal signal design in satin bowerbirds. *Journal of Animal Ecology* 75: 549–558.

70. Nordby, J. C., Campbell, S. E., and Beecher, M. D. 1999. Ecological correlates of song learning in song sparrows. *Behavioral Ecology* 10: 287–297.

71. Nottebohm, F. and Arnold, A. P. 1976. Sexual dimorphism in vocal control areas of songbird brain. *Science* 194: 211–213.

72. Nowicki, S., Peters, S., and Podos, J. 1998. Song learning, early nutrition, and sexual selection in birds. *American Zoologist* 38: 179–190.

73. Nowicki, S., Hasselquist, D., Bensch, S., and Peters, S. 2000. Nestling growth and song repertoire size in great reed warblers: Evidence for song learning as an indicator mechanism in mate choice. *Proceedings of the Royal Society of London B* 267: 2419–2424.

74. Nowicki, S., Searcy, W. A., and Peters, S. 2002. Brain development, song learning and mate choice in birds: A review and experimental test of the "nutritional stress hypothesis." *Journal of Comparative Physiology A* 188: 1003–1114.

75. Nowicki, S., Searcy, W. A., and Peters, S. 2002. Quality of song learning affects female response to male bird song. *Proceedings of the Royal Society of London B* 269: 1949–1954.

76. Pfaff, J. A., Zanetter, L., MacDougall-Shackleton, S. A., and MacDougall-Shackleton, E. A. 2007. Song repertoire size varies with HVC volume and is indicative of male quality in song sparrows (*Melospiza melodia*). *Proceedings of the Royal Society B* 274: 2035–2040.

77. Pitkow, L. J., Sharer, C. A., Ren, X. L., Insel, T. R., Terwilliger, E. F., and Young, L. J. 2001. Facilitation of affiliation and pair-bond formation by vasopressin receptor gene transfer into the ventral forebrain of a monogamous vole. *Journal of Neuroscience* 21: 7392–7396.

78. Poesel, A. and Nelson, D. A. 2012. Delayed song maturation and territorial aggression in a songbird. *Biology Letters* 8: 369–371.

79. Rowley, I. and Chapman, G. 1986. Cross-fostering, imprinting, and learning in two sympatric species of cockatoos. *Behaviour* 96: 1–16.

80. Searcy, W. A., Nowicki, S., Hughes, M., and Peters, S. 2002. Geographic song discrimination in relation to dispersal distances in song sparrows. *American Naturalist* 159: 221–230.

81. Semple, S., McComb, K., Alberts, S., and Altmann, J. 2002. Information content of female copulation calls in yellow baboons. *American Journal of Primatology* 56: 43–56.

82. Shenoy, K. and Crowley, P. H. 2011. Endocrine disruption of male mating signals: Ecological and evolutionary implications. *Functional Ecology* 25: 433–448.

83. Slabbekoorn, H. and den Boer-Visser, A. 2006. Cities change the songs of birds. *Current Biology* 16: 2326–2331.

84. Sockman, K. W., Sewall, K. B., Ball, G. F., and Hahn, T. P. 2005. Economy of mate attraction in the Cassin's finch. *Biology Letters* 1: 34–37.

85. Sockman, K. W., Salvante, K. G., Racke, D. M., Campbell, C. R., and Whitman, B. A. 2009. Song competition changes the brain and behavior of a male songbird. *Journal of Experimental Biology* 212: 2411–2418.

86. Soha, J. A., Nelson, D. A., and Parker, P. G. 2004. Genetic analysis of song dialect populations in Puget Sound white-crowned sparrows. *Behavioral Ecology* 15: 636–646.

87. Spencer, K. A., Wimpenny, J. H., Buchanan, K. L., Lovell, P. G., Goldsmith, A. R., and Catchpole, C. K. 2005. Developmental stress affects the attractiveness of male song and female choice in the zebra finch (*Taeniopygia guttata*). *Behavioral Ecology and Sociobiology* 58: 423–428.

88. Streatfeild, C. A., Mabry, K. E., Keane, B., Crist, T. O., and Solomon, N. G. 2011. Intraspecific variability in the social and genetic mating systems of prairie voles, *Microtus ochrogaster*. *Animal Behaviour* 82: 1387–1398.

89. Suh, A., Paus, M., Kiefmann, M., Churakov, G., Franke, F. A., Brosius, J., Kriegs, J. O., and Schmitz, J. 2011. Mesozoic retroposons reveal parrots as the closest living relatives of passerine birds. *Nature Communications* 2: doi:10.1038/ncomms1448.

90. Tinbergen, N. 1963. On the aims and methods of ethology. *Zeitschrift für Tierpsychologie* 20: 410–433.

91. Tobias, J. A., Aben, J., Brumfield, R. T., Derryberry, E. P,, Halfwerk, W., Slabbekoorn, H., and Seddon, N. 2010. Song divergence by sensory drive in Amazonian birds. *Evolution* 64: 2820–2839.

92. Visscher, P. K. 2003. How self-organization evolves. *Nature* 421: 799–800.

93. Wang, N., Braun, E. L., and Kimball, R. T. 2012. Testing hypotheses about the sister group of the Passeriformes using an independent 30-locus data set. *Molecular Biology and Evolution* 29: 737–750.

94. Wilbrecht, L., Crionas, A., and Nottebohm, F. 2002. Experience affects recruitment of new neurons but not adult neuron number. *Journal of Neuroscience* 22: 825–831.

95. Wolff, J. O. and Macdonald, D. W. 2004. Promiscuous females protect their offspring. *Trends in Ecology and Evolution* 19: 127–134.

96. Woolley, S. C. and Doupe, A. J. 2008. Social context-induced song variation affects female behavior and gene expression. *PLoS Biology* 6(3): e62. doi:10.1371/journal.pbio.0060062.

97. Young, K. A., Gobrogge, K. L., Liu, Y., and Wang, Z. X. 2010. The neurobiology of pair bonding: Insights from a socially monogamous rodent. *Frontiers in Neuroendocrinology* 32: 53–69.

98. Young, L. J. and Wang, Z. 2004. The neurobiology of pair bonding. *Nature Neuroscience* 7: 1048–1054.

99. Ziegler, H. P. and Marler, P. (Eds.). 2008. *Neuroscience of Birdsong*. Cambridge University Press, New York, NY.

Chapter 11

1. Anstey, M. L., Rogers, S. M., Ott, S. R., Burrows, M., and Simpson, S. J. 2009. Serotonin mediates behavioral gregarization underlying swarm formation in desert locusts. *Science* 323: 627–630.

2. Arnold, S. J. 1980. The microevolution of feeding behavior. In A. Kamil and T. Sargent (Eds.), *Foraging Behavior: Ecology, Ethological and Psychological Approaches*. Garland STPM Press, New York, NY.

3. Balda, R. P. 1980. Recovery of cached seeds by a captive *Nucifraga caryocatactes*. *Zeitschrift für Tierpsychologie* 52: 331–346.

4. Balda, R. P. and Kamil, A. C. 1992. Long-term spatial memory in Clark's nutcracker, *Nucifraga columbiana*. *Animal Behaviour* 44: 761–769.

5. Ben-Shahar, Y., Robichon, A., Sokolowski, M. B., and Robinson, G. E. 2002. Influence of gene action across different time scales on behavior. *Science* 296: 741–744.

6. Bentley, D. and Hoy, R. R. 1974. The neurobiology of cricket song. *Scientific American* 231 (Aug.): 4–44.

7. Berthold, P. 1991. Genetic control of migratory behaviour in birds. *Trends in Ecology and Evolution* 6: 254–257.

8. Berthold, P. and Pulido, F. 1994. Heritability of migratory activity in a natural bird population. *Proceedings of the Royal Society of London B* 257: 311–315.

9. Berthold, P. and Querner, U. 1995. Microevolutionary aspects of bird migration based on experimental results. *Israel Journal of Zoology* 41: 377–385.

10. Brakefield, P. M. 2011. Evo-devo and accounting for Darwin's endless forms. *Philosophical Transactions of the Royal Society B* 366: 2069–2075.

11. Brown, J. R., Ye, H., Bronson, R. T., Dikkes, P., and Greenberg, M. E. 1996. A defect in nurturing in mice lacking the immediate early gene *fosB*. *Cell* 86: 297–309.

12. Burke, M. K., Dunham, J. P., Shahrestani, P., Thornton, K. R., Rose, M. R., and Long, A. D. 2010. Genome-wide analysis of a long-term evolution experiment with *Drosophila*. *Nature* 467: 587–590.

13. Burmeister, S. S., Jarvis, E. D., and Fernald, R. D. 2005. Rapid behavioral and genomic responses to social opportunity. *PLoS Biology* 3(11): e363. doi:10.1371/journal.pbio.0030363.

14. Carroll, S. 2005. *Endless Forms Most Beautiful*. W. W. Norton, New York, NY.

15. Carroll, S. B. 2005. Evolution at two levels: On genes and form. *PLoS Biology* 3(7): e245. doi:10.1371/journal.pbio.0030245.

16. Carroll, S. B., Grenier, J. K., and Weatherbee, S. D. 2005. *From DNA to Diversity: Molecular Genetics and the Evolution of Animal Design*. Blackwell Publishing, Malden, MA.

17. Chandrasekaran, S., Ament, S. A., Eddy, J. A., Rodriguez-Zas, S. L., Schatz, B. R., Price, N. D., and Robinson, G. E. 2011. Behavior-specific changes in transcriptional modules lead to distinct and predictable neurogenomic states. *Proceedings of the National Academy of Sciences* 108: 18020–18025.

18. Chen, A., Kramer, E. F., Purpura, L., Krill, J. L., Zars, T., and Dawson-Scully, K. 2011. The influence of natural variation at the foraging gene on thermotolerance in adult *Drosophila* in a narrow temperature range. *Journal of Comparative Physiology A* 197: 1113–1118.

19. Collins, J. P. and Cheek, J. E. 1983. Effect of food and density on development of typical and cannibalistic salamander larvae in *Ambystoma tigrinum nebulosum*. *American Zoologist* 23: 77–84.

20. Cutler, D. M., Miller, G., and Norton, D. M. 2007. Evidence on early-life income and late-life health from America's Dust Bowl era. *Proceedings of the National Academy of Science* 104: 13244–13249.

21. Dawkins, R. 1982. *The Extended Phenotype*. W. H. Freeman, San Francisco, CA.

22. de Belle, J. S. and Sokolowski, M. B. 1987. Heredity of *rover/sitter*: Alternative foraging strategies of *Drosophila melanogaster* larvae. *Heredity* 59: 73–83.

23. de Belle, J. S., Hilliker, A. J., and Sokolowski, M. B. 1989. Genetic localization of *foraging* (*for*): A major gene for larval behavior in *Drosophila melanogaster*. *Genetics* 123: 157–163.

24. de Kort, S. R. and Clayton, N. S. 2006. An evolutionary perspective on caching by corvids. *Proceedings of the Royal Society B* 273: 417–423.

25. Donaldson, Z. R. and Young, L. J. 2008. Oxytocin, vasopressin, and the neurogenetics of sociality. *Science* 322: 900–904.

26. Dunlap, A. S., Chen, B. B., Bednekoff, P. A., Greene, T. M., and Balda, R. P. 2006. A state-dependent sex difference in spatial memory in pinyon jays, *Gymnorhinus cyanocephalus*: Mated females forget as predicted by natural history. *Animal Behaviour* 72: 401–411.

27. Fadool, D. A., Tucker, K., Perkins, R., Fasciani, G., Thompson, R. N., Parsons, A. D., Overton, J. M., Koni, P. A., Flavell, R. A., and Kaczmarek, L. K. 2004. Kv1.3 channel

gene-targeted deletion produces "super-smeller mice" with altered glomeruli, interacting scaffolding proteins, and biophysics. *Neuron* 41: 389–404.

28. Ferguson, J. N., Young, L. J., Hearn, E. F., Matzuk, M. M., Insel, T. R., and Winslow, J. T. 2000. Social amnesia in mice lacking the oxytocin gene. *Nature Genetics* 25: 284–288.

29. Fernald, R. D. 1993. Cichlids in love. *The Sciences* 33: 27–31.

30. Fernald, R. D. 2011. Systems biology meets behavior. *Proceedings of the National Academy of Sciences* 108: 17861–17862.

31. Francis, R. C., Soma, K. K., and Fernald, R. D. 1993. Social regulation of the brain-pituitary-gonadal axis. *Proceedings of the National Academy of Sciences* 90: 7794–7798.

32. Gamboa, G. J. 2004. Kin recognition in eusocial wasps. *Annales Zoologici Fennici* 41: 789–808.

33. Garcia, J. and Ervin, F. R. 1968. Gustatory-visceral and telereceptor-cutaneous conditioning: Adaptation in internal and external milieus. *Communications in Behavioral Biology* (A) 1: 389–415.

34. Garcia, J., Hankins, W. G., and Rusiniak, K. W. 1974. Behavioral regulation of the milieu interne in man and rat. *Science* 185: 824–831.

35. Gaskett, A. C. and Herberstein, M. E. 2008. Orchid sexual deceit provokes pollinator ejaculation. *American Naturalist* 171: E206–E212.

36. Gaulin, S. J. C. and FitzGerald, R. W. 1989. Sexual selection for spatial-learning ability. *Animal Behaviour* 37: 322–331.

37. Göth, A. and Evans, C. S. 2004. Social responses without early experience: Australian brush-turkey chicks use specific visual cues to aggregate with conspecifics. *Journal of Experimental Biology* 207: 2199–2208.

38. Harlow, H. F. and Harlow, M. K. 1962. Social deprivation in monkeys. *Scientific American* 207 (Nov.): 136–146.

39. Harlow, H. F., Harlow, M. K., and Suomi, S. J. 1971. From thought to therapy: Lessons from a primate laboratory. *American Scientist* 59: 538–549.

40. Hitchcock, C. L. and Sherry, D. F. 1990. Long-term memory for cache sites in the black-capped chickadee. *Animal Behaviour* 40: 701–712.

41. Holmes, W. G. and Sherman, P. W. 1982. The ontogeny of kin recognition in two species of ground squirrels. *American Zoologist* 22: 491–517.

42. Holmes, W. G. and Sherman, P. W. 1983. Kin recognition in animals. *American Scientist* 71: 46–55.

43. Holmes, W. G. 1986. Identification of paternal half-siblings by captive Belding's ground squirrels. *Animal Behaviour* 34: 321–327.

44. Huang, Z.-Y. and Robinson, G. E. 1992. Honeybee colony integration: Worker-worker interactions mediate hormonally regulated plasticity in division of labor. *Proceedings of the National Academy of Sciences* 89: 11726–11729.

45. Ijichi, N., Shibao, H., Miura, T., Matsumoto, T., and Fukatsu, T. 2004. Soldier differentiation during embryogenesis of a social aphid, *Pseudoregma bambucicola*. *Entomological Science* 7: 141–153.

46. Jones, C. M. and Healy, S. D. 2006. Differences in cue use and spatial memory in men and women. *Proceedings of the Royal Society B* 273: 2241–2247.

47. Kamakura, M. 2011. Royalactin induces queen differentiation in honeybees. *Nature* 473: 478–483.

48. Kannisto, V., Christensen, K., and Vaupel, J. W. 1997. No increased mortality in later life for cohorts born during famine. *American Journal of Epidemiology* 145: 987–994.

49. Kasumovic, M. M. and Andrade, M. C. B. 2006. Male development tracks rapidly shifting sexual versus natural selection pressures. *Current Biology* 16: R242–R243.

50. Kaun, K. R., Chakaborty-Chatterjee, M., and Sokolowski, M. B. 2008. Natural variation in plasticity of glucose homeostasis and food intake. *Journal of Experimental Biology* 211: 3160–3166.

51. Kent, C. F., Daskalchuk, T., Cook, L., Sokolowski, M. B., and Greenspan, R. J. 2009. The *Drosophila foraging* gene mediates adult plasticity and gene-environment interactions in behaviour, metabolites, and gene expression in response to food deprivation. *PLoS Genetics* 5(8): e1000609. doi:10.1371/journal.pgen.1000609.

52. Kerverne, E. B. 1997. An evaluation of what the mouse knockout experiments are telling us about mammalian behaviour. *Bioessays* 19: 1091–1098.

53. King, A. P. and West, M. J. 1983. Epigenesis of cowbird song—A joint endeavour of males and females. *Nature* 305: 704–706.

54. Koscik, T., O'Leary, D., Moser, D. J., Andreasen, N. C., and Nopoulos, P. 2009. Sex differences in parietal lobe morphology: Relationship to mental rotation performance. *Brain and Cognition* 69: 451–459.

55. Kroodsma, D. E. and Canady, R. A. 1985. Differences in repertoire size, singing behavior, and associated neuroanatomy among marsh wren populations have a genetic basis. *Auk* 102: 439–446.

56. Leoncini, I., Le Conte, Y., Costagliola, G., Plettner, E., Toth, A. L., Wang, M., Huang, Z., Bécard, J.-M., Crauser, D., Slessor, K. N., and Robinson, G. E. 2004. Regulation of behavioral maturation by a primer pheromone produced by adult worker honey bees. *Proceedings of the National Academy of Sciences* 101: 17559–17564.

57. Levine, S. and Mullins, R. F. 1966. Hormonal influences on brain organization in infant rats. *Science* 152: 1585–1592.

58. Lindauer, M. 1961. *Communication among Social Bees*. Harvard University Press, Cambridge, MA.

59. Lore, R. and Flannelly, K. 1977. Rat societies. *Scientific American* 236 (May): 106–116.

60. Lorenz, K. Z. 1952. *King Solomon's Ring*. Crowell, New York, NY.

61. Maret, T. J. and Collins, J. P. 1994. Individual responses to population size structure: The role of size variation in controlling expression of a trophic polyphenism. *Oecologia* 100: 279–285.

62. Mateo, J. M. and Holmes, W. G. 1997. Development of alarm-call responses in Belding's ground squirrels: The role of dams. *Animal Behaviour* 54: 509–524.

63. Mateo, J. M. and Johnston, R. E. 2000. Kin recognition and the "armpit effect": Evidence of self-reference phenotype matching. *Proceedings of the Royal Society of London B* 267: 695–700.

64. Mateo, J. M. 2002. Kin-recognition abilities and nepotism as a function of sociality. *Proceedings of the Royal Society of London B* 269: 721–727.

65. Mateo, J. M. 2006. The nature and representation of individual recognition odours in Belding's ground squirrels. *Animal Behaviour* 71: 141–154.

66. Mery, F., Belay, A. T., So, A. K. C., Sokolowski, M. B., and Kawecki, T. J. 2007. Natural polymorphism affecting learning and memory in *Drosophila*. *Proceedings of the National Academy of Sciences* 104: 13051–13055.

67. Milius, S. 2004. Where'd I put that? *Science News* 165: 103–105.

68. Moffat, S. D., Hampson, E., and Hatzipantelis, M. 1998. Navigation in a "virtual" maze: Sex differences and correlation with psychometric measures of spatial ability in humans. *Human Behavior and Evolution* 19: 73–87.

69. Möller, A., Pavlick, B., Hile, A. G., and Balda, R. P. 2001. Clark's nutcrackers *Nucifraga columbiana* remember the size of their cached seeds. *Ethology* 107: 451–461.

70. Morley, R. and Lucas, A. 1997. Nutrition and cognitive development. *British Medical Bulletin* 53: 123–124.

71. Olson, D. J., Kamil, A. C., Balda, R. P., and Nims, P. J. 1995. Performance of four seed-caching corvid species in operant tests of nonspatial and spatial memory. *Journal of Comparative Psychology* 109: 173–181.

72. Osborne, K. A., Robichon, A., Burgess, E., Butland, S., Shaw, R. A., Coulthard, A., Pereira, H. S., Greenspan, R. J., and Sokolowski, M. B. 1997. Natural behavior polymorphism due to a cGMP-dependent protein kinase of *Drosophila*. *Science* 277: 834–836.

73. Peakall, R. 1990. Responses of male *Zaspilothynnus trilobatus* Turner wasps to females and the sexually deceptive orchid it pollinates. *Functional Ecology* 4: 159–167.

74. Pennisi, E. 2000. Fruit fly genome yields data and a validation. *Science* 287: 1374.

75. Pfennig, D. W. and Collins, J. P. 1993. Kinship affects morphogenesis in cannibalistic salamanders. *Nature* 362: 836–838.

76. Pfennig, D. W., Sherman, P. W., and Collins, J. P. 1994. Kin recognition and cannibalism in polyphenic salamanders. *Behavioral Ecology* 5: 225–232.

77. Pfennig, D. W., Rice, A. M., and Martin, R. A. 2007. Field and experimental evidence for competition's role in phenotypic divergence. *Evolution* 61: 257–271.

78. Picciotto, M. R. 1999. Knock-out mouse models used to study neurobiological systems. *Critical Reviews in Neurobiology* 13: 103–149.

79. Pravosudov, V. V., and de Kort, S. R. 2006. Is the western scrub-jay (*Aphelocoma californica*) really an underdog among food-caching corvids when it comes to hippocampal volume and food caching propensity? *Brain, Behavior and Evolution* 67: 1–9.

80. Pulido, F. and Berthold, P. 2010. Current selection for lower migratory activity will drive the evolution of residency in a migratory bird population. *Proceedings of the National Academy of Sciences* 107: 7341–7346.

81. Rasmussen, K. M. 2001. The "fetal origins" hypothesis: Challenges and opportunities for maternal and child nutrition. *Annual Review of Nutrition* 21: 73–95.

82. Ratcliffe, J. M., Fenton, M. B., and Galef, B. G. 2003. An exception to the rule: Common vampire bats do not learn taste aversions. *Animal Behaviour* 65: 385–389.

83. Robinson, G. E. 1998. From society to genes with the honey bee. *American Scientist* 86: 456–462.

84. Robinson, G. E. 2004. Beyond nature and nurture. *Science* 304: 397–399.

85. Seeley, T. D. 1995. *The Wisdom of the Hive*. Harvard University Press, Cambridge, MA.

86. Sen Sarma, M., Rodriguez-Zas, S. L., Gernat, T., Nguyen, T., Newman, T., and Robinson, G. E. 2010. Distance-responsive genes found in dancing honey bees. *Genes, Brain and Behavior* 9: 825–830.

87. Sherry, D. F. 1984. Food storage by black-capped chickadees: Memory of the location and contents of caches. *Animal Behaviour* 32: 451–464.

88. Sherry, D. F., Forbes, M. R. L., Khurgel, M., and Ivy, G. O. 1993. Females have a larger hippocampus than males in the brood-parasitic brown-headed cowbird. *Proceedings of the National Academy of Sciences* 90: 7839–7843.

89. Simpson, S. J., Despland, E., Hagele, B. F., and Dodgson, T. 2001. Gregarious behavior in desert locusts is evoked by touching their back legs. *Proceedings of the National Academy of Sciences* 98: 3895–3897.

90. Simpson, S. J., Sword, G. A., and Lo, N. 2011. Polyphenism in insects. *Current Biology* 21: R738–R749.

91. Skinner, B. F. 1966. Operant behavior. In W. Honig (Ed.), *Operant Behavior*. Appleton-Century-Crofts, New York, NY.

92. Slagsvold, T. 1998. On the origin and rarity of interspecific nest parasitism in birds. *American Naturalist* 152: 264–272.

93. Slagsvold, T. and Hansen, B. T. 2001. Sexual imprinting and the origin of obligate brood parasitism in birds. *American Naturalist* 158: 354–367.

94. Slagsvold, T., Hansen, B. T., Johannessen, L. E., and Lifjeld, J. T. 2002. Mate choice and imprinting in birds studied by cross-fostering in the wild. *Proceedings of the Royal Society of London B* 269: 1449–1455.

95. Stein, Z., Susser, M., Saenger, G., and Marolla, F. 1972. Nutrition and mental performance. *Science* 178: 708–713.

96. Stoutamire, W. P. 1974. Australian terrestrial orchids, thynnid wasps and pseudocopulation. *American Orchid Society Bulletin* 43: 13–18.

97. Sullivan, J. P., Jassim, O., Fahrbach, S. E., and Robinson, G. E. 2000. Juvenile hormone paces behavioral development in the adult worker honey bee. *Hormones and Behavior* 37: 1–14.

98. Susser, M. and Stein, Z. 1994. Timing in prenatal nutrition: A reprise of the Dutch famine study. *Nutrition Reviews* 52: 84–94.

99. Toth, A. L. and Robinson, G. E. 2007. Evo-devo and the evolution of social behavior. *Trends in Genetics* 23: 334–341.

100. Warner, R. R. 1984. Mating behavior and hermaphroditism in coral reef fishes. *American Scientist* 72: 128–136.

101. West-Eberhard, M. J. 2003. *Developmental Plasticity and Evolution*. Oxford University Press, New York, NY.

102. White, S. A., Nguyen, T., and Fernald, R. D. 2002. Social regulation of gonadotropin-releasing hormone. *Journal of Experimental Biology* 205: 2567–2581.

103. Whitfield, C. W., Cziko, A. M., and Robinson, G. E. 2003. Gene expression profiles in the brain predict behavior in individual honey bees. *Science* 302: 296–299.

104. Young, K. A., Gobrogge, K. L., Liu, Y., and Wang, Z. X. 2010. The neurobiology of pair bonding: Insights from a socially monogamous rodent. *Frontiers in Neuroendocrinology* 32: 53–69.

Chapter 12

1. Allman, J., 1999. *Evolving Brains*. Scientific American Library, New York, NY.

2. Babineau, D., Lewis, J. E., and Longtin, A. 2007. Spatial acuity and prey detection in weakly electric fish. *PLoS Computational Biology* 3(3): e38. doi:10.1371/journal.pcbi.0030038

3. Bass, A. H. 1996. Shaping brain sexuality. *American Scientist* 84: 352–363.

4. Boulcott, P. D., Walton, K., and Braithwaite, V. A. 2005. The role of ultraviolet wavelengths in the mate-choice decisions of female three-spined sticklebacks. *Journal of Experimental Biology* 208: 1453–1458.

5. Brower, L. P. 1996. Monarch butterfly orientation: Missing pieces of a magnificent puzzle. *Journal of Experimental Biology* 199: 93–103.

6. Burkhardt, R. W. 2004. *Patterns of Behavior: Konrad Lorenz, Niko Tinbergen, and the Founding of Ethology*. University of Chicago Press, Chicago, IL.

7. Carter, R. 2010. *Mapping the Mind*. University of California Press, Berkeley, CA.

8. Catania, K. C. and Kaas, J. H. 1996. The unusual nose and brain of the star-nosed mole. *BioScience* 46: 578–586.

9. Catania, K. C. and Kaas, J. H. 1997. Somatosensory fovea in the star-nosed mole: Behavioral use of the star in relation to

innervation patterns and cortical representation. *Journal of Comparative Neurology* 387: 215–233.

10. Catania, K. C. 2000. Cortical organization in Insectivora: The parallel evolution of the sensory periphery and the brain. *Brain, Behavior and Evolution* 55: 311–321.

11. Catania, K. C. and Remple, M. S. 2002. Somatosensory cortex dominated by the representation of teeth in the naked mole-rat brain. *Proceedings of the National Academy of Sciences* 99: 5692–5697.

12. Catania, K. C. and Remple, F. E. 2005. Asymptotic prey profitability drives star-nosed moles to the foraging speed limit. *Nature* 433: 519–522.

13. Catania, K. C. and Henry, E. C. 2006. Touching on somatosensory specializations in mammals. *Current Opinion in Neurobiology* 16: 467–473.

14. Catania, K. C. 2008. Worm grunting, fiddling, and charming: Humans unknowingly mimic a predator to harvest bait. *PLoS ONE* 3(10): e3472. doi:3410.1371/journal.pone.0003472.

15. Clarac, F. and Pearlstein, E. 2007. Invertebrate preparations and their contribution to neurobiology in the second half of the 20th century. *Brain Research Reviews* 54: 113–161.

16. Conner, W. E. and Corcoran, A. J. 2012. Sound strategies: The 65-million-year-old battle between bats and insects. *Annual Review of Entomology* 57: 21–39.

17. Corcoran, A. J., Barber, J. R., Hristov, N. I., and Conner, W. E. 2011. How do tiger moths jam bat sonar? *Journal of Experimental Biology* 214: 2416–2425.

18. Davies, N. B. 2000. *Cuckoos, Cowbirds and Other Cheats.* T & AD Poyser, London, England.

19. Dawson, J. W., Dawson-Scully, K., Robert, D., and Robertson, R. M. 1997. Forewing asymmetries during auditory avoidance in flying locusts. *Journal of Experimental Biology* 200: 2323–2335.

20. De Waal, F. B. M. 1982. *Chimpanzee Politics: Power and Sex among Apes.* Harper & Row, New York, NY.

21. Delhey, K., Peters, A., Johnsen, A., and Kempenaers, B. 2007. Fertilization success and UV ornamentation in blue tits *Cyanistes caeruleus*: Correlational and experimental evidence. *Behavioral Ecology* 18: 399–409.

22. Drea, C. M. and Carter, A. N. 2009. Cooperative problem solving in a social carnivore. *Animal Behaviour* 78: 967–977.

23. Dunbar, R. I. M. 2003. The social brain: Mind, language, and society in evolutionary perspective. *Annual Review of Anthropology* 32: 163–181.

24. Frost, W. N., Hoppe, T. A., Wang, J., and Tian, L. M. 2001. Swim initiation neurons in *Tritonia diomedea*. *American Zoologist* 41: 952–961.

25. Froy, O., Gotter, A. L., Casselman, A. L., and Reppert, S. M. 2003. Illuminating the circadian clock in monarch butterfly migration. *Science* 300: 1303–1305.

26. Fullard, J. H., Dawson, J. W., and Jacobs, D. S. 2003. Auditory encoding during the last moment of a moth's life. *Journal of Experimental Biology* 206: 281–294.

27. Getting, P. A. 1983. Mechanisms of pattern generation underlying swimming in *Tritonia*. II. Network reconstruction. *Journal of Neurophysiology* 49: 1017–1035.

28. Getting, P. A. 1989. A network oscillator underlying swimming in *Tritonia*. In J. W. Jacklet (Ed.), *Neuronal and Cellular Oscillators*. Dekker, New York, NY.

29. Goris, R. C. 2011. Infrared organs of snakes: An integral part of vision. *Journal of Herpetology* 45: 2–14.

30. Gracheva, E. O., Cordero-Morales, J. F., González-Carcacía, J. A., Ingolia, N. T., Manno, C., Aranguren, C. I., Weissman, J. S., and Julius, D. 2011. Ganglion-specific splicing of TRPV1 underlies infrared sensation in vampire bats. *Nature* 476: 88–91.

31. Griffin, D. R. 1958. *Listening in the Dark.* Yale University Press, New Haven, CT.

32. Hare, B., Brown, M., Williamson, C., and Tomasello, M. 2002. The domestication of social cognition in dogs. *Science* 298: 1634–1636.

33. Hare, B., Rosati, A., Kaminski, J., Brauer, J., Call, J., and Tomasello, M. 2010. The domestication hypothesis for dogs' skills with human communication: A response to Udell et al. 2008 and Wynne et al. 2008. *Animal Behaviour* 79: E1–E6.

34. Hawryshyn, C. W., Ramsden, S. D., Betke, K. M., and Sabbah, S. 2010. Spectral and polarization sensitivity of juvenile Atlantic salmon (*Salmo salar*): Phylogenetic considerations. *Journal of Experimental Biology* 213: 3187–3197.

35. Heinrich, B. 1999. *Mind of the Raven: Investigations and Adventures with Wolf-Birds.* HarperCollins, New York, NY.

36. Holland, R. A. 2011. Differential effects of magnetic pulses on the orientation of naturally migrating birds. *Journal of the Royal Society Interface* 7: 1617–1625.

37. Holzhaider, J. C., Sibley, M. D., Taylor, A. H., Singh, P. J., Gray, R. D., and Hunt, G. R. 2011. The social structure of New Caledonian crows. *Animal Behaviour* 81: 83–92.

38. Hopkins, C. D. 1998. Design features for electric communication. *Journal of Experimental Biology* 202: 1217–1228.

39. Hunt, S., Cuthill, I. C., Bennett, A. T. D., and Griffiths, R. 1999. Preferences for ultraviolet partners in the blue tit. *Animal Behaviour* 58: 809–815.

40. Kalko, E. K. V. 1995. Insect pursuit, prey capture and echolocation in pipistrelle bats (Microchiroptera). *Animal Behaviour* 50: 861–880.

41. Keeton, W. T. 1969. Orientation by pigeons: Is the sun necessary? *Science* 165: 922–928.

42. Kell, C. A., von Kriegsterin, K., Rosler, R., Kleinschmidt, A., and Laufs, H. 2005. The sensory cortical representation of the human penis: Revisiting somatotopy in the male homunculus. *Journal of Neuroscience* 25: 5984–5987.

43. Kemp, D. J. 2008. Female mating biases for bright ultraviolet iridescence in the butterfly *Eurema hecabe* (Pieridae). *Behavioral Ecology* 19: 1–8.

44. Kruuk, H. 2004. *Niko's Nature: The Life of Niko Tinbergen and His Science of Animal Behavior.* Oxford University Press, Oxford, England.

45. Lappin, A. K., Brandt, Y., Husak, J. F., Macedonia, J. M., and Kemp, D. J. 2006. Gaping displays reveal and amplify a mechanically based index of weapon performance. *American Naturalist* 168: 100–113.

46. Marasco, P. D. and Catania, K. C. 2007. Response properties of primary afferents supplying Eimer's organ. *Journal of Experimental Biology* 210: 765–780.

47. May, M. 1991. Aerial defense tactics of flying insects. *American Scientist* 79: 316–329.

48. Menzel, R. and Giurfa, M. 2001. Cognitive architecture of a mini-brain: The honeybee. *Trends in Cognitive Sciences* 5: 62–71.

49. Miller, L. A. and Surlykke, A. 2001. How some insects detect and avoid being eaten by bats: Tactics and countertactics of prey and predator. *BioScience* 51: 570–581.

50. Moiseff, A., Pollack, G. S., and Hoy, R. R. 1978. Steering responses of flying crickets to sound and ultrasound: Mate attraction and predator avoidance. *Proceedings of the National Academy of Sciences* 75: 4052–4056.

51. Newman, E. A. and Hartline, P. H. 1982. The infrared "vision" of snakes. *Scientific American* 20 (Mar.): 116–127.

52. Nolen, T. G. and Hoy, R. R. 1984. Phonotaxis in flying crickets: Neural correlates. *Science* 226: 992–994.

53. O'Connell-Rodwell, C. E. 2007. Keeping an "ear" to the ground: Seismic communication in elephants. *Physiology* 22: 287–294.

54. Pérez-Barbería, F. J., Shultz, S., and Dunbar, R. I. M. 2007. Evidence for coevolution of sociality and relative brain size in three orders of mammals. *Evolution* 61: 2811–2821.

55. Pike, T. W., Blount, J. D., Bjerkeng, B., Lindstrom, J., and Metcalfe, N. B. 2007. Carotenoids, oxidative stress and female mating preference for longer lived males. *Proceedings of the Royal Society B* 274: 1591–1596.

56. Pike, T. W., Bjerkeng, B., Blount, J. D., Lindstrom, J., and Metcalfe, N. B. 2011. How integument colour reflects its carotenoid content: A stickleback's perspective. *Functional Ecology* 25: 297–304.

57. Reppert, S. M., Zhu, H. S., and White, R. H. 2004. Polarized light helps monarch butterflies navigate. *Current Biology* 14: 155–158.

58. Robert, D., Amoroso, J., and Hoy, R. R. 1992. The evolutionary convergence of hearing in a parasitoid fly and its cricket host. *Science* 258: 1135–1137.

59. Roeder, K. D. and Treat, A. E. 1961. The detection and evasion of bats by moths. *American Scientist* 49: 135–148.

60. Roeder, K. D. 1963. *Nerve Cells and Insect Behavior.* Harvard University Press, Cambridge, MA.

61. Roeder, K. D. 1970. Episodes in insect brains. *American Scientist* 58: 378–389.

62. Roth, G. and Dicke, U. 2005. Evolution of the brain and intelligence. *Trends in Cognitive Sciences* 9: 250–257.

63. Rutowski, R. L. 1998. Mating strategies in butterflies. *Scientific American* 279 (July): 64–69.

64. Sakai, S. T., Arsznov, B. M., Lundrigan, B. L., and Holekamp, K. E. 2011. Brain size and social complexity: A computed tomography study in Hyaenidae. *Brain Behavior and Evolution* 77: 91–104.

65. Saul-Gershenz, L. and Millar, J. G. 2006. Phoretic nest parasites use sexual deception to obtain transport to their host's nest. *Proceedings of the National Academy of Sciences* 103: 14039–14044.

66. Sheehan, M. J. and Tibbetts, E. A. 2010. Selection for individual recognition and the evolution of polymorphic identity signals in *Polistes* paper wasps. *Journal of Evolutionary Biology* 23: 570–577.

67. Sheehan, M. J. and Tibbetts, E. A. 2011. Specialized face learning is associated with individual recognition in paper wasps. *Science* 334: 1272–1275.

68. Shultz, S. and Dunbar, R. I. M. 2010. Social bonds in birds are associated with brain size and contingent on the correlated evolution of life-history and increased parental investment. *Biological Journal of the Linnean Society* 100: 111–123.

69. Sisneros, J. A. and Bass, A. H. 2003. Seasonal plasticity of peripheral auditory frequency sensitivity. *Journal of Neuroscience* 23: 1049–1058.

70. Sisneros, J. A. 2007. Saccular potentials of the vocal plainfin midshipman fish, *Porichthys notatus*. *Journal of Comparative Physiology A* 193: 413–424.

71. Soltis, J. 2010. Vocal communication in African elephants (*Loxodonta africana*). *Zoo Biology* 29: 192–209.

72. Stapley, J. and Whiting, M. J. 2005. Ultraviolet signals fighting ability in a lizard. *Biology Letters* 2: 169–172.

73. Stoutamire, W. P. 1974. Australian terrestrial orchids, thynnid wasps and pseudocopulation. *American Orchid Society Bulletin* 43: 13–18.

74. Stumpner, A. and Lakes-Harlan, R. 1996. Auditory interneurons in a hearing fly (*Therobia leonidei*, Ormiini, Tachinidae, Diptera). *Journal of Comparative Physiology A* 178: 227–233.

75. Surlykke, A. 1984. Hearing in notodontid moths: A tympanic organ with a single auditory neuron. *Journal of Experimental Biology* 113: 323–334.

76. Taylor, A. H., Elliffe, D., Hunt, G. R., and Gray, R. D. 2010. Complex cognition and behavioural innovation in New Caledonian crows. *Proceedings of the Royal Society B* 277: 2637–2643.

77. Tinbergen, N. and Perdeck, A. C. 1950. On the stimulus situations releasing the begging response in the newly hatched herring gull (*Larus argentatus* Pont.). *Behaviour* 3: 1–39.

78. Tinbergen, N. 1951. *The Study of Instinct.* Oxford University Press, New York, NY.

79. Topál, J., Gergely, G., Erdőhegyi, A., Csibra, G., and Miklósi, A. 2009. Differential sensitivity to human communication in dogs, wolves, and human infants. *Science* 325: 1269–1272.

80. Udell, M. A. R., Dorey, N. R., and Wynne, C. D. L. 2008. Wolves outperform dogs in following human social cues. *Animal Behaviour* 76: 1767–1773.

81. Udell, M. A. R., Dorey, N. R., and Wynne, C. D. L. 2010. What did domestication do to dogs? A new account of dogs' sensitivity to human actions. *Biological Reviews* 85: 327–345.

82. Vereecken, N. J. and Mahé, G. 2007. Larval aggregations of the blister beetle *Stenoria analis* (Schaum) (Coleoptera: Meloidae) sexually deceive patrolling males of their host, the solitary bee *Colletes hederae* Schmidt & Westrich (Hymenoptera: Colletidae). *Annales de la Société Entomologique de France* 43: 493–496.

83. Walcott, C. 1972. Bird navigation. *Natural History* 81: 32–43.

84. Wehner, R., Lehrer, M., and Harvey, W. R. 1996. Navigation: Migration and homing. *Journal of Experimental Biology* 199: 1–261.

85. White, D. J., Ho, L., and Freed-Brown, S. G. 2009. Counting chicks before they hatch: Female cowbirds assess temporal quality of nests for parasitism. *Psychological Science* 20: 1140–1145.

86. Wickler, W. 1968. *Mimicry in Plants and Animals.* World University Library, London, England.

87. Willows, A. O. D. 1971. Giant brain cells in mollusks. *Scientific American* 224 (Feb.): 68–75.

88. Yack, J. E., Kalko, J. E. V., and Surlykke, A. 2007. Neuroethology of ultrasonic hearing in nocturnal butterflies (Hedyloidea). *Journal of Comparative Physiology A* 193: 577–590.

89. Yager, D. D. and May, M. L. 1990. Ultrasound-triggered, flight-gated evasive maneuvers in the flying praying mantis, *Parasphendale agrionina*. II. Tethered flight. *Journal of Experimental Biology* 152: 41–58.

Chapter 13

1. Acharya, L. and McNeil, J. N. 1998. Predation risk and mating behavior: The responses of moths to bat-like ultrasound. *Behavioral Ecology* 9: 552–558.

2. Adkins-Regan, E. 2005. *Hormones and Animal Social Behavior.* Princeton University Press, Princeton, NJ.

3. Balthazart, J., Baillien, M., Charlier, T. D., Cornil, C. A., and Ball, G. F. 2003. The neuroendocrinology of reproductive behavior in Japanese quail. *Domestic Animal Endocrinology* 25: 69–82.

4. Baylies, M. K., Bargiello, T. A., Jackson, F. R., and Young, M. W. 1987. Changes in abundance or structure of the *Per* gene-product can alter periodicity of the *Drosophila* clock. *Nature* 326: 390–392.

5. Benkman, C. W. 1990. Foraging rates and the timing of crossbill reproduction. *Auk* 107: 376–386.

6. Chapman, T., Bangham, J., Vinti, G., Lung, O., Wolfner, M. F., Smith, H. K., and Partridge, L. 2003. The sex peptide of

Drosophila melanogaster. Female post-mating responses analyzed by using RNA interference. *Proceedings of the National Academy of Sciences* 100: 9923–9928.

7. Cheng, M. Y., Bullock, C. M., Li, C. Y., Lee, A. G., Bermak, J. C., Belluzzi, J., Weaver, D. R., Leslie, F. M., and Zhou, Q. Y. 2002. Prokineticin 2 transmits the behavioural circadian rhythm of the suprachiasmatic nucleus. *Nature* 417: 405–410.

8. Colwell, C. S. 2012. Linking neural activity and molecular oscillations in the SCN. *Nature Reviews Neuroscience* 12: 553–569.

9. Crews, D. and Greenberg, N. 1981. Function and causation of social signals in lizards. *American Zoologist* 21: 273–294.

10. Crews, D. 1984. Gamete production, sex hormone secretion, and mating behavior uncoupled. *Hormones and Behavior* 18: 22–28.

11. Crews, D. 1991. Trans-seasonal action of androgen in the control of spring courtship behavior in male red-sided garter snakes. *Proceedings of the National Academy of Sciences* 88: 3545–3548.

12. Davis-Walton, J. and Sherman, P. W. 1994. Sleep arrhythmia in the eusocial naked mole-rat. *Naturwissenschaften* 81: 272–275.

13. De Dreu, C. K. W., Greer, L. L., Van Kleef, G. A., Shalvi, S., and Handgraaf, M. J. J. 2011. Oxytocin promotes human ethnocentrism. *Proceedings of the National Academy of Sciences* 108: 1262–1266.

14. DeCoursey, P. J. and Buggy, J. 1989. Circadian rhythmicity after neural transplant to hamster third ventricle: Specificity of suprachiasmatic nuclei. *Brain Research* 500: 263–275.

15. Deviche, P. and Sharp, P. J. 2001. Reproductive endocrinology of a free-living, opportunistically breeding passerine (White-winged crossbill, *Loxia leucoptera*). *General and Comparative Endocrinology* 123: 268–279.

16. Edery, I. 2000. Circadian rhythms in a nutshell. *Physiological Genomics* 3: 59–74.

17. Farner, D. S. 1964. Time measurement in vertebrate photoperiodism. *American Naturalist* 95: 375–386.

18. Farner, D. S. and Lewis, R. A. 1971. Photoperiodism and reproductive cycles in birds. *Photophysiology* 6: 325–370.

19. Follett, B. K., Mattocks, P. W., Jr., and Farner, D. S. 1974. Circadian function in the photoperiodic induction of gonadotropin secretion in the white-crowned sparrow, *Zonotrichia leucophrys gambellii. Proceedings of the National Academy of Sciences* 71: 1666–1669.

20. Froy, O., Gotter, A. L., Casselman, A. L., and Reppert, S. M. 2003. Illuminating the circadian clock in monarch butterfly migration. *Science* 300: 1303–1305.

21. Gangestad, S. W., Simpson, J. A., Cousins, A. J., Garver-Apgar, C. E., and Christensen, P. N. 2004. Women's preferences for male behavioral displays change across the menstrual cycle. *Psychological Science* 15: 203–207.

22. Gangestad, S. W. and Thornhill, R. 2008. Human oestrus. *Proceedings of the Royal Society B* 275: 991–1000.

23. Gettler, L. T., McDade, T. W., Feranil, A. B., and Kuzawa, C. W. 2012. Longitudinal evidence that fatherhood decreases testosterone in males. *Proceedings of the National Academy of Sciences* 108: 16194–16199.

24. Ghalambor, C. K. and Martin, T. E. 2001. Fecundity-survival trade-offs and parental risk-taking in birds. *Science* 292: 494–497.

25. Granados-Fuentes, D., Tseng, A., and Herzog, E. D. 2006. A circadian clock in the olfactory bulb controls olfactory responsivity. *Journal of Neuroscience* 26: 12219–12225.

26. Greives, T. J., McGlothlin, J. W., Jawor, J. M., Demas, G. E., and Ketterson, E. D. 2006. Testosterone and innate immune function inversely covary in a wild population of breeding dark-eyed juncos (*Junco hyemalis*). *Functional Ecology* 20: 812–818.

27. Gwinner, E. and Dittami, J. 1990. Endogenous reproductive rhythms in a tropical bird. *Science* 249: 906–908.

28. Gwinner, E. 1996. Circannual clocks in avian reproduction and migration. *Ibis* 138: 47–63.

29. Hahn, T. P. 1995. Integration of photoperiodic and food cues to time changes in reproductive physiology by an opportunistic breeder, the red crossbill, *Loxia curvirostra* (Aves: Carduelinae). *Journal of Experimental Zoology* 272: 213–226.

30. Hahn, T. P., Wingfield, J. C., Mullen, R., and Deviche, P. J. 1995. Endocrine bases of spatial and temporal opportunism in arctic-breeding birds. *American Zoologist* 35: 259–273.

31. Hahn, T. P. 1998. Reproductive seasonality in an opportunistic breeder, the red crossbill, *Loxia curvirostra. Ecology* 79: 2365–2375.

32. Hahn, T. P., Pereyra, M. E., Sharbaugh, S. M., and Bentley, G. E. 2004. Physiological responses to photoperiod in three cardueline finch species. *General and Comparative Endocrinology* 137: 99–108.

33. Hahn, T. P., Cornelius, J. M., Sewall, K. B., Kelsey, T. R., Hau, M., and Perfito, N. 2008. Environmental regulation of annual schedules in opportunistically-breeding songbirds: Adaptive specializations or variations on a theme of white-crowned sparrow? *General and Comparative Endocrinology* 157: 217–226.

34. Hammock, E. A. D. and Young, L. J. 2006. Oxytocin, vasopressin and pair bonding: Implications for autism. *Philosophical Transactions of the Royal Society B* 361: 2187–2198.

35. Hamner, W. M. 1964. Circadian control of photoperiodism in the house finch demonstrated by interrupted-night experiments. *Nature* 203: 1400–1401.

36. Hau, M., Wikelski, M., Soma, K. K., and Wingfield, J. C. 2000. Testosterone and year-round territorial aggression in a tropical bird. *General and Comparative Endocrinology* 117: 20–33.

37. Hau, M., Ricklefs, R. E., Wikelski, M., Lee, K. A., and Brawn, J. D. 2010. Corticosterone, testosterone and life-history strategies of birds. *Proceedings of the Royal Society B* 277: 3203–3212.

38. Hedwig, B. 2000. Control of cricket stridulation by a command neuron: Efficacy depends on the behavioral state. *Journal of Neurophysiology* 83: 712–722.

39. Helm, B., Schwabl, I., and Gwinner, E. 2009. Circannual basis of geographically distinct bird schedules. *Journal of Experimental Biology* 212: 1259–1269.

40. Husak, J. F., Irschick, D. J., Meyers, J. J., Lailvaux, S. P., and Moore, I. T. 2007. Hormones, sexual signals, and performance of green anole lizards (*Anolis carolinensis*). *Hormones and Behavior* 52: 360–367.

41. Jenssen, T. A., Lovern, M. B., and Congdon, J. D. 2001. Field-testing the protandry-based mating system for the lizard, *Anolis carolinensis*: Does the model organism have the right model? *Behavioral Ecology and Sociobiology* 50: 162–171.

42. Johnson, C. H. and Hasting, J. W. 1986. The elusive mechanisms of the circadian clock. *American Scientist* 74: 29–37.

43. Ketterson, E. D. and Nolan, V., Jr. 1999. Adaptation, exaptation, and constraint: A hormonal perspective. *American Naturalist* 154(Suppl.): S4–S25.

44. Klein, S. L. 2000. The effects of hormones on sex differences in infection: From genes to behavior. *Neuroscience and Biobehavioral Reviews* 24: 627–638.

45. Klose, S. M., Welbergen, J. A., and Kalko, E. K. V. 2009. Testosterone is associated with harem maintenance ability in free-ranging grey-headed flying-foxes, *Pteropus poliocephalus. Biology Letters* 5: 758–761.

46. Krohmer, R. W. 2004. The male red-sided garter snake (*Thamnophis sirtalis parietalis*): Reproductive pattern and behavior. *ILAR Journal* 45: 65–74.

47. Krohmer, R. W., Boyle, M. H., Lutterschmidt, D. I., and Mason, R. T. 2010. Seasonal aromatase activity in the brain of the male red-sided garter snake. *Hormones and Behavior* 58: 485–492.

48. Lang, A. B., Kalko, E. K.V., Romer, H., Bockholdt, C., and Dechmann, D. K. N. 2006. Activity levels of bats and katydids in relation to the lunar cycle. *Oecologia* 146: 659–666.

49. Langmore, N. E., Cockrem, J. F., and Candy, E. J. 2002. Competition for male reproductive investment elevates testosterone levels in female dunnocks, *Prunella modularis*. *Proceedings of the Royal Society of London B* 269: 2473–2478.

50. Lema, S. C. and Nevitt, G. A. 2004. Variation in vasotocin immunoreactivity in the brain of recently isolated populations of a Death Valley pupfish, *Cyprinodon nevadensis*. *General and Comparative Endocrinology* 135: 300–309.

51. Lemaster, M. P. and Mason, R. T. 2001. Evidence for a female sex pheromone mediating male trailing behavior in the red-sided garter snake, *Thamnophis sirtalis parietalis*. *Chemoecology* 11: 149–152.

52. Levine, J. D. 2004. Sharing time on the fly. *Current Opinion in Cell Biology* 16: 1–7.

53. Lincoln, G. A., Guinness, F., and Short, R. V. 1972. The way in which testosterone controls the social and sexual behavior of the red deer stag (*Cervus elaphus*). *Hormones and Behavior* 3: 375–396.

54. Lockard, R. B. and Owings, D. H. 1974. Seasonal variation in moonlight avoidance by bannertail kangaroo rats. *Journal of Mammalogy* 55: 189–193.

55. Lockard, R. B. 1978. Seasonal change in the activity pattern of *Dipodomys spectabilis*. *Journal of Mammalogy* 59: 563–568.

56. Loher, W. 1972. Circadian control of stridulation in the cricket *Teleogryllus commodus* Walker. *Journal of Comparative Physiology* 79: 173–190.

57. Loher, W. 1979. Circadian rhythmicity of locomotor behavior and oviposition in female *Teleogryllus commodus*. *Behavioral Ecology and Sociobiology* 5: 383–390.

58. Loher, W., Weber, T., and Huber, F. 1993. The effect of mating on phonotactic behavior in *Gryllus bimaculatus* (DeGeer). *Physiological Entomology* 18: 57–66.

59. Macrae, C. N., Alnwick, K. A., Milne, A. B., and Schloerscheidt, A. M. 2002. Person perception across the menstrual cycle: Hormonal influences on social-cognitive functioning. *Psychological Science* 13: 532–536.

60. Mak, G. K., Enwere, E. K., Gregg, C., Pakarainen, T., Poutanen, M., Huhtaniemi, I., and Weiss, S. 2007. Male pheromone-stimulated neurogenesis in the adult female brain: Possible role in mating behavior. *Nature Neuroscience* 10: 1003–1011.

61. Marler, C. A., Walsberg, G., White, M. L., and Moore, M. C. 1995. Increased energy-expenditure due to increased territorial defense in male lizards after phenotypic manipulation. *Behavioral Ecology and Sociobiology* 37: 225–231.

62. McGlothlin, J. W., Jawor, J. M., and Ketterson, E. D. 2007. Natural variation in a testosterone-mediated trade-off between mating effort and parental effort. *American Naturalist* 170: 864–875.

63. Mendonca, M. T., Daniels, D., Faro, C., and Crews, D. 2003. Differential effects of courtship and mating on receptivity and brain metabolism in female red-sided garter snakes (*Thamnophis sirtalis parietalis*). *Behavioral Neuroscience* 117: 144–149.

64. Moore, M. C. and Kranz, B. 1983. Evidence for androgen independence of male mounting behavior in white-crowned sparrows (*Zonotrichia leucophrys gambelii*). *Hormones and Behavior* 17: 414–423.

65. Muehlenbein, M. P. and Watts, D. P. 2010. The costs of dominance: Testosterone, cortisol and intestinal parasites in wild male chimpanzees. *BioPsychoSocial Medicine* 4(21): doi:10.1186/1751-0759-4-21.

66. Neal, J. K. and Wade, J. 2007. Courtship and copulation in the adult male green anole: Effects of season, hormone and female contact on reproductive behavior and morphology. *Behavioural Brain Research* 177: 177–185.

67. O'Donnell, R. P., Shine, R., and Mason, R. T. 2004. Seasonal anorexia in the male red-sided garter snake, *Thamnophis sirtalis parietalis*. *Behavioral Ecology and Sociobiology* 56: 413–419.

68. Packer, C., Swanson, A., Ikanda, D., and Kushnir, H. 2011. Fear of darkness, the full moon and the nocturnal ecology of African lions. *PLoS ONE* 6(7): e22285. doi:10.1371/journal.pone.0022285.

69. Page, T. L. 1985. Clocks and circadian rhythms. In G. A. Kerkut and L. I. Gilbert (Eds.), *Comprehensive Insect Physiology, Biochemistry, and Pharmacology*. Pergamon Press, New York, NY.

70. Pengelley, E. T. and Asmundson, S. J. 1974. Circannual rhythmicity in hibernating animals. In E. T. Pengelley (Ed.), *Circannual Clocks*. Academic Press, New York, NY.

71. Pereyra, M. E., Sharbaugh, S. M., and Hahn, T. P. 2005. Interspecific variation in photo-induced GnRH plasticity among nomadic cardueline finches. *Brain, Behavior and Evolution* 66: 35–49.

72. Perrigo, G., Bryant, W. C., and vom Saal, F. S. 1990. A unique neural timing system prevents male mice from harming their own offspring. *Animal Behaviour* 39: 535–539.

73. Pillsworth, E. G., Haselton, M. G., and Buss, D. M. 2004. Ovulatory shifts in female sexual desire. *Journal of Sex Research* 41: 55–65.

74. Ralph, M. R., Foster, R. G., Davis, F. C., and Menaker, M. 1990. Transplanted suprachiasmatic nucleus determines circadian rhythm. *Science* 247: 975–978.

75. Reed, W. L., Clark, M. E., Parker, P. G., Raouf, S. A., Arguedas, N., Monk, D. S., Snadjr, E., Nolan, V., Jr., and Ketterson, E. D. 2006. Physiological effects on demography: A long-term experimental study of testosterone's effects on fitness. *American Naturalist* 167: 667–683.

76. Rodriguez-Zas, S. L., Southey, B. R., Shemsh, Y., Rubin, E. B., Cohen, M., Robinson, G. E., and Bloch, G. 2012. Microarray analysis of natural socially regulated plasticity in circadian rhythms of honey bees. *Journal of Biological Rhythms* 27: 12–24.

77. Roeder, K. D. 1963. *Nerve Cells and Insect Behavior*. Harvard University Press, Cambridge, MA.

78. Roff, D. A. and Fairbairn, D. J. 2007. The evolution and genetics of migration in insects. *BioScience* 57: 155–164.

79. Runfeldt, S. and Wingfield, J. C. 1985. Experimentally prolonged sexual-activity in female sparrows delays termination of reproductive activity in their untreated mates. *Animal Behaviour* 33: 403–410.

80. Safran, R. J., Adelman, J. S., McGraw, K. J., and Hau, M. 2008. Sexual signal exaggeration affects male physiological state in barn swallows. *Current Biology* 18: R461–R462.

81. Schneider, J. S., Stone, M. K., Wynne-Edwards, K. E., Horton, T. H., Lydon, J., O'Malley, B., and Levine, J. E. 2003. Progesterone receptors mediate male aggression toward infants. *Proceedings of the National Academy of Sciences* 100: 2951–2956.

82. Schwartz, C. C., Cain, S. L., Podruzny, S., Cherry, S., and Frattaroli, L. 2010. Contrasting activity patterns of sympatric and allopatric black and grizzly bears. *Journal of Wildlife Management* 74: 1628–1638.

83. Sinervo, D., Miles, D. B., Frankino, W. A., Klukowski, M., and DeNardo, D. F. 2000. Testosterone, endurance, and Darwinian fitness: Natural and sexual selection on the physiological bases of alternative male behaviors in side-blotched lizards. *Hormones and Behavior* 38: 222–233.

84. Skals, N., Anderson, P., Kanneworff, M., Löfstedt, C., and Surlykke, A. 2005. Her odours make him deaf: Crossmodal modulation of olfaction and hearing in a male moth. *Journal of Experimental Biology* 208: 595–601.

85. Small, T. W., Sharp, P. J., and Deviche, P. 2007. Environmental regulation of the reproductive system in a flexibly breeding Sonoran Desert bird, the rufous-winged sparrow, *Aimophila carpalis*. *Hormones and Behavior* 51: 483–495.

86. Soma, K. K., Tramontin, A. D., and Wingfield, J. C. 2000. Oestrogen regulates male aggression in the non-breeding season. *Proceedings of the Royal Society of London B* 267: 1089–1092.

87. Strand, C. R., Small, T. W., and Deviche, P. 2007. Plasticity of the rufous-winged sparrow, *Aimophila carpalis*, song control regions during the monsoon-associated summer breeding period. *Hormones and Behavior* 52: 401–408.

88. Striepens, N., Kendrick, K. M., Maier, W., and Hurlemann, R. 2011. Prosocial effects of oxytocin and clinical evidence for its therapeutic potential. *Frontiers in Neuroendocrinology* 32: 426–450.

89. Toh, K. L., Jones, C. R., He, Y., Eide, E. J., Hinz, W. A., Virshup, D. M., Ptácek, L. J., and Fu, Y.-H. 2001. An hPer2 phosphorylation site mutation in familial advanced sleep phase syndrome. *Science* 291: 1040–1043.

90. Tökölyi, J., McNamara, J. M., Houston, A. I., and Barta, Z. 2012. Timing of avian reproduction in unpredictable environments. *Evolutionary Ecology* 26: 25–42.

91. Toma, D. P., Bloch, G., Moore, D., and Robinson, G. E. 2000. Changes in *period* mRNA levels in the brain and division of labor in honey bee colonies. *Proceedings of the National Academy of Sciences* 97: 6914–6919.

92. Trainor, B. C., Bird, I. M., Alday, N. A., Schlinger, B. A., and Marler, C. A 2003. Variation in aromatase activity in the medial preoptic area and plasma progesterone is associated with the onset of paternal behavior. *Neuroendocrinology* 78: 36–44.

93. Turek, F. W., McMillan, J. P., and Menaker, M. 1976 . Melatonin: Effects of the circadian rhythms of sparrows. *Science* 194: 1441–1443.

94. Wade, J. 2005. Current research on the behavioral neuroendocrinology of reptiles. *Hormones and Behavior* 48: 451–460.

95. Wingfield, J. C. and Moore, M. C. 1987. Hormonal, social and environmental factors in the reproductive biology of free-living male birds. In D. Crews (Ed.), *Psychobiology of Reproductive Behavior: An Evolutionary Perspective*. Prentice Hall, Englewood Cliffs, NJ.

96. Wingfield, J. C. and Ramenofsky, M. 1997. Corticosterone and facultative dispersal in response to unpredictable events. *Ardea* 85: 155–166.

97. Wingfield, J. C., Lynn, S. E., and Soma, K. K. 2001. Avoiding the "costs" of testosterone: Ecological bases of hormone-behavior interactions. *Brain, Behavior and Evolution* 57: 239–251.

98. Winkler, S. M. and Wade, J. 1998. Aromatase activity and regulation of sexual behaviors in the green anole lizard. *Physiology & Behavior* 64: 723–731.

99. Xu, X., Coats, J. K., Yang, C. F., Wang, A., Ahmed, O. M., Alvarado, M., Isumi, T., and Shah, N. M. 2012. Modular genetic control of sexually dimorphic behaviors. *Cell* 148: 596–607.

100. Yapici, N., Kim, Y.-J., Ribiero, C., and Dickson, B. J. 2008. A receptor that mediates the post-mating switch in *Drosophila* reproductive behaviour. *Nature* 451: 33–38.

101. Young, M. W. 2000. Marking time for a kingdom. *Science* 288: 451–453.

102. Zera, A. J., Zhao, Z., and Kaliseck, K. 2007. Hormones in the field: Evolutionary endocrinology of juvenile hormone and ecdysteroids in field populations of the wing-dimorphic cricket *Gryllus firmus*. *Physiological and Biochemical Zoology* 80: 592–606.

103. Zhu, H., Sauman, I., Yuan, A., Emery-Le, M., Emery, P., and Reppert, S. M. 2008. Cryptochromes define a novel circadian clock mechanism in monarch butterflies that may underlie sun compass navigation. *PLoS Biology* 6(1): e4. doi:10.1371/journal.pbio.0060004.

104. Zucker, I. 1983. Motivation, biological clocks and temporal organization of behavior. In E. Satinoff and P. Teitelbaum (Eds.), *Handbook of Behavioral Neurobiology: Motivation*. Plenum Press, New York, NY.

105. Zuk, M., Johnsen, T. S., and MacLarty, T. 1995. Endocrine-immune interactions, ornaments and mate choice in red jungle fowl. *Proceedings of the Royal Society of London B* 260: 205–210.

Chapter 14

1. Alcock, J. 2001. *The Triumph of Sociobiology*. Oxford University Press, New York, NY.

2. Alexander, G. M. and Hines, M. 2002. Sex differences in response to children's toys in nonhuman primates (*Cercopithecus aethiops sabaeus*). *Evolution and Human Behavior* 23: 467–479.

3. Alexander, R. D. 1979. *Darwinism and Human Affairs*. University of Washington Press, Seattle, WA.

4. Allen, L., Beckwith, B., Beckwith, J., Chorover, S., Culver, D., Daniels, N., and Dorfman, D. 1975. Against "sociobiology." *New York Review of Books* 22 (Nov. 13): 43–44.

5. Allen, L., Beckwith, B., Beckwith, J., Chorover, S., Culver, D., Duncan, M., and Gould, S. J. 1976. Sociobiology: Another biological determinism. *BioScience* 26: 182–186.

6. Allison, T., Ginter, H., McCarthy, G., Nobre, A. C., Puce, A., Luby, M., and Spencer, D. D. 1994. Face recognition in human extrastriate cortex. *Journal of Neurophysiology* 71: 821–825.

7. Allison, T., Puce, A., and McCarthy, G. 2000. Social perception from visual cues: The role of the STS region. *Trends in Cognitive Sciences* 4: 267–278.

8. Alvergne, A., Faurie, C., and Raymond, M. 2007. Differential facial resemblance of young children to their parents: Who do children look like more? *Evolution and Human Behavior* 28: 135–144.

9. Alvergne, A., Faurie, C., and Raymond, M. 2009. Father-offspring resemblance predicts paternal investment in humans. *Animal Behaviour* 78: 61–69.

10. Alvergne, A., Faurie, C., and Raymond, M. 2010. Are parents' perceptions of offspring facial resemblance consistent with actual resemblance? Effects on parental investment. *Evolution and Human Behavior* 31: 7–15.

11. Alvergne, A. and Lummaa, V. 2010. Does the contraceptive pill alter mate choice in humans? *Trends in Ecology and Evolution* 25: 171–179.

12. Angier, N. 1999. *Woman: An Intimate Geography*. Houghton Mifflin Company, New York, NY.

13. Apicella, C. L. and Marlowe, F. W. 2004. Perceived mate fidelity and paternal resemblance predict men's investment in children. *Evolution and Human Behavior* 25: 371–378.

14. Apicella, C. L., Feinberg, D. R., and Marlowe, F. W. 2007. Voice pitch predicts reproductive success in male hunter-gatherers. *Biology Letters* 3: 682–684.

15. Atkinson, Q. D. 2011. Phonemic diversity supports a serial founder effect model of language expansion from Africa. *Science* 332: 346–349.

16. Beecher, M. D. and Beecher, I. M. 1979. Sociobiology of bank swallows: Reproductive strategy of the male. *Science* 205: 1282–1285.

17. Begley, S. 2009, June 29. Why do we rape, kill and sleep around? *Newsweek* (http://www.newsweek.com/id/202789).

18. Belot, M. and Fancesconi, M. 2006. Can anyone be "the one"? Evidence on mate selection from speed dating. *CEPR Discussion Papers* 5926.

19. Betzig, L. (Ed.) 1997. *Human Nature: A Critical Reader*. Oxford University Press, New York, NY.

20. Boone, J. L., III. 1986. Parental investment and elite family structure in preindustrial states: A case study of late medieval-early modern Portuguese genealogies. *American Anthropologist* 88: 859–878.

21. Borgerhoff Mulder, M. 1987. Resources and reproductive success in women with an example from the Kipsigis of Kenya. *Journal of Zoology* 213: 489–505.

22. Boyd, R. S. and Silk, J. B. 2011. *How Humans Evolved* (6th ed.). W. W. Norton, Los Angeles, CA.

23. Bressler, E. R., Martin, R. A., and Balshine, S. 2006. Production and appreciation of humor as sexually selected traits. *Evolution and Human Behavior* 27: 121–130.

24. Brooks, R., Shelly, J. P., Fan, J., Zhai, L., and Chau, D. K. P. 2010. Much more than a ratio: Multivariate selection on female bodies. *Journal of Evolutionary Biology* 23: 2238–2248.

25. Brown, J. L. and Eklund, A. 1994. Kin recognition and the major histocompatibility complex: An integrative review. *American Naturalist* 143: 435–461.

26. Brownmiller, S. 1975. *Against Our Will.* Simon and Schuster, New York, NY.

27. Brownmiller, S. and Merhof, B. 1992. A feminist response to rape as an adaptation in men. *Brain and Behavioral Sciences* 15: 381–382.

28. Bryant, G. A. and Haselton, M. G. 2009. Vocal cues of ovulation in human females. *Biology Letters* 5: 12–15.

29. Bryant, J. I. 2010. Dialect. In R. L. Jackson II (Ed.), *Encyclopedia of Identity* (pp. 219–220). Sage Publications, Urbana, IL.

30. Burch, R. L. and Gallup, G. G. 2004. Pregnancy as a stimulus for domestic violence. *Journal of Family Violence* 19: 243–247.

31. Buss, D. M. 1989. Sex differences in human mate preferences: Evolutionary hypotheses tested in 37 cultures. *Behavioral and Brain Sciences* 12: 1–14.

32. Buss, D. M., Larsen, R. J., Westen, D., and Semmelroth, J. 1992. Sex differences in jealousy: Evolution, physiology, and psychology. *Psychological Science* 3: 251–255.

33. Buss, D. M. and Schmitt, D. P. 1993. Sexual strategies theory: An evolutionary perspective on human mating. *Psychological Review* 100: 204–232.

34. Buss, D. M. 2003. *The Evolution of Desire* (2nd ed.). Basic Books, New York, NY.

35. Buss, D. M. and Shackelford, T. K. 2008. Attractive women want it all: Good genes, economic investment, parenting proclivities, and emotional commitment. *Evolutionary Psychology* 6: 134–146.

36. Buss, D. M. and Pinker, S. 2009. Pop psychology probe. *Scientific American* 300: 10–11.

37. Buss, D. M. and Duntley, J. D. 2011. The evolution of intimate partner violence. *Aggression and Violent Behavior* 16: 411–419.

38. Buss, D. M. 2012. *Evolutionary Psychology: The New Science of the Mind* (4th ed.). Pearson, Upper Saddle River, NJ.

39. Buston, P. M. and Emlen, S. T. 2003. Cognitive processes underlying human mate choice: The relationship between self-perception and mate preference in Western society. *Proceedings of the National Academy of Sciences* 100: 8805–8810.

40. Carroll, S. B. 2005. Evolution at two levels: On genes and form. *PLoS Biology* 3: 1159–1166.

41. Case, A., Lubotsky, D., and Paxson, C. 2002. Economic status and health in childhood: The origins of the gradient. *American Economic Review* 92: 1308–1334.

42. Cialdini, R. B. 2001. The science of persuasion. *Scientific American* 284 (Feb.): 76–81.

43. Clark, R. D. and Hatfield, E. 1989. Gender differences in receptivity to sexual offers. *Journal of Psychology and Human Sexuality* 2: 39–55.

44. Colarelli, S. M. and Dettmann, J. R. 2003. Intuitive evolutionary perspectives in marketing practices. *Psychology & Marketing* 20: 837–865.

45. Daly, M., Wilson, M., and Weghorst, S. T. 1982. Male sexual jealousy. *Ethology and Sociobiology* 3: 11–27.

46. Daly, M. and Wilson, M. 1983. *Sex, Evolution and Behavior* (2nd ed.). Willard Grant Press, Boston, MA.

47. Daly, M. and Wilson, M. 1985. Child abuse and other risks of not living with both parents. *Ethology and Sociobiology* 6: 197–210.

48. Daly, M. and Wilson, M. 1992. The man who mistook his wife for a chattel. In J. Barkow, L. Cosmides, and J. Tooby (Eds.), *The Adapted Mind*. Oxford University Press, New York, NY.

49. Daly, M. and Wilson, M. 2008. Is the "Cinderella effect" controversial?: A case study of evolution-minded research and critiques thereof. In C. Crawford and K. Dennis (Eds.), *Foundations of Evolutionary Psychology*. Lawrence Erlbaum Associates, New York, NY.

50. Davey, G. C. L. 2011. Disgust: The disease-avoidance emotion and its dysfunctions. *Philosophical Transactions of the Royal Society B* 366: 3453–3465.

51. Diamond, J. M. 1992. *The Third Chimpanzee*. HarperCollins, New York, NY.

52. Dunbar, R. 1996. *Grooming, Gossip, and the Evolution of Language*. Harvard University Press, Cambridge, MA.

53. Durante, K. M. and Li, N. P. 2009. Oestradiol level and opportunistic mating in women. *Biology Letters* 5: 179–182.

54. Egan, J. 2006, March 19. Wanted: A few good sperm. *New York Times Magazine*.

55. Ellis, A. W. and Young, A. W. 1996. *Human Cognitive Neuropsychology*. Psychology Press, East Sussex, England.

56. Feinberg, D. R. 2008. Are human faces and voices ornaments signaling common underlying cues to mate value? *Evolutionary Anthropology* 17: 112–118.

57. Felson, R. B. and Cundiff, P. R. 2011. Age and sexual assault during robberies. *Evolution and Human Behavior* 33: 10–16.

58. Fink, B., Neave, N., and Seydel, H. 2007. Male facial appearance signals physical strength to women. *American Journal of Human Biology* 19: 82–87.

59. Fischer, J., Semple, S., Fickenscher, G., Jürgens, R., Kruse, E., Heistermann, M., and Amir, O. 2011. Do women's voices provide cues of the likelihood of ovulation? The importance of sampling regime. *PLoS ONE* 6(9): e24490. doi:10.1371/journal.pone.0024490.

60. Folstad, I. and Karter, A. J. 1992. Parasites, bright males, and the immunocompetence handicap. *American Naturalist* 139: 603–622.

61. Frenda, S. J., Nichols, R. M., and Loftus, E. F. 2011. Current issues and advances in misinformation research. *Current Directions in Psychological Science* 20: 20–23.

62. Gallup, A. C., White, D. D., and Gallup, G. G. 2007. Handgrip strength predicts sexual behavior, body morphology, and aggression in male college students. *Evolution and Human Behavior* 28: 423–429.

63. Gangestad, S. W. and Thornhill, R. 2008. Human oestrus. *Proceedings of the Royal Society B* 275: 991–1000.

64. Garver-Apgar, C. E., Gangestad, S. W., Thornhill, R., Miller, R. D., and Olp, J. J. 2006. Major histocompatibility complex alleles, sexual responsivity, and unfaithfulness in romantic couples. *Psychological Science* 17: 830–835.

65. Gaulin, S. J. C. and Boster, J. S. 1990. Dowry as female competition. *American Anthropologist* 92: 994–1005.

66. Gaulin, S. J. C. and McBurney, D. H. 2003. *Psychology: An Evolutionary Approach* (2nd ed.). Prentice Hall, Upper Saddle River, NJ.

67. Geary, D. C. 2010. *Male, Female: The Evolution of Human Sex Differences* (2nd ed.). American Psychological Association, Washington, DC.

68. Gillespie, D. O. S., Russell, A. F., and Lummaa, V. 2008. When fecundity does not equal fitness: Evidence of an offspring quantity versus quality trade-off in pre-industrial humans. *Proceedings of the Royal Society B* 275: 713–722.

69. Gobbini, M. I. and Haxby, J. V. 2006. Neural systems for recognition of familiar faces. *Neuropsychologia* 45: 32–41.

70. Gottschall, J. A. and Gottschall, T. A. 2003. Are per-incident rape-pregnancy rates higher than per-incident consensual pregnancy rates? *Human Nature* 14: 1–20.

71. Gould, S. J. and Lewontin, R. C. 1979. The spandrels of San Marco and the Panglossian paradigm: A critique of the adaptationist programme. *Proceedings of the Royal Society of London B* 205: 581–598.

72. Gould, S. J. 1996. The diet of worms and the defenestration of Prague. *Natural History* 105: 18–24.

73. Grammar, K., Fink, B., Møller, A. P., and Thornhill, R. 2003. Darwinian aesthetics: Sexual selection and the biology of beauty. *Biological Reviews* 78: 385–407.

74. Green, C. S., Benson, C., Kersten, D., and Schrater, P. 2010. Alterations in choice behavior by manipulations of world model. *Proceedings of the National Academy of Sciences* 107: 16401–16406.

75. Greengross, G. and Miller, G. 2011. Humor ability reveals intelligence, predicts mating success, and is higher in males. *Intelligence* 39: 188–192.

76. Griskevicius, V., Tybur, J. M., and Van den Bergh, B. 2010. Going green to be seen: Status, reputation and conspicuous conservation. *Journal of Personality and Social Psychology* 98: 392–404.

77. Gueguen, N. 2011. Effects of solicitor sex and attractiveness on receptivity to sexual offers: A field study. *Archives of Sexual Behavior* 40: 915–919.

78. Harcourt, A. H., Harvey, P. H., Larson, S. G., and Short, R. V. 1981. Testis weight, body weight and breeding system in primates. *Nature* 293: 55–57.

79. Hartung, J. 1982. Polygyny and inheritance of wealth. *Current Anthropology* 23: 1–12.

80. Hartung, J. 1995. Love thy neighbor: The evolution of ingroup morality. *Skeptic* 3: 86–99.

81. Haselton, M. G. 2003. The sexual overperception bias: Evidence of a systematic bias in men from a survey of naturally occurring events. *Journal of Research in Personality* 37: 34–47.

82. Heath, K. M. and Hadley, C. 1998. Dichotomous male reproductive strategies in a polygynous human society: Mating versus parental effort. *Current Anthropology* 39: 369–374.

83. Hobaiter, C. and Byrne, R. W. 2011. The gestural repertoire of the wild chimpanzee. *Animal Cognition* 14: 745–767.

84. Holveck, M. J. and Riebel, K. 2011. Low-quality females prefer low-quality males when choosing a mate. *Proceedings of the Royal Society B* 277: 153–160.

85. Hurst, J. A., Baraitser, M., Auger, E., Graham, F., and Norell, S. 1990. An extended family with a dominantly inherited speech disorder. *Developmental Medicine & Child Neurology* 32: 352–355.

86. Huxley, T. H. 1910. *Lectures and Lay Sermons.* E. P. Dutton, New York, NY.

87. Irons, W. 1979. Cultural and biological success. In N. A. Chagnon and W. Irons (Eds.), *Evolutionary Biology and Human Social Behavior: An Anthropological Perspective.* Duxbury Press, North Scituate, MA.

88. Jasiénska, G., Ziomkiewicz, A., Ellison, P. T., Lipson, S. F., and Thune, I. 2004. Large breasts and narrow waists indicate high reproductive potential in women. *Proceedings of the Royal Society of London B* 271: 1213–1217.

89. Jones, B. C., DeBruine, L. M., Perrett, D. I., Little, A. C., Feinberg, D. R., and Smith, M. J. L. 2008. Effects of menstrual cycle phase on face preferences. *Archives of Sexual Behavior* 37: 78–84.

90. Jones, O. D. 1999. Sex, culture, and the biology of rape: Toward explanation and prevention. *California Law Review* 87: 827–942.

91. Kanazawa, S. 2003. Can evolutionary psychology explain reproductive behavior in the contemporary United States? *Sociological Quarterly* 44: 291–302.

92. Kaplan, H. and Hill, K. 1985. Hunting ability and reproductive success among male Ache foragers: Preliminary results. *Current Anthropology* 26: 131–133.

93. Karremans, J. C., Frankenhuis, W. E., and Arons, S. 2010. Blind men prefer a low waist-to-hip ratio. *Evolution and Human Behavior* 31: 182–186.

94. Kates, C. A. 2004. Reproductive liberty and overpopulation. *Environmental Values* 13: 51–79.

95. Kellogg, W. N. and Kellogg, L. A. 1933. *The Ape and the Child.* Hafner Publishing Company, New York, NY.

96. Kenrick, D. T., Sadalla, E. K., Groth, G., and Trost, M. R. 1990. Evolution, traits, and the stages of human courtship: Qualifying the parental investment model. *Journal of Personality* 58: 97–116.

97. Kenrick, D. T. and Keefe, R. C. 1992. Age preferences in mates reflect sex differences in reproductive strategies. *Behavioral and Brain Sciences* 15: 75–133.

98. Kenrick, D. T. 1995. Evolutionary theory versus the confederacy of dunces. *Psychological Inquiry* 6: 56–61.

99. Kenrick, D. T., Keefe, R. C., Gabrielidis, C., and Cornelius, J. S. 1996. Adolescents' age preferences for dating partners: Support for an evolutionary model of life-history strategies. *Child Development* 67: 1499–1511.

100. Kitcher, P. 1985. *Vaulting Ambition.* MIT Press, Cambridge, MA.

101. Kuhl, P. K. 2010. Brain mechanisms in early language acquisition. *Neuron* 67: 713–727.

102. Laeng, B., Mathisen, R., and Johnsen, J. A. 2007. Why do blue-eyed men prefer women with the same eye color? *Behavioral Ecology and Sociobiology* 61: 371–384.

103. Lalumiére, M. L., Chalmers, L. J., Quinsey, V. L., and Seto, M. C. 1996. A test of the mate deprivation hypothesis of social coercion. *Ethology and Sociobiology* 17: 299–318.

104. Lassek, W. D. and Gaulin, S. J. C. 2008. Waist-hip ratio and cognitive ability: Is gluteofemoral fat a privileged store of neurodevelopmental resources? *Evolution and Human Behavior* 29: 26–34.

105. Law-Smith, M. J., Perrett, D. I., Jones, B. C., Cornwell, R. E., Moore, F. R., Feinberg, D. R., Boothroyd, L. G., Durrani, S. J., Stirrat, M. R., Whiten, S., Pitman, R. M., and Hillier, S. G. 2006. Facial appearance is a cue to oestrogen levels in women. *Proceedings of the Royal Society B* 273: 135–140.

106. Leonard, W. R., Robertson, M. L., Snodgrass, J. J., and Kuzawa, C. W. 2003. Metabolic correlates of hominid brain evolution. *Comparative Biochemistry and Physiology A* 136: 5–15.

107. Lewkowicz, D. J., and Hansen-Tift, A. M. 2012. Infants deploy selective attention to the mouth of a talking face when learning speech. *Proceedings of the National Academy of Sciences* 109: 1431–1436.

108. Li, N. P., Bailey, J. M., Kenrick, D. T., and Linsenmeier, J. A. W. 2002. The necessities and luxuries of mate preferences: Testing the tradeoffs. *Journal of Personality and Social Psychology* 82: 947–955.

109. Li, N. P., Valentine, K. A., and Patel, L. 2011. Mate preferences in the US and Singapore: A cross-cultural test of the mate preference priority model. *Personality and Individual Differences* 50: 291–294.

110. Lie, H. C., Simmons, L. W., and Rhodes, G. 2010. Genetic dissimilarity, genetic diversity, and mate preferences in humans. *Evolution and Human Behavior* 31: 48–58.

111. Lieberman, P. 2007. The evolution of human speech: Its anatomical and neural bases. *Current Anthropology* 48: 39–66.

112. Lipson, S. F. and Ellison, P. T. 1996. Comparison of salivary steroid profiles in naturally occurring conception and non-conception cycles. *Human Reproduction* 11: 2090–2096.

113. Little, A. C., Burt, D. M., Penton-Voak, I. S., and Perrett, D. I. 2001. Self-perceived attractiveness influences human female preferences for sexual dimorphism and symmetry in male faces. *Proceedings of the Royal Society of London B* 268: 39–44.

114. Little, A. C., Jones, B. C., and Burriss, R. P. 2007. Preferences for masculinity in male bodies change across the menstrual cycle. *Hormones and Behavior* 51: 633–639.

115. Little, A. C., Jones, B. C., and Burt, D. M. 2007. Preferences for symmetry in faces change across the menstrual cycle. *Biological Psychology* 76: 209–216.

116. Little, A. C., Saxton, T. K., Roberts, S. C., Jones, B. C., DeBruine, L. M., Vukovic, J., Perrett, D. I., Feinberg, D. R., and Chenore, T. 2010. Women's preferences for masculinity in male faces are highest during reproductive age-range and lower around puberty and post-menopause. *Psychoneuroendocrinology* 35: 912–920.

117. Little, A. C., Jones, B. C., and DeBruine, L. M. 2011. Facial attractiveness: Evolutionary based research. *Philosophical Transactions of the Royal Society B* 366: 1638–1659.

118. Lummaa, V. 2003. Early developmental conditions and reproductive success in humans: Downstream effects of prenatal famine, birthweight, and timing of birth. *American Journal of Human Biology* 15: 370–379.

119. Lutz, W., Testa, M. R., and Penn, D. J. 2006. Population density is a key factor in declining human fertility. *Population and Environment* 28: 69–81.

120. Mace, R. 1998. The coevolution of human fertility and wealth inheritance strategies. *Philosophical Transactions of the Royal Society of London B* 353: 389–397.

121. Maguire, E. A., Gadian, D. G., Johnsrude, I. S., Good, C. D., Ashburner, J., Frackowiak, R. S. J., and Frith, C. D. 2000. Navigation-related structural change in the hippocampi of taxi drivers. *Proceedings of the National Academy of Sciences* 97: 4398–4403.

122. Maguire, E. A., Wollett, K., and Spiers, H. J. 2006. London taxi drivers and bus drivers: A structural MRI and neuropsychological analysis. *Hippocampus* 16: 1091–1101.

123. Massaro, D. W. and Stork, D. G 1998. Speech recognition and sensory integration. *American Scientist* 86: 236–244.

124. McCandliss, B. D., Cohen, L., and Dehaene, S. 2003. The visual word form area: Expertise for reading in the fusiform gyrus. *Trends in Cognitive Sciences* 7: 293–299.

125. McGurk, H. and Macdonald, J. 1976. Hearing lips and seeing voices. *Nature* 264: 746–748.

126. Mesoudi, A. 2007. Biological and cultural evolution: Similar but different. *Biological Theory* 2: 119–123.

127. Miller, G. F. 2000. *The Mating Mind.* Doubleday, New York, NY.

128. Miller, G. F., Tybur, J., and Jordan, B. 2008. Ovulatory cycle effects on tip earnings by lap-dancers: Economic evidence for human estrus? *Evolution and Human Behavior* 28: 375–381.

129. Miller, G. F. 2009. *Spent: Sex, Evolution and Consumerism.* Penguin Group, New York, NY.

130. Miller, R. M., Sanchez, K., and Rosenblum, L. D. 2010. Alignment to visual speech information. *Attention, Perception and Psychophysics* 72: 1614–1625.

131. Miller, S. L. and Maner, J. K. 2010. Scent of a woman: Men's testosterone responses to olfactory ovulation cues. *Psychological Science* 21: 276–283.

132. Moorad, J. A., Promislow, D. E. L., Smith, K. R., and Wade, M. J. 2011. Mating system change reduces the strength of sexual selection in an American frontier population of the 19th century. *Evolution and Human Behavior* 32: 147–155.

133. Mueller, U. and Mazur, A. 2001. Evidence of unconstrained directional selection for male tallness. *Behavioral Ecology and Sociobiology* 50: 302–311.

134. Muller, M. N., Thompson, M. E., and Wrangham, R. W. 2006. Male chimpanzees prefer mating with old females. *Current Biology* 16: 2234–2238.

135. Murdock, G. P. 1967. *Ethnographic Atlas.* Pittsburgh University Press, Pittsburgh, PA.

136. Pallett, P. M., Link, S., and Lee, K. 2010. New "golden" ratios for facial beauty. *Vision Research* 50: 149–154.

137. Palmer, C. T. 1991. Human rape: Adaptation or by-product? *Journal of Sex Research* 28: 365–386.

138. Pawłowski, B. and Dunbar, R. I. M. 1999. Impact of market value on human mate choice decisions. *Proceedings of the Royal Society of London B* 266: 281–285.

139. Penn, D. J. 2003. The evolutionary roots of our environmental problems: Toward a Darwinian ecology. *Quarterly Review of Biology* 78: 275–301.

140. Penton-Voak, I. S., Perrett, D. I., Castles, D. L., Kobayashi, T., Burt, D. M., Murray, L. K., and Minamisawa, R. 1999. Menstrual cycle alters face preference. *Nature* 399: 741–742.

141. Perusse, D. 1993. Cultural and reproductive success in industrial societies: Testing the relationship at the proximate and ultimate levels. *Behavioral and Brain Sciences* 16: 267–283.

142. Pinker, S. 1994. *The Language Instinct.* W. Morrow & Co., New York, NY.

143. Pinker, S. 2003. Language as an adaptation to the cognitive niche. In S. Kirby and M. Christiansen (Eds.), *Language Evolution: States of the Art.* Oxford University Press, New York, NY.

144. Plomin, R., Fulker, D. W., Corley, R., and DeFries, J. C. 1997. Nature, nurture, and cognitive development from 1 to 16 years: A parent-offspring adoption study. *Psychological Science* 8: 442–447.

145. Prokop, P. and Fedor, P. 2011. Physical attractiveness influences reproductive success of modern men. *Journal of Ethology* 29: 453–458.

146. Prokosch, M. D., Coss, R. G., Scheib, J. E., and Blozis, S. A. 2009. Intelligence and mate choice: Intelligent men are always appealing. *Evolution and Human Behavior* 30: 11–20.

147. Provine, R. R., Cabrera, M. O., Brocato, N. W., and Krosnowski, K. A. 2011. When the whites of the eyes are red: A uniquely human cue. *Ethology* 117: 395–399.

148. Provost, M. P., Quinsey, V. L., and Troje, N. F. 2008. Differences in gait across the menstrual cycle and their attractiveness to men. *Archives of Sexual Behavior* 37: 598–604.

149. Rees, W. E. 2002. An ecological economics perspective on sustainability and prospects for ending poverty. *Population and Environment* 24: 15–46.

150. Reid, R. L. and van Vugt, D. A. 1987. Weight related change in reproductive function. *Fertility and Sterility* 48: 905–913.

151. Rivas, E. 2005. Recent use of signs by chimpanzees (*Pan troglodytes*) in interactions with humans. *Journal of Comparative Psychology* 119: 404–417.

152. Roberts, S. C., Havlicek, J., Flegr, J., Hruskova, M., Little, A. C., Jones, B. C., Perrett, D. I., and Petrie, M. 2004. Female facial attractiveness increases during the fertile phase of the menstrual cycle. *Proceedings of the Royal Society of London B* 271: S270–S272.

153. Roberts, S. C., Gosling, L. M., Carter, V., and Petrie, M. 2008. MHC-correlated odour preferences in humans and the use of oral contraceptives. *Proceedings of the Royal Society B* 275: 2715–2722.

154. Rose, M. 1998. *Darwin's Spectre*. Princeton University Press, Princeton, NJ.

155. Rose, N. A., Deutsch, C. J., and Le Boeuf, B. J. 1991. Sexual behavior of male northern elephant seals: III. The mounting of weaned pups. *Behaviour* 119: 171–192.

156. Rupp, H. A., James, T. W., Ketterson, E. D., Sengelaub, D. R., Janssen, E., and Heiman, J. R. 2009. Neural activation in the orbitofrontal cortex in response to male faces increases during the follicular phase. *Hormones and Behavior* 56: 66–72.

157. Sacks, O. W. 1985. *The Man Who Mistook His Wife for a Hat and Other Clinical Tales*. Summit Books, New York, NY.

158. Sahlins, M. 1976. *The Use and Abuse of Biology*. University of Michigan Press, Ann Arbor, MI.

159. Schacter, D. L., Guerin, S. A., and St. Jacques, P. L. 2011. Memory distortion: An adaptive perspective. *Trends in Cognitive Sciences* 15: 467–474.

160. Scharff, C. and Petri, J. 2011. Evo-devo, deep homology and FoxP2: Implications for the evolution of speech and language. *Philosophical Transactions of the Royal Society B* 366: 2124–2140.

161. Schmitt, D. P., Shackelford, T. K., Duntley, J., Tooke, W., and Buss, D. M. 2001. The desire for sexual variety as a key to understanding basic human mating strategies. *Personal Relationships* 8: 425–455.

162. Senghas, A., Kita, S., and Özyürek, A. 2004. Children creating core properties of language: Evidence from an emerging sign language in Nicaragua. *Science* 305: 1779–1782.

163. Shackelford, T. K., Buss, D. M., and Weekes-Shackelford, V. A. 2003. Wife killings committed in the context of a lovers triangle. *Basic and Applied Social Psychology* 25: 137–143.

164. Shettleworth, S. J. 2010. *Cognition, Evolution and Behavior* (2nd ed.). Oxford University Press, New York, NY.

165. Singh, D. 1993. Adaptive significance of female physical attractiveness: Role of the waist-to-hip ratio. *Journal of Personality and Social Psychology* 65: 293–307.

166. Singh, D. and Bronstad, P. M. 2001. Female body odour is a potential cue to ovulation. *Proceedings of the Royal Society of London B* 268: 797–801.

167. Smith, E. A., Borgerhoff Mulder, M., and Hill, K. 2001. Controversies in the evolutionary social sciences: A guide for the perplexed. *Trends in Ecology and Evolution* 16: 128–135.

168. Sorenson, L. G. 1994. Forced extra-pair copulation in the white-cheeked pintail: Male tactics and female responses. *Condor* 96: 400–410.

169. Swaddle, J. P. and Reierson, G. W. 2002. Testosterone increases perceived dominance but not attractiveness in human males. *Proceedings of the Royal Society of London B* 269: 2285–2289.

170. Symons, D. 1979. *The Evolution of Human Sexuality*. Oxford University Press, New York, NY.

171. Thompson, M. E., Jones, J. H., Pusey, A. E., Brewer-Marsden, S., Goodall, J., Marsden, D., Matsuzawa, T., Nishida, T., Reynolds, V., Sugiyama, Y., and Wrangham, R. W. 2007. Aging and fertility patterns in wild chimpanzees provide insights into the evolution of menopause. *Current Biology* 17: 2150–2156.

172. Thornhill, R. and Thornhill, N. W. 1983. Human rape: An evolutionary analysis. *Ethology and Sociobiology* 4: 137–173.

173. Thornhill, R. and Gangestad, S. W. 1999. The scent of symmetry: A human sex pheromone that signals fitness? *Evolution and Human Behavior* 20: 175–201.

174. Thornhill, R. and Gangestad, S. W. 1999. Facial attractiveness. *Trends in Cognitive Sciences* 3: 452–460.

175. Thornhill, R. and Palmer, C. T. 2000. *A Natural History of Rape: The Biological Bases of Sexual Coercion*. MIT Press, Cambridge, MA.

176. Tovée, M. J., Maisey, D. S., Emery, J. L., and Cornelissen, P. L. 1999. Visual cues to female physical attractiveness. *Proceedings of the Royal Society of London B* 266: 211–218.

177. Townsend, J. M. 1989. Mate selection criteria: A pilot study. *Ethology and Sociobiology* 10: 241–253.

178. Trivers, R. L. 2011. *The Folly of Fools: Deceit and Self-Deception in Human Affairs*. Basic Books, New York, NY.

179. Tybur, J. M., Miller, G. F., and Gangestad, S. W. 2007. Testing the controversy: An empirical examination of adaptationists'attitudes toward politics and science. *Human Nature* 18: 313–328.

180. Ujhelyi, M. 1996. Is there any intermediate stage between animal communication and language? *Journal of Theoretical Biology* 180: 71–76.

181. Vargha-Khadem, F., Gadian, D. G., Copp, A., and Mishkin, M. 2005. *FOXP2* and the neuroanatomy of speech and language. *Nature Reviews Neuroscience* 6: 131–138.

182. Vaughan, A. E. 2003. The association between offender socioeconomic status and victim-offender relationship in rape offences—revised. *Sexualities, Evolution & Gender* 5: 103–105.

183. Vining, D. R., Jr. 1986. Social versus reproductive success: The central theoretical problem of human sociobiology. *Behavioral and Brain Sciences* 9: 167–187.

184. Waynforth, D. and Dunbar, R. I. M. 1995. Conditional mate choice strategies in humans: Evidence from lonely hearts advertisements. *Behaviour* 132: 755–779.

185. Wedekind, C., Seebeck, T., Bettens, F., and Paepke, A. J. 1995. MHC-dependent mate preferences in humans. *Proceedings of the Royal Society of London B* 260: 245–249.

186. West, S. A., El Mouden, C., and Gardner, A. 2011. Sixteen common misconceptions about the evolution of cooperation in humans. *Evolution and Human Behavior* 32: 231–262.

187. Wiederman, M. W. and Allgeier, E. R. 1992. Gender differences in mate selection criteria: Sociobiological or socioeconomic explanation? *Ethology and Sociobiology* 13: 115–124.

188. Wiederman, M. W. and Kendall, E. 1999. Evolution, sex, and jealousy: Investigation with a sample from Sweden. *Evolution and Human Behavior* 20: 121–128.

189. Williams, G. C. 1996. *The Pony Fish's Glow*. Basic Books, New York, NY.

190. Wilson, E. O. 1975. *Sociobiology: The New Synthesis*. Harvard University Press, Cambridge, MA.

191. Wilson, M. I. and Daly, M. 1997. Life expectancy, economic inequality, homicide, and reproductive timing in Chicago neighborhoods. *British Medical Journal* 314: 1271–1274.

192. Wilson, M. I., Daly, M., and Gordon, S. 1998. The evolved psychological apparatus of decision-making is one source of environmental problems. In T. M. Caro (Ed.), *Behavioral Ecology and Conservation Biology*. Oxford University Press, New York, NY.

193. Winking, J., Kaplan, H., Gurven, M., and Rucas, S. 2007. Why do men marry and why do they stray? *Proceedings of the Royal Society B* 274: 1643–1649.

194. Wood, B. and Harrison, T. 2011. The evolutionary context of the first hominins. *Nature* 470: 347–352.

Illustration Credits

Chapter 1 Opener: © Mark Moffett/Minden Pictures. 1.5: © Heini Wehrle/AGE Fotostock. 1.7: © George Schaller/Photoshot.

Chapter 2 Opener: © Mark Moffett/Minden Pictures. 2.5: From The Far Side by Gary Larson. Reproduced by permission of Chronicle Features, San Francisco. 2.6: © Sacramento Bee/Zuma Press. 2.18A: © 2002 by the Nature Publishing Group. 2.18B: © 1999 by the Royal Society of London. 2.21: © Eric Tourneret/Visuals Unlimited, Inc.

Chapter 3 Opener: © Scott Camazine/Alamy. 3.2: © 1986 by the Ecological Society of America. 3.4: © Reinhard Dirscherl/Alamy. 3.13A: © Franz Christoph Robiller/imagebroker/Alamy. 3.14: © Gerrit de Vries/Shutterstock. 3.18: © xpixel/Shutterstock. 3.22: © J. Jarvis/Visuals Unlimited, Inc. 3.23: Courtesy of Andrew D. Sinauer.

Chapter 4 Opener: David McIntyre. 4.5: © Jane Burton/Naturepl.com. 4.13: © 1987 by Prentice-Hall Inc., Englewood Cliffs, N.J. 4.14: © Nick Upton/Naturepl.com. 4.15: © Andrew Parkinson/Naturepl.com. 4.20B: © Desmette Frede/Wildlife Pictures/AGE Fotostock. 4.31A: © Merlin D. Tuttle/Bat Conservation International/Photo Researchers, Inc. 4.34A: © Andrew Darrington/Alamy. 4.35 bunting: © Gertjan Hooijer/Shutterstock. 4.35 blackbird: © Karel Brož/Shutterstock. 4.35 chaffinch: © Andrew Howe/istock.

Chapter 5 Opener: © Staffan Widstrand/Naturepl.com. 5.2: © birdpix/Alamy. 5.3: © gary forsyth/istock. 5.4: © Glen Bartley/All Canada Photos/Alamy. 5.11: © European Pressphoto Agency/Alamy. 5.15 individual: © David Dohnal/Shutterstock. 5.15 flock: © Mike Lane/AGE Fotostock. 5.17B: © Michael Tweedie/Photo Researchers, Inc. 5.21A: David McIntyre. 5.23: © 1987 by the American Association for the Advancement of Science. 5.24: © Steve Bloom Images/Alamy. 5.27: © Terry Wall/Alamy. 5.29: © 7877074640/Shutterstock. 5.31: © Rui Saraiva/Shutterstock. 5.33: © Jane Burton/Naturepl.com. 5.34 shark: © Ian Scott/Shutterstock. 5.34 dugong: © Kennet Havgaard/Aurora Photos/Corbis. 5.35: © Byron Jorjorian/Alamy.

Chapter 6 Opener: © Suzi Eszterhas/Naturepl.com. 6.1: © All Canada Photos/Alamy. 6.2: © visceralimage/Shutterstock. 6.6: © Rick & Nora Bowers/Alamy. 6.8: © Christian Musat/Shutterstock. 6.14: © John Kirinic/Shutterstock. 6.15: © Shari L. Morris/AGE Fotostock. 6.19 map: © 2004 by Blackwell Publishing and the Society for Conservation Biology. 6.19 thrush: © All Canada Photos/Alamy. 6.21: © Prisma Bildagentur AG/Alamy. 6.23: © Jonathan Blair/National Geographic. 6.24: © Rick & Nora Bowers/Alamy. 6.25: © Tom Vezo/Minden Pictures/Corbis. 6.26: © 2002 by the Royal Society of London. 6.28: © Karel Brož/Shutterstock. 6.29 map: © 1997 by the Cooper Ornithological Society. 6.29 fox sparrow: Courtesy of James C. Leupold/U.S. Fish and Wildlife Service.

Chapter 7 Opener: Photograph by John Mitani. 7.2A: © Phil Savoie/Naturepl.com. 7.6: Photograph by David M. Phillips/The Population Council. 7.10: © 2000 by Academic Press Ltd. 7.13: © Gregory Dimijian/Photo Researchers, Inc. 7.17: © 1996 by the Royal Society of London. 7.22: © Morales/AGE Fotostock. 7.25B: © 1979 by the American Association for the Advancement of Science. 7.28B: © Juniors Bildarchiv/AGE Fotostock. 7.33A: © 2002 by the Royal Society of London. 7.33 canary: © ene/istock. 7.34: © Jason Gallier/Alamy. 7.36: © blickwinkel/Alamy. 7.37: Courtesy of DiBgd at en.wikipedia. 7.38: © John Kirinic/Shutterstock. 7.44A: © Andrew Syred/Photo Researchers, Inc.

Chapter 8 Opener: © Winfried Wisniewski/Foto Natura/Minden Pictures/Corbis. 8.2: Courtesy of Stephen Childs/Creative Commons Attribution License. 8.4: © Dr. Paul A. Zahl/Photo Researchers, Inc. 8.7: © 2000 by Academic Press Ltd. 8.10B: © Bengt Lundberg/Naturepl.com. 8.11: © Txanbelin/Shutterstock. 8.16: © Dave Watts/Alamy. 8.25: © Joe Gough/Shutterstock. 8.26: © Merlin D. Tuttle/Photo Researchers, Inc. 8.27A: © Eduardo Rivero/Shutterstock. 8.27B: © Humberto Olarte Cupas/Alamy. 8.32: © Rick & Nora Bowers/Alamy. 8.34: © J. Norman Reid/Shutterstock. 8.35: © Hugh Maynard/Naturepl.com. 8.37: © Ocean/Corbis. 8.38 maps: © 2000 by Springer Verlag. 8.38 grouse: Photograph by Marc Dantzker. 8.39: © Michael Zysman/Shutter-

Index

Page numbers in *italic* denote entries that are included in a figure.

A

A1 and A2 auditory receptors, 364–370, 379–380
Abedus, 264
Abortion, 442, 447
  and reproductive success in meerkats, *63*
Aché of Paraguay, 434, 437
Acorn woodpeckers, 50
Action potentials, 365
Adaptation, 6, 13
  behavioral development in, 342–355, 356
  bird songs in, 311–315
  comparative method of hypothesis testing on, 106–108, 136
  cost–benefit analysis of, 103, 122–123, 136
  current utility of, 105
  deceptive communication in, 90
  definition of, 103–104, 105
  developmental homeostasis in, 342–345, 356
  of digger bees, 294
  female mate preference in, 433
  language and speech in, 429–432, 452–453
  in learning, 349–355
  mobbing behavior in, 104–108
  monogamy of males in, 218–225, 296
  optimality theory on, 122–132
  and proximate mechanisms of behavior, 391
  switch mechanisms in, 345–349, 356
  as ultimate cause of behavior, 294, 321
  value of, 103
Adaptation and Natural Selection (Williams), 18
Adélie penguins, 113, *114*
Adolescents, dating preferences of, 440
Adoption
  of genetic strangers, 200, 268, 270–281
  and indirect selection, *21*

Affiliation function of speech, 430
African cichlid fish
  frequency-dependent selection in, 133–134
  resource defense polygyny of, 243–244
African hedgehogs, *383*
African topi, 209
African tree frogs, 232, *233*
Against Our Will (Brownmiller), 446
Aggression
  bird song repertoire matching in, 315
  in territorial behavior, 144, 145
  testosterone levels in, 414–415, 416, 417, 418
Alarm signals, 95–96, *97*
  of Belding's ground squirrels, 53, 343
  false, 93
  of gazelles, 121
Alexander, Richard, 451
Alleles, 6–7, 20
  and coefficient of relatedness, 22
  in Hymenoptera, 26
Allomyrina dichotomus, 182
Alpine accentor, 245
Alston's singing mouse, 201
Altizer, Sonia, 166
Altruism, 15–41
  as adaptive behavior, 24–25, 63
  and coefficient of relatedness, 22, 41
  definition of, 23
  facultative, 63, 65
  and fitness, 24–25
  and group selection, 18–20
  haplodiploidy hypothesis of, 25–29
  of helpers at nests and burrows, 54–62
  and history of behavioral traits, 29–37
  and indirect selection, 21–23
  intelligent design theory of, 17–18
  and kin selection, 37–40, 41, 63
  maladaptive, *46*
  obligate, 63, 65
  reciprocal, *46*, 50–53, 65
  in vertebrates and insects, comparison of, 62–64

Amargosa River pupfish, 419
American redstarts, 145–146
American Sign Language, 425, 431
Amoebae, 25
Amoroso, John, 380
Amphipods, female defense polygyny in, 243
Analysis, levels of, 294
Andersson, Malte, 202
Andrade, Maydianne, 198
Androgens
  and pseudopenis in hyenas, 68–69, 77–78
  and siblicide, 286
  testosterone. *See* Testosterone
Angier, Natalie, 448
Animal Dispersion in Relation to Social Behaviour (Wynne-Edwards), 18
Annual cycles of behavior, 401–409
Anole lizards, 121, *122*
  brown, 237
  green, 414–415
Anolis
  A. cristatellius, *122*
  push-up displays of, 121, *122*
Antelopes, 112
  false alarm signals of, 93
  lek polygyny of, 252
  mating success of, 248
  monogamy of, 224
  sexual conflicts of, 209
  stotting of, 120–121
Anthropology, evolutionary, 454
Antipredator behavior, 101–136. *See also* Predation
Antler size of deer, 83
Antlered flies, 83, 243
Ants
  butterfly larvae raised in colonies of, 88
  conflicts in colonies of, 38
  female defense polygyny of, 243
  food-induced polyphenisms of, *346*
  haplodiploidy hypothesis on altruism of, 26–29
  mating in groups of, 110
  monogamy and altruism of, 29–31
  proportion of males in colonies of, *39*
  self-sacrificing behavior of, 16

Aphelocoma
  A. californica, 57
  A. coerulescens, 57–59
  A. ultramarina, 57–58
Aphids
  clones of, 64, 177
  soldier caste of, *346*
  territories and reproductive success of,
    152, *153*
Apini, *36*, 37
Apis, 34
  A. dorsata, 237
  A. florea, 34, 36, 237
  A. mellifera, 34, 35, 36, 37
  dances of, 32–37
  polyandry of, 237–238
Arbitrary contest resolution, 146–148
Arctic ground squirrels, 103, 146
Arctic moths, 103
Arctic skuas, 105
Area X in bird songs, *306, 307*
Arginine vasotocin, 419
Arnold, Steve, 337–338
Arnqvist, Göran, 240, 241
Artificial selection, 336
Asian honey bees, 34, 36, 237
Assassin bugs, 90, 112, *113*
Associated reproductive pattern, 414
Astatotilapia burtoni, *346*, 347–348
Auditory receptors, 389
  in fish, 380
  in moths, 364–370, 379–380
  and stimulus filtering, 379–380
Augrabies flat lizards, 189
Auklets, least, response to artificial sig-
    nals, 73
Australian antlered flies, 83
Australian brush turkeys, 343
Australian hawks, 124–125
Australian slender crayfish, 85, *86*
Australian thorny devil lizards, *116*
avpr1a gene, and monogamy in prairie
    voles, 297–298
Axons, 365

B
Baboons
  cooperation of, 184
  mating behavior of, 182–184, 194, 299
  paternal care of offspring, 270
  reciprocal grooming of, *51*
Balda, Russ, 332
Banana slugs, 337, 356
Banded mongoose, 152
Bank swallows, 108, 268
Banner-tailed kangaroo rats, 404
Baptista, Luis, 302–303
Bar-tailed godwit migration, 156
Barbary macaques, 230
Barking gecko, 82
Barking treefrogs, 253
Barn swallows
  gape color and growth of nestlings, 288
  mate choice of, 200
  offspring recognition of, 268–269
  survival cost of testosterone in, 417
  tail length of, 200

Barrow's goldeneye ducks, 275
Basolo, Alexandra, 74–75
Bats, 103
  and cricket response to ultrasound,
    375–376
  echolocation by, 364, 367
  female defense polygyny of, *242*
  fringe-lipped, 93, 94
  hammer-headed, 248, *249*
  as illegitimate receivers of communica-
    tion signals, 93, 94
  Mexican free-tailed, 267–268
  and moth antipredator adaptations,
    69–70, 109, 363–371
  parental care for offspring of, 267–268
  reciprocal cooperation of, 52–53
  taste aversion learning of, 355
  territorial songs of, 320
  testes and brain size of, 185
  vampire, 52–53, 355
Bears, 156, 404
Bedbugs, 212–213
Beecher, Michael, 314, 315
Bees
  altruism of, 26–31
  circadian rhythms of, 399–400, 401
  copulation of, *5*, 202, 359–360
  dances of, 32–37, 295
  digger bees. *See* Digger bees
  haplodiploidy hypothesis on, 26–29
  honey bees. *See* Honey bees
  interactive theory of development,
    324–328
  monogamy of, 29–31, *30*, 217, 218
  parental favoritism of, 281–282
  polyandry of, 29, 217, 237–238
  proportion of males in colonies of, *39*
  proximate and ultimate causes of behav-
    ior, 294
  queen production of, 39–40
  sexual behavior of, 4–5, 7–8, 172, 193,
    196, 217
  sleeping in clusters, 112, *113*
  social behavior of, 44, *45*
  sting as self-sacrificing behavior, 16, *17*
  swarming to new hive, 295
  ultraviolet perception of, 371
Beetles
  blister, *119*
  blue milkweed, 194–195
  buprestid, 103
  burying, 220, *221*, 282–283
  copulation of, 363
  dung, *182*, 196, *197*
  harlequin, 144–145
  horned scarab, 185
  Oriental, 203
  rhinoceros, *182*
Begging behavior of nestlings, 84, 94, *95*
  of brood parasites, 272–273, 303, *304*
  gape coloration in, 288
  of gull chicks, 361, 362
  honest signals in, 287–288
  offspring recognition by parents in,
    268–269
Behavioral biology, science of, 11–13

Behavioral development, 323–356
  adaptive, 342–355, 356
  evolution of, 341–355, 356
  genetic and environmental factors in,
    324–341, 356
  homeostasis in, 342–345
  individual differences in, 331–341, 356
  interactive theory of, 324–329, 356
  nature-versus-nurture controversy on,
    324–332
  polyphenisms in, 345–349
  switch mechanisms in, 345–349, 356
Behavioral ecology, 4–10, 11, 452, 454
  of altruism, 15–41
  of communication, 77–96
  of digger bees, 4–5, 7–8
Behavioral strategies or traits, 24
  definition of, 23
  descent with modification theory on,
    29, 32
  haploidiploidy hypothesis on, 28
  history of, 29–37
Belding's ground squirrels
  alarm calls of, 53, 343
  dispersal of, 153–154, *155*, 156
  olfactory cues in kin recognition,
    333–334
Bellbirds, 157, *158*, 258
Belostoma, 264
Belostomatidae, 265, 266
Benkman, Craig, 407
Berthold, Peter, 335
Biesmeijer, Jacobus, 34
Bighorn sheep, *186*, 241
Biological clocks, 395–410, 420
  in crickets, 395, 396–398
  and daily changes in behavior, 395–401
  genetics of, 399–400
  of homing pigeons, 374
  of monarch butterflies, 373
  and seasonal or annual cycles of behav-
    ior, 401–409
  suprachiasmatic nucleus in, 398–401
Biology, behavioral, science of, 11–13
Bird of paradise, *173*
Birds. *See also specific birds.*
  alarm calls of, 93, 95–96, *97*
  begging behavior of nestlings. *See*
    Begging behavior of nestlings
  brood parasites, 272–281. *See also* Brood
    parasites
  circannual rhythms of, 402–403, 405–409
  cognitive skills of, 175, 384–385
  cooperation of, 48–49, 50, 64
  dispersal of, 153, *154*
  egg retrieval routine of, 361
  female defense polygyny of, 241–242
  flying in V-formation, 160, *161*
  foraging decisions of, 124–130, 134–135
  helpers at nests of, 54–61
  hippocampus of, *353*
  ideal free distribution theory on habitat
    selection of, 140–141
  imprinting in, 329–330
  learning abilities of, 329–332
  lek polygyny of, 249–250

loud calling of ravens, 86–88
mate choice of, 48, 199–208, 231, 374, 436
mate guarding of, 195–196
mating systems of, 217, 218
migration of, 156–163, 166–168, 334–336
mobbing behavior of, 53, *54*, 101–107
monogamy of, 64, 222, 224–225, 228–230
nest sharing of, 50
optimal flock size of, 122, 123, *124*
parental care for offspring of, 220, 224–225, 257, 258–259, 284–289
photosensitivity of, 405–409
polyandry of females, 228–230, 232–234, 236, 238, 240
reciprocity of helpful behavior, 51
reproductive anatomy of, 192, *193*
reproductive behavior of, 171–177, 192, 409, 415–416
resource defense polygyny of, 244–246
selfish herds of, 114
siblicide of, 284–287
songs of, 299–320, 321. *See also* Bird songs
spatial learning and memory of, 330–332, 350–351, 352–353
survival cost of testosterone in, 417, 418
territorial behavior of, 18, *19*, 139, 140, 142, 145, 148–149, 151
testosterone and aggression of, 416
threat displays of, 81
ultraviolet-reflecting feathers of, 374
Bird songs, 299–320, 321
adaptation of, 311–315
benefits of, 311–312
control system in, 305–308, *407*
dialects in, 299–301, 304, 305, 311, *312,* 313, 316–317, 320, 332–333
individual differences in, 332–333
late acquisition hypothesis on, 313
learning of, 299–308, 311–314, 350
and mate choice, 200–201, 205, 316–319, 436
nature and nurture in development of, 324, 356
note complex and trill of, 312, *313*
proximate causes of, 299–308, 319–320, 323
repertoire matching in, 315–316
seasonal changes in, 407
selective attrition hypothesis on, 313
ultimate causes of, 308–320
Birkhead, Tim, 6
Biston betularia, 115–117
Black bears, 404
Blackbirds, *97,* 166–167
red-winged. *See* Red-winged blackbirds
Blackcap warblers, 140, 142, 334–336
Black-capped chickadees, 330–331
Black grouse, 208, 250, *251,* 262
Black-headed gulls, 101, *102,* 104–105, 106
Black-legged kittiwake gulls, 107
Blackpoll warblers, 162
Black redstarts, 336
Black-throated blue warblers, *140,* 167
Black-tipped hangingflies, 197
Black widow spiders, 198

Black-winged damselflies
copulation and sperm competition in, 191–192
polygyny of, 241, 243
territorial conflicts of, *149*
The Blind Watchmaker (Dawkins), 17
Blister beetles, *119*
Blueband goby, *194*
Blue-footed boobies, 285
Bluegill sunfish
cooperation and mutual defense of, 46, *47*
mating tactics of, 190
nesting colonies of, 114
paternal care of, 270
Blue jays, 354
butterfly toxins affecting, *119*
response to cryptically colored moths, 117, *118*
Blue milkweed beetles, 194–195
Bluethroats, 231, 236
Blue tits
in brood parasite experiment, 276
cross-fostering and imprinting in, 330
mate choice of, 199–200, 374
parental favoritism of, 286
polyandry of, 234
Blue whales, 127
Blue-winged warblers, 139
Bombini, *36*
Bonobos, 425
Boobies, 285, 286, 287
Boomsma, Jacobus, 25
Borgia, Gerald, 175
Bourke, Andrew, 23
Bowerbirds
mating systems of, 217, 241
reproductive behavior of, 171–177, 181, 200, 209, 217
Bradbury, Jack, 248
Brain
and bird songs, 302, 304–310, 317–318, 320, *407*
of bower-building birds, 174–175
and cognitive skills, 384–388
of nurse bees and foragers, 325–326, *328*
social brain hypothesis, 385–386
in speech and language, 426–430
of star-nosed moles, 381–383
stimulus filtering and cortical magnification in, 381–384
superior temporal sulcus of, 428, *429*
vasopressin receptors in, and monogamy in prairie voles, 296–297, 298
Brazilian harvestman, 243
Bretagnolle, Vincent, 249
Broca's area, 427, 428
Brockmann, Jane, 186
Bronstad, Matthew, 439
Brontops robustus, 182
Brontothere, 182
Bronze-cuckoos, 272–273, 278, 280–281
Brood parasites, 272–281
acceptance of eggs, 278–281
code breaking by, 362
cost–benefit analysis of, 278–281
interspecific, 273, 274–277, 330

mafia hypothesis on, 279–280
obligate, 276–277
song learning by, 303
spatial learning and memory of, 352–353
Brower, Lincoln, 164, 165
Brown, Jerram, 23
Brown anole lizards, 237
Brown bats, 364
Brown bears, 156
Brown boobies, *285*
Brown-headed cowbirds, 280, 352–353
Brownmiller, Susan, 446, 447
Brush turkeys, 343
Buchanan, Katherine, 317
Bufo bufo, 81–82
Buller, David, 447
Bumblebees, 35, *36,* 37
Buntings
indigo, 245
lark, *288*
lazuli, 48–49
reed, *97*
snow, 225
Buprestid beetles, 103
Burger, Joanna, 110
Burrowing owls, 135
Burying beetles, 220, *221,* 282–283
Buss, David, 434, 435, 442
Buston, Peter, 435
Butterflies
biological clock of, 395
conspicuous behavior of, 118–120
deceptive behavior of, 88
dilution effect of social defenses, 110, *111*
hearing ability of, 109
migration of, 156, 164–166, 372–373, *374,* 395
polyandry of, 239
puddling of, 110, *111*
territorial contests of, 147–148, 150, 283
ultraviolet reflecting patches on, 371
By-product hypothesis on helpers at nests, 56–58

C

Cab drivers, 429
Caecilian amphibians, parental care of offspring, 259
California ground squirrels, 108, *109*
California mouse, 222–223, 411
Calopteryx maculata, 191–192
Calorie maximization hypothesis on foraging decisions, 128–129, 130
Calvert, William, 164
Camouflage and cryptic coloration, 115–118
Canaries
mate choice of, 200–201, 205
songs of, 305
Cane toads, 124–125
Cannibalism, 198, 214
of praying mantis, 393
of spiders, 198, 259
of tiger salamanders, 346–347
Cape gannets, 128
Cardinal, Sophie, 37
Cardiocondyla, 243

Carey, Michael, 245
Caro, Tim, 121
Carotenoids
    and gape color of nestlings, 288
    and mate choice, 72–73, 199, 371–372
Carrion crows, 58, 104
Carroll, Sean, 342
Carter, Allisa, 386
Cartwheel display of long-tailed mana-
        kins, 49
Cassin's finches, 316
Catania, Kenneth, 383
Catbirds, green, 217
Catchpole, Clive, 317
Catharus, 158, *159*, 162, *163*
Catocala relicta, 117
Cattle egrets, 286–287
Caudomedial neostriatum in bird songs,
        306, 310
    ventral, 305
Central pattern generators, 378, 379
Centris pallida. See Digger bees
Ceratocephalus grayanus, 182
Ceratopsid dinosaur, 182
Cerebral neuron 2 (C2), 378
Chaffinches, *97*, 196, 299
Chamaeleo montium, 182
Chameleon, 182
Chapman, Tracey, 212
Chase-away selection theory on mate
        choice, 204, 210–213, 215
Cheetahs, 120–121
Chickadees, 330–331, 332
Chickens, 175, 187, 192
Child abuse, 449
Chimpanzees
    communication of, 424–425
    mate choice of, 196, 439, *440*
    mate guarding of, 446
    sexual conflicts of, 210
    territorial behavior of, 142–143
    testosterone levels and parasitic infec-
        tions of, 417
Chipping sparrows, *278*
Chromosomes
    in diploid and haploid cells, 26, *27*
    X chromosomes, 23
Cicadas, 88–89
Cichlid fish
    foraging decisions of, 129
    frequency-dependent selection of,
        133–134
    gonadotropin-releasing hormone of, 348
    polyphenisms of, *346*, 347–348
    resource defense polygyny of, 243–244
    sensory exploitation in spawning of, 71
Circadian rhythms, 396–401, 420
    in cricket calling behavior, 396–397
    master clock mechanism in, 398, 400
    suprachiasmatic nucleus in, 398–401
Circannual rhythms, 401–409, 420
Clark's nutcrackers, 331–332, 350–351
Claw size of crayfish, 85, *86*
Clethrionomys gapperi, 297
Cliff-nesting gulls, 106–107

Cliff swallows
    offspring recognition of, 268–269
    social behavior of, 44
Climate change, 11–12, 140
Clitoris, and pseudopenis of female spot-
        ted hyenas, 67–69, 77–80, 98, 295
Clock mechanisms, 395–410. *See also*
        Biological clocks
Clonal organisms, 64, 177
Clown shrimp, 218, *219*
Clusia, 72
Cockatoos, 303, *304*
Cockroaches, 370
Code breaking, 362
Coefficient of relatedness, 22, 24
    in haplodiploid system of sex determi-
        nation, 26
    in helpful behavior, 55, 64
    in parental favoritism, 281
    and proportion of males in colony, 39–40
Coercive sex, 175, 210, 446–449, 455
Coevolution, 103
Cognitive abilities, 384–388, 389
    in birds, 175, 384–385
Coho salmon, 176
Collared flycatchers, 192
Collared lizards, 83, 142, 372
Colletes hederae, 360
Coloration
    conspicuous, in toxicity, 118–120
    cryptic, 115–118
    of eyes, 445
    of gape, 288
    and mate choice, 72–73, 199, 371–372
Command centers, 391–410
    in bird songs, 305–308
    and central pattern generators, 378, 379
Communication, 67–98
    alarm signals in, 53, 93, 95–96, *97*, 121,
        343
    begging behavior in. *See* Begging behav-
        ior of nestlings
    behavioral ecology of, 77–96
    bird songs in, 299–320, 321. *See also* Bird
        songs
    deceptive, 88–93, 98, 120
    honest signals in, 82–85, 98, 121, *122*,
        204–205
    illegitimate receivers in, 93–96
    illegitimate signalers in, 90–93
    language and speech in, 424–432
    mobbing calls in, 53, *54*
    in parent recognition of offspring,
        268–269
    pseudopenis of spotted hyenas in, 67–69,
        77–80, 98, 295
    sensory exploitation in, 70–77
    stotting in, 120–121
    threat displays in, 74, *75*, 81–85, 98, 372
    ultrasonic, by moths, 69–70
Comparative method of hypothesis test-
        ing, 106–108, 136
Competition
    in arbitrary contest resolution, 146–148
    in female defense polygyny, 241–243
    game theory on, 113
    between human groups, 451

    in mate choice, 181–190, 214
    in nonarbitrary contest resolution,
        148–150
    in resource defense polygyny, 243–246
    in scramble competition polygyny, 241,
        246–247
    in selfish herd, 113–114
    sperm, 185, 190–196, 211, *233*, 446
    for territory, 140–152, 169
Conditional strategies
    in foraging decisions, 135
    in mating behavior, 185–188
    in migration, 166–168, 169
Conditioning, operant, 117, *118*, 353–355
Conflicts, 37–40
    arbitrary resolution of, 146–148
    fighting ability in, 181, *182*
    honest signals in, 84
    in mate competition, 181–190
    nonarbitrary resolution of, 148–150
    parent-offspring, 284
    policing behavior in, 38, *39*
    sexual, 38, 208–214, 215, 441–449
    siblicide in, 284–287
    threat displays in, 81–85, 98
Confusion effect hypothesis on stotting,
        121
Conspicuous behavior, 118–122
Convergent evolution, 136
    in male weaponry, *182*
    in mobbing behavior of gulls, 107–108
Coolidge effect, 175, 236
Cooperation, 46
    of baboons, 184
    costs and benefits of, 46–48, 50
    of females in sperm competition, 192
    of helpers at nests and burrows, 54–62
    postponed, *46*, 50
    reciprocal, 50–53
    of spotted hyenas, 80, 386
    of unrelated individuals, 46–53
Coots
    as brood parasites, 274, 275
    chick ornaments and parental care,
        288–289
Copulation
    of baboons, 182–184, 194, 299
    of bedbugs, 212–213
    coercive or forced, 175, 210, 446–449, 455
    Coolidge effect in, 175
    of digger bees, *5*, 202, 359–360
    extra-pair, 192, 231, 232–234, 442, 444
    maladaptive behavior in, 359–360, 363
    male dominance and mating success,
        182–183
    male performance in, and female mate
        choice, 202–203
    of marine iguanas, 184
    and mate choice in humans, 437
    and postcopulatory activities, 193–196
    of praying mantis, 393–394
    sperm competition in, 185, 190–196, 211,
        233, 446
Cordylochernes scorpioides, 144–145
Cornwallis, Charlie, 64
Cortical magnification, 381–384

Cost–benefit analysis
  of adaptation, 103, 122–123, 136
  of brood parasites, 278–281
  of conspicuous behavior, 118–119
  of dispersal, 153–156, 169
  of helpers at nests, 54–55
  of hormonal regulation, 416–419
  of language and speech, 429–430
  of learning, 350
  of migration, 159–166
  of mobbing behavior, 101–102, 104, 106
  of monogamy, 296
  of nesting colonies, 44
  of nest sharing, 50
  of parental care, 257, 258–266
  of postcopulatory activities, 193–196
  of social behavior, 44, 46–48
  of social defenses, 110–114
  of territorial behavior, 142–152, 169
Courtship behavior
  of bowerbirds, 171–172
  of empid flies, 77
  of manakins, 49, 50, 201–202, 248
  and mate choice, 200–213
  of moths, 70
  of parthenogenetic lizards, 76, 77
  and paternal care, 199
  and polygyny at display grounds,
    247–254
  of water mites, 70–72
Covey size of northern bobwhite quail,
    123, 124
Cowbirds
  behavioral development in, 343
  as brood parasites, 275, 278, 279, 280,
    352–353, 362, 385
  cognitive skills of, 385
  spatial learning and memory of, 352–353
Coyotes, 404
Crayfish threat displays, 85, 86
Creel, Scott, 113
Cresswell, Will, 114
Crickets
  aggression and acoustical signals of, 84
  behavioral development of, 343
  biological clock of, 395, 396–398
  chirping calls of, 93, 394
  mate choice of, 204, 234
  mate guarding and assistance of,
    219–220
  nervous system of, 396, 397, 398
  parasitoid flies of, 93, 380–381
  polyandry of, 230
  ultrasound response of, 375–376
Crockford, Catherine, 194
Crossbills, 127, 407–409
Crows
  and alarm calls of great tits, 96
  cognitive skills of, 384–385
  foraging decisions of, 125–127
  helpful behavior of, 58
  and mobbing behavior of gulls, 104
Crying by human infants, 84
Cryptic coloration, 115–118
Cryptic female choice, 203
Cuckoos, as brood parasites, 272–273, 274,
    278, 279–280, 281, 362

Cultural stereotypes on gender roles, 432
Cumulative selection, 7

D

Dale, Svein, 246
Daly, Martin, 449, 450
Damselflies, 149
  copulation and sperm competition of,
    191–192
  mate choice of, 200
  mate guarding of, 194
  polygyny of, 241, 243
Dances of honey bees, 32–37
  in swarming to new hive, 295
Danforth, Bryan, 37
Dark-eyed juncos, 232, 233, 417
Darwin, Charles, 3
  descent with modification theory of, 29,
    32, 68, 97, 293, 341
  imperfection principle of, 77
  natural selection theory of, 5–6, 13. See
    also Natural selection
  The Origin of Species, 3, 21
  sexual selection theory of, 172, 173, 214
Darwinian puzzles, 7, 41, 98
  in altruism, 16, 17
  in antipredator and foraging behaviors,
    135, 136
  in brood parasites, 278
  in conspicuous behavior, 118–121
  in cooperation among competitors, 48
  in deception, 88–93
  in eavesdropping by illegitimate receiv-
    ers, 93–96
  in helpers at nests and burrows, 54–62
  in human behavior, 448–449
  in loud calling by ravens, 86
  in male decisions on egg consumption,
    290
  in monogamy and polyandry, 227, 254,
    296
  in parental care, 271, 290, 291
  in reproductive behavior, 177, 193, 196
  in termite behavior, 16
  in territorial behavior, 139
  in threat displays, 81–82
  in V-formation of flying birds, 160
Dating preferences of male teenagers, 440
Davies, Nick, 81, 147, 148, 244, 274
Dawkins, Richard, 7, 17
Dear enemy effect, 151–152
Deception, 88–93, 98
  by illegitimate signalers, 90, 120
  by orchids, 90–92, 350
Decision making
  in foraging, optimality theory on,
    124–132
  game theory on, 113
De Dreu, Carsten, 413
Deer
  antler size of, 83
  lek polygyny of, 251
  seasonal reproductive behavior of, 414
Degu, 268
Descent with modification theory of
    Darwin, 29, 32, 68, 76, 97, 293, 341

Development
  behavioral, 323–356. See also Behavioral
    development
  bird songs as indicator of, 317–319
Developmental homeostasis, 342–345, 356
Dialects
  in bird songs, 299–301, 304, 305, 311, 312,
    313, 316–317, 320, 332–333
  in human speech, 430
Dickinson, Janis, 194–195
Dictyostelium discoideum, 25
Diet. See Nutrition
Dietary generalists and specialists, 355
Digger bees
  behavioral ecology of, 4–8
  copulation of, 5, 202, 359–360
  mate finding behavior of, 4–5, 11, 172
  mating systems of, 217
  olfactory system of, 359
  parental care for offspring of, 257
  postcopulatory behavior of, 193, 196
  proximate and ultimate causes of
    behavior, 294
  sexual selection theory on, 172, 181
Dik-diks, 224
Dilution effect, 110–111
Dimetrodon, 206, 207
  D. grandis, 207
Dinoponera quadriceps, 38
Dinosaurs, 182
Diplodiploid system of sex determination,
    26
Diploid eggs, 26, 27
Direct fitness, 24
  definition of, 23
  in helpful behavior, 47, 50–51, 55, 56, 58,
    63, 65
  in reciprocity, 50–51
Direct selection
  definition of, 23
  in helpful behavior, 53
  and warning coloration in toxicity, 120
Disease, as a cost for social species, 44
Dispersal, 153–168
  cost–benefit analysis of, 153–156, 169
  delayed, helpful behavior in, 56–58,
    59–60
  in migration. See Migration
  of pseudoscorpions, 144–145
  quality of territory affecting, 59–60
Display grounds, 248
  and lek polygyny, 247–254
Displays
  in courtship, 171–175
  pseudopenis of spotted hyenas in, 67–69,
    77–80, 98, 295
  push-up displays of lizards in, 85,
    121–122
  of threat, 74, 75, 81–85, 98, 372
Distraction hypothesis on mobbing
    behavior, 101, 104
Divergent evolution, 136
  and mobbing behavior of gulls, 107, 108
Djungarian hamsters, 222
DNA, 31
  and environment interactions, 324–328

microsatellite analysis of, 226–227
in social and sexual monogamy, 225, 226
Dobzhansky, Theodosius, 3
Dogs
    domestication of, 386
    intelligence of, 386–387
Domestic violence, 445
Domestication of dogs, 386
Dominance hierarchy
    of cichlid fish, 347–348
    in competition for mates, 181–184, 187
    of spotted hyenas, 78–80, 183
Dorsal flexion neurons, 377–378
Dorsal ramp interneurons, 377–378
Dorsal swim interneurons, 378
Doucet, Stéphanie, 174
Dragonflies, 110, 191
Drea, Christine, 386
Drongos, fork-tailed, 93
Drosophila
    biological clocks of, 399, *400*
    D. melanogaster, 211, 338–340
    foraging behavior of, *326*, 338–340
    for gene of, 339–340, 342
    mating systems of, 218
    reproductive behavior of, 211–212
    sitting and roving behaviors of, 338–340
Ducks
    as brood parasites, 274–275
    parental care of genetic strangers, 268, 274–275
Dugongs, 130–131
Dung beetles, *182, 196, 197*
Dunnocks
    polyandry of, 239, 244
    resource defense polygyny of, 244–245
    sperm competition and female cooperation of, *192*
    testosterone and aggression of, 416

E

Eared grebes, *178*
Ears
    of crickets, 375
    of moths, 69, 109, 110, 364–370
    of *Ormia ochracea* flies, 380–381
Earthworms, 362–363
Earwigs, 261, 283
Eastern moles, *383*
Eavesdropping by illegitimate receivers of communication signals, 93–96, 98
Eberhard, Bill, 202
Echolocation by bats, 364, 367
Eclectus parrots, 284
Ecology, behavioral, 4–10, 11, 452, 454
    of altruism, 15–41
    of communication, 77–96
    of digger bees, 4–5, 7–8
Eggert, Anne-Katrin, 220
Eggfly butterflies, *150,* 283
Eggs
    of brood parasites, 275, 278–281
    destruction in policing by social insects, 38
    male brooding of, 264–266
    male consumption of, 290
    and mate choice, 208

mobbing behavior for protection of, 101–107
    in nest sharing, 50
    parental guarding of, 259, *260, 261*
    and sexual differences theory on reproductive behavior, 175–177
Egrets, 284, 286–287
Eimer's organs, 382
Elasmotherium sibiricum, 182
Elbow orchids, 90–92
Elk, 113, 131
Emerald coral goby, 221
Emlen, Steve, 224, 435
Empid flies, 77, 179
Emu-wren, southern, *19*
Endocrine system. *See* Hormones
Energy minimization hypothesis, 131
English sparrow, *276*
Entrainment, 397
Environmental influences
    on individual behavioral differences, 332–334
    in interactive theory of development, 324–329, 356
    on language and speech, 334, 426, 428
    in learning, 329–332
    in nature versus nurture controversy, 324–332, 356
Epigaulus, 182
erg-1 gene, 348
Ervynck, A., 128
Escape behavior
    of crickets, 375–376
    of moths, 69–70, 109, 363–371, 379–380
    of sea slugs, 377–378
17β-Estradiol, 412
Estrildidae, 276
Estrogen
    and facial appearance, 439
    and female attractiveness, 438, 439, 440
    measurement of, 415
    and sexual motivation of Japanese quail, 412
Ethology, 361
Ethyl oleate, and foraging behavior of honey bees, 327–328
Euglossini, *36, 37*
Eurasian oystercatchers, 128–129
European blackbird migration, 166–167
European cuckoos, 272, 362
European earwigs, 261
European toads with deep-pitched croaks, 81–82
European warbler songs, 317
Eusocial behavior
    haplodiploidy hypothesis of, 25–29
    of honey bees, 16
    and monogamy, 29–31
Evans, Christopher, 343
Evolution
    and behavioral development, 341–355, 356
    of bird songs, 308–319, 320
    of cognitive skills, 384–388
    convergent, 107–108, 136
    divergent, 107, *108,* 136
    group selection in, 18–20

of human behavior, 423–455
    of mating systems, 217–255, 432–441
    of migration, 156–159
    natural selection in, 5–10, 13. *See also* Natural selection
    of nervous systems and behavior, 359–389
    of parental care, 257–291
    of reproductive behavior, 171–215
    sexual selection in. *See* Sexual selection
    as ultimate cause of behavior, 294
Evolutionarily stable strategy, 146–147
Evolutionary anthropology, 454
Evolutionary psychology, 452, 454
Exploitation, sensory, 70–77, 97
Explosive breeding assemblage, 246–247
Extramarital affairs, 437, 442, 444
    in Amazon culture, 240
    jealousy in, 445
Extra-pair copulation, 192, 442, 444
    genetic heterozygosity of offspring in, *236*
    in polyandry, 227–241
Eye colors, 445

F

Face recognition, 429
    by paper wasps, 387–388
Facial appearance
    as honest visual signal of dominance in paper wasps, 82, *83*
    mate choice based on, 432, *433,* 438, *441*
    in ovulation, 439
Facial fusiform area, 429
Fadool, Debi, 338
Fairy-wrens
    brood parasites of, 272–273, 280–281
    helpers at nests of, 56, 58–59
    microsatellite analysis of, 226–227
    polyandry of, 233, *234*
    sperm of, 176
Falcons, *310*
Fallow deer, 251
Farley, Claire, 131
Farner, Donald, 406
Favoritism, parental, 281–289
Feathers, 105
    and courtship displays, 172, *173*
    and mate choice, 48
    and parental care of coot chicks, 288–289
    reproductive benefit of bright colors in, 48
    ultraviolet-reflecting, 374
Feeding behavior
    foraging in. *See* Foraging behavior
    genetic influences on, 337–338
Female defense polygyny, 241–243, 255
Female-enforced monogamy, 220
Female mate preference
    adaptive value of, 433
    in birds, 316–319, 436
    in humans, 432–437, 455
    in lek polygyny, 249, 252–253
Fertility
    and body shape, 438
    and family income, 437

and insurance hypothesis on polyandry, 231, 232
and mate choice, *183*, 438–439, 440, *441*, 455
and mate guarding, *183*
Fertilization
and reproductive anatomy in birds, 192, *193*
sperm competition in, 185, 190–196, 211, *233*, 446
Fewell, Jennifer, 237
Fifteen-spined sticklebacks, 199
Fighting
arbitrary contest resolution in, 146–148
evolution of abilities in, 181, *182*
honest signals of competitive ability in, 84
in mate competition, 181
Finches
as brood parasites, 274, 276
foraging decisions and reproductive success of, 127
mate choice of, 199, 208, 436
photosensitivity of, 406
response to artificial signals, 73, *74*
songs of, *301, 302*, 303, 305, *306*, 316, 318
Fireflies, *89*, 90
communication flashes of, 88, 94
mate choice of, 197–198
parental investment, *178*
scramble competition polygyny of, 246
Fish. *See also specific fish.*
auditory receptors of, 380
female-enforced monogamy of males, 221
mate choice of, 72–73, 199, 371–372
mate guarding of, *194*
mating tactics of, 190
parental care for offspring of, 220, 260, *261*, 263–264, 270, 290
polyphenisms of, *346*, 347–348, 349
predator response of, 370
reproductive behavior of, 179, 180, 181
resource defense polygyny of, 243–244
sensory exploitation in communication of, 71, 74–76
songs of plainfin midshipman, 378–379
ultraviolet perception of, 95
Fitness
in adaptation, 103, 122–123, 136
and communication, 90, 97
definition of, 22
direct. *See* Direct fitness
in frequency-dependent selection, 133
in game theory, 115
haplodiploidy hypothesis on, 28
in helpful behavior, 47, 50–51, 55–56, 58–59, 63, 65
inclusive, 23, 24, 41, 452
indirect. *See* Indirect fitness
in mobbing behavior, 101–102, 105
in rape, 447–448
in reciprocal altruism, 50–51
in social interactions, *46*
in territorial behavior, 143
FitzGerald, Randall, 351

Fixed action pattern, 362
Flagfish, 290
Flatfish, 88
Flies
antlered, 83, 243
empid, 77, 179
Ormia ochracea, 93, 380–381
parasitoid, 380–381
tephritid, 120
Florida scrub jays
helpers at nests of, 57–59, 60
monogamy of, 225
Flycatchers, 192
polygyny of, 246
reciprocity of helpful behavior, 51
Flying foxes, gray-headed, 416–417
Food. *See also* Nutrition
foraging for. *See* Foraging behavior
hereditary differences in preferences for, 337–338, 356
and polyphenisms, 345–347
and taste aversion learning, 354–355
for gene, 328, 339–340, 342
Forager bees, 325–328
Foraging behavior
calorie maximization hypothesis on, 128–129, 130
of cliff swallows, 44
conditional strategies in, 135
dances of honey bees in, 32–37, *33*
and *Drosophila* genes, *326*, 338–340
and frequency-dependent selection, 133–134
game theory on, 132–135
genetic and environmental influences on, 325–328
in lunar cycle, 404–405
optimality theory on, 124–132
predation risk and daily activity patterns in, 404–405
Ford, E. B., 119
Fork-tailed drongos, 93
fosB gene, 340
Fox sparrow migration, 167–168
FOXP2 gene, 425–426, 428
Free-running cycles, 396
Frequency-dependent selection, 133–134
Fretwell, Steve, 140
Friendships of male baboons, 184
Fringe-lipped bats, 93, 94
Frogs
African tree frogs, 232, *233*
barking treefrogs, 253
illegitimate receivers of communication signals of, 93–94
lek polygyny of, 253
monogamy of, 222
parental investment, *178*
polyandry of, 232, *233*
scramble competition polygyny of, 246–247
sperm competition of, 190
Fruit flies
biological clocks of, 399, *400*
foraging behaviors of, 132–133, *326*, 338–340
for gene of, 339–340, 342

mating systems of, 218
per gene of, 399, *400*
reproductive behavior of, 211–212, 413–414
sitting and roving behaviors of, 338–340
sperm size of, 176
territorial behavior of, 149–150
Fullard, James, 370
Fusiform gyrus, 429

G
Galahs, 303, *304*
Galápagos hawks, 228
Galls, aphid, 152, *153*
Game theory, 136
in arbitrary contest resolution, 146
in feeding behavior, 132–135
and ideal free distribution theory, 140
in nonterritorial bird pairs, 149
in social defenses, 113–115
Gametes, sexual differences theory on, 175–177, 214
Ganglion, 392–393
Gape coloration, 288
Garcia, John, 354
Garter snakes
hereditary differences in food preferences of, 337–338, 356
hormones and reproductive behavior of, *414*, 418–419
Gaulin, Steve, 351
Gazelles, 120–121, 164
Gecko, barking, 82
Geese
egg retrieval routine of, 361
imprinting in, 329
Geffen, Eli, 276
Gender differences
in cultural stereotypes, 432
and female mimicry by males, 189
in gametes and reproductive behavior, 175–177, 214
in mate choice of humans, 432–442, 444
in parental investment, 177–180
in sexual behavior of humans, 442–446
in spatial learning and memory, 351–353
Gender roles
cultural stereotypes on, 432
reversal of, 179–180
Genetics
and adaptive value of traits, 103
and altruism, 24–25
and behavioral development, 324–341
and behavioral strategies, 23, 24
and biological clocks, 399–400
and bird songs, 299, 301, 304–305, 309–310, 320, 323
and coefficient of relatedness, 22
and compatibility in polyandry, 232, 234–236
and fitness, 23, 24
and foraging behavior, 325–328, 338–340
and gender differences in reproductive behavior, 177
and gonadotropin-releasing hormone in cichlid fish, 348

and good genes theory on mate choice, 204–205, 207, 208, 215
and group selection, 18, 19, 20
and haplodiploidy hypothesis on altruism, 25–29
homeobox genes, 341–342
and inbreeding avoidance, 154–155
and indirect selection, 21–23
and individual behavioral differences, 334–341
in interactive theory of development, 324–329, 356
and kin selection, 41
and learning, 329–332
in lek polygyny, 254
and mate choice, 441
microsatellite analysis of, 226–227
in monogamy, 225, 226–227, 297–298
in nature versus nurture controversy, 324–332, 356
and parental care of genetic strangers, 200, 268, 270–281, 291
and parental investment, 178, 262
and pleiotropy, 103
in polyandry, 231–237, 238, 240
as proximate cause of behavior, 294, 321, 323–324
and runaway selection theory on mate choice, 205–206
and spatial intelligence, 427
in speech and language, 425–426, 428
variations in, 6–7
Genetic strangers, parental care for, 200, 268, 270–281, 291
Genitalia and pseudopenis of spotted hyenas, 67–69, 77–80, 98, 295
Genocide, 451
Genotypes
in interactive theory of development, 325, 356
in polyandry, 235
Gerygones, 278
Ghalambor, Cameron, 258
Giant rhinoceros, 182
Gibbons, 223, 230
Gill, Frank, 132
Giraffes, 181
Global warming, 11–12, 140
Goby fish
blueband, 194
emerald coral, 221
two-spotted, 180, 181
Gochfeld, Michael, 110
Goldberg, Rube, 76
Golden-collared manakins, 201–202
Golden egg bugs, 263
Goldeneye ducks, 268, 275
Golden-mantled ground squirrels, 402
Golden-winged sunbirds, 132
Gonadotropin-releasing hormone, 348, 417
Good genes theory
on mate choice, 204–205, 207, 208, 215
on polyandry, 232–234, 235
Good parent theory of mate choice, 199–200, 214
Gorillas, 241
Göth, Ann, 343

Gould, Stephen Jay, 68, 77, 104, 295, 451
Grackles, 353
Grasser, Andreas, 34
Gray-headed flying foxes, 416–417
Gray mouse lemurs, 237
Great egrets, 284
Great Plains toads, 186
Great skuas, 105
Great snipes, 250
Great spotted cuckoos, 279–280
Great tits
in brood parasite experiment, 276
cross-fostering and imprinting in, 330
gape color and growth of nestlings, 288
mobbing and alarm calls of, 95–96, 97
songs of, 311
Great white pelicans, 160, 161
Greater ani, 50
Grebes, eared, 178
Green anoles, 414–415
Green catbirds, 217
Greene, Erick, 48
Greeting ceremony of spotted hyenas, 79–80, 98
Grenade ants, 16
Greylag geese
egg retrieval routine of, 361
imprinting in, 329
Griffin, Donald, 364
Griffith, Simon, 241
Grim, Thomas, 281
Grizzly bears, 156, 404
Grooming behavior, reciprocal, 51
Ground-nesting gulls
mobbing behavior of, 101–107, 108
parental care of genetic strangers, 271
Ground squirrels
arctic, 103, 146
Belding's. See Belding's ground squirrels
California, 108, 109
circannual rhythms in, 402
dispersal of, 153–154, 155, 156
golden-mantled, 402
olfactory cues in kin recognition, 333–334
scramble competition polygyny of, 246, 247
Group selection, 18–20, 23, 221
Grouse
black, 208, 250, 251, 262
ruffed, 153, 154
sage, 173, 251
Gryllus
G. bimaculatus, 394
G. campestris, 219
G. firmus, 397
Guarding of mate. See Mate guarding
Guinea pigs, 232
Gulls
begging behavior of chicks, 361, 362
black-headed, 101, 102, 104–105, 106
cliff-nesting, 106–107
divergent evolution of, 107
mobbing behavior of, 101–107
parental care of genetic strangers, 271–272

songs of, 310
Gunnison's prairie dogs, 231
Guppies
mate preferences of, 72–73
pike cichlid as predator of, 129

H
Habitat
and bird songs, 311–312
climate change affecting, 140
Habitat selection, 140–141, 169
and ideal free distribution theory, 140–141, 169
of monarch butterflies, 164–166
and territorial defense, 142–152
Hahn, Thomas, 407–408, 409
Halliday, Tim, 81
Hamadryas baboons, 51
Hamilton, W. D., 26, 37, 113, 204–205
Hamilton's rule, 24, 26, 63
Hammer-headed bats, 248, 249
Hamner, William, 406
Hamsters
circadian rhythms of, 399
monogamy of, 222
sperm and eggs of, 176
Handgrip strength, 433
Hangingflies, 210, 238
Hanuman langurs
infanticide of, 9–10, 11, 18–19, 240
polyandry of, 240
Haplodiploidy, 25–29
Haploid eggs, 26, 27, 40
Hare, Brian, 386–387
Hare, Hope, 28
Harlequin beetles, 144–145
Harlow, Harry, 343–344
Harlow, Margaret, 343–344
Harpegnathos saltator, 38
Hartung, John, 284
Harvester ants, 110
Harvestman
mate guarding of, 194, 243
parental care for offspring of, 261, 264
resource defense polygyny of, 243
Haselton, Martie, 442
Haskell, David, 94
Hausfater, Glen, 182
Hawks
and alarm calls of great tits, 95–96, 97
foraging decisions of, 124–125
mating systems of, 228
reactions of parental birds to, 258
and selfish herds of redshanks, 114
Healthy mates theory, 204, 207, 215
bird songs in, 316–318
Hearing
A1 and A2 auditory receptors in, 364–370
of crickets, 375–376
of moths, 69–70, 109, 110, 364–370, 379–380
of Ormia ochracea flies, 380–381
of predator and prey, 96, 109
of sparrows, and song learning, 300–301, 303
Hecatesia exultans, 69–70

Hedgehogs, African, *383*
Heeb, Philipp, 288
Heinrich, Bernd, 86–88
Helpful behavior, 45–62
 altruistic, 15–41. *See also* Altruism
 cooperation in, 46
 kin or indirect selection in, 53, 63
 at nests and burrows, 54–62
 as nonadaptive by-product, 56–58
 reciprocal, 50–53, 65
 of unrelated individuals, 46–53
 of vertebrates and insects, comparison
  of, 62–64
Hemipepsis ustulata, 150–151
Heredity, 6–7. *See also* Genetics
Herring gulls, 361, 362
Heterozygosity, genetic, in polyandry, 236
High vocal center (HVC) neurons in bird
 songs, 306, 307–308
Hilara sartor, 77
Hippocampus
 in navigation by humans, 429
 in spatial learning, 353
Hippocampus whitei, 220
Hippopotamus, 124
Hirundidae, 108
Histocompatibility complex, major, 236,
 237, 441
Hobson, Keith, 167
Holland, Brett, 210
Homeobox genes, 341–342
Homeostasis, adaptive developmental,
 342–345, 356
Home range, 144
Homicide, 445, 451
Homing pigeons, 374, *375*
Honest signals, 82–85, 98, 121, *122*
 in begging behavior of nestlings,
  287–288
 in mate choice, 204–205
Honey bees
 brain size of, 384
 circadian rhythms of, 399–400, 401
 dances of, 32–37, 295
 for gene of, 342
 interactive theory of development,
  324–328
 monogamy of, 218, 221
 nurse workers and foragers, 400, 401
 pheromones of queens, 85
 polyandry of, 29, 217, 237–238
 sting as self-sacrificing behavior, 16, 17
Hoover, Jeffrey, 280
Hori, Michio, 134
Hormones, 411–419, 420
 and circadian rhythms of crickets, 397
 costs in behavioral effects of, 416–419
 estrogen. *See* Estrogen
 and foraging behavior of honey bees,
  327
 in helpers at nests, 57–58
 and mate choice, 433, 438, 439, 440, *441*
 and monogamy of prairie voles, 296–297
 and paternal behavior, 411
 and photosensitivity of birds, 406, 407,
  408

 as proximate cause of behavior, 294, 321
 and pseudopenis in hyenas, 68–69, 77–78
 and reproductive behavior, 411–419
 seasonal changes in, 407
 testosterone. *See* Testosterone
Horned pig, *182*
Horned rodents, *182*
Horned scarab beetles, 185
Horns
 and antler size of deer, 83
 and evolution of male weaponry, 181,
  182
 and testes size, 185
Horses, energy consumption at different
 speeds, 131
Horseshoe crabs
 eggs as fuel for migrant birds, 159–160
 explosive breeding assemblage of, 246,
  247
 mating behavior of, 185–186, 187
 scramble competition polygyny of, 246,
  247
Horsfield's bronze-cuckoos, 272–273,
 280–281
Hotshot hypothesis on lek polygyny,
 249–250, *251*
Hotspot hypothesis on lek polygyny,
 249–252
Houbara bustard, 204, *205*
House finches, 406
House mice, 409–410
House sparrows, 205
House wrens, 218
Hox genes, 341–342
Hoy, Ronald, 375, 380
Hrdy, Sarah, 10, 11
Hughes, William, 29
Humans
 adaptationist approach to behavior of,
  452–453, 454, 455
 brain size of, 384, 430
 cognitive skills of, 384
 crying by infants, 84
 cultural stereotypes on gender roles of,
  432
 Darwinian puzzles in behavior of,
  448–449
 developmental homeostasis of, 344–345
 evolution of behavior, 423–455
 extramarital affairs of, 240, 437, 442, 444,
  445
 face recognition by, 429
 gene-environmental interactions in, 453
 irrational behavior of, 449
 language and speech of, 320, 424–432
 lipreading by, 428
 mate choice of, 431, 432–441, 455
 navigation skills of, 429
 nutrition affecting intellectual development of, 344–345, 431
 oxytocin affecting behavior of, 413
 parental care of offspring, 257, 283–284
 population growth and resource consumption, 449–450
 reproductive behavior of, 415, 432–449
 sensory analysis in, 384

 sexual conflicts of, 441–449
 spatial learning and memory of, 352,
  426, *427*
 sperm and eggs of, 176
Hummingbirds
 migration of, 156
 songs of, 308, *309*, 310, 311
Humor, 431
Huxley, T. H., 452
Hyenas
 dispersal of, 155, 156
 pseudopenis of females, 67–69, 77–80,
  98, 295
 spotted. *See* Spotted hyenas
Hymenocera picta, 218, *219*
Hymenoptera
 ants in. *See* Ants
 bees in. *See* Bees
 haplodiploidy and altruism in, 25–29
 monogamy and eusociality in, 29–31, *30*
 polyandry of, 29, *30*, 237–238
 wasps in. *See* Wasps
Hypoglossal nucleus in bird songs, 306
Hypothesis testing, 11–12, 13
 comparative method of, 106–108, 136

I

Ichthyoxenus fushanensis, 214
Ideal free distribution theory, 140–141, 169
Iguanas
 lek polygyny of, 252
 mating behavior of, 184
Illegitimate receivers of communication
 signals, 93–96
Illegitimate signalers, deception by, 90–93
Immune system
 in polyandry, 230–231, 236
 testosterone affecting, 417, 433
Imprinting, 329–330
Inbreeding, 154–155, 156, 235
Inclusive fitness, 24, 40, 41
 definition of, 23
 in helpers at nests, 55, 56, 58
 in social interactions, 46
Income
 and fertility, 437
 and mate preferences, 434–435, 437
Indigo buntings, 245
Indirect fitness, 24, 28
 definition of, 23
 in helpful behavior, 55, 56, 58–59, 65
 in social interactions, 46
Indirect selection, 21–23
 in helpful behavior, 53
Infanticide
 as adaptive behavior, 10, 11
 and group selection, 18–19
 of Hanuman langurs, 9–10, 11, 18–19, 240
 hormone levels in, 411
 of lions, *10*
 mate guarding and assistance in, 223
 of mice, 409–410, 411
 and natural selection, 9–10, 11, 18–19
 and polyandry of females, 240
Innate releasing mechanism, 362
Insectivores, sensory analysis in, 381–383

Insects
altruistic behavior of, 15–17, 20, 21, 23–29, 65
direct fitness benefits of helpful behavior, 46–47
evasive behavior in bat avoidance, 368
female defense polygyny of, 243
haplodiploidy hypothesis on, 25–29
mate choice of, 197–198
monogamy of, 25–26, 29, 30, 31
policing behavior of, 38, 39
polyandry of, 237–238, 239
proportion of males in colonies, 38–39
self-sacrificing behavior of, 15–17, 20, 21, 23–24, 62–64
sterile, 15–16, 21, 23–24
Instincts, 361–362
Intelligence
and cognitive abilities, 175, 384–388, 389
and mate choice, 444
nutrition affecting development of, 344–345, 431
Intelligent design theory, 17–18
Interactive theory of development, 324–329, 356
Interneurons, 365
dorsal ramp interneurons, 377–378
dorsal swim interneurons, 378
int-1 sensory interneurons, 375–376
ventral swim interneurons, 378
int-1 sensory interneurons, 375–376
Iporangaia pustulosa, 264
Irs4 gene, 413
Isolation experiments on development of social behavior, 343–344
Isopods, *182,* 188–190, 214
Ivy bees, 360, 362
Izzo, Amanda, 82

J

Jacanas, wattled, 228
Jamieson, Ian, 56
Japanese damselflies, 200
Japanese quail, 411–412
Jays, 350–351, 354
butterfly toxins affecting, *119*
helpers at nests of, 57–59, 60
mobbing calls and kinship of, 53, *54*
monogamy of, 225
pinyon, 350–351
plush-capped, 258
response to cryptically colored moths, 117, *118*
Steller's, 258
Jealousy, sexual, 445
Jiguet, Frédéric, 249
Juncos, 224, 232, 233, 417
Juvenile hormone
and circadian rhythms in crickets, 397
and foraging behavior of honey bees, 327

K

Kaas, Jon, 383
Kafue lechwe, *252*
Kamil, Alan, 117
Kangaroo rats, 152, 404, 405

Katydids
deceptive behavior of, 88–89
Mormon crickets, 179–180
parasitoid flies of, 381
parental investment, *178*
sex roles of, 179–180
Kellogg, Luella, 424
Kellogg, Winthrop, 424
Kemp, Darrell, 147–148
Kenrick, Doug, 444
Kessel, E. L., 77
Ketterson, Ellen, 417
Kettlewell, H. B. D., 115
Kingfishers, 54–56
Kin selection, 20, 23, 24, 41, 230
in helpful behavior, 53, 63
and social conflict, 37–40
Kipling, Rudyard, 104
Kirchner, Wolfgang, 34
Kirkpatrick, Mark, 205–206, 240, 241
Kirk's dik-dik, 224
Kitcher, Philip, 454
Kittiwake gulls, 107, *108*
Klug, Hope, 290
Komdeur, Jan, 59, 195
Krill, 127
Kroodsma, Donald, 316
Krüger, Oliver, 274
Kruuk, Hans, 101–106
Kubanochoerus gigas, 182
Kv1.3 protein, 338

L

Lacewings, 109, *368*
Lacey, Eileen, 146
Lande, Russell, 205–206
Language and speech, 424–432
adaptive value of, 429–432, 452–453
affiliation function of, 430
cost–benefit analysis of, 429–430
critical period in development of, 428
dialects in, 430
environmental factors in, 334, 426, 428
genetic factors in, 425–426, 428
history of, 425–426
learning of, 320, 334, 424–426, 428, 430
and mate choice, 431
neurophysiology of, 426–429
phonemes in, 425
sexual selection hypothesis on, 430–431
Langur monkeys
infanticide of, 9–10, 11, 18–19, 240
polyandry of, 240
Laridae, 108
Lark buntings, *288*
Lasioglossum, 44, 45
Late acquisition hypothesis on bird songs, 313
Lateral magnocellular nucleus of the anterior nidopallium (LMAN) in bird songs, 306–307
Lazuli buntings, 48–49
Leadbeater, Ellouise, 47
Leal, Manuel, 121
Learning
adaptive value of, 349–355
of bird songs, 299–308, 311–314, 320, 350

cost–benefit analysis of, 350
genetic and environmental factors in, 329–332
of language and speech, 320, 334, 424–426, 428, 430
olfactory cues in, 333–334
and spatial memory, 330–332, 350–353. *See also* Spatial learning and memory
of taste aversion, 354–355
Least auklets, 73
Leeches, 124
Lek polygyny, 241, 247–254, 255, 262
Lemmings, 20, *21*
Lemmiscus, 296
L. curtatus, 297
Lemurs, 69
Lethocerus, 264–266
Leuzinger, Lucas, 15
Levels of analysis, 294
Levey, Douglas, 157
Levine, Jon, 411
Lewontin, Richard, 104
Li, Norm, 434
Life expectancy, 450–451
Lill, Alan, 248
Lin, Chung-Ping, 259
Lindauer, Martin, 34–37, *35*
Lions
dispersal of, 156
hunting success during lunar cycle, 404–405
infanticide of, *10*
prides of, 50, 241
Lipreading, 428
Little bustard, 249–250
Lizards
cryptic coloration of, 116
female mimicry by males, 189
honest visual signals of, 83, 121–122
mate guarding of, 194
parthenogenetic, *76,* 77
polyandry of, 237
push-up displays of, 85, 121–122
side-blotched, 85, *417*
survival cost of testosterone in, *417*
territorial behavior of, 151
threat displays of, 74, *75,* 83, 85, 372
ultraviolet-reflecting patches of, 372
Lloyd, Jim, 246
Lockard, Robert, 404
Locusts, 345–346
Long-billed marsh wrens, 350
Long-tailed finches, *74*
Long-tailed manakins, 49, 50
Long-tailed widowbirds, 202, 203
Lorenz, Konrad, 329, 361
Lunar cycle and nocturnal foraging, 404–405
Lundberg, Arne, 244
Luteinizing hormone, 406, 407
Lyon, Bruce, 288

M

Madden, Joah, 174
Mafia hypothesis on brood parasites, 279–280
Magpies, 279–280

Majerus, Michael, 115, 116
Major histocompatibility complex, 236, 237, 441
Male mate choice, 438–441, 455
  and dating preferences of male teenagers, 440
  fertility hypothesis on, 438–439
Mallard ducks
  eggs of, 208
  mate choice of, 231
  predator response while nesting, 290
Mammals, monogamy of, 222–224, 442
Manakins
  courtship behavior of, 49, 50, 201–202, 248
  golden-collared, 201–202
  lek polygyny of, 248, 251
  long-tailed, 49, 50
  migration of, 167
  white-ruffed, 167
Mandrills, 236
Mangrove gerygone, 278
Margay cats, 88
Marine iguanas
  lek polygyny of, 252
  mating behavior of, 184
"Market value" of men, 435
Marler, Catherine, 143
Marler, Peter, 299, 301, 302
Marsh wrens, 350
Martin, Tom, 258
Masked boobies, 285
Masked shrews, *383*
Master clock in circadian rhythms, 398, 400
Mate assistance hypothesis on monogamy, 219–220, 222, 223, 224–225
Mate choice, 196–214, 215
  bird songs in, 200–201, 205, 316–319, 436
  chase-away selection theory on, 204, 210–213, 215
  of chimpanzees, 196, 439, *440*
  competition in, 181–190, 214
  and copulatory success, 437
  courtship behavior in, 200–213
  cryptic, 203
  facial appearance as factor in, 432, *433*, 438, *441*
  of females, 432–437. *See also* Female mate preference
  fertility hypothesis on, 438–439, 455
  of fish, 72–73, 199, 371–372
  good genes theory on, 204–205, 207, 208, 215
  good parent theory on, 199–200, 214
  healthy mates theory on, 204, 207, 215
  of humans, 431, 432–441, 444, 455
  intelligence as factor in, 444
  of males, 438–441, 455
  and "market value," 435
  null model on, 206
  nuptial gifts in, 179, 197–198
  in polyandry, 227, 234
  runaway selection theory on, 204, 205–206, 207, 208, 215
  and self-rating on attractiveness, 435–436

  speech and language in, 431
  ultraviolet reflectance in, 371–372
Mate guarding, 194–196, 214, 446
  of baboons, *183*
  in monogamy, 218, 219–220, 223, 296
  in resource defense polygyny, 243
Mateo, Jill, 333–334
Maternal care, 259–262, 290
  genetic influences on, 340
  and surrogate mothers in social deprivation experiments, 343–344
Mating behavior. *See* Reproductive behavior
Mating systems, 217–255
  diversity of, 217–218
  monogamy in, 217, 218–230. *See also* Monogamy
  polyandry in, 227–241. *See also* Polyandry
  polygyny in, 217, 241–254, 255. *See also* Polygyny
Matthews, L. Harrison, 68
Mattila, Heather, 238
May, Mike, 376
Mayflies, 110–111
McCracken, Gary, 267–268
McDonald, David, 49
Meadow voles, *297*, 351–352
Mediterranean fruit flies, 149–150
Meerkats, 62, *63*, 183
Meire, P. M., 128
Melatonin, 400
Melipona, 35, 36, 39–40
  colony kin structure and queen production, *40*
  M. beecheii, 40
  M. favosa, 40
  M. quadrifasciata, 40
  M. subnitida, 40
Meliponini, *36*, 37
Membracinae, 259, *260*
Memory
  and social amnesia in knockout mutations, 341
  spatial. *See* Spatial learning and memory
Menstrual cycle, 415
  fertility and mate choice during, *183*, 439, 440, *441*
  mate guarding in, *183*
Mexican free-tailed bats, 267–268
Mexican jays, 57–58, 350–351
Mexican tetra, 95
Mexican whiptail lizards, 194
Meyer, Alex, 76
Mice
  circadian rhythms of, 400, *401*
  hormone regulation of behavior, 412–413
  infanticide of, 409–410, 411
  knockout experiments on, 340–341, 411
  mate choice of, 201
  maternal behavior of, 340
  olfactory processing by, 338, 400
  paternal care of, 222–223
  reproductive behavior of, 412–413
  single-gene effects in, 340–341
  social amnesia in, 341
Microsatellite analysis, 226–227

Microtus, 296
  M. californicus, 297
  M. montanus, 297
  M. ochrogaster, 296–298
  M. pennsylvanicus, 297
  M. pinetorum, 297
  spatial learning and memory of, 351–352
Midshipman fish, 378–379, 380
Migration, 153–168, 169
  conditional strategies in, 166–168, 169
  cost–benefit analysis of, 159–166
  departure date from wintering grounds in, 145–146
  and developmental switch mechanisms in locusts, 345–346
  genetic factors in, 334–336
  history of, 156–159
  of monarch butterflies, 372–373, *374*, 395
  multistrategy hypothesis on, 168
  seasonal differences in speed of travel in, 157
  short-range, 157, 158
Milkweed beetles, 194–195
Miller, Geoffrey, 430–431, 439
Miller, Jeremy, 198
Millipedes, 203
Mimicry of females by males, 189
Mobbing behavior, 101–108
  as adaptation, 104–108
  of California ground squirrels, 108, *109*
  comparative method of hypothesis testing on, 106–108
  cost–benefit analysis of, 101–102, 104, 106
  distraction hypothesis on, 101, 104
  of great tits, 95–96, *97*
  of gulls, 101–107, *108*
  of Siberian jays, 53, *54*
Mole rats, 80
  activity patterns of, 401, *402*
  altruism of, 61–62
  queen, 61–62
  sensory analysis of, 384
Moles
  eastern, *383*
  star-nosed, 381–383
Møller, Anders, 200
Mollusks, crows foraging on, 125–127
Monarch butterflies
  biological clock of, 395
  conspicuous coloration and toxicity of, 118–120
  migration of, 156, 164–166, 372–373, *374*, 395
Mongoose, 152
Monkeys
  infanticide of Hanuman langurs, 9–10, 11, 18–19, 240
  margay cats imitating calls of, 88
  mate choice of, 236
  parental care for offspring of, 262
  polyandry of, 240
  social deprivation experiments on, 343–344
Monodon monoceros, 182
Monogamy, 217, 218–230, 254
  adaptive value for males, 218–225, 296
  and altruism, 25–26, 29–31, *30*, 64

of birds, 64, 222, 224–225, 228–230
cost–benefit analysis of, 296
female-enforced, 220, *221*
genetic analysis in, 225, 226–227
of humans, 442
of mammals, 222–224, 442
mate assistance hypothesis on, 219–220, 222, 223, 224–225
mate guarding in, 218, *219*, 219–220, 223, 296
and polyandry of female mates, 227–230
of prairie voles, 296–298, 321
social, 225, 226–227
Montane voles, *297*, 342
Montezuma oropendola, 241–242
Montgomerie, Bob, 174, 176
Mooney, Richard, 307
Moore, Frank, 161
Moore, Michael, 143
Moral issues, 451
in rape, 448
in social injustice and inequality, 453
Mormon crickets, 179–180
Mormons, 444
Morocconites malladoides, 182
Moths, 364–370
acoustic stimuli triggering diving behavior of, 364
antipredator adaptations of, 69–70, 109, 110, 363–371, 391
attraction to artificial lights, 103
cryptic coloration of, 115–118
ears of, 69, 109, 110, 364–370
neural networks of, 365, 391
sex pheromones of, 391
stimulus filtering of, 379–380
ultrasonic communication of, 69–70, 109, 110, 364, 366, 371
Motor nuclei, 379
Multistrategy hypothesis on migration, 168
Mumme, Ronald, 59
Mussels, oystercatchers feeding on, 128–129, 130

N

Naked mole rats, 80
activity patterns of, 401, *402*
altruism of, 61–62
queen, 61–62
sensory analysis of, 384
Narwhal, *182*
Naso annulatus, 182
Natural selection, 5–10, 13, 293
and adaptation, 391
and communication systems, 97
and group selection, 18–19
and human behavior, 451, 452, 454
and infanticide, 9–10, 11, 18–19
and pseudopenis in hyenas, 77–80
and runaway selection theory on mate choice, 206
and sexual selection, 172, 173
as ultimate cause of behavior, 294
Nature versus nurture controversy, 324–332, 356

Navigation
by bats, ultrasound in, 364
hippocampus in, 429
by homing pigeons, 374, *375*
by monarch butterflies, 372–373
NCM (caudomedial neostriatum) in bird songs, *306*, 310
Neal, Jennifer, 415
Nectar
and foraging decisions of birds, 132
honey bee dance on source of, 32–37
Neff, Bryan, 270
Neostriatum, caudomedial, in bird songs, *306*, 310
ventral, 305
Nephila pilipes, 72
Nepoidea, 265
Nervous system, 359–389
brain in. *See* Brain
central pattern generators in, 378, 379
and cognitive skills, 384–388
command centers in, 391–410
and complex responses to simple stimuli, 360–370
cortical magnification in, 381–384
of crickets, 396, 397, *398*
and moth avoidance of bats, 363–370
neurons in. *See* Neurons
of praying mantis, 392–393
response to relayed messages in, 376–379
selective relay of sensory inputs in, 374–376
in speech and language, 426–429
stimulus filtering in, 379–383, 389
Nestlings
begging behavior of. *See* Begging behavior of nestlings
and brood parasites, 272–281
gape coloration of, 288
helpers in care of, 54–61
honest signals of, 84, *288*
parental care of, 258–259, 284–289
reproductive value of, 288
Nests
brood parasites in, 272–281
in colonies, 44
helpers at, 54–61
male protection of, 199
sharing of, 50
Net benefit theory on deceptive communication, 89, 90
Net stance, 70
Neumania papillator, 71
Neurons
action potentials of, 365
and adaptive behavior, 391
and begging behavior of gull chicks, 361
and bird songs, 305–308
cerebral neuron 2 (C2), 378
cortical, 383
dorsal flexion, 377–378
and evasive behavior of moths, 366, 369
and proximate causes of behavior, 294
structure and function of, 365, 366
ventral flexion, 377–378

Neurotransmitters, 365
New Caledonian crows, 384–385
Nicrophorus
N. defodiens, 220
N. vespilloides, 282–283
Noctuid moths, 364–370
antipredator adaptations of, 109, 110, 364–370, 371, 391
ears of, 109, 110, 364–370
neural networks of, 365, 391
sex pheromones of, 391
stimulus filtering by, 379–380
Nolan, Val, 245
Northern bobwhite quail, optimal covey size for, 123, *124*
Northern swordtail, 95
Norway rats, 353–355
Note complex of bird songs, 312, *313*
Notodontid moths, 369
Novel environment theory on deceptive communication, 89
Nowicki, Steve, 317, 318
Nucleus
hypoglossal, 306
lateral magnocellular nucleus of the anterior nidopallium in bird songs, 306–307
motor, 379
robust nucleus of the arcopallium in bird songs, 306, *307*
suprachiasmatic, 398–401
Null model on mate choice, 206
Nunn, Charles, 230–231
Nuptial gifts, 179, 197–198, 447
Nurse honey bees, 325–326, *328*
Nurture versus nature controversy, 324–332, 356
Nutcrackers, 331–332, 350–351
Nutrition
and bird songs, 317–318
of dietary generalists and specialists, 355
and food-induced polyphenisms, 345–347
foraging for. *See* Foraging behavior
and hereditary differences in food preferences, 337–338, 356
and intellectual development of humans, 344–345, 431
long-term effects of early deficits in, 431
in pregnancy of humans, 344–345
and taste aversion learning, 354–355

O

Obesity, 103
Obligate altruism, 63, 65
Occam's razor, 106–107
Octopus, 88
Odor cues. *See* Olfactory cues
Oldfield mice, *155*
Oldroyd, Benjamin, 237
Olfactory bulb, 400
Olfactory cues
for digger bees, 359
genetic factors in processing of, 338
in recognition behavior, 333–334

Onthophagus
　O. nigriventris, 185
　O. raffrayi, *182*
Operant conditioning, 117, *118*, 353–355
Operational sex ratio, 178, 179, 180
　in lek polygyny, 252
Optic lobe of crickets, 397, *398*
Optimality theory, 122–132, 136
　and antipredator behavior, 122–124
　and foraging decisions, 124–132
Orangutans, 210
Orb-weaving spiders
　reproductive behavior of, 193
　web ornaments of, 135
Orchid bees, *36, 37*
Orchid pollination by thynnine wasps,
　90–92, 350
Orians, Gordon, 141
Oriental beetles, 203
The Origin of Species (Darwin), 3, 21
Oring, Lew, 224
Ormia ochracea, 93, 380–381
Osmia rufa, 281–282
Ovary development, and policing behav-
　ior of colony mates, 38
Ovulation
　and mate choice, 415, 439, 440
　and mate guarding, *183*
　and sexual activity in humans, 415
Owings, Donald, 404
Owls, 404
　burrowing, 135
　songs of, 309, *310*
Oxt gene, 341
Oxytocin, 341, 413
Oyamel firs, and migration of monarch
　butterflies, 164–166
Oystercatchers, foraging decisions of,
　128–129, 130

P

Pacemaker in circadian rhythms, 398, 400
Packer, Craig, 404
Panda principle, 76–77, 266
Pandas, thumb of, 76–77
Panorpa, 187–188, 447
Paper wasps, *24*
　face recognition by, 387–388
　honest visual signals of, 82, *83*
　reaction to foreign workers, 333
　social behavior of, 37–38, 46–47
Paracerceis sculpta, 188–190
Parasites
　and bird songs, 316–317, 319
　in cliff swallow colonies, 44
　and display structures of bowerbirds,
　　174, 204
　honest signals on, 204–205
　of monarch butterflies, 166
　and testosterone levels, 417
Parasitic birds, 272–281. *See also* Brood
　parasites
Parasitoid flies, 380–381
Parental care, 257–291
　for brood parasites, 272–281
　cost–benefit analysis of, 257, 258–266,
　　290

discriminating, 267–281
effort in, 287
egg-guarding in, 259, *260,* 261
by fathers. *See* Paternal care
favoritism in, 281–289
by fish, 220, 260, *261,* 263–264, 270, 290
for genetic strangers, 200, 268, 270–281,
　291
and good parent theory of mate choice,
　199–200, 214
by humans, 257, 283–284
infanticide in. *See* Infanticide
investment in offspring in, 177–180, 260,
　262, 287
in monogamy, 220, 222–223, 224–225,
　227–229
by mothers, 259–262, 290, 340
recognition of offspring in, 267–269, 291
resemblance of offspring in, 444
and siblicide, 284–287
social deprivation experiments on,
　343–344
testosterone affecting, 411, 417
and value of offspring, 287–289
Parrots
　parental favoritism of, 284
　songs of, 303, *304,* 308, *309,* 310, 311
Parthenogenetic lizards, 76, 77
Passeriformes, 308
Passerine songbirds, *309, 310*
Paternal care, 263–266
　compared to maternal care, 259–262
　cost–benefit analysis of, 262, 263–264,
　　290
　and female mate choice, 199–200
　in fish, 220, 260, *261,* 263–264, 270, 290
　infanticide in, 409–410, 411
　in monogamy, 220, 222–223, 224–225,
　　227–229
　resemblance of offspring affecting, 444
　testosterone affecting, 411, 417
Payoff asymmetry hypothesis, 150–151, 169
Peacock butterflies, 118
Peacocks
　courtship display of, 172
　lek polygyny of, 251
　mate choice by, 203, 207–208
Pelicans, 160, *161*
Penguins, 113, *114*
Penis
　of bedbugs, 212, *213*
　of black-winged damselflies, 191–192
　and female mate choice, 203
　and pseudopenis of female spotted
　　hyenas, 67–69, 77–80, 98, 295
Peppered moths, coloration of, 115–117
per gene, 399–400
Perching position, and cryptic coloration,
　117–118
Perdita, 8
Perissodus microlepis, 133–134
Peromyscus californicus, 411
Perrigo, Glenn, 410
Perusse, Daniel, 437
Petrie, Marion, 207–208
Petrinovich, Lewis, 303

Phalaropes
　red, 228
　red-necked, 229
Phenotypes, 334, 356
　in interactive theory of development,
　　325, 328–329
　and polyphenisms, 345–349
Pheromones
　of burying beetles, 220, *221*
　food sources of bees marked with, 35
　illegitimate receivers of, 93
　of noctuid moths, 391
　of queen honey bees, 85
　of thynnine wasps, 92, 350
Phonemes, 425
Photinus, 88, 89, 90, *178*
　scramble competition polygyny of, 246
Photoperiods and circannual rhythms,
　402–409, 420
Photosensitivity and breeding behavior of
　sparrows, 405–406
Photuris, 88, 89, 94
Phylogenetic trees, 29, 31
　on honey bee dances, *36, 37*
　on *Lasioglossum* bees, 44, *45*
　on song learning in birds, 308–309
Phytalmia, *83*
Pied babblers, 93
Pied flycatchers
　polygyny of, 246
　reciprocity of helpful behavior, 51
Pied kingfishers, altruism of, 54–56
Pietrewicz, Alexandra, 117
Pigeons, 309, *310,* 354, 374, *375*
Pike cichlids, foraging decisions of, 129
Pin-tailed whydah, *276*
Pine voles, *297*
Pink cockatoos, 303, *304*
Pinyon jays, 350–351
Pipefish, 179, 220
PKG enzyme, 328
Plainfin midshipman fish, 378–379
Platyfish, *75*
Pleiotropy, 103
Plomin, Robert, 426
Plumage. *See* Feathers
Plush-capped jays, 258
Poecilimon veluchianus, 381
Pogonomyrmex, *110*
Policing behavior in social insects, 38, *39*
Polistes, *24*
　face recognition by, 387–388
　P. dominulus, 46–47, *83*
　P. fuscatus, 387–388
　P. metricus, 387–388
　reactions to foreign workers, 333
　social behavior of, 37–38, 46–47
Pollination of orchids by thynnine wasps,
　90–92, 350
Pollution affecting mate attraction and
　choice, 319
Polyandry, 217, 227–241, 254
　benefits of, 230–241
　convenience, 241
　fertility insurance hypothesis of, 231,
　　232

genetics in, 231–237, 238, 240
of Hymenoptera, 29, *30*, 237–238
and monogamy of males, 227–230
Polybia, 112
Polygyny, 217, 241–254, 255
female defense, 241–243, 255
of humans, 442, 444
lek, 241, 247–254, 255, 262
and parental favoritism, 284
resource defense, 241, 243–246, 255
scramble competition, 241, 246–247, 255
threshold hypothesis on, 245
Polyphenisms, 345–349
Poplar galls, 152, *153*
Pornography, 442
Possum, rock-haunting, 223
Postcopulatory activities, 193–196
Potbellied seahorses, 198
Prairie dogs, polyandry of, 231
Prairie voles
mate guarding and assistance of, 223, 342
monogamy of, 296–298, 321
outbreeding of, 155
spatial learning and memory of, 352
Praying mantis, 109, *368*, 392–394
Predation, 101–136
and adaptation of moths, 69–70, 109, 110, 363–371
arms race between predator and prey in, 103, 104
conspicuous behavior in, 118–122
cost–benefit analysis of behavior in, 101–102
cryptic coloration in, 115–118
and daily activity patterns, 404–405
deceptive signals in, 88–89, 90
and frequency-dependent selection, 133–134
by illegitimate receivers of communication signals, 93–96
and mating success of redback spiders, 198
mobbing behavior in, 101–108
optimality theory on, 122–132
and parental care of offspring, 258, 259
polyphenisms in, *346*, 347
and reproductive value, 290
social defenses in, 110–114
Pregnancy, nutrition in, 344–345
Priapella, 74, *75*
P. olmecae, 74, 76
Pribil, Stanislav, 245
Primates
polyandry of, 230–231
reciprocal grooming behavior of, 51
Principles of Social Evolution (Bourke), 23
Prisoner's dilemma, 52
Problem-solving behavior, 387
Proctor, Heather, 70
Progesterone and paternal behavior in mice, 411
Prokineticin 2 (PK2), 400–401
Prolactin levels in helpers at nests, 57–58
Prostitution, 442
Prothonotary warblers, 279, 280
Protoceratid ungulate, *182*

Protocerebral ganglion in praying mantis, 393
Proximate causes of behavior, 293–308, 319–321, 323–324, 420
and adaptive outcomes, 391
hormones in, 411–419, 420
neural mechanisms in, 359, 389
in sexual assault, 447
Prum, Richard, 206
Pseudopenis of spotted hyenas, 67–69, 77–80, 98, 295
Pseudoscorpions, 144–145, 259
polyandry of, 235–236
Psychology, evolutionary, 452, 454
Purple-crowned fairy wrens, 56
Purple martins, 157
Push-up displays of lizards, 85, 121–122

Q

Quail
Japanese, 411–412
northern bobwhite, optimal covey size for, 123, *124*
Queen insects
behavioral development in queen bees, 328
and coefficient of relatedness, 24, 39–40
and conflicts in ant colonies, 38
and control of offspring production, 28
and direct fitness benefits of helpful behavior, 46–47
and female defense polygyny, 243
and haplodiploidy, 26, 28, 32
monogamous, 25, 29
pheromones of queen honey bees, 85
and policing of reproductive behavior, 38
polyandry of, 217, 237–238
production of queen bees, 39–40
in termite colonies, 15–16
Queen naked mole rats, 61–62
Queller, David, 261, 263
Quetzal, *173*
Quinn, John, 114

R

Raccoons, 94
Racey, P. A., 68–69
Rana temporaria, 190
Rape, 446–449
RA (robust nucleus of the arcopallium) in bird songs, 306, *307*
Ratnieks, Francis, 39
Rats
circadian rhythms of, 399, 401
kangaroo rats, 152, 404, 405
mole rats. *See* Mole rats
operant conditioning of, 353–355
white, 353–355, 399, 401
Ravens, 46
loud calling of, 86–88
Reciprocity, 50–53, 65
in altruistic behavior, 50–53, 65
fitness consequences of, *46*
Redback spiders
developmental flexibility of, 349
sexual suicide of, 198
Red-backed voles, 296, *297*

Red crossbills, 407–409
foraging behavior and mate selection of, 127
Red damselflies, *194*
Red deer, 414
Red-eyed vireos, 160–161
Red knots, 141, 142, 153
migration of, 159–160, 164
Red mason bees, 281–282
Red-necked phalaropes, 229
Red phalaropes, 228
Redshanks, 114
optimal flock size for, 124, *125*
Red-shouldered widowbirds, 148–149
Red-sided garter snakes, *414*, 418–419
Redstarts, 145–146, 336
Red-winged blackbirds
habitat selection of, 141
hippocampus of, *353*
polyandry of, 231, 238
polygyny of, 245–246
territorial behavior of, 142, 150
Reed buntings, *97*
Reed warblers, *272*, 278, 281
Rejection behavior, 209
Relatedness, coefficient of. *See* Coefficient of relatedness
Repertoire matching in bird songs, 315–316
Reproductive behavior, 171–215
in associated reproductive pattern, 414
of baboons, 182–184, 194, 299
of bees, 4–5, 7–8, 11, 193, 196, 202, 359–360
of bowerbirds, 171–177, 181
coercive or forced, 175, 210, 446–449, 455
coexistence of alternative strategies in, 188–190
conditional strategies in, 185–188
conflicts in, 38, 208–214, 215, 441–449
courtship in. *See* Courtship behavior
desire for sexual variety in, 442, *443*, 444
in dissociated reproductive pattern, *414*, 418–419
environmental cues in timing of, 407–409
gender differences in, 175–177, 432–446
of guppies, 72–73
hormones affecting, 411–419
of humans, 415, 432–449
jealousy in, 445
of lizards, *76*, 77, 194
mate choice in, 72–73, 196–214, 215. *See also* Mate choice
mate competition in, 181–190
mate guarding in, 194–196. *See also* Mate guarding
monogamy in, 218–230. *See also* Monogamy
nuptial gifts in, 179, 197–198, 447
operational sex ratio in, 178, 179, 180
and parental investment, 177–180
perceptions on sexual intent in, 442–443
and photosensitivity of birds, 405–409
polyandry in, 227–241. *See also* Polyandry
polygyny in, 241–254. *See also* Polygyny
polyphenisms in, 349

postcopulatory activities in, 193–196
of praying mantis, 393–394
rejection in, 209
of satellite males, 185–186, 188
seasonal changes in, 414–415
sex role reversal in, 179–180
sexual conflicts in, 208–214, 215, 441–449
sexual selection in, 172–215
social conditions affecting, 409–410
and sperm competition, 185, 190–196, 211, *233*, 446
of white-fronted bee-eaters, 60–61
Reproductive success
of bowerbirds, 175, *176*
in brightly colored plumage, 48
and courtship behavior, 174
of dominant males, 182–183
and evolutionary change, 6
and female mate choice, 433–434
and fitness, 23, 24
and foraging decisions, 127
and habitat selection, 140, 141
in helpful behavior, 47, 55, 56, 58–59, 65
and infanticide of Hanuman langurs, 9–10
measures of, 105
in mobbing behavior, 101, 105
of monogamous males, 218, 225
persistence of hereditary traits in, 18
in polyandry, 235
of satellite males, 185–186
and territories of gall-forming poplar aphids, 152, *153*
winter territoriality affecting, 146
Reproductive value, 290
of nestlings, 288
residual, 283
Residual reproductive value, 283
Resource defense polygyny, 241, 243–246, 255
Resource-holding power, 145–146, 148–149, 169
Reyer, Uli, 54, 55, 56
Rhesus monkeys, social deprivation experiments on, 343–344
Rhinoceros, giant, *182*
Rhinoceros beetles, *182*
Rice, Bill, 210
Ring-billed gulls, *271, 272*
RNA, messenger, 325, 327, *328*
Robert, Daniel, 380
Robins
parental care for offspring of, 258
territorial behavior of, 151
Robinson, Gene, 328
Robinson, Scott, 280
Robust nucleus of the arcopallium (RA) in bird songs, 306, *307*
Rock-haunting possum, 223
Rodenhouse, Nicholas, 140
Roeder, Kenneth, 363–364, 365, 366, 368, 370, 371, 392–393
Roosters, 187, 192
and Coolidge effect, 175
Rough-winged swallows, *108,* 268
Round dance of honey bees, 32
Royalactin, 341

Rubenstein, Dustin, 236
Ruddy turnstones, foraging decisions of, 134–135
Ruff, 188
Ruffed grouse, 153, *154*
Rufous-bellied thrush, 258
Rufous-winged sparrows, 407
Runaway selection theory on mate choice, 204, 205–206, 207, 208, 215
Ryan, Mike, 93

S

Sac-winged bats, 320
Sagebrush voles, *297*
Sage grouse, *173, 251*
Sahlins, Marshall, 453
Saino, Nicola, 288
St. Peter's fish, 263–264
Sakaluk, Scott, 220
Salamanders, 346–347
Salmon, sperm and eggs of, 176
Sand crickets, 397
Sandberg, Ronald, 161
Sandpipers
foraging decisions of, 134–135
mating strategies of, 188
mating systems of, 217, 229–230
parental care for offspring of, 260
selfish herds of, 114
Sardell, Rebecca, 234
Satellite males, 185–186, 188
Satin bowerbirds
mating systems of, 217, 241
reproductive behavior of, 171–177, 200, 217
Savanna baboons
cooperation of, 184
mating behavior of, 182–184
Sceloporus virgatus, 74, *75*
Schmitt, David, 442
Scientific research, hypothesis testing in, 11–12, 13
Scorpionflies, 187–188, 447
Scramble competition polygyny, 241, 246–247, 255
Scrub jays, 350–351
helpers at nests of, 57–59, 60
monogamy of, 225
Sea anemones, 64
Sea slugs, 377–378
Sea stars, 377
Sea turtles, 103
Seahorses, 198, 220
Seals, 268
Searcy, William, 245, 319
Seasonal cycles of behavior, 401–409, 420
hormone changes and reproductive behavior in, 414–415
Sedge warblers, 205, 317
Sedge wrens, 316
Seeley, Thomas, 34, 238
Selective attrition hypothesis on bird songs, 313
Selfish behavior
and altruistic behavior, 22, 37
and social defenses of herds, 112, 113–114

Selfish herd, 112, 113–114
Self-sacrificing behavior
altruistic, 15–17
in parental care of offspring, 259
in sexual suicide of spiders, 198, 218
Sensory exploitation, 70–77, 97
Sensory interneurons, 375–376
Serins, 318
Serotonin, 346
Serracutisoma proximum, 261
Sex determination
diplodiploid system of, 26
haploidiploid system of, 26–29
Sex differences. *See* Gender differences
Sex hormones, 413
estrogen. *See* Estrogen
and pseudopenis in hyenas, 68–69
testosterone. *See* Testosterone
Sex peptide hormone and receptor, 414
Sex roles
cultural stereotypes on, 432
reversal of, 179–180
Sexual behavior. *See* Reproductive behavior
Sexual conflicts, 38, 208–214, 215
in humans, 441–449
Sexual deception by orchids, 90–92, 350
Sexual selection, 172–215, 455
and bowerbird behavior, 172–177
Darwin on, 172, 173, 214
in lek polygyny, 254
and male monogamy, 222
and male traits harmful to females, 211–212
and mate choice, 196–214, 215
and mate competition, 181–190
null model of, 206
and parental investment, 177–180, 262
sex role reversal in, 179
and speech in humans, 430–431
and sperm competition, 190–196
Sexually transmitted diseases in polyandry, 230–231
Seychelles warblers
helpers at nests of, 59–60
mate guarding of, 195–196
Shackelford, Todd, 435
Sharks, and foraging decisions of dugongs, 130–131
Sherman, Paul, 53
Sherry, David, 330–331
Shier, Debra, 152
Shrews, *383*
Shrimp, clown, 218, *219*
Shuster, Steve, 189–190
Siberian jays, 53, *54*
Siblicide, 284–287, 291
Side-blotched lizards, 85, *417*
Sign language, 424–425, 426, 431
Sign stimulus, 362
Silver gulls, 101, *102*
Singh, Devendra, 439
Skinner, B. F., 69, 353
Skinner boxes, 353–354
Skuas, 105
Slagsvold, Tore, 276
Smith, Bob, 265, 266

Snakes
  and alarm calls of great tits, 96
  and defensive reaction of arctic ground
    squirrels, 103
  hereditary differences in food prefer-
    ences of, 337–338, 356
  hormones and reproductive behavior of,
    414, 418–419
  and mobbing behavior of California
    ground squirrels, 108, *109*
  and push-up displays of lizards, 121, *122*
  response of fish to, 370
Snow buntings, 225
Social amnesia in knockout mutations, 341
Social behavior, 43–65
  altruistic, 15–41. *See also* Altruism
  costs and benefits of, 44, 46–48
  defensive, 101–115. *See also* Social de-
    fenses
  helpful, 45–62. *See also* Helpful behavior
  of insects. *See* Social insects
  isolation experiments on, 343–345
Social bonding hypothesis on pseudopenis
  in hyenas, 80
Social brain hypothesis, 385–386
Social cohesion hypothesis on stotting, 121
Social conflicts, 37–40
Social defenses, 101–115
  cost–benefit analysis of, 110–114
  and game theory, 113–115
  mobbing behavior in, 101–108
Social environment
  and bird song development, 302–303
  and task specialization of worker honey
    bees, 327
Social insects
  altruistic behavior of, 15–17, 20, 21,
    23–29, 65
  direct fitness benefits of helpful behav-
    ior, 46–47
  haplodiploidy hypothesis on, 25–29
  monogamy of, 25–26, 29, *30*, 31
  policing behavior of, 38, *39*
  polyandry of, 237–238
  proportion of males in colonies, 38–39
Sociobiology, 452, 453, 454, 455
Sociobiology: The New Synthesis (Wilson),
  452
Soldier aphids, 64
Somatosensory cortex, 382–383, *384*
Songbirds, 299–320, *321*. *See also specific*
  birds.
  alarm calls of, 95–96, *97*
  honest signals of, 84
  imprinting in, 330
  mate choice of, 200–201
  migration of, 156–163, 166–168
  parental care for offspring of, 257,
    258–259
  polyandry of, 232, 236, 239
  resource defense polygyny of, 244–245
  songs of. *See* Bird songs
  territorial behavior of, *19*
Songs
  of birds, 299–320, *321*. *See also* Bird songs
  of plainfin midshipman fish, 378–379

Song sparrows
  polyandry of, 234
  social conditions affecting reproductive
    behavior of, 409
  songs of, 302, 303, 314, 315, 318, 319
  territorial behavior of, 150
Sooty shearwater migration, 156–157
Sparrowhawks, *96*, 114
Sparrows
  brood parasites of, *278*
  hormones and reproductive behavior of,
    415–416
  migration of, 405
  photosensitivity and breeding behavior
    of, 405–407
  social conditions affecting reproductive
    behavior of, 409
  songs of, 299–308, 311–315, 316–318, 319,
    320, 323. *See also* Bird songs
Spatial learning and memory
  adaptive value of, 350–353
  in birds, 330–332, 350–351
  in scramble competition polygyny, 246,
    247
  and verbal abilities, 426, *427*
Spear-nosed bat, *242*
Speckled wood butterflies, 147–148
Speech and language, 424–432. *See also*
  Language and speech
Sperm
  and chase-away selection theory,
    211–212
  chemicals transferred with, 211–212
  competition in fertilization success of,
    185, 190–196, 211, *233*, 446
  and mate guarding, 446
  and sexual differences theory on repro-
    ductive behavior, 175–177
Spermatophores, *178*, 179, 180
  in polyandry, 239
Spiders
  bright spots and stripes of, 72
  developmental flexibility of, 349
  feeding behavior of, 130
  mate choice of, 198
  nocturnal prey captured by, 72
  nuptial gifts of, 198
  orb-weaving, 135, 193
  parental care for offspring of, 259
  redback, 198, 349
  reproductive behavior of, 193–194, 210
  response to web movements, 90
  sexual suicide of, 198, 218
  and signal deception by tephritid flies,
    120
  tropical jumping, 130
  web ornaments of, 135
Spiteful behavior, *46*
Spotted hyenas
  brain and cognitive abilities of, 386
  competition for food, *78*
  cooperation of, 80, 386
  dispersal of, 155
  dominant females, 78–79, 183
  greeting ceremony of, 79–80, 98
  pseudopenis of females, 67–69, 77–80,
    98, 295

  siblicide of, 286
  social bonding hypothesis on, 80
  submission hypothesis on, 79–80
Spotted sandpipers, 217, 229–230, 260
Springboks, *120*
Squirrels, ground. *See* Ground squirrels
Star-nosed moles, 381–383
Starlings
  extra-pair matings of, 236
  mate choice of, 205
  monogamy of, 222, 224, 225
  paternal care of, 224, 225
  polyandry of, 239
  songs of, *303, 305, 307*
Stegodyphus lineatus, 259
Steller's jays, 258
Stephens' kangaroo rats, 152
Sterile insects
  monogamy of, 25
  self-sacrificing behavior of, 15–16, 21,
    23–24
Sticklebacks
  mate choice of, 199, 371–372
  parental care for offspring of, 263
Stiles, Gary, 157
Stimulus filtering, 379–383, 389
Sting of honey bee as self-sacrificing
    behavior, 16, *17*
Stingless tropical bees, 35–36, 37
Stitchbirds, 194, 210
Stonechats, circannual rhythms in,
    402–403, 405
Storks, 309, *310*
Stotting, 120–121
Strawberry finches, *302, 303*
Styracosaurus albertensis, 182
Subantarctic fur seals, 268
Subesophageal ganglion in praying man-
    tis, 393
Submission hypothesis on pseudopenis in
    spotted hyenas, 79–80
Sundews, deceptive signaling by, 91
Superb fairy-wrens
  brood parasites of, 272–273, 280–281
  polyandry of, 233, *234*
Superb starlings, 236, 239
Superior temporal sulcus, 428, *429*
Suprachiasmatic nucleus, 398–401
Surrogate mothers in social deprivation
    experiments, 343–344
Swainson's thrush, 162, *163*, 167
Swaisgood, Ronald, 152
Swallows
  bank, 108, 268
  barn. *See* Barn swallows
  begging calls of, 94
  cliff, 44, 268–269
  gape color and growth of nestlings, 288
  mate choice of, 200
  nest colonies of, 44
  offspring recognition by, 268–269
  rough-winged, *108*, 268
  social behavior of, 44
  survival cost of testosterone in, 417
  tail length of, 200
Swamp sparrow songs, 308, 317–318
Sweeney, Bernard, 111

Swifts, 309, *310*
Switch mechanisms in behavioral development, 345–349, 356
Swordtails, 74–76, 95
Synapses, 365
Syngnathus typhle, 179
Synthetoceras, 182
Sytl4 gene, 413

T
Tamarin monkeys, 88
Tarantula hawk wasps, 150–151
Taste aversion learning, 354–355
tau gene, 399
Taxicab drivers, 429
Taylor, Richard, 131
Tegrodera aloga, 446
Teleogryllus, 395, 396
   T. oceanicus, 93, 204, 375–376
Temporal sulcus, superior, 428, *429*
Tephritid flies, 120
Terminal investment, 283
Termites
   haplodiploidy and eusocial behavior
      of, 32
   monogamy of, 230
   sterile, self-sacrificing behavior of,
      15–16, *17,* 23–24
Terns, *112*
Territorial behavior, 169
   arbitrary contest resolution in, 146–148
   of Australian songbirds, *19*
   bird songs in, 311–312, 314–316, 319
   coexistence on home range in, 144
   cost–benefit analysis of, 142–152, 169
   dear enemy effect in, 151–152
   and dispersal, 59–60, 144–145, 153–168
   evolutionarily stable strategy in, 146–147
   foraging decisions in, 132
   group selection theory on, 18
   and ideal free distribution theory on
      habitat selection, 140–141
   loud calling by ravens in, 86–88
   nonarbitrary contest resolution in,
      148–150
   payoff asymmetry hypothesis on,
      150–151, 169
   and polyandry of females, 228, 238
   and polygyny, 241, 243–246
   and reproductive success, 146
   and resource defense polygyny, 243–246
   and resource-holding power, 145–146,
      148–149, 169
   of warblers, 139, 140, 145
   in wintering grounds, 145
Testes growth, and photosensitivity of
   birds, 406, 408
Testosterone
   and aggression, 414–415, 416, 417, 418
   costs of behavioral effects, 416–419
   and immune function, 417, 433
   and male attractiveness, 433, *441*
   measurement of, 415
   and paternal behavior, 411, 417
   and pseudopenis in hyenas, 68–69, 77–78
   and reproductive behavior, 411–412,
      414–419

and territoriality of lizards, 143
Thamnophis elegans, 337–338
Thirteen-lined ground squirrels, 246, *247*
Thomson's gazelle, 120–121
Thornhill, Nancy, 446
Thornhill, Randy, 187–188, 197, 446
Thorny devil lizard, *116*
Threat displays, 81–85, 98
   of lizards, 74, *75,* 83, 85, 372
Threat level hypothesis on dear enemy
   effect, 152
Three-wattled bellbird migration, 157, *158*
Threshold hypothesis on polygyny, 245
Thrushes
   migration of, 158, *159,* 162, *163,* 166–167
   parental care for offspring of, 258
Thumb of pandas, 76–77
Thynnine wasps in orchid pollination,
   90–92, 350
Tibbetts, Elizabeth, 82
Tiger moths, 371
Tiger salamanders, 346–347
Tiger sharks, and foraging decisions of
   dugongs, 130–131
Tinbergen, Niko, 101, 293, 298, 361, *362*
Toads
   deep-pitched croaks of, 81–82
   hawks foraging on, 124–125
   mating behavior of, *186*
Tobias, Joe, 151
Topi antelopes
   false alarm signals of, 93
   mating success at leks, 248
   sexual conflicts of, 209
Toxicity, warning coloration in, 118–120
Transcription factors, 326, 327
Tree frogs
   African, 232, *233*
   barking, 253
Treehoppers, 259, *260*
Tree swallows, begging calls of, 94
Trigona, 35
Trill of bird songs, 312, *313*
Trilobite, 182
Trioceros montium, 182
Tritonia diomedea, 377–378
Trivers, Robert, 28, 50, 177, 283
Tropical harvestman, 194
Tropical jumping spiders, 130
Tropical stingless bees, 35–36, *37*
Trypoxylus dichotomus, 182
Tsimane of Bolivia, 442
Túngara frogs, 93–94
Turkeys, social behavior of, 343
Turnstones, foraging decisions of, 134–135
Two-spotted goby fish, 180, *181*
Tympanum, *364*
   auditory receptors A1 and A2 in,
      364–370

U
Udell, Monique, 386
Uganda kob, 252, *253*
Ultimate causes of behavior, 293–298,
   308–321

Ultrasound
   in bat navigation, 364
   cricket response to, 375–376
   in moth courtship signals, 70
   moth detection of, 69–70, 109, 110, 364,
      366, 371, 379–380
Ultraviolet radiation
   animal perception of, 95
   and circadian rhythms, 395–401
   and circannual rhythms, 401–409
   and cricket calling cycles, 396–397
   and monarch butterfly migration,
      372–373, *374,* 395
   reflection patterns of, 95, 371–372, *374*
Unicornfish, *182*
Urban areas, bird songs in, 312

V
V-formation of flying birds, 160, *161*
Vampire bats
   as dietary specialists, 355
   reciprocal cooperation of, 52–53
Vannote, Robin, 111
Variation, and evolutionary change, 5, 6
Vasopressin, 342, 413
   and monogamy in prairie voles, 296–297,
      342
Ventral caudomedial neostriatum in bird
   songs, 305
Ventral flexion neurons, 377–378
Ventral swim interneurons, 378
Verbal and spatial abilities, 426, *427*
Verbal courtship, 430–431
Vertebrates, altruism in, compared to
   insects, 62–64
Viduidae, 276
Vireo migration, 160–161
Voles
   evolutionary relationships of, *297*
   mate guarding and assistance of, 223,
      296, 342
   meadow, *297,* 351–352
   monogamy of, 296–298, 321
   montane, *297,* 342
   outbreeding of, 155
   prairie. *See* Prairie voles
   red-backed, 296
   spatial learning and memory of, 351–352
von Frisch, Karl, 32

W
Waage, Jon, 192
Wade, Juli, 415
Wade, Michael, 189–190
Waggle dance of honey bees, 32, *33,* 35, 36
Walnut sphinx caterpillar, 118
Warblers
   begging calls of nestlings, 94, *95,* 272
   brood parasites of, 278, 279, 280, 281
   habitat selection by, 140
   helpers at nests of, 59–60
   mate guarding of, 195–196
   migration of, 158–159, 162, 167, 334–336
   sedge, 205, 317
   songs of, 308, 317
   territorial behavior of, 139, 140, 142, 145
   yellow, 118, 139, 279

Wasps
  direct fitness benefits of helpful behavior, 46–47
  face recognition by, 387–388
  haplodiploidy hypothesis on altruism of, 26–29
  honest visual signals of, 82, *83*
  monogamy and altruism of, 29–31
  paper wasps. *See* Paper wasps
  proportion of males in colonies of, *39*
  reaction to foreign workers, 333
  social behavior of, 37–38, 46–47
  social defenses of, 111, *112*
  territorial behavior of, 150–151
  thynnine, in orchid pollination, 90–92, 350
Water bugs, parental care for offspring of, 264–266
Water mites, 70–72
Water supply as factor in migration, 164
Wattled jacana, 228
Waxbill, *276*
Webs
  ornaments of orb-weaving spiders, 135
  spider response to movements of, 90
Weidinger, Karel, 140
Wenseleers, Tom, 39
Wernicke's area, 427
West, Stuart, 20
Western scrub jays, 57
Whales
  foraging decisions of, 127
  migration of, 163
Whelks, crows foraging on, 125–127
Whiptail lizards, *76, 77,* 194
Whistling moths, 69–70
White-bearded manakins, 248
White-crowned sparrows
  hormones and reproductive behavior of, 415–416
  migration of, 405
  photosensitivity and breeding behavior of, 405–406

songs of, 299–306, 308, 311, 312, 313, 316–317, 320, 323. *See also* Bird songs
White-fronted bee-eaters, 60–61
White-handed gibbons, 223
White rats, 353–355
  circadian rhythms of, 399, 401
White-ruffed manakins, 167
White-winged crossbills, 407, *408*
White-winged fairy-wrens, microsatellite analysis of, 226–227
Whitfield, Charles, 325
Whitfield, Philip, 134
Whitham, Tom, 152
Widowbirds
  as brood parasites, 276
  long-tailed, 202, 203
  red-shouldered, 148–149
Wieczorek, John, 146
Wigby, Stuart, 212
Wiklund, Christer, 147–148, 239
Wildebeests, 164
Willard, Dan, 283
Williams, George C., 18, 19
Wilson, David Sloan, 20
Wilson, E. O., 452
Wilson, Margo, 449, 450
Winking, Jeffrey, 442
Wintering grounds
  departure date from, 145–146
  territorial behavior in, 145
Wolanski, Eric, 164
Wolf, Larry, 132
Wolf spiders, 210
Wolff, Jerry, 296
Wolves, 43, 46
  and foraging decisions of elk, 131
  intelligence of, 386–387
Wood ducks as brood parasites, 274–275
Wood frogs, 246–247
Woodpeckers, 50, *310*
Wood thrushes, 157
Wood warbler migration, 158–159

Worker insects
  interactive theory of development in honey bee workers, 324–328
  reaction to foreign workers of paper wasps, 333
  self-sacrificing behavior of, 15–17
Worms, 362–363
Wrens
  fairy-wrens. *See* Fairy-wrens
  helpers at nests of, 56, 58–59
  mating systems of, 218
  response to brood parasites, 272–273
  songs of, 316
Wynne-Edwards, V. C., 18, 20

X
X chromosomes, 23
Xiphophorus, 74–76
  X. helleri, 75
  X. maculatus, 75

Y
Yarrow's spiny lizard, 143, 144
Yellow-eyed juncos, 224
Yellowthroats, 139
Yellow-toothed cavy, 232, *233*
Yellow warblers, 118, 139, 279
Yoder, James, 153
Yom-Tov, Yoram, 276
Young, Larry, 296–297, 298

Z
Zach, Reto, 125–127
Zebra finches, 274
  foraging decisions and reproductive success of, 127
  mate choice of, 199, 208, 436
  response to artificial signals, *74*
  songs of, *301,* 305, *306,* 318
Zebras, 164
Zeh, Jeanne, 235
ZENK gene, 305, 309–310
Zera, Anthony, 397
Zuk, Marlene, 93, 204–205

About the Book

Editor: Sydney Carroll

Production Editor: Kathaleen Emerson

Copy Editor: Lou Doucette

Production Manager: Christopher Small

Photo Researcher: David McIntyre

Book Design and Production: Jefferson Johnson

Illustration Program: Elizabeth Morales

Indexer: Linda Hallinger

Cover and Book Manufacturer: World Print, Ltd.